WITHDRAWN

TREATISE ON INVERTEBRATE PALEONTOLOGY

Prepared under Sponsorship of
The Geological Society of America

The Paleontological Society   The Society of Economic Paleontologists and Mineralogists
The Palaeontographical Society   The Palaeontological Association

Directed and Edited by
RAYMOND C. MOORE

PART C

PROTISTA 2

SARCODINA
CHIEFLY "THECAMOEBIANS" AND FORAMINIFERIDA

By ALFRED R. LOEBLICH, JR., and HELEN TAPPAN
with some systematic descriptions of Foraminiferida by R. WRIGHT BARKER, W. STORRS
COLE, R. C. DOUGLASS, MANFRED REICHEL, and M. L. THOMPSON

VOLUME 2

THE GEOLOGICAL SOCIETY OF AMERICA
and
THE UNIVERSITY OF KANSAS PRESS
1964
REPRINTED 1978

© 1964 by The University of Kansas Press
AND
The Geological Society of America

All Rights Reserved

Library of Congress Catalogue Card
Number: 53-12913
ISBN 0-8137-3003-1

Composition by The University of Kansas Press
Lawrence, Kansas

Printing by The Meriden Gravure Company
Meriden, Connecticut

Binding by the Tapley-Rutter Company, Inc.
Moonachie, New Jersey

Address all communications to The Geological Society of America, 3300 Penrose Place, Boulder, CO 80301

PART C
PROTISTA 2

SARCODINA
CHIEFLY "THECAMOEBIANS" AND FORAMINIFERIDA

By ALFRED R. LOEBLICH, JR., and HELEN TAPPAN
with additions by others

VOLUME 2

Suborder ROTALIINA
Delage & Hérouard, 1896

[*nom. correct.* LOEBLICH & TAPPAN, 1961, p. 219 (*pro* Rotalidae DELAGE & HÉROUARD, 1896, p. 143)]——[In synonymic citations superscript numbers indicate taxonomic rank assigned by authors (^{1}subclass, ^{2}tribu, ^{3}division, ^{4}order, ^{5}suborder, ^{6}group); dagger(†) indicates *partim*——[=^{4}Polythalamacea and ^{4}Polythalamacés DE BLAINVILLE, 1825, p. 375; ^{4}Cellulacea and ^{4}Cellulacés DE BLAINVILLE, 1825, p. 368]——[=^{2}Nantilites LATREILLE, 1825, p. 165; =^{2}Polycyclica LATREILLE, 1825, p. 164; =^{2}Milleporita LATREILLE, 1825, p. 166]——[=^{4}Enallostèguest D'ORBIGNY in DE LA SAGRA, 1839, p. xxxix, 125 (*nom. neg.*); =^{4}Stichostèguest D'ORBIGNY in DE LA SAGRA, 1839, p. xxxvii, 5 (*nom. neg.*); =^{4}Helicostèguest D'ORBIGNY in DE LA SAGRA, 1839, p. xxxviii, 27 (*nom. neg.*); =^{4}Cyclostèguest D'ORBIGNY, 1851, p. 192 (*nom. neg.*); =^{6}Helicoideat SCHULTZE, 1854, p. 53; =Rhabdoidea SCHULTZE, 1854, p. 53; =Nautiloidea REUSS, 1860, p. 151; =Helicostegia REUSS, 1860, p. 151, 205; =Turbinoidea REUSS, 1860, p. 151]——[=^{4}Monostègues D'ORBIGNY in DE LA SAGRA, 1839, p. xxxvii, 1 (*nom. neg.*); =^{5}Monosomatia EHRENBERG, 1839, table opposite p. 120; =^{4}Monostega DIESING, 1848, p. 497; =Monothalamia† SCHULTZE, 1854, p. 52; =^{3}Monothalamiat MARRIOTT, 1878, p. 30; =^{4}Monothalamia HAECKEL, 1894, p. 164; =^{4}Monosomatia COPELAND, 1956, p. 183]——[=Foraminifera Monomera REUSS, 1862, p. 362; =Foraminifera Polymera REUSS, 1862, p. 365; =^{6}Vitrea CARPENTER, 1879, p. 375, 378; =Canaliculata MÖBIUS, 1880, p. 104; =Basistoma SCHUBERT, 1920, p. 148; =Telostoma SCHUBERT, 1920, p. 172; =Schizostoma† SCHUBERT, 1920, p. 179; =^{4}Flexostylidia CALKINS, 1926, p. 355; =Sektion Neohellenoideat WEDEKIND, 1937, p. 72, 84; =^{4}Hellenoidea WEDEKIND, 1937, p. 79; =^{5}Biloculinideat SIGAL in PIVETEAU, 1952, p. 157; =^{5}Pluriloculinideat SIGAL in PIVETEAU, 1952, p. 160]——[=^{5}Perforata CARPENTER, PARKER & JONES, 1862, p. 149; =^{2}Perforata CLAUS, 1872, p. 108; =^{6}Perforata CARPENTER, 1879, p. 375; =^{1}Perforata LANKESTER, 1885, p. 847; =Perforata (Foraminifera) HAECKEL, 1894, p. 164; =^{4}Perforida DELAGE & HÉROUARD, 1896, p. 135; =^{5}Perforina CALKINS, 1901, p. 108; =^{4}Orthostili (Perforata) SILVESTRI, 1937, p. 89]——[=^{4}Dentata HOFKER, 1951, p. 14; =^{5}Protoforaminata HOFKER, 1951, p. 42; =^{5}Biforaminata HOFKER, 1951, p. 306; =^{5}Conorbida HOFKER, 1951, p. 307; =^{5}Deuteroforaminata HOFKER, 1951, p. 412]——[=^{4}Lagenidea LANKESTER, 1885, p. 847; =^{5}Lagenidae DELAGE & HÉROUARD, 1896, p. 136; =^{4}Lagenaceae HARTOG in HARMER & SHIPLEY, 1906, p. 59; =^{4}Lagenida CALKINS, 1909, p. 39]——[=^{4}Nodosalidia CALKINS, 1926, p. 355; =^{4}Nodosaridia KÜHN, 1926, p. 135; =^{4}Nodosarioidea WEDEKIND, 1937, p. 86; =^{5}Cristellariaceat WEDEKIND, 1937, p. 93; =^{5}Lenticulinacea WEDEKIND, 1937, p. 99; =^{5}Polymorphinacea WEDEKIND, 1937, p. 103; =^{5}Robulinacea WEDEKIND, 1937, p. 104]——[=^{4}Buliminida FURSENKO, 1958, p. 24]——[=^{4}Chilostomellidea LANKESTER, 1885, p. 847; =^{5}Chilostomellidae DELAGE & HÉROUARD, 1896, p. 138; =^{4}Cheilostomellaceae HARTOG in HARMER & SHIPLEY, 1906, p. 59; =^{4}Chilostomellida CALKINS, 1909, p. 39]——[=^{4}Rotalidea LANKESTER, 1885, p. 847; =^{5}Rotalidae DELAGE & HÉROUARD, 1896, p. 143; =^{4}Rotaliaceae HARTOG in HARMER & SHIPLEY, 1906, p. 59; =^{4}Rotalida CALKINS, 1909, p. 39; =^{4}Rotaliaridia KÜHN, 1926, p. 152; =^{5}Rotaliacea WEDEKIND, 1937, p. 85, 115; =^{4}Rotaliida FURSENKO, 1958, p. 23]——[=^{4}Globigerinidea LANKESTER, 1885, p. 847; =^{5}Globigerinidae DELAGE & HÉROUARD, 1896, p. 141; =^{4}Globigerinidae HARTOG in HARMER & SHIPLEY, 1906, p. 59; =^{4}Globigerinida CALKINS, 1909, p. 39; =^{4}Heterohelicida FURSENKO, 1958, p. 24]——[=^{4}Nummulinidea LANKESTER, 1885, p. 848; =^{5}Nummulitidae DELAGE & HÉROUARD, 1896, p. 147; =^{5}Nummulitidea LISTER in LANKESTER, 1903, p. 146; =^{4}Nummulitaceae HARTOG in HARMER & SHIPLEY, 1906, p. 59; =^{4}Nummulitida CALKINS, 1909, p. 39; =^{5}Nummulitacea WEDEKIND, 1937, p. 119; =^{4}Nummulitinidea COPELAND, 1956, p. 188]——[=^{5}Tinoporinae CALKINS, 1901, p. 109; =^{4}Textulinida CALKINS, 1926, p. 356]

Wall calcareous, perforate. *Perm.-Rec.*

Superfamily NODOSARIACEA
Ehrenberg, 1838

[*nom. correct.* LOEBLICH & TAPPAN, 1961, p. 295 (*pro* superfamily Nodosariidea NØRVANG, 1957, p. 23, *nom. transl. ex* family Nodosarina EHRENBERG, 1838)]——[In synonymic citations superscript numbers indicate taxonomic rank assigned by authors (^{1}superfamily, ^{2}group, ^{3}family group) and a dagger(†) indicates *partim*]——[=^{2}Lagenidae BÜTSCHLI in BRONN, 1880, p. 196; =Titanostichostegia EIMER & FICKERT, 1899, p. 676; =^{1}Enclinostegia† EIMER & FICKERT, 1899, p. 682 (*nom. nud.*); =^{3}Nodosalidia† RHUMBLER in KÜKENTHAL & KRUMBACH, 1923, p. 86; =^{1}Lagenidea GLAESSNER, 1945, p. 126; =^{1}Lagenicae EASTON, 1960, p. 65, 78]

Wall of finely perforate, radial laminated calcite; chambers planispirally coiled or uncoiled, or straight, or coiled about longitudinal axis; aperture peripheral or terminal, typically radiate, or may be slitlike or rounded. *Perm.-Rec.*

Family NODOSARIIDAE Ehrenberg, 1838

[*nom. correct.* LISTER in LANKESTER, 1903, p. 144 (*pro* family Nodosarina EHRENBERG, 1838, p. 200)]——[All names of family rank, a dagger(†) indicates *partim*]——[=Polystoma† LATREILLE, 1825, p. 161 (*nom. nud.*); =Polythalama† LATREILLE, 1825, p. 161 (*nom. nud.*); =Helicostèguest D'ORBIGNY, 1826, p. 268 (*nom. nud., nom. neg.*); =Stichostèguest D'ORBIGNY, 1826, p. 251 (*nom. nud., nom. neg.*); =Stichostegia† REUSS, 1860, p. 151, 178]——[=Equilateralidae† D'ORBIGNY in DE LA SAGRA, 1839, p. xxxvii, 11 (*nom. nud.*); =Aequilateralidae† D'ORBIGNY, 1846, p. 28]——[=Nautiloidae† D'ORBIGNY in DE LA SAGRA,

(C511)

Protista—Sarcodina

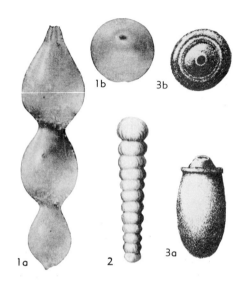

Fig. 400. Nodosariidae (Nodosariinae; *1-3, Nodosaria*) (p. C512).

1839, p. xxxviii, 38 (*nom. nud.*); =Nautiloida† SCHULTZE, 1854, p. 53; =Nautiloidea REUSS, 1860, p. 151 (*nom. nud.*)]
——[=Rhizopodes† DUJARDIN, 1841, p. 126, 240 (*nom. nud., nom. neg.*); =Rhabdoidea† REUSS, 1862, p. 365 (*nom. nud.*); =Rhabdoina† BÜTSCHLI in BRONN, 1880, p. 197 (*nom. nud.*)]——[=Frondicularidae REUSS, 1860, p. 151, 191; =Frondicularideae GÜMBEL, 1870, p. 53]——[=Vaginulinidae REUSS, 1860, p. 151; =Vaginulinideae GÜMBEL, 1868, p. 53]——[=Dentalinoidea SCHWAGER, 1877, p. 18; =Les Lenticulacées LAMARCK, 1809, p. 322 (*nom. neg.*); =Lenticulinidae CHAPMAN, PARR & COLLINS, 1934, p. 554; =Robulinidae WEDEKIND, 1937, p. 104; =Marginulinellidae WEDEKIND, 1937, p. 94; =Marginulinidae WEDEKIND, 1937, p. 99; =Hydromylinidae DE WITT PUYT, 1941, p. 54]——[=Lagenidea REUSS, 1862, p. 305; =Lagenida CARPENTER, PARKER & JONES, 1862, p. 154; =Lagenideae GÜMBEL, 1870, p. 28; =Lagene SCHWAGER, 1876, p. 476; =Lagenoidea SCHWAGER, 1877, p. 18; =Lagenideae SCHULZE, 1877, p. 29; =Lagenina LANKESTER, 1885, p. 847; =Lagenetta HAECKEL, 1894, p. 164; =Lageninae DELAGE & HÉROUARD, 1896, p. 137; =Lagénidos GADEA-BUISÁN, 1947, p. 18 (*nom. neg.*)]——[=Nodosarida SCHULTZE, 1854, p. 53; =Nodosaridae REUSS, 1860, p. 151, 178; =Nodosarideae GÜMBEL, 1870, p. 30; =Nodosarie SCHWAGER, 1876, p. 476; =Nodosaretta HAECKEL, 1894, p. 164; =Nodosarinae DELAGE & HÉROUARD, 1896, p. 137; =Arnodosaridia RHUMBLER, 1913, p. 342 (*nom. van.*); =Nodosariellidae WEDEKIND, 1937, p. 93; =Plectofrondiculariidae MONTANARO GALLITELLI, 1957, p. 143]——[=Orthocerata, and Orthocérés DE BLAINVILLE, 1825, p. 376; =Orthoceratidae† BRODERIP, 1839, p. 321; =Radiolata† CROUCH, 1827, p. 41 (*nom. nud.*); =Radiolididae† BRODERIP, 1839, p. 321; =Orthocerinida SCHMARDA, 1871, p. 165; =Cristaceat & Cristacés DE BLAINVILLE, 1825, p. 383; =Spherulacea and Sphérulacés DE BLAINVILLE, 1825, p. 369]

Test free, one or more chambers in planispiral, biserial, uncoiling, curved or straight series; aperture simple, slitlike or radiate, peripheral in coiled forms, terminal in straight forms, may have apertural chamberlet, or may have elongate neck. *Perm.-Rec.*

Subfamily NODOSARIINAE Ehrenberg, 1838

[*nom. correct.* CHAPMAN, 1900, p. 30 (*pro* subfamily Nodosaridea REUSS, 1862, p. 334, *nom. transl. ex* family Nodosarina EHRENBERG, 1838)]——[All names of subfamily rank] ——[=Vaginulinidea REUSS, 1862, p. 366; =Frondicularidea REUSS, 1862, p. 307, 335, 366, 395; =Dentalinidae SCHWAGER, 1877, p. 18; =Lageninae BRADY, 1881, p. 44; =Nodosarinae BRADY, 1884, p. 69; =Glandulonodosariinae SILVESTRI, 1901, p. 109; =Frondiculariinae GALLOWAY, 1933, p. 235; =Robulinae GALLOWAY, 1933, p. 250; =Lenticulininae CHAPMAN, PARR & COLLINS, 1934, p. 554; =Marginulinae NØRVANG, 1957, p. 83 (*nom. imperf.*); =Lenticulinae NØRVANG, 1957, p. 93 (*nom. imperf.*)]

Test with one or more chambers arranged in straight, arcuate or enrolled series; aperture terminal, rounded or radiate. *Perm.-Rec.*

Nodosaria LAMARCK, 1812, *1087, p. 121 [*Nautilus radicula* LINNÉ, 1758, *1140, p. 711; SD (SM) LAMARCK, 1816, *1089, pl. 465] [=*Orthocera* MODEER in SOLDANI, 1789, *1809, p. 41 (obj.); SD MELVILLE, 1959, *1253, p. 21, *nom. reject.* ICZN pending, see MELVILLE, 1959, *1253; *Orthocera* LAMARCK, 1799, *1083, p. 80 (type, *Nautilus raphanus* LINNÉ, 1758, *1140, p. 711) (*non Orthocera* MODEER, 1789, *nom. reject.* ICZN pending, see MELVILLE, 1959, *1253); *Nodosarina* PARKER & JONES, 1859, *1417a, p. 477 (type, *Nautilus raphanus* LINNÉ, 1758); *Pyramidulina* COSTA in FORNASINI, 1894, *731, p. 224 (type, *Pyramidulina eptagona* COSTA, 1894); *Herrmannia* ANDREAE, 1895, *20, p. 172 (*nom. nud.*); *Nodosariopsis* RZEHAK, 1895, *1605, p. 228 (type, *Nodosaria perforata* SEGUENZA, 1880, *1713, p. 332, SD LOEBLICH & TAPPAN, herein); *Lagena (Cidaria)* GRZYBOWSKI, 1896, *835, p. 267, 292 (type, *Lagena (Cidaria) cidarina* GRZYBOWSKI, 1896, SD LOEBLICH & TAPPAN, herein) (*non Cidaria* TREITSCHKE, 1825); *Glandulonodosaria* SILVESTRI, 1900, *1751, p. 4 (type, *Nodosaria ambigua* NEUGEBOREN, 1856, *1351, p. 71; *Pseudoglandulina* CUSHMAN, 1929, *442, p. 87 (type, *Nautilus comatus* BATSCH, 1791, *102, pl. i, fig. 2a,b; *Nodosariella* WEDEKIND, 1937, *2041, p. 93 (type, *Nautilus raphanus* LINNÉ, 1758)]. Test free, multilocular, rectilinear, rounded in section, sutures distinct and commonly perpendicular to axis of test, surface smooth, costate, striate, hispid or tuberculate; aperture terminal, central basically radiate, may be produced on neck. *Perm.-Rec.*, cosmop.——FIG. 400,*1*. **N. radicula* (LINNÉ), U. Plio., Italy; *1a,b,* side, top views, ×40 (*7).——FIG. 400,*2. N. ambigua* NEUGEBOREN, Mio., Rumania; ×25 (*700).——FIG. 400,*3. N. cidarina* (GRZYBOWSKI), L.Oligo., Pol.; *3a,b,* side, top views, ×45 (*835).

Alfredosilvestris ANDERSEN, 1961, *18, p. 71 [**A. levinsoni*; OD]. Test free, uniserial, chambers of microspheric form and early chambers of megalospheric form arched and compressed with chevron-shaped sutures, as in *Lingulina*, later chambers rounded in section, with straight and horizontal sutures; aperture terminal, radiate. [*Alfredosilvestris* resembles *Lingulina* in the compressed early stage, but differs in having a radiate instead of slitlike aperture.] *Rec.*, USA (La.).——FIG. 401,*8*. **A. levinsoni*; *8a,b,* side, edge views of megalospheric holotype, ×66 (*18).

Foraminiferida—Rotaliina—Nodosariacea

Amphicoryna SCHLUMBERGER in MILNE-EDWARDS, 1881, *1285, p. 881 [*Marginulina falx JONES & PARKER, 1860, *998, p. 302; SD (SM) BRADY, 1884, *200, p. 556, =Nautilus scalaris BATSCH, 1791, *102, p. 1, 4] [=Plesiocorine SCHLUMBERGER in MILNE-EDWARDS, 1882, *1286, p. 31

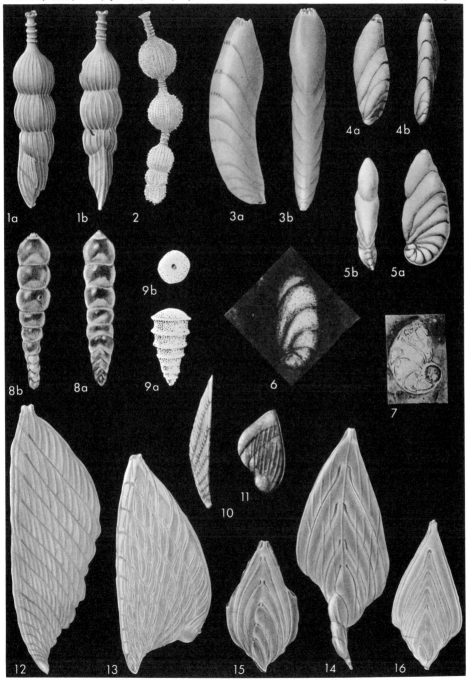

FIG. 401. Nodosariidae (Nodosariinae; 1,2, Amphicoryna; 3-7, Astacolus; 8, Alfredosilvestris; 9, Austrocolomia; 10-13, Citharina; 14-16, Citharinella) (p. C512-C516).

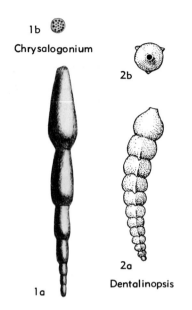

Fig. 402. Nodosariidae (Nodosariinae; *1, Chrysalogonium; 2, Dentalinopsis*) (p. C514, C516).

(type, *P. edwardsi* SCHLUMBERGER, 1882); *Plesiocoryna* SCHLUMBERGER in MILNE-EDWARDS, 1882, *1286, p. 31 *(nom. null.)*; *Amphicoryne* BRADY, 1884, *200, p. 556 *(nom. van.)*; *Amphycorina* DE FOLIN & PERIER, 1887, *727Ab, p. 159 *(nom. null.)*; *Lagenonodosaria* SILVESTRI, 1900, *1751, p. 3 (type, *Nodosaria scalaris* var. *separans* BRADY, 1884, *200, p. 510); *Nodosariopsis* SILVESTRI, 1902, *1755, p. 52 (type, *Marginulina falx* JONES & PARKER, 1860, SD LOEBLICH & TAPPAN, herein (obj.) (*non Nodosariopsis* RZEHAK, 1895; =?*Vaginuloglandulina* SILVESTRI, 1906, *1764, p. 24 (type, *V. laevigata*)]. Test free, elongate, early chambers compressed, in microspheric form arranged in loose coil as in *Astacolus*, later development uniserial; sutures oblique and flush in early stages, later constricted and horizontal; wall smooth or longitudinally costate; aperture terminal, radiate, at end of distinct neck. Mio.-Rec., cosmop.——FIG. 401,1. **A. scalaris* (BATSCH), Rec., Syra Arch., Medit.; *1a,b*, side, dorsal views, ×46 (*2117).——FIG. 401,2. *A. separans* (BRADY), Rec., Pac.; ×17 (*2117).

Astacolus DE MONTFORT, 1808, *1305, p. 262 [**Astacolus crepidulatus* DE MONTFORT, 1808, =*Nautilus crepidulus* FICHTEL & MOLL, 1798, *716, p. 64; OD] [=*Chrysolus* DE MONTFORT, 1808, *1305, p. 26 (obj.); *Periples* DE MONTFORT, 1808, *1305, p. 270 (type, *P. elongatus* DE MONTFORT, 1808); *Crepidulina* DE BLAINVILLE, 1824, *141a, p. 188 (type, *C. astacolus* DE BLAINVILLE, 1824, =*Nautilus crepidulus* FICHTEL & MOLL, 1798, SD LOEBLICH & TAPPAN, herein) (obj.);

Cochlidion ZALESSKY, 1926, *2099, p. 92 (type, *C. alexandrae* ZALESSKY, 1926); *Cochlea* ZALESSKY, 1926, *2099, p. 93 (type, *C. sapracolli* ZALESSKY, 1926, SD LOEBLICH & TAPPAN, herein) (*non Cochlea* DA COSTA, 1778; *nec* MARTYN, 1784; *nec* HITCHCOCK, 1888); *Polymorphinella* CUSHMAN & HANZAWA, 1936, *504, p. 46 (type, *P. vaginulinaeformis* CUSHMAN & HANZAWA, 1936); *Polymorphinoides* CUSHMAN & HANZAWA, 1936, *504, p. 48 (type, *P. spiralis* CUSHMAN & HANZAWA); *Sacculariella* WEDEKIND, 1937, *2041, p. 102 (type, *S. ensis* WEDEKIND, 1937); *Gladiaria* WEDEKIND, 1937, *2041, p. 105 (*nom. nud.*) (*non* WICK, 1939); *Gladiaria* WICK, 1939, *2059, p. 479 (type, *Cristellaria hermanni* ANDREAE, 1896, *21, p. 298); *Gladiaria* THALMANN, 1941, *1897e, p. 652 (type, *Cristellaria decorata* REUSS, 1855, *1544, p. 269) (*non Gladiaria* WICK, 1939); *Enantiovaginulina* MARIE, 1941, *1215, p. 160, 255 (type, *Cristellaria recta* D'ORBIGNY, 1840, *1394, p. 28)]. Test free, elongate, arcuate, compressed; chambers numerous, low, broad, added along slightly curved axis; sutures oblique, highest at outer margin, curved, straight or sinuate; aperture radiate, terminal, at peripheral angle. Perm.-Rec.——FIG. 401,3. **A. crepidulus* (FICHTEL & MOLL), Plio., Italy; *3a,b*, side, face views, ×33 (*2117).——FIG. 401,4. *A. vaginulinaeformis* (CUSHMAN & HANZAWA), Pleist., Ryukyu Is.; *4a,b*, side, dorsal views, ×33 (*504).——FIG. 401,5. *A. spiralis* (CUSHMAN & HANZAWA), Pleist., Ryukyu Is.; *5a,b*, side, face views, ×22 (*504).——FIG. 401,6. *A. alexandrae* (ZALESSKY), Jur., USSR; ×73 (*2099).——FIG. 401,7. *A. sapricolli* ZALESSKY), Jur., USSR; ×73 (*2099).

[*Astacolus* differs from *Vaginulina* in having oblique sutures and a more distinctly curved axis. It differs from *Lenticulina* in having a curved axis, rather than a closely enrolled test, and in later chambers being added so as to touch only the chamber immediately preceding, and in not being involute. Slightly irregular forms have been described as *Enantiovaginulina*, *Polymorphinella*, and *Polymorphinoides*, but as some specimens of most nodosariid genera may show irregular chamber development, this is not regarded as of generic or even specific importance.]

Austrocolomia OBERHAUSER, 1960, *1384, p. 37 [**A. marschalli*; OD]. Similar to *Nodosaria*, but with rounded aperture and no neck; chambers considerably overlapping and in type-species with elevated "sutures"; wall single-layered. U.Trias. (Carn.), Aus.——FIG. 401,9. **A. marschalli*; *9a,b*, side, top views, ×45 (*1384).

Chrysalogonium SCHUBERT, 1907 (separate of 1908), *1687, p. 243 [**Nodosaria polystoma* SCHWAGER, 1866, *1703, p. 217; OD (M)]. Test similar to *Nodosaria* but with series of pores taking place of radial apertural slits of *Nodosaria*. U. Cret.-Rec., Pac.-N. Am.-Eu.-Atl.-Carib.——FIG. 402,1. **C. polystoma* (SCHWAGER), U.Tert., India (Kar Nicobar); *1a,b*, side, top views, ×22.5 (*700).

Citharina D'ORBIGNY in DE LA SAGRA, 1839, *1611, p. xxxvii [**Vaginulina (Citharina) strigillata* REUSS, 1846, *1538, p. 106; SD LOEBLICH & TAP-

PAN, 1949, *1156, p. 259] [=*Cytharina* D'ARCHAIC, 1843, *36, p. 333 (*nom. null.*); *Hybridina* KÜBLER & ZWINGLI, 1866, *1060, p. 8

(type, *H. obliqua* KÜBLER & ZWINGLI, 1866, SD LOEBLICH & TAPPAN, herein); *Pseudovaginulina* WEDEKIND, 1937, *2041, p. 95 (type, *P. oxyacan-*

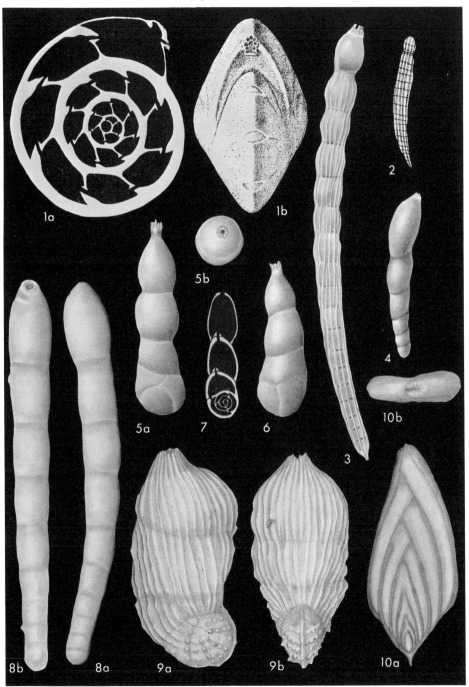

FIG. 403. Nodosariidae (Nodosariinae; *1, Cribrorobulina; 2-4, Dentalina; 5-7, Dimorphina; 8, Dentalinoides; 9, Marginulinopsis; 10, Dyofrondicularia*) (p. C516, C521-C522).

tha WEDEKIND, 1937); *Saccularia* WEDEKIND, 1937, *2041, p. 95 (type, *Marginulina inaequistriata* TERQUEM, 1864, *1885, p. 401); *Pseudocitharina* PAYARD, 1947, *1432, p. 118 (type, *Marginulina colliezi* TERQUEM, 1866, *1886, p. 430)]. Test flattened, subtriangular in outline, may be keeled; chambers numerous, extending nearly to base at inner margin; wall smooth, striate or costate; aperture radiate, at outer margin. *L.Jur.-Paleoc.*, cosmop.——FIG. 401,*10*. **C. strigillata* (REUSS), Cret., Boh.; enlarged (*700).——FIG. 401,*11*. *C. inaequistriata* (TERQUEM), L.Jur. (Lias.), Ger.; ×20 (*92).——FIG. 401,*12*. *C. colliezi* (TERQUEM), L.Jur.(U.Lias.), Fr.; lectotype here designated and refigured (specimen in TERQUEM Coll., Muséum Natl. Hist. Nat., Paris, *1886, pl. 17, fig. 10), ×48 (*2117).——FIG. 401,*13*. *Citharina discors* (KOCH), L.Cret.(Gault), Eng.; ×46 (*2117).

Citharinella MARIE, 1938, *1214, p. 99 [**Flabellina karreri* BERTHELIN, 1880, *133, p. 62; OD]. Test free, flattened, lanceolate to flabelliform, chambers low, broad, uniserial early ones arranged as in *Citharina*, extending backward toward ovate or fusiform proloculus at one side, later chambers chevron-shaped and symmetrical, as in *Frondicularia*; surface may be smooth, costate or striate; aperture terminal, slightly produced, radial. *Jur.-Cret.*, Eu.-N.Am.——FIG. 401,*14*. **C. karreri* (BERTHELIN), L.Cret.(Alb.), Eng.; ×100 (*2117). ——FIG. 401,*15,16*. *C. tarrantensis* (LOEBLICH & TAPPAN), L.Cret.(Alb.), USA(Tex.); *15,16*, megalospheric and microspheric tests, ×44 (*2117).

Cribrorobulina THALMANN, 1947, *1897g, p. 372 [**Robulina serpens* SEGUENZA, 1880, *1713, p. 143; OD] [=*Cribrorobulina* SELLI, 1941, *1716, p. 90 *(nom. nud.)*]. Test like *Lenticulina*, but aperture consisting of numerous small round openings instead of being radiate. *Mio.-Rec.*, Eu. ——FIG. 403,*1*. **C. serpens* (SEGUENZA), L.Plio., Italy; *1a,b*, sec. and idealized apert. view, ×66 (*1716).

Dentalina RISSO, 1826, *1579a, p. 16 [**Nodosaria (Dentaline) cuvieri* D'ORBIGNY, 1826, *1391, p. 255, OD (M)] [=Les Dentalines D'ORBIGNY, 1826, *1391, p. 254 *(nom. neg.)*; *Svenia* BROTZEN, 1937, *238, p. 66 (type, *Nodosaria laevigata* NILSSON, 1826, *1358, p. 342); *Dentalinella* WEDEKIND, 1937, *2041, p. 94 (type, *D. cuneata* WEDEKIND, 1937); *Enantiodentalina* MARIE, 1941, *1215, p. 149, 255 (type, *Nodosaria (Dentaline) communis* D'ORBIGNY, 1826, *1391, p. 254)]. Test elongate, arcuate, uniserial; sutures commonly oblique; aperture radiate, terminal, may be eccentric or nearly central. [Differs from *Nodosaria* in being asymmetrical.] *Perm.-Rec.*, cosmop.——FIG. 403,*2,3*. **D. cuvieri*, Rec., Adriatic (*2*), Rec., Gulf Mex. (*3*); *2*, enlarged (*700); *3*, ×22 (*2117).——FIG. 403,*4*. *D. trujilloi* LOEBLICH & TAPPAN [*nom. nov. pro Dentalina intermedia* REUSS, 1860, *1548, p. 186 (*non Dentalina intermedia* CORNUEL, 1848; *nec* HANTKEN, 1875)], U.Cret.(Cenom.), USA(Tex.); ×48 (*2117).

Dentalinoides MARIE, 1941, *1215, p. 207, 256 [**D. canulina*; OD]. Test elongate, straight or slightly arcuate, uniserial, circular in section; sutures horizontal; wall calcareous, perforate; aperture large, rounded, slightly to one side of center and opening toward concave side of arcuate test. *U.Cret.*, Eu.-N.Am.——FIG. 403,*8*. **D. canulina*, Senon., Fr.; *8a,b*, ×216 (*2117).

[This genus was originally placed in the Ellipsoidinidae (=Pleurostomellidae) because of the eccentric rounded aperture, but that family consists of perforate granular-walled forms with internal siphons between chambers, neither of which have been demonstrated for *Dentalinoides*. It is here placed with the Nodosariidae, differing from *Dentalina* in the rounded, rather than radiate aperture.]

Dentalinopsis REUSS, 1860, *1547, p. 81 [**D. semitriquetra*; OD (M)]. Test free, elongate, uniserial, straight or arcuate, early chambers angled or triangular in section, later rounded; aperture terminal, rounded. *L.Cret.*, cosmop.——FIG. 402,*2*. **D. semitriquetra*, Apt., Ger.; *2a,b*, side, top views, enlarged (*762).

[Placed in the family Buliminidae (subfamily Uvigerininae) by CUSHMAN (*486) and in the Uvigerinidae (subfamily Angulogerininae) by GALLOWAY (*762), the genus is here believed closely related to the Nodosariidae. It cannot be an end member of the above-mentioned subfamilies, as it is found only in the Lower Cretaceous, whereas these subfamilies are largely Cenozoic. The absence of phialine lip and internal tube also indicates that it is not related to these forms. Jurassic species previously placed here should be referred to *Tristix*.].

Dimorphina D'ORBIGNY, 1826, *1391, p. 264 [**D. tuberosa*; OD (M)] [=*Glandulodimorphina* A. SILVESTRI, 1901, *1752, p. 17 (type, *Dimorphina tuberosa* D'ORBIGNY, 1826, SD LOEBLICH & TAPPAN, herein) (obj.)]. Test free, elongate, early portion close-coiled, later uncoiling and uniserial, as in *Marginulina*; aperture terminal, radiate, produced on neck, at the outer margin. [*Dimorphina* differs from *Marginulina* in having an enrolled early stage. Regarded previously as having an initial polymorphine stage (*486), it is now known to be lenticuline in early development (*1717).] *Jur.-Rec.*, cosmop.——FIG. 403,*5-7*. **D. tuberosa*, Plio., Italy *(5,6)*, Adriatic *(7)*; *5a,b*, side, top views; *6*, side view; all ×44 (*2117); *7*, sec. showing early coil, ×24 (*1717).

Dyofrondicularia ASANO, 1936, *46, p. 330 [**D. nipponica*; OD]. Test free, elongate, flattened, early chambers equitant, uniserially arranged, later broad, low chambers biserially arranged; aperture radiate. *Plio.*, Japan.——FIG. 403,*10*. **D. nipponica*; *10a,b*, side, top views of holotype, refigured, ×48 (*2117).

Flabellinella SCHUBERT, 1900, *1680, p. 551 [**Frondicularia tetschensis* MATOUSCHEK, 1895, *1235, p. 143; OD (M)]. Early stage as in *Vaginulina*, later chambers equitant as in *Frondicularia*; aperture radiate. *U.Cret.*, Eu.——FIG. 404,*4*. *F. zitteliana* (EGGER), U.Cret., Bavaria, *4a,b*, side, top views, ×44 (*2117).

Frondicularia DEFRANCE in D'ORBIGNY, 1826, *1391,

Foraminiferida—Rotaliina—Nodosariacea

p. 256 [*Renulina complanata* DEFRANCE, 1824, *141a, p. 178, SD CUSHMAN, 1913, *404c, p. 81] [=*Pleiona* FRANZENAU, 1888, *744, p. 146, 203 (type, *P. princeps* FRANZENAU, 1888) (*non Pleiona* DEYROLLE, 1864; *nec* PAETEL, 1875); *Frondovaginulina* SCHUBERT, 1912, *1691, p. 179 (type,

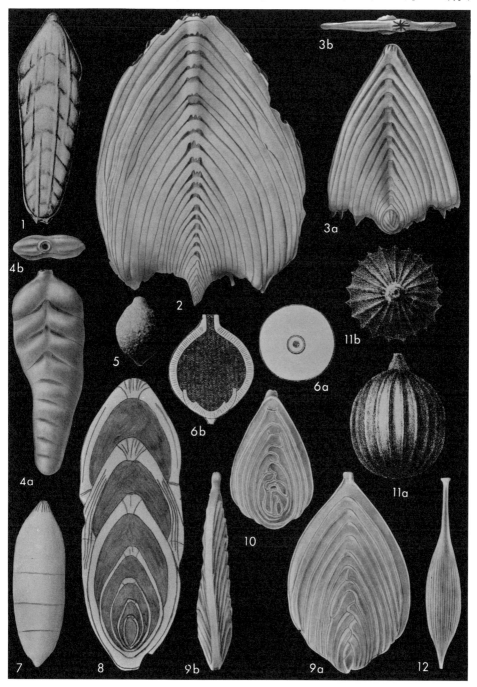

FIG. 404. Nodosariidae (Nodosariinae; *1-3, Frondicularia; 4, Flabellinella; 5,6, Lagenoglandulina; 7,8, Involutaria; 9,10, Kyphopyxa; 11,12, Lagena*) (p. C516-C518).

FIG. 405. Nodosariidae (Nodosariinae; *1, Lankesterina*) (p. C518).

Frondicularia REUSS, 1844, *1537, p. 211, SD LOEBLICH & TAPPAN, herein); *Ichthyolaria* WEDEKIND, 1937, *2041, p. 93 (type, *Frondicularia bicostata* D'ORBIGNY, 1850, *1397b, p. 242); *Pseudofrondicularia* WEDEKIND, 1937, *2041, p. 94 (type, *Frondicularia carinata* BURBACH, 1886, *253a, p. 47); *Annulofrondicularia* KEIJZER, 1945, *1030, p. 196 (type, *Frondicularia annularis* D'ORBIGNY, 1846, *1395, p. 59)]. Test free, elongate or palmate, flattened; chambers low, broad, and equitant; sutures strongly arched or angled at center of test; aperture terminal, radiate, may be produced on short neck. *Perm.-Rec.*——FIG. 404,*1*. *F. bicostata* D'ORBIGNY, L.Jur.(M. Lias.), Fr.; side view of holotype, ×36 (*1198).——FIG. 404,*2,3*. **F. complanata* (DEFRANCE), M.Plio.(Piacenz.), Italy; *2*, microspheric form, ×10; *3a,b*, megalospheric form, ×10 (*2117).

Involutaria GERKE, 1957, *778, p. 33 [**I. triassica*; OD]. Test elongate, chambers uniserially arranged, similar to *Nodosaria* or *Pseudonodosaria*, but with early chambers (wall and chamber cavity) completely overlapping and only few final chambers not enclosing all previous ones; wall calcareous, finely perforate, hyaline, radial; aperture terminal, radiate. *U.Trias.*, USSR(Krasnodar).——FIG. 404,*7,8*. **I. triassica*; *7*, side view of holotype, ×45; *8*, long. sec., ×83 (*778).

Kyphopyxa CUSHMAN, 1929, *440, p. 1 [**Frondicularia christneri* CARSEY, 1926, *282, p. 41; OD]. Test palmate, early chambers citharine in microspheric form, followed by biserial stage which occupies about half of test, final chambers uniserial, equitant, and strongly overlapping, random chambers may even envelop early stage and be cyclical; sutures commonly thickened and elevated; aperture terminal, radiate. *U.Cret.*, N.Am.——FIG. 404,*9*. **K. christneri* (CARSEY), USA (Tex.); *9a,b*, side, edge views of topotype, ×28; *10*, side view, ×28 (*2117).

Lagena WALKER & JACOB in KANMACHER, 1798, *1011, p. 634 [**Serpula (Lagena) sulcata* WALKER & JACOB, 1798; SD PARKER & JONES, 1859, *1417b, p. 337] [=*Serpula (Lagena)* BOYS & WALKER, 1784 (publ. rejected, ICZN Op. 558, 1959); *Vermiculum* MONTAGU, 1803, *1298, p. 517 (type, *V. perlucidum* MONTAGU, 1803); *Lagenula* DE MONTFORT, 1808, *1305, p. 311 (type, *L. floscula* DE MONTFORT, 1808); *Amphorina* D'ORBIGNY, 1849, *1396, p. 666 (type, *A. gracilis* COSTA, 1856, *392, p. 121), *non Lagena gracilis* WILLIAMSON, 1848, =*Amphorina costai* ANDERSEN, 1961, *18, p. 78) (*non Amphorina* DE QUATREFAGES, 1844); *Phialina* COSTA, 1856, *392, p. 122 (type, *P. piriformis* COSTA, 1856, SD LOEBLICH & TAPPAN, herein) (*non Phialina* BORY DE ST. VINCENT, 1827); *Tetragonulina* SEGUENZA, 1862, *1712, p. 53 (type, *T. prima* SEGUENZA, 1862); *Capitellina* MARSSON, 1878, *1228, p. 122 (type, *C. multistriata* MARSSON, 1878); *Ectolagena* SILVESTRI, 1900, *1751, p. 4 (type, *Serpula (Lagena) sulcata* WALKER & JACOB, 1798, SD LOEBLICH & TAPPAN, herein) (obj.); *Procerolagena* PURI, 1954, *1487, p. 104 (type, *Lagena gracilis* WILLIAMSON, 1848, *2064, p. 13)]. Test unilocular, rarely 2 or more chambers; surface variously ornamented; aperture on elongate neck which may have phialine lip, not radiate. [Differences in chamber shape are here regarded as of specific, not generic, value, hence the elongate forms (e.g., "*Amphorina*," "*Procerolagena*," Fig. 404,*12*), are considered congeneric.] *Jur.-Rec.*, cosmop.——FIG. 404,*11*. **L. sulcata* (WALKER & JACOB), Rec., S.Pac.; *11a,b*, side, top views, ×80 (*200).——FIG. 404,*12*. *L. mollis* CUSHMAN, Rec., Baffin Is.; ×102 (*2117).

Lagenoglandulina SILVESTRI, 1923, *1775, p. 12 [**Glandulina subovata* STACHE, 1865, *1825, p. 185; OD (M)]. Test free, subovate, similar to *Pseudonodosaria* in development, but with final chamber completely overlapping earlier uniserial chambers, which are apparent only in section; aperture terminal, rounded. *Eoc.-Rec.*, N.Z.-Eu.-C.Am.——FIG. 404,*5,6*. **L. subovata* (STACHE), Eoc., Italy; *5*, ext., ×15; *6a,b*, outline view of top and long. sec. showing strong overlap of uniserial chambers resulting in unilocular appearance, ×34 (*1775).

Lankesterina LOEBLICH & TAPPAN, 1961, *1181, p. 219 [**Bolivina frondea* CUSHMAN, 1922, *417, p. 126; OD]. Test free, small, symmetrically biserial throughout, with flattened sides and truncate margins; chambers low and broad, as in later stage of *Dyofrondicularia*, but without early uniserial stage; wall calcareous, finely perforate; aperture terminal, radial. *Oligo.*, N.Am.——FIG. 405,*1*. **L. frondea* (CUSHMAN), USA; *1a,b*, side, top views, ×80 (*514).

[Differs from *Polymorphina* in being completely symmetrical throughout and in having truncate margins, similar to the other palmate genera of the Nodosariinae, but differs from these in being biserial throughout. *Polymorphina* is somewhat asymmetrical, particularly in its early development, and may show traces of a sigmoid development.]

Lenticulina LAMARCK, 1804, *1085a, p. 186 [**Lenti-

culites rotulata LAMARCK, 1804; SD CHILDREN, 1823, *337, p. 153] [=*Lenticulites* LAMARCK, 1804, *1085a, p. 187 (obj.); *Phonemus* DE MONT-FORT, 1808, *1305, p. 11 (type, *Nautilus vortex* FICHTEL & MOLL, 1798, *716, p. 33); *Pharamum* DE MONTFORT, 1808, *1305, p. 34 (type, *Nautilus*

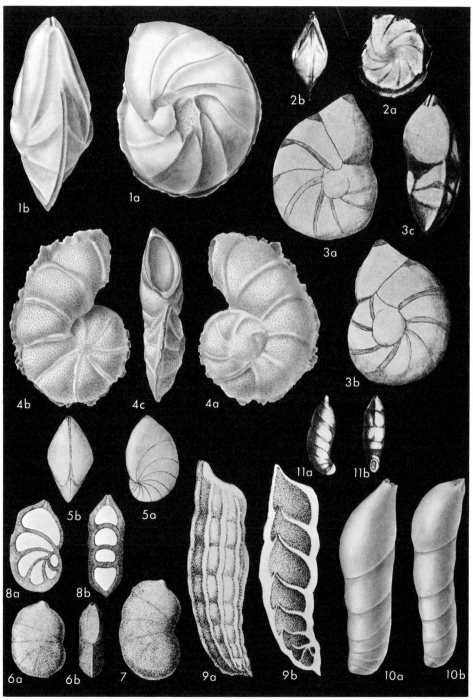

FIG. 406. Nodosariidae (Nodosariinae; *1-8, Lenticulina; 9-11, Marginulina*) (p. C518-C521).

calcar LINNÉ, 1758, *1140, p. 709); *Antenor* DE MONTFORT, 1808, *1305, p. 70 (type, *A. diaphaneus* DE MONTFORT, 1808); *Oreas* DE MONTFORT, 1808, *1305, p. 94 (type, *O. subulatus* DE MONTFORT, 1808, =*Nautilus acutauricularis* FICHTEL & MOLL, 1798, *716, p. 102) (non *Oreas* HUEBNER, 1807); *Robulus* DE MONTFORT, 1808, *1305, p. 214 (type, *R. cultratus* DE MONTFORT, 1808); *Patrocles* DE MONTFORT, 1808, *1305, p. 218 (type, *P. querelans* DE MONTFORT, 1808); *Spincterules* DE MONTFORT, 1808, *1305, p. 222 (type, *Nautilus costatus* FICHTEL & MOLL, 1798, *716, p. 47, non *Nautilus (Orthoceras) costatus* BATSCH, 1791); *Clisiphontes* DE MONTFORT, 1808, *1305, p. 226 (type, *C. calcar* DE MONTFORT, 1808); *Herion* DE MONTFORT, 1808, *1305, p. 231 (type, *H. rostratus* DE MONTFORT, 1808); *Rhinocurus* DE MONTFORT, 1808, *1305, p. 234 (type, *R. araneosus* DE MONTFORT, 1808); *Macrodites* DE MONTFORT, 1808, *1305, p. 238 (type, *M. cucullatus* DE MONTFORT, 1808); *Lampas* DE MONTFORT, 1808, *1305, p. 242 (type, *L. trithemus* DE MONTFORT, 1808) (non *Lampas* MEUSCHEN, 1787); *Scortimus* DE MONTFORT, 1808, *1305, p. 250 (type, *S. navicularis* DE MONTFORT, 1808); *Linthuris* DE MONTFORT, 1808, *1305, p. 254 (type, *L. cassidatus* DE MONTFORT, 1808); *Robulina* D'ORBIGNY, 1826, *1391, p. 282, 283, 287 (type, *Robulus cultratus* DE MONTFORT, 1808); *Soldania* D'ORBIGNY, 1826, *1391, p. 281 (type, *S. carinata* D'ORBIGNY, 1826; SD LOEBLICH & TAPPAN, herein); *Nautilina* COSTA, 1856, *392, p. 370 (type, *N. puteolana* COSTA, 1856) (non *Nautilina* STEIN, 1850); *Clisophontes* SCUDDER, 1882, *1709a, p. 77 (nom. van.); *Linthurus* SHERBORN, 1893, *1731a, p. 181, 182 (nom. van. pro *Linthuris* DE MONTFORT, 1808); *Cristellariopsis* RZEHAK, 1895, *1605, p. 227 (type, *C. punctata* RZEHAK, 1895); *Darbyella* HOWE & WALLACE, 1932, *972, p. 23 (type, *D. danvillensis* HOWE & WALLACE, 1932); *Perisphinctina* WEDEKIND, 1937, *2041, p. 105 (type, *Robulina depauperata* REUSS, 1851, *1541, p. 70) (erroneously cited as *R. pauperata* REUSS, 1851, by THALMANN, 1941, *1897e, p. 658); *Perisphinctina* WICK, 1939, *2059, p. 482 (type, *Cristellaria (Robulina) articulata* REUSS, 1863, *1553, p. 53, non *Cristellaria articulata* TERQUEM, 1862); *Enantiocristellaria* MARIE, 1941, *1215, p. 162, 255 (type, *Cristellaria navicula* D'ORBIGNY, 1840, *1394, p. 27); *Hydromylina* DEWITT PUYT, 1941, *2069, p. 54 (type, *H. rutteni* DEWITT PUYT, 1941); *Rimalina* PÉRÉBASKINE, 1946, *1444, p. 359 (type, *R. pinatensis* PÉRÉBASKINE, 1946); *Eoflabellina* PAYARD, 1947, *1432, p. 101 (type, *Peneroplis d'orbignyi* ROMER, 1839, *1582, p. 47); *Darbyellina* HARRIS & SUTHERLAND, 1954, *882, p. 207 (type, *D. hempsteadensis* HARRIS & SUTHERLAND, 1954)]. Test free, planispiral or rarely slightly trochoid, lenticular, biumbonate, periphery angled or keeled; chambers increasing gradually in size, in general of greater breadth than height; sutures radial, straight or curved and depressed, flush or elevated; surface may be variously ornamented with thickened, elevated sutures, bosses or sutural nodes; aperture radial at peripheral angle. Trias.-Rec., cosmop.——FIG. 406,*1*. *L. rotulata* (LAMARCK), U.Cret.(Senon.), Fr.; *1a,b*, side, face views, ×30 (*2117).——FIG. 406,*2*. *L. cultrata* (DE MONTFORT), L. Plio., Italy; *2a,b*, side, face views, ×27 (*7).——FIG. 406,*3*. *L. danvillensis* (HOWE & WALLACE), U.Eoc.(Jackson.), USA(La.); *3a-c*, opposite sides and face views, ×40 (*972).——FIG. 406,*4*. *L. hempsteadensis* (HARRIS & SUTHERLAND), Paleoc.(Midway.), USA(Ark.); *4a-c*, opposite sides and face view of holotype, refigured, ×47 (*2117).——FIG. 406,*5*. *L. pinatensis* (PÉRÉBASKINE), U.Cret., Fr.; *5a,b*, side, face views, ×33 (*1444).——FIG. 406,*6-8*. *L. punctata* (RZEHAK), L.Tert., Aus.; *6a,b*, side, face views; *7*, side view; *8a,b*, long. secs. in plane of coiling and perpendicular to this plane, showing radial laminated wall characteristic of Nodosariidae, but described by RZEHAK as characterizing *Cristellariopsis*; all ×28 (*1605).

[*Robulus* is regarded as a synonym of *Lenticulina*, as considerable gradation in length of the radial apertural slits may occur. *Darbyella* is merely an abnormal asymmetrical form of *Lenticulina*, and in large assemblages of any species of this genus random asymmetrical, twinned, or even partially uncoiled specimens may be obtained. Similarly *Darbyellina* is represented by an abnormal specimen showing both a slight asymmetrical development and a final chamber which fails to reach the earlier whorl, and thus appears to be uncoiling. These aberrant forms do not warrant distinct generic assignments.]

Marginulina D'ORBIGNY, 1826, *1391, p. 258 [**M. raphanus* D'ORBIGNY, 1826, non *Nautilus raphanus* LINNÉ, 1758; SD DESHAYES, 1830, *590, p. 416 (LOEBLICH & TAPPAN, 1961,*1179, p.77)] [=*Buccinina* COSTA, 1861, *393, p. 53 (type, *B. subrecta* COSTA, 1861, SD LOEBLICH & TAPPAN, herein); *Hemicristellaria* STACHE, 1865, *1825, p. 222 (type, *H. procera* STACHE, 1865); *Ellipsomarginulina* A. SILVESTRI, 1923, *1774, p. 265 (type, *Marginulina raphanus* D'ORBIGNY, 1826, *1391, p. 258, SD LOEBLICH & TAPPAN, herein) (obj.); *Marginulinella* WEDEKIND, 1937, *2041, p. 94 (type, *Nautilus (Orthoceras) costatus* BATSCH, 1791, *102, pl. i, fig. 1a-g); *Enantiomarginulina* MARIE, 1941, *1215, p. 163, 255 (type, *E. d'orbignyi* MARIE, 1941); *Enantioamphicoryna* MARIE, 1956, *1221, p. B243 (type, *E. obesa* MARIE, 1956)]. Early portion slightly coiled but not completely enrolled, as in *Marginulinopsis*, later rectilinear; sutures oblique, especially in early portion; aperture of dorsal angle, somewhat produced. Trias.-Rec., cosmop.——FIG. 406,*9*. **M. raphanus*; *9a,b*, side view and long. sec., enlarged (*1391).——FIG. 406,*10*. *M. glabra* D'ORBIGNY, Plio., Italy; *10a,b*, side views, ×60 (*2117).——FIG. 406,*11*. *M. procera* (STACHE), L.Tert., N.Z.; *11a,b*, ×13 (*700).

[*Marginulina glabra* D'ORBIGNY, 1826, was cited as type of the genus by CUSHMAN (1913, *404c, p. 79), despite the fact that the type had previously been fixed by DESHAYES (1830, *590, p. 416) as *Nautilus raphanus* LINNÉ (=*Mar-*

ginulina raphanus D'ORBIGNY, *590, p. 418). The status of the genus as based on the type-species is discussed by LOEBLICH & TAPPAN (1961, *1179).]

Marginulinopsis A. SILVESTRI, 1904, *1760, p. 253

[*M. densicostata* THALMANN, 1937; SD THALMANN, 1937, *1899a, p. 348]. Test with early stage as in *Lenticulina,* later uncoiling and rec-

FIG. 407. Nodosariidae (Nodosariinae; *1,2, Tentifrons; 3-5, Palmula; 6, Neoflabellina; 7,8, Planularia; 9, Orthomorphina*) (p. C522, C524).

tilinear as in *Marginulina;* aperture terminal, radiate. [*Marginulinopsis* is similar to *Dimorphina,* but differs in having a keeled or angular periphery in the coiled portion.] *Jur.-Rec.,* cosmop.——FIG. 403,9. *M. densicostata* THALMANN, Rec., Challenger Sta. 24, off Culebra Is., W.Indies, 390 fathoms; holotype (BMNH-ZF 1808) refigured, originally described as *Marginulina costata* BATSCH by BRADY (*200, pl. 65, fig. 11), 9a,b, ×50 (*2117).

Neoflabellina BARTENSTEIN, 1948, *90, p. 122 [*Flabellina rugosa* D'ORBIGNY, 1840, *1394, p. 23; SD CUSHMAN, *433, p. 189] [=*Flabellina* D'ORBIGNY in DE LA SAGRA, 1839, *1611, p. 42 (obj.) (*non* VOIGHT, 1834; *nec* FORBES & HANLEY, 1851; *nec* DE GREGORIO, 1930)]. Test large, palmate, similar to *Palmula* but with flattened, parallel sides, and angular or keeled margins, thickened and elevated sutures, surface commonly highly ornamented with ribs, reticulations or nodes. *U.Cret.-Paleoc.,* cosmop.——FIG. 407,6. *N. rugosa* (D'ORBIGNY), U.Cret., Fr.; lectotype (MNHN) here designated and refigured, ×48 (*2117).

Orthomorphina STAINFORTH, 1952, *1833, p. 8 [*Nodogenerina havanensis* CUSHMAN & BERMUDEZ, 1937, *491, p. 14; OD]. Test rectilinear, uniserial; chambers inflated; wall calcareous, perforate, surface smooth or costate; aperture terminal, rounded, and may have slight neck or everted rim. [Differs from *Nodosaria* in having rounded, rather than radiate, aperture, and from *Siphonodosaria* in lacking apertural tooth. Originally placed in the Heterohelicidae, this form seems to have no relation to those planktonic genera and is here transferred to the Nodosariidae.] *Eoc.-Rec.,* Carib.-Eu.-N. Am.-Pac.-Asia-Atl.——FIG. 407,9. *O. havanensis* (CUSHMAN & BERMUDEZ), Eoc., Cuba; 9a,b, side, top views of paratype, ×44 (*2117).

Palmula LEA, 1833, *1099, p. 219 [*P. sagittaria;* OD (M)] [=*Planularia* NILSSON, 1826, *1358, p. 342 (type, *P. elliptica* NILSSON, 1826, SD LOEBLICH & TAPPAN, herein) (*non Planularia* DEFRANCE, 1826); *Frondiculina* VON MÜNSTER in ROEMER, 1838 (*non* LAMARCK, 1816), *1581, p. 382 (type, *F. obliqua* VON MÜNSTER, 1838, SD LOEBLICH & TAPPAN, herein); *Falsopalmula* BARTENSTEIN, 1948, *90, p. 124, 127 (type, *Flabellina tenuistriata* FRANKE, 1936, *741, p. 93); *Phalsopalmula* AGALAROVA, 1960, *3A, p. 79 (*nom. van.*)]. Test free, flattened, elongate or palmate, early portion planispirally coiled in microspheric forms, or arcuate in megalospheric forms, later becoming uncoiled and rectilinear, with low, broad, arched, and equitant chambers, as in *Frondicularia;* sutures radial in early portion, later strongly arched or angled at center of test; aperture terminal, radiate. [Certain of the geologically older species were separated as *Falsopalmula,* being somewhat smaller, and considered to be more closely related to ancestral *Lenticulina.* The differences are here regarded as specific only, as early forms of most nodosariid genera show their close interrelationship.] *L.Jur.-Rec.,* N.Am.-Eu.——FIG. 407,3,4. *P. sagittaria,* Paleoc., USA (N.J.); 4a,b, side and top views, ×5; 3, early portion of test partially acid-treated to show coil, ×22 (*2117).——FIG. 407,5. *P. tenuistriata* (FRANKE), L.Jur.(U.Lias.), Ger.; 5a,b, side, top views of topotype, ×65 (*2117).

Pandaglandulina LOEBLICH & TAPPAN, 1955, *1167, p. 7 [*P. dinapolii;* OD] [=*Pandoglandulina* GERKE, 1957, *778, p. 36 (*nom. null.*)]. Test free, uniserial, chambers strongly overlapping, and with slightly arcuate axis; sutures very slightly radiate in early portion, later horizontal, may be slightly depressed; aperture terminal, radiate. *Mio.-Rec.,* Eu.——FIG. 408,1,2. *P. dinapolii,* L.Plio., Italy; *1,* paratype; 2a,b, side, top views of holotype; all ×45 (*2117).

Planularia DEFRANCE in DE BLAINVILLE, 1826, *141c, p. 244 (*non* NILSSON, 1826) [*Peneroplis auris* DEFRANCE in DE BLAINVILLE, 1824, *141a, p. 178; OD (M)] [=*Megathyra* EHRENBERG, 1843, *672, p. 409 (type, *M. planularia,* SD LOEBLICH & TAPPAN, herein)]. Similar to *Astacolus,* but with compressed sides and carinate margins. *Mio.-Rec.,* cosmop.——FIG. 407,7,8. *P. auris* (DEFRANCE), Plio., Italy; 7,8a, side views; 8b, edge view; all ×33 (*2117).

Pseudarcella SPANDEL, 1909, *1823, p. 199 [*P. rhumbleri;* OD] [=*Arpseudarcelloum* RHUMBLER, 1913, *1572b, p. 349 (*nom. van.*) (obj.)]. Test free, consisting of single conical or plano-convex chamber; wall calcareous, finely perforate, lamellar character and microstructure unknown, surface smooth or reticulate; aperture a large round opening in center of flat to concave surface of test. [The systematic position is doubtful. Because of the calcareous wall it is not considered to be related to the pseudochitinous Arcellidae. Petrographic and X-ray studies of the test wall are needed to aid in its placement, but none have been made to date. At least a superficial similarity to the tests of the Nodosariidae has been noted, and as all known perforate calcareous unilocular hyaline foraminifers are currently placed in the Nodosariacea, the present genus is also tentatively included.] *Eoc.-Mio.,* Eu.(Fr.-Ger.-Belg.-Italy)-Carib.(Puerto Rico).——FIG. 409,1. *P. rhumbleri,* M.Oligo., Ger.; 1a-c, side and apert. views and axial sec., approx. ×55 (*1823).——FIG. 409, 2. *P. jeugueuri* Y. LE CALVEZ, Eoc., Belg.; 2a,b, oblique side and apert. views, ×90 (*1115).——FIG. 409,3. *P. campanula* Y. LE CALVEZ, Eoc., Belg.; 3a,b, oblique side and apert. views, ×84 (*1115).——FIG. 409,4. *P. patella* GALLOWAY & HEMINWAY, U.Oligo., Carib.(Puerto Rico); 4a,b, side and apert. views, ×56 (*764).

Pseudonodosaria BOOMGAART, 1949, *173, p. 81 [*Glandulina discreta* REUSS, 1850, *1540, p. 366;

Foraminiferida—Rotaliina—Nodosariacea C523

OD] [=*Rectoglandulina* LOEBLICH & TAPPAN, 1955, *1167, p. 3 (type, *R. appressa* LOEBLICH & TAPPAN, 1955)]. Test free, uniserial and rectilinear throughout, chambers embracing strongly, at least in early portion, later chambers may be inflated and less embracing; sutures horizontal;

FIG. 408. Nodosariidae (Nodosariinae; *1,2, Pandaglandulina; 3-7, Pseudonodosaria; 8, Tribrachia; 9,10, Pseudotristix, 11,12, Saracenaria*) (p. C522-C524).

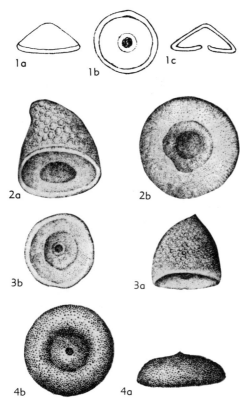

Fig. 409. Nodosariidae (Nodosariinae; *1-4*, *Pseudarcella*) (p. C522).

aperture terminal, radiate. *Perm.-Rec.*, Eu.-N.Am.-Australia-Asia-Pac.-Atl.——Fig. 408,*3,4*. **P. discreta* (Reuss), U.Tert., Java; *3a,b*, side, top views, ×62; *4*, side view, ×53 (*2117).——Fig. 408, *5,6*. *P. obesa* (Loeblich & Tappan), U.Cret., USA(Ark.); microspheric and megalospheric specimens, ×116 (*2117).——Fig. 408,*7*. *P. appressa* (Loeblich & Tappan), U.Cret., USA(Ark.); *7a,b*, side, top views, ×107 (*2117).

Pseudotristix K. V. Miklukho-Maklay, 1960, *1279, p. 156 [**Tristix (P.) techerdynzevi*; OD] [=*Pseudotristix* K. V. Miklukho-Maklay, 1958, *1278, p. 481, 484 *(nom. nud.)*; *Tristix (P.)* K. V. Miklukho-Maklay, 1960, *1279, p. 156 (obj.)]. Test uniserial, chambers low, gradually enlarging, trilobate in section, not overlapping; sutures straight, horizontal; wall calcareous; aperture terminal, radiate. *U.Perm.(Kazan.)*, Russian Platform.——Fig. 408,*9,10*. **P. tcherdynzevi*; *9a,b*, side, apert. views, ×66; *10*, side view, ×71 (*1279).

Saracenaria Defrance in de Blainville, 1824, *141a, p. 176 [**S. italica* Defrance, 1824; OD (M)] [=*Hemirobulina* Stache, 1865, *1825, p. 227 (type, *H. arcuatula* Stache, 1865); *Saracenella* Franke, 1936, *741, p. 87 (type, *Marginulina trigona* Terquem, 1866, *1886, p. 435)]. Test free, planispiral in early stage, later with tendency to uncoil; triangular in section, with broad flat apertural face, the outer margin and 2 angles of face may be acute and keeled to somewhat rounded; aperture at peripheral angle, radiate. *Jur.-Rec.*, cosmop.——Fig. 408,*11*. **S. italica*, Rec., Carib.; *11a,b*, side, face views, ×35 (*2117).——Fig. 408,*12*. *S.* sp., Rec., Gulf Mex.; ×44 (*2117).

Tentifrons Loeblich & Tappan, 1957, *1172, p. 225 [**T. barnardi* Loeblich & Tappan, 1957; OD]. Test free in early stages, with chambers in citharine arrangement, loosely coiled, becoming uniserial, flattened and palmate, with smooth and centrally excavated chevron-shaped chambers; attached in later stages with equitant chambers slightly inflated, extremely papillose and fistulose; sutures raised and thickened in early portion, slightly depressed in irregular attached portion; aperture as in *Citharinella* in early stages, later stage with numerous apertures at ends of fistulose extensions. *U.Cret.*, Eu.——Fig. 407,*1,2*. **T. barnardi*, Senon., Eng., *1*, paratype, ×57; *2*, holotype, ×21 (*1172).

Tribrachia Schubert, 1912, *1691, p. 183 [**T. inelegans* Loeblich & Tappan, 1950; SD Loeblich & Tappan, 1950, *1157, p. 15]. Test free, elongate, tapering, chambers triangular to trifoliate in section, low, broad, extending backward toward proloculus at angles, strongly arched upward on concave faces of test; sutures distinct, strongly arched on sides of test, curving downward at angles; aperture terminal, radiate, may be produced on neck. *M.Jur.-Cret.*, N.Am.-Eu.——Fig. 408,*8*. **T. inelegans*, M.Jur.(Callov.), Wyo.; *8a,b*, side, top views, ×48 (*1157).

Vaginulina d'Orbigny, 1826, *1391, p. 257 [**Nautilus legumen* Linné, 1758, *1140, p. 711; SD Cushman, 1913, *404c, p. 80] [=*Vaginulinella* Kaptarenko-Chernousova, 1956, *1017, p. 68 *(nom. null. pro Vaginulina)*; *Vaginula* Risso, 1826, *1579a, p. 16 (obj.)]. Test straight to arcuate as in *Dentalina*, but compressed or ovate in section; aperture at dorsal angle, radiate. *Trias.-Rec.*, cosmop.——Fig. 410,*1,2*. **V. legumen* (Linné), Rec., Adriatic; *1a,b*, side, edge views; *2*, side view; all ×15 (*2117).

Vaginulinopsis Silvestri, 1904, *1760, p. 251 [**Vaginulina soluta* Silvestri var. *carinata* Silvestri, 1898, *1750, p. 166; =*Vaginulinopsis inversa* (Costa) var. *carinata* (Silvestri), 1904, =*Vaginulinopsis carinata* (Silvestri); SD Thalmann, 1937, *1899a, p. 347]. Test close-coiled, as in *Lenticulina*, in early stage, later uncoiling, slightly compressed as in *Vaginulina*, aperture at dorsal angle, radiate. [The type-species was not fixed by original designation, as was erroneously stated by Thalmann (1937, *1899a, p. 347). *Vaginulina soluta* Silvestri, 1898, was stated to

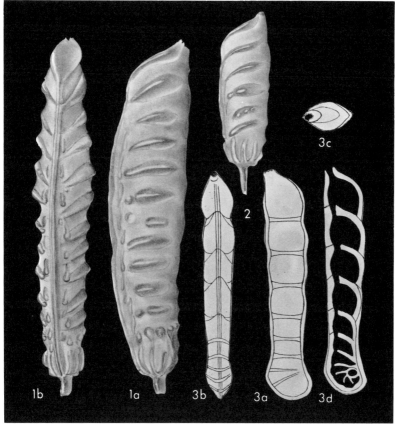

Fig. 410. Nodosariidae (Nodosariinae; *1,2, Vaginulina; 3, Vaginulinopsis*) (p. C524-C525).

be a synonym of *Marginulina inversa* Costa (1856, *392, p. 183) and the variety *carinata* was transferred to *Vaginulinopsis inversa* (Costa) by Silvestri, 1904. *M. inversa* was a homonym of *M. inversa* Neugeboren, 1851, hence the specific name *soluta* should be retained for the noncarinate species described by Silvestri. If regarded solely as a subspecies, the present typespecies should be referred to *V. soluta* subsp. *carinata*. However, as the types of *V. soluta* do not show the early coil, we regard the present form as a distinct species]. *Trias.-Rec.,* cosmop.——Fig. 410,*3*. *V. carinata*, Mio., Sicily; *3a-d,* side, edge, and top views and long. sec., ×30 (*1899a).

Subfamily PLECTOFRONDICULARIINAE Cushman, 1927

[Plectofrondiculariinae Chapman & Parr, 1936, p. 143 (*nom. correct.* pro Plectofrondicularinae Cushman, 1927, p. 62)]

Test biserial to uniserial; aperture terminal, dentate or cribrate. *Eoc.-Rec.*

Plectofrondicularia Liebus, 1902, *1134, p. 76 [*P. concava*; SD Cushman, 1928, *439, p. 238] [=*Parafrondicularia* Asano, 1938, *49, p. 187, 189 (type, *P. japonica*)]. Test elongate, compressed, biserial in early stage, later uniserial, sutures limbate; aperture terminal, radial with elevated margin at outer edge, projecting laminae between grooves of aperture may fuse centrally, as in *Amphimorphina*, so that aperture consists of one or more small, irregularly distributed, elliptical openings. [As shown by Montanaro Gallitelli (1957, *1303, p. 144), this genus does not have an early coiled stage and no internal apertural modifications and is not related to the Heterohelicidae or Buliminidae.] *Eoc.-Rec.*, Eu.-N. Am.-N. Z.-Japan-S. Am.-Carib.-Sumatra-Cyprus.——Fig. 411,*1*. *P. floridana* Cushman, U.Oligo., Dominican Republic; *1a,b,* side, top views of microspheric form, ×65 (*1303).——Fig. 411,*2*. *P. concava*, Tert., Ger.; *2a-d,* side and edge views, long. and transv. secs., ×44 (*1134).——Fig. 411,*3*. *P. japonica* (Asano), Plio., Japan; *3a,b,* side, top views of holotype, ×48 (*2117).

Amphimorphina Neugeboren, 1850, *1349, p. 125 [*A. haueriana*; OD (M)] [=*Amphimorphinella* Keijzer, 1953, *1031, p. 274 (type, *A. butonensis*)]. Test elongate, early stage may be compressed, uniserial in megalospheric form, with

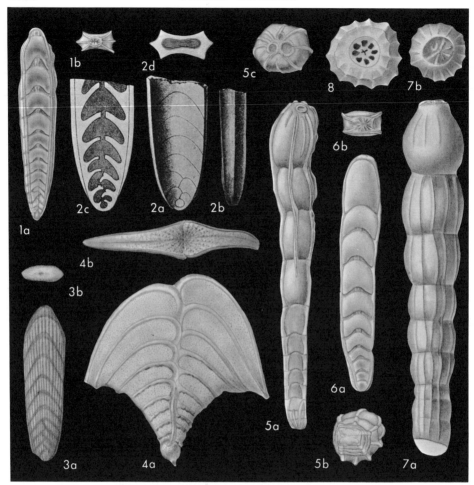

Fig. 411. Nodosariidae (Plectofrondiculariinae; *1-3, Plectofrondicularia; 4, Bolivinella; 5-8, Amphimorphina*) (p. C525-C528).

6 to 10 biserially arranged chambers in microspheric form, chambers equitant in early stage, then may be inflated; aperture in early stage radial, ribs between radial grooves converging in later growth to meet centrally, leaving 3 to 6 pores open between strong radial costae, forming cribrate aperture; apertural chamberlet may be present, as in other Nodosariidae. *M.Eoc.-Rec.,* Eu.-N.Am.-Carib.——Fig. 411,*5,6*. **A. haueriana,* Mio., Hung. *(5),* L.Mio., Fr. *(6); 5a-c,* side, basal and top views, ×52; *6a,b,* side and apert. views of megalospheric form, ×74 (*1303).——Fig. 411,*7,8*. *A. butonensis* (KEIJZER), Mio.-Plio., Malay Arch.; *7a,b,* side, top views of holotype; *8,* top view of broken paratype showing intercameral openings; all ×47 (*1031).

Bolivinella CUSHMAN, 1927, *428, p. 79 [**Textularia folium* PARKER & JONES, 1865, *1418, p. 370, 420; OD]. Test compressed, flabelliform, biserial throughout, with no trace of coiling present; chambers broad, low, sutures may be limbate; aperture indistinct, but apparently basal and cribrate, apertural face obscured by numerous papillae commonly aligned in series radiating from apertural area. *Eoc.-Rec.,* Australia-Carib.-N.Am.-Eu.-Pac.——Fig. 411,*4*. **B. folia* (PARKER & JONES), Rec., Fiji; *4a,b,* side and apert. views, ×130 (*1303).

[*Bolivinella* was placed by GALLOWAY (*762) and CUSHMAN (*486) near *Bolivinitella* in the Bolivinitinae. SIGAL in PIVETEAU (*1458) placed it in the Heterohelicidae (superfamily Buliminidea). POKORNÝ (*1478) assigned it to the superfamily Buliminidea but in the subfamily Plectofrondiculariinae, which MONTANARO GALLITELLI (*1303) elevated to family rank. The genus is here transferred to the Nodosariidae, since no trace of internal apertural modifications are seen, for example, internal tubes or tooth plates such as are characteristic of the Buliminacea. According to MONTANARO GALLITELLI (1957, *1303, p. 144), "the aperture in some specimens seems to consist of a cribrose lamina, with four or six minute openings and is covered by numerous papillae, sometimes hirsute and aligned in radiating rows. . . . An open elongate aperture, as described by Cushman and figured by Parker and Jones is only visible when the specimen is damaged. . . ." The original types of the genotype species in the PARKER & JONES collection in the British Museum (Natural History) were

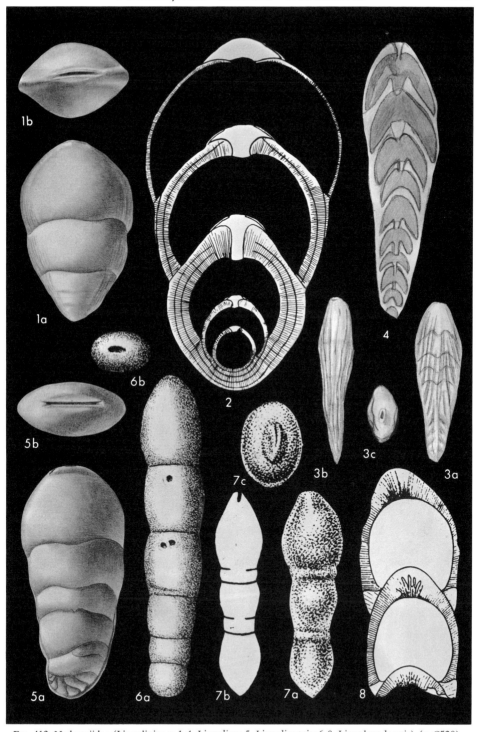

FIG. 412. Nodosariidae (Lingulininae; *1-4, Lingulina; 5, Lingulinopsis; 6-8, Lingulonodosaria*) (p. C528).

isolated by us in 1953. A lectotype was selected (BMNH-ZF3595, ex 94.4.3.1521) and paratypes isolated (BMNH-ZF3594), all from beach sand, Melbourne, Australia.]

Subfamily LINGULININAE Loeblich & Tappan, 1961

[Lingulininae LOEBLICH & TAPPAN, 1961, p. 298]

Test multilocular, chambers arranged in straight or arcuate series; aperture terminal, single elongate slit. *Perm.-Rec.*

Lingulina D'ORBIGNY, 1826, *1391, p. 256 [**L. carinata*; SD CUSHMAN, 1913, *404c, p. 61] [=*Frondicularia (Frondiculina)* GERKE, 1957, *778, p. 43 (type, *F. (F.) dubiella*) (*non Frondiculina* LAMARCK, 1816; *nec* MUENSTER, 1835); *Frondiculinita* GERKE, 1961, *782, p. 74 (*nom. nov.* pro *Frondicularia (Frondiculina)* GERKE, 1957)]. Test free, elongate, uniserial and compressed, with succeeding chambers strongly overlapping, as in *Pseudonodosaria*; aperture an elongate terminal slit in plane of compression. *Perm.-Rec.*, cosmop.——FIG. 412,*1,2*. **L. carinata*, Rec., Carib. (*1*), Rec., Sicily (*2*); *1a,b*, side, top views, ×15 (*2117); *2*, long. sec., ×48 (*700).——FIG. 412,*3,4*. *L. dubiella* (GERKE), L.Jur.(M.Lias.), USSR; *3a-c*, side, edge, and top views, ×68; *4*, sec., ×124 (*778).

Berthelinella LOEBLICH & TAPPAN, 1957, *1172, p. 225 [**Frondicularia paradoxa* BERTHELIN, 1879, *132, p. 33; OD]. Test free, elongate palmate, flattened; proloculus followed by reduced biserial stage of 1 or 2 pairs of chambers, later chambers uniserial and equitant; aperture slitlike. [*Berthelinella* resembles *Plectofrondicularia* in chamber arrangement and *Lingulina* in the slitlike aperture.] *Jur.*, Fr.-Alaska.——FIG. 413,*1,2*. **B. paradoxa* (BERTHELIN), L.Jur.(L.Pliensbach.), Fr.; *1*, side view; *2a,b*, side and top views; all ×137 (*2117).

Daucinoides DE KLASZ & RÉRAT, 1962, *1043, p. 181 [**D. circumtegens*; OD]. Test uniserial, subcircular in section, elongate proloculus followed by completely enveloping uniserial chambers, each succeeding one enclosing all previously formed; wall calcareous, finely perforate, microstructure not described, surface may be finely striate; aperture terminal, commonly a rectilinear slit or more rarely irregular in form. *L.Mio.*, W.Afr.(Gabon-Cameroon-Nigeria).——FIG. 413,*3,4*. **D. circumtegens*, Cameroon; *3a*, side view of holotype; *3b-d*, apert. views of different specimens; *4*, median sec. showing overlapping chambers; all ×27 (*1043).

[Originally placed in the Ellipsoidinidae (=Pleurostomellidae), the genus is here classed in the Lingulininae of the family Nodosariidae, because of the absence of an internal tube connecting successive apertures. As topotype specimens examined by us are pyritized, no evidence as to the wall structure is available. *Daucinoides* is similar to *Involutaria*, but differs in having a slitlike, rather than radial, aperture.]

Ellipsocristellaria SILVESTRI, 1920, *1773, p. 57 [**Lingulinopsis sequana* BERTHELIN, 1880, *133, p. 63; OD (M)]. Test enrolled as in *Lenticulina*, but with slitlike terminal aperture, as in *Lingulina*. *L.Cret.*, Fr.——FIG. 413,*5*. **E. sequana* (BERTHELIN), Alb., Fr.; *5a-c*, side, edge, and top views, ×80 (*133).

Gonatosphaera GUPPY, 1894, *843, p. 651 [**G. prolata*; OD] [=*Linguloglandulina* SILVESTRI, 1903, *1756, p. 49 (type, *L. laevigata* SILVESTRI, 1903)]. Test free, uniserial, with strongly overlapping chambers, chambers circular in section, but with bilaterality shown in some species by development of marginal keel which extends from proloculus up sides of test to merge into apertural lips at apex of test; aperture a terminal, elongate, narrow slit, with distinctly projecting apertural lips or flanges which pass laterally into marginal keel when present. [Differs from *Lingulina* in being rounded in section, rather than compressed. Placed in the Pleurostomellinae by CUSHMAN (*431), it differs in having a perforate radial wall and a symmetrical aperture.] *Eoc.-Mio.*, Carib.-S.Am.-Eu.——FIG. 413,*7*. **G. prolata*, Mio., Trinidad; *7a-c*, side, edge, and top views, ×40 (*2117).——FIG. 413,*8*. *G. laevigata* (SILVESTRI), Rec., Sicily; *8a,b*, top view and long. sec., ×29, ×32 (*700).

Lingulinopsis REUSS, 1860, *1545, p. 23 [**Lingulina bohemica* REUSS, 1846, *1538, p. 108; OD (M)]. Early stage enrolled as in *Lenticulina*, later uniserial as in *Lingulina*, compressed to slightly ovate in section; aperture a single terminal elongate slit in plane of compression. *U.Cret.-Rec.*, Eu.-S.Pac.——FIG. 412,*5*. *L. carlofortensis* BORNEMANN, Rec., Ki Is.; *5a,b*, side, top views, ×20 (*2117).

Lingulonodosaria SILVESTRI, 1903, *1756, p. 48 [**Lingulina nodosaria* REUSS, 1863, *1554, p. 59; SD GALLOWAY, 1933, *762, p.252] [=*Lingulinella* GERKE, 1952, *777, fide GERKE, 1960, *780 (type, *L. arctica*)]. Test elongate, uniserial, ovate in section, with very little overlap of chambers; aperture a terminal slit; differs from *Lingulina* as *Nodosaria* does from *Pseudonodosaria*. *L.Perm.-L.Cret.*, Eu.-N.Am.-Sib.——FIG. 412,*6*. **L. nodosaria* (REUSS), L.Cret.(Gault), Eng.; *6a,b*, side, top views, ×120 (*311).——FIG. 412,*7,8*. *L. arctica* (GERKE), Perm., Sib.; *7a,b*, side, edge views; *7c*, apert. view of holotype; all ×100; *8*, long. sec., ×132 (*780).

Mucronina EHRENBERG, 1839, *667, table opposite p. 120 [**Nodosaria (Mucronine) hasta* D'ORBIGNY, 1826, *1391, p. 256; SD (SM) PARKER, JONES & BRADY, 1865, *1419, p. 27] [=*Les Mucronines* D'ORBIGNY, 1826, *1391, p. 256 (*nom. neg.*); *Nodosaria (Mucronina)* PARKER, JONES & BRADY, 1865, *1419, p. 27; *Staffia* SCHUBERT, 1911, *1689b, p. 78 (type, *Nodosaria tetragona* COSTA, 1855, *391, p. 116); *Nodomorphina* CUSHMAN, 1927, *428, p. 80 (type, *Nodosaria compressiuscula* NEUGEBOREN, 1852, *1350, p. 59)]. Test elongate, narrow, uniserial, strongly carinate margins, later chambers becoming increasingly compressed;

aperture a terminal slit; differs from *Lingulonodosaria* in compressed sides and keeled margins. *Mio.-Rec.*, Eu.——FIG. 414,*1*. *M. tetragona* (COSTA), Plio., Italy; *1a*, side view; *1b-f*, secs. of successive test stages, ×15 (*1899a).

Rimulina D'ORBIGNY, 1826, *1391, p. 257 [*R.

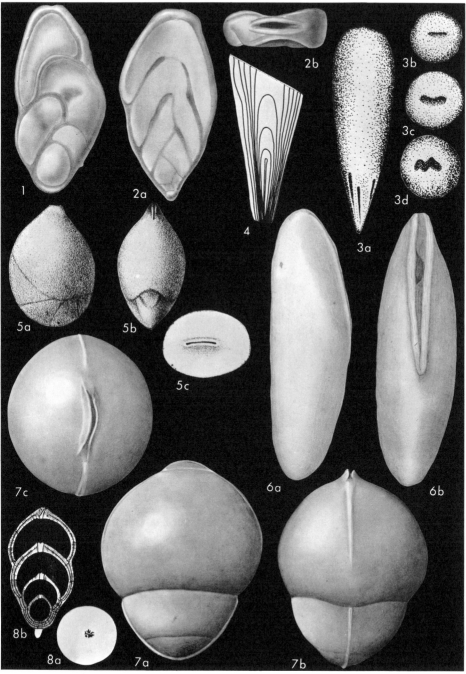

FIG. 413. Nodosariidae (Lingulininae; *1,2, Berthelinella; 3,4, Daucinoides; 5, Ellipsocristellaria; 6, Rimulina; 7,8, Gonatosphaera*) (p. C528-C530).

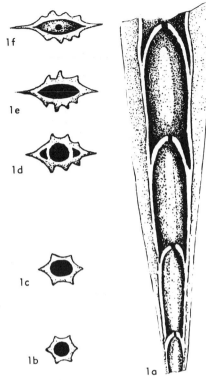

Fig. 414. Nodosariidae (Lingulininae; *1, Mucronina*) (p. C528-C529).

glabra; OD (M)]. Test elongate single chamber with elongate slit aperture extending from apex about half length of one edge. *Rec.*, Adriatic.——Fig. 413,6. **R. glabra; 6a,b,* side, edge views of holotype, refigured, ×77 (*2117).

Family POLYMORPHINIDAE d'Orbigny, 1839

[Polymorphinidae D'ORBIGNY in DE LA SAGRA, 1839, p. xxxix, 131]——[All names of family rank and a dagger(†) indicates *partim*]——[=Polymorphinidées D'ORBIGNY, 1840, p. 9 *(nom. neg.)*; =Polymorphinideae REUSS, 1860, p. 230; =Polymorphinidea REUSS, 1860, p. 151; =Polymorphinida JONES in GRIFFITH & HENFREY, 1875, p. 320; =Polymorphinidee SCHWAGER, 1876, p. 479; =Polymorphinina BÜTSCHLI in BRONN, 1880, p. 200; =Polymorphinae DELAGE & HÉROUARD, 1896, p. 138]——[=Énallostèguest D'ORBIGNY, 1826, p. 260 *(nom. nud., nom. neg.)*; =Turbinoidae D'ORBIGNY in DE LA SAGRA, 1839, p. xxxviii, 71 *(nom. nud.)*; =Uvellina EHRENBERG, 1839, table opposite p. 120 *(nom. nud.)*; =Enantiomorphinidae MARIE, 1941, p. 142]——[=Ramulinina LANKESTER, 1885, p. 847; =Ramulinae DELAGE & HÉROUARD, 1896, p. 138; =Ramulinidae LISTER in LANKESTER, 1903, p. 145]

Test multilocular, chambers in spiral or sigmoidal coil about longitudinal axis of growth, or biserial or uniserial, typically somewhat overlapping; aberrant forms may be irregular and attached; apertures all develop in same direction, terminal, radiate. *Trias.-Rec.*

Subfamily POLYMORPHININAE d'Orbigny 1839

[*nom. transl.* BRADY, 1881, p. 44 *(ex* family Polymorphinidae D'ORBIGNY, 1839]——[=Enantiomorphininae LOEBLICH & TAPPAN, 1961, p. 298]

Test free, chambers arranged in spiral, sigmoidal, biserial or asymmetrically alternating series; aperture terminal, radiate. *Trias.-Rec.*

Polymorphina D'ORBIGNY, 1826, *1391, p. 265 [**P. burdigalensis;* SD GALLOWAY & WISSLER, 1927, *766, p. 53] [*=Les Polymorphines* D'ORBIGNY, 1826, *1391, p. 265 *(nom. neg.)*; *Rostrolina* VON SCHLICHT, 1870, *1648, pl. 25, 26 (type, *Polymorphina burdigalensis* D'ORBIGNY, 1826, *1391, p. 265, SD LOEBLICH & TAPPAN, herein) (obj.); *Glandulopolymorphina* A. SILVESTRI, 1901, *1752, p. 17 (type, *Polymorphina burdigalensis* D'ORBIGNY, 1826, SD LOEBLICH & TAPPAN, herein) (obj.)]. Test elongate, somewhat compressed, commonly twisted; chambers biserial, early ones may be somewhat sigmoid. *Paleoc.-Rec.,* cosmop. ——Fig. 415,*1.* **P. burdigalensis,* Mio.(Burdigal.), Fr.; *1a,b,* ×49 (*2117).

Enantiomorphina MARIE, 1941, *1215, p. 144 [**E. lemoinei;* OD]. Test elongate, ovate to subcylindrical with chambers overlapping in alternating series, unequally inclined on longitudinal axis, although not completely biserial; sutures flush; aperture terminal, radiate. *U.Cret.(Senon.),* Eu. ——Fig. 415,*5.* **E. lemoinei,* Fr.; *5a,b,* opposite sides; *5c,* edge view, ×87 (*2117).

Eoguttulina CUSHMAN & OZAWA, 1930, *514, p. 16 [**E. anglica;* OD]. Test with chambers added in elongate spiral series in planes less than 90° apart, each succeeding chamber farther from base. *Jur.-U.Cret.,* Eu., N.Am.——Fig. 415,*2.* **E. anglica.* U.Cret.(Cenoman.), Eng.; *2a-c,* opposite sides and base of holotype, ×90 (*2117).

Falsoguttulina BARTENSTEIN & BRAND, 1949, *94, p. 671 [**F. wolburgi;* OD]. Test with chambers arranged in low spiral series, in planes approximately 120° apart; aperture a simple curved slit, not radiate. *L.Cret.(Valangin.),* Ger.——Fig 415,*9.* **F. wolburgi; 9a-d,* opposite sides, top, and basal views, ×156 (*2117).

Glandulopleurostomella SILVESTRI, 1903, *1757, p. 217 [**Polymorphina subcylindrica* HANTKEN, 1875, *863, p. 60; OD (M)] [*=Paleopolymorphina* CUSHMAN & OZAWA, 1930, *514, p. 12, 112 (type, *Polymorphina pleurostomelloides* FRANKE, 1928, *740, p. 121)]. Test elongate, early chambers spiral, later ones biserially arranged. *Jur.-Oligo.,* Eu.-N.Am.——Fig. 415,*3.* **G. subcylindrica* (HANTKEN), L.Oligo., Hung.; ×20 (*863).——Fig. 415,*4. G. pleurostomelloides* (FRANKE), U.Cret.(Cenoman.), Ger.; *4a-c,* side, edge, and top views, ×56 (*2117).

Globulina D'ORBIGNY in DE LA SAGRA, 1839, *1611, p. 134 [**Polymorphina (Globuline) gibba* D'ORBIGNY, 1826, *1391, p. 266; SD CUSHMAN,

Foraminiferida—Rotaliina—Nodosariacea

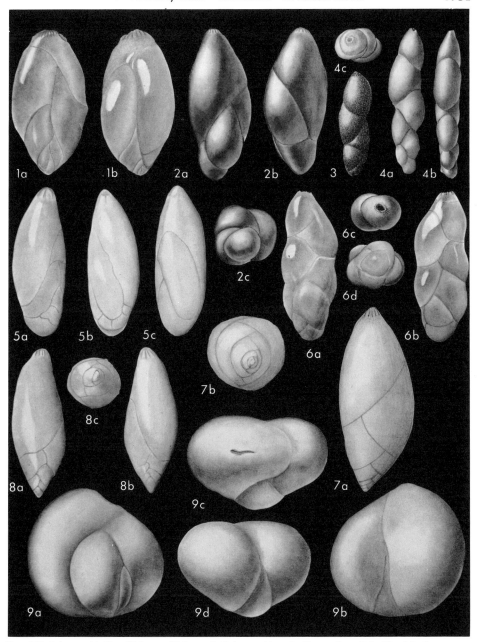

FIG. 415. Polymorphinidae (Polymorphininae; *1, Polymorphina; 2, Eoguttulina; 3,4, Glandulopleurostomella; 5, Enantiomorphina; 6, Pseudopolymorphina; 7,8, Pyrulinoides; 9, Falsoguttulina*) (p. C530, C533).

1927, *433, p. 189] [=*Polymorphina (Les Globulines)* D'ORBIGNY, 1826, *1391, p. 266 *(nom. van.)*; *Guttulina (Globulina)* D'ORBIGNY in DE LA SAGRA, 1839, *1611, p. 134 (obj.); *Aulostomella* ALTH, 1850, *13, p. 263 (type, *A. pediculus* ALTH, 1850, SD LOEBLICH & TAPPAN, herein)]. Test globular to ovate, chambers strongly overlapping, added in planes approximately 144° apart; sutures flush, not depressed, aperture radiate, but commonly obscured by fistulose growth. *U.Jur.-Rec.*, cosmop.——FIG. 416,*1.* *G. gibba*, Mio. (Torton.), Aus.; *1a,b*, side, basal views, ×45 (*514).

Guttulina D'ORBIGNY in DE LA SAGRA, 1839, *1611,

p. 132 [*Polymorphina (Guttuline) communis* d'Orbigny, 1826, *1391, p. 266; SD Galloway & Wissler, 1927, *766, p. 56] [=*Polymorphina (Les Guttulines)* d'Orbigny, 1826, *1391, p. 266 *(nom. neg.)*; *Guttulina (Guttulina)* d'Orbigny in de la Sagra, 1839, *1611, p. 132 (obj); *Sig-*

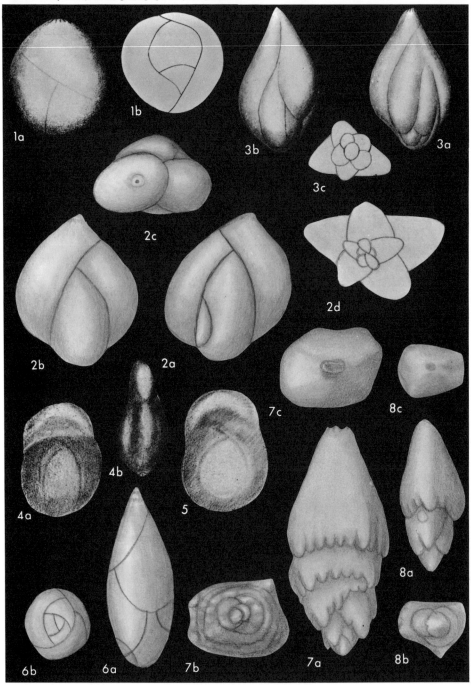

Fig. 416. Polymorphinidae (Polymorphininae; *1, Globulina; 2,3, Guttulina; 4,5, Pseudopolymorphinoides; 6, Pyrulina; 7,8, Sagoplecta*) (p. C530-C533).

momorpha CUSHMAN & OZAWA, 1928, *513, p. 17 (type, *S. sadoensis* CUSHMAN & OZAWA, 1928)]. Test ovate to elongate; inflated chambers added in quinqueloculine spiral series, in planes 144° apart, each successive chamber extending farther from base but strongly overlapping; sutures depressed; aperture radiate. *Jur.-Rec.*, cosmop.——FIG. 416,2. **G. communis* (D'ORBIGNY), Plio., Italy; *2a-d,* opposite sides, top view, and diagram. sec., enlarged (*1611).——FIG. 416,3. *G. sadoensis* (CUSHMAN & OZAWA), U.Plio., Japan; *3a-c,* opposite sides and basal view, ×45 (*514).

Paradentalina UCHIO, 1960, *1961, p. 60 [**Enantiodentalina muraii* UCHIO, 1953, *1960, p. 152; OD]. Like *Dentalina*, but with early chambers definitely biserial. [The Cretaceous species placed in *Enantiodentalina* are not congeneric with the Recent *Dentalina communis* D'ORBIGNY, which was selected as type of *Enantiodentalina*. As *Enantiodentalina* is thus a synonym of *Dentalina*, *Paradentalina* was proposed for species with an early biserial stage]. *Cret.-Rec.*, N.Am.-Eu.-Japan.——FIG. 417,1,2. **P. muraii* (UCHIO), Plio.-Pleist., Japan; *1,* holotype, ×70; *2a,b,* side, face views of paratype, ×65 (*1960).

Pseudopolymorphina CUSHMAN & OZAWA, 1928, *513, p. 15 [**P. hanzawai*; OD]. Test elongate; early chambers in quinqueloculine arrangement, later biserial; chambers high and overlapping only slightly; sutures depressed; aperture radiate. *Jur.-Rec.,* Japan-Pac.-Australia-Atl.-N.Am.-S.Am.-Eu.-Carib.——FIG. 415,6. **P. hanzawai,* Plio., Japan; *6a-d,* opposite sides, top, and basal views, ×15 (*2117).

Pseudopolymorphinoides VAN BELLEN, 1946, *113, p. 41 [**P. limburgensis*; OD]. Early stage inflated, with chambers in quinqueloculine arrangement, final chamber terminal and compressed; sutures flush; aperture an elongate slit. [Differs from *Falsoguttulina* in being quinqueloculine, rather than triloculine, in early stage]. *M.Eoc.*, Eu.(Neth.).——FIG. 416,4,5. **P. limburgensis; 4a,b,* side, edge views of holotype; *5,* side view of paratype; all ×35 (*113).

Pyrulina D'ORBIGNY in DE LA SAGRA, 1839, *1611, p. 107 [**Polymorphina (Pyruline) gutta* D'ORBIGNY, 1826, *1391, p. 267, 310; OD (M)] [=*Polymorphina (Les Pyrulines)* and *Polymorphina (Pyruline)* D'ORBIGNY, 1826, *1391, p. 267, 310 (subgeneric names=*nom. neg.*); *Pirulina* BRONN & ROEMER, 1853, *214a, p. 88 *(nom. van.)*; *Pyrulinella* CUSHMAN & OZAWA, 1928, *513, p. 16 (type, *Polymorphina lanceolata* REUSS, 1851, *1541, p. 83)]. Test fusiform; early chambers arranged in spiral series approximately 120° apart, later chambers biserial; sutures flush; aperture radiate. *Jur.-Rec.,* cosmop.——FIG. 416,6. **P. gutta* (D'ORBIGNY), Plio., Italy; *6a,b,* side, basal views, enlarged (*1391).

Pyrulinoides MARIE, 1941, *1215, p. 169, 255 [**Pyrulina acuminata* D'ORBIGNY, 1840, *1394,

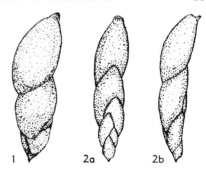

FIG. 417. Polymorphinidae (Polymorphininae; *1,2, Paradentalina*) (p. C533).

p. 43; OD]. Test free, elongate, fusiform; chambers biserially arranged, much embracing; sutures oblique, flush; aperture terminal, radiate. [*Pyrulinoides* differs from *Pyrulina* in being biserial throughout, and in lacking the early spiral stage.] *Trias.-U.Cret.*, Eu.-N.Am.——FIG. 415,7,8. **P. acuminata* (D'ORBIGNY), U.Cret.(Senon.), Fr.; *7a,b,* lectotype (MNHN), side, and basal views, ×36 (*2117); *8a-c,* opposite sides and basal view of hypotype, ×48 (*2117).

Sagoplecta TAPPAN, 1951, *1873, p. 14 [**S. goniata;* OD]. Test free, elongate, early portion biserial, later portion uniserial and quadrate or flattened with later chambers equitant and chevron-shaped, overhanging earlier chambers at angles of test; wall calcareous; aperture terminal, radiate. [*Sagoplecta* differs from *Spirofrondicularia* in having a distinctly biserial, rather than tetraloculine, early stage.] *U.Trias.*, N.Am.(Alaska).——FIG. 416, 7,8. **S. goniata; 7a-c,* side, basal, and top views of microspheric holotype, ×95 (*1873); *8a-c,* side, basal, and top views of megalospheric paratype, ×95 (*1873).

Sigmoidella CUSHMAN & OZAWA, 1928, *513, p. 18 [**S. kagaensis;* OD] [=*Sigmoidella (Sigmoidina)* CUSHMAN & OZAWA, 1928, *513, p. 18 (type, *S. (S.) pacifica* CUSHMAN & OZAWA, 1928)]. Test compressed, chambers arranged in sigmoid series, those on each side reaching to base and covering earlier chambers on one side. *M.Eoc.-Rec.,* Japan-Formosa-Indon.-N.Am.-N.Z.——FIG. 418,1. **S. kagaensis,* U.Plio., Japan; *1a,b,* opposite sides of paratype, ×49 (*2117).——FIG. 418,2. *S. pacifica,* Rec., Philip.; *2a-c,* opposite sides and basal view, ×24 (*514).

Sigmomorphina CUSHMAN & OZAWA, 1928, *513, p. 17 [**Sigmomorpha (Sigmomorphina) yokoyamai;* OD] [=*Sigmomorpha (Sigmomorphina)* CUSHMAN & OZAWA, 1928, *513, p. 17 (obj.); *Ellisina* LALICKER, 1950, *1082, p. 18 (type, *Ellisina spatula* LALICKER, 1950) *(non Ellisina* NORMAN, 1903); *Pealerina* LALICKER in THALMANN, 1950, *1902, p. 43, *nom. subst.* pro *Ellisina* LALICKER, 1950 *(non Ellisina* NORMAN, 1903); *Sigmomorphina (Sigmomorphinoides)* ROUVIL-

LOIS, 1960, *1589, p. 62 (type, *Sigmomorphina (Sigmomorphinoides) parisiensis* ROUVILLOIS, 1960)]. Test elongate, compressed, chambers added in planes slightly less than 180° apart, forming sigmoid series, each chamber farther removed from base but strongly overhanging at

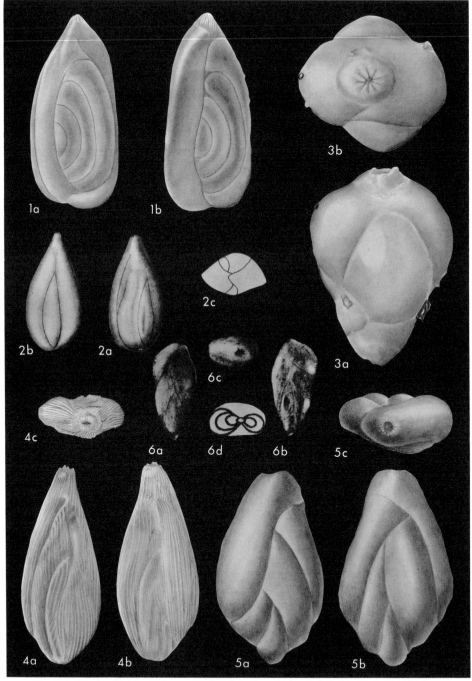

FIG. 418. Polymorphinidae (Polymorphininae; *1,2, Sigmoidella; 3, Spirofrondicularia; 4-6, Sigmomorphina*) (p. C533-C535).

edges of test; sutures depressed; aperture radiate. [*Sigmomorphinoides* was separated by the presence of 2 apertures on the final chamber, apparently an accidental occurrence in abnormal specimens and not here regarded as of generic importance.] *Jur.-Rec.*, Japan-Eu.-N. Am.-S. Am.-Cuba-Trinidad-N.Z.-Australia-Antarctic. —— FIG. 418,*4*. **S. yokoyamai,* Plio., Japan; *4a-c,* opposite sides and top view, ×61 (*2117).——FIG. 418,*5*. *S. spatula* (LALICKER), Jur., USA(Mont.); *5a-c,* opposite sides and top view of holotype, ×119 (*2117).——FIG. 418,*6*. *S. parisiensis* ROUVILLOIS, L.Eoc.(Thanet.), Fr.; *6a-d,* opposite sides, top, and basal view of holotype, ×30 (*1589).

Spirofrondicularia SCHUBERT, 1902, *1681, p. 16 [**Polymorphina frondicularioides* CHAPMAN, 1894, *310, p. 716; SD GALLOWAY, 1933, *762, p. 262] [=*Quadrulina* CUSHMAN & OZAWA, 1930, *514, p. 12, 18 (type, *Polymorphina rhabdogonioides* CHAPMAN, 1894, *310, p. 716)]. Test with early chambers tetraloculine, added in planes 90° apart; sutures depressed; aperture terminal, radiate. *L. Jur.-L.Cret.,* Eu.——FIG. 418,*3*. **S. frondicularioides* (CHAPMAN), L.Cret.(Apt.), Eng.; *3a,b,* side, top views, ×192 (*2117).

Tobolia DAIN in N. K. BYKOVA *et al.,* 1958, *265, p. 39 [**T. veronikae*=*T. veronica* E. V. BYKOVA, DAIN & FURSENKO in RAUZER-CHERNOUSOVA & FURSENKO, 1959, *1509, p. 17 *(nom. van.);* OD]. Test globular, chambers added in planes 140° apart, as in *Guttulina,* strongly overlapping, sutures flush to slightly depressed; slitlike aperture somewhat produced. *U.Cret.(Maastricht.),* Sib. ——FIG. 419,*1*. **T. veronikae*; *1a-c,* opposite sides, edge, and basal views, ×72 (*265).

Subfamily WEBBINELLINAE Rhumbler, 1904

[Webbinellinae Rhumbler, 1904, p. 224] [=Arwebbina RHUMBLER, 1913, p. 346 *(nom. van.)*]

Test attached, one or more chambers connected by stolons, early portion may be globular or polymorphine, with attachment rounded or irregularly spreading. *Jur.-Rec.*

Webbinella RHUMBLER, 1904, *1569, p. 228 [**Trochammina (Webbina) irregularis hemisphaerica* JONES, PARKER & BRADY, 1865, *1002, p. 26, =*Webbina hemisphaerica* JONES, PARKER & BRADY, 1865, *1002, p. 27; SD CUSHMAN, 1918, *411a, p. 61] [=*Arwebbinum* RHUMBLER, 1913, *1572b, p. 346 (obj.) *(nom. van.)*]. Test attached, early multilocular polymorphine or pyruline stage surrounded by flangelike chamber spreading on surface of substratum; wall calcareous, perforate, no apparent aperture. [Restudy of the holotype of the type-species showed it to be a calcareous perforate polymorphinid and not an attached arenaceous single-chambered form (*1172). *L.Cret.-Rec.,* cosmop.——FIG. 420,*7*. **W. hemisphaerica* (JONES, PARKER & BRADY), Plio.(L.Crag), Eng.; ×48 (*1172).

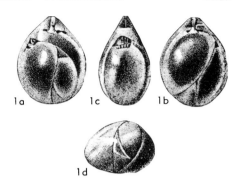

FIG. 419. Polymorphinidae (Polymorphininae; *1, Tobolia*) (p. *C*535).

Bullopora QUENSTEDT, 1856, *1495, p. 292 [**B. rostrata* QUENSTEDT, 1857; SD (SM) QUENSTEDT, 1857, *1495, p. 580]. [=*Arperneroum* RHUMBLER, 1913, *1572b, p. 444 (type, *Webbina irregularis* D'ORBIGNY, 1850, *1397b, p. 111); *Placopsum* RHUMBLER, 1913, *1572b, p. 445 (type, *Webbina breoni* TERQUEM & PIETTE in TERQUEM, 1862, *1883, p. 458); *Arplacopsum* RHUMBLER, 1913, *1572b, p. 445 (type, *Webbina breoni* TERQUEM & PIETTE in TERQUEM, 1862) *(nom. van.)*]. Test attached, composed of single series of hemispherical chambers, rounded to ovate in outline, earlier chambers may be closely appressed, later ones connected by more or less well-developed stoloniferous necks as in type-species; in microspheric forms chambers increase rapidly in size, but chambers may all be of approximately equal size in megalospheric forms; wall calcareous, perforate; aperture at open end of stolon-like neck. *Jur.-Cret.,* cosmop.——FIG. 420,*1*. **B. rostrata,* U.Jur.(Malm alpha), Ger.; *1a-d,* ×19 (*2117).——FIG. 420,*2,3*. *B. breoni* (TERQUEM & PIETTE), L.Jur.(Lias.), Fr.; *2,3a,b,* ×10 (*1572b).——FIG. 420,*4*. *B. irregularis* (D'ORBIGNY), U.Cret., Czech.; *4a,* side view, ×10; *4b,* view of detached specimen, ×22; *4c,* long. sec., ×28 (*1445).

[*Bullopora* was originally named and figured (QUENSTEDT, 1856, fasc. 2, p. 292, pl. 41, fig. 26 and 1856, fasc. 3, p. 554, pl. 72, fig. 35) with no species named. In 1857, (fasc. 4, p. 580, pl. 73, fig. 28) *B. rostrata* was named, automatically becoming the type of the genus by subsequent monotypy. Much confusion concerning the type-species is found, for it has been variously regarded as a calcareous imperforate form (*1200, p. 25), considered to be a senior synonym of *Nubeculinella* and *Nodobacularia* (*1200, p. 27), belonging to the Ophthalmidiidae (*1200, p. 25, *1478, p. 254), Nubeculariidae (*64, p. 838), Nodosinellidae (*762, p. 167) or as a calcareous perforate form belonging to the Polymorphinidae (*486, p. 230, *1509, p. 264) and including *Vitriwebbina* as a junior synonym. LOEBLICH & TAPPAN in 1954, with Drs. E. BUCH & K. FEIFEL collected at the type locality, which is an erosional slope exposing the Upper Jurassic (Malm alpha), middle *Impressa* Mergel, in the valley of Fils, between Unter Böhringer and Reichenbach i T., northeast of Reichenbach, Wurttemberg, Germany. The type-species was clearly stated by QUENSTEDT to be from the lower Weisse Jura alpha (explanation of pl. 73), not the Oberer Lias, Zeta zone, *Aalensis* Mergel, as reported by ELLIS & MESSINA (*700). The *Bullopora* recorded from the Lias Zeta

(p. 292, pl. 41, fig. 26) and from the Brauner Jura Zeta (p. 554, pl. 72, fig. 35) was never given a specific name, and does not show the stoloniferous necks which QUENSTEDT stated to be characteristic of the species *B. rostrata*.]

Histopomphus LOEBLICH & TAPPAN, 1949, *1156, p. 262 [***Globulina redriverensis* TAPPAN, 1943, *1872, p. 505; OD]. Test large, early portion

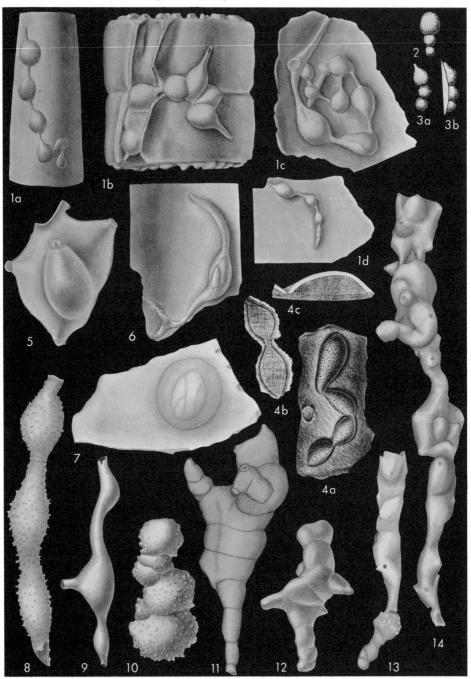

FIG. 420. Polymorphinidae (Webbinellinae; *1-4, Bullopora; 5, Vitriwebbina; 6, Histopomphus; 7, Webbinella;* Ramulininae; *8,9, Ramulina; 10, Ramulinella; 11,12, Washitella; 13,14, Sporadogenerina*) (p. C535-C537).

polymorphine, later attached portion consisting of branching or bifurcating undivided tubular chamber; wall calcareous, perforate; aperture rounded or low arch at ends of tubular chamber. [Differs from *Vitriwebbina* in possessing a multilocular early polymorphine stage, followed by an elongate branching tubular attached stage, and from *Webbinella* in having an irregular branching attachment, instead of the circular disclike attachment.] L.Cret., N.Am.——Fig. 420,6. *H. redriverensis (TAPPAN), L.Cret.(Alb.), USA(Okla.); ×20 (*2117).

Vitriwebbina CHAPMAN, 1892, *309, p. 52, 53 [*V. sollasi CHAPMAN, 1892; SD CUSHMAN, 1927, *433, p. 189]. Differs from *Bullopora* in having central initial chamber surrounded by broad flangelike chamber rather than uniserial series of simple chambers; may also have additional chambers after bilocular beginning; apertures at open ends of tubular projections from flange. [Lectotype of *V. sollasi* here designated, specimen figured by CHAPMAN (*309, pl. 2, fig. 1). CHAPMAN'S specimen of fig. 3 is a *Bullopora*.] Cret., Eu.-N.Am.——Fig. 420,5. *V. sollasi, L.Cret.(Gault), Eng.; specimen broken free of substratum to which it had been attached, ×70 (*2117).

Subfamily RAMULININAE Brady, 1884

[Ramulininae BRADY, 1884, p. 71]

Test free, with one or more chambers connected by stolons. *Jur.-Rec.*

Ramulina JONES in WRIGHT, 1875, *2079, p. 88 [*R. laevis; OD]. Test consisting of globular or irregular chambers loosely connected by stolonlike necks, or by straight or branching tube with local irregular chamber-like swellings; apertures rounded, at open ends of tube or stoloniferous necks. [Because of confusion concerning the generic status of *R. aculeata* (D'ORBIGNY), which has been referred to both *Dentalina* and *Ramulina*, even in a single publication (*484, p. 67, 100), it was restudied by us in Paris and found to represent a true *Ramulina*.] *Jur.-Rec.*, cosmop.——Fig. 420,9. *R. laevis, U.Cret., Ire., ×17 (*2079).——Fig. 420,8. R. aculeata (D'ORBIGNY), U.Cret., Fr.; lectotype, here designated and refigured (MNHN), ×20 (*2117).

Ramulinella PAALZOW, 1932, *1405, p. 135 [*R. suevica; OD (M)]. Similar to *Ramulina* but with closely appressed irregularly arranged chambers and without intercameral stolons. U.Jur.(Oxford.), Eu.-N.Am.(USA).——Fig. 420,10. *R. suevica, Ger.; side view, ×42 (*1405).

Sporadogenerina CUSHMAN, 1927, *430, p. 95 [*S. flintii CUSHMAN, 1927 (=*Ramulina proteiformis FLINT, 1899, *723, p. 321); OD]. Test elongate, with irregular early portion and later uniserial or branching stage; chambers inflated, somewhat globular; aperture radiate, terminal in early stage, later with multiple radiate apertures, irregularly placed. Rec., Gulf Mex.——Fig. 420,13,14. *S. proteiformis (FLINT); 13, side view of holotype of S. flintii; 14, side view of hypotype; both ×25 (*2117).

Washitella TAPPAN, 1943, *1872, p. 515 [*W. typica; OD]. Test free, consisting of well-defined but very irregularly arranged chambers, which may be in linear or slightly coiled series or variously branched; apertures simple, rounded, at ends of series of chambers, commonly more than one per chamber. [*Washitella* differs from *Sporadogenerina* in having rounded, rather than radiate, apertures and more regular chambers.] L.Cret. (Alb.)-U.Cret.(Cenoman.), USA(Okla.-Tex.).——Fig. 420,11,12. *W. typica, L.Cret., Tex. (11), Okla. (12); 11, hypotype, ×75; 12, holotype, ×75 (*2117).

Family GLANDULINIDAE Reuss, 1860

[Glandulinidae REUSS, 1860, p. 151] [=Stichostègues D'ORBIGNY, 1826, p. 251 *(partim)* *(nom. nud., nom. neg.)*; =Ovulinida HAECKEL, 1894, p. 185 *(nom. nud.)*]

Test unilocular or with chambers in biserial, uniserial or polymorphine arrangement; aperture terminal, radial or slitlike, with simple, straight or curved internal (entosolenian) tube. *Jur.-Rec.*

Subfamily GLANDULININAE Reuss, 1860

[*nom. correct.* LOEBLICH & TAPPAN, 1961, p. 299 *(pro* subfamily Glandulinidea REUSS, 1862, p. 307), *nom. transl. ex* family Glandulinidae REUSS, 1860] [=Glandulinea HANTKEN, 1875, p. 41]

Test biserial, uniserial or polymorphine; aperture terminal, radial or slitlike, with internal tube. *Jur.-Rec.*

Glandulina D'ORBIGNY in DE LA SAGRA, 1839, *1611, p. 12 [*Nodosaria (Glanduline) laevigata D'ORBIGNY, 1826, *1391, p. 252; SD CUSHMAN, 1927, *433, p. 189] [=Psecadium NEUGEBOREN, 1856, *1351, p. 99 (type, P. ellipticum); Encorycium EHRENBERG, 1858, *683, p. 12 (type, E. nodosaria); Atractolina VON SCHLICHT, 1870, *1648, p. 69 (type, Nodosaria (Glanduline) laevigata D'ORBIGNY, 1826, SD LOEBLICH & TAPPAN, herein) (obj.)]. Test free, elongate, circular in section, early portion biserial, later uniserial; chambers strongly overlapping and increasing in size; sutures distinct, flush; aperture terminal, central, radiate, with entosolenian tube. [Although superficially resembling *Pandaglandulina*, type material of the type-species of *Psecadium* was stated by CUSHMAN (*486, p. 228) to be biserial in the early stage and thus belongs with *Glandulina*.] Paleoc.-Rec., cosmop.——Fig. 421,1,2. *G. laevigata (D'ORBIGNY), Rec., Can. (1), Greenl. (2); 1, side view; 2, specimen showing internal tube, ×49 (*1162).

Dainita LOEBLICH & TAPPAN, herein [*nom. nov. pro Mariella* DAIN in N. K. BYKOVA, *et al.*, 1958, *265, p. 41 (*non* NOWAK, 1916; *nec* MÖRCH, 1865, *nom.*

null. pro *Mariaella* GRAY, 1855)] [*Mariella sibirica* DAIN in N. K. BYKOVA et al., 1958, *265, p. 41, here designated as type-species)]. Similar to *Siphoglobulina* but with later stage biserial; aperture radiate, with tube attached to one wall of final chamber. *L.Cret.(Hauteriv.)-U.Cret.(Maas-*

FIG. 421. Glandulinidae (Glandulininae; *1,2, Glandulina; 3,4, Esosyrinx; 5, Siphoglobulina; 6-8, Tristix; 9, Laryngosigma; 10, Globulotuba*) (p. C537, C539-C540).

Foraminiferida—Rotaliina—Nodosariacea C539

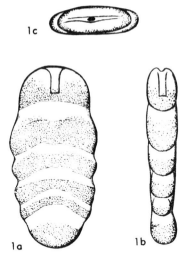

Fig. 422. Glandulinidae (Glandulininae; *1*, *Entolingulina*) (p. C539).

tricht.), Eu.-Sib.——Fig. 423,*1*. **D. sibirica* (DAIN), U.Cret.(Maastricht.), Sib.; *1a-d*, opposite sides, edge, and basal views, ×47 (*265).

Entolingulina LOEBLICH & TAPPAN, 1961, *1181, p. 220 [**Lingulina aselliformis* BUCHNER, 1942, *250, p. 121; OD]. Test free, elongate, compressed, of 2 or more chambers in rectilinear series, commonly with considerable overlap of earlier chambers; wall calcareous, finely perforate, hyaline; aperture ovate or elongate slit, with distinct entosolenian tube projecting into final chamber. Rec., Eu.-Antarctic.——Fig. 422,*1*. **E. aselliformis* (BUCHNER), Rec., Italy; *1a-c*, side, edge, and top views, ×200 (*250).

Esosyrinx LOEBLICH & TAPPAN, 1953, *1162, p. 85 [**Pseudopolymorphina curta* CUSHMAN & OZAWA, 1930, *514, p. 105; OD]. Test free, chambers biserially arranged throughout and in single plane; aperture terminal, radiate, with internal tube. [*Esosyrinx* differs from *Pseudopolymorphina* in being biserial throughout and in having an internal tube, and from *Laryngosigma* in having chambers in a single plane rather than a sigmoid series.] Rec., Atl.——Fig. 421,*3,4*. **E. curta* (CUSHMAN & OZAWA); *3a,b*, side and basal views of holotype, *4*, side view of hypotype; all ×48 (*1162).

Globulotuba COLLINS, 1958, *375, p. 385 [**G. entoseleniformis*; OD] Test ovate, circular in section; chambers in triloculine arrangement, sutures flush; aperture radiate, with short, free, internal entosolenian tube. Rec., Australia.——Fig. 421,*10*. **G. entoseleniformis*; *10a,b*, side and basal views, ×150 (*375).

Laryngosigma LOEBLICH & TAPPAN, 1953, *1162, p. 83 [**L. hyalascidia*; OD]. Test free, somewhat compressed; chambers biserially arranged, added in planes slightly less than 180° apart, forming sigmoid series with each succeeding chamber farther removed from base; aperture terminal, radiate, with entosolenian tube. Rec., Atl.-Arctic-Antarctic-Australia.——Fig. 421,*9*. **L. hyalascidia*, Alaska; *9a-c*, opposite sides and basal view, ×100 (*1162).

[*Laryngosigma* is similar to *Sigmomorphina* but differs in possessing an entosolenian tube within the aperture. It differs from *Esosyrinx* in being sigmoid and biserial, and from *Siphoglobulina* in being biserial rather than triserial, and in having a free entosolenian tube which is not attached to the interior chamber wall.]

Oolitella MAKIYAMA & NAKAGAWA, 1941, *1206, p. 242, 243 [**O. irregularis*, OD]. Test with irregularly arranged inflated chambers; wall thin, finely perforate; aperture terminal, rounded, with entosolenian tube. Pleist., Japan.——Fig. 424,*1-3*. **O. irregularis*; *1*, holotype, showing entosolenian tube; *2,3*, paratypes; all ×100 (*1206).

Siphoglobulina PARR, 1950, *1429, p. 332 [**S. siphonifera*; OD]. Test elongate-ovate to subfusiform; chambers in triloculine series, strongly overlapping but each farther removed from base; aperture radiate, with entosolenian tube extending downward along inner wall of final chamber and opening to exterior in short slit at its lower end, relict slits of earlier chambers remaining visible. L.Tert.-Rec., Australia-Antarctic.——Fig. 421,*5*. **S. siphonifera*, Mio., Australia; *5a-c*, side, face, and basal views, ×44 (*2117).

Tristix MACFADYEN, 1941, *1200, p. 54 [**Rhabdogonium liasinum* BERTHELIN, 1879, *132, p. 35; OD] [=*Tricarinella* TEN DAM & SCHIJFSMA, 1945, *558, p. 233 (type, *Rhabdogonium excavatum* REUSS, 1863, *1554, p. 91; *Quadratina*

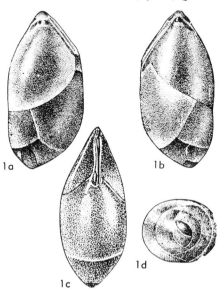

Fig. 423. Glandulinidae (Glandulininae; *1*, *Dainita*) (p. C537-C539).

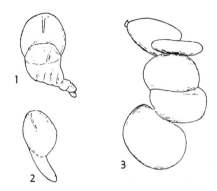

Fig. 424. Glandulinidae (Glandulininae; *1-3, Oolitella*) (p. C539).

TEN DAM, 1946, *552, p. 65 (type, *O. depressula* TEN DAM, 1946)]. Test free, uniserial, generally triangular in section, but rarely quadrate; wall calcareous, hyaline; aperture terminal, rounded to radiate, with entosolenian tube in at least some species. [Differs from *Glandulina* in being uniserial throughout, angular in section and with less overlapping chambers.] L.Jur.-Eoc., Eu.-N.Am.——FIG. 421,6. *T. liasina* (BERTHELIN), L.Jur.(L.Pleinsbach.), Fr.; side view, ×146 (*2117).——FIG. 421,7,8. *T. reesidei* LOEBLICH & TAPPAN, U.Jur., USA; *7a,b,* side and top views of normal triangular form, ×48; *8a,b,* side and top views of rarer quadrate form, ×64 (*2117).

Subfamily SEABROOKIINAE Cushman, 1927

[Seabrookiinae CUSHMAN, 1927, p. 86]

Test compressed, early stage with proloculus and 2 chambers to whorl, later chambers added 180° from preceding and completely enveloping earlier formed chambers; aperture terminal, oval to slitlike, commonly with thickened lip. *U.Cret.-Rec.*

Seabrookia BRADY, 1890, *202, p. 570 [**S. pellucida*; OD (M)] [=?*Cerviciferina* GODDARD & JENSEN, 1907, *799, p. 305 (type, *C. hilli* GODDARD & JENSEN, 1907)]. Test free, elongate ovate, compressed, early stage with 3 chambers to whorl, rapidly enlarging chambers 2 per coil in later stages, completely involute, aperture of successive chambers at opposite ends of test, as in miliolids; wall calcareous, perforate, radial in structure, may have peripheral keel, in type-species aboral end ornamented with small blunt spines along keel; aperture a terminal slit bordered by distinct lip. *U.Cret.-Rec.,* Eu.-Atl.-Pac.-Cuba.——FIG. 425,*1.* *S. pellucida,* Rec., Pac.; *1a-c,* opposite sides and apert. view, ×140 (*2117).

[*Seabrookia* has been included in the Chilostomellidae (*486, *762, *1458) but has a perforate radial wall, whereas tests of chilostomellid genera are granular. It resembles some of the Miliolidae in alternation of the aperture to opposite ends of the test in successive chambers, but differs from these in having a perforate radial wall. This wall character and the entosolenian tube places *Seabrookia* in the family Glandulinidae.]

Subfamily OOLININAE Loeblich & Tappan, 1961

[Oolininae LOEBLICH & TAPPAN, 1961, p. 299]

Test unilocular, with slitlike or radiate aperture and entosolenian tube. *Jur.-Rec.*

Oolina D'ORBIGNY, 1839, *1393, p. 18 [**O. laevigata*; SD GALLOWAY & WISSLER, 1927, *766, p. 50] [=*Ovulina* EHRENBERG, 1845, *675, p. 358 (*non Ovulina* SCHULTZE, 1854; *nec* GRUBER, 1884) (nom. van. pro *Oolina* D'ORBIGNY, 1839) (obj.); *Cenchridium* EHRENBERG, 1845, *675, p. 357 (type, *C. sphaerula* EHRENBERG, 1845); *Entosolenia* WILLIAMSON, 1848, *2064, p. 16 (type, *E. lineata* WILLIAMSON, 1848); *Entosalenia* PARKER & JONES, 1857, *1416, p. 278 (nom. van.) (obj.); *Obliquina* SEGUENZA, 1862, *1712, p. 75 (type, *O. acuticosta* SEGUENZA, 1862); *Lagenulina* TERQUEM, 1876, *1888, p. 67 (type, *L. sulcata* TERQUEM, 1876, SD LOEBLICH & TAPPAN, herein); *Entolagena* SILVESTRI, 1900, *1751, p. 4 (type, *Vermiculum globosum* MONTAGU, 1803, *1298, p. 523); *Lagena (Reussoolina)* COLOM, 1956, *376, p. 71 (type, *Oolina apiculata* REUSS, 1851, *1542, p. 22)]. Test single globular to ovate chamber, rarely somewhat asymmetrical; surface may be smooth or ornamented with striae, reticulations or costae; aperture rounded and may have radiating grooves surrounding aperture on exterior, internally provided with entosolenian tube; mononucleate; at least some species ectoparasitic on other foraminifers, having reproductive cycle reduced to only asexual generation, with small size and single nucleus suggesting that haploid stage is represented. *Jur.-Rec.,* cosmop.——FIG. 425,*2. O. lineata* (WILLIAMSON), Rec., Alaska; *2a,b,* side and top views, ×75 (*1162).——FIG. 425,*3. O. striatopunctata* (PARKER & JONES), Rec., Alaska; chamber broken, showing entosolenian tube, ×75 (*1162). ——FIG. 425,*4.* *O. laevigata,* Rec., Falk. Is.; *4a,b,* side and top views of holotype (MNHN), ×58 (*2117).——FIG. 425,*5. O. apiculata* REUSS, U. Cret., Pol.; side view, ×54 (*700).——FIG. 425, *6. O. acuticosta* (SEGUENZA), Mio., Sicily; *6a-c,* side and opposite edges, ×30 (*700).

[*Oolina marginata* is an ectoparasite on *Discorbis,* and during its reproductive stage moves to margin of the host, constructs a chitinoid cyst around the aperture into which the protoplasm moves after dissolution of the entosolenian tube. The protoplasm and nucleus then divide asexually into 2 to 6 parts, each reorganizes, secretes a calcareous test, leaves the cyst, and returns to the host (*1109).]

Fissurina REUSS, 1850, *1540, p. 366 [**F. laevigata;* OD (M)] [=*Hyaleina* COSTA, 1856, *392, p. 366 (type, *Fissurina laevigata* REUSS, 1850, SD LOEBLICH & TAPPAN, herein) (obj.); *Trigonulina* SEGUENZA, 1862, *1712, p. 74 (*non* D'ORBIGNY, 1846) (type, *T. oblonga* SEGUENZA, 1862); *Ellipsolagena* A. SILVESTRI, 1923, *1774, p. 265, 268 (type, *Lagena acutissima* FORNASINI, 1890, *729, p. 1; *Ellip-

Foraminiferida—Rotaliina—Nodosariacea

Fig. 425. Glandulinidae (Seabrookiinae; *1, Seabrookia;* Oolininae; *2-6, Oolina; 7,8, Fissurina; 9,10, Parafissurina*) (p. C540-C543).

sofissurina A. Silvestri, 1923, *1774, p. 265 (type, *Fissurina laevigata* Reuss, 1850, SD Loeblich & Tappan, herein) (obj.)]. Test rounded to ovate in outline; compressed, trigonal or tetragonal in section, and may be keeled; surface smooth, costate, beaded, pitted or reticulate; aperture slitlike to oval or rounded, in center of fissure-like cavity at one end of test; entosolenian tube projecting inward from aperture into chamber cavity. [*Ellipsolagena* is a synonym of *Fissurina* (*1428)

with the type-species *Lagena acutissima* Fornasini, 1890, by monotypy, and not with *Lagena ventricosa* Silvestri, 1904, as type by subsequent designation of Cushman (1927, *431, p. 72).] *Cret.-Rec.,* cosmop.——Fig. 425,7. *F. marginata* (Montagu), Rec., Alaska; *7a-c,* side, edge, and top views, ×75 (*1162).——Fig. 425,8. *F. laevigata* Reuss, Tert., Ger.; *8a,b,* side, apert. views, ×60 (*1540).

Parafissurina Parr, 1947, *1428, p. 123 [**Lagena*

ventricosa SILVESTRI, 1904, *1758, p. 10; OD]. Test single ovate chamber, commonly compressed; surface smooth or rarely keeled; **aperture arched** or crescentic subterminal opening at one side of test, with overhanging hoodlike extension of wall; entosolenian tube as in *Oolina* and *Fissurina*.

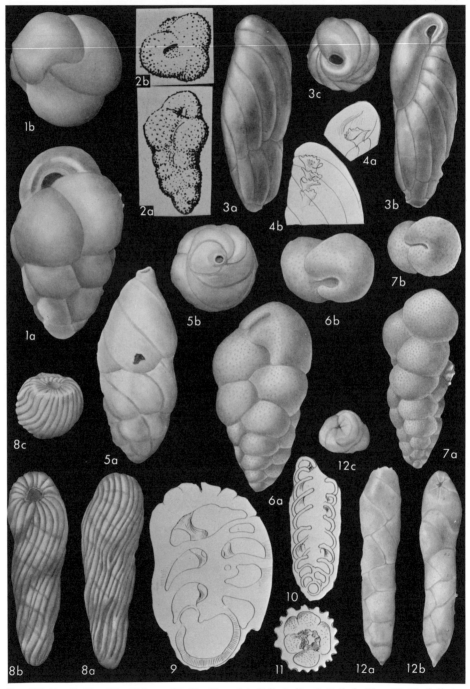

FIG. 426. Turrilinidae (Turrilininae; *1,2, Turrilina; 3,4, Buliminella; 5, Buliminellita; 6,7, Neobulimina; 8-12, Buliminoides*) (p. C543-C545).

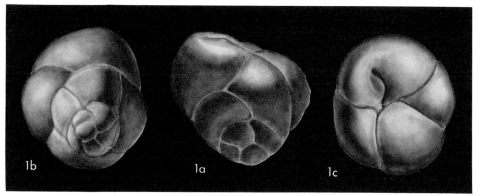

FIG. 427. Turrilinidae (Turrilininae; *1, Baggatella*) (p. C543).

[The hooded aperture is reminiscent of the Pleurostomellidae, but the radially built wall shows relationship with the Oolininae.] *M.Eoc.-Rec.,* cosmop.——FIG. 425,*9,10*. *P. *ventricosa* (SILVESTRI), Mio., Italy; *9a*, side view showing hooded aperture; *9b*, opposite side showing entosolenian tube; *9c*, top view; all ×111 (*2117); *10a,b*, profile and cross sec., ×55 (*1758).

Superfamily BULIMINACEA Jones, 1875

[*nom. correct.* LOEBLICH & TAPPAN, 1961, p. 299 (*pro* superfamily Buliminidea GLAESSNER, 1945, p. 134, and Buliminicae EASTON, 1960, p. 65, 79)]——[In synonymic citations superscript numbers indicate taxonomic rank assigned by authors (¹superfamily, ²family group); dagger (†) indicates *partim*]——[=¹Enclinostegia† EIMER & FICKERT, 1899, p. 682 (*nom. nud.*); =²Textulinidia† RHUMBLER in KÜKENTHAL & KRUMBACH, 1923, p. 88]

Test high trochospiral or modified to biserial or uniserial; wall finely or coarsely perforate, of radially built calcite; aperture primary, basal slit, or in apertural face, or terminal; may have internal tooth plate or tube, and aperture may be on neck. *U. Trias.-Rec.*

Family TURRILINIDAE Cushman, 1927

[*nom. transl.* LOEBLICH & TAPPAN, 1961, p. 300 (*ex* subfamily Turrilininae CUSHMAN, 1927)] [=Buliminellidae HOFKER, 1951, p. 121]

Test high trochospiral, with more than 3 chambers to whorl, or may be reduced to biserial; wall of radially lamellar calcite; apertural face poreless, formed by outgrowth from tooth plate, may be radially grooved. *M.Jur.-Rec.*

Subfamily TURRILININAE Cushman, 1927

[Turrilininae CUSHMAN, 1927, p. 65] [=Buliminellinae N. K. BYKOVA in RAUZER-CHERNOUSOVA & FURSENKO, 1959, p. 323; =Baggatellinae N. K. BYKOVA in RAUZER-CHERNOUSOVA & FURSENKO, 1959, p. 325]

Test high-spired, with 3 or more chambers to whorl; aperture loop-shaped, in face of last-formed chamber. *M.Jur.-Rec.*

Turrilina ANDREAE, 1884, *19, p. 120 [**T. alsatica*; OD (M)] [=*Corrosina* NYIRÖ, 1954, *1382, p. 68, 71, 73 (type, *C. pupoides*)]. Test free, elongate, high-spired, 3 or more chambers to whorl; wall calcareous, finely perforate, monolamellar, microstructure unknown, surface smooth or roughened; aperture a small, basal arch in final chamber, presence or absence of internal tooth plate unknown. [Originally *Corrosina* was placed in the Heterohelicidae, as related to *Guembelitria*, but more prismatic in form. Both *Turrilina* and *Corrosina* were first described from the Oligocene of western and central Europe, respectively. More information is needed as to wall structure and the presence or absence of an internal tooth plate.] *Eoc.(Ypres.)-U.Oligo.,* Eu.——FIG. 426,*1*. **T. alsatica*, M.Oligo., Fr.; *1a,b*, side, top views, ×235 (*2117).——FIG. 426,*2*. *T. pupoides* (NYIRÖ), U.Oligo.(Chatt.), Fr.; *2a,b*, side, apert. views of holotype, ×115 (*1382).

Baggatella HOWE, 1939, *971, p. 79 [**B. inconspicua*; OD]. Test free, tiny, with relatively low spire, 4 or 5 chambers to whorl; aperture loop-shaped, extending up face of final chamber. *M. Eoc.-U.Oligo.,* N.Am.-Carpathians.——FIG. 427,*1*. **B. inconspicua*, M.Eoc.(Cook Mountain), USA (La.); *1a-c*, side, basal, and apert. views, ×300 (*2117).

Buliminella CUSHMAN, 1911, *404b, p. 88 [**Bulimina elegantissima* D'ORBIGNY, 1839, *1393, p. 51; OD]. Test free, elongate, with high close spiral formed by numerous very high, narrow chambers, commonly with many chambers to whorl and few whorls; wall calcareous, perforate, radial in structure, apertural face just above aperture poreless to sharp angle of apertural ridge, surface smooth to striate, rarely spinose; aperture loop-shaped, with upper end relatively broad, internal tooth plate connecting aperture with that of previous chamber. [Early Cretaceous species referred to *Buliminella* belong to *Praebulimina* or *Caucasina*.] *U.Cret.(Maastricht.)-Rec.,* cosmop. ——FIG. 426,*3,4*. **B. elegantissima* (D'ORBIGNY), Rec., Brazil (*3*), Peru (*4*); *3a-c*, opposite sides and apert. view, ×208 (*2117); *4a*, optical sec.

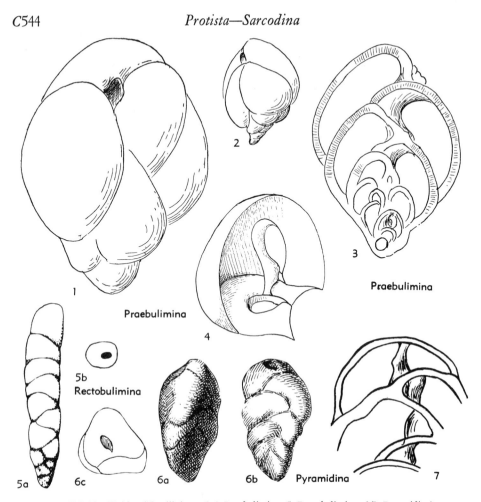

Fig. 428. Turrilinidae (Turrilininae; *1-4, Praebulimina; 5, Rectobulimina; 6,7, Pyramidina*) (p. C545-C546).

showing tooth plate in final chamber from apert. side, ×333; *4b,* successive tooth plates in optical sec. from opposite side, ×333 (*928c).

Buliminellita CUSHMAN & STAINFORTH, 1947, *526, p. 78 [*B. mirifica*; OD]. Test elongate, chambers arranged in high trochospiral coil, approximately 3 to 5 high, narrow chambers to whorl; aperture in early stage as in *Buliminella* but terminal and rounded in adult and produced on neck. *U.Eoc.-Mio.,* Ecuad.-Afr.——FIG. 426,5. *B. mirifica,* U.Eoc., Ecuad.; *5a,b,* side, top views of holotype, ×116 (*2117).

Buliminoides CUSHMAN, 1911, *404b, p. 90 [*Bulimina williamsoniana* BRADY, 1881, *196c, p. 56; OD] [=*Elongobula* FINLAY, 1939, *717c, p. 321 (type, *E. chattonensis*)]. Test free, elongate, early chambers in low trochospiral coil, then spire increasing rapidly in height with coiling around open umbilicus, about 5 chambers to whorl, aligned oblique to axis, septal walls partially resorbed internally so that chambers open into umbilical hollow; wall calcareous, perforate radial in structure; surface smooth or with prominent longitudinal costae which cross sutures obliquely and obscure structure externally; aperture umbilical, with simple tooth plate. [As *Elongobula chattonensis* differs only in the absence of ornamentation from typical *Buliminoides*, the genus is here regarded as synonymous. The Upper Cretaceous *Elongobula creta* FINLAY apparently belongs to *Buliminella.*] *Oligo.-Rec.,* Indo-Pac.-W.trop. Atl. ——FIG. 426,8-11. *B. williamsoniana* (BRADY), Rec., Fiji *(8),* Indon. *(9-11); 8a-c,* opposite sides and apert. view, ×94 (*2117); *9,* long. sec. showing tooth plates, ×210; *10,* long. sec. showing resorbed internal walls, ×150; *11,* transv. sec. showing chambers around hollow umbilical axis, ×150 (*928c).——FIG. 426,*12. B. chattonensis* (FINLAY), L.Oligo.(Duntroon.), N.Z.; *12a-c,* opposite sides and apert. view, ×94 (*2117).

Neobulimina CUSHMAN & WICKENDEN, 1928, *541, p. 12 [*N. canadensis*; OD]. Test free, elongate,

Fig. 429. Turrilinidae (Turrilininae; *1, Tosaia; 2, Quadratobuliminella; 3, Sporobuliminella*) (p. C546-C547).

early stage triserial, later biserial, not compressed; chambers inflated; aperture loop-shaped opening extending up terminal face. *L.Cret.(Alb.)-U.Cret. (Maastricht.)*, cosmop.——Fig. 426,6,7. **N. canadensis*, U.Cret., Can.; *6a,7a*, side views; *6b,7b*, apert. views; all ×208 (*2117).

Praebulimina Hofker, 1953, *939, p. 27 [*Bulimina ovulum Reuss, 1844, *1537, p. 215 (*non Bulimina ovula* d'Orbigny, 1839) =*Bulimina reussi* Morrow, 1934, *1319, p. 195; OD] [=*Praebulimina* Hofker, 1951, *928c, p. 144, *935, p. 6 (*nom. nud.*); *Praebulimina* Thalmann, 1952, *1897j, p. 979 (type, *Praebulimina* sp. Hofker, 1951, *928c, p. 145, *nom. nud.*)]. Test flaring, inflated, chambers triserially arranged, externally similar to *Bulimina*; wall calcareous, perforate, thick and opaque in appearance; aperture loop-shaped, with simple internal tooth plate, instead of complex projecting one of *Bulimina*. *M.Jur. (Bathon.)-U.Cret.(Maastricht.)*, cosmop.——Fig. 428,*1-3*. *P. reussi* (Morrow), U.Cret.(U.Turon.), Sweden; *1*, ext. of megalospheric form, ×268; *2*, ext. of microspheric test, ×43; *3*, long. sec. of microspheric test, showing successive tooth plates, ×268 (*935).——Fig. 428,*4*. *P.* sp., U.Cret., Neth.; opened final chamber showing simple toothplate bordering side of apert. opening and extending to margin of previous septal foramen, ×220 (*928c).

[Although a generic description was given for *Praebuli-*

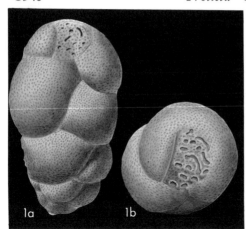

Fig. 430. Turrilinidae (Turrilininae; *1*, *Sporobulimina*) (p. C546).

mina in 1951 (*928c) the genus was a *nomen nudum* as no type-species was designated, though various species of "*Bulimina*" were discussed under the generic heading. Another publication in the same year (*935) discussed only *P. ovula*, but did not state it to be the type-species. THALMANN (1952, *1897j) cited the *Praebulimina* sp. figured by HOFKER (*928c) as type-species, but as this was not a valid named species, *Praebulimina* remained a *nomen nudum* until the designation by HOFKER in 1953 of *Bulimina ovulum* REUSS, 1844, as type-species. As *B. ovulum* REUSS was a homonym of *B. ovula* D'ORBIGNY, the former had been renamed by MORROW, 1934, as *B. reussi*, which is thus the valid name for the European species. HOFKER (1957, *948, p. 184, 187) recognized both *Praebulimina ovulum* (REUSS) and *P. reussi* (MORROW), including the original reference of REUSS in both synonymies, but regarding the American species as distinct from the European one. If so, the American species would require a new name, as *B. ovulum* REUSS cannot be resurrected for the European species and *B. reussi* MORROW was proposed only as a nom. nov. for *B. ovulum* REUSS. However, HOFKER regarded *Bulimina brevis* FRANKE [=*B. brevis* D'ORBIGNY?] as a synonym of *B. reussi* MORROW from the Niobrara formation and the *B. reussi* from the American Gulf Coast as synonymous with *B. ventricosa* BROTZEN, all species being transferred to *Praebulimina*. Many of the Cretaceous species previously placed in *Bulimina*, *Buliminella*, and *Reussella* should be referred to *Praebulimina* or *Pyramidina*.]

Pyramidina BROTZEN, 1948, *241, p. 62 [*Bulimina? curvisuturata* BROTZEN, 1940, *239, p. 29; OD] [=*Pyramidina* BROTZEN, 1940, *239, p. 29 (nom. nud.)]. Test free, flaring, subtriangular in section, chambers broad, low, triserially arranged and subangular; wall calcareous, finely perforate, surface may be somewhat nodose; aperture a high loop-shaped opening which has tendency to close at basal part, remaining only as more or less defined suture connecting subterminal aperture to base of chamber. U.Cret.(Santon.)-Paleoc.(Dan.), Eu.-N.Am.——FIG. 428,6. *P. curvisuturata* (BROTZEN), Paleoc.(U.Dan.); *6a-c,* opposite sides and apert. view, ×100 (*239).——FIG. 428,7. *P. cushmani* (BROTZEN), U.Cret.(L.Campan.), Ger.; apert. portion of long. sec. showing tooth plates, ×160 (*948).

[In 1940, *Bulimina? curvisuturata* was described by BROTZEN (*239) with the statement that it did not wholly agree with that genus because of a tendency to terminal development of the aperture, and that it probably should be placed in a new genus, *Pyramidina*. In the discussion he referred to this species as "*Bulimina (Pyramidina) curvisuturata*," but also discussed "*Reussella (Pyramidina) cushmani*," and since no type-species was designated and 2 species were discussed, the generic name proposed was invalid until 1948, when type designation was made. Although the main generic features given by BROTZEN were the subangular test shape and tendency for the loop-shaped aperture to close at the lower part, with only a suture connecting the opening to the chamber base, the same apertural characters were shown in *Praebulimina* sp. of HOFKER (*928c) from the Upper Cretaceous of the Netherlands. In the diagnosis of *Praebulimina* (*928c, p. 144) HOFKER stated that he included "those Buliminidae found in the Upper Cretaceous of Sweden (Brotzen) and the Netherlands," and cited the publication in which *Bulimina? curvisuturata* was described. In 1957 HOFKER (*948) regarded *Pyramidina* as a synonym of *Reussella*, discussing BROTZEN's *Reussella (Pyramidina) cushmani*, but did not mention the type-species. As noted by HOFKER (*948, p. 202), the Cretaceous species are finely perforate and the tooth plate less complex, in contrast to the more coarsely perforate true *Reussella* of the Cenozoic. *Pyramidina* is therefore here recognized for the subangular finely perforate species particularly characteristic of the Upper Cretaceous, differing from the more coarsely perforate, sharply angular or keeled Cenozoic *Reussella*. It differs from *Praebulimina* in its low, broad and angular rather than rounded or inflated chambers. *Pseudouvigerina* differs in having a distinctly terminal aperture in the adult.]

Quadratobuliminella DE KLASZ, 1953, *1041, p. 435 [*Q. pyramidalis*; OD]. Test similar to *Buliminella* but quadrate in section, chambers elongate, quadriserially arranged; aperture low and umbilical as in *Buliminoides*. *Paleoc.(Dan.),* Bav.-Fr.——FIG. 429,2. *Q. pyramidalis*, Bav.; *2a,b,* side, top views, ×174 (*2117).

Rectobulimina MARIE, 1956, *1221, p. B249 [*R. carpentierae*; OD]. Test similar to *Siphogenerina* in being triserial in early stage, later biserial and finally uniserial; wall calcareous, perforate; aperture terminal, rounded to oval, flush with surface and not produced into phialine lip, presence or absence of internal tooth plates not known. [*Rectobulimina* is tentatively placed in the Turrilinidae, but information as to the internal structure is lacking.] U.Cret.(Maastricht.), Belg.——FIG. 428,5. *R. carpentierae*; *5a,b,* side, apert. views of holotype, ×77.5 (*1221).

Sporobulimina STONE, 1949, *1842, p. 82 [*S. perforata*; OD]. Test elongate, triserial, wall calcareous, perforate, primary aperture narrow elongate slit extending from base of chamber about half distance up apertural face, supplementary apertures consist of numerous irregular openings in face of chamber at one side and adjacent to primary aperture. U.Cret., Peru.——FIG. 430,1. *S. perforata*; *1a,b,* side and apert. views of holotype, ×82 (*2117).

Sporobuliminella STONE, 1949, *1842, p. 81 [*S. stainforthi*; OD]. Test tightly coiled in low spire; with about 4 inflated chambers to whorl; primary aperture low interiomarginal opening with narrow lip, with numerous small supplementary apertures over nodose or pustulose roughly circular area or pore plate extending up terminal face from primary aperture. U.Cret., Peru.——FIG. 429,3. *S. stainforthi*; *3a,b,* opposite sides of holotype, ×93 (*2117).

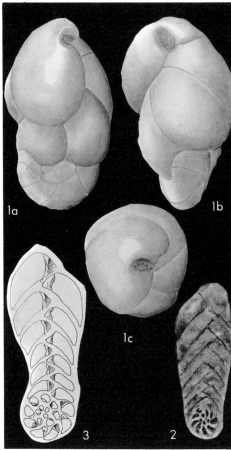

FIG. 431. Turrilinidae (Lacosteininae; *1, Lacosteina; 2,3, Spirobolivina*) (p. C547).

Tosaia TAKAYANAGI, 1953, *1862, p. 30 [*T. hanzawai*; OD]. Test free, small, flaring, early stage obscure, triserial completely or through most of development, rarely with last few chambers in biserial arrangement; wall calcareous, smooth, finely perforate, microstructure not known; aperture basal, relatively small, with a narrow bordering lip. [Originally regarded as belonging to the Heterohelicidae, *Tosaia* was later questionably referred to the Buliminidae by MONTANARO GALLITELLI (*1303). Additional information is required as to wall character and presence or absence of apertural tooth plate.] *Plio., Japan.*——FIG. 429,*1*. **T. hanzawai*; *1a-c*, side, basal and apert. views, ×99 (*1303).

Subfamily LACOSTEININAE Sigal, 1952

[Lacosteininae SIGAL in PIVETEAU, 1952, p. 220]

Early portion planispirally coiled, later changing abruptly to elongate growth axis with 2, 3, or 4 chambers to whorl; aperture loop-shaped, in face of final chamber. *U. Cret.-U.Eoc.*

Lacosteina MARIE, 1945, *1216, p. 295 [*L. gouskovi*; OD]. Test free, elongate; early portion in planispiral coil of few chambers, later changing direction of coiling and forming high spire of about 2 volutions with 3 or 4 chambers to whorl, chambers inflated; sutures distinct, depressed; wall calcareous, finely perforate, surface smooth; aperture loop-shaped, at inner margin of final chamber. *U.Cret.(Campan.),* Morocco-USA (Alaska-Calif.).——FIG. 431,*1*. **L. gouskovi*, Morocco; *1a-c*, side, edge, and apert. views, ×166 (*2117).

[*Lacosteina* differs from *Bulimina* and *Praebulimina* in having the early planispiral coil perpendicular to the plane of coiling of its later high-spired part of the test. MARIE (1945, *1216, p. 295) stated that the genus resembled *Bulimina* in the later stage and the Heterohelicidae in its initial stage, and accordingly suggested that *Lacosteina* represents the ancestral genus of the Buliminidae, which was therefore derived from a planispiral ancestry, rather than from the high-spired *Terebralina*, as CUSHMAN had earlier concluded. Although the ontogeny might suggest such an ancestry, the geological record does not bear out this relationship, since the earliest Buliminidae occur in the Jurassic. *Lacosteina* is apparently a specialized offshoot occurring in the Upper Cretaceous.]

Spirobolivina HOFKER, 1956, *945, p. 915 [*Bolivinopsis pulchella* CUSHMAN & STAINFORTH, 1947, *526, p. 78; OD]. Test free, elongate, with early planispiral stage of about 1.5 volutions, later biserial, compressed; wall thin, calcareous, finely perforate; aperture a loop-shaped opening, with small internal tooth plate similar to *Bolivina*, tooth plates of successive chambers differing in orientation by 180°. [*Spirobolivina* was proposed for calcareous perforate species with internal tooth plate, previously placed erroneously in *Bolivinopsis*, which is an agglutinated form.] *Paleoc.-U. Eoc.,* S.Am.-N.Am.——FIG. 431,*2,3*. **S. pulchella* (CUSHMAN & STAINFORTH), U.Eoc., S.Am. (Ecuad.); *2*, holotype, side view, ×80 (*526); *3*, long. sec., ×120 (*945).

Family SPHAEROIDINIDAE Cushman, 1927

[nom. transl. LOEBLICH & TAPPAN, 1961, p. 300 (ex subfamily Sphaeroidininae CUSHMAN, 1927)] [=Uvellina EHRENBERG, 1839, table opposite p. 120 (partim) (nom. nud.)]

Early portion trochospiral, later streptospiral, with chambers embracing most of preceding ones; aperture interiomarginal, with rounded tooth, or with later secondary sutural openings. *U.Cret.-Rec.*

Sphaeroidina D'ORBIGNY, 1826, *1391, p. 267 [*S. bulloides*; OD (M)] [=*Sexloculina* CŽJŽEK, 1848, *545, p. 138 (type, *S. haueri*); ?*Bolbodium* EHRENBERG, 1872, *687, p. 276 (type, *B. sphaerula*)]. Test subglobular, coiling variable, depending on fluctuation in position of aperture; chambers hemispherical, few, number depending on changes of apertural position and relative size and placement of chambers, each placed centrally about previous aperture, strongly embracing, me-

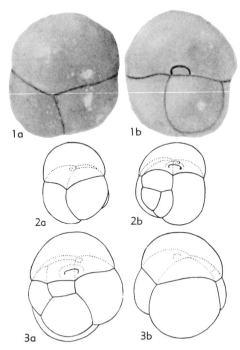

FIG. 432. Sphaeroidinidae; *1-3, Sphaeroidina* (p. C547-C548).

dian apertural planes of later chambers diverging from those of earlier ones alternating to left and right, or constantly to one side, or irregularly to right and left by angle up to 180°, commonly 90°, in latter case resulting in relatively regular spiral; wall of calcite, very finely perforate, radial in structure, surface smooth or faintly roughened near aperture, which is crescentic slit near suture and may occur above junction of 3 chambers, bordered by lip, also may have simple or bifid tooth. [Placed in the Chilostomellidae by CUSHMAN (1948, *486, p. 321), it was transferred to the Cassidulinidae by VAŠÍČEK (1956, *1983, p. 160). However, as both the Chilostomellidae and Cassidulinidae are characterized by a perforate granular wall structure, the radially built Sphaeroidininae have been elevated to a separate family by us (*1177, p. 300) and placed in the superfamily Buliminacea.] *U.Eoc.-Rec.*, cosmop.——FIG. 432, *1-3. *S. bulloides*, Rec., Italy (*1*), Mio.(Torton.), Czech.(Morav.) (*2,3*); *1a,b*, opposite sides of topotype, ×73 (*530); *2a,b, 3a,b*, diagram. figures showing chamber arrangement, opposite sides seen through sides of final chamber with preceding whorl indicated by dotted lines, ×44 (*1983).

Pullenoides HOFKER, 1951, *935, p. 10 [*P. senoniensis*; OD]. Test free, subglobular, early stage trochospiral, later chambers planispiral and embracing, with tendency to biseriality; wall calcareous, very finely perforate, as in *Sphaeroidina*, opaque, surface smooth, microstructure not determined; aperture a loop-shaped opening in early stage, later with numerous, small supplementary sutural openings, no internal tooth plate. *U.Cret. (U.Senon.)*, Neth.——FIG. 433,*1-3. *P. senoniensis*; *1a-c*, opposite sides and edge view, ×74 (*2117); *2,3*, horiz. and transv. secs., ×53 (*935).

Family BOLIVINITIDAE Cushman, 1927

[*nom. transl.* GLAESSNER, 1936, p. 127 (*ex* subfamily Bolivinitinae CUSHMAN, 1927, p. 61)] [=Bolivininae GLAESSNER, 1937, p. 420; Bolivinidae HOFKER, 1951, p. 48]

Test biserial at least in young stage, aperture comma-shaped, parallel to compression of test, basal or terminal, with internal tooth plate. *U.Trias.-Rec.*

Bolivinita CUSHMAN, 1927, *429, p. 90 [*Textilaria quadrilatera* SCHWAGER, 1866, *1703, p. 253; OD]. Test free, compressed, broad sides flat to concave, rectangular in transverse section, 4 angles of test with strongly developed axial costae; chambers biserial throughout, gradually increasing in relative breadth, proloculus may have one or more spines; sutures straight, depressed on lateral edges, oblique and may be limbate on broader faces; wall thin, calcareous, perforate radial in structure, completely covered by minute pores and sporadic larger ones, surface of early portion may be spinose or vertically costate; aperture basal, subcircular, elliptical, perpendicular to suture and with bordering lip, tooth plate may project slightly, somewhat arched at upper surface, flaring and curved internally and may be spatulate at free lower end, those of successive chambers alternating in direction. *Mio.-Rec.*, Atl.-Pac.-Kar Nicobar-N.Z.-N. Am.-Java-Sumatra-Australia.——FIG. 434,*1-3. *B. quadrilatera* (SCHWAGER), Rec., Philip.; *1a,b*, side, edge views of microspheric form; *2a,b*, side, edge views of megalospheric form; *3*, edge view of megalospheric form with portion of final chamber removed to show tooth plate; all ×65 (*1303).

[*Bolivinita* closely resembles *Bolivina* in chamber arrangement and apertural features, differing in its marginal keels, quadrate section, and absence of retral processes. Although regarded as a synonym of *Bolivina* by HOFKER (1951, *928c, p. 106), *Bolivinita*, as here understood, has a more restricted geologic occurrence, and therefore its retention seems to be useful. Such a taxonomic modification, with specialized morphology and limited geologic occurrence, may be afforded generic or subgeneric status by different workers, but is here regarded as of generic status.]

Altistoma DE KLASZ & RÉRAT, 1962, *1043, p. 180 [*A. scalaris*; OD]. Test biserial, strongly overlapping chambers with lobulate lower margin, sutures depressed; wall calcareous, finely perforate, surface smooth; aperture large, high symmetrical arch bordered by thickened lip, in laterally compressed apertural face. *Eoc.-L.Mio.*, W.Afr. (Gabon).——FIG. 434,*4. *A. scalaris*, L.Mio.; *4a-c*, side, edge, and apert. views of holotype, ×133 (*1043).

Foraminiferida—Rotaliina—Buliminacea C549

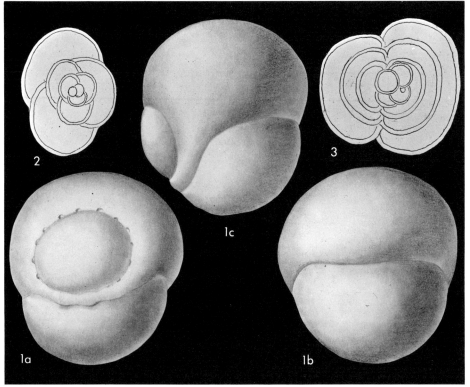

FIG. 433. Sphaeroidinidae; *1-3, Pullenoides* (p. C548).

Bolivina D'ORBIGNY, 1839, *1393, p. 60 [*B. plicata; SD CUSHMAN, 1911, *404b, p. 31] [=Grammostomum EHRENBERG, 1839, *667, table opposite p. 120 (type, G. tenue); Proroporus EHRENBERG, 1844, *673, p. 75 (type, P. lingua); Clidostomum EHRENBERG, 1845, *675, p. 358 (type, C. polystigma); Afrobolivina REYMENT, 1959, *1557, p. 19 (type, A. afra)]. Test elongate, may be somewhat compressed; chambers broad, low, biserially arranged throughout, basal margins of chambers with retral processes or backward directed chamber overlaps; wall calcareous, perforate, radial in structure, smooth, striate, or costate and may have marginal keel; aperture a narrow, elongate loop up chamber face, one margin ending blindly or bent upward as collar, opposite border attached to one side of doubly folded internal tooth plate (U-shaped in section), attached half of tooth plate projecting inward to coalesce with free half of tooth plate of previous foramen, free half of tooth plate projecting through aperture at one extremity and bisecting it, narrowing rapidly inward, tooth plate thus being trough-shaped structure with concave portion alternately turning from one side to opposite in successive chambers. *U. Cret.-Rec.,* cosmop.——FIG. 434,7. *B. plicata,* Rec., Panama; *7a,b,* side, apert. views, ×99 (*2117).——FIG. 434,*8,9. B. afra* (REYMENT),
U.Cret.(Maastricht.), Nigeria; *8a,b,* side and top views of microspheric test showing surface ribs and chamber overlaps, ×40; *9,* dissected final chamber showing rear side of tooth plate *(t)*, intercameral foramen *(f)*, lip of preceding tooth plate *(l)*, apertural depression *(d)*, crenulated terminal wall of penultimate chamber *(c)* and crenulations from interior *(cr),* ×147 (*1557).

[Although it has been stated that *Bolivina* and *Virgulina* [=*Fursenkoina*] are intergradational (*472), *Bolivina,* as all Buliminacea, has a perforate radial wall structure, and *Fursenkoina* has a perforate granular wall structure. HOFKER (*928c), REYMENT (*1557) and others have regarded *Bolivinita* and *Bolivinoides* as synonyms of *Bolivina,* but they are here considered to be distinct, although all are biserial in chamber arrangement and possess internal tooth plates. The differing geologic ranges of these distinct morphologic types seem to indicate their generic validity. *Bolivina* is therefore restricted to include biserial species with internal tooth plates, basal aperture, radially built perforate hyaline walls, and chamber retral projections or overlaps, varying from a few broad lobes, as in the type-species, to the numerous smaller projections, as in *Afrobolivina afra.* As the so-called secondary vertical septa described for *Afrobolivina* are merely internal indentations of the wall between chamber overlaps, *Afrobolivina* is regarded as a synonym of *Bolivina.* Species without chamber overlaps, commonly keeled and strongly compressed, are placed by us in *Brizalina.*]

Bolivinoides CUSHMAN, 1927, *429, p. 89 [*Bolivina draco MARSSON, 1878, *1228, p. 157; OD]. Test free, rhomboidal, flaring, compressed; chambers low and broad, biserially arranged throughout; septa thick, sutures oblique, obscured externally by strong ornamentation; wall calcareous, single-layered, lamellar, finely perforate, radial in structure, interior tuberculate, exterior surface with strong longitudinal costae and tuberculate; aperture elongate, loop-shaped, basal, extending up face of final chamber with bordering lip and in-

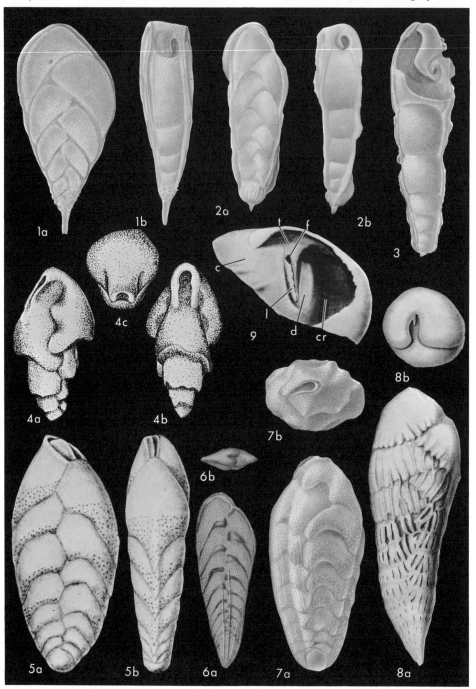

Fig. 434. Bolivinitidae; *1-3, Bolivinita; 4, Altistoma; 5,6, Brizalina; 7-9, Bolivina* (p. C548-C549, C552).

Foraminiferida—Rotaliina—Buliminacea C551

ternal tooth plate. [*Bolivinoides* may have an internal tuberculate wall, but does not show the exterior chamber overlaps or retral processes found in *Bolivina*, as here restricted. It is also char-

Fig. 435. Bolivinitidae; *1,2, Bolivinoides; 3, Grimsdaleinella; 4, Tappanina; 5, Unicosiphonia; 6,7, Gabonella* (p. C549-C555).

acterized by heavy longitudinal ornamentation.]
U. Cret.(U. Santon.)-Paleoc., Eu.-N. Am.-Carib.-S. Am.-Australia-N.Z.-Indon.——Fig. 435,*1,2*. **B. draco* (Marsson), U.Cret.(Campan.), Ger.; *1a-c,* side, edge, and top views, ×97; *2,* dissected specimen showing thick septa, internally tuberculate wall, and apert. tooth plate in later chambers, ×100 (*1303).

Brizalina Costa, 1856, *392, p. 296 [**B. aenariensis*; OD (M)]. Test elongate, tapering, commonly compressed and laterally carinate, biserial throughout, lacking basal chamber lobes, crenulations or retral processes of *Bolivina* but having straight or curved, commonly limbate sutures; wall calcareous, perforate, radially built, with ornamentation consisting of variously arranged pores, longitudinal costae, carinae, and marginal or apical chamber spines; aperture loop-shaped, extending up from base of final chamber, with tooth plate as in *Bolivina. U.Trias.-Rec.,* cosmop.——Fig. 434,*5;* 436,*1*. **B. aenariensis,* Rec., Ire. (434,*5*), Plio., Italy (436,*1*); 434,*5a,b,* side and apert. views, ×75 (*472); 436,*1a,b,* holotype, side, edge views, approx. ×60 (*700).——Fig. 434,*6*. *B.* sp. cf. *B. vadescens* (Cushman), Rec., Sweden; *6a,b,* side view and edge view showing projecting tooth plate, ×140 (*924).——Fig. 436,*2*. *B. pseudopunctata* (Höglund), Rec., Sweden; *2a,* optical sec. of apert. end showing internal tooth plate in alternating arrangement; *2b,* transv. sec. of final chamber through aperture showing U-shaped sec. of tooth plate fastened at one border to chamber wall, ×500 (*924).

[*Brizalina,* as here emended, includes many species previously placed in *Bolivina* that do not show retral chamber processes or crenulations, such as are found in *Bolivina plicata.* The original description of *Brizalina* erroneously described the presence of a neck; this was on the basis of a broken specimen in which only the axis and tooth plates of the final pair of chambers were preserved. Similar preservation has been noted in many specimens of the type-species.]

Gabonella de Klasz, Marie & Meijer, 1960, *1042, p. 167 [**G. elongata* de Klasz & Meijer; OD]. Test free, elongate, biserial, chambers broad and low, plane of biseriality somewhat twisted; sutures strongly depressed, commonly with strong reentrant toward center of chamber margins; wall calcareous, finely perforate, radial in structure; aperture hook-shaped, extending upward from base of final chamber, then curving sharply to run nearly parallel to suture, with narrow bordering lip. [*Gabonella* differs from *Grimsdaleinella* in its distinctly twisted test, low hook-shaped aperture, and small tooth, instead of high comma-shaped aperture. It differs from *Bolivina* in lacking crenulated sutures or retral chamber processes and distinctive tooth plate.] *U.Cret.(Santon.-Maastricht.), ?Paleoc.(Dan.),* Afr.——Fig. 435, *6,7*. **G. elongata,* U.Cret.(Maastricht.), Gabon; *6a,b,* side and edge views showing twisted test and deeply incised sutures; *7a-c,* opposite sides and edge of specimen in which twisting results

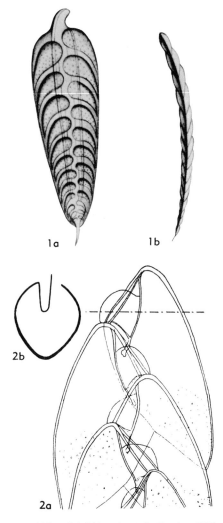

Fig. 436. Bolivinitidae; *1,2, Brizalina* (p. C552).

in nearly quadrate test, apert. tooth visible; all ×62 (*2117).

Grimsdaleinella Bolli, 1959, *162, p. 1 [**G. spinosa;* OD]. Test free, chambers biserially arranged, inflated, and laterally produced into spinelike extensions; wall calcareous, finely perforate, surface smooth, hispid or striate; aperture an asymmetrical arch or slit extending up face, presence or absence of tooth plate unknown. *U.Cret. (Turon.-Coniac.),* Trinidad.——Fig. 435,*3*. **G. spinosa; 3a-c,* opposite sides and top view of holotype, ×73 (*162).

[Originally regarded as belonging to the Heterohelicidae, and differing from *Chiloguembelina* in having lateral spines, the genus is here judged to belong probably to

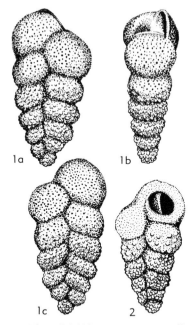

Fig. 437. Bolivinitidae; *1,2, Laterostomella* (p. C553).

the Bolivinitidae, as indicated by its loop-shaped aperture and broad low chambers, although no information is available as to the presence of an internal tooth plate. The original generic description stated "biserially arranged throughout or planispiral in early stage," but the description of the type-species of the monotypic genus stated "biserial throughout," further suggesting the possibility of a relationship with the Bolivinitidae.]

Laterostomella DE KLASZ & RÉRAT, 1962, *1043, p. 177 [*L. guembeliniformis*; OD]. Test elongate, biserial, chambers inflated; sutures depressed; wall calcareous, finely to coarsely perforate, with rugose or striate surface; aperture elongate, with bordering lip, situated in cavity at one side of apertural face, outer margin somewhat flaring and infolded to form tooth plate. *Mio.*, W.Afr.(Gabon).——FIG. 437,*1*. *L. guembeliniformis*, Burdigal.; *1a-c*, opposite sides and edge view of holotype; *2*, dissected specimen showing tooth plate; all ×133 (*1043).

Loxostomoides REISS, 1957, *1528a, p. 241 [*Bolivina applini* PLUMMER, 1927, *1461, p. 69 (recte= *B. applinae*); OD] [=*Bolivina (Loxostomoides)* REYMENT, 1959, *1557, p. 16 (obj.)]. Test free, narrow, elongate, oval in section; chambers biserial, with tendency to become uniserial in later stage; sutures with retral processes or crenulations of base of chambers; wall calcareous, perforate, radial in structure; aperture an elongate loop extending from base of chamber in early stages, becoming areal in later stages although never central and completely terminal, internal tooth plate present. *U.Cret.(Senon.)-Paleoc.*, N.Am.-Israel.——FIG. 438,*1*. *L. applinae* (PLUMMER), Paleoc. (Midway.), USA(Tex.); *1a-c*, side, edge, and apert. views, ×50 (*472).

Rectobolivina CUSHMAN, 1927, *431, p. 68 [*Sagrina bifrons* BRADY, 1881, *196c, p. 64; OD] [=*Geminaricta* CUSHMAN, 1936, *468, p. 61 (type, *Bolivinella virgata* CUSHMAN, 1929, *441, p. 33)]. Test elongate, may be slightly compressed or circular in section, in early stage biserial, later uniserial, biserial stage much reduced in megalospheric generation and may consist of only slightly eccentric second chamber; wall calcareous, finely perforate, radial in structure, surface smooth, nodose, or more commonly longitudinally costate; aperture terminal, rounded to elongate, with internal twisted tooth plate, those of successive chambers alternating in position in planes 180° apart. *M.Eoc.-Rec.*, cosmop.——FIG. 438,*2-5*. *R. bifrons* (BRADY), Rec., Pac.; *2a,b*, side and apert. views of microspheric test; *3a,b*, side and apert. views of megalospheric test, ×65 (*2117); *4*, sec. showing tooth plates; *5a,b*, side and edge views of isolated tooth plate, enlarged (*928c).——FIG. 438,*6-8*. *R. virgata* (CUSHMAN), Mio., Fr.; *6a,b*, side and apert. views of holotype with only biserial stage; *7a,b*, *8a,b*, side and apert. views of hypotypes, ×102 (*2117).——FIG. 438, *9-11*. *R. raphana* (PARKER & JONES), Rec., Ind.O.; *9-11a,b*, side and apert. views of paratypes showing variation in degree of biserial development, ×55 (*2117).

[As shown by HOFKER (1951, *928c), many species previously have been placed incorrectly in *Siphogenerina*, *Loxostomum*, and *Bifarina* that should be assigned to *Rectobolivina*, as they are unlike the type-species of those genera. The present generic definition of *Rectobolivina* also includes the type-species of *Geminaricta*, hence the latter is a junior synonym. The type-specimens of *Uvigerina (Sagrina) raphanus* PARKER & JONES were examined by us in 1953 in the British Museum (Natural History); a lectotype was selected and has been designated (BMNH-ZF3582), together with paratypes (BMNH-ZF3581), on Recent *Chama hippopus* from the Indian Ocean. As the type-specimens of this species have up to 22 ribs and are either biserial in the early stage or uniserial throughout, they are regarded as specifically and generically distinct from *Siphogenerina costata* SCHLUMBERGER for which the species had been considered a senior synonym. *S. costata* has 5 or 6 costae, an early triserial microspheric stage, and a biserial early stage in the megalospheric form. It was also regarded as a *Rectobolivina* by HOFKER (1951, *928c, p. 62). The "double aperture" of *Geminaricta* was illustrated only on broken specimens. Complete specimens show only an elongate aperture. Probably an erroneous interpretation of the fragmentary remains of the tooth plate led to separating *Geminaricta* as a distinct genus.]

Tappanina MONTANARO GALLITELLI, 1955, *1301, p. 190 [*Bolivinita selmensis* CUSHMAN, 1933, *459, p. 58; OD]. Test biserial, flaring, sides flattened, resulting in transverse section; chambers cuneiform, apparently concave on broad sides, more or less inflated laterally, with well-developed horizontal or arched rib across chambers and along zigzag suture and lateral margins; sutures depressed, straight or arched; wall calcareous, finely perforate, surface appearing rough owing to development of ridges; aperture narrow, elongate, at base of final chamber, with tooth plate as in *Bolivina*. [*Tappanina* is characterized by its strong horizontal carinae, narrow incised sutures, and degeneration into discontinuous thickenings of the 4 axial lamellar sutural costae which are characteristic of *Loxostomum*, but it differs in the bolivine character of the aperture.] *U.Cret.-Paleoc.*, N.Am.-Eu.——FIG. 435,*4*. **T. selmensis* (CUSH-

FIG. 438. Bolivinitidae; *1, Loxostomoides; 2-11, Rectobolivina* (p. C553).

Foraminiferida—Rotaliina—Buliminacea C555

MAN), U.Cret., USA(Tenn.); *4a,b,* side, apert. views of holotype, ×130 (*1302).

Unicosiphonia CUSHMAN, 1935, *465, p. 81 [*U. *crenulata*; OD]. Test similar to *Rectobolivina* but chambers with basal crenulations or retral processes as in *Bolivina* and *Loxostomoides*; aperture ter-

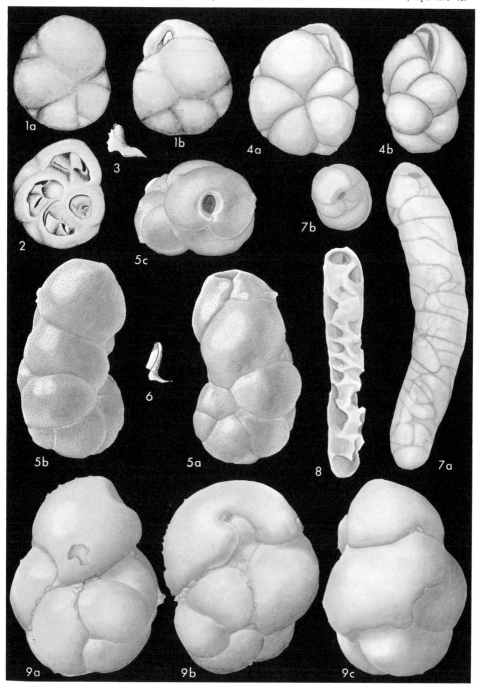

FIG. 439. Islandiellidae; *1-4, Islandiella; 5,6, Cassidulinoides; 7,8, Orthoplecta; 9, Stichocassidulina* (p. C556).

minal, rounded, with internal columellar process or tooth plate. *Tert.,* Atl.(Georges Bank).——Fig. 435,5. **U. crenulata*; 5a,b, side, top views of holotype, ×64 (*2117).

Family ISLANDIELLIDAE Loeblich & Tappan, n.fam.

Test with biserially arranged enrolled chambers, at least in early stage, or derived from such, later stage may uncoil; as in Cassidulinidae, but with calcareous, perforate, radiate fibrous wall and primary aperture provided with internal tooth plate extending inward from aperture to previous foramen. *?U.Cret., Paleoc.-Rec.*

Islandiella NØRVANG, 1958, *1361, p. 26 [**Cassidulina islandica* NØRVANG, 1945, *1359, p. 41; OD] [=*Cassilamellina* VOLOSHINOVA, 1960, *2020, p. 59 (type, *Cassidulina californica* CUSHMAN & HUGHES, 1925, *508, p. 12)]. Test relatively large, lenticular to subglobular, periphery rounded, umbilicus closed; chambers biserially arranged and planispirally enrolled, sutures slightly depressed; wall calcareous, thick, perforate, lamellar, radiate, fibrous in microstructure, surface smooth; aperture elongate, interiomarginal opening, with internal platelike tooth extending from posterior edge of aperture to anterior corner of preceding foramen and with free margin projecting from aperture and partially closing it. [*Islandiella* differs from *Cassidulina* in having a radiate, instead of granular, wall structure and in the presence of an internal tooth plate.] *?U.Cret., Paleoc.-Rec.,* cosmop.——Fig. 439,*1-3.* **I. islandica* (NØRVANG), Rec., Iceland; *1a,b,* opposite sides, showing elongate aperture and projecting tooth plate; *2,* partially dissected specimen showing free tongue of internal platelike tooth at base of open chambers; all ×33 (*1361); *3,* isolated tooth plate, enlarged (*928c).——Fig. 439,*4. I. californica* (CUSHMAN & HUGHES), Pleist., USA(Calif.); *4a,b,* side, edge views, ×37 (*766).

Cassidulinoides CUSHMAN, 1927, *431, p. 84 [**Cassidulina parkeriana* BRADY, 1881, *196c, p. 59; OD]. Test free, elongate, robust, early stage subglobular with chambers biserially arranged and enrolled as in *Cassidulina*, later uncoiling, but continuing biserial development; wall calcareous, perforate, radial in structure; aperture in adult loop-shaped, extending upward from base of chamber into rounded opening at its summit. *U.Eoc.-Rec.,* Atl.-Pac.-N. Am.-Australia-S. Am.-Carib.-Indon.-Japan-Eu.——Fig. 439,*5,6.* **C. parkeriana* (BRADY), Rec., Falk. Is.; *5a-c,* opposite sides and top view, ×153 (*2117); *6,* isolated tooth plate, enlarged (*928c).

Orthoplecta BRADY, 1884, *200, p. 355, 428 [**Cassidulina (Orthoplecta) clavata*; OD (M)] [=*Cassidulina (Orthoplecta)* BRADY, 1884, *200, p. 355, 428 (obj.)]. Test free, elongate, narrow, slightly arcuate, of nearly equal diameter throughout, no regular chamber arrangement, but with spiraling internal column, which gives extremely irregular septation as it spirals and in places touches exterior wall; wall calcareous, finely perforate, radial in structure; aperture subterminal, ovate, just above sutural junction. [Although originally considered a subgenus of *Cassidulina*, *Orthoplecta* has a perforate radial rather than a granular wall structure, and is neither cassiduline nor biserial in the early stage.] *Rec.,* Pac.——Fig. 439,*7,8.* **O. clavata*; *7a,b,* side, top views of holotype; *8,* dissected hypotype, ×146 (*1166).

Stichocassidulina STONE, 1946, *1841, p. 59 [**S. thalmanni*; OD]. Test subglobular, periphery rounded; chambers inflated, biserial and enrolled as in *Islandiella*, involute; sutures depressed; wall calcareous, finely perforate, microstructure unknown; aperture large loop-shaped opening in apertural face, perpendicular to basal suture, partially covered by toothlike plate, numerous small, secondary, sutural openings also occurring along all sutures of test. [*Stichocassidulina* is here placed with *Islandiella* because of the presence of the apertural tooth. Confirmation of its placement requires information as to the wall microstructure.] *U.Eoc.,* S.Am.——Fig. 439,*9.* **S. thalmanni,* Peru; *9a-c,* opposite sides and edge view of holotype, ×80 (*2117).

Family EOUVIGERINIDAE Cushman, 1927

[*nom. transl.* LOEBLICH & TAPPAN, 1961, p. 300 (*ex* subfamily Eouvigerininae CUSHMAN, 1927, p. 63)]——[=Stilostomellinae FINLAY, 1947, p. 275]

Test biserial in young, later may become uniserial; aperture terminal, with internal siphon, and may have everted phialine lip. *L.Cret.-Rec.*

Eouvigerina CUSHMAN, 1926, *424, p. 4 [**E. americana* (=*Loxostomum aculeatum* EHRENBERG, 1854, *680, p. 22); OD] [=*Zeauvigerina* FINLAY, 1939, *717a, p. 541 (type, *Z. zelandica*)]. Test biserial throughout, but may be slightly twisted, final chamber nearly central in position; sutures depressed; wall calcareous, finely perforate, surface may be smooth, carinate or hispid; aperture terminal, with neck and phialine lip, commonly with crenulated margin, internally with thin columellar tooth plate. [Although the presence of an internal apertural tooth plate has not been demonstrated in the type-species of *Zeauvigerina*, owing to unfavorable preservation and lack of sufficient material for sectioning, it is here regarded as congeneric with *Eouvigerina* because of the similarity in chamber arrangement and apertural characters, including the crenulated phialine lip.] *L.Cret.(Alb.)-U.Eoc.,* N.Am.-N.Z.-Eu.——Fig. 440,*1-3.* **E. aculeata* (EHRENBERG),

Foraminiferida—Rotaliina—Buliminacea C557

U.Cret.(Campan.), USA (Tex.) *(1,2)*, Neth. or Ger. *(3)*; *1a,b*, side, apert. views of holotype of *E. americana*, ×162; *2*, dissected specimen showing internal tooth plate, ×162 (*1303); *3a,b*, vert. secs. through breadth and thickness showing character of tooth plates, ×106 (*948).——FIG. 440,

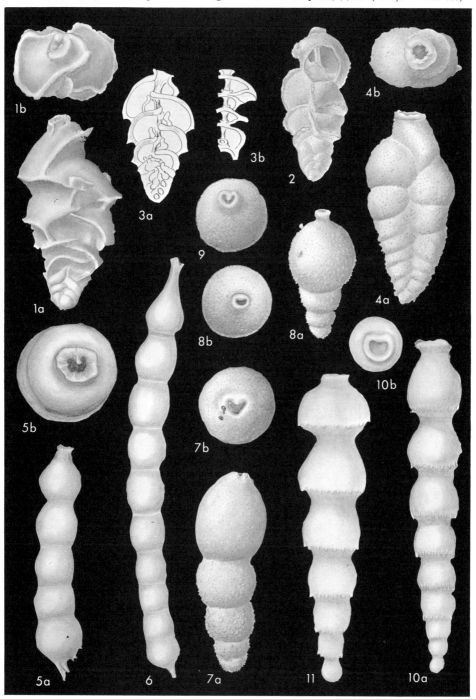

FIG. 440. Eouvigerinidae; *1-4, Eouvigerina; 5,6, Siphonodosaria; 7-11, Stilostomella* (p. C556-C559).

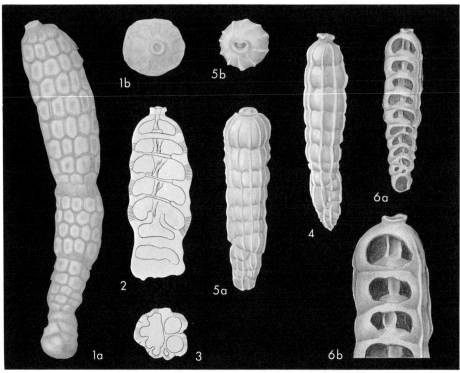

Fig. 441. Eouvigerinidae; *1-3, Millettia; 4-6, Siphogenerinoides* (p. C558).

4. *E. zelandica* (Finlay), Up.M.Eoc., N.Z.; *4a,b,* side, top views of paratype, ×180 (*1303).

Millettia Schubert, 1911, *1689b, p. 89 [*non* Sherborn, 1893, *1731a, p. 206, *nom. correct. pro Milletia* Wright, 1889, *2080, p. 448 (*nom. nud.*); *non Milletia* Duncan, 1889] [*Sagrina? tessellata* Brady, 1884, *200, p. 85; SD Schubert, 1911, *1689a, p. 320] [=*Schubertia* A. Silvestri, 1912, *1772, p. 68 (*non Schubertia* Gistl, 1848) (*nom. subst. pro Millettia* Schubert, 1911) (obj.)]. Test free, elongate, narrow, arcuate, very early portion biserial, later portion consisting of few elongate subcylindrical chambers. rapidly increasing in height and subdivided into chamberlets by vertical and horizontal partitions in honeycomb pattern, chamberlets arranged in regular transverse rows; wall calcareous, perforate radial in structure; surface marked into hexagonal patterns by junction of chamberlet walls with outer wall; aperture terminal, rounded with slight lip and internal tooth plate. *Rec.,* Pac.——Fig. 441, *1-3.* *M. tessellata* (Brady), Admiralty Is. *(1),* Indon. *(2,3); 1a,b,* side, top views of lectotype (BMNH-ZF2359), ×146 (*1166); *2,* long. sec., showing tooth plates, ×160; *3,* horiz. sec. showing vertical partitions and chamberlets, ×210 (*928c).

[Hofker (1951, *928c, p. 67) incorrectly restricted the genus *Sagrina* to *S. tessellata,* but the type of *Sagrina* is *S. pulchella* by monotypy. Furthermore, *S. tessellata* had been made the type-species for *Millettia* and *Schubertia. Schubertia* was proposed as a replacement for *Millettia* Schubert (*non* Sherborn; *non Milletia* Wright; *nec Milletia* Duncan), but *Schubertia* is also preoccupied by the molluscan genus *Schubertia* Gistl, 1848. Furthermore, *Millettia* Sherborn (*nom. correct.*) and *Milletia* Wright were both *nomina nuda,* hence have no standing in zoological nomenclature. *Milletia* Duncan, 1889, is an echinoid genus, but according to the Rules of Nomenclature (Art. 56) a difference in spelling of even one letter is sufficient to prevent generic homonymy; hence *Millettia* Schubert is here reinstated.]

Siphogenerinoides Cushman, 1927, *431, p. 63 [*Siphogenerina plummeri* Cushman, 1926, *422, p. 18; OD]. Test elongate, biserial in early stage in both microspheric and megalospheric forms, later uniserial, with straight, nearly horizontal sutures; wall calcareous, perforate, surface with numerous longitudinal costae; aperture terminal, elliptical or reniform, with internal tooth plate of spoutlike shape, those of successive chambers with concave side facing in alternate directions, each apertural foramen except that of final chamber connecting to terminal end of tooth plate of its own chamber and also to that of succeeding chamber, convex surface of both sections being oriented toward opening. *U.Cret.-Paleoc.,* N.Am.-Afr.——Fig. 441,*4-6.* *S. plummeri* (Cushman), U.Cret.(Maastricht.), USA(Tex.); *4,* side view of microspheric form, ×66 (*2117); *5a,b,* side, top views of megalospheric form, ×66; *6a,b,* long. sec. showing tooth plates, ×66 and ×133 (*1303).

Siphonodosaria A. SILVESTRI, 1924, *1779, p. 18 [*Nodosaria abyssorum BRADY, 1881, *196c, p. 63; SD (SM) CUSHMAN, 1927, *431, p. 67] [=Sagrinnodosaria JEDLITSCHKA, 1931, *985, p. 125 (type, Nodosaria abyssorum BRADY, 1881, SD LOEBLICH & TAPPAN, herein) (obj.)]. Test free, narrow, elongate, uniserial, straight to arcuate; chambers subglobular, proloculus may have basal spines; sutures constricted; wall calcareous, thick, perforate, radial in structure; aperture rounded, produced on slight neck, bordered with phialine lip, and with distinct teeth projecting into aperture. Eoc.-Rec., Eu.-N.Am.-S.Am.-Carib.-Atl.-Pac.——FIG. 440,5,6. *S. abyssorum (BRADY), Rec., S.Pac.; 5a,b, side, apert. views of lectotype, here designated (BRADY, 1884, *200, pl. 63, fig. 8) (BMNH-ZF3649), 5a, ×22, 5b, ×48; 6, paratype (BMNH-ZF1926), one of unfigured syntypes, ×22 (*2117).

[GALLOWAY (1933, *762, p. 376) regarded Nodogenerina as a synonym of Siphonodosaria and STAINFORTH (1952, *1833, p. 7) also stated that "no difference is readily apparent between Siphonodosaria SILVESTRI and Nodogenerina CUSHMAN." Siphonodosaria is here restricted to forms with completely crenulate or dentate phialine lip, in addition to the distinct apertural tooth, whereas Stilostomella (including Nodogenerina) has a simple lip and single tooth.]

Stilostomella GUPPY, *843, p. 649 [*S. rugosa; OD] [=Nodogenerina CUSHMAN, 1927, *428, p. 79 (type, N. bradyi =Sagrina virgula BRADY, 1884, *200, p. 583, partim)]. Test free, elongate, uniserial and rectilinear, with gradually enlarging subglobular chambers; wall calcareous, hyaline, finely perforate, surface may be spinose, or spines may be restricted to lower chamber margin; aperture terminal, may be produced on neck, with phialine lip and slight indentation at one side owing to surface reflection of internal spatulate tooth. Cret.-Rec., Pac.-Atl.-Carib.-N.Z.——FIG. 440,7-9. *S. rugosa, Mio., Trinidad; 7-8a,b, 9, side and top views, of paratypes, ×33 (*2117).——FIG. 440,10,11. S. bradyi (CUSHMAN), Rec., Brazil(off Pernambuco); 10a,b, side, top views of lectoytpe of Sagrina virgula BRADY (1884, *200, pl. 76, fig. 8) here designated, BMNH-ZF2363; 11, megalospheric paratype, ×146 (*2117).

[Stilostomella was regarded as unrecognizable by CUSHMAN (1948, *486, p. 277) and he placed Nodogenerina in the Heterohelicidae. Stilostomella was regarded as a valid genus in the Pleurostomellidae by GALLOWAY (1933, *762, p. 384) and Nodogenerina was considered a synonym of Siphonodosaria in the Uvigerinidae. FINLAY (1947, *717e, p. 273) regarded Nodogenerina and Siphonodosaria both as junior synonyms of Stilostomella, placing the latter genus in the Lagenidae [=Nodosariidae] in a new subfamily Stilostomellinae. Stilostomella is here separated from Siphonodosaria in having a single tooth or indentation of the phialine lip, whereas Siphonodosaria has a more prominent tooth and the entire inner margin of the lip is crenulate or dentate. The prominent apertural tooth, instead of a radial or slit aperture, separates it from the Nodosariidae.]

Family BULIMINIDAE Jones, 1875

[nom. correct. EIMER & FICKERT, 1899, p. 680 (pro family Buliminida JONES in GRIFFITH & HENFREY, 1875, p. 320)]—— [All names cited are of family rank; dagger (†) indicates partim]——[=Stichostèguest D'ORBIGNY, 1826, p. 251 (nom. neg.; nom. nud.); =Hélicostèguest D'ORBIGNY, 1826, p. 268 (nom. neg.; nom. nud.); =Uvellinat EHRENBERG, 1839, table opposite p. 120 (nom. nud.); =Helicosorinat EHRENBERG, 1839, table opposite p. 120 (nom. nud.); =Equilateralidaet D'ORBIGNY IN DE LA SAGRA, 1839, p. xxxvii, 11 (nom. nud.); =Turbinoidaet D'ORBIGNY in DE LA SAGRA, p. xxxviii, 71 (nom. nud.); =Aequilateralidaet D'ORBIGNY, 1846, p. 28 (nom. nud.); =Uvellidaet REUSS, 1860, p. 225 (nom. nud.)]——[=Buliminidee SCHWAGER, 1876, p. 479; =Buliminidea SCHWAGER, 1877, p. 19; = Bulimina LANKESTER, 1885, p. 847; =Buliminae DELAGE & HÉROUARD, 1896, p. 140]——[=Pavoninidae EIMER & FICKERT, 1899, p. 678; =Globobuliminidae HOFKER, 1956, p. 908; =Hyalovirgulinidae HOFKER, 1956, p. 45 (nom. nud.)]

Test high trochospiral, with not more than 3 chambers to whorl, may reduce to biserial; aperture a loop in apertural face, with platelike internal tooth connecting successive chambers, or aperture may be indistinct and represented only by pores in terminal chamber face. Paleoc.-Rec.

Subfamily BULIMININAE Jones, 1875

[nom. correct. BRADY, 1881, p. 44 (pro subfamily Buliminidae SCHWAGER, 1877, p. 19)] [=Buliminae RHUMBLER, 1895, p. 89; =Globobulimininae HOFKER, 1951, p. 248]

Test triserial throughout; aperture loop-shaped, with distinctive tooth plate. Paleoc.-Rec.

Bulimina D'ORBIGNY, 1826, *1391, p. 269 [*B. marginata; SD CUSHMAN, 1911, *404b, p. 76]. Test triserial in early stage, may tend to reduce to uniserial in later portion; wall calcareous, finely to coarsely perforate, radial in structure; aperture extending up from base of apertural face, with free border that may have elevated rim and fixed border attached to internal folded tooth plate, which with fixed shank is attached to internal chamber wall below aperture, with free shank that may be dentate or smooth, flaring or enrolled and subtubular. [Bulimina differs from Praebulimina in having a tooth plate with developed border, and from Globobulimina in one shank of the tooth plate free, instead of both fixed, and in lacking strongly embracing chambers.] Paleoc.-Rec., cosmop.——FIG. 442,1-3. *B. marginata, Rec., Italy (1), Rec., Sweden (2); 1a,b, side, apert. views, ×50 (*519); 2a,b, apert. end showing tooth plate (t), aperture lip (l), and free shank of tooth plate (s), ×105 (*924); 3, isolated tooth plate, enlarged (*928c).

Globobulimina CUSHMAN, 1927, *431, p. 67 [*G. pacifica; OD] [=Bulimina (Desinobulimina) CUSHMAN & PARKER, 1940, *518, p. 19 (type, Bulimina auriculata BAILEY, 1851, *65, p. 12)]. Test globular to ovate, chambers triserially arranged, strongly overlapping earlier ones; wall calcareous, thin, finely perforate, radial in structure, surface smooth; aperture loop-shaped, with tendency to become terminal, tooth plate doubly folded pillar-like trough joined to apertural border at one side, upper part with projecting fanlike tip, lower portion extending into chamber cavity as arched trough, then curving forward, free shank coalescing with free border of aperture, lower part of tooth plate touching projected tip of tooth plate of preceding chamber. [The modified definition of the genus by Höglund (*924) based on apertural features, includes *Desinobulimina*.] *Paleoc.(Dan.)-Rec.*, cosmop.——FIG. 442,4. **G. pacifica*, Rec., Pac.; *4a-c*, opposite

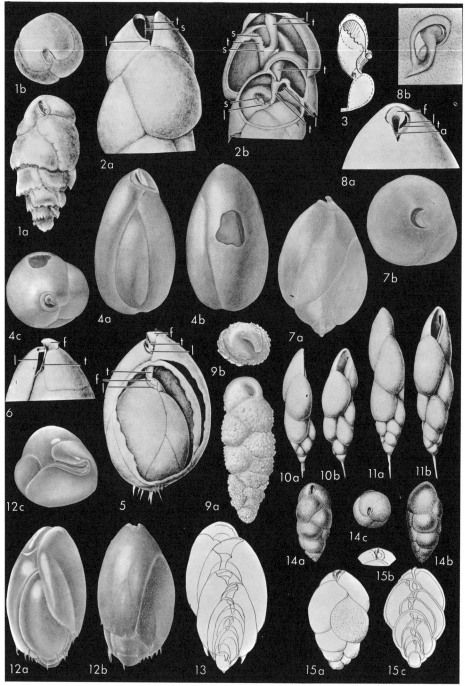

FIG. 442. Buliminidae (Bulimininae; *1-3, Bulimina; 4-8, Globobulimina; 9, Virgulopsis; 10-11, Stainforthia; 12-15, Praeglobobulimina*) (p. C559-C561).

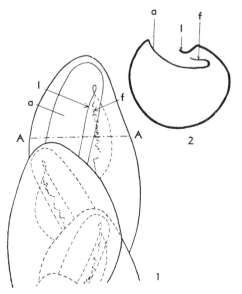

Fig. 443. Buliminidae (Bulimininae; *1,2*, *Stainforthia*) (p. C561).

sides and apertural views of holotype, ×56 (*2117).——Fig. 442,*5*. *G. turgida* (BAILEY), Rec., Sweden; dissected specimen showing tooth plate (*t*), fanlike tip (*f*), and apert. lip (*l*) in 2 successive chambers, ×70 (*924).——Fig. 442,*6*. *G.* sp., Rec., Gulf Mex.; apert. region, as in fig. 5, ×22 (*924).——Fig. 442,*7*. *G. auriculata* (BAILEY), Rec., N.Atl.; *7a,b*, side, apert. views, ×51 (*2117).——Fig. 442,*8*. *G. auriculata gullmarensis* HÖGLUND; *8a,b*, apert. region showing tooth plate attachment (*a*) and apert. features as in fig. 5, *8a*, ×70, *8b*, ×105 (*924).

Praeglobobulimina HOFKER, 1951, *928c, p. 248 [*Bulimina pyrula* var. *spinescens* BRADY, 1884, *200, p. 400; OD] [=*Protoglobobulimina* HOFKER, 1951, *928c, p. 252 (type, *Bulimina pupoides* D'ORBIGNY, 1846, *1395, p. 185)]. Test free, elongate, triserial with strongly overlapping chambers; wall calcareous, perforate radial in structure; aperture elongate, extending up from base of final chamber toward its apex, successive chambers connected internally by tooth plate with simple fold and fixed shank joined to anterior wall below aperture and wide free shank joined to chamber wall only at anterior end of aperture, free shank may be dentate and basal part reduced, small tip protruding through aperture. Paleoc.-Rec., cosmop. ——Fig. 442,*12,13*. *P. spinescens* (BRADY), Rec., Pac.; *12a-c*, opposite sides and apert. view of topotype, ×62 (*2117); *13*, sec. showing tooth plates, ×125 (*928c).——Fig. 442,*14,15*. *P. pupoides* (D'ORBIGNY), Mio., Aus. (*14*), Rec., Italy (*15*); *14a-c*, opposite sides and apert. view of microspheric specimen, ×25 (*516); *15a,b*, side view and opposite side of apert. region showing tooth plate, *15c*, long. sec. showing successive tooth plates, all ×83 (*928c).

[HAYNES (1954, *886) emended the original diagnosis but incorrectly cited the type-species as *Praeglobobulimina spinescens* HOFKER. Since the type-species was designated by HOFKER as *P. spinescens* (BRADY) this cannot be changed. HOFKER regarded *Praeglobobulimina* as characterized by elongate pores, and *Protoglobobulimina* as having elongate pores with fine pores between. The apertural tooth plate has a flaring free tip in the type (and only species) originally placed in *Praeglobobulimina*, and it is straight and collar-like in *Protoglobobulimina*. As shown by HAYNES (1954, *886, p. 185) generic separations based on pore size and distribution do not agree in many cases with those based on apertural features and (p. 188) the observation that pores are of greater length (through the wall) than their diameter may result in ovate appearance when seen on the convex test surface. Thus, the feature of pore shape does not seem to be valid for generic distinction, although pore patterns may have specific value. The actual proportions of the tooth plates also are here regarded as specific rather than generic in importance, and Recent species with flaring free tip are not regarded as generically distinct from those with less complex free tip.]

Stainforthia HOFKER, 1956, *945, p. 908 [*Virgulina concava* HÖGLUND, 1947, *924, p. 257; OD]. Test narrow, elongate, early stage triserial, at least in microspheric generation, later with twisted biserial development; chambers inflated, laterally overlapping; wall calcareous, hyaline, finely perforate, radial in structure, surface smooth or longitudinally costate, may have one or more apical spines; aperture loop-shaped in face, with narrow incurved lip at one side and broad tooth plate at opposite side bending under lip and partially closing opening, tooth plate with serrated free folded portion, lower portion of tooth plate attached to preceding chamber wall. Eoc.-Rec., Eu.-S.Am.——Fig. 442,*10,11*; 443,*1,2*. *S. concava* (HÖGLUND), Rec., Sweden; 442,*10a,b*, *11a,b*, side and edge views, ×93; 443,*1,2*, later portion in optical sec. and cross sec. showing apert. features with narrow incurved lip at one side (*l*), and folded tooth plate with serrate free shank (*f*), and attached opposite border (*a*), ×340 (*924).

Virgulopsis FINLAY, 1939, *717c, p. 321 [*V. pustulata*; OD]. Test free, elongate, early stage triserial, later biserial; wall calcareous, finely perforate, surface plicate or pustulose; aperture loop-shaped, extending up face, internal features unknown. [Details of the inner structure and character of the tooth plate are needed for accurate placement of this genus. It may prove to be a junior synonym of *Uvigerinella* or of *Neobulimina*, or a senior synonym of *Stainforthia*.] M. Mio., N.Z.——Fig. 442,*9*. *V. pustulata*; *9a,b*, side, apert. views, ×82 (*2117).

Subfamily PAVONININAE Eimer & Fickert, 1899

[nom. transl. CUSHMAN, 1927, p. 59 (ex family Pavoninidae EIMER & FICKERT, 1899)] [=Reussiinae CUSHMAN, 1927, p. 68 (pro Reussia SCHWAGER, 1877) (non Reussia M'COY, 1854); =Reussellinae CUSHMAN, 1933, p. 223 (nom. subst.)]

Test triserial in early stage, rarely biserial, later uniserial; aperture loop-shaped or represented by pores on terminal chamber face. [Simple forms such as *Reussella* have an apertural tooth plate, whereas other genera show a tooth plate in early stages (*Chrysalidinella*) and a majority apparently show no tooth plates but may have sec-

Fig. 444. Buliminidae (Pavonininae; *1-4, Pavonina; 5, Fijiella; 6, Acostina; 7-10, Chrysalidinella*) (p. C563).

ondary resorption of the terminal face to form irregular intercameral openings. Restudy of the ontogeny and internal morphology of all genera may later result in separating the forms with tooth plates (Reussellinae CUSHMAN, 1933) from the Pavoninīnae, but meanwhile they are retained together.] *Eoc.-Rec.*

Pavonina D'ORBIGNY, 1826, *1391, p. 260 [*P. flabelliformis*; OD (M)] [=*Bifarinella* CUSHMAN & HANZAWA, 1936, *504, p. 46 (type, *B. ryukyuensis*); *Valvopavonina* HOFKER, 1951, *928c, p. 35 (obj.)]. Test with reduced triserial stage of 3 chambers, later biserial, and finally uniserial, spreading with low broad arched chambers, strongly recurved at margins; wall calcareous, radial in structure, coarsely perforate; no distinct aperture, terminal face of final chamber merely coarsely perforate like remainder of test, large rounded openings originally reported and since mentioned by various authors not found to be present on well-preserved specimens, possibly due to secondary resorption. *Mio.-Rec.*, Pac.-Atl.-Afr.-N.Am.-Madag.——FIG. 444,*1-3*. *P. flabelliformis*, Rec., Kerimba Arch. (*1*), Rec., Mauritius (*2,3*); *1a,b*, side, top views, ×82; *2a,b*, side, top views, ×82 (*2117); *3a,b*, edge view, showing pores, and long. sec. of same specimen showing septal openings, large pores through walls and fine pores through one side of the proloculus, and reduced triserial stage, ×80 (*928c).——FIG. 444,*4*. *P. ryukyuensis* (CUSHMAN & HANZAWA), Pleist., Ryukyu Is.; *4a,b*, side and top views of holotype, showing early biserial stage, later uniserial stage with spreading test, and large septal perforations with intervening pillars, ×48 (*2117).

[The early triserial stage mentioned has not been confirmed by us, as all specimens of the type-species observed show only a biserial stage. It is possible that ornamentation may have been mistaken for additional chambers in some reports, the wall being secondarily thickened and laminar. Nevertheless, PARR found a distinctly triserial base in *P. triformis* (*1422) and HOFKER illustrated a reduced triserial stage in *P. flabelliformis* (*928c). Well-preserved specimens of *P. flabelliformis* examined at high magnification show numerous irregularly scattered pores on the terminal face, identical in appearance to those on sides of the test, but no large regularly aligned apertural pores have been seen by us in either wet or dry or stained tests. The systematic placement has also varied, PARR (*1422) placing *Pavonina* near *Reussella* and *Chrysalidinella* in the Buliminidae, although *Pavonina* has no apertural tooth plate, and HOFKER (*928c) placing it in the Valvulinidae because of absence of a tooth plate and presence of coarse pores in the wall, although other representatives of that family are agglutinated. *Bifarinella* was placed in the Virgulininae by CUSHMAN (1937, *472), a group with perforate granular walls and apertural tooth plate, and *Pavonina* was placed in the Reussellinae, a group with perforate radial walls and apertural tooth plate. Both the holotype of *B. ryukyuensis* (in the CUSHMAN collection) and paratype are broken specimens, no terminal face being preserved. The "slitlike aperture" and everted lip consist merely of fragments of the final chamber wall. Not previously mentioned is the fact that the final septum preserved has numerous pores, with only narrow bridges remaining across the test, so that it has the identical large septal pores found in *Pavonina*, and there described as a multiple aperture. The early biserial stage and later uniserial stage both occur in most specimens and species of *Pavonina*, the less flabelliform test of *B. ryukyuensis* being here regarded as only of specific value.]

Acostina BERMÚDEZ, 1949, *124, p. 152 [*Chrysalogonium piramidale* ACOSTA, 1940, *3, p. 4; OD]. Test elongate pyramidal, triangular in section, with carinate angles, chambers uniserial throughout; aperture terminal, consisting of numerous small pores in protruding portion of terminal face. *U.Oligo.-Rec.*, Cuba-Dominican Republic.——FIG. 444,*6*. **A. piramidale* (ACOSTA), Rec., Cuba; *6a,b*, side, apert. views of holotype, ×37 (*3).

Chrysalidinella SCHUBERT, 1908, *1687, p. 242 [*Chrysalidina dimorpha* BRADY, 1881, *196c, p. 54; OD (M)] [=*Chrysalidinoides* UCHIO, 1952, *1959, p. 154 (type, *C. pacificus*)]. Test elongate, commonly pyramidal, early portion triserial and triangular, later uniserial and triangular in section or rarely quadrangular; sutures arched; wall smooth, calcareous, coarsely perforate, radial in structure; aperture basal in early stage as in *Reussella* with small tooth plates, in uniserial stage consisting of numerous scattered pores on terminal face, without tooth plates, early septa showing some larger irregular openings, probably due to resorption. *Eoc.-Rec.*, Cuba-Kerimba Arch.-Pac.-N.Am.-Carib.-Indon.——FIG. 444,*7-9*. **C. dimorpha* (BRADY), Rec., Pac. (*7*), Rec., Sumatra (*8,9*); *7a,b*, side, top views, ×74 (*2117); *8*, outline view of septum, showing apert. pores and secondary irregular openings due to resorption, ×80; *9*, isolated tooth plate, enlarged (*928c). ——FIG. 444,*10*. *C. pacifica* (UCHIO), Rec., Japan; *10a-c*, lat. and terminal views of holotype, ×56 (*1959).

[*Chrysalidinoides* was based on a single specimen which became quadrate in the adult, although early development was triserial and triangular. As many triangular genera have occasional aberrant quadrate specimens (e.g., *Tristix*, *Triplasia*) the present form is regarded as adventitious.]

Fijiella LOEBLICH & TAPPAN, 1962, *1185, p. 109 [*Trimosina simplex* CUSHMAN, 1929, *443, p. 158; OD]. Test triserial and triangular throughout; wall calcareous, coarsely perforate, surface smooth, lateral margins carinate and may be spinose; primary aperture a narrow elongate basal slit with terminal supplementary cribrate openings. [*Fijiella* differs from *Reussella* and *Trimosina* in having the supplementary cribrate terminal aperture, and from *Chrysalidinella* in lacking a uniserial stage.]. *Rec.*, Pac.——FIG. 444,*5*. **F. simplex* (CUSHMAN), Fiji; *5a,b*, side, apert. views, ×60 (*476).

Mimosina MILLETT, 1900, *1284e, p. 547 [*M. histrix*; SD CUSHMAN, 1927, *433, p. 190]. Test in early stage triserial, later biserial, each chamber in later stage ornamented with spine; wall calcareous, surface ornamented with very fine longitudinal ridges with fine pores between, radially built; aperture in 2 parts, one nearly terminal, second marginal and tending to be more oval in outline. *Rec.*, Malay Arch.-Tropical Pac.-Kerimba Arch.-Atl.-Medit.——FIG. 445,*1,2*. **M. histrix*, Malay Arch.; *1a,b*, *2a,b*, side and top views, ×104 (*2117).

Reussella GALLOWAY, 1933, *762, p. 360 [*pro*

Reussia SCHWAGER, 1877, *1705, p. 21 (non M'COY, 1854)] [*Verneuilina spinulosa REUSS, 1850, *1540, p. 374; OD]. Test triserial and triangular throughout, gradually enlarging; wall calcareous, coarsely perforate; aperture basal in final chamber, with internal tooth plate. [Reussella is

FIG. 445. Buliminidae (Pavonininae; *1,2, Mimosina; 3-5, Reussella; 6, Trimosina; 7, Tubulogenerina; 8,9, Valvobifarina*) (p. C563-C565).

restricted here to include only sharply angular species, commonly with carinate or spinose angles, coarsely perforate wall, and complex tooth plate. Upper Cretaceous species that have been previously referred to *Reussella* are here regarded as belonging to *Pyramidina*, differing in their less angular margins, finely perforate walls, and simpler tooth plate.] *M.Eoc.(Lutet.)-Rec.*, cosmop.——FIG. 445,*3-5*. **R. spinulosa* (REUSS), Mio., Aus.; *3*, side view, ×100; *4a,b*, side, apert. views, ×94 (*2117); *5*, apertural tooth plate, magnified (*928c).

Trimosina CUSHMAN, 1927, *431, p. 64 [**T. milletti=Mimosina spinulosa* var. MILLETT, 1900, *1284e, p. 548; OD]. Test triserial, similar to *Mimosina* but without later biserial development; wall calcareous, perforate, radial in structure; aperture an elongate slit, in face of final chamber and paralleling its base. *Rec.*, Indo-Pac.——FIG. 445,*6*. **T. milletti*, Malay Arch.; *6a,b*, side, top views, ×90 (*1284e).

Tubulogenerina CUSHMAN, 1929, *428, p. 78 [**Textularia (Bigenerina) tubulifera* PARKER & JONES, 1863, *1417e, p. 94; OD]. Test elongate, early stage triserial in microspheric form, followed by short biserial stage, later chambers uniserial and compressed or rounded in section; wall calcareous perforate, surface may be distinctly nodose or longitudinally costate; aperture a narrow, elongate, crescentic slit in terminal face, with internal tooth plate. *M.Eoc.(Lutet.)-Oligo.,?Mio.*, Eu.-N.Am.-Australia.——FIG. 445,*7*. **T. tubulifera* (PARKER & JONES), *M.Eoc.*(Lutet.), Fr.; *7a,b*, side, apert. views of topotype, ×109 (*2117).

Valvobifarina HOFKER, 1951, *928c, p. 39 [**Bifarina mackinnoni* MILLETT, 1900, *1284d, p. 281; OD]. Test in early portion triserial, triangular in section, later changing abruptly to twisted biserial arrangement of cuneate chambers; wall ornamented with numerous calcareous knobs, each with large pore and commonly with spines at chamber margins; aperture terminal, narrow and elongate, occupying width of chamber and surrounded by everted rim. *Rec.*, Malay Arch.-Timor Sea.——FIG. 445,*8,9*. **V. mackinnoni* (MILLETT), Timor Sea (*8*), Macassar Straits (*9*); *8a,b*, *9a,b*, side and top views, ×74 (*2117).

[HOFKER (1951, *928c, p. 42) originally placed *Valvobifarina* in the agglutinated family Valvulinidae, because of its scattered large pores and knobs of "somewhat arenaceous chalky matter." He regarded both this genus and *Bolivinitella* (=*Loxostomum*) as closely related to *Siphogaudryina*. As correctly stated by HOFKER, neither *Loxostomum* nor *Valvobifarina* are related to the "Bolivininae," the former having a granular wall and lacking a tooth plate and *Valvobifarina* having a triserial, rather than biserial, early development.]

Family UVIGERINIDAE Haeckel, 1894

[*nom. correct.* GALLOWAY & WISSLER, 1927, p. 74 (*pro* family Uvigerinida HAECKEL, 1894, p. 185)]——[In synonymic citations superscript numbers indicate taxonomic rank assigned by authors (¹family, ²subfamily); dagger (†) indicates *partim*]——[¹Uvellina† EHRENBERG, 1839, table opposite p. 120 (*nom. nud.*); =²Turbinoidea† D'ORBIGNY in DE LA SAGRA, 1839, p. xxxviii, 71 (*nom. nud.*); =²Angulogerininae GALLOWAY, 1933, p. 377; =²Uvigerininae CUSHMAN, 1913, p. 91]

Test triserial to biserial in early stage, later may become biserial or uniserial; aperture terminal, with neck and internal tooth plate connecting apertures of successive chambers. *U.Cret.-Rec.*

Uvigerina D'ORBIGNY, 1826, *1391, p. 268 [**U. pygmea*; SD PARKER, JONES & BRADY, 1865, *1419, p. 36] [=*Uvigerina (Uhligina)* SCHUBERT, 1899, *322, p. 222 (type, *U. (U.) uhligi*) (*non Uhligina* YABE & HANZAWA, 1922); *Aluvigerina* HOFKER, 1951, *928c, p. 201 (*nom. nud.*); *Aluvigerina* THALMANN, 1952, *1897j, p. 970 (obj.); *Miniuva* VELLA, 1961, *2002, p. 480 (type, *M. minima*)]. Test elongate, triserial, rounded in section, chambers inflated, wall calcareous, perforate, surface smooth, hispid or costate; aperture terminal, rounded with nonperforate neck and may have phialine lip, internal tooth plate with distinct wing at one side. [The type-species was spelled *pigmea* in the text (*1391, p. 269) but *pygmea* on the plate explanation (*1391, pl. 12, p. 310). *Miniuva* was separated for an extremely small costate species with short neck, features here regarded as of specific value.] *Eoc.-Rec.*, cosmop.——FIG. 446, *1,2*. **U. pygmea*, Plio., Italy (*1*), Rec., Italy (*2*); *1a,b*, side, apert. views, ×94 (*2117); *2*, sectioned specimen, showing tooth plates with wings (shaded portion), ×104 (*928c).

Clavelloides DE KLASZ & RÉRAT, 1962, *1043, p. 182 [**C. tenuistriata*; OD]. Test elongate, tapering, with broad, low, slightly enveloping, uniserially arranged chambers; sutures horizontal, slightly depressed; wall calcareous, microstructure unknown, surface longitudinally striate; aperture terminal, in slight depression, subelliptical; interior with columellar process connecting foramina of adjacent chambers. *L.Eoc.-M.Eoc.*, W.Afr.(Gabon).——FIG. 446,*3,4*. **C. tenuistriata*; *3*, ext. holotype, ×27; *4*, long. sec., ×27 (*1043).

[This genus was originally placed in the Ellipsoidinidae (=Pleurostomellidae), but differs from characteristic genera of that family in the ornate surface and very large size. It is here tentatively referred to the Uvigerinidae, though the wall microstructure is unknown. If granular, it should be placed with the Pleurostomellidae; if radial, the present position would be correct. Additional details as to the character of the columellar process or tooth plate would aid in determining the systematic position.]

Compressigerina BERMÚDEZ, 1949, *124, p. 219 [**Uvigerina coartata* D. K. PALMER, 1941, *1410b, p. 304 (=*U. compressa* PALMER, 1941, *1410a, p. 182) (*non U. compressa* CUSHMAN, 1925); OD]. Test free, small, with early stage triserial, later biserial with twisted axis as in *Sigmavirgulina* and finally tending to become uniserial, peripheral margins angled or keeled; wall calcareous, finely perforate, radial in structure, may have longitudinal carinae and fine spines at chamber angles; aperture terminal, ovate, produced on slight neck, with internal tooth plate. *Oligo.-Rec.*, Carib.——FIG. 446,

6. *C. coartata (D. K. Palmer), M.Mio., Dominican Republic; 6a-c, side, edge, and top views, ×143 (*2117).

Euuvigerina Thalmann, 1952, *1897j, p. 974 [*Uvigerina aculeata d'Orbigny, 1846, *1395, p. 191; OD] [=Euuvigerina Hofker, 1951, *928c,

Fig. 446. Uvigerinidae; 1,2, Uvigerina; 3,4, Clavelloides; 5, Kolesnikovella; 6, Compressigerina; 7, Hopkinsina; 8-14, Orthokarstenia (p. C565-C568).

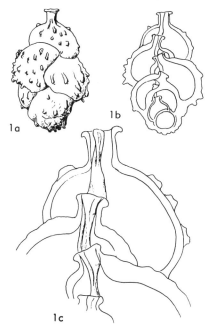

Fig. 447. Uvigerinidae; *1, Euuvigerina* (p. *C566-C567*).

p. 217 (nom. nud.); *Hofkeruva (Hofkeruva)* VELLA, 1961, *2002, p. 473 (type, *H. (H.) mata*); *Hofkeruva (Laminiuva)* VELLA, 1961, *2002, p. 474 (type, *H. (L.) tutamoea*); *Hofkeruva (Tereuva)* VELLA, 1961, *2002, p. 475 (type, *Uvigerina paeniteres* FINLAY, 1939, *717b, p. 103); *Hofkeruva (Trigonouva)* VELLA, 1961, *2002, p. 476 (type, *H. (T.) zeacuminata*)]. Test similar in appearance to *Uvigerina*, with chambers triserially arranged, rounded in section; thick-walled and finely perforate, with surface commonly spinose and apertural neck nonperforate; with simple straight, narrow tooth plate, base of which is attached to margins of previous foramen, lacking broad wing of tooth plate of *Uvigerina*. Eoc.-Rec., cosmop.——FIG. 447,1. *E. aculeata* (D'ORBIGNY), Rec., Indon.; *1a,b*, ext. and long. sec. showing simple tooth plates, ×32; *1c*, apert. area, ×153 (*928c).

[VELLA (1961, *2002) subdivided the uvigerine forms on the basis of surface ornamentation into many genera and subgenera. Although distinct lineages of costate or spinose species can be demonstrated, we do not regard them to require generic separation, for which characters recognized are those of chamber arrangement and apertural position and modifications, including tooth plates. VELLA stated (*2002, p. 473) that *Hofkeruva* and its subgenera have tooth plates identical to those of *Euuvigerina aculeata*; hence, they are here regarded as congeneric.]

Hopkinsina HOWE & WALLACE, 1932, *972, p. 61 [**H. danvillensis*; OD]. Test elongate, early stage triserial, later biserial, wall calcareous, perforate, surface smooth or more commonly longitudinally striate or costate; aperture terminal, with lip or may be slightly produced on neck, apertural tooth plate not described in type-species. [Differs from *Uvigerina* in its later biserial stage, from *Uvigerinella* in its terminal aperture, and from *Trifarina* in being rounded in section and in being biserial rather than uniserial in the adult.] Eoc.-Rec., N.Am.-Eu.——FIG. 446,7. **H. danvillensis*, U.Eoc.(Jackson.), USA(La.); *7a,b*, side, top views of topotype, ×130 (*2117).

Kolesnikovella N. K. BYKOVA, 1958, *265, p. 68 [**Tritaxia elongata* HALKYARD, 1918, *861, p. 45; OD]. Test similar to *Trifarina*, triserial in early stage with tendency to become uniserial, lower chamber margins with retral processes and sinuous margin; wall calcareous perforate; aperture terminal, rounded, produced on neck, commonly with phialine lip. Eoc., W.Eu.-USSR-Cuba-USA(Calif.)-W.Indies.——FIG. 446,5. *K. elongata* (HALKYARD), Eoc., Fr.; *5a-c*, opposite sides and apert. view, ×97 (*861).

Orthokarstenia DIETRICH, 1935, *597, p. 80 [**Orthocerina ewaldi* KARSTEN, 1856, *1025, p. 114; OD] [=*Siphogenerita* FURRER, 1961, *757, p. 271 (type, *Siphogenerinoides clarki* CUSHMAN & CAMPBELL, 1936, *499, p. 91)]. Test free, elongate, gradually enlarging from rounded base, early stage of microspheric form triserial, then short biserial stage, megalospheric form with proloculus followed by biserial stage, adult uniserial in both generations; adult chambers subcylindrical, somewhat inflated, lower margin commonly with re-entrants, resulting in appearance of lobulate sutures, sutures distinct, straight, depressed; wall calcareous, finely perforate, surface smooth or with ornamentation of longitudinal costae or striae; aperture terminal, elliptical to reniform, with short neck and distinct lip, internally provided with a spoutlike columellar process, semicylindrical spout arising from apertural lip and extending inward with concave side toward opening, those of successive chambers discontinuous, orientation of convex side changing from 120° to 180° in successive chambers, spout may terminate in small circular opening adjacent to concave side of true aperture but is not connected to it. U.Cret.(Turon.-Maastricht.), N.Am.-S.Am.-Afr.——FIG. 446,8-10. **O. ewaldi* (KARSTEN), Turon., S.Am.(Colom.); *8*, ext., approx. ×17; *9a,b*, vert. and cross secs. showing internal semicylindrical siphon, approx. ×30 (*1025); *10a,b*, ext. and vert. sec. showing early triserial stage, approx. ×20 (*597).——FIG. 446,11,12. **O. clarki* (CUSHMAN & CAMPBELL), U.Cret.(Campan.), USA(Calif.); *11*, microspheric test, ×21; *12a,b*, megalospheric test, side, and apert. views, ×21; *12c*, basal view showing chamber arrangement, ×41 (*757).——FIG. 446,13,14. *O. whitei* (CHURCH), U.Cret.(Maastricht.), USA(Calif.); *13a,b*, side and apert. views of megalospheric test, ×21; *14a,b*, dissected test showing

Fig. 448. Uvigerinidae; *1-4, Pseudouvigerina; 5,6, Rectuvigerina; 7-11, Sagrina* (p. C568-C569).

alternating position of internal process in successive chambers, ×48 (*757).

Pseudouvigerina CUSHMAN, 1927, *428, p. 81 [*Uvigerina cristata MARSSON, 1878, *1228, p. 150; OD] [=*Praeuvigerina* HOFKER, 1951, *928c, p. 188 (type, *Uvigerina westfalica* FRANKE, 1912, *738, p. 280)]. Test small, triserial throughout, rounded, triangular or trihedral in section; wall calcareous, finely perforate, surface may be smooth or tuberculate, angles of test may have double vertical costae; aperture circular or subelliptical, with short neck provided internally with narrow columellar tooth plate. *U.Cret.,* Eu.-N.Am.—— FIG. 448,*1-3.* **P. cristata* (MARSSON), Maastricht., Ger.; *1a,b,* side, apert. views of topotype, U.Cret., Ger., ×111; *2,3,* acid-dissected specimens showing internal tooth plate, ×107 (*1303).—— FIG. 448,*4. P. westfalica* (FRANKE), U.Cret. (Senon.), Neth.; *4a,b,* side view and sec. of apert. region showing tooth plates, ×125 (*928c).

[*Pseudouvigerina* may have arisen from early *Pyramidina* and given rise to *Trifarina.* HOFKER (1957, *948, p. 220) regarded *Pseudouvigerina* as synonymous with *Reussella*, but recognized the younger name as the valid one. *Pseudouvigerina* is here separated on the basis of its terminal aperture, whereas *Pyramidina* and *Reussella* have basal apertures.]

Rectuvigerina MATHEWS, 1945, *1234, p. 590, 598, 601 [*Siphogenerina multicostata* CUSHMAN & JARVIS, 1929, *509, p. 14; OD] [=*Rectuvigerina (Rectuvigerina)* MATHEWS, 1945, *1234, p. 590, 598, 601 (obj.); *Rectuvigerina (Transversigerina)* MATHEWS, 1945, *1234, p. 599 (type, *Siphogenerina raphanus* (PARKER & JONES) var. *transversus* CUSHMAN, 1918, *409, p. 64); *Ruatoria* VELLA, 1961, *2002, p. 480 (type, *R. ruatoria*); *Ciperozea* VELLA, 1961, *2002, p. 481 (type, *Siphogenerina ongleyi* FINLAY, 1939, *717b, p. 111)]. Similar to *Siphogenerina* but with triserial to uniserial chamber arrangement in both megalospheric and microspheric generations, whereas *Siphogenerina* has biserial to uniserial megalospheric generation. *M.Eoc.-Rec.*, cosmop.——FIG. 448,5. *R. multicostata* (CUSHMAN & JARVIS), Mio. (originally recorded as Eoc.), Trinidad; 5a,b, side, apertural views of holotype, ×44 (*2117).——FIG. 448,6. *R. transversa* (CUSHMAN), Oligo., Panama C.Z.; 6a,b, side, apert. views of holotype, ×49 (*2117).

[*Ruatoria* was stated to differ from *Rectuvigerina* in being smaller, with "staggered" terminal chambers and broad neck. *Ciperozea* was stated to have a more elongate triserial portion and cuneate, rather than truly rectilinear, chambers and low longitudinal ribs. Although slightly cuneate in *Ruatoria* and *Ciperozea*, these terminal chambers are nevertheless uniserial and they are regarded as synonymous with *Rectuvigerina*.]

Sagrina D'ORBIGNY in DE LA SAGRA, 1839, *1611, p. 149 [*S. pulchella*; OD (M)] [=*Sagraina* BRONN & ROEMER, 1853, *214a, p. 92 (*nom. van.*); *Bitubulogenerina* HOWE, 1934, *970, p. 420 (type, *B. vicksburgensis*); *Tritubulogenerina* HOWE, 1939, *971, p. 69 (type, *T. mauricensis*, =*Bitubulogenerina mauricensis* HOWE, 1934, *970, p. 421)]. Test free, elongate, circular to ovate in section, triserial in early stage, later biserial; chambers commonly with angular lower margin; wall calcareous, hyaline, coarsely perforate, radial in structure, surface variously ornamented with longitudinal costae, prominent nodes; aperture elongate, bordered with distinct lip, extending up face from base of chamber beyond middle of chamber, outer portion of penultimate aperture visible also, but partially filled by tooth plate; flaring, folded tooth plate extending upward from border of previous foramen, then flaring back through chamber to attach at one side of aperture and in part forming apertural border overlapping previous chamber, entire inner margin of tooth plate serrated, forming fringed border to interior of aperture. *M.Eoc.-Rec.*, Cuba-Carib.-N.Am.-Atl. ——FIG. 448,7-9. *S. pulchella*, Rec., Cuba (7); Atl. (8), W.Indies (9); 7a,b, side, apert. views of lectotype (MNHN, Paris), ×109 (*2117); 8a-c, side, basal, apert. views, ×107 (*2117); 9, long. sec. of megalospheric form showing tooth plates, ×117 (*946).——FIG. 448,10. *S. vicksburgensis* (HOWE), Oligo., USA(Miss.); 10a,b, side, apert. views, ×130 (*2117).——FIG. 448, 11. *S. mauricensis* (HOWE), M.Eoc., USA(La.); 11a-c, side, basal, apert. views, ×227 (*2117).

[*Sagrina* described by D'ORBIGNY, 1839, was monotypic, including only *S. pulchella*. PARKER & JONES (1863, *1417e, p. 95) incorrectly emended the genus on the basis of the arenaceous *S. rugosa* D'ORBIGNY, 1840, adding that "the other *Sagrina* (*S. pulchella* d'Orb . . .) (biserial, ribbed and not sandy) is a *Uvigerina*." BRADY (1884, *200, p. 580) stated, "The generic term *Sagrina* was introduced by d'Orbigny for a biserial or Textulariform variety of *Uvigerina* with longitudinal costae." As D'ORBIGNY later also included an arenaceous species, BRADY concluded that while it would have been better to allow the name to lapse, it had been revived by PARKER & JONES for a "group of dimorphous *Uvigerinae*, usually biserial in the arrangement of their early segments and Nodosariform in their later growth, and it is to this particular set of forms that the genus is now restricted." CUSHMAN (1928, *439, p. 249) placed *Sagrina* in the synonymy of *Bolivina* but did correctly consider the type-species to be *S. pulchella*. GALLOWAY (1933, *762, p. 348) recognized *Sagrina* as a distinct genus, placing it in the Heterohelicidae. HOWE (1934, *970, p. 420) defined *Bitubulogenerina*, comparing it to the similar *Tubulogenerina*, which has a uniserial adult stage but without mention of *Sagrina*. HOFKER (1951, *928c, p. 67) incorrectly restricted *Sagrina* to *S. tesselata* (which is the type-species of *Millettia*) and placed *S. pulchella* in *Bitubulogenerina*. As *Sagrina* has priority, *Bitubulogenerina* is the junior synonym. A lectotype of *S. pulchella*, here designated and redrawn, and paratypes were selected by us from the D'ORBIGNY collection in the Muséum National d'Histoire Naturelle, Paris, France. They are Recent, off Cuba. The monotypic *Tritubulogenerina* was based on a small completely triserial form, *T. mauricensis*, which was described from the same strata and locality as *Bitubulogenerina mauricensis* HOWE, 1934, a form with early triserial stage and later biserial development. As *Tritubulogenerina mauricensis* appears merely to represent a young form or the megalospheric generation of the earlier-described species, it is a junior synonym of *Bitubulogenerina* and of *Sagrina*.]

Siphogenerina SCHLUMBERGER in MILNE-EDWARDS, 1882, *1286, p. 51 [*Siphogenerina costata* SCHLUMBERGER, 1883, *1650, p. 26; SD CUSHMAN, 1927, *433, p. 190] [=*Ellipsosiphogenerina* A. SILVESTRI, 1902, *1754, p. 101 (type, *Siphogenerina costata* SCHLUMBERGER, 1883, *1650, p. 26, SD LOEBLICH & TAPPAN, herein) (obj.); *Ellipsosiphongenerina* A. SILVESTRI, 1923, *1774, p. 265 (*nom. null.*)]. Test free, early stage biserial, later uniserial, or rarely with early triserial stage (probably microspheric); wall calcareous, hyaline, finely perforate, radial in structure, surface smooth or variously ornamented with longitudinal costae, striations or pits; aperture terminal, rounded with short neck or rim and phialine lip; apertural tooth plates projecting inward, those of successive chambers added in planes 120° apart. *Eoc.-Rec.*, cosmop.——FIG. 449,1-4. *S. costata*, Rec., Tahiti (1), Fiji (2,3), W.Indies (4); 1, holotype, ×28 (*1650); 2a,b, side, apert. views of microspheric hypotype; 3, side view of megalospheric hypotype; all ×32 (*476); 4, sec. of apert. end showing tooth plates in 2 successive chambers, probably ×160 (originally stated to be ×240, but magnification of figures and measurements of specimens do not agree) (*946).

[*Siphogenerina* was originally defined (1882, *1286) without included species. A subsequent article by the same author (1883, *1650) included 3 species (*S. glabra, S. costata, S. ocracea*), but none was designated as type-species. CUSHMAN (1913, *404c, p. 104) incorrectly designated *Uvigerina (Sagrina) raphanus* PARKER & JONES as the type-species of *Siphogenerina*, although this was not one of the 3 species originally included. Later CUSHMAN (1927, *433,

p. 190) corrected this, designating *S. costata* SCHLUMBERGER as the type-species, although stating that it was a synonym of *S. raphanus*. MATHEWS (1945, *1234, p. 589) cited the type as *S. costata* and regarded it as specifically distinct from *S. raphanus*. BANDY & BURNSIDE (1951, *76, p. 14) stated that CUSHMAN was in error in designating *S. costata* as type and that *S. glabra* as the first species included should be the type. However, any of the 3 species de-

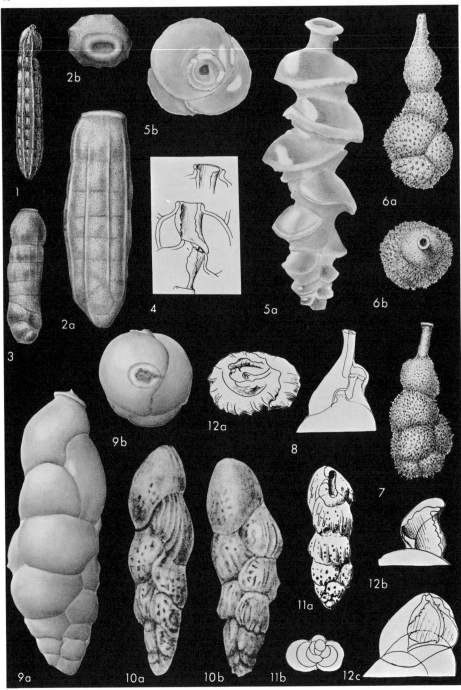

FIG. 449. Uvigerinidae; *1-4, Siphogenerina; 5-8, Siphouvigerina; 9, Uvigerinella; 10-12, Virgulinopsis* (p. C569-C572).

Foraminiferida—Rotaliina—Buliminacea

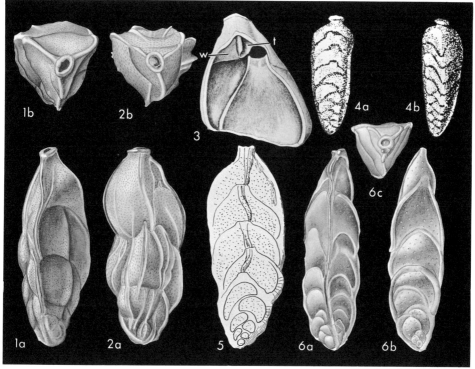

Fig. 450. Uvigerinidae; *1-6, Trifarina* (p. C571-C572).

scribed in 1883 was available for designation as type. BANDY (1952, *72, p. 17) later agreed with CUSHMAN that *S. costata* was a synonym of *S. raphanus* and stated that the latter was therefore the type-species. The type area given for *S. costata* by SCHLUMBERGER was Tahiti and New Caledonia. BANDY stated (1952, *72, p. 18) that some specimens of *S. costata* from Tahiti have a triserial early stage and others have a biserial early stage. We examined the types of *S. raphanus* in the British Museum (Natural History), and regard it as distinct from *S. costata*. *Siphogenerina* is here regarded as including only species which are triserial in the microspheric early stage and biserial in the megalospheric stage. As *S. raphanus* is biserial to uniserial in the microspheric form and only uniserial in the megalospheric stage it has been transferred to *Rectobolivina*.]

Siphouvigerina PARR, 1950, *1429, p. 342 [*Uvigerina porrecta* BRADY var. *fimbriata* SIDEBOTTOM, 1918, *1741, p. 147; OD] [=*Neouvigerina* HOFKER, 1950, *932, p. 67 *(nom. nud.)*; *Neouvigerina* HOFKER, 1951, *928c, p. 206 *(nom. nud.)*; *Neouvigerina* THALMANN, 1952, *1897j, p. 977 (type, *Uvigerina asperula* var. *ampullacea* BRADY, 1884, *200, p. 579)]. Test elongate, chambers triserial and closely appressed in the early stage, later tending to uniseriality, with chambers loosely attached and sutures deeply incised; wall calcareous, surface hispid or with granulations which may coalesce into costae; aperture terminal, rounded, with perforate neck and phialine lip, tooth plate straight and simple, attached to outer margin of previous foramen. *Oligo.-Rec.* cosmop.——FIG. 449,5. *S. fimbriata* (SIDEBOTTOM), Rec., Australia; *5a,b,* side, top views, ×185 (*2117).——FIG. 449,6-8. *S. ampullacea* (BRADY), Rec., S.Atl. *(6,7),*

Sumatra *(8); 6a,b,* side, apert. views, ×64; *7,* side view, ×64 (*200); *8,* optical sec. of terminal portion showing tooth plates, approx. ×100 (*928c).

Trifarina CUSHMAN, 1923, *411d, p. 99 [*T. bradyi*; OD] [=*Angulogerina* CUSHMAN, 1927, *431, p. 69 (type, *Uvigerina angulosa* WILLIAMSON, 1858, *2065, p. 67); *Candela* N. K. BYKOVA, in N. K. BYKOVA et al., 1958, *265, p. 70 (type, *Trifarina labrum* SUBBOTINA, 1953, *1846, p. 247 *(non Candela* HERRMANNSEN, 1846); *Dymia* N. K. BYKOVA, 1962, *264, p. 22 *(nom. subst. pro Candela* N. K. BYKOVA, 1958, *non* HERRMANNSEN, 1846); *Norcottia* VELLA, 1961, *2002, p. 478 (type, *Hopkinsina mioindex* FINLAY, 1947, *717e, p. 282)]. Test free, elongate, triangular in section; chambers triserially arranged, early ones closely appressed, later ones more loosely appressed and tending to become uniserial; wall calcareous, finely perforate, radial in structure, commonly with longitudinal costae; aperture terminal, ovate, on short neck with thickened rim, tooth plate with wing at dorsal side. *Eoc.-Rec.,* cosmop.——FIG. 450,*1-3. T. angulosa* (WILLIAMSON), Rec., Br.I.; *1a,b,* side, apert. views of paratype, ×83 (*2117); *2a,b,* side, apert. views, ×99 (*2117); *3,* apert. end dissected to show tooth plate *(t)* and its wing *(w),* ×123 (*924).——FIG. 450,*4. T. labrum* (SUBBOTINA), U.Eoc., Ukraine; *4a,b,* opposite sides of holotype, ×96 (*1509).——FIG. 450,*5,6.* **T. bradyi,* Rec., Indon. *(5),* Atl. *(6);*

5, long. sec. showing tooth plates, ×160 (*928c); 6a-c, opposite sides and apert. view of paratype, ×94 (*2117).

[*Trifarina* differs from *Uvigerina* in being angular in section and in the tendency to become uniserial in the adult. The synonymy of *Angulogerina* with *Trifarina* was shown by HOFKER (1956, *946, p. 77), although he recognized the junior name as valid. The original types of *Uvigerina angulosa* WILLIAMSON were studied by us in the British Museum (Natural History), and a lectotype here designated (BMNH-ZF3576) with paratypes (BMNH-ZF3575) (all ex 96.8.13.32, WILLIAMSON collection, Recent. off Great Britain). *Norcottia* was proposed to include a finely costate Miocene species, as VELLA utilized surface ornamentation for generic separation. In chamber arrangement and other features it resembles *Trifarina*; hence *Norcottia* is here considered a synonym of *Trifarina*.]

Uvigerinella CUSHMAN, 1926, *426, p. 58 [*Uvigerina (Uvigerinella) californica*; OD] [=*Uvigerina (Uvigerinella)* CUSHMAN, 1926, *426, p. 58 (obj.)]. Test similar to *Uvigerina*, triserial; wall calcareous, perforate, surface smooth or longitudinally costate; aperture slitlike, extending up face of final chamber, rather than terminal in position, and may have elevated rim or collar, but without neck and phialine lip, character of tooth plate not described in type-species. [Additional study is needed of the type-species of this and other genera. Possibly *Virgulopsis* or *Virgulinopsis* may be synonyms of the present genus, but evidence for determining this is insufficient as yet.]. *Eoc.-Rec.*, N.Am.-W.Indies.——FIG. 449, 9. **U. californica*, Mio., USA(Calif.); *9a,b*, side, apert. views of paratype, ×97 (*2117).

Virgulinopsis HOFKER, 1956, *946, p. 47 [*Bolivina cubana* BERMUDEZ, 1935, *117, p. 196; OD]. Test with short triserial early stage, later biserial; wall calcareous, finely perforate, surface commonly striate or costate, apertural face poreless; aperture elongate, nearly terminal in position, with flaring tooth plate, attached portion folded, and irregularly lobed, free folded part narrow, with fimbriate margin, occurring in an excavation of upper apertural margin. *Rec.*, Carib.——FIG. 449, 10-12. *V. cubana* (BERMUDEZ), Cuba; *10a,b*, opposite sides, ×120 (*472); *11a,b*, edge view showing aperture and basal view showing triserial base, ×160 (*946); *12a*, top view showing aperture, ×210; *12b,c*, views of tooth plate, one showing chamber outline, ×210 (*946).

[*Virgulinopsis* differs from *Bolivina* in its early triserial development, and from *Stainforthia* in its coarser perforations, longitudinal ornamentation, and more highly developed tooth plate. It is possibly intermediate between *Stainforthia* and *Sagrina*. The magnifications here given are corrected, as the figures given by HOFKER (*946) apparently were reduced to two-thirds the size stated in the figure explanations.]

Superfamily DISCORBACEA Ehrenberg, 1838

[*nom. correct.* LOEBLICH & TAPPAN, herein (*pro* Discorbidea SMOUT, 1954, p. 81)]——[In synonymic citations superscript numbers indicate taxonomic rank assigned by authors (¹superfamily, ²family group); dagger (†) indicates *partim*]——[=¹Orthoklinostegia† EIMER & FICKERT, 1899, p. 685 (*nom. nud.*); =²Rotaliaridia† RHUMBLER in KÜKENTHAL & KRUMBACH, 1923, p. 88; =¹Discorbidea SMOUT, 1954, p. 81; =¹Monolamellidea REISS, 1957, p. 128 (*nom. nud.*)]——[=Asterigerinacea LOEBLICH & TAPPAN, 1961, p. 302]

Test trochospiral or derived from such; wall of radial laminated calcite, perforate, noncanaliculate, single walls and septa; aperture interiomarginal or areal, or derived from such. *M.Trias.-Rec.*

Family DISCORBIDAE Ehrenberg, 1838

[*nom. correct.* GLAESSNER, 1945, p. 145 (*pro* Discorbina EHRENBERG, 1838, p. 200)]——[All names cited are of family rank; dagger (†) indicates *partim*]——[=Polystomat† LATREILLE, 1825, p. 161 (*nom. nud.*); =Cristaceat†, and Cristacést† DE BLAINVILLE, 1825, p. 383 (*nom. nud.*); =Hélicostégues† D'ORBIGNY, 1826, p. 268 (*nom. nud.*); =Uvellinat† EHRENBERG, 1839, table opposite p. 120 (*nom. nud.*); =Turbinoidaet† D'ORBIGNY in DE LA SAGRA, 1839, p. xxxviii, 71 (*nom. nud.*); =Valvulineriidae BROTZEN, 1942, p. 16; =Laticarinidae HOFKER, 1951, p. 307; =Valvulineridae HOFKER, 1951, p. 484; =Marginolamellidaet† HOFKER, 1951, p. 485 (*nom. nud.*); =Discorbididae POKORNÝ, 1954, p. 215 (*nom. van.*); =Conorbinidae HOFKER, 1954, p. 167; =Discorbinidae HOFKER, 1954, p. 167; =Pseudoparrellidae SUBBOTINA in RAUZER-CHERNOUSOVA & FURSENKO, 1959, p. 272; =Discorbiidae HORNIBROOK, 1961, p. 97 (*nom. van.*)]

Test free, trochospiral; chambers simple; wall calcareous, perforate, radial in structure, monolamellid; aperture basal or areal. *M.Trias.-Rec.*

Subfamily DISCORBINAE Ehrenberg, 1838

[*nom. correct.* GALLOWAY, 1933, p. 285 (*pro* subfamily Discorbisinae CUSHMAN, 1927, p. 75; *nom. transl. ex* family Discorbina EHRENBERG, 1838)] [=Discorbininae SCHUBERT, 1921, p. 156; =Pseudoparellinae VOLOSHINOVA in VOLOSHINOVA & DAIN, 1952, p. 81; =Discorbinellinae SIGAL in PIVETEAU, 1952, p. 228; =Discorbidinae POKORNÝ, 1954, p. 215 (*nom. van.*); =Discorbiinae HORNIBROOK, 1961, p. 97 (*nom. van.*)]

Test free, trochospiral, low- to highspired, umbilical region open; aperture basal, umbilical. *M.Trias.-Rec.*

Discorbis LAMARCK, 1804, *1085a, p. 182 [*Discorbites vesicularis*; OD (M)] [=*Discorbites* LAMARCK, 1804, *1085a, p. 182 (obj.); *Discorbitus* RAFINESQUE, 1815, *1496, p. 140 (*nom. van.*); Les Discorbes D'ORBIGNY, 1826, *1391, p. 274 (*nom. neg.*); Les Trochulines D'ORBIGNY, 1826, *1391, p. 274 (*nom. neg.*); *Trochulina* D'ORBIGNY in EHRENBERG, 1839, *667, chart following p. 120 (type, *Rotalia turbo* D'ORBIGNY, 1826, *1391, p. 274); *Cyclodiscus* EHRENBERG, 1839, *667, chart opposite p. 120 (*nom. subst. pro Discorbis* LAMARCK, 1804) (obj.); *Allotheca* EHRENBERG, 1843, *672, p. 407 (type, *A. megathyra*); *Aristerospira* EHRENBERG, 1858, *683, p. 11 (type, *A. isoderma*); *Discorbina* PARKER & JONES in CARPENTER, PARKER & JONES, 1862, *281, p. 200, 203 (type, *Rotalia turbo* D'ORBIGNY, 1826, *1391, p. 274); *Rotorbinella* BANDY, 1944, *69, p. 372 (type, *R. colliculus*); *Biapertorbis* POKORNÝ, 1956, *1477, p. 262 (type, *B. biaperturata*)]. Test free, trochospiral, plano-convex, flattened on umbilical side, periphery angled; all chambers visible on umbonate spiral side, only chambers of final whorl visible on umbilical side, with a flap extending from basal portion of each chamber toward umbilical region, opening extending along proximal side of each radial umbilical flap, connecting through cavity beneath flaps to interior of chambers themselves; primary aperture an interiomarginal, extraumbilical arch, secondary sutural openings at opposite side of chamber flap remaining open as later chambers are formed; biflagellate gametes occur. *Eoc.-Rec.*, Eu.-N.Am.-Pac.-N.Z.-Australia-Atl.——Fig. 451,*1-3*. **D. vesicularis*,

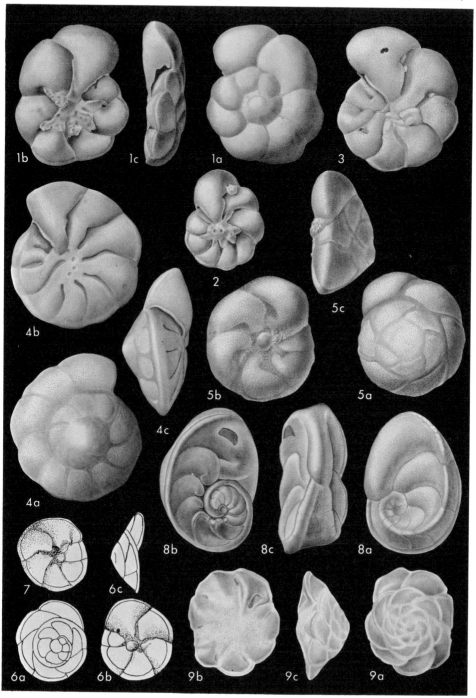

Fig. 451. Discorbidae (Discorbinae; *1-7, Discorbis; 8, Bronnimannia; 9, Buccella*) (p. C572-C575).

M.Eoc.(Lutet.), Fr.; *1a-c,* opposite sides and edge view of well-preserved topotype, showing flange-like umbilical flaps with openings at each extremity, ×17; *2,* umbilical side of smaller topotype, with less well-preserved flaps coalescing centrally and with a few central perforations, ×17; *3,* umbilical side of somewhat abraded topotype, umbilical flaps being destroyed and secondary apertures showing as sutural slits, ×25 (*2117).——FIG. 451,*4. D. turbo* (D'ORBIGNY), M.Eoc. (Lutet.), Fr.; *4a-c,* opposite sides and edge view of specimen compared with and nearly identical to lectotype in Paris, ×28 (*2117).——FIG. 451, *5. D. colliculus* (BANDY), Eoc., USA(Ore.); *5a-c,* opposite sides and edge view, ×74 (*2117).——FIG. 451,*6,7. D. biaperturata* (POKORNÝ), U.Eoc., Czech.; *6a-c,* opposite sides and edge view of paratype, ×85; *7,* umbilical view of holotype, ×85 (*1477).

[Many dissimilar forms have been placed in *Discorbis,* as, unfortunately, no characteristic illustrations were available for the type-species until approximately 1950. The original figure given by LAMARCK was extremely diagrammatic, the vesicular portion of the umbilical side being indicated only by the slightly angled sutures. CUSHMAN (1927, *432, pl. 24, fig. 1a-c) illustrated a topotype specimen; although the spiral and edge views given are recognizable, the drawing of the umbilical side does not show the "alar projections" referred to in the discussion. Y. LE CALVEZ (1949, *1112, pl. 3, figs. 36-38) has illustrated the central vesicular area much better. The preservation and degree of abrasion of the specimens cause a considerable degree of variation in the lateral extent of these projections, from narrow radial projections to approximately continuous flaps which almost overlap. Most texts have copied either the original figures of *Discorbis* or the misleading illustrations of CUSHMAN just cited; others have illustrated different species, some of which are not even congeneric with the type-species. Many unrelated forms thus have been placed in *Discorbis,* some of them completely lacking essential features of the genus. For these reasons, during the last 15 years there has been a great proliferation of generic names proposed for various discorbine species. Many of these are currently recognized, whereas others are here regarded as synonyms of one genus or another. In some instances the same species has been cited as type or placed within 3 or 4 different generic taxa. The species *Rotalia* (Trochuline) *turbo* D'ORBIGNY, 1826, has been designated as the type-species of *Trochulina* D'ORBIGNY, 1839, by subsequent monotypy, BASSET (1885, *101, p. 162); also it has been defined as the type-species of *Discorbina* PARKER & JONES, 1862, was included by BROTZEN (1936, *237, p. 141) in *Conorbina* but by HOFKER at various times in *Discopulvinulina* (1951, *928c) and *Rotorbinella* (1954, *942, p. 34). *Trochulina* was named by D'ORBIGNY in EHRENBERG, 1839, but no species were cited, although 3 species had been mentioned by D'ORBIGNY in 1826 (*1391, p. 274) under the French vernacular subgeneric term "Les Trochulines." Two were *nomina nuda* but the third, *Rotalia* (Trochuline) *turbo,* was valid. The latter name was first used in combination with the Latin subgeneric name *Trochulina* by BASSETT, 1885, thus automatically becoming the type of *Trochulina* by subsequent monotypy. This type designation thus validated the generic name *Trochulina,* which therefore takes precedence over the later name *Discorbina* PARKER & JONES, 1862. HORNIBROOK & VELLA (1954, *960, p. 26) discussed the genus *Discorbina* (type, *Rotalia turbo*) and considered *Rotorbinella* BANDY, 1944, to be a synonym, stating (p. 27), "The main diagnostic features of *Rotorbinella* are the prominent umbilical plug and channeled ventral sutures, characters that are strongly indicated in d'Orbigny's figure of *Rotalia* (Trochuline) *turbo.* Specimens of what we believe to be *Rotalia turbo,* from the Paris Basin Lutetian, are very close to *Discorbis finlayi* Dorreen, 1948, which Bermúdez regards as a typical *Rotorbinella. Rotorbinella* is thus a synonym of *Discorbina,* and moreover is not far removed from *Discorbis* in the strict sense." HOFKER (1951, *928c) included *R. turbo* in his new genus *Discopulvinulina* but later (1954, *942, p. 34) stated that *turbo* should be placed in *Rotorbinella,* adding, "*Rotorbinella turbo* (d'Orbigny) does not occur in the Lutetian of the Paris Basin, as

Hornibrook and Vella believe; the species which they had at hand must have been *Rotorbinella perovalis* (Terquem) . . . ." He also stated that perhaps the species should be called *Conorbina turbo.* Apparently D'ORBIGNY did not illustrate *R. turbo,* but included it in his modèles (No. 73). A figure of this was given by PARKER, JONES & BRADY (1865, *1419, pl. 2, fig. 68). However, this figure does not agree with D'ORBIGNY'S specimens. Commonly D'ORBIGNY'S models, and his illustrations as well, were not intended to portray a type-specimen exactly, but instead were a composite, much-generalized illustration which sometimes combined features of more than a single species. This fact makes reference to his type-specimens absolutely imperative, and the only reliable basis for systematic work. In 1954, we examined the D'ORBIGNY types in the Muséum National d'Histoire Naturelle in Paris, among them several specimens of *Rotalia turbo,* the type-specimens of which are from the Paris Basin Lutetian. HOFKER was thus mistaken in stating that *R. turbo* does not occur in the Lutetian of the Paris Basin. One of D'ORBIGNY'S original specimens of *R. turbo* is here designated as lectotype and this specimen is now so labeled on a separate slide in the Museum in Paris. The specimen of *R. turbo* here figured is from the Lutetian at the classic locality of Chaussey, Seine-et-Oise, France, and was compared to the lectotype in Paris, and found to be identical in all features. It is a true *Discorbis.* Of the genera to which *R. turbo* has been referred, *Conorbina* is regarded as a valid genus, on the basis of its type-species; *Discopulvinulina* is a synonym of *Discorbinella;* and *Trochulina, Discorbina,* and *Rotorbinella* are regarded as synonyms of *Discorbis.* In addition, *Biapertorbis* is regarded as a synonym of *Discorbis,* the type-species showing the umbilical flap separating the 2 apertures characteristic of *Discorbis,* and an umbilical "plug" like that found in some species of *Discorbis* but varying considerably in the degree of development.]

Aboudaragina NAKKADY, 1955, *1345, p. 261 [**A. eponidelliformis;* OD]. Test trochospiral, ventrally umbilicate; wall calcareous, finely perforate, microstructure and lamellar character unknown; aperture a large, rounded, interiomarginal equatorial opening in depressed terminal face. [This genus and type-species are known only from the original publication, in which the figures are generalized, and at least the spiral view apparently incorrect, as it does not show a trochospiral coil, but concentric whorls.] M.Jur., Egypt.——FIG. 452,*1. *A. eponidelliformis,* U.Dogger(Bathon.); *1a-c,* opposite sides and edge view of holotype, ×54 (*700).

Bronnimannia BERMÚDEZ, 1952, *127, p. 39 [**Discorbis palmerae* BERMÚDEZ, 1935, *117, p. 207; OD]. Test free, auriculate in outline, planispiral, evolute on both sides, plano-convex to nearly biconcave, umbilical region open on apertural side, closed on opposite side with sharp, acute-angled peripheral ridge and truncate peripheral margin sloping sharply to marginal keel; umbilical flap near aperture of each chamber, those of earlier chambers of final whorl remaining visible; sutures arched on apertural side and curved backward at periphery, sigmoid on opposite side, curving backward from umbonal area, abruptly angled at sharp dorsal angle, and curving again to peripheral keel; wall coarsely perforate on umbonal, ridged side, finely perforate on apertural side; aperture opening beneath umbilical chamber flaps and connecting laterally along spiral suture to openings of earlier chambers of final whorl. [*Bronnimannia* differs from *Planulinoides* in having a slitlike aperture beneath the ventral umbilical chamber flaps, whereas in *Planulinoides* the aperture is peripheral and consists of an oblique ovate open-

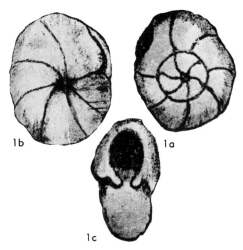

Fig. 452. Discorbidae (Discorbinae; *1, Aboudaragina*) (p. C574).

ing surrounded by a lip, in addition to the ventral openings beneath rudimentary umbilical flaps.] *Rec.*, Gulf Mex.-Atl.-Pac.——Fig. 451,*8*. **B. palmerae* (Bermúdez), Cuba (Bahia de Matanzas); lectotype, here designated (Coll. U.S.N.M.) from original syntypes of Bermúdez; *8a-c*, opposite sides and edge view, ×78 (*2117).

Buccella Andersen, 1952, *17, p. 143 [*Eponides hannai Phleger & F. L. Parker, 1951, *1455, p. 21; OD]. Test trochospiral, planoconvex to biconvex, periphery keeled; umbilical region and inner part of last whorl of chambers partially obscured by granular or pustulose coating on umbilical side; primary aperture interiomarginal, midway between umbilicus and periphery, supplementary sutural apertures at posterior margin of each chamber, near periphery on umbilical side. *Oligo.-Rec.*, N. Am.-Atl.-Pac.-Carib.-Japan-Eu.-S.Am.——Fig. 451,*9*. **B. hannai* (Phleger & F. L. Parker), Rec., Gulf Mex.; *9a-c*, opposite sides and edge view of holotype, ×80 (*17).

Conorbina Brotzen, 1936, *237, p.141 [**C.marginata*; OD]. Test plano-convex, trochospiral; chambers crescentic, increasing in breadth as added, so that final whorl has relatively few; sutures oblique and curved on spiral side, nearly radial on opposite side; aperture a low slit at base of final chamber, in slight re-entrant of chamber margin, near periphery on umbilical side. [*Conorbina* differs from *Glabratella* in lacking open umbilicus with surrounding radial ornamentation, and in having a suturally placed aperture.] *L.Cret.(Alb.)-U. Cret.(Senon.)*, Eu.-N.Am.——Fig. 453,*1*. **C. marginata*, L.Senon., Sweden; *1a-c*, opposite sides and edge view, ×190 (*2117).

Diplotremina Kristan-Tollman, 1960, *1059, p. 64 [**D. astrofimbriata*; OD]. Test free, trochospiral, margin of large open umbilicus deeply lobed; chambers increasing gradually in size; wall calcareous, perforate, microstructure and lamellar character unknown; primary aperture interiomarginal, about midway between umbilicus and periphery, umbilical chamber flap separating it from secondary umbilical opening, both apertures with crenulated margins. *M.Trias.*, Aus.——Fig. 454,*1*. **D. astrofimbriata*; *1a-c*, opposite sides and edge view of holotype, ×125 (*1059).

Discorbinella Cushman & Martin, 1935, *512, p. 89 [**D. montereyensis*; OD] [=*Discopulvinulina* Hofker, 1951, *936, p. 359 (type, *Rosalina bertheloti* d'Orbigny in Barker-Webb & Berthelot, 1839, *86, p. 135)]. Test free, plano-convex, compressed to scalelike, spiral side convex, nearly involute, only small portion of earlier whorls visible centrally, opposite side flattened to slightly concave, umbilicate, but nearly involute, with very little of previous coil visible at center, periphery carinate; aperture an interiomarginal arch, nearly peripheral on umbilical side, with supplementary opening at opposite margin of umbilical chamber flap; gametes biflagellate (in *D. bertheloti*). *Rec.*, Atl.-Pac.-Gulf Mex.——Fig. 453,*2*. **D. montereyensis*, USA(Calif., Monterey Bay); *2a-c*, opposite sides and edge of holotype, ×115 (*2117).——Fig. 453,*3*. *D. bertheloti* (d'Orbigny), Gulf Mex.; *3a-c*, opposite sides and edge view, ×68 (*2117).

[*Discorbinella* differs from *Discorbis* in its spiral side being only partially evolute, in having very simple umbilical chamber flaps, and in having a nearly peripheral primary aperture. Hofker (1951, *936, p. 359; *928c, p. 448) proposed the name *Discopulvinulina* to include a variety of forms previously placed in *Discorbis*, *Discorbina*, *Pulvinulina*, *Rotalia*, *Cibicidoides*, and *Rosalina*. If species originally included by Hofker under *Discopulvinulina* were in reality congeneric, his proposed name would be preoccupied by no less than six other valid generic names. However, on the basis of the type-species of these genera, *Discopulvinulina* is distinct from those, but a junior synonym of *Discorbinella*. d'Orbigny recorded *Rosalina bertheloti* (type-species of *Discopulvinulina*) from the Canary Islands, in marine sands at Teneriffe. Our figured specimen from the Gulf of Mexico is one of the hypotypes originally figured by Flint (1899, *723, pl. 72, fig. 4).]

Discorbitura Bandy, 1949, *70, p. 99 [**D. dignata*; OD]. Test free, trochospiral, concavo-convex, all chambers visible on convex spiral side; only last whorl visible on flat to concave umbilical side, periphery keeled; chambers with slight re-entrant at their irregular posterior margin on umbilical side, sufficiently pronounced as to nearly subdivide chamber into peripheral and umbilical lobe, and may have series of grooves branching out from sutures, umbilical region may be filled with nodes and pustules; sutures somewhat thickened on spiral side, depressed on opposite side: aperture peripheral, round areal opening at short distance above base of final chamber face, secondary openings which may be filled appear beneath posterior umbilical margin of chambers, rarely one or more of these remaining open after later chambers are added. *Oligo.*, N.Am.——Fig. 453,*4*. **D. dignata*, USA(Ala.); *4a-c*, opposite sides and edge views, ×139 (*2117).

[The secondary apertures on the umbilical side were not mentioned in the original description, nor was the infolding of the posterior chamber margins, although BANDY stated (*70, p. 99) that the sutures were "usually channeled with re-entrants." The holotype, paratypes, and metatypes have been examined by us and when examined at a sufficiently high magnification, all specimens show the features described above. *Discorbitura* resembles *Discorbinella*, but differs in having an areal aperture, less distinct umbilical supplementary apertures, in being involute rather than partially evolute on the umbilical side, and in possessing umbilical nodes and branching sutural grooves.]

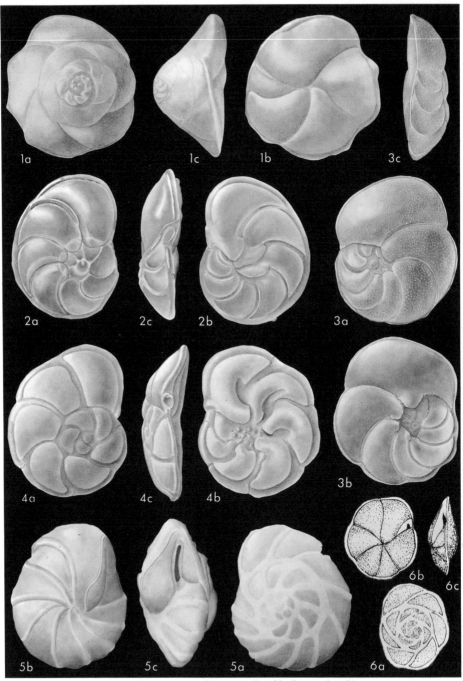

FIG. 453. Discorbidae (Discorbinae; *1, Conorbina; 2,3, Discorbinella; 4, Discorbitura; 5,6, Epistominella*) (p. C575-C576, C578).

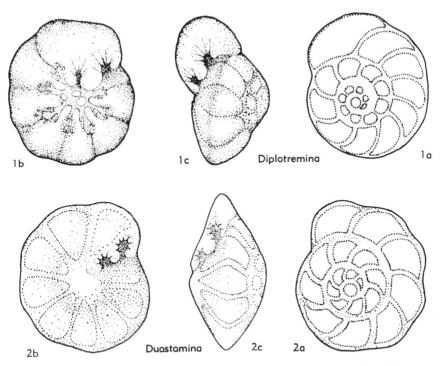

FIG. 454. Discorbidae (Discorbinae; *1, Diplotremina; 2, Duostomina*) (p. C575, C577).

Duostomina KRISTAN-TOLLMANN, 1960, *1059, p. 68 [*D. biconvexa*; OD]. Test free, trochospiral, chambers enlarging gradually, wall calcareous, perforate, microstructure and lamellar character unknown; similar to *Diplotremina* but with closed umbilicus and both apertures and intervening flap at forward margin of final chamber, instead of one being umbilical in position. *M.Trias.*, Aus.——FIG. 454,2. **D. biconvexa*; *2a-c,* opposite sides and edge view of holotype, ×125 (*1059).

Earlmyersia RHUMBLER, 1938, *1576, p. 209 [*"*Pulvinulina punctulata* (D'ORBIGNY)", HERON-ALLEN & EARLAND, 1913, *909, p. 134 (*non Rotalia punctulata* D'ORBIGNY, 1826) (=*Earlmyersia punctulata liliputana* RHUMBLER, 1938); OD]. Test trochospiral, plano-convex to concavo-convex, all whorls visible on spiral side and most of final 2 whorls visible on umbilical side, sutures thickened, strongly curved; wall calcareous, finely perforate, radial in structure, umbilical side with fine papillae; aperture obscure, an interiomarginal slit midway between periphery and umbilicus; growth or reproductive cysts may occur, during which agglutinated material temporarily covers protoplasm. *Rec.,* Ire.-Helgoland.——FIG. 455,*1-3. *E. liliputana,* Helgoland *(1),* W.Ire.(Clare Is.) *(2,3); 1a-c,* opposite sides and edge view, ×200 (*1576); *2,3,* spiral and umbilical sides of different specimens, ×120 (*909).

[The original definition of *Earlmyersia* stressed the flattened test, finely perforate wall, and the presence of pustulose ornamentation on the umbilical side. Both the descriptions by HERON-ALLEN & EARLAND and by RHUMBLER also mentioned specimens attached by the umbilical surface and surrounded by agglutinated material, comparable to the growth and reproductive cysts such as have been described for other Discorbinae. The type-species seems close to *Discorbinella* but as described, it differs in the apertural characters. *Discorbinella* has a primary peripheral interiomarginal aperture, a distinct umbilical chamber flap and a smaller opening behind this flap. A restudy of the type-species of *Earlmyersia* would show whether these features are also present therein, but meanwhile the genus is recognized tentatively as originally described. The type-species for *Earlmyersia* was originally designated by RHUMBLER (*1576, p. 209) as "*Pulvinulina punctulata* (d'Orb.) bei Heron-Allen und Earland in: Proc. roy. irish Acad., V. 31, Pt. 64, 1913, p. 134, T 4, fig. 20, 21." On the following page RHUMBLER (*1576, p. 210) described the specimens of HERON-ALLEN & EARLAND as "*Earlymyersia punctulata* (d'ORBIGNY) forma: *liliputana* nom. nov.!," and the above reference, plate and figures again were cited. Taxa proposed as *forma* remain available if proposed before 1961 [ICZN Art. 17(9)], hence the type-species is "*P. punctulata* (d'ORBIGNY)" HERON-ALLEN & EARLAND, 1913 (*non Rotalia punctulata* D'ORBIGNY, 1826) =*Earlmyersia punctulata liliputana* RHUMBLER, 1938.]

Eoeponidella WICKENDEN, 1949, *2060, p. 81 [*E. linki*; OD (M)] [=*Heminwayina* BERMÚDEZ, 1951, *126, p. 325 (type, *Discorbis multisectus* GALLOWAY & HEMINWAY, 1941, *764, p. 384)]. Test free, plano-convex to nearly biconvex, umbilical side may be slightly depressed centrally, all chambers visible on convex spiral side, only final whorl visible on umbilical side where each chamber has supplementary chamber along its forward

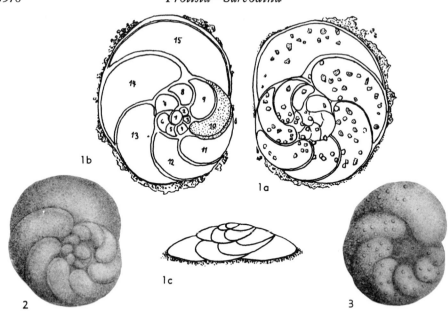

Fig. 455. Discorbidae (Discorbinae; *1-3, Earlmyersia*) (p. C577).

margin near umbilical region, supplementary chambers apparently formed after main chamber; wall calcareous, rather coarsely perforate, microstructure not known; primary aperture a broad high interiomarginal umbilical arch, but in specimens possessing final supplementary chamber against final chamber, only smaller aperture remains near proximal chamber margin. [The large open-arched aperture leading into the umbilical area, coarsely perforate test, and simple radial chambers, place *Discorbis multisectus* GALLOWAY & HEMINWAY, the type-species of *Heminwayina*, in the genus *Eoeponidella*.] *U.Cret.-Oligo.*, N.Am.-Carib.——Fig. 456,*1*. **E. linki*, U.Cret., Can.(Sask.); *1a-c*, opposite sides and edge view, ×242 (*2117).——Fig. 456,*2*. *E. multisecta* (GALLOWAY & HEMINWAY), Mio., Puerto Rico; *2a-c*, opposite sides and edge view of holotype, ×93 (*764).

Epistominella HUSEZIMA & MARUHASI, 1944, *974, p. 397 [**E. pulchella*; OD] [=*Pulvinulinella* CUSHMAN, 1926, *426, p. 62 (type, *P. subperuviana*) (non *Pulvinulinella* EIMER & FICKERT, 1899); *Pseudoparrella* CUSHMAN & TEN DAM, 1948, *502, p. 49 (type, *Pulvinulinella subperuviana* CUSHMAN, 1926, *426, p. 63)]. Test trochospiral; all chambers visible on spiral side, only those of last whorl visible on umbilical side; sutures oblique on spiral side, nearly radial on umbilical side; wall calcareous, perforate, radial in structure and monolamellid; aperture an elongate vertical slit in face, near and parallel to peripheral keel. *U.Cret.-Rec.*, Japan-N. Am.-Pac. - Gulf Mex.-Eu.——Fig. 453,*6*. **E. pulchella*, Plio., Japan; *6a-c*, opposite sides and edge views, ×70 (*52b).——Fig. 453,

5. *E. subperuviana* (CUSHMAN), Mio., USA (Calif.); *5a-c*, opposite sides and edge view of holotype, ×125 (*2117).

Eurycheilostoma LOEBLICH & TAPPAN, 1957, *1172, p. 228 [**E. altispira*; OD]. Test free, trochospiral, high-spired, umbilical side excavated, earliest whorl with 4 to 6 chambers, which increase in breadth as added, so that in adults only 3 or 4 chambers occur in each whorl, final chamber occupying most of umbilical side, extending around both sides of open umbilicus, final whorl may abruptly attain greater diameter, resulting in flaring test; aperture a broad arch at inner margin of last chamber, opening into umbilicus, and partially covered by broad umbilical flap which may have serrate margin. [*Eurycheilostoma* differs from *Neoconorbina* in being high-spired, rather than low, scalelike, and in having a rounded periphery. The apertural characters are similar, the broad umbilical flap with apertural re-entrants at the extremities occurring in both genera.] *L.Cret.*, N.Am.——Fig. 456,*3*. **E. altispira*, Alb., USA(Tex.); *3a-c*, opposite sides and edge view of holotype, ×192 (*1172).

Gavelinopsis HOFKER, 1951, *928c, p. 485 [**Discorbina praegeri* HERON-ALLEN & EARLAND, 1913, *909, p. 122; OD] [=*Gavelinopsis* HOFKER, 1951, *936, p. 359 (nom. nud.)]. Test free, planoconvex or biconvex, periphery keeled, all chambers visible on convex spiral side, only those of final whorl visible from flat to slightly convex umbilical side, which has prominent umbilical plug; sutures curving backward at periphery on spiral side, nearly radial on umbilical side; wall calcareous, hyaline, finely perforate; aperture a low interiomarginal slit at short distance from periphery on umbilical side, with slight lip above.

Rec., Atl.-Pac.——FIG. 456,4. *G. praegeri (HERON-ALLEN & EARLAND), Ire.; 4a-c, opposite sides and edge view, ×111 (*2117).

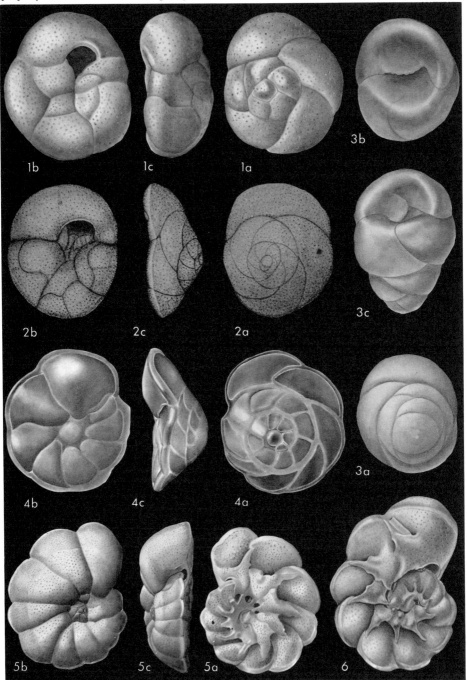

FIG. 456. Discorbidae (Discorbinae; 1,2, Eoeponidella; 3, Eurycheilostoma; 4, Gavelinopsis; 5,6, Lamellodiscorbis) (p. C577-C580).

[*Gavelinopsis* differs from *Conorbina* in having a distinct umbilical plug and in having a more ovate aperture bordered by a distinct lip. It differs from *Discorbis* in having an umbilical plug, instead of an umbilicus covered only by the highly developed umbilical chamber flaps of *Discorbis*. HOFKER (*936, p. 359) introduced this generic name citing *Gavelinopsis atlantica* HOFKER as type, but no description was given and the species was a *nomen nudum*. Later (*928c, p. 485) the genus was described and *Gavelinopsis praegeri* (HERON-ALLEN & EARLAND) (=*Discorbina praegeri* HERON-ALLEN & EARLAND, 1913) was cited as type. In this paper HOFKER again referred to the undescribed *Gavelinopsis atlantica* as occurring in the West Indies. BERMÚDEZ (1952, *127, p. 150) considered that HOFKER referred to the same species by both names, but *G. atlantica* was described by HOFKER (1956, *946, p. 212) as a new species from off Frederiksted, Santa Cruz. It is probably not congeneric, having prominent umbilical flaps and should be placed in *Rosalina*. Neither genus is considered by us to be related to the gavelinellids or anomalinids.]

Helenina SAUNDERS, 1961, *1634, p. 148 [**Pseudoeponides anderseni* WARREN, 1957, *2039, p. 39; OD] [=*Helenia* SAUNDERS, 1957, *1632, p. 374 (obj.) (*non* WALCOTT, 1889)]. Test free, trochospiral, biconvex, periphery rounded; chambers numerous, all visible from spiral side, only those of final whorl visible on umbilical side, final chamber with umbilical flap; sutures depressed, radial on umbilical side, curved to sinuate on spiral side with sutural slits on both spiral and umbilical sides, opening into chambers; wall calcareous, finely perforate; aperture an interiomarginal slit, extending from umbilicus across peripheral margin onto spiral side where it follows suture line 0.5 to 0.7 of distance to periphery, additional sutural slit occurring on umbilical side, extending from umbilical flap of chamber toward periphery. *Rec.*, N.Am.-W.Indies(Trinidad).——FIG. 457,*1*. **H. anderseni* (WARREN), Trinidad; *1a-c*, opposite sides and edge views, ×118 (*1632).

[Differs from *Pseudoeponides* in having the supplementary slits sutural in position, whereas those of *Pseudoeponides* are nearly perpendicular to the sutures on the umbilical side and those of the spiral side are areal in position in the chamber walls. *Epistomaria* resembles *Helenina* in possessing sutural slits on both spiral and umbilical sides but differs in having supplementary chamberlets on the umbilical side, which also are bordered with slits, and an areal aperture in the face of the final chamber in addition to the interiomarginal aperture.]

Lamellodiscorbis BERMÚDEZ, 1952, *127, p. 39 [**Discorbina dimidiata* JONES & PARKER in CARPENTER, PARKER & JONES, 1862, *281, p. 201; OD]. Test free, plano-convex, periphery sharply angled and keeled, with inflated chambers around umbonal boss on spiral side, umbilical surface flattened, somewhat evolute, with alar projections on inner part of proximal margins of chambers, with opening on their umbilical side and leaving opening both in front of and behind flaps just before they attach to test at their outer ends, flaps usually coalescing at their inner margins so as to form continuous ring or spiral around open umbilicus; sutures depressed on both sides, somewhat limbate on umbilical side; wall calcareous, coarsely perforate, spiral side commonly with secondary coating that covers inner two-thirds of chambers and partially fills pores; aperture an arch at periphery, extending short distance past keel on spiral side and about one-third of distance to umbilicus on opposite side, although it may merge with opening under chamber flaps so that a definite umbilical exent cannot be delineated, aperture bordered above by narrow lip. *Rec.*, Australia.——FIG. 456,*5,6*. **L. dimidiata* (JONES & PARKER); *5a-c*, opposite sides and edge view of paratype; *6*, umbilical side of larger paratype showing well-developed alar projections and apertures, ×26 (*2117).

[Differs from *Discorbis* in its evolute umbilical side and relatively involute spiral side, in having a distinct open umbilicus, umbonal plug on the spiral side, and in extension of the aperture somewhat onto the spiral side. The type-species superficially resembles *Discorbis vesicularis* LAMARCK and, in fact, the description on the plate legend of PARKER & JONES (1865, *1418, p. 422) stated that it was "merely *D. vesicularis* modified by being sharp-edged, and flat, and even scooped on the under face (opposite to that which is flat in *Truncatulina*)." All whorls are visible spirally in *D. vesicularis* and only the final whorl visible on the umbilical side, the opposite being true in *Lamellodiscorbis dimidiata*. The chamber flaps are also better developed and are perforate to a greater extent in the present species. The illustrations given by BERMÚDEZ (1952, *127, pl. 4, figs. 4a-c) are not of this species, or genus, but as noted by HORNIBROOK & VELLA (1954, *960, p. 27) are a copy of the figures of "*Discorbina vesicularis* (Lamarck)" given by BRADY (1884, *200, pl. 87, figs. 2a-c), whose figures show the convex evolute dorsal side and involute, somewhat flattened ventral side, typical of *Discorbis*, although it is not *D. vesicularis* LAMARCK. ¶In 1953, we studied the types of JONES & PARKER in the British Museum (Natural History). As no holotype had been selected for *D. dimidiata*, one of the original specimens is here designated as lectotype (BMNH-ZF 3651), the remainder of the syntypes becoming paratypes (BMNH-ZF 3650). All are from Recent sponge sands near Melbourne, Australia. The generic description and comparisons here given are based upon these original specimens of JONES & PARKER.]

Laticarinina GALLOWAY & WISSLER, 1927, *767, p. 193 [*pro Carinina* GALLOWAY & WISSLER, 1927, *766, p. 51 (*non* HUBRECHT, 1887)] [**Pulvinulina repanda* var. *menardii* subvar. *pauperata* PARKER & JONES, 1865, *1418, p. 395; OD] [=*Parvicarinina* FINLAY, 1940, *717d, p. 467 (type, *Truncatulina tenuimargo* var. *alto-camerata* HERON-ALLEN & EARLAND, 1922, *911, p. 209)]. Test free, planispiral, broad peripheral keel may show growth lines; chambers saddle-shaped, anterior margin of keel forming separation between 2 lobes of next-developed chamber, lobes larger on one side and closely appressed, final chambers may be irregular in outline and rarely small, irregularly placed, supplementary chambers may appear on side where lobes are larger, lobes small and less closely appressed on opposite side, interconnected by small tubular necks, final chamber commonly with broad attachment flange somewhat loosely attached at posterior umbilical margin, leaving opening beneath which connects to chamber interior, wide scarlike whitish area may occur around final 2 or 3 chambers on side with larger lobes; wall calcareous, finely perforate, keel apparently imperforate, although small irregularly spaced lines, "bubbles," and tubules may appear, possibly due to parasitic organisms; peripheral aperture at one side of keel, low slit perpendicular to periphery may be slightly produced in large specimens, this peripheral aperture being absent in some specimens and entire forward margin tightly closed, with supplementary openings beneath posterior umbilical margin of smaller lobes of later chambers suggesting apertures beneath umbilical chamber flaps. *Paleoc.-Rec.*, Atl.-Pac.-Carib.-N.Z.

Afr.-Eu.——Fig. 457,2,3. **L. pauperata* (Parker & Jones), Rec., Carib.; *2*, apert. or umbilical side; *3a,b*, opposite sides of another specimen, ×19 (*2117).——Fig. 457,*4*. *L. altocamerata* (Heron-

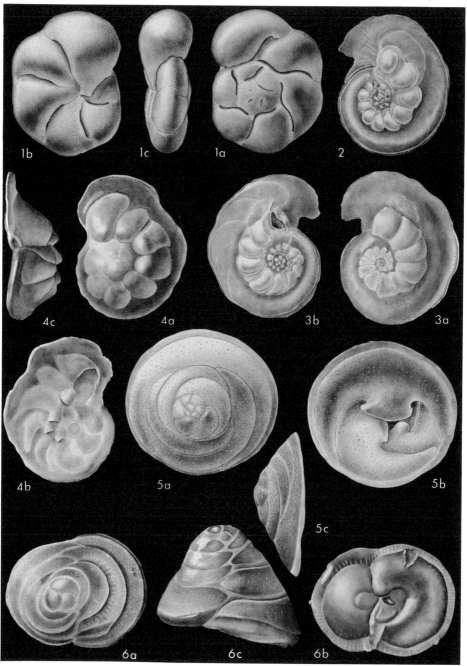

Fig. 457. Discorbidae (Discorbinae; *1, Helenina; 2-4, Laticarinina; 5, Neoconorbina; 6, Patellinella*) (p. C580-C582).

ALLEN & EARLAND), L.Mio., N.Z.; *4a-c,* opposite sides and edge view, showing peripheral and umbilical apertural openings, ×73 (*2117).

[BERMÚDEZ (1952, *127, p. 40) cited the type-species of *Parvicarinina* as *P. alatocamerata* (HERON-ALLEN & EARLAND, =*Truncatulina tenuimargo* HERON-ALLEN & EARLAND). This species was described as *alto-camerata* and was so designated by FINLAY. The umbilical openings described above are found in the type-species of both *Laticarinina* and *Parvicarinina*, although not previously reported for *Laticarinina*. We have examined the original types of PARKER & JONES in the British Museum (Natural History) and isolated a lectotype that is here designated (BMNH-ZF 3574 *ex* 94.4.3.319) for *Pulvinulina repanda* var. *menardii* subvar. *pauperata* PARKER & JONES. FINLAY (1940, *717d, p. 467) had regarded *Laticarinina* as having only the peripheral aperture and *Parvicarinina* as having only the umbilical openings. However, the type-species of *Laticarinina* has both types of apertures well developed. Furthermore, the type-species of *Parvicarinina* also has the forward peripheral aperture; hence, *Parvicarinina* is regarded as a junior synonym.——¶CUSHMAN & TODD (1941, *527, p. 105) regarded *Laticarinina* as closely related to *Cibicides*, stating that "the aperture in the adult is on the dorsal side on the inner margin of the last-formed chamber, low and elongate, similar to that in many species of *Cibicides*." The genus was placed in the Anomalinidae by CUSHMAN (1948, *486, p. 334). The aperture is unlike that of *Cibicides*, however, and the test is not coarsely perforate, nor perforate granular in structure, as in the Anomalinidae. GALLOWAY placed the genus in the Nonionidae, stating (1933, *762, p. 264), "*Laticarinina* evolved from *Nonion* by developing a peripheral flange. Free specimens are planispiral and symmetrical, but attached specimens are distorted and on that account bear some slight resemblance to the Rotaliidae." However, none of the Nonionidae show umbilical flaps with supplementary openings, and the Nonionidae have a perforate granular wall structure, whereas that of *Laticarinina* is perforate radial.——¶BERMÚDEZ (1952, *127, p. 18) placed *Parvicarinina* in the subfamily Discorbisinae [=Discorbinae], family Rotaliidae, and placed *Laticarinina* in the subfamily Planulininae (*127, p. 21), family Anomalinidae. *Planulina* has radial perforate walls, as does *Laticarinina*, but the apertural characters are quite distinct. As *Parvicarinina* is a synonym of *Laticarina*, "both" must be placed in the same family.]

Neoconorbina HOFKER, 1951, *936, p. 357 [*Rosalina orbicularis* TERQUEM, 1876, *1888, p. 75 (*non Rosalina orbicularis* D'ORBIGNY, 1850) (=*Discorbina terquemi* RZEHAK, 1888, *1602, p. 228); OD]. Test free, trochospiral, conical, concavo-convex, periphery acutely angled and carinate; early chambers subglobular, increasing very rapidly in breadth on spiral side and very little in height as added, so that final chamber occupies much of periphery and is much broader than high, chambers on umbilical side with distinct flap at midline and apertural re-entrant on either side; wall calcareous, of calcite, by X-ray powder diffraction film; aperture in forward re-entrant of chamber on umbilical side, covered by succeeding chambers to remain as intercameral opening, supplementary aperture occurring in other re-entrant of final chamber, those of earlier chambers of final whorl remaining open. *Rec.,* Atl.O.-Pac.O.——FIG. 457, 5. **N. terquemi* (RZEHAK), Atl.; *5a-c,* opposite sides and edge view, ×111 (*2117).

[HOFKER described *Neoconorbina* with *N. orbicularis* (TERQUEM) (=*Rosalina orbicularis* TERQUEM, 1876) designated as type-species (*936, p. 357). The Siboga monograph (*928c) was mentioned (*936, p. 360) as being in press. *Neoconorbina* was described in detail in the Siboga paper (1951, *928c, p. 433) but in it HOFKER stated, "The type of the species [*sic*] is *Neoconorbina pacifica* Hofker." Undoubtedly THALMANN (1952, *1897j, p. 977) considered the Siboga paper as the original reference for the genus and therefore erroneously listed *Neoconorbina pacifica* HOFKER as the type-species. The genus was defined and the type-species fixed by original designation and monotypy, however, in the earlier paper cited above. BERMÚDEZ (1952, *127, p. 34) regarded *Neoconorbina* as a synonym of *Rosalina* D'ORBIGNY. However, *Rosalina* differs from *Neoconorbina* in the presence of sutural slits, which are the remnants of earlier apertures. *Neoconorbina* also has a conical form, lunate chambers, and an overlapping final chamber on the umbilical side. It differs from *Conorboides* in having 2 distinct apertures, one at each side of the umbilical flap.]

Patellinella CUSHMAN, 1928, *436, p. 5 [**Textularia inconspicua* BRADY, 1884, *200, p. 357; OD]. Test free, conical, trochoid, plano-convex, earliest whorl may have more than 2 chambers, test later biserial, all whorls visible dorsally, only final pair visible ventrally; wall calcareous, finely perforate, radial in structure; aperture ventral, broad arch opening into umbilicus, not covered by next following chamber. *Rec.,* S.Pac.O.(Tasm.).——FIG. 457,6. **P. inconspicua* (BRADY); *6a-c,* opposite sides and edge view, ×183 (*2117).

[Differs from *Patellinoides* in having more than 2 chambers in the early whorl, not a simple spiraling tube, and in having a less complex apertural region. WOOD (1949, *2073, p. 250) noted that the type of *Patellinella inconspicua* shows a perforate radial wall structure, whereas *Spirillina*, *Patellina*, and *Patellinoides* all have a test composed of a single crystal of calcite. This has been verified by us. Furthermore, the absence of an early undivided spire in the present genus, such as is found in the Patellininae, substantiates their separation shown by COLLINS (1958, *375, p. 400), who placed *Patellinella* in the "Discorbisinae." HOFKER (1951, *936, p. 358) described *Discobolivina* as including the earlier genera "*Patellina*, *Patellinoides*, etc.," citing *Discobolivina corrugata* (WILLIAMSON) (=*Patellina corrugata* WILLIAMSON) as type-species. In the Siboga monograph (1951, *928c, p. 422) HOFKER also included *Patellinella* in *Discobolivina*, giving the generic description as an original description in this publication without citing a type-species.——¶THALMANN (1952, *1897j, p. 973) stated that HOFKER had not designated a type-species for *Discobolivina*, hence he selected *Patellinoides conica* HERON-ALLEN & EARLAND. As HOFKER'S designation of *Patellina corrugata* WILLIAMSON as type was earlier, the designation by THALMANN was invalid and *Discobolivina* is a junior objective (isogenotypic) synonym of *Patellina*. HOFKER'S discussion and figures of "*Discobolivina inconspicua* (Brady)" are of the species BRADY described as *Textularia jugosa*, and specimens belonging to BRADY'S species *inconspicua* are included by HOFKER under *Discobolivina conica*. As BRADY well described and figured the 2 species, it is impossible to alter the names applied to them.——¶The species *Textularia jugosa* does not belong to *Patellinella*; hence HOFKER'S discussion of this form, although under the name *inconspicua*, has no bearing on the present genus. *Textularia jugosa* shows an open umbilical region into which open the apertures of final pair of chambers, and in *Patellinella* the apertures of the two final chambers are distinctly separated. Typical *Patellinella* is not characterized by a strongly ornamented test, as in *T. jugosa* BRADY.]

Pijpersia THALMANN, 1954, *1904, p. 153 [*pro Ruttenia* PIJPERS, 1933, *1457, p. 30 (*non* RODHAIN, 1924)] [**Bonairea coronaeformis* PIJPERS, 1933, *1456, p. 72; OD] [=*Bonairea* PIJPERS, 1933, *1456, p. 72 (obj). (*non* BURRINGTON BAKER, 1924); *Pseudoruttenia* Y. LE CALVEZ, 1959, *1115, p. 92 (type, *P. diadematoides*)]. Test free, trochospiral, spiral side ornamented by tubercles and keels, umbilical side flat to concave and may show radial grooves; chambers inflated to angular, strongly overlapping on umbilical side; wall microstructure unknown; aperture umbilical, with broad umbilical flap. [Similar to *Glabratella* in the commonly ornamented spiral side and radially ornamented umbilical side, but differs in having a prominent umbilical flap, similar to *Conorboides*.] *Eoc.,* W.Indies(Bonaire-Trinidad)-

C.Am.(Panama)-Eu.——FIG. 458,*1*. **P. coronaeformis* (PIJPERS), Eoc.,Bonaire; *1a-c*, opposite sides and back edge (not apert.) view of topotype, ×163 (*2117).——FIG. 458,*2*. *P. diadematoides* (Y. LE CALVEZ), Cuis., Fr.; *2a-c*, opposite sides and edge view of holotype, ×90 (*1115).

Planodiscorbis BERMÚDEZ, 1952, *127, p. 40 [**Discorbina rarescens* BRADY, 1884, *200, p. 651, OD].

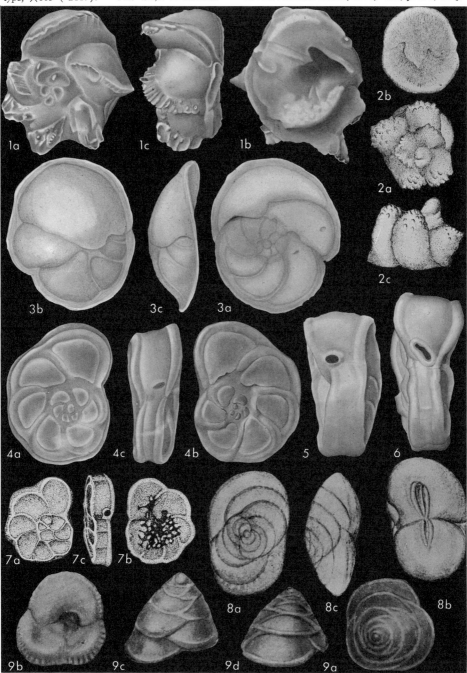

FIG. 458. Discorbidae (Discorbinae; *1,2, Pijpersia; 3, Planodiscorbis; 4-7, Planulinoides; 8, Pseudopatellinella; 9, Pseudopatellinoides*) (p. C582-C584).

Test free or possibly attached during life, planoconvex or concavo-convex, with spiral side flattened or concave and all chambers visible, umbilical side convex with only chambers of final whorl visible around closed umbilicus, periphery angled with broad keel, chambers increasing rapidly in size, final chamber occupying much of umbilical side; aperture a small arch in slight re-entrant at basal margin of final chamber on flattened spiral side, about halfway between periphery and umbilical region. *Rec.*, Pac.O.——Fig. 458,3. **P. rarescens* (Brady); *3a-c,* opposite sides and edge view of lectotype, ×79 (*2117).

[*Planodiscorbis* is very similar in character to *Discorbinella,* but is completely involute on the umbilical side, rather than partially evolute on both sides. It differs from *Discorbis* in having the spiral side flattened, possessing the aperture, with the umbilical side convex and involute. In *Discorbis* the umbilical side is flat and involute and contains the aperture, and the spiral side is evolute. *Planodiscorbis* also lacks the characteristic umbilical alar extensions of the chambers found in *Discorbis*.——¶Brady's types of *Discorbina rarescens* in the British Museum (Natural History) were examined by us and we here designate as lectotype the specimen figured by Brady (1884, *200, pl. 90, fig. 2) (BMNH-ZF3648, from *Challenger* station 185, off Raine Island, Torres Straits, at a depth of 155 fathoms). The remaining original syntypes are now designated as paratypes (BMNH-ZF1414).]

Planulinoides Parr, 1941, *1424, p. 305 [**Discorbina biconcava* Jones & Parker in Carpenter, Parker & Jones, 1862, *281, p. 201; OD] [=*Discotruncana* Shirai, 1960, *1734, p. 539 (type, *D. japonica*)]. Test free, biconcave, nearly planispiral, evolute, with broad truncate double-keeled periphery, evolute on spiral side, partially evolute on opposite side; primary areal aperture peripheral and somewhat oblique toward umbilical side and surrounded by lip, supplementary apertures on umbilical side at inner margin of chambers, under rudimentary umbilical flap. Plio.-Rec., Australia-Japan.——Fig. 458,4-6. **P. biconcava* (Jones & Parker), Rec., Victoria; *4a-c,* opposite sides and edge view; *5,6,* edge views of additional specimens showing variation in peripheral aperture, ×115 (*2117).——Fig. 458,7. *P. japonica* (Shirai), Plio., Japan; *7a-c,* opposite sides and edge view, ×47 (*1734).

[Parr originally stated that "the aperture is peripheral instead of being situated on the under surface, as in *Discorbis*." Apparently the openings on the umbilical side were not observed, although they are shown in the figures of Parker & Jones (1865, *1418, pl. 19, fig. 10b) and Brady (1884, *200, pl. 91, fig. 2b). *Planulinoides* was considered to be a synonym of *Discorbinella* Cushman & Martin by Cushman (1948, *486, p. 288) but it differs in being biconcave, bicarinate, and in having a truncate periphery, whereas *Discorbinella* is plano-convex, with a single keel, and more prominent umbilical flaps. *Planulinoides* differs from *Bronnimannia* Bermúdez in having a peripheral aperture and a double keel. A lectotype for *Discorbina biconcava* Jones & Parker was isolated by us in 1953 and is here designated (BMNH-ZF3646, with paratype ZF3645), both from Recent shore sand, Melbourne, Australia.]

Pseudopatellinella Takayanagi, 1960, *1863, p. 121 [**P. cretacea*; OD]. Test free, trochospiral, spiral side convex and evolute, umbilical side flattened; early chambers subglobular, rapidly increasing in breadth and becoming crescentic in spiral view, with only 2 chambers to whorl; wall calcareous, perforate, microstructure and lamellar character unknown, inner surface of wall undulating, but without septula; aperture a narrow slit on umbilical side, extending up center of chamber face. [Although superficially resembling *Patellina,* this genus does not have the nonseptate coiled stage such as is characteristic of the Spirillinidae.] *U.Cret.*, Japan.——Fig. 458,8. **P. cretacea*; *8a-c,* opposite sides and edge view of holotype, ×140 (*1863).

Pseudopatellinoides Krasheninnikov, 1958, *1051, p. 241 [**P. primus*; OD]. Test free, small, conical, trochospirally coiled with highly convex spiral side and flattened, centrally umbilicate opposite side, periphery angled and carinate; chambers few, commonly 3 to whorl, broad, low, semilunate, and all visible on spiral side, only 3 of last whorl visible on umbilical side where each occupies approximately one-third of test; sutures strongly oblique, thickened and flush on spiral side, radial, curved and depressed on umbilical side; wall calcareous, hyaline, finely perforate, radial in structure; aperture an interiomarginal, umbilical slit or slight arch, which does not extend to periphery. [*Pseudopatellinoides* differs from *Patellinella* in having 3 chambers to whorl throughout development.] *Mio.(U.Torton.),* USSR.——Fig. 458,9. **P. primus*; *9a-d,* spiral, umbilical, and edge views, ×100 (*1051).

Rosalina d'Orbigny, 1826, *1391, p. 271 [**R. globularis*; SD Galloway & Wissler, 1927, *766, p. 62] [=*Turbinolina* d'Orbigny in de la Sagra, 1839, *1611, p. 89 (type, *Rosalina globularis* d'Orbigny, 1826, *1391, p. 271; SD Loeblich & Tappan, herein); *Semirosalina* Hornibrook, 1961, *959, p. 103 (type, *S. inflata*)]. Test planoconvex, free or attached by flattened umbilical surface, all chambers visible from convex spiral side, only those of final whorl visible around open umbilicus on umbilical side; aperture a low interiomarginal arch at base of final chamber near periphery on umbilical side, with broad chamber flap just beneath aperture extending into open umbilicus, secondary sutural opening at opposite side of flap, those of previous chambers also remaining open. *Rec.*, Atl.O.-Pac.O.-Antarctic.——Fig. 459,1. **R. globularis,* Antarctic(Ross Sea); *1a-c,* opposite sides and edge view, ×74 (*2117).——Fig. 460,1. *R. inflata* (Hornibrook), L.Mio., N.Z.; *1a-c,* opposite sides and edge view of holotype, ×100 (*959).

[Cushman (1948, *486, p. 286) & Galloway (1933, *762, p. 286) considered *Rosalina* a synonym of *Discorbis.* Bermúdez (1952, *127, p. 34) considered it a valid genus but placed *Neoconorbina* Hofker in the synonymy of *Rosalina.* All 3 are here considered to be distinct, *Rosalina* being intermediate, but lacking the pronounced ventral chamber flaps and closed umbilicus of *Discorbis* and differing from *Neoconorbina* in the presence of sutural slits which are remnants of earlier apertures. It differs from *Conorbina* in having an open umbilicus and a more extensive aperture nearer the umbilicus, with the proximal portions of earlier apertures remaining as sutural secondary openings. No definite locality was given in the original reference, the species being merely noted to occur on all ocean coasts. The specimen here figured was compared by us with

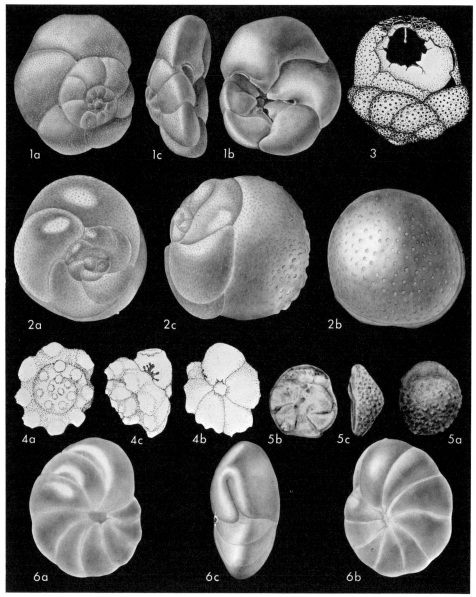

Fig. 459. Discorbidae (Discorbinae; *1, Rosalina; 2,3, Tretomphalus; 4, Variostoma; 5, Vernonina; 6, Stetsonia*) (p. C584-C586).

D'ORBIGNY's type-specimen in the Muséum National d'Histoire Naturelle, Paris, France.]

Stetsonia F. L. PARKER, 1954, *1414, p. 534 [*S. minuta*; OD]. Test small, lenticular, slightly trochospiral but involute on both sides, periphery narrowly rounded, chambers increasing gradually in size, low and broad; sutures radial, curved, slightly depressed; wall thin, calcareous, finely perforate, radial in structure, lamellar character not described; aperture an elongate slit extending from base of final chamber in equatorial position up face in slightly diagonal line on umbilical side, with narrow lip. *Rec.*, Gulf Mex.——FIG. 459,6. *S. minuta*; *6a-c*, opposite sides and edge view, ×325 (*2117).

Tretomphalus MÖBIUS, 1880, *1293, p. 67, 99 [*Rosalina bulloides* D'ORBIGNY in DE LA SAGRA, 1839, *1611, p. 98; OD (M)]. Test with early benthonic stage similar to *Discorbis*, reproductive cycle with alternation of generations, asexually

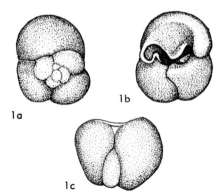

Fig. 460. Discorbidae (Discorbinae; *1, Rosalina*) (p. C584).

produced megalospheric individuals becoming encysted before development of gametes, developing large globular float chamber over umbilical region, small entosolenian tube extending inward from distal surface of float, through which ectoplasm may protrude, large gas bubble then developing within float chamber, added buoyancy allowing test to float to surface after breaking free from growth cyst, commonly 2 or more pelagic parent individuals then approaching closely by means of extended pseudopodia so as to insure fusion of maximum number of gametes, equally biflagellate gametes escaping through areal pores, those from different parent individuals fusing in pairs to form benthonic microspheric tests which in turn give rise asexually to megalospheric generation; test calcareous, with pseudochitinous inner membrane, wall microstructure unknown (if monolamellid, as in *Discorbis*, it is the only pelagic one); aperture umbilical, as in *Discorbis* in benthonic stage, and consisting of areal pores on pelagic float chamber. *Rec.*, Australia-Medit.-Gulf Mex.-tropical and subtropical Atl.O.-Pac.O.-Ind.O.-Medit. Sea-Red Sea.——Fig. 459,2; 461,1. **T. bulloides* (D'Orbigny), *Rec.*, USA(Fla.), (459,2), *Rec.*, USA(Calif.) (461,1); 459,*2a-c*, opposite sides and edge view of hypotype, ×135 (*2117); 461,*1a*, sectioned decalcified specimen showing large central gas bubble, internal tube of float chamber, and gametes *in situ*, ×300; 461,*1b*, biflagellate gametes below and left, fused gametes at right and zygote in center, ×2,000 (*1341).——Fig. 459,3. *T. myersi* Cushman, *Rec.*, USA(Calif.); side view of dissected specimen showing perforated float chamber exposing internal float with entosolenian tube, ×170 (*1341).

Variostoma Kristan-Tollmann, 1960, *1059, p. 55 [**V. spinosum*; OD]. Test free, trochospiral, may be high-spired; all chambers visible on spiral side, opposite side involute, deeply umbilicate, with lobulate umbilical margin; wall calcareous-perforate, granular in structure, lamellar character unknown; aperture interiomarginal, extraumbilical, with lobulate margin. *M.Trias.*, Eu. (Aus.).——Fig. 459,4. **V. spinosum*; *4a-c*, opposite sides and edge view of paratype, ×25 (*1059).

Vernonina Puri, 1957, *1488, p. 124 [**V. tuberculata*; OD]. Test trochospiral, hemispherical, with convex spiral side, covered with numerous rounded granules, flattened umbilical side with central plug or granules; sutures oblique but obscured by surface ornamentation on spiral side, radial and depressed on umbilical side; wall calcareous, perforate, microstructure and lamellar character not described; aperture interiomarginal, on umbilical side about half distance between umbilicus and periphery. *U.Eoc.*, USA(Fla.).——Fig. 459,5. **V. tuberculata*; *5a-c*, opposite sides and edge view of paratype, ×40 (*1488).

Subfamily BAGGININAE Cushman, 1927

[Baggininae Cushman, 1927, p. 77]——[Dagger (†) indicates *partim*]——[=Praerotalininae† Hofker, 1933, p. 125 (*nom. nud.*); =Cancrisinae Chapman, Parr & Collins, 1934, p. 567; =Valvulineriinae Brotzen, 1942, p. 17; =Cancrininae Sigal in Piveteau, 1952, p. 228 (*nom. van.*)]

Test free, trochospiral, umbilical area closed, with clear thin imperforate area adjacent to umbilicus; aperture basal. *L. Cret.-Rec.*

Baggina Cushman, 1926, *426, p. 63 [**B. californica*; OD]. Test free, subglobular, trochospiral, chambers few, rapidly enlarging and somewhat overlapping on spiral side, with closed umbilicus on opposite side; wall calcareous, perforate, radial in structure; aperture a broad umbilical opening below clear, nonperforate lunate area in face of final chamber. [*Baggina* differs from *Cancris* in having an open aperture without a lip, and in being somewhat involute on the spiral side.] *Cret.-Rec.*, cosmop.——Fig. 462,1. **B. californica*, Mio., USA(Calif.); *1a-c*, opposite sides and edge view of paratype, ×56 (*2117).

Cancris de Montfort, 1808, *1305, p. 267 [**C. auriculatus* (=*Nautilus auriculus* Fichtel & Moll, 1798, *716, p. 108); OD] [=*Carcris* Deshayes, 1830, *590, p. 191 (*nom. null.*); *Pulvinulinella* Eimer & Fickert, 1899, *692, p. 628 (obj.) (*non* Cushman, 1926)]. Test free, trochospiral, biconvex, commonly elongate and auriculate in shape, spiral side evolute, opposite side may have slightly open umbilicus; chambers rapidly enlarging, relatively low and broad; wall calcareous, perforate, radial in structure, may have peripheral keel; aperture on umbilical side, broad apertural lip extending over opening and projecting into umbilicus. [*Cancris* differs from *Baggina* in being more elongate, evolute on the spiral side, keeled, and in having an open umbilicus and an apertural lip. It resembles *Baggina* in having a broad nonperforate area above the aperture.] *Eoc.-Rec.*, cosmop.——Fig. 462,3. **C. auriculus* (Fichtel & Moll), Plio., Italy; *3a-c*, opposite sides and edge view, ×45 (*2117).

289 [*Sphaeroidina corticata* HERON-ALLEN & EARLAND, 1915, *910b, p. 681; OD]. Test free, small, consisting of 4 subglobular chambers arranged in apposed pairs, all visible externally, perhaps representing much reduced trochospiral coiling; wall calcareous; perforate, radial in structure, surface covered with numerous irregular knobs and ridges, presenting extremely rugose appearance, lamellar character unknown; aperture consisting of large pores between pairs of chambers on umbilical side, separated by pillar-like extensions from final chamber. *Rec.*, SE.Afr. (Moz.).——FIG. 462,4. **R. corticata* (HERON-ALLEN & EARLAND); *4a-c,* opposite sides and edge view of lectotype, ×79 (*2117).

[*Rugidia* differs from *Physalidia* in having a rugose exterior and multiple apertural openings between pillars along margins of the final chamber. A lectotype was selected by us at the British Museum (Natural History), and is here designated (BMNH-ZF3623, *910b, pl. 51, fig. 14) with paratypes (BMNH-ZF3621) from Kerimba Station 11, Manangoroshi to Lurio Points, Kerimba Archipelago, off Mozambique.]

Valvulineria CUSHMAN, 1926, *426, p. 59 [**V. californica*; OD] [=*Rotamorphina* FINLAY, 1939, *717c, p. 325 (type, *R. cushmani* FINLAY, 1939, *717c, p. 325 (non *Valvulineria cushmani* CORYELL & EMBICH, 1937) (=*Valvulineria teuriensis* LOEBLICH & TAPPAN, nom. nov., herein)]. Test free, trochospiral, umbilicate, periphery rounded; chambers increasing gradually in size; sutures radial, thickened; wall calcareous, finely perforate, radial in structure, monolamellid, surface smooth; aperture interiomarginal, extraumbilical-umbilical, with broad thin apertural flap projecting over the umbilicus. *L.Cret. (Alb.)-Rec.*, cosmop.——FIG. 462,5-7; 463. **V. californica*, Mio., USA(Calif.); *462,5a-c*, opposite sides and edge view; *462,6,7*, umbilical sides showing more extensive umbilical flaps; all ×49 (*2117); 463, horiz. sec. showing monolamellar radial structure, ×100 (*1529).——FIG. 462,8. *V. teuriensis* LOEBLICH & TAPPAN, nom. nov., U. Cret.(Teurian), N.Z.; *8a-c,* opposite sides and edge view, ×44 (*2117).

Family GLABRATELLIDAE
Loeblich & Tappan, n.fam.

Test trochospiral, low to high-spired, umbilical side flattened; wall calcareous; hyaline, perforate, radial in structure; aperture umbilical in position; in Recent forms reproduction plastogamic, with specimens attaching in pairs by umbilical surfaces, gametes triflagellate; habitat commonly littoral. [The genera here included are distinct from the Discorbidae in having an umbilical aperture, flattened to concave and radially striate or grooved umbilical side and a plastogamic reproductive cycle with triflagellate rather than biflagellate gametes.] *Eoc.-Rec.*

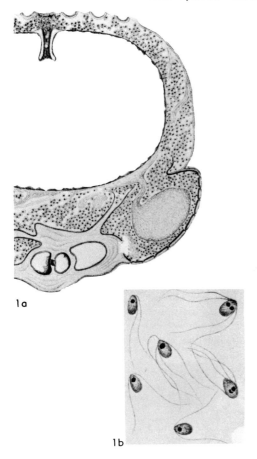

FIG. 461. Discorbidae (Discorbinae; *1, Tretomphalus*) (p. C585-C586).

Physalidia HERON-ALLEN & EARLAND, 1928, *913, p. 288 [**P. simplex*; SD GALLOWAY, 1933, *762, p. 337]. Test free, ovate or reniform in outline, composed of few (2 to 4) subglobular chambers arranged in apposition; wall calcareous, hyaline, radial in structure, very thin, coarsely perforate, with perforations produced into very thin tubules, lamellar character unknown; wall imperforate for short distance just beneath aperture on chamber opposite; aperture slitlike, at base of final chamber near its junction with earlier chambers, with slight lip on upper border. [Although previously placed with the Pegidiidae, the imperforate region near the aperture and lack of a distinct canal system and thickened lamellar wall suggest that *Physalidia* does not belong with the Rotaliacea. No specimens were available for sectioning, hence the present placement is tentative.] *Rec.*, Pac.O. ——FIG. 462,2. **P. simplex*, S.Pac.O.(Cook Is.); *2a,b,* side, edge views of holotype, ×79 (*2117).

Rugidia HERON-ALLEN & EARLAND, 1928, *913, p.

Glabratella DORREEN, 1948, *610, p. 294 [*G. crassa; OD] [=Conorbella HOFKER, 1951, *928c, p. 448, 466 (type, Discorbina pulvinata BRADY, 1884, *200, p. 650); Pileolina BERMÚDEZ, 1952, *127, p. 38 (type, Valvulina pileolus D'ORBIGNY, 1839, *1393, p. 47)]. Test hemispherical, all

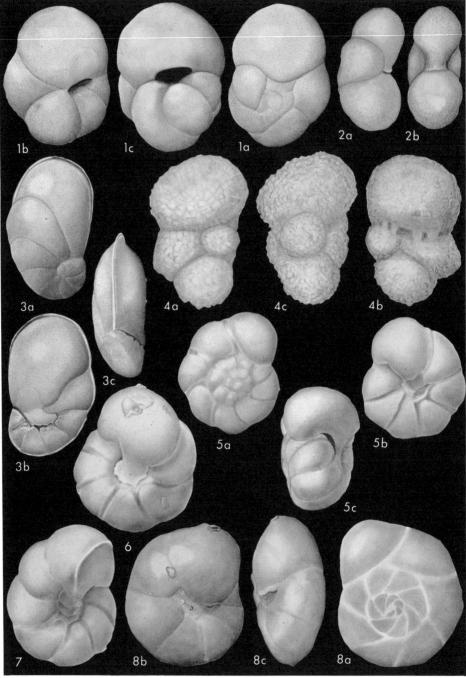

FIG. 462. Discorbidae (Baggininae; 1, Baggina; 2, Physalidia; 3, Cancris; 4, Rugidia; 5-8, Valvulineria) (p. C586-C587).

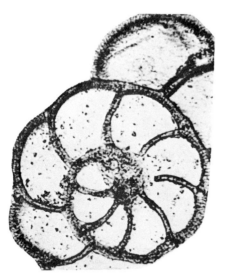

Fig. 463. Discorbidae (Baggininae; *Valvulineria*) (p. C587).

chambers visible from convex spiral side, only those of final whorl visible on flattened umbilical side, periphery rounded; schizont forms generally larger and flatter, gamont more high-spired; chambers relatively few, enlarging rapidly as added, sutures arcuate on spiral side, radial on opposite side; wall calcareous, hyaline, perforate, spiral surface generally ornamented with pustules, umbilical side with radial ornamentation, consisting of fine grooves or tiny, radially arranged pustules; aperture a small rounded opening restricted to open umbilicus; sexual reproduction plastogamic, with 2 specimens attaching by their umbilical surfaces, gametes triflagellate, habitat of plastogamic forms always littoral. *U.Eoc.-Rec.*, N.Z.-Pac.O.-Atl.-Australia-E.Afr.(Kerimba Arch.)-Medit. Sea-Eu.——Fig. 464,*1*. **G. crassa*, U.Eoc., N.Z.; *1a-c*, opposite sides and edge view of paratype, ×119 (*2117).——Fig. 464,*2*. *G. pulvinata* (Brady), Rec., S.Australia; *2a-c*, opposite sides and edge view, ×188 (*2117).——Fig. 464,*3*. **G. pileolus* (d'Orbigny), Rec., S.Am.(Chile); *3a-c*, opposite sides and edge view, enlarged (*127).——Fig. 465. *G. mediterranensis* (d'Orbigny), Rec., Medit.; living triflagellate gamete, ×1500 (*1109).

[*Glabratella* resembles *Discorbis* in having a flat, involute umbilical side and convex, evolute spiral side, but differs in lacking the umbilical alar extensions of the chambers, in having the typical umbilical radial ornamentation, and in having as an aperture only the open umbilical area. It differs from *Angulodiscorbis* in being low-spired, and in having relatively few chambers in each whorl, and in having the open umbilical aperture, instead of a sutural aperture at the base of the final chamber. *Glabratella* was defined by Dorreen with *G. crassa* as type, but also included in the genus was *Discorbina pulvinata* Brady, which Hofker later (1951, *928c) selected as type-species of *Conorbella* without reference to the prior *Glabratella*.——¶Bermúdez (1952, *127, p. 36, 37) recognized both genera, considering *Glabratella* to have a central umbilical aperture and *Conorbella* to have a slitlike interiomarginal aperture. However, the sutural aperture described by Hofker and cited by Bermúdez for *Conorbella* is lacking in the type-species, hence species showing this feature should be separated, possibly placed in *Angulodiscorbis* which Hofker and Bermúdez had considered to be a synonym of *Conorbella*. *Discorbina pulvinata* is very similar to *G. crassa* except in minor features of ornamentation and degree of convexity, which we regard as only of specific importance, hence *Conorbella*, as based on the type-species, is classed as a synonym of *Glabratella*, as it had been by Hornibrook & Vella (1954, *960, p. 25). *Pileolina* is also regarded as a synonym.——¶ J. Le Calvez (1952, *1110) studied some plastogamic species of "*Discorbis*" (=*Glabratella*) and noted that they have triflagellate gametes, and that the schizont generation is commonly larger and flatter than the gamont, the 2 generations commonly having been given distinct specific names. Synonymies of many of these were noted by Le Calvez. *Discorbina pulvinata* Brady (type-species of *Conorbella* and also originally included in *Glabratella*), *Valvulina pileolus* d'Orbigny (type-species of *Pileolina*) and *Discorbis patelliformis* and *D. opercularis* (included later in *Conorbella* by Bermúdez and Hofker) all were among the plastogamic species studied by Le Calvez. These similarities in reproductive habits substantiate the congeneric status of *Glabratella*, *Conorbella* and *Pileolina* and their separation from *Discorbis*.]

Angulodiscorbis Uchio, 1953, *1960, p. 156 [**A. quadrangularis*; OD]. Test free, spiral side extremely high-spired, with all chambers visible, and may be somewhat angular in section, opposite side flat to convex, umbilicate, with only chambers of final whorl visible; chambers numerous, crescentic, broad, low, with considerable overlap; spiral surface commonly with vertical ornamentation, resulting in angular test, or with vertical ribs or very fine striae, with pores of wall aligned in fine striae, ornamentation of umbilical side also with many fine radial striae; aperture a low slit at base of final chamber. *Rec.*, Pac.O.——Fig. 466,*1*. **A. quadrangularis*, Ifaluk Atoll; *1a-c*, spiral, umbilical, and edge views, ×148 (*2117).

[As had been noted for nearly all similar conical, high-spired species, pairs of specimens are frequently found attached by their umbilical surfaces, which are resorbed by these plastogamic species during the reproductive process. Other specimens may later become detached, and with much of the ventral surface dissolved, appear to have an oversized umbilicus. Hofker (1951, 928c, p. 466) considered some of the high-spired species to belong to *Conorbella* and Bermúdez (1952, *127, p. 37) considered *Angulodiscorbis* a synonym of *Conorbella*. On the basis of their type-species, *Conorbella* is here regarded as a synonym of *Glabratella*, and *Angulodiscorbis* is available for the high-spired rotaliiform species with a slitlike aperture at base of the final chamber.]

Bueningia Finlay, 1939, *717b, p. 122 [**B. creeki*; OD] [=*Ruttenella* Keyzer, 1953, *1031, p. 279 (type, *R. butonensis*) (non *Ruttenella* van den Bold, 1946); *Lamarckinita* Keyzer, 1955, *1032, p. 119 (nom. subst. pro *Ruttenella* Keyzer, 1953 non van den Bold, 1946)]. Test small, inflated, both sides involute, umbilical side flattened, with distinct peripheral keel and deep umbilicus, opposite side convex; wall calcareous, finely perforate except for keel, microstructure and lamellar character unknown; aperture umbilical, with small apertural lip. *L.Mio.-Plio.*, N.Z.-W.Indies(Indon.).——Fig. 464,*5*. *B. butonensis* (Keyzer), Mio.-Plio., Indon.; *5a,b*, opposite sides, ×111 (*2117).——Fig. 464,*6*. **B. creeki*, L.Mio., N.Z.; *6a-c*, opposite sides and edge view, ×115 (*2117).

Heronallenia Chapman & Parr, 1931, *324, p. 236

C590 Protista—Sarcodina

[*Discorbina wilsoni Heron-Allen & Earland, 1922, *911, p. 206; OD]. Test trochospiral, compressed, plano-convex, periphery carinate but rounded; chambers increasing rapidly in breadth as added, in few whorls, umbilical side with broad open umbilicus; sutures thickened on spiral side; wall calcareous; finely perforate, radial in structure, lamellar character unknown, surface

Fig. 464. Glabratellidae; *1-3, Glabratella; 4, Heronallenia; 5-6, Bueningia* (p. C588-C591).

Foraminiferida—Rotaliina—Discorbacea

Fig. 465. Glabratellidae; *Glabratella* (p. C588-C589).

radially grooved; aperture a large ovate opening into umbilicus. *Eoc.-Rec.*, Antarctic-Australia-N. Am.-Carib.——Fig. 464,4. **H. wilsoni* (HERON-ALLEN & EARLAND), Rec., Antarctic; *4a-c*, opposite sides and edge view, ×93 (*2117).

Schackoinella WEINHANDL, 1958, *2043, p. 141 [**S. sarmatica*; OD]. Test trochospiral, with inflated chambers, open umbilicus and single thick spine projecting from each chamber on spiral side; wall finely perforate, microstructure and lamellar character not described; aperture apparently basal and umbilical in position. [*Schackoinella* was defined as belonging to the Hantkeninidae, but it differs from that group in having a trochospiral coil. No information is available as to the lamellar character of the type-species, but the general appearance strongly suggests its placement with the Glabratellidae. Additional study is needed of its internal characters.] *Mio.(Sarmat.)*, Eu.(Aus.).——Fig. 467,1. **S. sarmatica*; *1a-c*, opposite sides and edge view, ×68 (*2043).

Family SIPHONINIDAE Cushman, 1927

[*non. transl.* N. K. BYKOVA, VASILENKO, VOLOSHINOVA, MYATLYUK & SUBBOTINA in RAUZER-CHERNOUSOVA & FURSENKO, 1959, p. 270 (*ex* subfamily Siphonininae CUSHMAN, 1927, p. 77)]

Test trochospiral or may become uncoiled or biserial, periphery commonly with fimbriate keel; aperture oval, bordered by distinct lip and projecting on neck. *Eoc.-Rec.*

Siphonina REUSS, 1850, *1540, p. 372 [**S. fimbriata*, =*Rotalina reticulata* ČŽJŽEK, 1848, *545, p. 145; OD (M)]. Test free, biconvex, trochospiral, lenticular, periphery wih fimbriate keel; umbilicus closed; wall calcareous, coarsely perforate, radial in structure, monolamellar, surface may be ornamented with radial striae or pustules; sutures oblique on spiral side, radial on umbilical side; aperture areal, elliptical, nearly equatorial, with short neck and phialine lip. *Eoc.-Rec.*, Eu.-N.Am.-Carib.-Australia-Pac.O.-Atl.O.-S.Am.-Afr.——Fig. 468,1. **S. reticulata* (ČŽJŽEK), Mio., Eu.(Aus.); *1a-c*, opposite sides and edge view of holotype of *S. fimbriata* REUSS, approx. ×47 (*1540).

Siphonides FERAY, 1941, *714, p. 174 [**S. biserialis*; OD]. Test free, tiny, early stage as in *Siphonina*, later chambers uncoiled and biserially arranged, periphery with fimbriate keel; aperture subterminal, with neck and phialine lip. *M.Eoc.*, USA (Tex.).——Fig. 468,7. **S. biserialis*; *7a-c*, opposite sides and edge view of topotype, ×218 (*2117).

Siphoninella CUSHMAN, 1927, *431, p. 77 [**Truncatulina soluta* BRADY, 1884, *200, p. 670; OD]. Test similar to *Siphonina* in early stage, later chambers uncoiling and rectilinear; aperture terminal with neck and phialine lip. *M.Eoc.-Rec.*, Carib.-N.Am.——Fig. 468,2. **S. soluta* (BRADY), Rec., W.Indies; *2a-c*, opposite sides and edge view, ×100 (*200).

Siphoninoides CUSHMAN, 1927, *431, p. 77 [**Planorbulina echinata* BRADY, 1879, *196b, p. 283; OD]. Test subglobular, irregularly trochospiral, few chambers to whorl, involute; wall calcareous, coarsely perforate, surface commonly spinose or tuberculate; aperture circular, with neck and phialine lip. *Mio.-Rec.*, Australia-Pac.O.-Ind.O.——Fig. 468,3-6. **S. echinata* (BRADY), Rec., W.Pac.O.(Admiralty Is.) (*3,4*), Hawaii (*5,6*); *3,4*, side and edge views of different speci-

Fig. 466. Glabratellidae; *1, Angulodiscorbis* (p. C589).

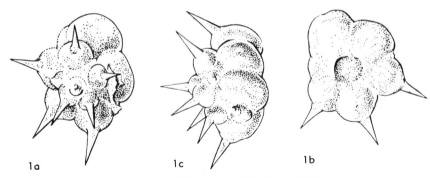

Fig. 467. Glabratellidae; *1, Schackoinella* (p. C591).

mens, ×100; *5,6,* apert. view and optical sec. showing chamber arrangement, ×100 (*200).

ASTERIGERINIDAE
By R. W. Barker
[Shell Development Company, Houston, Texas]

Family ASTERIGERINIDAE d'Orbigny, 1839

[Asterigerinidae d'Orbigny in de la Sagra, 1839, p. xxxix, 116] [=Helicotrochina Agassiz, 1844, p. 2 *(nom. nud.) (partim);* Asterigerinida Copeland, 1956, p. 187 *(nom. van.)*]

Test free, simple, calcareous, unequally biconvex, dorsal side usually more elevated; numerous chambers arranged in flat turbinoid spiral, with oblique sutures; dorsal chambers all visible in simple spiral, those on ventral side with less oblique sutures and alternating with small secondary chambers arranged in rosette form around umbilical plug; surface smooth; primary chambers showing slit aperture on inner side of ventral face of last chamber, secondary chambers with loop-shaped aperture leading into primaries, growth of these two series thus alternating; no canal system (*241, *553, *762, *1392). [Warm, shallow water; probably linked to *Discorbis* and perhaps to the Ceratobuliminidae, according to Brotzen.] *Cret.-Rec.*

Asterigerina d'Orbigny in de la Sagra, 1839, *1611, p. 117 [*A. carinata*; SD Cushman, 1927, *433, p. 190]. Test simple, 3 to 5 whorls visible dorsally; ventrally secondary chamberlets form star-shaped rosette around umbilical plug. *Cret.-Rec.,* cosmop.——Fig. 469,2. **A. carinata*, Rec., W. Indies(Barbados); *2a-c,* dorsal., lat., vent. sides, ×60 (*2110).

Asterigerinata Bermúdez, 1949, *124, p. 266 [*A. dominicana*; OD]. Differs from *Asterigerina* in having more convex dorsal side and ventral side almost flat; aperture shorter and more elliptical; test approaching *Discorbis* in general form, vitreous, compressed, secondary chambers smaller and more globular than in *Asterigerina* (*124). *Oligo.-Rec.,* cosmop.——Fig. 469,1. **A. dominicana*, U.Oligo., W.Indies(Santo Domingo); *1a-c,* dorsal, lat., vent. sides of topotype, ×55 (*2110).

Asterigerinella Bandy, 1949, *70, p. 118 [*A. gallowayi*; OD]. Similar to *Asterigerina* but planispiral, tending to become evolute; spire visible on both sides; periphery lobulate or carinate; chambers numerous, closely appressed and enlarging gradually; surface smooth or papillate (*70). *Eoc.*, N.Am.——Fig. 470,1. **A. gallowayi*, U. Eoc., USA(Miss.); *1a-c,* dorsal, lat., vent. sides, ×30 (*2110).

Asterigerinoides Bermúdez, 1952, *127, p. 61 [*Discorbina gürichi* Franke, 1912, *739, p. 29; OD]. Many-chambered trochoidal test similar to *Asterigerina* but possessing prominent spheroidal umbo on ventral side; differs from *Asterigerinata* in having more numerous chambers and long, narrow, slitlike aperture on inner edge of last chamber (*127). *Oligo.*, Eu.(Fr.-Belg.-Ger.-Neth.)-N.Am.(USA).——Fig. 471,1. **A. guerichi* (Franke), Neth.; *1a-c,* vent., lat., dorsal sides, enlarged (*557, *127).

Family EPISTOMARIIDAE Hofker, 1954

[Epistomariidae Hofker, 1954, p. 166]

Test trochospiral, supplementary chamberlets on umbilical side; interiomarginal primary aperture, and supplementary sutural and areal apertures. *U.Cret.-Rec.*

Epistomaria Galloway, 1933, *762, p. 286 [pro *Epistomella* Cushman, 1928, *436, p. 6 *(non* Zittel, 1878)] [*Discorbina rimosa* Parker & Jones in Carpenter, Parker & Jones, 1862, *281, p. 205); OD]. Test free, trochospiral, biconvex, early whorls visible on spiral side, chambers enlarging rapidly as added, with complex system of internal partitions junction of which with outer wall give appearance of supplementary chamberlets around umbilicus and occupy much of umbilical side; sutures depressed; radial, curved; wall calcareous, perforate, but wall microstructure and lamellar character unknown; primary aperture a low interiomarginal slit, extending from periphery nearly to umbilicus, second aperture in face of final chamber, and series of slitlike accessory apertures paralleling peripheral margin, one at suture formed by attachment of internal plate

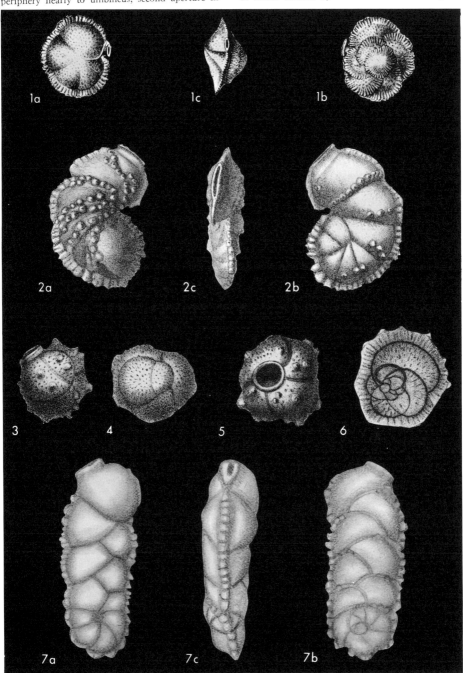

FIG. 468. Siphoninidae; *1, Siphonina; 2, Siphoninella; 3-6, Siphoninoides; 7, Siphonides* (p. C591-C592).

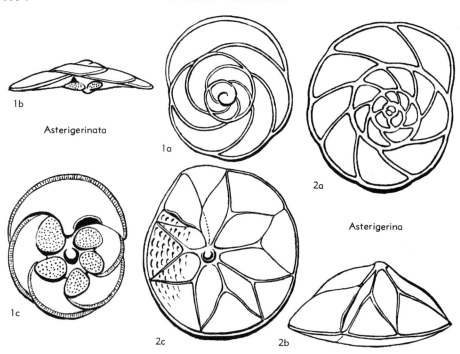

FIG. 469. Asterigerinidae; *1, Asterigerinata; 2, Asterigerina* (p. C592).

on each chamber on umbilical side, and additional supplementary apertures along sutures on both spiral and umbilical sides. [A lectotype for *Discorbina rimosa* PARKER & JONES was selected and isolated in the British Museum (Natural History) by us and is here designated (BMNH-P41670), also paratypes (BMNH-P41669), from the Eocene, Hauteville, France.] *Eoc.*, Eu.——FIG. 472,*1-3*. **E. rimosa* (PARKER & JONES); M. Eoc.(Lutet.), Fr. *(1,2)*, USSR(Ukraine) *(3)*; *1a-c,* opposite sides and edge view of paratype, ×40; *2,* edge view of another paratype, ×40 (*2117); *3,* sec. showing internal partitions, ×33 (*1509).

Elphidioides CUSHMAN, 1945, *482, p. 7 [**E. americanus*; OD]. Test free, trochospiral, biconvex, all whorls visible from spiral side, umbilical side involute with umbilicus covered by extension of final chamber, periphery rounded; chambers numerous, gradually increasing in size; sutures radial, nearly straight, slightly depressed, with sutural pores and retral processes; wall calcareous, coarsely perforate, microstructure and lamellar character unknown; aperture an interiomarginal slit, midway between periphery and umbilicus on umbilical side, with supplementary, curved, slitlike, oblique areal opening. *U.Eoc.* (Jackson.), USA(Ga.).——FIG. 472,*4,5*. **E. americanus;* *4a-c,* opposite sides and edge view of paratype; *5,* edge view of additional paratype; all ×111 (*2117).

Epistomaroides UCHIO, 1952, *1959, p. 158 [**Discorbina polystomelloides* PARKER & JONES, 1865, *1418, p. 421; OD] [=*Epistomarioides* THALMANN, 1953, *1897k, p. 866 (*nom. null. pro Epistomaroides* UCHIO, 1952)]. Test free, trochospiral but nearly equally biconvex, all whorls visible on spiral side, umbilical side with supplementary chambers formed by transverse internal partition as in *Eponidella* and *Epistomaria;* sutures deeply incised with shell material bridging them as in Elphidiidae; wall calcareous, thin, coarsely perforate, surface with granulose ornamentation which forms network of ridges in very large specimens and extending over sutures as sutural bars; microstructure and lamellar character unknown; primary aperture a low interiomarginal arch extending from peripheral margin to umbilicus, opening into supplementary chambers present at edge of sutural incision, internal extension from secondary chamberlets opening into areal aperture on final chamber. [A lectotype for *Discorbina polystomelloides* PARKER & JONES was chosen by us and is here designated (BMNH-ZF3603) with paratypes (BMNH-ZF3602) all from "Juke's No. 2, at 14 fathoms, north of Sir C. Hardy's inside reefs, northeast coast of Australia."] *Rec.*, Australia-N.Guinea-Japan-E.Afr.(Kerimba Arch.).——FIG. 473,*1-3*. **E. polystomelloides* (PARKER & JONES); Australia (Lord Howe Is.) *(1,2),* Kerimba Arch *(3); 1a-c,* opposite sides showing incised sutures and bars, and edge view showing basal and areal apertures,

Foraminiferida—Rotaliina—Discorbacea C595

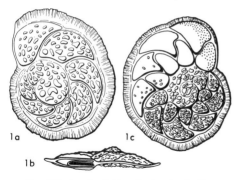

Fig. 470. Asterigerinidae; *1, Asterigerinella* (p. C592).

×23 (*2117); *2a-c,* opposite sides and edge view of larger specimen with more prominent network of ornamentation and sutural bridges, ×26 (*2117); *3a,* optical sec. of specimen in balsam, spiral side with primary chambers shown in outline, secondary chambers shaded; *3b,* same from umbilical side, showing extensions to apert. openings, ×23 (*910a).

Eponidella CUSHMAN & HEDBERG, 1935, *506, p. 13 [*E. libertadensis*; OD] [=*Paranonion* LOGUE & HAAS, 1943, *1189, p. 177 (type, *P. venezuelanum*)]. Test free, trochospiral but nearly biconvex, supplementary umbilical series of chambers appearing to be result of transverse chamber partition that extends from aperture across umbilical side of chambers and attaches to previous septum, but not reaching inner wall of spiral side of test; wall calcareous, coarsely perforate, with pseudochitinous inner layer, that of supplementary portion being thinner walled; lamellar character and microstructure unknown; aperture interiomarginal, extending in loop up peripheral apertural face, lower portion closed secondarily so that intercameral foramina consist only of areal openings. *Mio.-Rec.,* S. Am. (Venez.)-USA-Carib.——Fig. 472,6,7. **E. libertadensis*, Mio., Venez.; *6a-c,* opposite sides and edge view, ×168 (*2117); *7a-d,* edge views of holotype and paratypes showing septal foramen and fragments of internal partition, ×100 (*506).——Fig. 472,8,9. *E. venezuelana* (LOGUE & HAAS), U.Mio., Venez.; *8a-d,* edge views showing apertural development for comparison with *E. libertadensis*, ×100 (*1189); *9a-c,* opposite sides and edge view of holotype, ×122 (*2117).

[*Eponidella* appears closely related to *Palmerinella* but has less complex apertural and septal foramina. Details of the wall structure and internal features of the secondary partitions need additional study. *Paranonion* is a synonym of *Eponidella*, but its type-species does not show the suture of the internal partition as well as *E. libertadensis*. This may be a result of a difference in preservation, however.]

Nuttallides FINLAY, 1939, *717a, p. 520 [**Eponides trumpyi* NUTTALL, 1930, *1371, p. 287; OD]. Test trochospiral, lenticular, with poreless peripheral keel; chambers broad, low; sutures oblique on spiral side, radial and gently curved on umbilical side; umbilicus closed by poreless umbonal boss; internal plate extending diagonally from septal foramen toward peripheral apertural notch but not connecting to opposite wall so as to form supplementary chamberlets; wall calcareous, perforate, radial in structure, septa monolamellid, imperforate; aperture interiomarginal, extending from umbilical boss nearly to peripheral keel, with small notch parallel to plane of coiling. *Eoc.,* Mex.-N.Z.——FIG. 473,7,8. **N. trumpyi* (NUTTALL), Mex.; *7a-c,* opposite sides and edge view of lectotype, here designated (USNM CUSHMAN Coll. 59492), ×65 (*2117); *8,* interior of final chamber showing internal plate and marginal notch, enlarged (*108).

[The internal partitions in *Nuttallides* were described by BELFORD (1958, *108, p. 93, who regarded the genus as belonging to the Epistominidae, but possibly intermediate between *Alabamina* and *Epistomina*. As *Alabamina* has a granular, rather than radiate, wall structure, and *Epistomina* has an aragonite, rather than calcite, test, *Nuttallides* is not regarded as close to either of these genera.]

Nuttallinella BELFORD, 1959, *109, p. 20 [*pro Nuttallina* BELFORD, 1958, *108, p. 96 (*non* DALL,

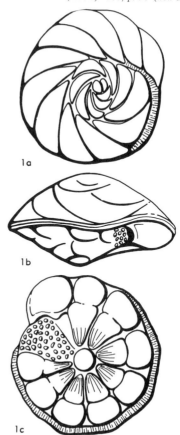

Fig. 471. Asterigerinidae; *1, Asterigerinoides* (p. C592).

C596 *Protista—Sarcodina*

1871)] [*Nuttallina coronula* BELFORD, 1958, *108, p. 97; OD]. Test trochospiral, plano-convex, with flattened spiral side, periphery with broad flangelike imperforate keel; all chambers visible from spiral side; umbilical side with small open umbilicus; sutures radial, straight to sinuate; wall

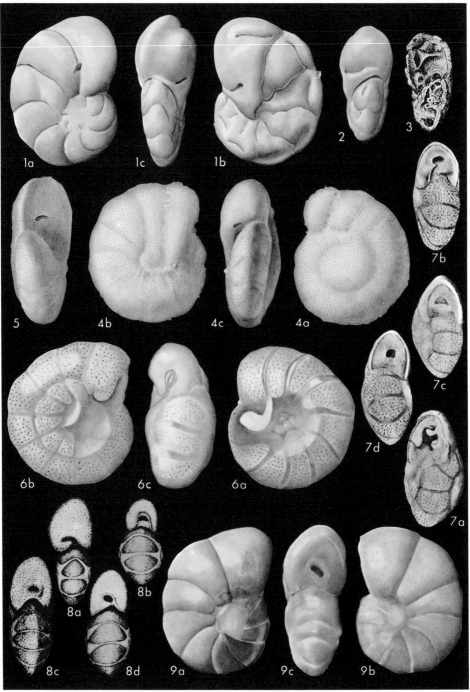

FIG. 472. Epistomariidae; *1-3, Epistomaria; 4,5, Elphidioides; 6-9, Eponidella* (p. C592-C595).

Foraminiferida—Rotaliina—Discorbacea

calcareous, perforate, radial in structure, septal walls single, monolamellid, imperforate; aperture elongate, interiomarginal on umbilical side, with narrow lip, and may have small flap over umbilicus, internal tooth plate extending diagonally across chamber from near periphery back to pre-

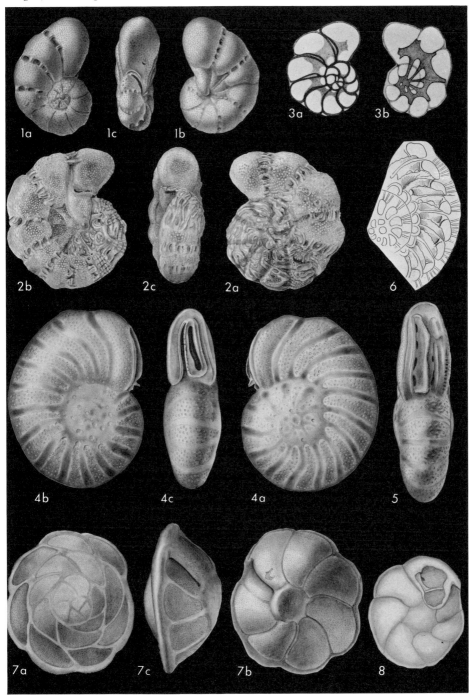

Fig. 473. Epistomariidae; *1-3, Epistomaroides; 4-6, Palmerinella; 7,8, Nuttallides* (p. C594-C595, C598).

vious septal foramen, as in *Nuttallides,* but tooth plate of *Nuttallinella* has strongly folded upper edge. [Differs from *Nuttallides* in having an open umbilicus, a more extensive aperture, which extends to the umbilicus, and folded upper margin of the tooth plate.] *U.Cret.(U.Santon.-U. Campan.),* Australia.——FIG. 474,2-4. **N. coronula* (BELFORD), Santon., W.Australia; *2a-c,* opposite sides and edge view of holotype, ×66; *3,* axial section showing tooth plate in final chamber at right, ×72 (*108); *4,* paratype with dissected penultimate chamber showing tooth plate, enlarged (*108).

Palmerinella BERMÚDEZ, 1934, *116, p. 83 [**P. palmerae*; OD]. Test free, discoidal, adult nearly planispiral and evolute, with low, broad chambers and small secondary chambers at umbilical margin of primary chambers on umbilical side, central portion of both sides of test with clear secondary shell material which may also obscure secondary chamberlets; secondary shell material pierced by few large pores; sutures gently curved; wall calcareous, coarsely perforate; microstructure and lamellar character unknown; aperture a broad open arch in terminal face with narrow raised bordering lip that extends somewhat to umbilical side in final chamber, this portion of earlier apertures being covered by secondary chamberlets; vertical internal partition subdividing aperture, curving at lower border to form continuous S-shaped ridge and leaving 2 elongate slits open in final chamber, these being closed by secondary plate in earlier chambers, plate containing vertical row of large perforations. [The asymmetrical aperture suggests that the nearly planispiral development is modified from a trochospiral ancestry.] *Rec.,* Carib.-N.Am.(USA).——FIG. 473, *4-6.* **P. palmerae,* Cuba; *4a-c,* opposite sides and edge view of lectotype showing aperture; *5,* edge view of paratype with broken final chamber showing septal partition and pores, ×90 (*1632); *6,* equat. sec. showing internal partitions and chamberlets, ×80 (*949).

Pseudoeponides UCHIO in KAWAI et al., 1950, *1027, p. 190 and UCHIO, 1951, *1957, p. 38 [**P. japonica*; OD] [=*Epistomaria (Epistomariella)* KUWANO, 1950, *1071, p. 315 (type, *E. (E.) miurensis,* =*P. japonicus* UCHIO, 1950)]. Test free, lenticular, trochospiral, chambers numerous, broad, semilunar in outline, with strongly oblique sutures on spiral side but nearly radial on umbilical side; wall calcareous, very finely perforate, wall microstructure and lamellar character not described; aperture a low interiomarginal opening midway between umbilicus and periphery, supplementary slitlike openings parallel to spiral suture near mid-point of each chamber on spiral side at junction of spiral and septal sutures, surrounded by poreless area of chamber wall, additional hook-shaped supplementary slits at posterior margin of each chamber on umbilical side extending perpendicularly from near mid-point of previous suture and curving toward the anterior margin, thus suggesting presence of internal tooth plate, which extends vertically through chamber to attach at supplementary opening on spiral side. *Plio.-Rec.,* Japan-Neth.-Carib.——FIG. 474,5,6. **P. japonicus,* Plio., Japan; *5a-c,* opposite sides and edge view of paratype, ×148 (*2117); *6a,* axial sec. showing tooth plates extending from aperture on umbilical side through test to attach to wall of spiral side proximal to supplementary openings; *6b,* partial sec. showing tooth plates, ground from umbilical side through center of test so as to cut final whorl of chambers; both ×107 (*950).

[*Pseudoeponides* was stated by UCHIO (1951, *1957) to be related to *Mississippina* or *Epistomina,* because of the supplementary apertures. KUWANO (1950, *1071) considered it to be a subgenus of *Epistomaria.* Later UCHIO (1953, *1960) included *Rotalina umbonata* REUSS in *Pseudoeponides,* regarded the genus as related to *Eponides,* and placed it in the "Rotaliinae." HOFKER (1958, *950) and REISS (1960, *1533) regarded *Pseudoeponides* to have a true rotaliid genus (double septa), related to *Ammonia.* HOFKER (1956, *945) considered *P. japonica* to have a "very highly developed toothplate and toothplate foramina at the dorsal side of each chamber." *Epistomina* has an aragonitic wall, the Rotaliinae are characterized by double septa and a canal system, *Eponides,* including *Rotalina umbonata* (=*Eponides*) is bilamellid, without supplementary openings and tooth plate.]

Torresina PARR, 1947, *1427, p. 129 [**T. haddoni*; OD]. Test free, trochospiral, compressed, chambers increasing gradually in size, few to whorl, chamber interior divided by secondary partitions projecting inward from peripheral margin; wall calcareous, perforate, microstructure and lamellar character unknown; aperture peripheral, short slit in plane of coiling and inclined toward umbilical side, second opening interiomarginal on umbilical side, and may have an umbilical chamber flap, as in *Discorbinella.* *U.Tert.-Rec.,* Australia.——FIG. 474,1. **T. haddoni,* Rec., Torres Straits; *1a-c,* opposite sides and edge view of topotype, ×133 (*2117).

Superfamily SPIRILLINACEA Reuss, 1862

[*nom. correct.* LOEBLICH & TAPPAN, 1961, p. 317 (*pro* superfamily Spirillinoidea CHAPMAN, PARR & COLLINS, 1934, p. 554, and superfamily Spirillinidea POKORNÝ, 1958, p. 311)] [=family group Archi-Monothalamidia RHUMBLER in KÜKENTHAL & KRUMBACH, 1923, p. 85 (*partim*)]

Test planispiral to conical, simple forms with proloculus followed by enrolled tubular second chamber, nonseptate or with septa in later stages, advanced forms with septa throughout, becoming biserial, later may develop annular chambers; wall perforate, calcareous, may consist optically of single crystal of calcite; amoeboid gametes in plastogamic reproductive cycle; quadrinucleate. *?Trias.,Jur.-Rec.*

Family SPIRILLINIDAE Reuss, 1862

[*nom. correct.* RHUMBLER, 1895, p. 85 (*pro* family Spirillinidea REUSS, 1862, p. 364)]——[All names cited of family

Foraminiferida—Rotaliina—Spirillinacea C599

rank]——[=Spirillinina LANKESTER, 1885, p. 847; =Spirillinida HAECKEL, 1894, p. 185; =Spirillinae DELAGE & HÉROUARD, 1896, p. 144]

Proloculus followed by nonseptate enrolled tubular second chamber which may be septate in later stages, becoming biserial and may develop annular chambers; wall

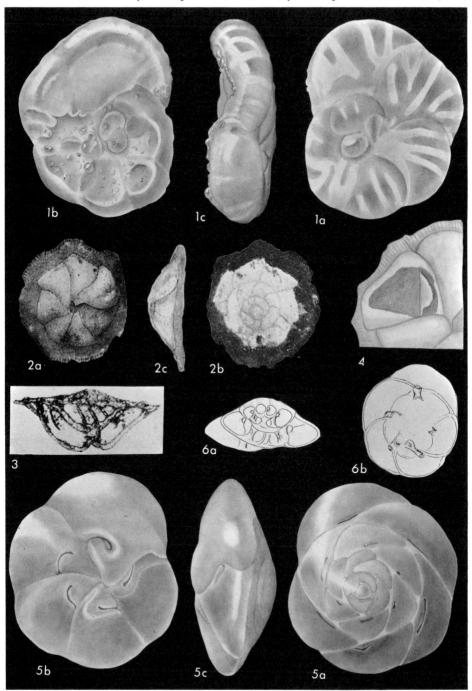

FIG. 474. Epistomariidae; *1, Torresina; 2-4, Nuttallinella; 5,6, Pseudoeponides* (p. C595-C598).

consisting optically of single crystal of calcite. ?Trias.,Jur.-Rec.

Subfamily SPIRILLININAE Reuss, 1862

[nom. transl. BRADY, 1884, p. 72 (ex family Spirillinidea REUSS, 1862)]——[All names cited of subfamily rank]——[=Arspirillinia RHUMBLER, 1913, p. 388 (nom. van.); =Turrispirillininae CUSHMAN, 1927, p. 73; =Terebralininae CUSHMAN, 1927, p. 65]

Test with proloculus and enrolled tubular, nonseptate second chamber only; aperture simple, single, at open end of tube. ?Trias.,Jur.-Rec.

Spirillina EHRENBERG, 1843, *672, p. 402 [*S. vivipara; OD (M)] [=Mychostomina BERTHELIN, 1881, *134, p. 557 (type, Spirillina vivipara revertens RHUMBLER, 1906, *1571, p. 32, SD GALLOWAY, 1933, *762, p. 88); Arspirillinum RHUMBLER, 1913, *1572b, p. 388 (nom. van.)]. Test free, planispiral, or with one side slightly concave, proloculus followed by closely appressed, spirally wound, undivided tubular second chamber, proloculus of "megalospheric" generation (agamont) smaller than that of "microspheric" generation; wall calcareous, hyaline, composed of single crystal of calcite (more rarely 2 or 3), deposited over pseudochitinous membrane, finely to coarsely perforate, although, according to SOLLAS (*1811, p. 207), the wall has "pseudopores" rather than true perforations, commonly with one side more coarsely perforate than the opposite; aperture terminal, peripheral, somewhat crescentic where final whorl lies against previous whorl, or final portion of tubular chamber may be somewhat turned inward to one side of periphery and directed toward umbilicus. ?Trias.,Jur.-Rec., cosmop.——FIG. 475,1,2. *S. vivipara, Rec., soft muddy white sand at 10 fathoms, Dry Tortugas, Fla. (Station 23), between Middle Ground and White Shoal (1), Gulf Mex. (2); 1a-c, opposite sides and edge view of neotype (Cushman Coll. 10186, U.S. Natl. Mus.), here designated, ×150; 2, hypotype showing mychostomine appearance of agamont form, ×150 (*2117).——FIG. 475, 3,4. S. revertens RHUMBLER, Rec., W.Pac.O. (Caroline Is.); 3a-c, opposite sides and edge view of hypotype with incurved chamber (agamont); 4a,b, opposite sides of typical spirilline form (gamont); all ×148 (*2117).

[EHRENBERG's types, originally in Berlin, were reportedly destroyed during the last war. As was noted by WOOD (1949, p. 245) the type-species of Spirillina, as generally understood, has a test composed of a single crystal of calcite, but EHRENBERG's original description stated that acid had no effect on the shell. CUSHMAN (1931, *451, p. 4) stated: "Ehrenberg originally described this species from off the Coast of Mexico, near Vera Cruz. I examined the type in the Ehrenberg collection in Berlin and the drawing given by Ehrenberg is an excellent one of the type specimen. The species is a fairly common one in the West Indian region . . . ." CUSHMAN stated of his own illustrated specimens: "The figures given show the typical form and appearance of this species in the West Indian region from which it was described." As EHRENBERG stated that the test was insoluble in acid, this would imply that he was concerned with a siliceous or agglutinated form, such as Ammodiscus. However, no noncalcareous species of similar appearance is known from the type area. As CUSHMAN's types were from the same general area, Gulf of Mexico, as the original of EHRENBERG and as he had seen the original types (now lost) and stated that his specimens were typical, we are here designating as neotype the specimen illustrated by CUSHMAN (1931, *451, pl. 1, figs. 4a-c), here refigured. The original description was probably in error in stating that the test is not soluble in acid and by designating a neotype upon which to base the emended generic definition, the nomenclature can be stabilized.——¶Mychostomina BERTHELIN, 1881, was defined without citation of species, and Spirillina vivipara var. revertens RHUMBLER, 1906, was designated as type-species by GALLOWAY, 1933 (subsequent monotypy). CUSHMAN in his various classifications of the foraminifers (1933, 1940, 1948) considered Mychostomina as a synonym of Spirillina, whereas GALLOWAY (1933, *762, p. 88) stated that Turrispirillina CUSHMAN may be a synonym of Mychostomina. BERMÚDEZ (1952, *127, p. 18) recognized all three genera. We have examined specimens of Spirillina revertens from the Caroline Islands. In this species only slight overlap of the umbilical region by the distal end of the tubular chamber is seen and even this is difficult to determine in some specimens. In addition, very similar specimens of a typical planispiral Spirillina are associated with S. revertens.——¶MYERS (1936, *1337, p. 123), in his study of living cultures of Spirillina vivipara and their ontogenetic development and reproduction, stated, "In the final stage of the agamont test the distal end of the spiral chamber is usually turned inward, so that the aperture is directed toward the umbilicus." Thus, the agamont (sexually produced) generation may show the "generic" character of Mychostomina, with a recurved distal end of the spiral chamber, and the gamont test would show the typical Spirillina-like planispiral coil. PHLEGER & PARKER (1951, pl. 13, figs. 3a, b) also figured a specimen of Spirillina vivipara from the Gulf of Mexico which shows the recurved distal end of the spiral chamber. We have here refigured it for comparison with the 2 forms of the type-species of Mychostomina. The genus Gyrammina EIMER & FICKERT, 1899, with its type-species, Trochammina annularis BRADY, 1876, was placed in the synonymy of Spirillina by GALLOWAY (1933, *762, p. 85).——¶The types of this species in the BRADY collection in the British Museum (Natural History) in London were examined by us and the species found to be unrecognizable on the basis of the type material. S. vivipara is one of the best-known species of all foraminifers and many details have been published as to its morphological characters (MYERS, 1936, *1337, p. 123), shell composition (WOOD, 1949, *2073, p. 245), ontogenetic development, reproductive process (MYERS, 1936, *1337, p. 125), cytology (MYERS, 1936, *1337, p. 126), and ecology (MYERS, 1936, *1337, p. 122). It has been widely recorded from Recent oceans.]

Alanwoodia LOEBLICH & TAPPAN, 1955, *1166, p. 26 [*Patellina campanaeformis BRADY, 1884, *200, p. 634; OD]. Test free, conical, high-spired, ventrally flattened or slightly excavated, consisting of proloculus and long, undivided, broad and low tubular chamber in high, open conical spire, central area being filled with clear or laminated calcite, tiny pores around exterior spiral suture, wall calcareous, test composed of single calcite crystal; aperture ventral, at open end of spiraling tube. Rec., Pac.O.——FIG. 476,1,2. *A. campanaeformis (BRADY); 1, long. sec. of holotype showing clear central filling; 2a-c, opposite sides and edge of paratype; all ×146 (*1166).

Conicospirillina CUSHMAN, 1927, *431, p. 73 [*Spirillina trochoides BERTHELIN, 1879, *132, p. 37; OD]. Test free, conical, consisting of proloculus and undivided tubular spiraling second chamber, spiral side convex and evolute, umbilical side concave and nearly completely involute, final whorl nearly or completely overlapping all previous whorls, rarely leaving open umbilicus; aperture at open end of tube on umbilical side. [Differs from Spirillina in being ventrally involute and in being conical in form. It differs

Foraminiferida—Rotaliina—Spirillinacea C601

from *Turrispirillina* in its involute ventral side with the final whorl occupying the entire ventral side.] Jur.-Rec., Eu.-Atl.O.-Pac.O.——FIG. 475,5.

**C. trochoides* (BERTHELIN), L.Jur.(L.Pliensbach.), Eu.(Fr.); *5a-c*, opposite sides and edge view of topotype, ×257 (*2117).

FIG. 475. Spirillinidae (Spirillininae; *1-4, Spirillina; 5, Conicospirillina; 6,7, Miliospirella; 8,9, Planispirillina*) (p. C600-C602).

Miliospirella GRIGELIS in N. K. BYKOVA et al., 1958, *265, p. 75 [*M. lithuanica; OD]. Test with proloculus followed by enrolled nonseptate tubular second chamber with plane of coiling changing regularly, so that successive whorls are approximately 120° apart, giving pseudotriloculine appearance; wall calcareous, coarsely perforate; aperture simple, at open end of tube. [Differs from Spirillina in its pseudotriloculine coiling, from Triloculina in its nonseptate tube, simple aperture, and perforate wall, and from Agathammina in its perforate wall.] M.Jur.(U.Callov.), Eu. (Lith.).——FIG. 475,6,7. *M. lithuanica; 6a-c, side, edge, and top views of holotype, ×120; 7, sec. showing arrangement of successive whorls of nonsegmented tube, ×240 (*265).

Planispirillina BERMÚDEZ, 1952, *127, p. 26 [*Spirillina limbata BRADY var. papillosa CUSHMAN, 1915, *404e, p. 6; OD] [=Trochospirillina MITYANINA, 1957, *1290, p. 230 (type, T. granulosa)]. Test free, planispiral, periphery rounded to truncate, all whorls visible on spiral side, all whorls except last obscured on ventral side by secondary accumulation of nodes and pustules of clear calcite which completely fill central region; wall calcareous, hyaline, coarsely perforate dorsally, finely perforate ventrally; aperture at open end of tube. Jur.-Rec., Pac.O.-Eu.-Medit. Sea-Australia. ——FIG. 475,8. *P. papillosa (CUSHMAN), Rec., Pac.; 8a-c, opposite sides and edge view of holotype, ×75 (*2117).——FIG. 475,9. P. granulosa (MITYANINA), U.Jur.(L.Oxford.), Belorussian SSR; 9a-c, spiral, umbilical, and edge views of type-specimen(s), ×44 (*1290).

[Planispirillina differs from Spirillina in the presence of secondary granules on its umbilical side. The original figures of Trochospirillina granulosa (*1290) seem to refer to a single specimen but either they represent different specimens or one figure is reversed, as the aperture is shown to the left in both figures.]

Sejunctella LOEBLICH & TAPPAN, 1957, *1172, p. 228 [*S. earlandi; OD]. Test free, planispiral, discoidal, may have peripheral keel; globular to ovate proloculus followed by loosely wound, spiral, undivided, tubular second chamber that does not lie in contact with previous whorl but is separated from it by solid platelike area; wall calcareous, finely perforate, chamber wall and peripheral keel, when present, formed of single calcite crystal but intercalary plate between coils of tubular chamber not composed of single crystal but of secondary granular calcite; aperture a rounded opening at end of tubular chamber. Rec., Atl.O.——FIG. 477,4. *S. earlandi; side view of holotype, showing fimbriate peripheral keel and intercalary plate between whorls, composed of keels of earlier whorls with addition of secondary granular calcite, ×253 (*1172).

[Differs from Spirillina in the presence of its platelike intercalation between the planispiral whorls, a condition considered to be generically important, not only on external appearance but also because it differs in structure, being composed of granular calcite instead of a single crystal, as is the remainder of the test. The type-species has a peripheral keel on the final whorl, but this may be lacking in other species.]

FIG. 476. Spirillinidae (Spirillininae; 1,2, Alanwoodia) (p. C600).

Terebralina TERQUEM, 1866, *1887, p. 471, 473 [pro Spirigerina TERQUEM, 1866, *1886, p. 454 (non D'ORBIGNY, 1847)] [*Spirigerina antiqua TERQUEM, 1866, *1886, p. 353, 454, =Terebralina regularis TERQUEM, 1866, *1887, p. 473; OD (M)]. Test consisting of proloculus and undivided tubular second chamber in high trochospiral coil; wall calcareous, perforate; aperture at open end of tubular chamber. [Although previously placed in the Buliminidae, Terebralina is here placed in the Spirillinidae because of its nonseptate coil, simple aperture, and absence of tooth plate. It differs from Turrispirillina in being extremely high-spired.] L.Jur.(Lias.), Eu.(Fr.).——FIG. 477, 5. *T. antiqua (TERQUEM); side view, ×66 (*519).

Turrispirillina CUSHMAN, 1927, *431, p. 73 [*Spirillina conoidea PAALZOW, 1917, *1403, p. 217; OD] [=Turrispirrillina NEAVE, 1940, *1348d, p. 594 (nom. null.)]. Test free, conical, consisting of proloculus and spirally wound tubular second chamber, which forms hollow cone, all coils visible dorsally and ventrally; wall calcareous, finely perforate, dorsal surface somewhat roughened; aperture at open end of tube on ventral side of test. [Differs from Spirillina in its hollow conical spire rather than being planispiral.] Jur.-Rec., Eu.-N.Am.-Australia-Antarctic. ——FIG. 477,1. *T. conoidea (PAALZOW), U.Jur., Eu.(Ger.); 1a-c, opposite sides and edge view of topotype, ×90 (*2117).

Subfamily PATELLININAE Rhumbler, 1906

[Patellininae RHUMBLER, 1906, p. 35] [=Arpatellinia RHUMBLER, 1913, p. 390 (nom. van.)]

Proloculus and trochospirally coiled nonseptate chamber in early stage, followed by septate stage with 2 chambers to whorl, or chambers annular; aperture umbilical. L. Cret.-Rec.

Foraminiferida—Rotaliina—Spirillinacea C603

Patellina WILLIAMSON, 1858, *2065, p. 46 [*P. corrugata; OD (M)] [=*Arpatellum* RHUMBLER, 1913, *1572b, p. 391 *(nom. van.)*; *Discobolivina* HOFKER, 1951, *936, p. 358 (obj.)]. Test free, conical, spiral side elevated and evolute, umbilical side flat and involute, elliptical proloculus fol-

FIG. 477. Spirillinidae (Spirillininae; *1, Turrispirillina; 4, Sejunctella; 5, Terebralina;* Patellininae; *2,3, Patellinoides; 6,7, Patellina*) (p. C602-C604).

lowed by spirally wound tubular undivided second chamber of 1 to 3 whorls in microspheric form, proloculus continuous with spiral tube in megalospheric test, smaller in size that that of microspheric generation, later stage with 2 broad, low chambers to each whorl, primary chambers divided by numerous incomplete secondary transverse septa and commonly with intercalated shorter third series, these transverse septa giving typical cancellated appearance to test but extending only approximately width of chambers, as seen from the spiral side, not reaching across umbilical portion of chambers; wall calcareous, built as single calcite crystal, finely perforate; aperture a low arch under exterior margin of scroll-like median septum of final chamber at center of test, median septa of entire test arranged above each other to form columella. L.Cret.-Rec., cosmop.——
Fig. 477,6,7, *P. corrugata, Rec., Can. (6), Greenl. (7); 6a-c, opposite sides and edge view of microspheric hypotype; 7, megalospheric hypotype; all ×100 (*1162).

[Hofker (1951, *936, p. 358) stated "that all species known as Patellina and Patellinoides do not show in the initial part a spiral without septa, but that in contrary all genera and species observed show a more or less highly developed conorbine initial part, with fine, only in a clarifier visible, septa." He also added (*928c, p. 422) "that those records which mention an undivided first part of the test, are erroneous ones, or that this character is due to the insufficient state of fossilization."——¶However, the exacting and detailed work on Recent living specimens of Patellina corrugata by Myers (1935, *1336, pl. 13, fig. 18), definitely showed the presence of an undivided spire, and nuclear characters in camera lucida drawings of decalcified cytological preparations. Myers also noted that the microspheric tests had a distinct proloculus, followed by an undivided spire, whereas the megalospheric test showed no separation of the proloculus from the spirally wound tubular chamber. The proloculus of the microspheric generation of Patellina was also shown by Myers (1935, *1335, p. 399) to be larger than that of the megalospheric generation, so that "the terms megalospheric and microspheric, when applied to the dimorphic tests of this species, are not descriptive of the relative diameters of the initial chambers of these two stages . . . . The diameter of a megalospheric test having a given number of semilunar chambers is larger than that of a microspheric test having a similar number of chambers because of the larger diameter of the spiral stage of the megalospheric test . . . [p. 402]. The diameter of the initial chamber of a megalospheric test is influenced by the diameter of the nucleus involved, and may or may not depend on the amount of cytoplasm that surrounds the nucleus." Myers also studied the internal features of the test (*1335, p. 395, fig. 7, and p. 397) and the columella which forms the S-shaped ventral structure considered as a tooth plate by Hofker (probably this is the "previously unmentioned" feature Hofker considered a basis for his genus Discobolivina) and discussed the morphology of the secondary septa (not mentioned by Hofker).——¶Hofker (1951, *928c, p. 422) stated that the wall was without pores, but Myers (1935, *1335, p. 396) had shown the presence of a row of pores even in the microspheric proloculus of P. corrugata and several rows of pores in the dorsal wall of the chamberlets in later chambers. The conorbine initial stage in Patellina, reported by Hofker (1951, *936, p. 358) does not occur in P. corrugata, as Myers showed in cytological preparations. We have also examined this species in reflected light at magnifications higher than X200, by transmitted light with anise oil as a clarifier, and in oil immersion at X400, and no conorbine early portions were found, only a spiral nonseptate coil. Only the genus Patellinella has this early conorbine stage, and it lacks the secondary septa of Patellina. Williamson originally described Patellina corrugata from Arran, Skye, Shetland, Brixham, and Fowey, all from the British Isles, and from Hunde Island, in Davis Straits, Arctic Canada. He did not cite a holotype, and since all localities are represented on a single slide preserved in the British Museum (Natural History), London, England, it is impossible to state which is the type locality.]

Patellinoides Cushman, 1933, *461, p. 236 [*P. conica Heron-Allen & Earland, 1932, *916, p. 408; OD] [=Patellinoides Heron-Allen & Earland, 1932, *916, p. 407 (nom. nud.)]. Test free, tiny, conical, plano-convex, somewhat ovate in outline, trochoid, all chambers visible dorsally, only final pair visible ventrally; proloculus followed by simple undivided spiral tubular chamber of 1 or 2 volutions, then followed by chambers arranged biserially around internal S-shaped columella, as in Patellina, but lacking radial secondary septa which form partial chamberlets in Patellina; wall calcareous, perforate, composed of single calcite crystal, light reflections from fine pores sometimes giving radial pattern to exterior of test but no true internal secondary partitions present; aperture ventral, small arch near umbilicus. Rec., N.Atl.O.-S.Atl.O.——Fig. 477,2,3. *P. conica (Heron-Allen & Earland), S.Atl.; 2a-c, opposite sides and edge view of lectotype (here designated) (BMNH-ZF3568 from R.R.S. William Scoresby station WS 408, lat. 53°50'00" S., long. 62°10'00" W., Falk. Is., at 454 m.), ×200; 3a-c, opposite sides and edge view of hypotype, ×187 (*2117).

[Patellinoides was named by Heron-Allen & Earland (1932, *916, p. 407) as a new genus including P. conica, n. sp., and P. depressa, n. sp., neither of which was designated as type-species for the genus. Thus, it was a nomen nudum, with no status of availability or validity (Zool. Code, 1961, Art. 13(b)). The genus was validated when the type-species was designated by Cushman (1933, *461, p. 236), who also gave a description and figure for this genus; Cushman must be considered the author of the genus, therefore. The species names conica and depressa published by Heron-Allen & Earland in 1932 comply with the Rules in being accompanied by adequate "indications," and they are not nomina nuda because given in combination with an invalid generic name; their availability for designation of the specific taxa described is not affected by the status of Patellinoides as a nomen nudum in the original publication (Zool. Code, 1961, Art. 11(g)(ii), Art. 17(3)). In assigning these species to Patellinoides Cushman, the names of the authors Heron-Allen & Earland need to be enclosed by parentheses, just as if they had used the generic name Patellina with conica and depressa.]

Family ROTALIELLIDAE
Loeblich & Tappan, n.fam.

Test trochospiral, consisting of few crescentic to subglobular chambers; wall calcareous, finely perforate, radial in structure, monolamellar; aperture central on umbilical side; quadrinucleate, sexual reproduction with amoeboid gametes. [Because of the similarity in the reproductive cycle, with the amoeboid gametes and the quadrinucleate agamont form, this family is placed in the Spirillinacea.] Rec.

Rotaliella Grell, 1954, *818, p. 269 [*R. heterocaryotica; OD]. Test tiny, to 60μ diam., free, trochospiral, chambers inflated, subglobular to crescentic, 3 to whorl, increasing rapidly in size, proloculus followed by small hourglass-shaped

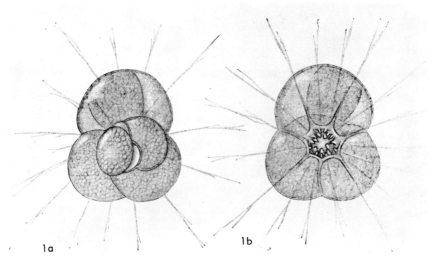

Fig. 478. Rotaliellidae; *1, Rotaliella* (p. C604-C605).

"intermediate" chamber preceding second normal chamber, adult gamont (megalospheric) with 5 chambers, microspheric test commonly with 6 or rarely 7 chambers; wall calcareous, very thin, transparent, finely perforate, radial in structure and monolamellar; aperture umbilical in position, apertural border with numerous small teeth projecting inward; pseudopodia few, relatively thin, with granular streaming; cytoplasm with 1 or 2 large yellow oil globules, that of agamont (microspheric) form greenish due to presence of large numbers of small *Chlamydomonas* cells, probably ingested as food, agamont heterokaryotic, with single vegetative or somatic nucleus and 3 generative nuclei, dividing in asexual reproduction to form 12 embryonic gamonts, adult mononucleate gamonts each producing 10 to 24 amoeboid gametes which form and may fuse with others within individual parent test (autogamy), zygote developing proloculus and reniform "intermediate" chamber of test before release from parent test. *Rec.*, Eu.(Yugosl.-W.Fr.).——Fig. 478, *1*. **R. heterocaryotica*, Yugosl.; *1a*, spiral side showing few globular chambers, with proloculus followed by narrow reniform "intermediate" chamber; *1b*, umbilical side with thin elongate pseudopodia radiating from umbilicus and umbilical margin with inward-pointing teeth, ×680 (*818).

Superfamily ROTALIACEA Ehrenberg, 1839

[*nom. correct.* LOEBLICH & TAPPAN, 1961, p. 303 (*pro* superfamily Rotalidea GLAESSNER, 1945, p. 143)]——[In synonymic citations superscript numbers indicate taxonomic rank assigned by authors (¹superfamily, ²family group); dagger(†) indicates *partim*]——[=¹Orthoklinostegia† EIMER & FICKERT, 1899, p. 685 (*nom. nud.*); =²Rotaliaridia RHUMBLER in KÜKENTHAL & KRUMBACH, 1923, p. 88; =²Rotaliformes BROTZEN, 1942, p. 9 (*nom. neg.*); =¹Rotaliidea SMOUT, 1954, p. 40; =¹Rotaliicae BRÖNNIMANN, 1958, p. 175]

Canaliculate, double walls and septa of radial laminated calcite secondarily formed; without primary aperture or large pores, or with pores on apertural face or elsewhere, and may have interiomarginal intercameral foramina. *U.Cret.-Rec.*

Family ROTALIIDAE Ehrenberg, 1839

[*nom. correct.* CHAPMAN, 1900, p. 10 (*pro* family Rotalina EHRENBERG, 1839, table opposite p. 120)]——[All names cited of family rank; dagger(†) indicates *partim*]——[=Polythalama† LATREILLE, 1825, p. 161 (*nom. nud.*); =Turbinacea† and Turbinacés DE BLAINVILLE, 1825, p. 390 (*nom. nud.*); =Hélicostègues† D'ORBIGNY, 1826, p. 268 (*nom. nud., nom. neg.*); =Radiolata† CROUCH, 1827, p. 41 (*nom. nud.*); =Radiolididea† BRODERIP, 1839, p. 321; =Turbinoidea† D'ORBIGNY IN DE LA SAGRA, 1839, p. xxxviii, 71 (*nom. nud.*); =Turbinoida† SCHULTZE, 1854, p. 52 (*nom. nud.*)]——[=Rotalideae REUSS, 1860, p. 221; =Rotalida SCHMARDA, 1871, p. 164; =Rotalidea HANTKEN, 1875, p. 80; =Rotalidee SCHWAGER, 1876, p. 479; =Rotalidae BRADY, 1881, p. 44; =Rotalinae DELAGE & HÉROUARD, 1896, p. 145; =Rotaliaridae RHUMBLER, 1913, p. 339; =Arrotalaridia RHUMBLER, 1913, p. 342 (*nom. van.*); =Rotaliformes THALMANN, 1945, p. 403 (*nom. neg.*); =Rotálidos GADEA BUISÁN, 1947, p. 19 (*nom. neg.*)]——[=Pegidiidae HERON-ALLEN & EARLAND, 1928, p. 283; =Pegidiida COPELAND, 1956, p. 188 (*nom. van.*)]——[=Chapmaniidae GALLOWAY, 1933, p. 316; =Chapmaniinidae THALMANN, 1938, p. 207; =Chapmaniida COPELAND, 1956, p. 187 (*nom. van.*)]

Test trochospiral throughout; with radial canals or fissures and intraseptal and subsutural canals. *U.Cret.-Rec.*

Subfamily ROTALIINAE Ehrenberg, 1839

[*nom. correct.* CHAPMAN, 1900, p. 11 (*pro* subfamily Rotalida SCHULTZE, 1854, p. 52)]——[All names cited of subfamily rank]——[=Rotalinae CARPENTER, PARKER & JONES, 1862, p. 198; =Rotalina JONES in GRIFFITH & HENFREY, 1875, p. 320; =Rotalidae SCHWAGER, 1877, p. 20; =Rotalininae HOFKER, 1933, p. 125]

Test trochospiral, all external openings, except perforations, on umbilical side; with radial canals or fissures or umbilical cavities, and commonly with intraseptal and subsutural canals. *U.Cret.-Rec.*

C606 Protista—Sarcodina

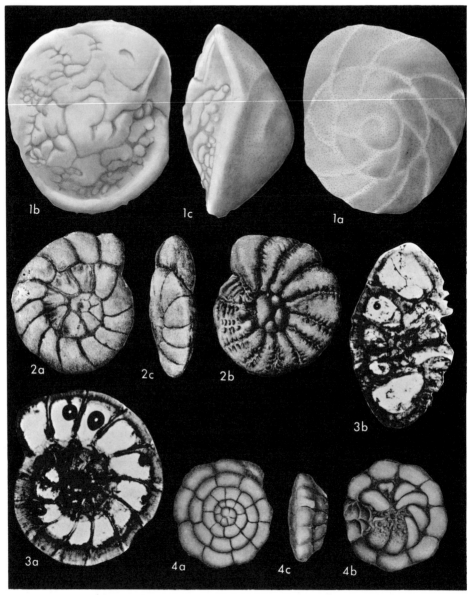

Fig. 479. Rotaliidae (Rotaliinae; *1, Rotalia; 2-4, Ammonia*) (p. C606-C607).

Rotalia LAMARCK, 1804, *1085a, p. 183 [***Rotalites trochidiformis* LAMARCK, 1804, *1085a; SD GALLOWAY & WISSLER, 1927, *766, p. 59] [=*Rotalina* DE BLAINVILLE, 1828, *143, p. 66 (*nom. van. pro Rotalia* LAMARCK, 1804)]. Test free, trochospiral, lenticular to plano-convex, 1-4 mm. diam., all whorls visible from spiral side, spire multilocular and single, direction of coiling random; chambers simple, 8 to 17 to whorl; septa primarily double, formed by upward bending of chamber floor; wall calcareous, coarsely perforate, of radially fibrous calcite; spiral side smooth, umbilical side with plug split by anastomosing fissures into numerous tubercles and pillars that crowd central portion of test, pillars not continuous from one whorl to next, as in *Dictyoconoides* and *Lockhartia*, but limited to each whorl, although they may fuse laterally to close fissures and form solid central mass, with umbilical canal beneath cortical chamber layer receiving tributary canals from umbilical slitlike apertures at inner side of chambers; in some species fissures or canals also present in septa. [The double septa have long been noted in *Rotalia* (CARPENTER, PARKER & JONES, 1862,

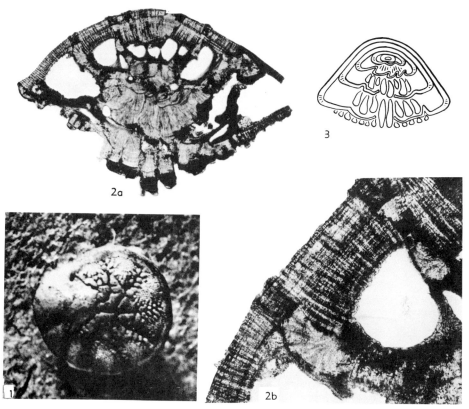

Fig. 480. Rotaliidae (Rotaliinae; *1-3, Rotalia*) (p. C606-C607).

*281, p. 214; ANDREAE, 1884, *19, p. 215) although only recently has use been made of this character in classification (SMOUT, 1954, *1803, p. 9).] *U.Cret.(Senon.)-Rec.,* cosmop.——FIG. 479, *1*; 480,*1-3*. **R. trochidiformis* LAMARCK, M.EOC. (Lutet.), Eu.(Fr.); 479,*1a-c,* opposite sides and edge view, ×31 (*2117); 480,*1,* lectotype, ×20; 480,*2a,* axial sec. showing radially built lamellar walls and umbilical pillars, which are not continuous from one whorl to next, ×25; 480,*2b,* portion of preceding sec., ×85; 480,*3,* diagram. sec. showing character of umbilical plugs (*561).

Ammonia BRÜNNICH, 1772. *248, p. 232 [***Nautilus beccarii* LINNÉ, 1758, *1140, p. 710; SD FRIZZELL & KEEN, 1949, *752, p. 106] [=*Hammonium* FICHTEL & MOLL, 1798, *716, p. 13, 15 (obj.); *Discorbula* LAMARCK, 1816, *1089, p. 14 (type, *D. ariminensis*); *Streblus* FISCHER DE WALDHEIM, 1817, *720, p. 449 (obj.); *Les Turbinulines* D'ORBIGNY, 1826, *1391, p. 275 *(nom. neg.)*; *Turbinulina* RISSO, 1826, *1579a, p. 18 (obj.); *Rolshausenia* BERMÚDEZ, 1952, *127, p. 63 (type, *Rotalia rolshauseni* CUSHMAN & BERMÚDEZ, 1946, *493, p. 119); *Rotalidium* ASANO, 1936, *48, p. 350 (type, *R. pacificum*)]. Test free, biconvex, low trochospiral coil of 3 or 4 volutions, sutures slightly curved, thickened, depressed on umbilical side, septa primarily double; wall calcareous, finely perforate, radial in structure; umbilical surface with irregular granules along suture and over umbilical region; umbilicus with open umbilical fissures and plug in young forms, which is broken up into numerous fused pillars and bosses in adult specimens, umbilical plugs extending inward to proloculus, no umbilical canal; aperture interiomarginal. [*Rotalidium* is regarded as a synonym of *Ammonia,* the "supplementary chamberlets" being the characteristic umbilical extensions, and the very rare type-species as a possible synonym of *Rotalia japonica* HADA, 1931, also described from Recent deposits along the Japanese coast.] *Mio.-Rec.,* cosmop.——FIG. 479,*2,3*. **A. beccarii* (LINNÉ), Rec., Italy *(2),* S.Fr. *(3); 2a-c,* opposite sides and edge view of topotype, ×27 (*437); *3a,b,* axial and equat. secs., ×50 (*358).——FIG. 479,*4*. *A. pacifica* (ASANO), Rec., Japan; *4a-c,* opposite sides and edge view of holotype, ×33 (*48).

Asanoina FINLAY, 1939, *717a, p. 541 [***Rotaliatina globosa* YABE & ASANO, 1937, *2087, p. 124; OD]. Test large, to 2 mm. diam., globose, relatively high trochospiral coil of 2 or more whorls,

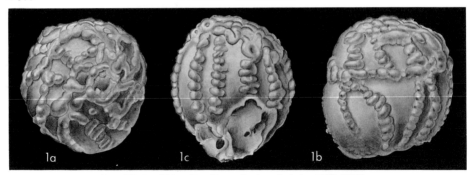

Fig. 481. Rotaliidae (Rotaliinae; *1, Asanoina*) (p. C607-C608).

with strongly convex spiral and umbilical sides, nonumbilicate; sutures raised and granulate; internal structure not described; aperture slitlike interiomarginal opening. *Plio.-Rec.*, Malay Arch. (Java).——Fig. 481,*1*. **A. globosa* (YABE & ASANO); *1a-c*, spiral and 2 edge views, intercameral foramen visible as rounded opening near umbilical region, ×33 (*2117).

Asterorotalia HOFKER, 1950, *932, p. 73, 76 [*Calcarina pulchella* D'ORBIGNY in DE LA SAGRA, 1839, *1611, p. 80, =*Rotalia trispinosa* THALMANN, 1933, *1895, p. 248; OD (M)]. Test free, trochospiral, biconvex, with 3 prominent slender spines radiating from test and continuous through all whorls from earliest, margin carinate; septa with intraseptal passages, opening as series of pores or fissures in and along sutures of umbilical side, partly covered by thin plates with distal openings; wall calcareous, perforate radial in structure, elongate spines formed by outer, main chamber lamellae, around stream of protoplasm emerging from intraseptal space, each spine containing tubular radial canal, surface of spiral side with irregular raised knobs and elevated sutures; interiomarginal aperture nearly equatorial in position, with strongly developed lips, which are fused in sutural region, posterior end of lip extended toward periphery, partly covering previous chamber and intraseptal fissure, leaving labial aperture in sutural position between lip and chamber, interior with strongly twisted tooth plate, intercameral foramina broadly elliptical in outline. [The type-species was originally described as *Calcarina pulchella* D'ORBIGNY, 1839, and transferred to *Rotalia* by BRADY (1884, *200, p. 710), an apparent synonym of *Rotalia pulchella* D'ORBIGNY (1826, *1391, p. 274). It was later transferred to *Pulvinulina* (=*Eponides*) by JONES, PARKER & BRADY (1866, *1002, pl. 2, fig. 25-27), and *Calcarina pulchella* was renamed *Rotalia trispinosa* by THALMANN (1933, *1895, p. 248). However, as the 2 species were originally described in distinct genera and are not now regarded as congeneric, the specific name *pulchella* is valid for the present type-species.] *Pleist.-Rec.*, Carib. (Cuba)-E.Indies(Indon.)-Pac.O.——FIG. 482,*1-4*. **A. pulchella* (D'ORBIGNY); Rec., Indon.; *1a,b*, opposite sides, ×50 (*200); *2*, umbilical side of young specimen showing aperture, sutural plates, and their distal openings; *3a*, portion of umbilical side of larger specimen; *3b*, final chamber, showing poreless but tuberculate apertural face, and aperture; all ×168 (*928c); *4*, horiz. sec. showing spines with central canal, septal flaps, and intraseptal passages, ×55 (*1534).

Dictyoconoides NUTTALL, 1925, *1367, p. 384 [*nom. subst. pro Conulites* CARTER, 1861, *287b, p. 53 (*non* FISCHER DE WALDHEIM, 1832; *nec* COZZENS, 1846] [**Conulites cooki* CARTER, 1861, *287b, p. 53; OD]. Test conical, with proloculus at apex, spiral side with thin imperforate lamina beneath which is layer of rectangular, spirally arranged chambers, in multiple spire, umbilical side with radiating pillars of shell matter extending out from apex and 0.1-0.15 mm. diam. at surface, with intervening spaces of nearly same size, spaces being divided by horizontal partitions; septa double, with median intraseptal canal and subsutural canal system; wall calcareous, umbilical side with granules, cavities in umbilical region separated by perforate plates and buttressing pillars; aperture multiple, umbilical, consisting of pores between pillars. *M.Eoc.*, Asia(India-Qatar Penin.)-Afr.(Somali.).——FIG. 483,*1-5*. **D. cooki* (CARTER), Somali. (*1*), India (*2-5*); *1a-c*, opposite sides and edge view, ×6 (*1788c); *2*, axial sec., ×8; *3*, tang. sec. through rectangular chambers, ×16; *4*, horiz. sec. through pillars, ×8 (*1367); *5*, axial sec. of lectotype, showing chambers near outer margin of conical test and prominent vertical pillars, ×10 (*561).

Dictyokathina SMOUT, 1954, *1803, p. 64 [**D. simplex*; OD]. Test trochospiral, with umbilical mass containing strong vertical radial canals, as in *Kathina*, but with spire repeatedly doubling in plane of coiling to form multiple spire, as in *Dictyoconoides*; wall calcareous, radially fibrous, finely perforate and laminated; intercameral foramen an interiomarginal slit, probably representing earlier aperture. *Paleoc.*, ?*L.Eoc.*, Arabia (Qatar Penin.)-Iraq.——FIG. 484,*1-4*. **D. simplex*, Paleoc., Qatar (*1,3,4*), Iraq (*2*); *1,2*, horiz. secs.

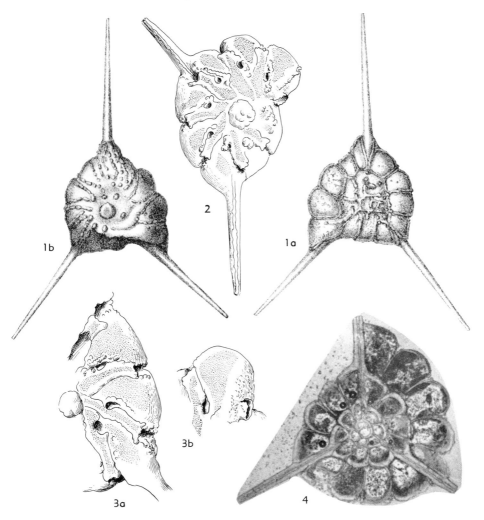

Fig. 482. Rotaliidae (Rotaliinae; *1-4, Asterorotalia*) (p. C608).

of megalospheric and microspheric forms, ×12, ×10; *3a-c,* opposite sides and edge view of paratype, ×16; *4,* nearly axial vert. sec. of megalospheric form, ×25 (*1803).

Kathina SMOUT, 1954, *1803, p. 61 [**K. delseata;* OD]. Test trochospiral, differing from *Dictyoconoides* in having chambers arranged in simple spire, umbilical side may have central plug with strong vertical canals; chambers simple, without supplementary chamberlets or umbilical extensions found in *Lockhartia* and *Sakesaria;* septa double, with intraseptal and subsutural canals but no definite sutural openings or retral processes, strong vertical canals opening as pores or slits on umbilical side; wall very finely perforate, of radially fibrous calcite, lamellar thickening pronounced, but no pustules or ornamentation; aperture an interiomarginal slit. *U.Cret.-Paleoc.,* Arabia (Qatar Penin.) - Carib.(Cuba).——FIG. 484,*5-8.* **K. delseata,* Paleoc., Qatar; *5a,b,* spiral and umbilical sides of paratype; *6,* decorticated holotype showing simple spire; *7,* paratype, umbilical view showing apertures at end of vertical canals; all ×12; *8,* axial sec. of paratype, ×25 (*1803).

Lockhartia DAVIES, 1932, *561, p. 406 [**Dictyoconoides haimei* DAVIES, 1927, *559, p. 280; OD]. Test conical to lenticular, trochospiral; chambers forming outer layer of cone, leaving wide umbilical area, chamber walls curving inward toward umbilicus leaving open only marginal slit which opens into cavity between outer wall laminae; local thickening and bending of umbilical laminae may result in irregular buttresses or pillars, which fill umbilical area, appearing as granules at umbilical surface, may be labyrinthic; wall calcareous, of laminated radially fibrous calcite, coarsely perforate, aperture an interiomarginal slit. [*Lockhartia* has numerous intercommunicating umbilical cavities into which the cortical chambers open, and which open to the exterior as large pores on the

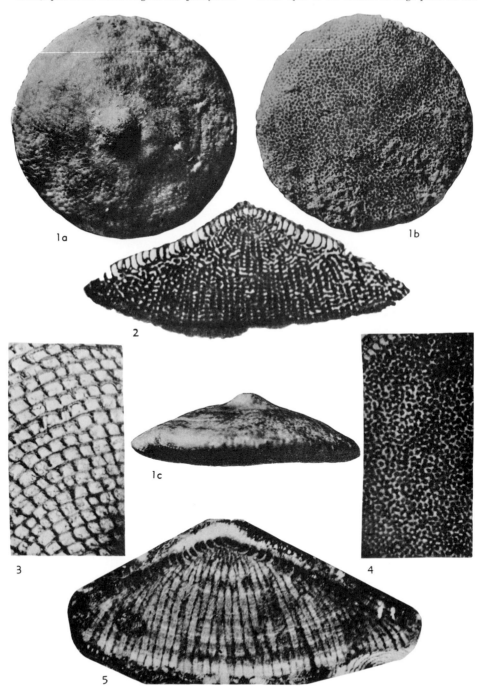

FIG. 483. Rotaliidae (Rotaliinae; *1-5, Dictyoconoides*) (p. C608).

umbilical side. *Rotalia* differs in having a solid umbilical plug or fissured one with a spiral canal beneath the chambers and tributary canals connecting to them. *Dictyoconoides* is similar to *Lockhartia*, but has intercalary whorls into the spire.] *Paleoc.-M.Eoc.*, Asia(India-Arabia-Iraq)-

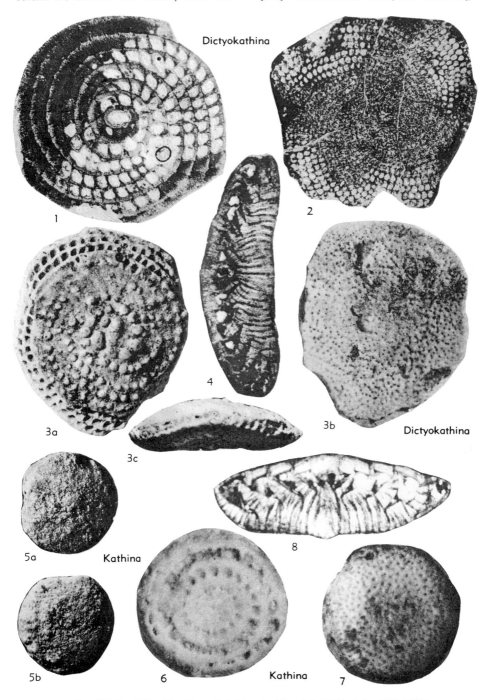

FIG. 484. Rotaliidae (Rotaliinae; *1-4, Dictyokathina; 5-8, Kathina*) (p. C608-C609).

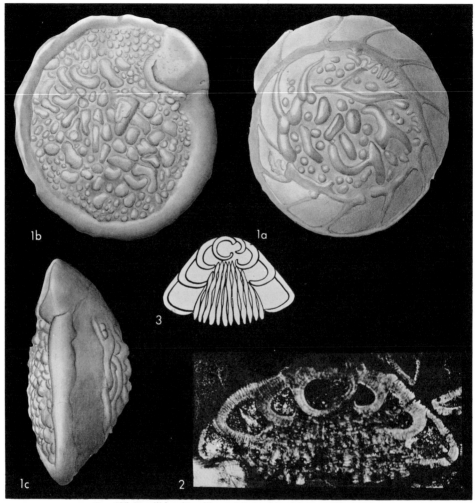

Fig. 485. Rotaliidae (Rotaliinae; *1-3, Lockhartia*) (p. C609-C612).

E. Afr.-S. Am.——Fig. 485,*1-3*. **L. haimei* (Davies), Paleoc., India; *1a-c*, opposite sides and edge view of topotype, ×26 (*2117); *2*, axial sec. showing umbilical pillars, ×30; *3*, diagram. sec. (*561).

Pararotalia Y. Le Calvez, 1949, *1112, p. 32 [**Rotalia inermis* Terquem, 1882, *1890, p. 68; OD] [=*Neorotalia* Bermúdez, 1952, *127, p. 75 (type, *Rotalia mexicana* Nuttall, 1928, *1370, p. 374); *Woodella* Haque, 1956, *876, p. 194 (type, *W. granosa*)]. Test free, trochospiral, plano-convex to biconvex, umbilicus filled by plug which may be broken out in preservation, chambers rounded to ovate in plan, may have smoothly rounded periphery or develop short, blunt peripheral spine on each chamber, umbilical region of each chamber partially covered by umbilical flap; wall calcareous, perforate, radially built, rotaliid in structure, smooth or variously ornamented with large solid spines or fine scattered spines or nodes; apertures on umbilical side, interiomarginal and extraumbilical-umbilical, with lip; internal "tooth plate" near umbilical and axial chamber wall, intercameral foramen narrow, elongate, comma-shaped or slitlike areal opening, consisting of portion of former aperture, roughly paralleling base of apertural face and restricted by tooth plate of following chamber. [*Woodella* is apparently synonymous with *Pararotalia* and the type-species *W. granosa* appears to be conspecific with *Rotalia capdevilensis* Cushman & Bermúdez.] U.Cret.(Coniac.)-Rec., cosmop.——Fig. 486,*1-3*. **P. inermis* (Terquem), M.Eoc.(Lutet.), Eu.(Fr.); *1a-c*, opposite sides and edge view, ×73; *2*, apert. region of dissected specimen showing tooth plate of final chamber attached to intercameral foramen of penultimate chamber, ×128 (*1171); *3*, equat. sec. showing double septa and intra-

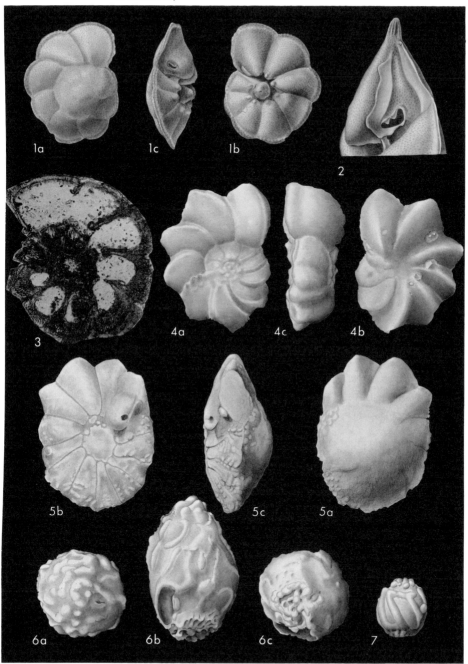

Fig. 486. Rotaliidae (Rotaliinae; *1-5, Pararotalia; 6,7, Sakesaria*) (p. C612-C614).

septal passages, ×65 (*1534).——Fig. 486,*4. P. nammalensis* (Haque), Paleoc., Asia(Pak.); *4a-c,* opposite sides and edge view of specimen originally described as *Woodella,* ×98 (*2117).——Fig. 486,*5. P. mexicana* (Nuttall), U.Eoc., Mex.; *5a-c,* opposite sides and edge view of lectotype, here designated (*1370, pl. 50, fig. 7), ×29 (*2117).

Pseudorotalia Reiss & Merling, 1958, *1534, p. 13 [*Rotalia schroeteriana* Carpenter, Parker & Jones, 1862, *281, p. 212; OD]. Test trochospiral, periphery acute, with imperforate keel;

chambers with imperforate umbilical lips confined to radial sector and with imperforate plate-like extensions formed by each succeeding chamber lamella covering umbilical area, those added by successive chambers with intervening cavities, imperforate plates may be pierced by few scattered large rounded openings, surrounded by thickened rims but without pillars or buttresses, opening of successive plates not aligned; septa secondarily doubled by septal flaps, which leave intraseptal passages that open to outside by means of double row of canals in alternating arrangement and sutural position, on both sides of test; wall lamellar, of radially fibrous calcite, coarsely perforate; cameral aperture interiomarginal on umbilical side, partly covered by narrow extension of apertural face, which is resorbed when new chambers are added and aperture becomes intercameral foramen, strongly developed and twisted tooth plate attached at angle, extending backward to close lower part of preceding intercameral foramen, apertural lip forming interiomarginal labial aperture at inner umbilical side of chamber, those of successive chambers remaining open. [*Pseudorotalia* differs from *Rotalia*, *Ammonia*, and *Lockhartia* in having sutural canals on both spiral and umbilical sides and in lacking umbilical labial apertures.] Plio.-Rec., E.Indies (Indon.-Borneo).——Fig. 487,1-5. *P. schroeteriana* (Carpenter, Parker & Jones), Rec., Borneo *(1-3,5)*, ?loc. *(4)*; *1*, horiz. sec. showing bifurcating sutural canals, ×55; *2*, vert. sec. showing tooth plates, umbilical lips and cavities, and sutural canals, ×55; *3*, horiz. sec. showing tooth plates and relationship to septal flap, ×55; *4*, oblique ext. view, enlarged; *5*, diagram of dissected chamber showing intercameral foramen in septal face, tooth plate attaching below it and labial aperture at umbilical end of chamber *(1-3,5,* *1534; *4, *281).

Sakesaria Davies in Davies & Pinfold, 1937, *563, p. 49 [*S. cotteri*; OD (M)]. Test similar in structure to *Lockhartia*, but differing in having elongate axis of coiling, more numerous whorls, and convex rather than flattened umbilical side; wall calcareous, coarsely perforate, surface commonly ornamented with raised and limbate sutures, pustules and bars. Paleoc.-L.Eoc., Asia(India-Arabia, Qatar Penin.)-Afr.(Somali.).——Fig. 486,6,7; 487,6,7. *S. cotteri*, L.Eoc., Qatar Penin. (486,6,7), India (487,6,7); 486,6a-c, opposite sides and edge view showing high spire and characteristic ornament, ×22 (*2117); 486,7, edge view of young specimen, ×22 (*2117); 487,6, axial sec., ×25 (*563); 487,7, axial sec., ×20 (*1803).

Smoutina Drooger, 1960, *631b, p. 306 [*S. cruysi*; OD]. Test trochospiral, biconvex, simple spire visible on spiral side, opposite side with central umbilical filling occupying about half of test diameter; chambers communicating with spiral canals at their umbilical end; septa double, with fissures on umbilical side that connect with branching spiral canal system in umbilical mass, which contains vertical canals opening as pores at surface; wall lamellar, of radially built calcite, finely perforate; aperture of final chamber not described, intercameral foramen elongate. [*Smoutina* differs from *Rotalia* in having a less completely fissured umbilical mass, and from *Kathina* in having a spiral canal system.] U.Cret.-M.Eoc., S. Am.(Fr.Guiana)-W.Indies(Cuba)-USA(Fla.).——Fig. 487,8-11. *S. cruysi*, Paleoc., Fr. Guiana; *8a-c*, opposite sides and edge view of holotype; *9*, axial half sec. showing vert. canals of umbilical mass; *10a,b*, horiz. half secs. near umbilical and spiral sides showing canal systems, double septa, and radial walls; *11*, peripheral view of broken specimen showing intraseptal and vert. canals and nearly basal intercameral foramina; all ×27 (*631b).

Subfamily CUVILLIERININAE
Loeblich & Tappan, n.subfam.

Test trochospiral to nearly planispiral, spiral and umbilical sides not differentiated in structure; canal system with subsutural and intraseptal canals and vertical canals or fissures, without differentiated marginal cord, spines or retral processes. U.Cret. (Campan.)-Mio.

Cuvillierina Debourle, 1955, *567b, p. 55 [*C. eocenica*, =*Laffitteina vallensis* Ruiz de Gaona, 1948, *1595, p. 87, =*L. vanbelleni* Grimsdale, 1952, *826, p. 232; OD] [=*Cuvillierina* Debourle, 1955, *567a, p. 19 *(nom. nud.)*]. Test free, planispiral, but slightly asymmetrical, exterior with reticulate ornamentation related to canal system, commonly with chevron pattern over sutures, open umbilical region with numerous pillars, and spongy with vertical and lateral canals present, as in *Notorotalia* and *Elphidium*, on both sides of test; septa double, rows of sutural canals connecting vertical grooves with intraseptal passages; septal flap "tooth plate" nearly equatorial but longitudinally folded, bending forward to coalesce with distal face of chambers and forming "spiral canal," which is not a true canal; wall calcareous, perforate, radially built; intercameral foramina comma-shaped, similar to those of *Pararotalia* and *Laffitteina*. [*Cuvillierina* was originally placed in the Nonionidae, but has a radially built rotaliid wall structure rather than granular wall structure. Because of the absence of retral processes and the planispiral coiling it was placed in the Miscellaneidae by Reiss, 1957, *1528b.] Eoc.(Ypres.), Eu. (Spain-Fr.)-Asia (Iraq.-Syria-Israel).——Fig. 488,1-4. *C. vallensis* (Ruiz de Gaona), Fr. *(1-3)*, Syria *(4)*; *1a,b*, side and apert. views, ×82 (*2117); *2*, equat. sec. showing double septa with intraseptal passages; *3*, axial sec., ×87 (*1534); *4*, portion of tang. sec. showing vert. canals in umbilical region and divergent canals over chambers of outer whorl, ×27 (*826).

Arnaudiella DOUVILLÉ, 1907, *618, p. 599 [**A. grossouvrei*; OD]. Test thin, lenticular, 5 to 7 mm. diam., planispiral, with approximately 4

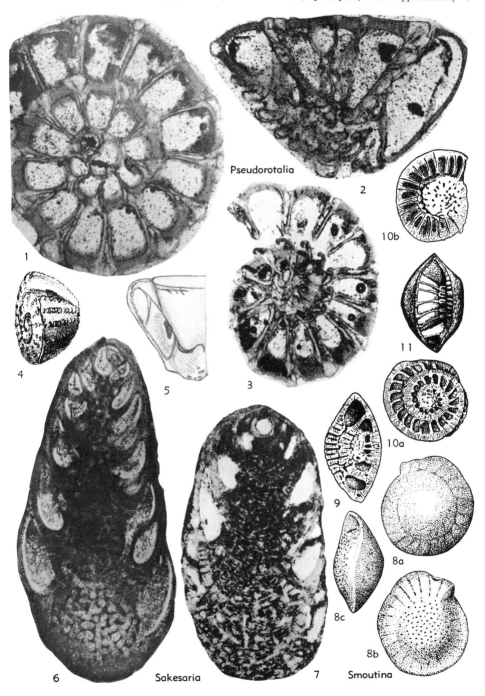

FIG. 487. Rotaliidae (Rotaliinae; *1-5, Pseudorotalia; 6,7, Sakesaria; 8-11, Smoutina*) (p. C613-C614).

whorls; chambers numerous, involute, with layers of vacuoles resembling lateral chamberlets; wall calcareous, lamellar, spiral septum strongly thickened, umbilical pillars appearing as nodes at surface. *U.Cret.(Campan.)*, Eu.(Fr.).——FIG. 489, *1-3.* **A. grossouvrei*; *1*, holotype, ×8 (*2118); *2*, oblique tang. sec. cutting chambers near center and showing thickened spiral septum containing

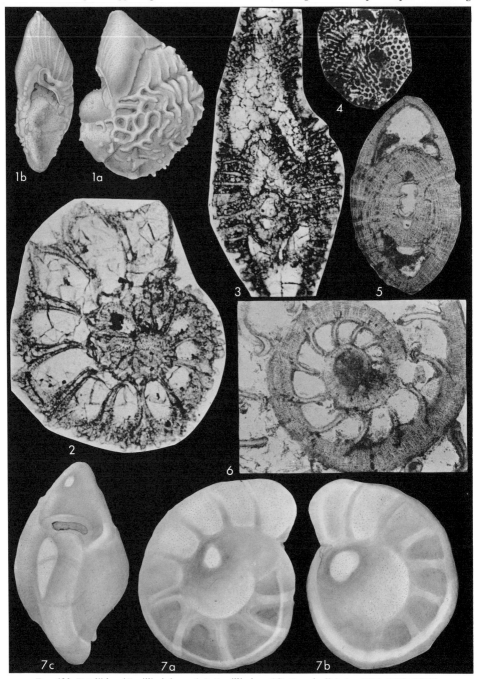

FIG. 488. Rotaliidae (Cuvillierininae; *1-4, Cuvillierina; 5-7, Crespinella*) (p. C614-C615, C617).

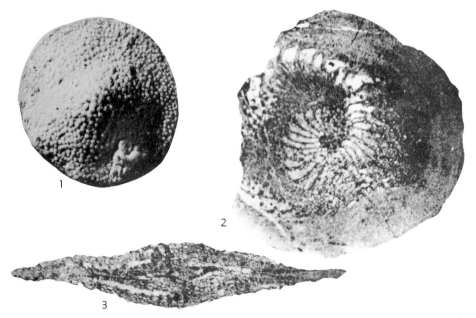

Fig. 489. Rotaliidae (Cuvillierininae; *1-3, Arnaudiella*) (p. C615-C616).

small vacuoles, ×13; *3*, axial sec. showing involute whorls and vacuolated spiral septum, ×13 (*618).

Crespinella PARR, 1942, *1426, p. 361 [*Operculina? umbonifera* HOWCHIN & PARR, 1938, *968, p. 309; OD]. Test free, early stage trochospiral, in adult biinvolute and nearly planispirally enrolled, biconvex and biumbonate, periphery subacute to rounded, chambers increasing gradually in size, numerous; sutures indistinct, radial and slightly curved; wall calcareous, thick, lamellar, microstructure unknown, distinctly perforate, apparently with interseptal canals and tubular passages in plane of coiling; aperture an interiomarginal equatorial or somewhat asymmetrical slit with projecting upper lip. *Mio.*, S.Australia.——FIG. 488, *5-7*. **C. umbonifera* (HOWCHIN & PARR); *5*, axial sec., ×35; *6*, central portion of equat. sec., ×47 (*1426); *7a-c*, opposite sides and edge view, ×40 (*2117).

Daviesina SMOUT, 1954, *1803, p. 66 [*D. khatiyahi*; OD] [=*Miscellanoides* SANDER, 1962, *1625A, p. 13 (type, *M. bramkampi*)]. Test operculine, biconvex to concavoconvex, but slightly asymmetrical, umbilical region with pillars, fissures, and vertical canals on both sides of test; septa double, with intraseptal canals; wall calcareous, lamellar, perforate, radially built; aperture not observed, intercameral foramen a basal slit. [*Miscellanoides* was described in 1962, but in a footnote the author stated that the genus had been described previously by SMOUT, 1954, as *Daviesina*.] *Paleoc.*, Arabia(Qatar Penin.).——FIG. 490,*1-4*. **D. khatiyahi*, M.Paleoc.; *1a-c*, opposite sides and edge of microspheric form; *2a-c*, megalospheric form; all ×17 (*2117); *3*, nearly axial sec. of microspheric form, ×28; *4*, equat. sec. of megalospheric form, ×17 (*1803).

Fissoelphidium SMOUT, 1955, *1804, p. 208 [*F. operculiferum*; OD]. Test planispiral, bilaterally symmetrical, chambers numerous; septa double and sutures fissured in dendritic pattern; umbilical region with fissured umbilical mass similar to that of *Rotalia* but occurring on both sides of test; wall calcareous, lamellar and radially fibrous, perforate; aperture a series of pores in somewhat protruding apertural plate in interiomarginal position, plate being resorbed when next chamber forms, leaving equatorial interiomarginal slitlike foramen. *U.Cret.(Maastricht.)*, Asia(Arabia-Iraq).——FIG. 490,*5*; 491,*1-3*. **F. operculiferum*, Qatar Penin; 490,*5a,b*, side and edge views showing fissured umbilical mass, dendritic fissured sutures, and perforated apertural plate, ×28 (*2117); 491,*1*, edge view showing intercameral slitlike foramen; 491,*2*, axial sec. showing umbilical thickening; 491,*3*, equat. sec. showing double septa; all ×30 (*1804).

Penoperculoides COLE & GRAVELL, 1952, *372, p. 714 [*P. cubensis*; OD] [=*Penoperculinoides* HANZAWA, 1962, *875, p. 140 *(nom. van.)*]. Test slightly asymmetrical, trochoid in early stages, adult nearly planispiral, involute; wall calcareous, laminated and finely tubulated; aperture an arched slit at base of last-formed chamber so arranged that it extends more on one side of median line than other. *M.Eoc.*, Carib.——FIG. 492,*1*. **P. cubensis*, Cuba; *1a-d*, ext. views, ×10; *1e*, axial sec., ×20; *1f,g*, equat. secs., ×20 (*372).

Pokornyellina LOEBLICH & TAPPAN, *nom. nov.* [*pro Pokornyella* LOEBLICH & TAPPAN, 1961, *1181, p. 220 (*nom. subst. pro Siderina* ABRARD, 1926) (*non Pokornyella* OERTLI, 1956)] [**Siderina douvillei* ABRARD, 1926, *2, p. 31, here designated as type-species] [=*Siderina* ABRARD, 1926, *2, p. 31 (*non* DANA, 1848) (obj.)]. Test large, to 7 mm. diam., discoidal, slightly asymmetrical, laterally compressed but with prominent umbilical thickening on both sides, consisting of pillars which appear

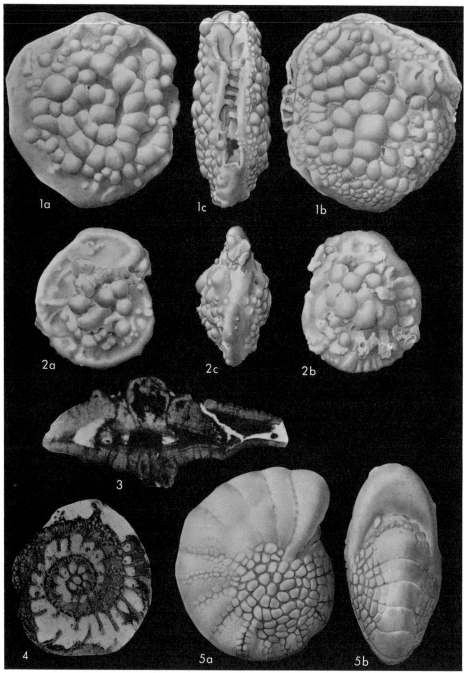

FIG. 490. Rotaliidae (Cuvillierininae; *1-4, Daviesina*; *5, Fissoelphidium*) (p. C617).

Fig. 491. Rotaliidae (Cuvillierininae; *1-3, Fissoelphidium*) (p. C617).

at umbilical surface as small nodes; chambers broad, low and numerous, planispirally coiled; aperture not described. *U.Cret.(Campan.)*, Eu. (Fr.).——Fig. 493,*1,2*. **P. douvillei* (ABRARD); *1*, ext., ×4.5; *2*, equat. sec., ×6 (*2).

[The original illustrations and description of *Siderina* ABRARD suggest that it may be congeneric with *Arnaudiella* or *Pseudosiderolites*. Until type material of all three typespecies can be re-examined, the present genus is tentatively recognized, and is renamed, inasmuch as *Siderina* ABRARD, 1926, and *Pokornyella* LOEBLICH & TAPPAN, 1961, are both homonyms.]

Pseudosiderolites SMOUT, 1955, *1804, p. 206 [**Siderolites vidali* DOUVILLÉ, 1907, *618, p. 599; OD]. Test lenticular, bilaterally symmetrical, planispirally coiled, with numerous radial canals, umbilical region with pillars, showing as nodes at surface; septa double, with intraseptal canals; walls perforate, of radially built calcite, lamellar and thickened particularly in marginal area; aperture not described. [*Pseudosiderolites* differs from *Arnaudiella* in having prominent radial canals and in lacking intralamellar vacuoles.] *U.Cret.*, Eu.(Spain-Fr.).——Fig. 493,*3-5*. **P. vidali* (DOUVILLÉ), Maastricht., Spain; *3*, ext., holotype, ×4 (*618); *4*, axial sec.; *5*, equat. sec. showing thickened marginal region, radial canals, and double septa; ×15 (*1450).

Pseudowoodella HAQUE, 1956, *876, p. 202 [**P. mamilligera*; OD]. Test free, trochospiral, biconvex, periphery broadly rounded; spiral side evolute but flat to slightly excavated centrally, umbilical side involute, nonumbilicate, sutures radial; wall calcareous, hyaline perforate, radial in structure, lamellar character unknown, surface with single short spine at center of each chamber on spiral side; aperture equatorial, interiomarginal. [The genus was originally placed in the Anomalinidae, but the spiny ornamentation is not characteristic of that group, which also differs in having a granular wall. The type-species needs restudy as to possible lamellar character of the wall and presence of a canal system.] *Paleoc.-L. Eoc.*, Asia(Pak.).——Fig. 493,*6*. **P. mamilligera*, Paleoc.; *6a-c*, opposite sides and edge view of holotype, ×115 (*876).

Storrsella DROOGER, 1960, *631a, p. 295 [**Cibicides haasteri* VAN DEN BOLD, 1946; *155, p. 125; OD]. Test trochospiral, similar to *Fissoelphidium*, fissured sutures on both sides of test, but with umbilical thickened mass only on umbilical side, as

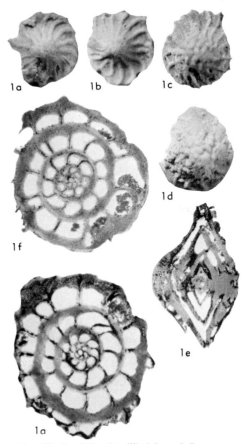

Fig. 492. Rotaliidae (Cuvillierininae; *1, Penoperculoides*) (p. C617).

in *Rotalia*; aperture of final chamber not described, intercameral foramen interiomarginal, subequatorial, somewhat toward umbilical side. *Paleoc.-L.Eoc.*, C. Am.(Guat.-Br.Hond.)-W. Indies (Cuba)-S.Am.(Fr. Guiana).——FIG. 493,7-9. **S. haasteri* (VAN DEN BOLD), Guat. (7), Fr. Guiana (8,9); 7a-c, opposite sides and edge view, showing fissures; 8, equat. half sec. showing double

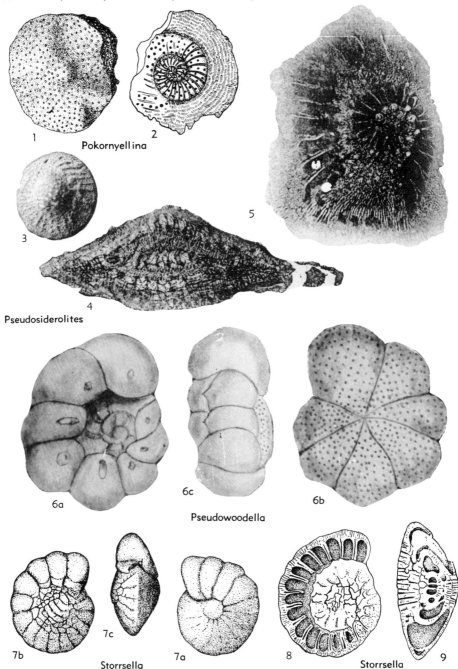

FIG. 493. Rotaliidae (Cuvillierininae; 1,2, *Pokornyellina*; 3-5, *Pseudosiderolites*; 6, *Pseudowoodella*; 7-9, *Storrsella*) (p. C618-C620).

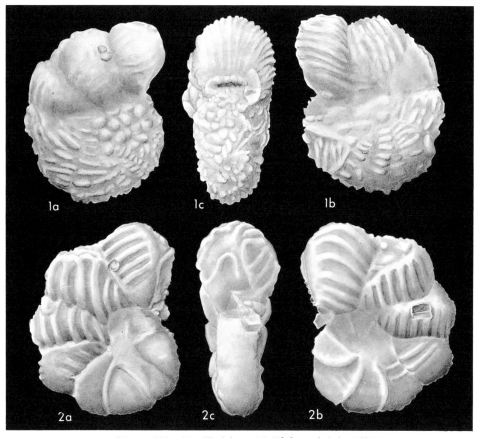

Fig. 494. Rotaliidae (Cuvillierininae; *1,2, Thalmannita*) (p. C621).

septa and fissured umbilical mass; *9*, axial half sec. showing fissured umbilical plugs on both sides of test; all ×45 (*631a).

Thalmannita BERMÚDEZ, 1952, *127, p. 76 [*Rotalia madrugaensis* CUSHMAN & BERMÚDEZ, 1947, *494, p. 24; OD] [=*Ornatanomalina* HAQUE, 1956, *876, p. 196 (type, *O. geei*)]. Test free, small, slightly trochoid in early stage, later planispiral, peripheral outline lobulate to angular, peripheral margin rounded, about 8 to 10 chambers to whorl; sutures radial to slightly curved; wall calcareous, perforate-radial in structure, surface ornamented with strong spiraling costae interrupted at sutures and may be interrupted by median ridges on chamber or broken into smaller nodes and ridges, similar nodes and pustules may occur in umbilical region; aperture a low equatorial, interiomarginal slit. [*Thalmannita* was originally referred to the Rotaliinae and *Ornatanomalina* to the Anomalinidae.] *Paleoc.-Oligo.*; W.Indies(Cuba-Puerto Rico)-Asia(Pak.).——FIG. 494,*1*. **T. madrugaensis* (CUSHMAN & BERMÚDEZ), Paleoc., Cuba; *1a-c*, opposite sides and edge view of holotype, ×86 (*2117).——FIG. 494,*2*. *T. hafeezi* (HAQUE), Paleoc., Pak.; *2a-c*, opposite sides and edge view of topotype, originally referred to *Ornatanomalina*, ×95 (*2117).

Subfamily CHAPMANININAE Thalmann, 1938

[*nom. transl.* FRIZZELL, 1949, p. 482 (*ex* family Chapmaninidae THALMANN, 1938)]

Test conical, early portion trochospiral, later uniserial; double walls and septa, with intraseptal spaces; septa invaginated into tubes or chamberlets; aperture consisting of tube openings. *M.Eoc.-Mio.*

Chapmanina A. SILVESTRI, 1931, *1784, p. 74 [*nom. subst. pro Chapmania* A. SILVESTRI & PREVER in SILVESTRI, 1904, *1759, p. 117 (*non* MONTICELLI, 1893; *nec* SPULER, 1910; *nec* DE MIRANDA RIBEIRO, 1920; *nec* BERNHAUER, 1933)] [*Chapmania gassinensis* A. SILVESTRI, 1905, *1762, p. 130=*Chapmania aegyptiensis* (CHAPMAN) A. SILVESTRI, 1904, (*sic*) *1759, p. 117 (*non Patellina egyptiensis* CHAPMAN, 1900) = *Archapmanoum gassinicoum* RHUMBLER, 1913, *1572b, p. 392 (*nom. van.*); OD (M), ICZN pending] [=*Archapmanoum* RHUMBLER, 1913, *1572b, p. 392 (obj.) (*nom. van.*); *Preverina* FRIZZELL, 1949, *751, p. 489 (type, *Chapmania galea* A. SILVESTRI, 1923, *1776,

p. 90]. Test conical, with early stage of few chambers trochospirally coiled, later whorls with small rectangular cortical chambers in widely flaring arrangement, possibly in multiple spire, umbilical region perforated with horizontal laminae and interlamellar pillars, similar to *Dictyo-*

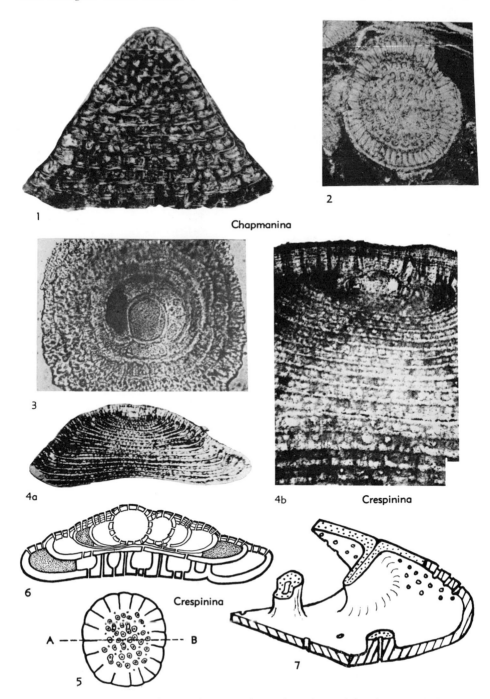

Fig. 495. Rotaliidae (Chapmanininae; *1,2, Chapmanina; 3-7, Crespinina*) (p. *C621-C624*).

Foraminiferida—Rotaliina—Rotaliacea

conoides, sutures fissured on umbilical side; septal walls invaginated from lower margin, resulting in double septa with intraseptal spaces; wall calcareous, perforate; aperture consisting of large pores in umbilical area, surrounded by tubelike pillars that extend from one umbilical lamina to

Fig. 496. Rotaliidae (Chapmanininae; *1-3, Chapmanina; 4-6, Sherbornina; 7,8, Crespinina*) (p. C621-C625).

the next, chambers connecting to interlamellar spaces by means of pores.*M.Eoc.-M.Mio.*, Eu.——Fig. 495,*1,2*; 496,*1,2*. **C. gassinensis* (SILVESTRI), Eoc., Italy (495,*1,2*; 496,*1*), Fr. (496,*2*); 495,*1*, axial sec. showing outer cortical layer of chambers and umbilical series of plates and pillars, ×37 (*1784); 495,*2*, sagittal sec., ×34 (*1780); 496,*1a-c*, spiral, umbilical, and edge views of topotype, ×28 (*2117); 496, *2a-c*, spiral, umbilical, and edge views showing fissured and perforated base and small rectangular cortical chambers, ×35 (*2117).——Fig. 496,*3*. *C. galea* (SILVESTRI), Mio., Italy; original figure of holotype and only specimen, ×40 (*1776).

[*Chapmania* 1904 was based on Italian specimens which were referred to the species *Patellina egyptiensis* CHAPMAN, 1900 (type-species of *Dictyoconus* BLANCKENHORN, 1900). In 1905 SILVESTRI noted that his specimens were neither conspecific nor congeneric and proposed the specific name *Chapmania gassinensis*. *C. gassinensis* has since then been regarded as the type-species of *Chapmania* SILVESTRI & PREVER. The generic name *Chapmanina* SILVESTRI, 1931, was proposed as a *nom. subst.* for *Chapmania* SILVESTRI & PREVER, 1904, a homonym of *Chapmania* MONTICELLI, 1893, and *C. gassinensis* has generally been regarded as its type-species. FRIZZELL, 1949, *751, noted that the type-species of *Chapmania*, by monotypy, was *Patellina egyptiensis* CHAPMAN and stated that a petition was being prepared for recognition of *C. gassinensis* as type-species, by use of the plenary powers of the ICZN. However, no petition was submitted (personal communication) and the generic status remained doubtful, hence the writers prepared such a petition in early 1963. *Chapmanina* was interpreted by FRIZZELL (*751) as having an early coil and later stage with low uniserial chambers, with secondary septa. It is here regarded as closely related to *Dictyoconoides* in structure, but has more widely spaced pillars and a longer axis. It differs from *Dictyokathina* in having a fissured base. *Preverina* was described by FRIZZELL from the figure and description of the type-species, *Chapmania galea* SILVESTRI. The type-species is known only from a drawing of a single vertical section, whose central part was replaced with crystalline calcite; the type-specimen is lost and no additional material referable to this species has been found at the type locality (*344A). According to FRIZZELL (*751, p. 489), "*Preverina* differs from *Chapmania* in the single wall and absence of intraseptal spaces. It is distinguished as well by the relatively larger initial spiral, and by the greater number of rows of chamberlets." The initial spire of the holotype is obscured by recrystallization and the monolamellar character is questionable, as the original figure (here reproduced) shows apparent single septa in part, but also shows apparent double septa in other parts of the section. No features are shown in the original figure that would preclude its assignment to *Chapmanina*, and *Preverina* is therefore regarded as a synonym.]

Crespinina WADE, 1955, *2026, p. 45 [**C. kingscotensis*; OD]. Test free, low and conical, megalospheric form with globular proloculus followed by embracing second chamber and annular undivided third chamber, microspheric form with planispiral stage with chambers increasing rapidly in length to become embracing, later annular chambers being subdivided by imperforate radial partitions, resulting in numerous rectangular chamberlets, all chambers visible from convex perforate spiral side, umbilical side partly imperforate, with perforate pillars extending from one horizontal lamina to next, but not continuous through test; wall calcareous, lamellar, septa double, formed by invagination of outer wall; intercameral connection by means of fine pores which open into chambers directly or may run through pillars, external large pores serving as apertures. [*Crespinina* is similar to *Dictyoconoides*, but dif-

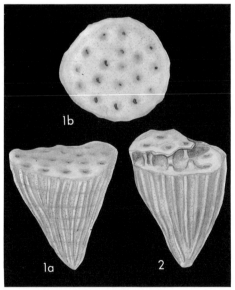

FIG. 497. Rotaliidae (Chapmanininae; *1,2*, Ferayina) (p. C624-C625).

fers in its pillars not being continuous throughout the test, and thus differs in much the way that *Rotalia* differs from *Lockhartia*. The multiple spiral chamber development is also not evident in *Crespinina*.] *U.Eoc.-L.Oligo.*, S.Australia.——Fig. 495,*3-7*; 496,*7,8*. **C. kingscotensis*, Eoc.; 495,*3*, horiz. sec. through early chambers of megalospheric form, ×140; 495,*4a,b*, vert. sec. through apex, ×40, ×140 (*2026); 495,*5*, diagram. view of umbilical side of small specimen showing marginal partitions, perforate pillars, and pores, ×60 (*2026); 495,*6*, axial sec. along line AB of 495,*5*, showing perforate protoconch and deuteroconch and later chambers with imperforate marginal partitions and central perforate pillars, ×175 (*2026); 495,*7*, diagram showing perforated pillars and pores in largely imperforate lower surface, perforate upper surface, and infolded double septa (*2026); 496,*7a-c*, opposite sides and edge view, ×30 (*2117); 496,*8*, diagram. figure of early whorls of microspheric test, ×300 (*2026).

Ferayina FRIZZELL, 1949, *751, p. 483, 492 [**F. coralliformis*; OD]. Test free, conical, proloculus followed by 3 or 4 tiny low chambers of undetermined arrangement, later with rapidly enlarging, low, uniserially arranged chambers; septa horizontal, flat, imperforate except for large rounded intercameral foramina, sutures indistinct at surface; wall of calcite (by X-ray analysis), finely perforate, radial in microstructure, surface with low longitudinal costae which increase by bifurcation and are thus equidistant throughout; aperture multiple, consisting of numerous rounded openings on terminal face, each provided in-

Fig. 498. Rotaliidae (Pegidiinae; *1, Pegidia; 2,3, Sphaeridia*) (p. C625-C627).

ternally with hollow pillar-like process, extending to previous septum. *M.Eoc.,* USA(Tex.-Calif.)-S.Am.(Ecuad.-Peru).——Fig. 497,*1,2.* **F. coralliformis,* Claiborne F., Tex.; *1a,b,* side, apertural views of topotype; *2,* side view of partially dissected specimen showing hollow pillar-like processes connecting adjacent septa; all ×105 (*2117).

[*Ferayina* was originally placed with the Chapmaniidae by FRIZZELL. HOFKER (1956, *945, p. 897) stated that the wall has an imperforate outer layer and contains embedded mineral particles; thus he considered the genus related to *Dictyoconus* and the valvulinids. The wall of topotypes of *Ferayina* was investigated by us and proved by X-ray and petrographic analysis to consist of radially built calcite; hence the genus is not regarded as related to the valvulinids.]

Sherbornina CHAPMAN, 1922, *322, p. 501 [**S. atkinsoni;* OD]. Test discoidal, thin, up to 2 mm. diam., early stage with nearly planispirally arranged chambers, megalospheric form with 4 to 10 enrolled chambers, microspheric form with 14 enrolled chambers, later with 3 or 4 more embracing chambers followed by cyclical chambers, all later chambers with corrugated margins near sutures, projections of successive chambers alternating in position; wall calcareous, coarsely perforate, radial in structure, lamellar, with well-developed canal system of septal canals in young stage and septal and radial canals in adult, with branches forming radial canals that open at surface as coarse pores, test perforations smaller than canal-system pores, surface may be pustulose; no visible aperture. *U.Eoc.-Mio.,* S.Pac.O.(Tasm.).——Fig. 496,*4-6.* **S. atkinsoni,* Oligo.; *5a,b,* side, edge views, ×33 (*2117); *4,* specimen split

in median plane showing corrugated septa and early embryonic coil, ×35 (*2028); *6,* vert. sec. showing lamellar walls, canals, and pores, ×100 (*2028).

Subfamily PEGIDIINAE Heron-Allen & Earland, 1928

[*nom. transl.* CHAPMAN & PARR, 1936, p. 144 (*ex* family Pegidiidae HERON-ALLEN & EARLAND, 1928)]

Trochospirally derived test, with chambers few and inflated, each successive chamber opposed to or partially enveloping that preceding, early chambers resorbed during growth; aperture a series of tubes which may pierce umbilical shell material. *Mio.-Rec.*

Pegidia HERON-ALLEN & EARLAND, 1928, *913, p. 290 [*Rotalia dubia D'ORBIGNY, 1826, *1391, p. 274, =*Pegidia papillata* HERON-ALLEN & EARLAND in HERON-ALLEN & BARNARD, 1918, *905, p. 90; OD] [*Pegidia* HERON-ALLEN & EARLAND in HERON-ALLEN & BARNARD, 1918, *905, p. 90 (*nom. nud.*)]. Test free, sublenticular, unequally biconvex, with 3 or 4 chambers arranged in apposition, early chambers may be resorbed as new ones form; calcareous wall and septa thick, perforate, radially built, lamellar character not described, surface of spiral side may be closely tuberculate, peripheral margin with broad, smooth keel, grooves radiating from umbilicus and tubular vertical canals piercing solid umbilical plug, opening at surface; no aperture other than openings of tubular canals. *Mio.-Rec.,* Eu.(Île de France) - Indian O. (Mauritius Is.) - Afr. (Kerimba Arch.)-E.Indies(Java-Philip. Is)-W.Indies-W.Pac.O. (Caroline Is., Ifaluk Atoll)-Eu.——FIG. 498,*1.* **P. dubia* (D'ORBIGNY), Rec., Mauritius; *1a-c,* opposite sides and edge view of topotype, ×33 (*2117).

Sphaeridia HERON-ALLEN & EARLAND, 1928, *913, p. 294 [**S. papillata*; OD]. Test free, 0.7-0.85 mm. diam., globular, chambers 3 or 4, increasing

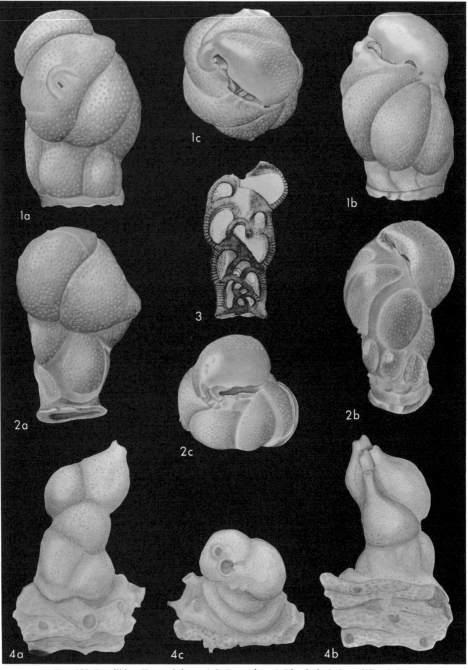

FIG. 499. Rotaliidae (Rupertininae; *1-3, Rupertina; 4, Biarritzina*) (p. C627-C628).

Foraminiferida—Rotaliina—Rotaliacea

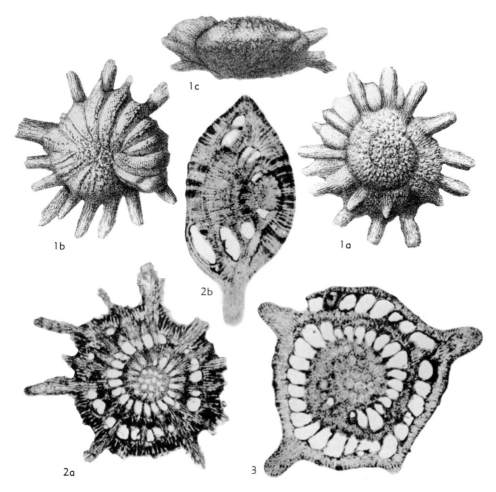

Fig. 500. Calcarinidae; *1-3, Calcarina* (p. C628-C629).

rapidly in size, arranged in apposition and strongly enveloping, probably resorbed as new chambers are formed, umbilical region filled by large solid plug that occupies about one-fourth surface of test and is perforated by series of bifurcating vertical tubular canals; wall thick, calcareous, perforate radial in structure, lamellar character not described, surface ornamented with beads or pustules of clear shell material; aperture consisting of pores at surface marking outlets of tubular canals. *Rec.*, Afr.(Kerimba Arch.)-Ind.O.(Mauritius Is.).——Fig. 498,*2,3*. **S. papillata*; *2a-c*, specimen from Ind.O. showing globular form, clear calcite pustules, umbilical plug, and bifurcating tubular canals, ×64 (*2117); *3*, broken specimen from Mauritius Is. showing thick wall, main septum, traces of resorbed earlier septa, and inner opening of tubular canals in center, ×33 (*913).

Subfamily RUPERTININAE
Loeblich & Tappan, 1961

[Rupertininae LOEBLICH & TAPPAN, 1961, p. 312 (*nom. subst. pro* Rupertiinae GALLOWAY, 1933, p. 302)] [=Rupertinae SILVESTRI, 1937, p. 143 (*nom. van.*)]

Test attached by basal disc, early chambers trochospiral, later extending upward from base in loose spiral; wall calcareous, coarsely perforate, radiate in structure, septa doubled as in Rotaliidae; aperture narrow, interiomarginal. *?Eoc., Mio.-Rec.*

Rupertina LOEBLICH & TAPPAN, 1961, *1177, p. 312 [*nom. subst. pro Rupertia* WALLICH, 1877, *2036, p. 502 (*non Rupertia* GRAY, 1865)] [**Rupertia stabilis* WALLICH, 1877, *2036, p. 502; OD]. Test attached by large prominent basal disc; chambers numerous, early ones in close coil, later vertically elongated and coiling in tall spire; wall calcareous, radiate in structure, coarsely

Fig. 501. Calcarinidae; *1-3, Baculogypsina;* 4, *Baculogypsinoides;* 5, *Siderolites* (p. C629-C631).

perforate, with rotaliid wall structure; aperture commonly narrow, slitlike, bordered above by prominent lip. ?*Eoc., Mio.-Rec.,* Atl.O.-S.Pac.O.-W.Pac.O.(Bismarck Arch.)-Ind.O.-USA-W.Indies (Carib.)-Eu.——Fig. 499,*1-3.* *R. *stabilis* (WALLICH), Rec., Atl.O.; *1a-c, 2a-c,* opposite sides and apert. views, ×40 (*2117); *3,* long. sec., ×27 (*200).

Biarritzina LOEBLICH & TAPPAN, *nom. subst.* herein [*pro Columella* HALKYARD, 1918, *861, p. 28 (*non* WESTERLUND, 1878)] [**Columella carpenteriaeformis* HALKYARD, 1918, *861, p. 28, here designated as type-species]. Test attached by flaring base, then growing upright; chambers few, inflated, early chambers trochospirally coiled, later chambers in loose, elevated spire, tending to become uniserial; sutures depressed; wall calcareous, with coarse perforations scattered between fine pores; aperture rounded, terminal, with distinct bordering lip or neck of nonperforate calcite. *Tert.-Rec.,* Eu.-Australia-Pac.O.(Philip. Is.)-W. Indies(Carib.)-Atl.O.——Fig. 499,*4.* **B. carpenteriaeformis* (HALKYARD), Eoc., Fr.; *4a-c,* opposite sides and apert. view of topotype attached to bryozoan, ×22 (*2117).

[The original type-specimens of the type-species, from the Auversian of Biarritz, France, deposited in the collection of Victoria University, Manchester, England, were destroyed during the war. *Columella* was regarded as a synonym of *Carpenteria* by GALLOWAY (1933, *762) but *Carpenteria* is here restricted to the low conical forms like its type-species. As *Columella* HALKYARD is a homonym, it is here renamed and the subcylindrical species previously placed in *Carpenteria* should be referred to *Biarritzina.*]

Family CALCARINIDAE Schwager, 1876

[*nom. correct.* EIMER & FICKERT, 1899, p. 703 (*pro* family Calcarine SCHWAGER, 1876, p. 481)]——[In synonymic citations superscript numbers indicate taxonomic rank assigned by authors ([1]family, [2]subfamily)] —— [=[1]Tinoporidea SCHWAGER, 1877, p. 21; =[2]Tinoporinae BRADY, 1884, p. 74; =[1]Tinoporidae DELAGE & HÉROUARD, 1896, p. 147; =[1]Tinoporidae LISTER in LANKESTER, 1903, p. 146; =[2]Tinoporininae HOFKER, 1933, p. 125 (*nom. van.*)]——[=[2]Calcarininae HOFKER, 1927, p. 42; =[1]Siderolitidae FINLAY, 1939, p. 525; =[2]Siderolitinae SIGAL in PIVETEAU, 1952, p. 250; =[1]Baculogypsinidae SMOUT, 1955, p. 205]

Test coiled, without differentiation into spiral and umbilical surfaces, advanced genera may become globular, large spines formed by thickenings, and not marginal projections of chambers; canal system diffuse and confused with perforations. *U.Cret-Rec.*

Calcarina D'ORBIGNY, 1826, *1391, p. 276 [**Nautilus spengleri* GMELIN, 1788, *798, p. 3371; SD PARKER & JONES, 1859, *1417a, p. 482]. Test large, 1 or 2 mm. diam., lenticular, biconvex, trochospiral throughout, chambers numerous, no later acervuline chambers present; sutures radial, depressed, but largely obscured by supplementary

Foraminiferida—Rotaliina—Rotaliacea C629

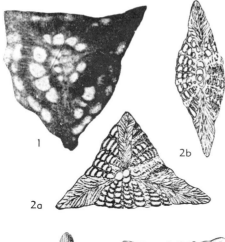

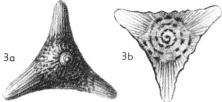

Fig. 502. Calcarinidae; *1-3, Baculogypsinoides* (p. C629).

lamellar calcite on umbilical side; intraseptal passages present, umbilical cavities interrupted by pillars and radial and lateral canals; wall calcareous, chamber roofs and floors with 2 layers, thin inner layer and coarsely perforate, thicker outer layer, surface thickly covered with tubercles; 6 to 30 thick, elongate, longitudinally striated, peripheral spines may bifurcate terminally, probably serving for anchorage on reef algae; aperture narrow and strongly indented, interiomarginal, intercameral foramina identical in form. [The type-species, discussed by LOEBLICH & TAPPAN (1962, *1186, p. 33, 34) is *Nautilus spengleri* GMELIN, by subsequent designation of PARKER & JONES, 1859. The type cannot be *Calcarina calcar,* either by tautonomy or subsequent designation, as that species was a *nomen nudum* in the original publication.] ?U.Cret., Rec., Pac.O.——Fig. 500, *1-3. *C. spengleri* (GMELIN), Rec., Admiralty Is. (*1*), Okinawa (*2*), Marshall Is. (*3*); *1a-c,* opposite sides and edge view, ×30 (*200); *2a,b,* horiz. and axial secs., ×20 (*531); *3,* horiz. sec., ×20 (*531).

Baculogypsina SACCO, 1893, *1607, p. 206 [*Orbitolina sphaerulata* PARKER & JONES, 1860, *1417d, p. 34; OD (M)] [=*Taurogypsina* SACCO, 1893, *1607, p. 205 (type, *T. taurobaculata*)]. Test free, periphery lobulated, with few coarse radial spines, early stage trochospiral, juvenarium or embryonic apparatus distinct, consisting of spherical proloculus followed by 1.5 whorls of planispirally arranged chambers, without canal system but with umbilical plugs on one side, later chambers arranged in radially disposed layers with numerous thin, conical pillars interspersed, ends of the pillars projecting at surface as tubercles; wall of chamber roofs and floors finely perforate, radial spines arising from juvenarium in its plane of coiling, of solid supplementary shell material, pierced by anastomosing canals and covered with several layers of chambers except at tip. Mio.-Rec., Eu.-Pac.O.——Fig. 501,*1-3.* *B. sphaerulata* (PARKER & JONES), Rec., Fiji Is. (*1*), Pleist., Saipan (*2,3*); *1,* side view of paratype, ×20 (*2117); *2,* equat. sec., ×53 (*364); *3,* axial sec., ×27 (*364).

[A lectotype for *Orbitolina sphaerulata* PARKER & JONES was chosen by us in the British Museum (Natural History) and is here designated (BMNH-ZF3599) and paratypes (BMNH-ZF3598) (all *ex* 94.4.3.1822) all from Recent deposits at Rewa Reef, Fiji. *Baculogypsina* was placed in the Cibicidinae by HANZAWA (1952, *872), in the Calcarinidae by CUSHMAN (1948, *486), and in the Baculogypsinidae by SMOUT (1955, *1804).]

Baculogypsinoides YABE & HANZAWA, 1930, *2093, p. 43 [*B. spinosus*; OD (M)] [=*Silvestriella* HANZAWA, 1952, *872, p. 17 (type, *Calcarina tetraedra* GÜMBEL, 1870, *840, p. 656)]. Early stage trochospiral, as in *Calcarina,* later chambers acervuline, lateral walls compact, peripheral wall coarsely perforate; wall calcareous, with coarse tubuli, commonly with 3 or 4 thick blunt spines, with anastomosing canal system, arising near proloculus and extending outward in plane of coiling, interior with numerous thin conical vertical pillars, which project at surface as tubercles. Eoc.-Rec., Philip. Is.-Eu.-China Sea(Ryukyu Is.).——Fig. 501,*4;* 502,*1.* *B. spinosus,* Rec., Philip. (501,*4*), Pleist., Ryukyu Is. (502,*1*); 501,*4,* lectotype, ×33 (*2117); 502,*1,* equat. sec., ×40 (*872).——Fig. 502,*2,3. S. tetraedra* (GÜMBEL), U.Eoc., Italy (*2*), Eoc., Aus. (*3*); *2a,b,* equat. and axial secs., ×7.5 (*872); *3a,b,* ext. and equat. sec., ×10 (*840).

[The type-specimens of *B. spinosus* were stated to be those figured by CUSHMAN (1919, *412, pl. 45) as *Siderolites? tetraedra* (GÜMBEL), which are not conspecific with GÜMBEL's form. A lectotype is here designated and redrawn (USNM 15364b, *412, pl. 45, figs. 2a,b, from *Albatross* Station D5179, Philippines). *Silvestriella* was shown by KÜPPER (1954, *1069) to be a synonym of *Baculogypsinoides.*]

Schlumbergerella HANZAWA, 1952, *872, p. 19 [*Baculogypsina floresiana* SCHLUMBERGER, 1896, *1657, p. 88; OD]. Test large, globular, to 3.5 mm. diam., with spines projecting slightly or forming tubercles; juvenarium of megalospheric form consisting of 3 chambers (of raspberry form, not coiled), microspheric form with early coil, later chambers undifferentiated acervuline, forming angle of about 60° to axis of spine, spines arising from juvenarium, containing radial and ramifying canals; pillars also present, similar to spines but smaller and with fewer canals, and different in structure from chamber walls, perhaps representing radial rows of calcified lateral chambers; wall calcareous, perforate; apertures consisting of rows of rounded openings in cham-

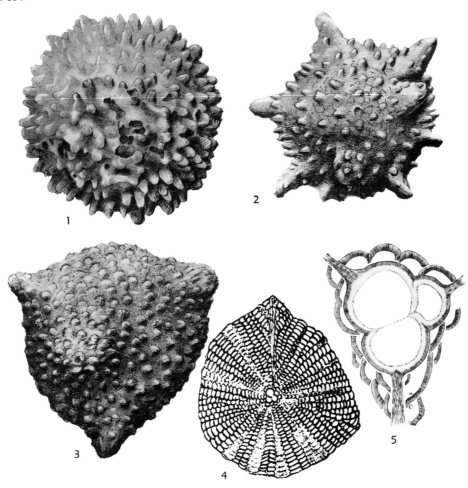

Fig. 503. Calcarinidae; *1-5, Schlumbergerella* (p. C629-C630).

ber roofs, with 2 to 4 apertures in row, smaller openings or perforations also connecting adjacent chambers through wall of roof, these openings being widest at outer surface, stolons also connecting acervuline chambers, one stolon opening into preceding chamber and 2 stolons opening into different later chambers, possibly with additional stolons. *Pleist.-Rec.*, E.Indies(Indon.).——Fig. 503,*1-5*. **S. floresiana* (SCHLUMBERGER), Rec.; *1,2*, microspheric specimens, ×16.5, ×20 (*928a); *3*, megalospheric specimen, ×20 (*928a); *4*, transv. sec. showing early juvenarium, ×21 (*872); *5*, central part of equat. sec. showing juvenarium, spines, and acervuline chambers, ×56 (*1069).

Siderolites LAMARCK, 1801, *1084, p. 376 [**S. calcitrapoides*; OD (M)] [=*Siderolithes* DE MONTFORT, 1808, *1305, p. 151 (obj.); *Siderolina* DEFRANCE, 1824, *579e, p. 180 (obj.); *Sideroporus* BRONN, 1825, *209, p. 30, 31 (type, *S. calcitrapa*, =*Sidérolite calcitrapoïde* FAUJAS, 1799, *712, p. 188); *Siderolithus* BRONN, 1838, *210, p. 711 (obj.)]. Test large, planispirally coiled throughout from globular proloculus, without raspberry type of embryonic apparatus and without supplementary acervuline chambers; wall of chamber roofs and floors of 2 layers, inner layer thin and finely perforate, outer layer thick and coarsely perforate, few large coarse spines originating near proloculus and radiating in plane of coiling, spines with ramifying canal system and commonly protruding somewhat at periphery, numerous conical pillars piercing successive spiral lamellae and appearing as tubercles at test surface. [Differs from *Calcarina* in being planispiral rather than trochospiral throughout.] *U.Cret.-L.Eoc.*, Eu.-Asia(India).——Fig. 501,*5*; 504,*1-3*. **S. calcitrapoides*, U.Cret.(Maastricht.), Neth.; 501,*5a,b*, side and edge views, ×11 (*2117); 504,*1,2*, equat. and axial secs., ×20 (*872); 504,*3*, de-

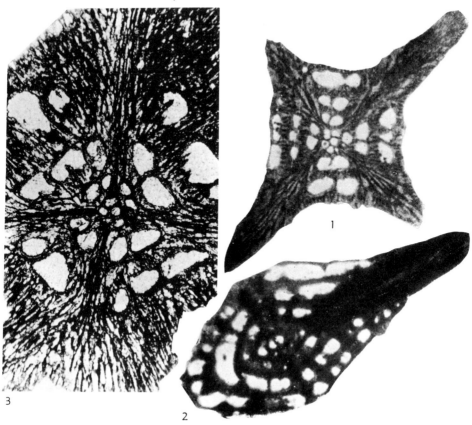

Fig. 504. Calcarinidae; *1-3, Siderolites* (p. C630-C631).

calcified equat. sec. in canada balsam preparation, ×20 (*1998).

Family ELPHIDIIDAE Galloway, 1933

[*nom. transl.* Sigal in Piveteau, 1952, p. 240 (*ex* subfamily Elphidiinae Galloway, 1933, p. 265)]——[In synonymic citations dagger(†) indicates *partim*]——[=Polythalama† Latreille, 1825, p. 161 (*nom. nud.*); =Hélicostègues† d'Orbigny, 1826, p. 268 (*nom. nud.; nom. neg.*); =Helicotrochina† Ehrenberg, 1839, table opp. p. 120 (*nom. nud.*); =Nautiloida† Schultze, 1854, p. 53 (*nom. nud.*); =Polystomellidea Reuss, 1862, p. 308, 388; =Polystomellida Schmarda, 1871, p. 165; =Polystomellina Lankester, 1885, p. 848; =Polystomellinae Delage & Hérouard, 1896, p. 150; =Polystomellidae Eimer & Fickert, 1899, p. 698; =Canaliferidae Krasheninnikov, 1953, p. 89 (*non* Canaliferidae Broderip, 1839)]

Test planispiral, trochospiral, or uncoiling; sutural canal system opening into single or double row of sutural pores; wall calcareous, perforate, radial in structure; aperture interiomarginal, single or multiple, or areal. *Paleoc.-Rec.*

Subfamily ELPHIDIINAE Galloway, 1933

[Elphidiinae Galloway, 1933, p. 265] [=Orbientina Marriott, 1878, p. 30 (*nom. nud.*); =Polystomellida Schultze, 1854, p. 53; =Polystomellina Jones in Griffith & Henfrey, 1875, p. 320; =Polystomellinae Brady, 1881, p. 44; =Cribroelphidiinae Voloshinova, 1958, p. 167]

Test free, planispiral and symmetrical, at least in adult, may uncoil in later stages, with sutural pores and sutural canal system, and retral processes projecting across sutures; aperture consisting of interiomarginal or areal pores or both. *Paleoc.-Rec.*

Elphidium de Montfort, 1808, *1305, p. 14 [*Nautilus macellus Fichtel & Moll var. β Fichtel & Moll, 1798, *716, p. 66; OD] [=Pelorus de Montfort, 1808, *1305, p. 22 (type, *Nautilus ambiguus* Fichtel & Moll, 1798, *716, p. 62); *Andromedes* de Montfort, 1808, *1305, p. 38 (type, *Nautilus strigillatus* Fichtel & Moll var. α Fichtel & Moll, 1798, *716, p. 49); *Sporilus* de Montfort, 1808, *1305, p. 42 (type, *Nautilus strigillatus* Fichtel & Moll, var. β Fichtel & Moll, 1798, *716, p. 49); *Themeon* de Montfort, 1808, *1305, p. 202 (type, *Nautilus crispus* Linné, 1758, *1140, p. 709, =*Themeon rigatus* de Montfort, 1808); *Geophonus* de Montfort, 1808, *1305, p. 18 (type, *Nautilus macellus* Fichtel & Moll var. α Fichtel & Moll, 1798, *716, p. 66); *Ceophonus* Bosc, 1816, *176, p. 491 (*nom. null.* pro *Geophonus* de Montfort,

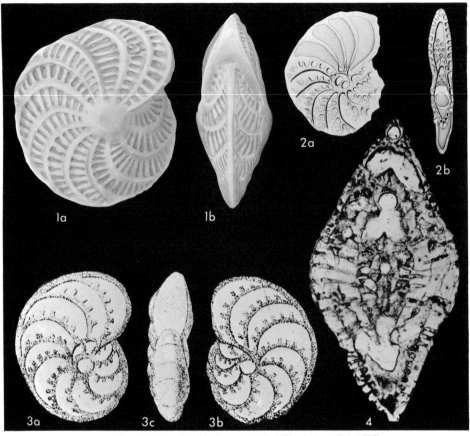

Fig. 505. Elphidiidae (Elphidiinae; *1-4, Elphidium*) (p. C631-C635).

1808); *Polystomella* Lamarck, 1822, *1090, p. 624 (type, *Nautilus crispus* Linné, 1758, *1140, p. 709); *Themeone* Berthold in Latreille, 1827, *1097A, p. 161 *(nom. van.)*; *Polystomatium* Ehrenberg, 1839, *667, table opp. p. 120 (type, *Nautilus strigillatus* Fichtel & Moll, 1798, *716, p. 49); *Geoponus* Ehrenberg, 1839, *667, p. 132 *(nom. van. pro Geophonus* de Montfort, 1808); *Planoelphidium* Voloshinova, 1958, *2019, p. 165 (type, *Polystomella laminata* Terquem, 1878, *1889, p. 16); *Faujasinella* Voloshinova, 1958, *2019, p. 162 (type, *Elphidium semiinvoluta* Myatlyuk in Dabagyan, Myatlyuk & Pishvanova, 1956, *547, p. 228); *Discorotalia* Hornibrook, 1961, *959, p. 141 (type, *Polystomella tenuissima* Karrer, 1865, *1020, p. 83)]. Test planispiral, bilaterally symmetrical, involute, chambers numerous, with numerous retral processes or internal chamber projections along septal borders, ending blindly against septal face in final chamber, but pierced by tiny pore formed by resorption of septum at base of retral process in earlier chambers, resulting in numerous tubular perforations connecting chambers; septa secondarily doubled, incomplete septal flap being formed against apertural face as succeeding chamber forms, leaving septum single-layered near center and base and double near outer edges where it encloses canal system, with prominent lamellar thickening of outer wall; canal system complex, spiral canal present along umbilical chamber margins leading to vertical umbilical canals through umbilical plug, and also giving rise to subsutural septal canals at each septum in intraseptal space between septal face and septal flap formed by succeeding chamber and lying below retral processes, communicating with surface by means of diverging canals; wall calcareous, finely perforate, radial in structure, surface commonly with grooves (fossettes) or ridges paralleling periphery (striped crenulation) and commonly coinciding with internal retral processes, or surface may be smooth or finely pustulose; aperture consisting of row of pores at base of septal face, earlier septa may also have areal foramina due to resorption; pseudopodia extremely numerous, long, and attenuated; alternation of asexual (producing up to 200 embryos) and sexual reproduction with development of inequally biflagellate gametes. [Habitat shallow water or tide pools on sandy or shelly bottoms, with

Foraminiferida—Rotaliina—Rotaliacea

algae, radiating pseudopodia binding together a mass of sand to prevent dislodging during moderate turbulence.] *L.Eoc.-Rec.,* cosmop.——FIG. 505,*1,2*. **E. macellum* (FICHTEL & MOLL), Rec., Eu.(Italy); *1a,b,* side, apert. views, ×68 (*2117); *2a,* horiz. sec. in canada balsam, showing canal system; *2b,* transv. sec. showing canal system and septal foramina, ×70 (*928a).——FIG. 505,*3*. *E. semiinvolutum* MYATLYUK, U.Eoc., Carpathians; *3a-c,* opposite sides and edge view showing faintly

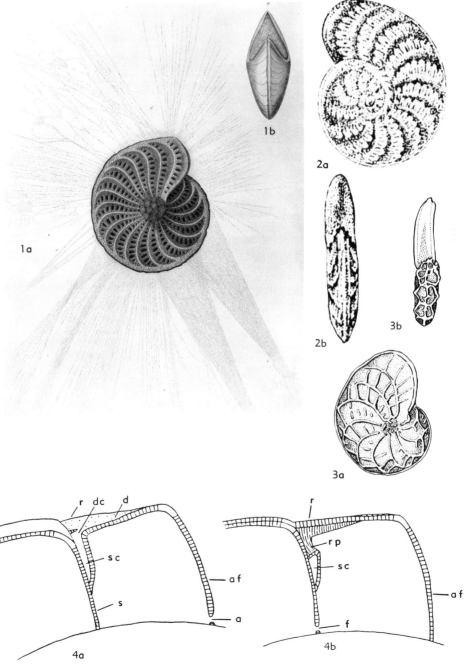

FIG. 506. Elphidiidae (Elphidiinae; *1-4, Elphidium*) (p. *C631-C635*).

asymmetrical form, ×100 (*2019).—FIG. 505, 4; 506,4. *E. crispum* (LINNÉ), Mio., Asia(Israel); 505,4, vert. sec., ×87 (*1534); 506,4a, diagram. sec. through surface depression or fossette; 506,4b, sec. through ridge between surface depressions showing aperture *(a)*, apertural face *(af)*, sur-

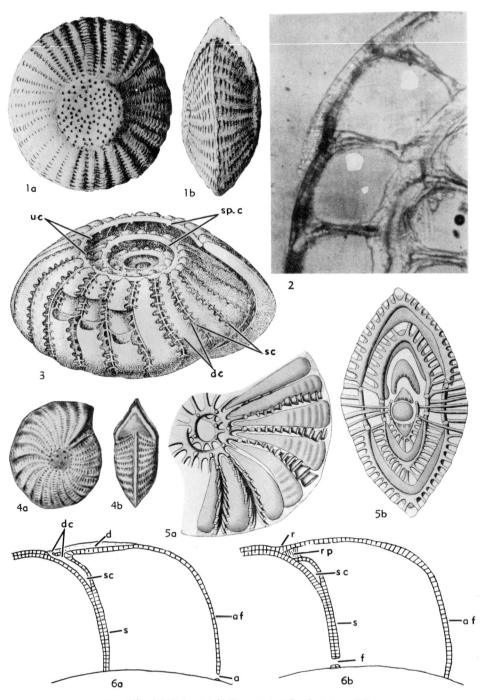

FIG. 507. Elphidiidae (Elphidiinae; *1-6, Cellanthus*) (p. C635).

face depression *(d)*, diverging canal *(dc)*, foramen *(f)*, surface ridge *(r)*, retral process *(rp)*, septum *(s)*, and septal canal *(sc)*, locality of specimens not given, ×300 (*2027).——FIG. 506, *1. E. strigillatum* (FICHTEL & MOLL), Rec., Eu. (Italy); *1a*, living specimen showing granular pseudopodia, which actually extend to a length 4 to 6 times diam. of test; *1b*, edge view, ×72 (*1695).——FIG. 506,*2. E. laminatum* (TERQUEM), Plio., Eu.(Albania); *2a,b*, side and apert. views, side view suggesting an evolute condition but arched chamber seen in edge view suggesting that this is a normally involute form, ×100 (*2019).——FIG. 506,*3. E. tenuissimum* (KARRER), Oligo.-Mio., N.Z.; *3a,b*, side, edge views, ×120 (*959).

[Numerous recent revisions of *Elphidium* have been undertaken, varying from the inclusive usage of CUSHMAN (1939, *473) to finer divisions variously based on apertural form and position, type of internal canals, and other characters. The present somewhat restricted usage of *Elphidium* is based on revisions of the genus by HOFKER (1956, *946), UJIIÉ (1956, *1964), WADE (1957, *2027), VOLOSHINOVA (1958, *2019), and KRASHENINNIKOV (1960, *1054). The previously synonymized *Elphidium*, *Cellanthus*, *Cribroelphidium*, and *Elphidiella* are recognized as distinct genera. *Planoelphidium* was described as being partially evolute, although this was not indicated in the original description of the type-species from the Pliocene of Rhodes, nor was this apparent except by dampening specimens (e.g., from Albania). TERQUEM reported *E. macellus* from the Middle Eocene (Lutetian) of the Paris Basin, but the form from these deposits was not conspecific with the type-species and was later named *Elphidium terquemianum* LE CALVEZ (1950, *1113). It appears that compressed tests result from very narrow highly arched chambers, and that the very thin lateral shell layers become transparent on dampening, erroneously suggesting an evolute condition. The Recent evolute species *E. subevolutum* CUSHMAN, mentioned by VOLOSHINOVA (*2019) as a probable *Planoelphidium* may be an incompletely developed *Ozawaia*, as both occur at Rotonga. *Planoelphidium* is therefore regarded as a synonym of *Elphidium*. *Faujasinella* was defined as differing from *Elphidium* in being slightly asymmetrical, although the aperture and canal system are as in *Elphidium*. It was separated from *Faujasina* because it is biconvex. As many species of *Elphidium* may have slightly asymmetrical specimens, this is not regarded as a generic character and *Faujasinella* is considered a synonym of *Elphidium*. *Discorotalia* was stated to differ from *Notorotalia* in being discoidal, with a cluster of areal pores as an aperture and having sporadic sutural pores. The tests of the 2 species included are only slightly evolute on one side; otherwise they seem referable to *Elphidium*, which also may have areal pores and narrow interseptal surface costae. *Discorotalia* is therefore also regarded as a synonym of *Elphidium*.]

Cellanthus DE MONTFORT, 1808, *1305, p. 206 [**Nautilus craticulatus* FICHTEL & MOLL, 1798, *716, p. 51; OD] [=*Vorticialis* LAMARCK, 1812, *1087, p. 122 (obj.); *Cellulia* AGASSIZ, 1844, *5, p. 6 (*nom. van.* pro *Cellanthus* DE MONTFORT, 1808); *Helicoza* MÖBIUS, 1880, *1293, p. 103 (obj.); *Carpenterella* KRASHENINNIKOV, 1953, *1050, p. 88 (*non* COLLENETTE, 1933; *nec* BERMÚDEZ, 1949) (obj.)]. Test large, planispiral, chambers numerous, with large umbilical plug on each side which may occupy over half diameter of test, chambers gradually enlarging but not involute, if umbilical plug is discounted; septa completely double and may enclose additional intraseptal canals; canal system similar to *Elphidium*, but more highly developed, spiral canal at umbilical chamber margin giving rise to straight unbranched canals that lead to surface of umbilical plugs and also to subsutural intraseptal canals which may branch into divergent canals near surface; wall calcareous, finely perforate, radial in structure, surface not highly ornamented as in *Elphidium*, but only with perforations of canal system; aperture single row of pores at base of apertural face. Plio.-Rec., Indo-Pac.reg.——FIG. 507,*1-6*. **C. craticulatus* (FICHTEL & MOLL), Rec., Tonga Is. (*4*), Indon. (*5*); *1a,b*, side and apert. views of microspheric adult, ×30; *2*, portion of equat. sec., showing septal canals between septal face and septal flap, and retral processes at peripheral margin, locality not given, ×200; *3*, schematic figure of internal cast of chambers and canals, with umbilical plug removed to show spiral canal *(spc)*, umbilical canal *(uc)*, septal or meridional canals *(sc)*, diverging canals *(dc)*, ×20 (*2019); *4a,b*, side, apert. views of megalospheric specimen, ×64 (*473); *5a*, schematic equat. sec. combining parts of both megalospheric and microspheric canal systems, 3 septal rows showing megalospheric canal system character, and *2* showing the forking canals of microspheric test, ×52.5; *5b*, axial sec., ×52.5 (*928a); *6a*, diagram. view of last chamber cut through surface depression; *6b*, same cut through ridge between depressions, showing aperture *(a)*, apertural face *(af)*, surface depression *(d)*, diverging canal *(dc)*, foramen *(f)*, surface ridge *(r)*, retral process *(rp)*, septum *(s)*, and septal canal *(sc)*, ×300 (*2027).

Cribroelphidium CUSHMAN & BRÖNNIMANN, 1948, *498, p. 18 [**C. vadescens*; OD] [=*Elphidionion* HOFKER, 1951, *936, p. 356 (type, *Polystomella poeyana* D'ORBIGNY in DE LA SAGRA, 1839, *1611, p. 55); *Cribroelphydium* TINOCO, 1955, *1935, p. 30 (*nom. van.*) (obj.); *Cribroelphidium (Rimelphidium)* VOLOSHINOVA, 1958, *2019, p. 173 (type, *Elphidium vulgare* var. *vulgare* VOLOSHINOVA in VOLOSHINOVA & DAIN, 1952, *2022, p. 53)]. Test free, planispiral and involute, commonly robust with rounded periphery and few chambers to whorl; sutures distinct, depressed and may be crossed by solid pillars or septal bars but without retral processes at chamber margins, large sutural pores may be present between septal bars leading to simplified sutural canal system; wall calcareous, coarsely perforate, radiate in microstructure; aperture multiple, with one or more pores at base of septal face and with one or more areal pores in addition. Mio.-Rec., cosmop.——FIG. 508,*1*. **C. vadescens*, Rec., W.Indies(Trinidad); *1a,b*, side, edge views, ×167 (*1940).——FIG. 508,*2. C. kugleri* CUSHMAN & BRÖNNIMANN, Rec., W.Indies(Trinidad); *2a,b*, side, edge views of holotype, ×174 (*2117).——FIG. 508,*3,4. C. poeyanum* (D'ORBIGNY), Rec., W.Indies(Cuba); *3a,b*, side, edge views of lectotype, here designated (MNHN, Paris), ×64 (*2117); *4a,b*, septum showing areal pores and simple canal, and septal canal from interior of chamber showing relation to incised suture, and solid nonperforate septal

C636 · Protista—Sarcodina

bridges, enlarged (*946).——FIG. 508,5. *C. vulgare* (VOLOSHINOVA), U.Mio., E.USSR(Sakhalin Is.); *5a,b,* side, edge views, ×66 (*2019).

[Previously considered a synonym of *Elphidium* by us (*1162, p. 105), *Cribroelphidium* is here recognized as differing from *Elphidium* in the absence of hollow retral processes, the presence of solid, nonperforate septal bridges, coarser pores in the wall, and a simpler canal system, which does not connect to the chamber interior through retral processes. It resembles *Cribrononion* in having solid septal bridges and simple canal system but differs in the presence of an areal aperture in addition to the pore or

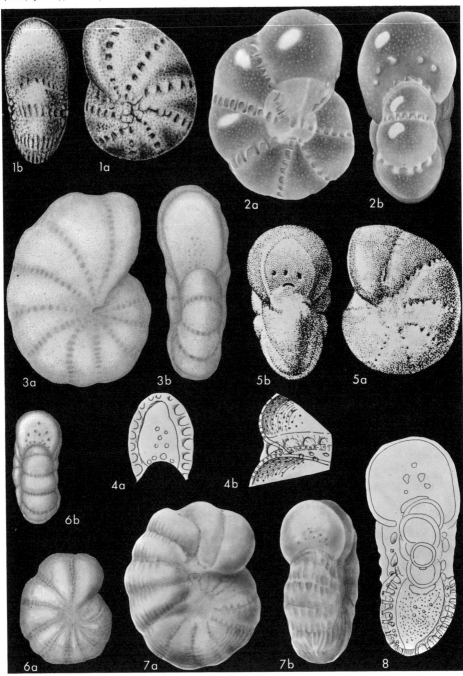

FIG. 508. Elphidiidae (Elphidiinae; *1-5, Cribroelphidium; 6-8, Elphidiella*) (p. C635-C639).

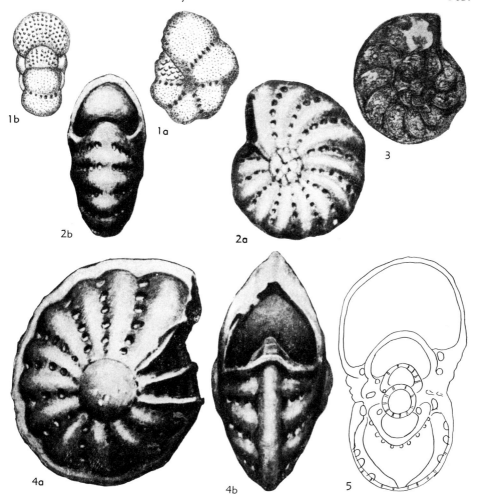

Fig. 509. Elphidiidae (Elphidiinae; *1-5, Cribrononion*) (p. C637-C638).

row of pores at base of the septal face. Some species previously placed in *Cribroelphidium* by reason of the presence of a multiple areal aperture belong to *Elphidium*, as shown by the presence of retral processes and a complex canal system, or to *Elphidiella*. It is regarded as belonging to the Elphidiidae, because of the canal system and radially built wall. A lectotype is here designated for *Polystomella poeyana* d'Orbigny, type-species of *Elphidiononion*. The lectotype (here redrawn) and paratypes (Recent, off Cuba) are in the d'Orbigny collection, Muséum Natl. Histoire Nat., Paris.]

Cribrononion Thalmann, 1947, *1899c, p. 312 [*Nonionina heteropora* Egger, 1857, *657, p. 300; OD] [=*Nonion (Cribrononion)* Thalmann, 1947, *1899c, p. 312 (obj.); *Canalifera* Krasheninnikov, 1953, *1050, p. 88 (type, *Elphidium eichwaldi*) (nom. nud.); *Canalifera (Canalifera)* Krasheninnikov, 1960, *1054, p. 59 (type, *Elphidium eichwaldi* Bogdanovich in Serova, 1955, *1719, p. 354); *Canalifera (Criptocanalifera)* Krasheninnikov, 1960, *1054, p. 60 (type, *C. (C.) clara*)]. Test planispiral, bilaterally symmetrical, involute, chambers simple; sutures ex-

cavated to open into intraseptal canal, connecting to spiral canal at each side in umbilical region, no retral processes, but solid and imperforate septal bridges may occur; wall calcareous, coarsely perforate, radial in structure; aperture a single opening or row of pores at base of apertural face, single slitlike foramen in earlier septa possibly due to later resorption. *Mio.-Rec.*, cosmop.——Fig. 509,*1*. **C. heteroporum* (Egger), Mio., Eu.(Bav.); *1a,b*, side, edge views, showing pustulose apertural face, not an areal aperture, ×60 (*700).——Fig. 509,*2,3*. *C. clarum* (Krasheninnikov), M.Mio.(U.Torton.), Eu.(Ukraine); *2a,b*, side and edge views of holotype showing arched slitlike foramen; *3*, sec., ×80 (*1054).——Fig. 509,*4*. *C. eichwaldi* (Bogdanovich), M.Mio.(U.Torton.), Eu.(Ukraine); *4a,b*, side, edge views, ×80 (*1054).——Fig. 509,*5*. *C. incertum* (Williamson), Rec., Arctic(Iceland); axial sec. showing

FIG. 510. Elphidiidae (Elphidiinae; *1,2, Laffitteina; 3,4, Ozawaia*) (p. C639-C640).

basal multiple foramen and canal system, enlarged (*946).

[Originally placed in the Nonionidae, this genus is here regarded as related to the Elphidiidae because of its canal system, radially built wall, and septal pores. These features are not characteristic of the Nonionidae. Although specimens of the type were not available for examination, other species here included (e.g., *Polystomella umbilicatula* var. *incerta* WILLIAMSON, 1858) have been found to be radially built, as are the Elphidiidae, and not granular, as in the Nonionidae. *Canalifera* was defined as having an aperture with a single row of pores and the subgenus *Criptocanalifera* as having a single arched basal slit. The single slit described in the type-species of the monotypic subgenus represented a foramen of an earlier septum, not a terminal aperture, and is probably due to resorption. *Cribrononion* also includes some species that have been placed in *Elphidiononion* (=*Cribroelphidium*). As here redefined, *Cribrononion* includes species with the shell morphology of *Cribroelphidium* but with basal aperture of one or more openings and without the multiple areal aperture of *Cribroelphidium*.]

Elphidiella CUSHMAN, 1936, *469, p. 89 [*Polystomella arctica* PARKER & JONES in BRADY, 1864, *186, p. 471; OD]. Test free, planispiral and involute, bilaterally symmetrical with equitant chambers commonly leaving axial umbilical plug

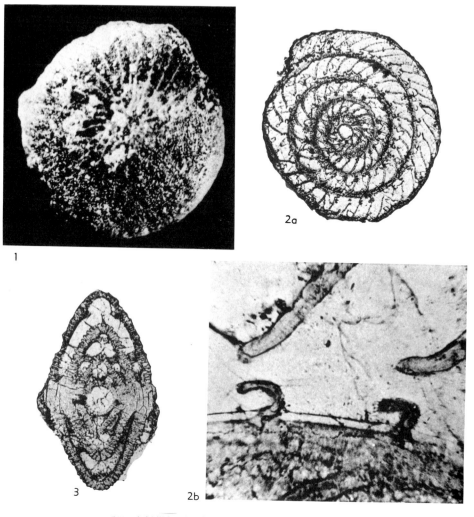

Fig. 511. Elphidiidae (Elphidiinae; *1-3, Pellatispirella*) (p. C640).

which lacks radial canals; sutures with openings to subsutural and vertical canals, generally forming double row of alternating pores along each radial suture, without retral processes, although striate surface ornamentation may be present, consisting of grooves originating at sutural pores and extending onto chamber walls; wall calcareous, radial and laminated in structure; aperture multiple, interioareal, consisting of scattered pores in apertural face. *Paleoc.(Dan.)-Rec.,* cosmop.——
Fig. 508,6-8. **E. arctica* (Parker & Jones), Rec., Arctic *(6),* Alaska *(7),* N.Atl. *(8); 6a,b,* side, apert. views, ×23 (*2117); *7a,b,* side, apert. views, ×26 (*1162); *8,* axial sec. showing septal canal, simple sutural septal foramen, and areal openings due to resorption, approx. ×50 (*946).

[*Elphidiella* differs from *Elphidium* in lacking retral processes. It differs from *Laffitteina* in being bilaterally symmetrical. Wade (1957, *2027) regarded the retral processes as unimportant and *Elphidiella* as a synonym of *Elphidium*. Smout (1955, *1804) considered these processes of family importance, and removed *Elphidiella* from the Elphidiidae. Ujiié (1956, *1964) regarded the double row of pores as a main criterion and included *Nautilus craticulatus* (the type-species of *Cellanthus*) in *Elphidiella*. If congeneric, the valid name for this group would then necessarily be *Cellanthus*. The 3 genera are here recognized on a similar basis to that used for grouping their type-species by Parker & Jones (1865, *1418, p. 400), *Polystomella arctica* (=*Elphidiella*) has a well-developed canal system, septal bridges, and apertural bars, but no retral processes; *P. craticulata* (=*Cellanthus*) was stated by Parker & Jones, 1865, to be characterized by a highly developed canal system, retral processes, septal bridges, and apertural bars, whereas *P. macella, P. strigillata* and *P. crispa* (=*Elphidium*) have retral processes, septal bridges, and apertural bars, but only a feebly developed canal system.]

Laffitteina Marie, 1946, *1217, p. 430 [**L. bibensis;* OD]. Test free, lenticular; chamber numerous, planispirally arranged, internally asymmetrical, with spiral lamella tending to adhere somewhat to one side of test; wall calcareous, septal walls double, with interseptal space enclosing part of canal system, which opens as double row of sutural pores and with vertical umbilical canals; aperture a basal peripheral slit. Paleoc.(Montian), Eu.(Fr.)-W.Afr.(Mauritania).——FIG. 510,1,2. *L. bibensis, Fr.; 1a,c, spiral side and edge of holotype showing double row of septal pores, ×21 (*2117); 1b, opposite side of holotype, showing numerous scattered pores in umbilical thickening, ×20 (*1217); 2a, equat. sec. showing chambers, interseptal canals and canals in supplementary shell material, ×26 (*1217); 2b, axial sec. of decorticated specimen showing asymmetrical chamber cavities and adherence of spiral lamella to lower side, canal system and apertures of chamber, ×22 (*1217).

[Laffitteina resembles Elphidiella in its double row of sutural pores, differing in internal asymmetry, and from asymmetrical Faujasina in its double row of sutural pores. Regarded as a synonym of Lockhartia by BERMÚDEZ (1952, *127), it differs from that genus in being planispiral and in having a sutural canal system and pores.]

Ozawaia CUSHMAN, 1931, *449, p. 80 [*O. tongaensis; OD]. Test similar to Elphidium in early stage, later uncoiling, with chambers becoming rounded in section, retral processes in both coiled and uncoiled stages; aperture in early stage series of pores at base of apertural face, as in Elphidium, cribrate in terminal face of adult. Rec., S.Pac.O.——FIG. 510,3,4. *O. tongaensis, Tonga Is.; 3a,b, side and edge views of young megalospheric specimen, ×95 (*2117); 4a,b, side and apert. views of microspheric holotype, ×95 (*2117).

Pellatispirella HANZAWA, 1937, *867, p. 114 [*Camerina matleyi VAUGHAN, 1929, *1991, p. 376; OD]. Test lenticular to compressed, 1-2.4 mm. diam., periphery rounded to subcarinate, biumbonate, with umbilical plugs, proloculus followed by numerous gradually enlarging planispiral and involute chambers, 20 to 40 in final whorl, no multilocular embryonic apparatus; septa may be slightly elevated and may bifurcate toward periphery; wall calcareous, finely perforate, surface smooth, septa double, walls solid, no marginal cord, umbilical plug perforated by vertical canals; primary aperture siphonate, equatorial and areal in position, with secondary smaller apertures at each side along base of septal face. M.Eoc., W. Indies-C.Am.——FIG. 511,1-3. *P. matleyi (VAUGHAN), W.Indies(Jamaica); 1, side view, ×20; 2a,b, equat. secs. showing character of siphonate primary aperture, ×20, ×230; 3, axial sec. of megalospheric specimen showing pectinate character of spiral lamella, ×40 (*362).

Protelphidium HAYNES, 1956, *887, p. 86 [*P. hofkeri; OD] [=Porosononion PUTRYA in VOLOSHINOVA, 1958, *2019, p. 135 (type, Nonionina subgranosa EGGER, 1857, *657, p. 299, =Nonionina tuberculata D'ORBIGNY, 1846, *1395, p. 108)]. Test planispiral and involute, similar to Nonion, but with perforate, radial wall structure, no sutural pores or retral processes but vertical canals pierc-

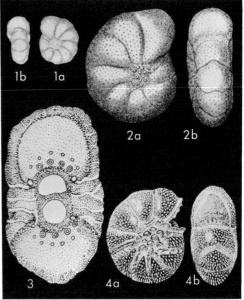

FIG. 512. Elphidiidae (Elphidiinae; 1-4, Protelphidium) (p. C640).

ing umbilical plug of secondary shell material; primary aperture not seen, possibly interiomarginal, secondary areal foramina and umbilical pores. Paleoc.-Plio., Eu.——FIG. 512,1. *P. hofkeri, Paleoc., Eng.; 1a,b, side, apert. views of holotype, ×50 (*887).——FIG. 512,2-4. P. tuberculatum (D'ORBIGNY), Mio., Aus. (2), Caucasus (3,4); 2a,b, side, edge views of topotype, ×90 (*473); 3, axial sec. showing cribrate foramina, spiral and vertical canals, ×150 (*2019); 4a,b, side, edge views, ×100 (*2019).

[Originally placed in the Nonionidae, it differs from these in its radially built, rather than granular perforate, wall. Porosononion was said to differ from Protelphidium in having a multiple aperture, but this is not present on the terminal face, only appearing in earlier septa by resorption. The type-species of Porosononion was also placed in Cribrononion by THALMANN (1947, *1899c) but that genus has sutural pores and canals.]

Subfamily FAUJASININAE Bermúdez, 1952

[Faujasininae BERMÚDEZ, 1952, p. 192] [=Notorotaliinae HORNIBROOK, 1961, p. 129]

Test trochospiral to planispiral, may have umbilical plug with anastomosing canals; sutural pores associated with well-developed sutural canal system; wall calcareous, surface with coalescing granules or narrow ribs connecting sutures; aperture of 1 or 2 rows or cluster of pores, near base of apertural face. M.Eoc.-Rec.

Faujasina D'ORBIGNY in DE LA SAGRA, 1839, *1611, p. 109 [*F. carinata; OD (M)] [=Faujassina TERQUEM, 1882, *1890, p. 48 (nom. van.)]. Test free, plano-convex, chambers numerous, low and broad, all visible on flat spiral side, only those of final whorl visible on convex umbilical side; spiral canal system well developed on umbilical side, rudimentary on spiral side, interseptal canals joining 2 spiral canals; sutures curved backward at periphery; wall calcareous, with regular, closely spaced, interseptal bars and grooves extending forward from sutures; aperture an interiomarginal row of pores. *Plio.*, Eu.-Japan.——FIG. 513,*1,2*;

FIG. 513. Elphidiidae (Faujasininae; *1,2, Faujasina; 3, Parrellina; 4-7, Polystomellina*) (p. C640-C643).

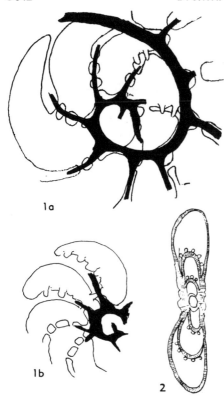

Fig. 514. Elphidiidae (Faujasininae; *1,2, Faujasina*) (p. C640-C642).

514,*1,2*. **F. carinata*, Plio., St. Erth, Eng.; 513, *1a-c*, opposite sides and edge view of hypotype, showing apert. pores, ×45 (*2117); 513,*2a-c*, opposite sides and edge view of lectotype (MNHN, Paris), here designated, stated (probably erroneously) to be from the Maastrichtian of Holland (Netherlands), ×48 (*2117); 514,*1a*, canal system of central portion of umbilical side, ×150; 514,*1b*, canal system of spiral side, ×350; 514,*2*, axial sec. showing foraminal pores, ×150 (*929).

[Originally reported to be from the Maastrichtian, this genus is undoubtedly solely a late Cenozoic one. Specimens identical to the original types occur in the Pliocene of St. Erth, as noted by CUSHMAN (1939, *473) and it is very probable that mislabeling or contamination of material resulted in the original "Maastrichtian" record of this genus. It was regarded as a post-Miocene genus by SMOUT (1955, *1804, p. 203). A lectotype is here designated and refigured; it and 4 paratypes are in the D'ORBIGNY collection, Muséum National d'Histoire Naturelle, Paris, labeled as from the Maastrichtian at Maastricht, Holland. It has not been reported since from this area or age. The figured hypotype is from the Pliocene of St. Erth and shows the apertural row of pores which are somewhat obscure in the original type-specimens.]

Parrellina THALMANN, 1951, *1899d, p. 224 [*nom. subst. pro Elphidioides* PARR, 1950, *1429, p. 373 (*non* CUSHMAN, 1945)] [**Polystomella imperatrix* BRADY, 1881, *196c, p. 66; OD]. Test free, bilaterally symmetrical, planispiral and involute; chambers numerous; sutures distinct, raised, with spiraling, irregular anastomosing ridges across chambers between sutures, costae roughly coinciding with retral processes; apertural face with vertical ridges extending up from its base; wall calcareous, perforate-radiate in structure, may be ornamented with few thick blunt peripheral spines, well-developed canal system with dendroid septal canals and diverging canals, septal pores small; aperture consisting of fine pores near base of apertural face, but may be obscured by ornamentation. [*Parrellina* is similar to *Polystomellina* in surface ornamentation, and to *Elphidium* in its symmetrical test, differing from both in the anastomosing canal system.] Oligo.-Rec., Tasm.-Australia (New S. Wales-Vict.).——FIG. 513,*3*. **P. imperatrix* (BRADY), Rec., Australia; *3a,b*, side, apert. views, ×37 (*2117).——FIG. 515,*1,2*. *P. craticulatiformis* WADE, L.Mio., S.Australia; *1*, nearly equat. sec. showing dendroid canal system; *2*, axial sec. showing radial and anastomosing umbilical canals, ×40 (*2027).

Polystomellina YABE & HANZAWA, 1923, *2089, p. 99 [**Polystomella (Polystomellina) discorbinoides*; OD (M)] [=*Polystomella (Polystomellina)* YABE & HANZAWA, 1923, *2089, p. 99 (obj.); *Notorotalia* FINLAY, 1939, *717a, p. 517 (type, *N. zelandica*)]. Test trochospiral, lenticular to planoconvex, periphery subangular to keeled, umbonal region of umbilical side with overlapping septal flaps or extensions of chambers, chambers with retral processes, intraseptal canal system, canals

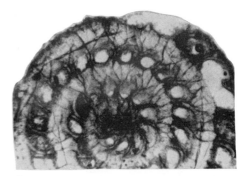

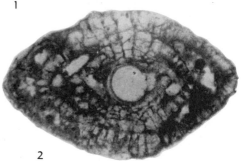

Fig. 515. Elphidiidae (Faujasininae; *1,2, Parrellina*) (p. C642).

Fig. 516. Elphidiidae (Faujasininae; *Polystomellina*) (p. C642-C643).

narrowing near surface to lead into diverging canals and tiny irregularly developed sutural pores which may occur on both sides of sutures, vertical umbilical canals also present; wall calcareous, perforate, surface typically ornamented with prominent spiraling or discontinuous ribs connecting elevated sutures; aperture 1 or 2 rows of pores near base of apertural face. *M.Eoc.-Rec.*, Japan-N.Z.-Australia-S.Am.-Antarct.——FIG. 513,4-6. **P. discorbinoides*, Plio., Japan; *4*, umbilical side, showing somewhat anastomosing ridges, and canaliculate umbonal plug, ×40 (*473); *5a-c*, opposite sides and edge view showing apert. pores, ×62 (*2117); *6*, axial sec. showing septal foramina and vertical umbilical and septal canals and pores, ×60 (*2089).——FIG. 513,7; 516. *P. zelandica* (FINLAY), M.Pleist., N.Z.; 513,*7a-c*, opposite sides and edge view of paratype showing characteristic ornamentation, ×33 (*2117); 516, axial sec. of topotype showing umbilical septal flaps, vertical canals, and apert. pores, ×60 (*959).

[*Polystomellina* was originally described as similar to *Faujasina* but with the umbilical side flattened and spiral side convex, and the original figures somewhat misleadingly suggested a conical test with septal pores but otherwise smooth surface. The type-species is from a limestone and preservation somewhat obscures the features, but when colored with a slight amount of dye a highly ornate surface is apparent, with ridges extending from suture to suture between septal pores on the spiral side and discontinuous and anastomosing ridges on the more flattened umbilical side, canal openings also appearing in the umbonal plug. The true characters of the species are better shown by CUSHMAN (*473, pl. 19, fig. 12a-c). CUSHMAN (1936, *469) also described 2 additional species of *Polystomellina*. FINLAY, 1939, defined *Notorotalia* for certain species previously placed in *Rotalia*, but also included CUSHMAN's 2 Australian species of *Polystomellina* (*469), stating that *Polystomellina* differed in being flat on the umbilical side rather than biconvex, and in having only porous radial sutures as ornament, lacking the characteristic reticulation of "*Notorotalia*." He did not comment on CUSHMAN's figures of the type-species of *Polystomellina* that correctly show the surface ornamentation. According to FINLAY, no visible aperture is present in *Notorotalia*, but HORNIBROOK (*959) showed that all species do have a row of small pores at the base of the apertural face, although these commonly are difficult to see. *Polystomellina* was stated to have a single low opening, but the type-species also shows a number of pores between strong ridges at the base of the apertural face, these also being more evident as foramina in earlier septa. Later workers have placed numerous species in *Notorotalia* but have left *Polystomellina* with only the type-species. As *Notorotalia* is identical in surface, apertural, and internal features, it is here suppressed as a junior synonym of *Polystomellina*. A lectotype for the previously unfigured *Notorotalia zelandica* was selected and illustrated by VELLA (1957, *2001).]

Porosorotalia VOLOSHINOVA, 1958, *2019, p. 167 [**Notorotalia clarki* VOLOSHINOVA in VOLOSHINOVA & DAIN, 1952, *2022, p. 56; OD] [=*Cribrorotalia* HORNIBROOK, 1961, *959, p. 138 (type, *Notorotalia tainuia* DORREEN, 1948, *610, p. 290)]. Test trochospiral, strongly biconvex, may have peripheral keel; 10 to 13 chambers in final whorl, retral processes present, prominent umbilical plug with labyrinthic canal system, internal spiral canal occurring on umbilical side and opening into sutural pores; wall calcareous, thick, lamellar, radially built, sculpture consisting of numerous granules especially well developed on umbilical side, where they may coalesce to form ribs or bars joining sutures; external aperture generally not visible, but may consist of row of very tiny pores, septal foramina of distinct row of openings at base of septum, enlarged by resorption. *Eoc.-Pleist.*, USSR(Sakhalin Is.)-N.Z.-N.Am.——FIG. 517,*1*. **P. clarki* (VOLOSHINOVA), Mio., Sakhalin Is.; *1a-c*, opposite sides and edge view of holotype showing septal foramina visible through broken final chamber, ×80 (*2019).——FIG. 517,*2,3*. *P. tainuia* (DORREEN), U.Eoc., N.Z.; *2a-c*, opposite sides and edge views of holotype lacking external aperture; *3*, edge view of paratype, final chamber broken, showing septal foramina; all ×93 (*610).——FIG. 517,*4*. *P. obesa* (HORNIBROOK), L.Mio., N.Z.; axial sec. showing canal system and foramina, ×40 (*959).

[*Porosorotalia* was originally placed by VOLOSHINOVA in the Cribroelphidiinae with *Cribroelphidium*, *Elphidiella*, and *Cellanthus*. *Cribrorotalia* was classed by HORNIBROOK in the new subfamily Notorotalinae [=Faujasininae] with *Notorotalia* (=*Polystomellina*), *Discorotalia* (=*Elphidium*), *Polystomellina*, *Faujasina*, and *Parrellina*. Both *Porosorotalia* and *Cribrorotalia* were independently separated from "*Notorotalia*" on the basis of their granular, rather than costate, ornamentation, differences in the canal system, and well-developed umbilical plug in *Porosorotalia*. Both authors included DORREEN's (*610) species in their new genera.]

NUMMULITIDAE
By W. STORRS COLE
[Cornell University]

Family NUMMULITIDAE de Blainville, 1825

[*nom. correct.* EIMER & FICKERT, 1899, p. 706 (*pro* family Nummulacea DE BLAINVILLE, 1825, p. 372)]—[All cited names are of family rank; dagger(†) indicates *partim*]——[=Nummulacés DE BLAINVILLE, 1825, p. 372 (*nom. neg.*); =Nummulitidea REUSS, 1862, p. 308; =Nummulinida CARPENTER, PARKER & JONES, 1862, p. 238; =Nummulitideae GÜMBEL, 1870, p. 84; =Nummuliti SCHWAGER, 1876, p. 477; =Nummulinidae SCHULZE, 1877, p. 29; =Nummulitina LANKESTER, 1885, p. 848; =Nummulinetta HAECKEL, 1894, p. 164; =Nummulitinae DELAGE & HÉROUARD, 1896, p. 152; =Nummulariidae WEDEKIND, 1937, p. 111; =Nummulitidos GADEA BUISÁN, 1947, p. 18 (*nom. neg.*); =Nummulitida HAECKEL, 1894, p. 185 (*nom. van.*)]——[=Polythalamat LATREILLE, 1825, p. 161 (*nom. nud.*); =Enthomostèguest D'ORBIGNY, 1826, p. 304 (*nom. nud.*; *nom. neg.*); =Helicosorinat EHRENBERG, 1839, table opp. p. 120 (*nom. nud.*); =Velellidaet AGASSIZ, 1844, p. 5 (*nom. nud.*); =Camerinidae MEEK & HAYDEN, 1865, p. 11]——[=Cycloclypeina LANKESTER, 1885, p. 848; =Cycloclypeida HAECKEL, 1894, p. 185; =Cycloclypeinae DELAGE & HÉROUARD, 1896, p. 152; =Cycloclypeidae GALLOWAY, 1933, p. 441] [Editor's Note—The author of this section has agreed here to use Nummulitidae and *Nummulites* in order to conform with editorial policy of the *Treatise* in accepting names legally fixed by ICZN.]

Test normally planispiral, but one terminal genus with annular ephebic cham-

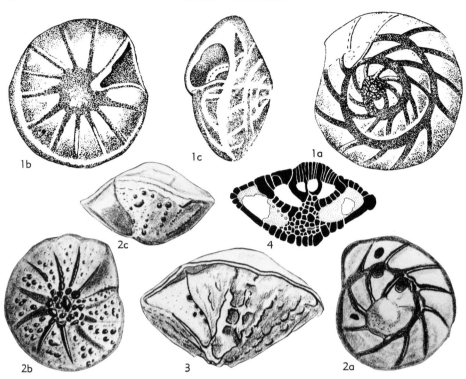

Fig. 517. Elphidiidae (Faujasininae; *1-4, Porosorotalia*) (p. C643).

bers; slightly asymmetrical to bilaterally symmetrical; involute or evolute; median chambers numerous, simple, or subdivided into chamberlets; with or without lateral chambers; complex canal system consisting of septal, marginal, and vertical canals; aperture typically an arched slit at the base of the septa. *U.Cret.-Rec.*

The nummulitids can be divided into 4 kinds on the development of the median layer: (1) those with planispirally coiled, simple chambers (*Nummulites,* Fig. 518,*1*); (2) those with planispirally coiled chambers subdivided into chamberlets (*Heterostegina,* Fig. 518,*4*); (3) those with planispirally coiled initial chambers associated with annular lateral chambers subdivided into chamberlets (*Cycloclypeus,* Fig. 518, *2*); and (4) those with a double median layer in the peripheral part of the test (*Biplanispira,* Fig. 518,*3*).

Transverse sections show additional features useful in generic and specific classification. In involute tests (Fig. 518,*1*) the chamber cavities extend to the axis of the test, producing elongate, V-shaped cavities (alar prolongations), whereas in evolute tests these prolongations do not appear (see Fig. 520,*3*). Lateral chambers may be present (*Spiroclypeus,* Fig. 518,*6*) or absent (*Heterostegina,* Fig. 518,*4*) in genera which have similar median sections. One genus (*Cycloclypeus,* Fig. 518,*2*) has the median layer covered on each side by walls made of laminellae, but others (*Pellatispira,* Fig. 518,*5*) have walls composed of coarse pillars between which numerous large vertical canals occur.

In the past, many generic names have been erected for nummulitids with undivided median chambers, based on the assumption that the type of coiling (involute or evolute), number of the coils, height of the coils, shape of the chambers, character of the spiral wall, and strength of the marginal cord are structures of constant nature within groups of species and accordingly usable for defining genera. These structures can be used to distinguish species from one another, even though they vary within limits between specimens of the same species. Thus, the structures mentioned are char-

Foraminiferida—Rotaliina—Rotaliacea C645

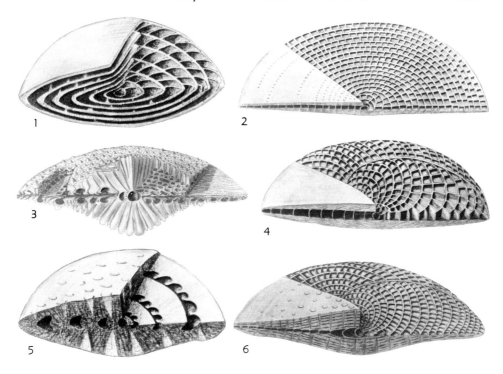

FIG. 518. Nummulitidae. Oblique views of representative genera showing internal structures revealed by transverse and equatorial sections (diagrammatic, not to scale).——*1. Nummulites.*——*2. Cycloclypeus.*——*3. Biplanispira.*——*4. Heterostegina.*——*5. Pellatispira.*——*6. Spiroclypeus* (*3,* *1969; others, *2121).

acteristic of species, not genera.

Sulcoperculina (U.Cret.) presumably was derived from a rotaliid ancestor, and, in turn it generated *Miscellanea* and *Nummulites.* The heterostegine kind of test was developed by subdivision of the median chambers into chamberlets. The more advanced *Spiroclypeus* has lateral chambers, although in median section it is identical with *Heterostegina.* The most advanced genus, *Cycloclypeus,* has an initial heterostegine stage and undoubtedly had a heterostegine ancestor.

In most species great size differentiation is observed between relatively small specimens of the megalospheric generation and specimens many times larger which represent the microspheric generation.

Subfamily NUMMULITINAE de Blainville, 1825

[*nom. transl.* BRADY, 1881, p. 44 (*ex* family Nummulacea DE BLAINVILLE, 1825)] [All cited names are of subfamily rank] [=Nummulinina JONES in GRIFFITH & HENFREY, 1875, p. 320; =Nummulitidae SCHWAGER, 1877, p. 19; =Camerininae CUSHMAN, 1928, p. 209; =Assilininae PURI, 1957, p. 97]

Median chambers numerous, simple, but in one genus occurring in double peripheral layer; without distinct lateral chambers, but vacuoles may develop in wall of spiral sheet. *U.Cret.-Rec.*

Nummulites LAMARCK, 1801, *1084, p. 101 [validated by ICZN under plenary powers (Opinion 192, 1945, p. 154)] [**Camerina laevigata* BRUGUIÈRE, 1792, *247, p. 399; SD ICZN, 1945] [=*Helicites* GESNER, 1758 (non-Linnean); *Camerina* BRUGUIÈRE, 1792, *247, p. 395 (type, *C. laevigata*); *Phacites* BLUMENBACH, 1799, *150a, pl. 40 (type, *P. fossilis*); *Lycophris* DE MONTFORT, 1808, *1305, p. 159 (type, *Lycophris lenticularis*); *Egeon* DE MONTFORT, 1808, *1305, p. 167 (type, *E. perforatus*); *Helicites* DE BLAINVILLE, 1824, *141a, p. 179 (type, not designated); *Nummulina* D'ORBIGNY, 1826, *1391, p. 295-296 (obj.); *Nummularia* SOWERBY & SOWERBY, 1826, *1820, p. 73 (obj.); *Operculina* D'ORBIGNY, 1826, *1391, p. 281 (type, *Lenticulites complanatus* DEFRANCE, 1822, *579c, p. 453); *Nummulita* FLEMING, 1828, *722, p. 233 (obj.); *Assilina* D'ORBIGNY in DE LA SAGRA, 1839, *1611, p. 48 (type, *Nummulites spira* DE ROISSY, 1805, *1584, p. 57; *Discospira* MORRIS in MANTELL, 1850, *1213, p. 142 (type, *Discospira* sp., =*Nummulites complanata* LAMARCK, 1804, *1085b, p. 242); *Monetulites* EHRENBERG, 1855, *681, p. 289 (type, not designated); *Cumerina*

C646 *Protista—Sarcodina*

SCUDDER, 1882, *1709a, p. 93 (*nom. null. pro Camerina* BRUGUIÈRE, 1792); *Discospora* SHERBORN, 1893, *1731a, p. 102 (*nom. null. pro Discospira* MORRIS, 1850); *Frilla* DE GREGORIO, 1894, *816A, p. 10 (type, *Operculina ammonea* LEYMERIE, 1846, *1132A, p. 359); *Gümbelia* PREVER, 1902, *1481,

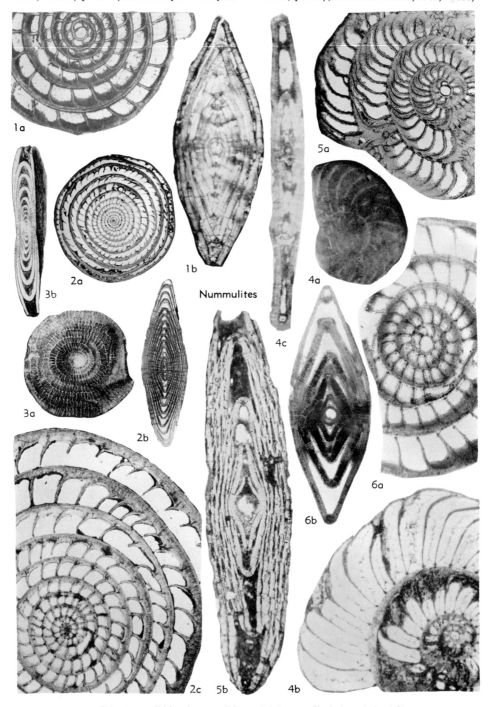

FIG. 519. Nummulitidae (Nummulitinae; *1-6, Nummulites*) (p. C645-C647).

p. 11 (type, *Nautilus lenticularis* FICHTEL & MOLL, 1798); *Bruguieria* PREVER, 1902, *1481, p. 11 (obj.); *Hantkenia* PREVER, 1902, *1481, p. 11 (*non* MUNIER-CHALMAS in FISCHER, 1885) (type, *Nummulites tchihatcheffi* D'ARCHIAC & HAIME, 1853, *38, p. 98; OD) (*non Nummulites complanata* LAMARCK, 1804, *1085b, p. 242, invalidly designated by GALLOWAY, *762, p. 416, because species not in originally included list assigned to genus and because type-species was otherwise fixed by OD); *Laharpia* PREVER, 1902, *1481, p. 11 (type, *Camerina tuberculata* BRUGUIÈRE, 1792, *247, p. 400); *Paronaea* PREVER, 1903, *1481A, p. 461 (type, *Nummulites tchihatcheffi* D'ARCHIAC & HAIME, 1853); *Paronia* PREVER in CHELUSSI, 1903, *330A, p. 74, =*Hantkenia* PREVER, 1902, obj. (*non Paronia* DIAMARE, 1900); *Verbeekia* A. SILVESTRI, 1908, *1770, p. 137 (type, *Amphistegina cumingii* CARPENTER, 1860, *271b, p. 32) (*non Verbeekia* FRITSCH, 1877, p. 90); *Palaeonummulites* SCHUBERT, 1908, *1686, p. 378 (type, *Nummulina pristina* BRADY, 1874, *191, p. 225); *Operculinella* YABE, 1908, *2084, p. 126 (type, *Amphistegina cumingii* CARPENTER, 1860, *271b, p. 32); *Operculinoides* HANZAWA, 1935, *866, p. 18 (type, *Nummulites willcoxi* HEILPRIN, 1883, *893, p. 191); *Pseudonummulites* A. SILVESTRI, 1937, *1787, p. 149 (type, *Amphistegina cumingii* CARPENTER, 1860, *271b, p. 32); *Paraspirocypeus* HANZAWA, 1937, *867, p. 116 (type, *Camerina chawneri* PALMER, 1934, *1408, p. 261); *Ranikothalia* CAUDRI, 1944, *304, p. 367 (type, *Nummulites nuttalli* DAVIES, 1927, *559, p. 266); *Nummulitoides* ABRARD, 1956, *2A, p. 489 (type, *Operculina (N.) tessieri*); *Planocamerinoides* COLE, 1957, *365, p. 262 (type, *Nummularia exponens* J. DE SOWERBY in SYKES, 1840, *1860, p. 719); *Eoassilina* SINGH, 1957, *1793A, p. 210 (type, *E. elliptica*); *Neooperculinoides* GOLEV, 1961, *807, p. 114 (type, *Nautilus ammonoides* GRONOVIUS, 1781, *828, p. 282)]. Test involute to evolute, spiral sheet with or without vacuoles. *Paleoc.-Rec.*, cosmop., trop.——FIG. 519,*2*. **N.* laevigata* (BRUGUIÈRE), Eoc., Eu.(Fr.); *2a,b,* med. and transv. secs., of microspheric specimens, ×3; *2c,* part of med. sec., ×12.5 (*2113c).——FIG. 519,*1*. *N. striatoreticulata* (RUTTEN), U.Eoc., C.Am.(Panama); *1a,b,* med. and transv. secs., ×12.5 (*2113c).——FIG. 519,*3*. *N. exponens* (SOWERBY), Eoc., Asia(India); *3a,* ext. view, ×1; *3b,* transv. sec., ×1.5 (*1860).——FIG. 519,*4*. *N. complanatus* (DEFRANCE), Mio., Japan; *4a,* ext. view, ×10; *4b,c,* med. and transv. secs., ×20 (*367).——FIG. 519,*6*. *N. willcoxi* (HEILPRIN), Eoc., N.Am.(Fla.); *6a,b,* med. and transv. secs., ×20 (*2113c).——FIG. 519,*5*. *N. chawneri* (D. K. PALMER), Mio., W.Indies(Cuba); *5a,b,* med. and transv. secs., ×20 (*365).

Biplanispira UMBGROVE, 1937, *1970, p. 309 [*nom. subst. pro Heterospira* UMBGROVE, 1936, *1969, p. 156 (*non* KOKEN, 1896)] [**Heterospira mirabilis* UMBGROVE, 1936, *1969, p. 157; OD]. Median layer single except in the wide peripheral flange where a double row of chambers occurs; covering layers thick, perforate. *Eoc.*, Indo-Pac.Reg.——FIG. 520,*1*. **B. mirabilis* (UMBGROVE), U.Eoc., Saipan Is.; *1a,b,* med. and transv. secs., ×20 (*2113c).

Miscellanea PFENDER, 1935, *1451, p. 230 [**Nummulites miscella* D'ARCHIAC & HAIME, 1854, *38, p. 345; OD] [=*Miscellanea* PFENDER, 1934, *1450, p. 80 (*nom. nud.*)]. Like *Nummulites* but with a coarsely perforate spiral sheet composed of closely spaced pillars. *Paleoc.*, Eu.-Asia (India).——FIG. 520,*3*. **M. miscella* (D'ARCHIAC & HAIME), India; *3a,* ext. view, ×10; *3b,* med. sec., ×12.5; *3c,* transv. sec., ×20 (*362).

Pellatispira BOUSSAC, 1906, *178, p. 91 [**P. douvillei* (=*Nummulites madaraszi* HANTKEN, 1876, *863, p. 75; OD] [=*Vacuolispira* TAN, 1936, *1869, p. 177 (type, *Pellatispira inflata* UMBGROVE, 1928, *1967, p. 63)]. Median layer single, composed of a loose coil of chambers separated by canaliferous shell material; covering walls thick, coarsely perforate. *Eoc.*, Eu.-Indo-Pac.Reg.——FIG. 520,*2*. **P. madaraszi* (HANTKEN), Italy; *2a,* ext. view, ×7; *2b,c,* med., transv. secs., ×11 (*178).

Sulcoperculina THALMANN, 1939, *1899b, p. 330 [**Camerina(?) dickersoni* PALMER, 1934, *1408, p. 243; OD]. Test involute, slightly asymmetrical, peripheral margin with radial plates forming a peripheral sulcus. *U.Cret.*, trop. Am.——FIG. 520, *4*. **S. dickersoni* (PALMER), Cuba; *4a,* ext. view, ×40 (*1408); *4b,c,* med., transv. secs., ×40 (*2113a); *4d,* schematic internal structure (*2111).

Subfamily CYCLOCLYPEINAE Bütschli, 1880

[*nom. correct.* BRADY, 1884, p. 76 (*pro* subfam. Cycloclypidae BÜTSCHLI in BRONN, 1880, p. 215] [=Cycloclypeina CALKINS, 1901, p. 109; Heteroclypeinae SCHUBERT, 1906, p. 640; Heterostegininae GALLOWAY, 1933, p. 421]

Median chambers subdivided into chamberlets; without lateral chambers, or with distinct lateral chambers, or with median layer covered by laminellated walls or by coarse pillars between which numerous large vertical canals occur. *Eoc.-Rec.*

Cycloclypeus W. B. CARPENTER, 1856, *271a, p. 555 [**C. mammilatus* CARTER, 1861, *287a, p. 461; SD (SM) CARTER, 1861, *287a, p. 461] [=*Heteroclypeus* SCHUBERT, 1906, *1683, p. 640 (type, *Heterostegina cycloclypeus* A. SILVESTRI, 1905, *1761, p. 126) (*non Heteroclypeus* COTTEAU, 1895]. Microspheric generation initially like *Heterostegina*; megalospheric generation initially with bilocular embryonic chambers followed by heterostegine-like periembryonic chambers; later chambers in both generations annular, divided into rectangular chamberlets. [CARPENTER validly described *Cycloclypeus* in 1856 but included it in no named species. When CARTER (1861) referred his new species *C. mammilatus* to *Cycloclypeus*, this was the first specific taxon assigned and thus

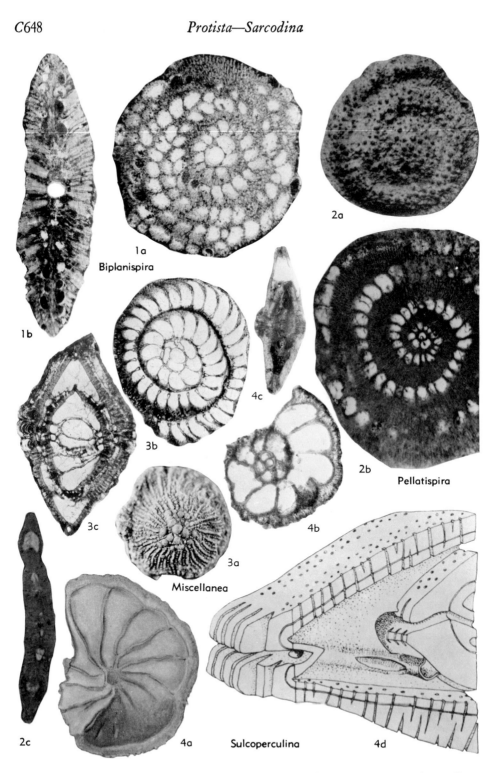

Fig. 520. Nummulitidae (Nummulitinae; *1, Biplanispira; 2, Pellatispira; 3, Miscellanea; 4, Sulcoperculina*) (p. *C647*).

Foraminiferida—Rotaliina—Rotaliacea C649

automatically was established as type-species.—Ed.] *Eoc.-Rec.*

C. (Cycloclypeus). Test circular without rays or marked concentric annular inflations. *Eoc.-Rec., Eu.-Indo-Pac.Reg.*——FIG. 521,*1*. *C. (C.) *car-penteri*, Rec., Bikini Atoll; *1a*, med. sec., ×32; *1b*, transv. sec., ×16 (*2113c).

C. (Katacycloclypeus) TAN, 1932, *1864, p. 39 [*C. *annulatus* MARTIN, 1880, *1229, p. 157; OD]. Test with marked concentric annular in-

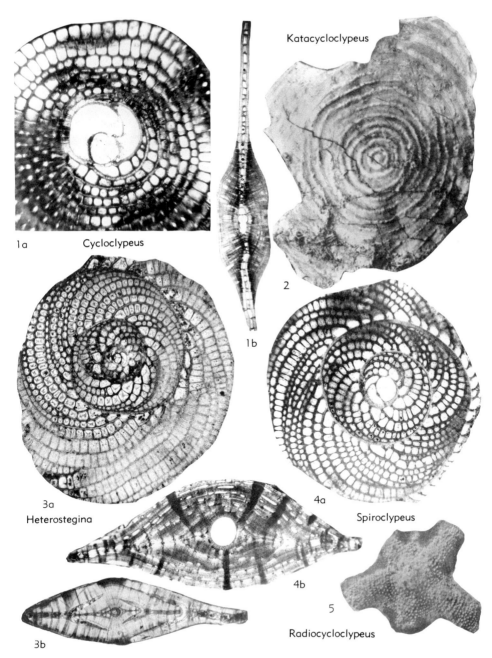

FIG. 521. Nummulitidae (Cycloclypeinae; *1*, *Cycloclypeus (Cycloclypeus)*; *2*, *C. (Katacycloclypeus)*; *5*, *C. (Radiocycloclypeus)*; *3*, *Heterostegina*; *4*, *Spiroclypeus*) (p. C647-C650).

flations. L.Mio., Indo-Pac.Reg.——Fig. 521,2. *C. (K.) annulatus Martin, Mio., Fiji(Lau Is.); ext. view, ×0.8 (*2116).

C. (Radiocycloclypeus) Tan, 1932, *1864, p. 39, 92 [*C. (R.) stellatus; OD]. Test with radiating rays. Mio., Indo-Pac.Reg.——Fig. 521,5. *C. (R.) stellatus, Malay Arch.(Borneo); ext. view, ×6.5 (*2033).

Heterostegina d'Orbigny, 1826, *1391, p. 304, 305 [*H. depressa; SD Parker, Jones & Brady, 1865, *1419, p. 36] [=?Heterosteginella Silvestri, 1937, *1787, p. 118 (nom. nud., type not designated); Grzybowskia Bieda, 1950, *137A, p. 167 (type, G. multifida)]. Like Nummulites but later chambers divided into rectangular chamberlets. Eoc.-Rec., cosmop., trop.——Fig. 521,3. H. antillea Cushman, U.Oligo., C.Am.(Panama); 3a,b, med., transv. secs., ×16 (*363).

Spiroclypeus Douvillé, 1905, *615, p. 458 [*S. orbitoideus; OD]. Like Heterostegina but lateral chambers developed on each side of the median layer. Eoc.-L.Mio., Indo-Pac.Reg.-Eu.; Oligo., W. Indies.——Fig. 521,4. S. tidoenganensis van der Vlerk, L.Mio., Saipan Is.; 4a,b, med. and transv. secs., ×16 (*2113c).

MIOGYPSINIDAE
By W. Storrs Cole
[Cornell University]

Family MIOGYPSINIDAE Vaughan, 1928

[nom. transl. Tan, 1936, p. 45 (ex Miogypsininae Vaughan in Cushman, 1928, p. 354)] [=Miogypsinoidinae Hanzawa, 1947, p. 262]

Test trigonal, suborbicular, or digitate, evenly or modified lenticular, composed of equatorial layer either with well-developed lateral chambers or with appressed laminae on each side; megalospheric generation with bilocular embryonic chambers with rude or well-developed spire of periembryonic chambers, situated apically, subapically, or subcentrally; microspheric generation with spire of chambers situated apically; spiral canal and intraseptal canal present; well-developed arcuate, rhombic, or elongate-hexagonal equatorial chambers interconnected by stolons (Fig. 522,1). M.Oligo.-L.Mio.

The primary differentiation of migypsinid genera is based on characteristics shown by vertical sections. Miogypsinoides lacks lateral chambers (Fig. 522,2), the equatorial layer being covered on each side by zones of appressed laminae, whereas Miogypsina has well-developed lateral chambers on each side of the equatorial layer (Fig. 522,1).

Miogypsina is subdivided into two subgenera: M.(Miogypsina) and M.(Miolepidocyclina). Inasmuch as their recognition is based on the position of the embryonic apparatus, oriented equatorial sections are needed. In M.(Miogypsina) the embryonic apparatus is apically situated, so that either the embryonic chambers or the periembryonic chambers are in contact with the peripheral zone of the test. In M.(Miolepidocyclina) the embryonic apparatus is separated from the peripheral zone of the test by one or more rows of equatorial chambers. Specific determinations within the subgenera are based on arrangements of the periembryonic chambers in relation to the embryonic chambers, and secondarily on shape of the equatorial chambers and characteristics of the lateral chambers.

Present evidence indicates that the miogypsinids are specialized, short-ranged descendants of some type of rotaliid. Some investigators place the miogypsinids in the Rotaliidae (*85).

Miogypsina Sacco, 1893, *1607, p. 205 [*Nummulites globulina Michelotti, 1841, *1256, p. 297; OD] [=Flabelliporus Dervieux, 1894, *588, p. 59 (type, F. dilatatus=Nummulites globulina Michelotti, 1841; SD herein); Lepidosemicyclina Rutten, 1911, *1596, p. 1135 (type, L. thecideaeformis; SD herein); Miogypsinopsis Hanzawa, 1940, *869, p. 773 (type, Miogypsina gunteri Cole, 1938, *356, p. 42)]. Lateral chambers present, well developed. U.Oligo.-L.Mio.

M. (Miogypsina). Megalospheric embryonic apparatus apically situated, without equatorial chambers between it and marginal fringe. U. Oligo.-L. Mio., Eu.-Indo-Pac. Reg.-N. Am.-S. Am. ——Fig. 522,3. M. (M.) antillea (Cushman), Oligo., C.Am.(Panama); 3a,b, equat., vert. secs., ×40 (*363).

M. (Miolepidocyclina) A. Silvestri, 1907, *1766, p. 80 [*Orbitoides (Lepidocyclina) burdigalensis Gümbel, 1870, *840, p. 719; OD] [=Heterosteginoides Cushman, 1918, *410, p. 97 (type, H. panamensis); Miogypsinita Drooger, 1952, *630, p. 58 (type, Miogypsina mexicana Nuttall, 1933, *1372, p. 175)]. Megalospheric embryonic apparatus subapically to subcentrally situated with normal equatorial chambers between it and marginal fringe. U.Oligo.-L.Mio., Eu.-Afr.-Indo-Pac. Reg.-N.Am.-S.Am.——Fig. 522,4. *M. (M.) burdigalensis (Gümbel), Burdigal., N. Afr.(Morocco); 4a, equat. sec., ×25 (*215); 4b, vert. sec., ×48 (*215).

Miogypsinoides Yabe & Hanzawa, 1928, *2092, p. 535 [*Miogypsina dehaartii van der Vlerk, 1924, *2012, p. 429; OD] [=Conomiogypsinoides Tan, 1936, *1866, p. 51 (type, Miogypsina abunensis Tobler, 1927, *1938, p. 328); Miogypsinella Hanzawa, 1940, *869, p. 765, 770, 775 (type,

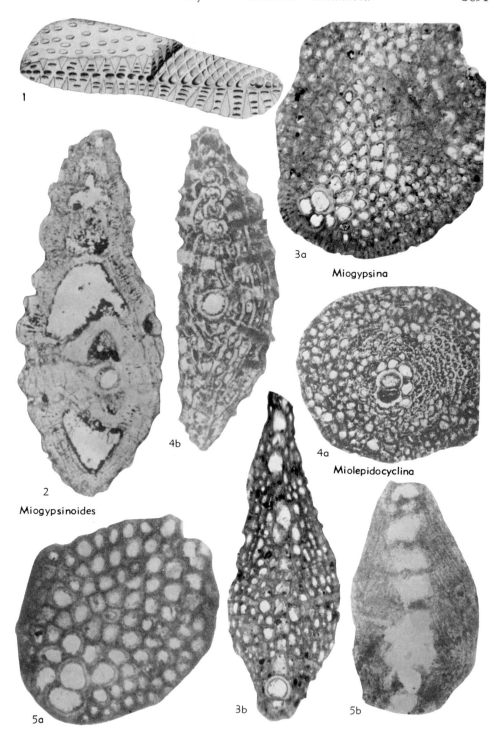

Fig. 522. Miogypsinidae; *1*, Diagrammatic illustration showing structure; *2,5, Miogypsinoides; 3, Miogypsina (Miogypsina); 4, M. (Miolepidocyclina)* (p. C650-C652).

M. borodinensis)]. Lateral chambers absent. *M. Oligo.-L.Mio.*, Eu.-Indo-Pac.Reg.-N.Am.——Fig. 522,5. **M. dehaartii* (VAN DER VLERK), L.Mio., Indonesia (Moluccas Is.); *5a,b*, equat., vert. secs., ×40 (*2012). [See also Fig. 522,*2*, *M. complanata* (SCHLUMBERGER), ×90.]

Superfamily GLOBIGERINACEA Carpenter, Parker & Jones, 1862

[*nom. correct.* LOEBLICH & TAPPAN, 1961, p. 307 (*pro* superfamily Globigerinidea MOROZOVA, 1957, p. 1110, and superfamily Globigerinaceae BANNER & BLOW, 1959, p. 4)]——[All cited names of superfamily rank; dagger(†) indicates *partim*]——[=Orthoklinostegia† EIMER & FICKERT, 1899, p. 685 (*nom. nud.*); =Bilamellidea† REISS, 1957, p. 127 (*nom. nud.*)]

Test enrolled, planispiral or trochospiral or modified from such; chambers basically globular, later may be compressed or variously modified; double walls of lamellar radial hyaline calcite, distinctly perforate, may have canaliculate keels; aperture primarily formed, interiomarginal, or may be modified to become areal or terminal, single, or more rarely multiple, and may have secondary or accessory openings, may have apertural lips. [Habit planktonic, with resultant modifications including fine elongate spines which support the frothy areolated ectoplasm.] *M.Jur.-Rec.*

Family HETEROHELICIDAE Cushman, 1927

[Heterohelicidae CUSHMAN, 1927, p. 59] [=Gümbelinidae WEDEKIND, 1937, p. 112; =Heterohelicida COPELAND, 1956, p. 188 (*nom. van.*)]

Early stage trochospiral, planispiral, biserial or triserial, later may show serial reductions or proliferations; aperture large, simple and interiomarginal, or terminal in uniserial forms, without internal columellar processes. *M.Jur.-Oligo.*

Subfamily GUEMBELITRIINAE Montanaro Gallitelli, 1957

[Gümbelitriinae MONTANARO GALLITELLI, 1957, p. 136]

Primitively trochospiral, triserial or quadriserial, later may develop proliferation of chambers; aperture simple. *M.Jur.-Eoc.*

Guembelitria CUSHMAN, 1933, *458, p. 37 [**G. cretacea*; OD] [=*Gümbelitria* CUSHMAN, 1933, *458, p. 37 (obj.)]. Test triserial throughout; chambers inflated, globular; sutures distinct, depressed; aperture an interiomarginal arch at base of last-formed chamber. *L.Cret.-Eoc.*, cosmop.——Fig. 523,*1*. **G. cretacea*, U.Cret., USA(Tex.); *1a,b*, side, top views of holotype, ×312 (*1303).

Gubkinella SULEYMANOV, 1955, *1852, p. 623 [**G. asiatica*; OD] [=*Globigerina (Conoglobigerina)* MOROZOVA in MOROZOVA & MOSKALENKO, 1961, *1318, p. 24 (type, *G. (C.) dagestanica*)]. Test free, high trochospiral; in type-species with 4 inflated chambers to whorl; aperture a low interiomarginal arch. *M.Jur.(Bajoc.-Callov.) - U.Cret.(Senon.)*, USSR-Eu.-N. Am.-W. Indies(Trinidad).——Fig. 524,*1*. **G. asiatica*, U.Cret.(Senon.), USSR(Kyzyl-Kumy); *1a,b*, side, top views, ×200 (*1852).——Fig. 524,*2*. *G. dagestanica* (MOROZOVA), M.Jur.(L.Bathon.), USSR(Dagestan); *2a-c*, holotype, ×100 (*1318).

[The original description of the genus indicated it to be quadriserial, belonging to the Heterohelicidae. *Globigerina graysonensis* TAPPAN, from the Albian-Cenomanian of North America seems most probably congeneric, having a similar high-spired test, low aperture, and 4 chambers in the final whorl, or as many as 5 or as few as 3 chambers in each whorl. N. K. BYKOVA, VASILENKO, VOLOSHINOVA, MYATLYUK & SUBBOTINA in RAUZER-CHERNOUSOVA & FURSENKO (1959, *1509, p. 267, 268) transferred *Gubkinella* to the family Discorbidae, subfamily Discorbinae, and illustrated a specimen of the type-species showing up to 5 chambers in an early whorl. Because of the extremely inflated chambers, the widespread occurrence of some of the species, and its association, we believe this genus to be planktonic in habit. The subgenus *Conoglobigerina* was recently described for an apparently congeneric high trochospiral Jurassic species.]

Guembelitriella TAPPAN, 1940, *1871, p. 115 [**G. graysonensis*; OD]. Test free, small, triserial in early stage, similar to *Guembelitria*, later becoming multiserial at top; chambers globular, increasing rapidly in size; sutures distinct, depressed; wall calcareous, finely perforate; aperture an interiomarginal arch at base of final chamber, rarely more than one. *U.Cret.(Cenoman.)*, USA.——Fig. 523,*2,3*. **G. graysonensis*; *2a,b*, side and top views of holotype; *3*, side view of paratype showing multiple apertures in final chamber, ×174 (*1303).

Woodringina LOEBLICH & TAPPAN, 1957, *1169, p. 39 [**W. claytonensis*; OD]. Test free, early stage with single whorl of 3 chambers, followed by biserial stage; chambers inflated; wall calcareous, radial in structure, finely perforate; aperture a low arched slit, bordered above by slight lip. [Differs from *Tosaia* in having a much-reduced early coil consisting of a single whorl of 3 chambers, whereas *Tosaia* has an early trochoid stage followed by a triserial and finally a reduced biserial stage.] *Paleoc.(Dan.)*, USA(Ala.).——Fig. 523,*4*. **W. claytonensis*; *4a,b*, holotype, opposite sides, ×187; *4c,d*, edge, and basal views, ×187 (*1169).

Subfamily HETEROHELICINAE Cushman, 1927

[*nom. subst.* CUSHMAN, 1927, p. 59 (*pro* Spiroplectinae CUSHMAN, 1911, p. 4)] [=Gümbelininae CUSHMAN, 1927, p. 59]

Early stage planispiral or biserial, later may develop chamber proliferation or serial reduction; aperture simple and interiomarginal, or terminal in uniserial forms. *L. Cret. Oligo.*

Heterohelix EHRENBERG, 1843, *672, p. 429 [**Spiroplecta americana* EHRENBERG, 1844, *673, p. 75; SD (SM) EHRENBERG, 1844, *673, p. 75] [=*Spiroplecta* EHRENBERG, 1844, *673, p. 75 (obj.); *Gümbelina* EGGER, 1899, *659, p. 31 (type, *Textularia globulosa* EHRENBERG, 1840) (non

Gümbelina KUNTZE, 1895)]. Test small, consisting of subglobular biserially arranged chambers, early portion of microspheric test commonly planispiral; surface smooth or striate; aperture large, interiomarginal, symmetrical. *L.Cret.(Apt.)-U.Cret.(Maastricht.)*, cosmop.——FIG. 523,5. *H.

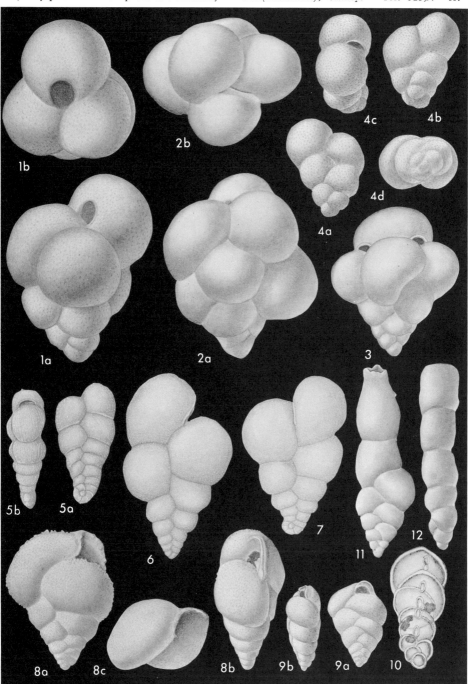

FIG. 523. Heterohelicidae (Guembelitriinae; *1, Guembelitria; 2,3, Guembelitriella; 4, Woodringina;* Heterohelicinae; *5-7, Heterohelix; 8,9, Chiloguembelina; 10-12, Bifarina*) (p. C652-C654).

americana (EHRENBERG), U.Cret.(L.Maastricht.), USA(Tex.); *5a,b,* side, edge views (holotype of *H. navarroensis*), ×146 (*1150).——FIG. 523, *6,7. H. globulosa* (EHRENBERG), U.Cret. (Maastricht.), USA(Tex.); *6,7,* megalospheric and microspheric tests, ×105 (*1150).

[The synonymic status of *Guembelina* and *Heterohelix* was discussed in detail by MONTANARO GALLITELLI (1957, *1303, p. 137) who showed that as most species of "*Guembelina*" have a microspheric coil, no valid morphologic distinction is found between it and *Heterohelix*; hence *Guembelina* was suppressed as a junior synonym. In addition, *Guembelina* EGGER, 1899, is a homonym of *Guembelina* KUNTZE, 1895, as recently shown by us (*1178). The original specimens of *Heterohelix* (*Spiroplecta americana* EHRENBERG) are from Upper Cretaceous chalk in northeastern Mississippi and the upper Missouri region. Specimens from the Selma chalk of Mississippi show *H. americana* and *H. navarroensis* to be synonymous.]

Bifarina PARKER & JONES, 1872, *1417g, p. 198 [**Dimorphina saxipara* EHRENBERG, 1854, *680, pl. 32; OD] [=*Tubitextularia* ŠULC, 1929, *1849, p. 148 (type, *Pseudotextularia bohemica*); *Rectoguembelina* CUSHMAN, 1932, *452, p. 6 (type, *R. cretacea*, =*Bifarina nodosaria* WHITE, 1929, *2055, p. 45)]. Early stage biserial, as in *Heterohelix*, later uniserial, with terminal, rounded aperture, which may be produced on short neck. *L.Cret. (U.Alb.)-Paleoc.,* N.Am.-Eu.——FIG. 523,*10.* *B. saxipara* (EHRENBERG), U.Cret.(Maastricht.), USA (Miss.); original specimen mounted in balsam, enlarged (*700).——FIG. 523,*11. B. nodosaria* WHITE (=paratype of *Rectoguembelina cretacea* CUSHMAN, ?U. Cret.(Maastricht.) or Paleoc. (Midway.), USA(Ark.); ×224 (*2117).——FIG. 523, *12. B. bohemica* (ŠULC), U.Cret.(Senon.), Czech.; topotype, ×148 (*1303).

[GLAESSNER (1936, *792, p. 108) and MONTANARO GALLITELLI (1957, *1303, p. 143) noted the synonymy of *Rectoguembelina* CUSHMAN, 1932, and *Tubitextularia* ŠULC, 1929. In addition, both are synonyms of *Bifarina* PARKER & JONES, as based on the type-species, *Dimorphina saxipara* EHRENBERG. CUSHMAN (1946, *484, p. 131) noted that *Bifarina saxipara* "may even possibly be a *Rectoguembelina, Nodosarella,* or another such form." The type-species of *Bifarina* cannot be placed in a later described genus, however. *Bifarina* has in the past included a number of quite distinct forms, similar only in their early biserial and later uniserial stages. Only the type-species is congeneric with forms previously placed in *Rectoguembelina* or *Tubitextularia* (which generic names are thus junior synonyms). Other species included in *Bifarina* by CUSHMAN (1937, *472) correctly should be placed in *Rectobolivina*, viz., *B. hungarica* VADAZ, *B. vicksburgensis* (CUSHMAN), *B. tombigbeensis* HADLEY, in *Valvobifarina*, viz., *B. elongata* (MILLETT), *B. mackinnoni* (MILLETT), and *B. mackinnoni* var. *robusta* (SIDEBOTTOM), and possibly in *Tubulogenerina*, viz., *B. reticulosa* CUSHMAN, *B. zanzibarensis* CUSHMAN, or *Loxostomum,* viz., *B. adelae* LIEBUS, *B. millepunctata* (TUTKOWSKY). GALLOWAY (1933, *762, p. 354) regarded *Rectobolivina* as a synonym of *Bifarina,* and placed *Bifarina nodosaria* WHITE in the genus *Rectoguembelina,* although correctly noting the synonymy of *Rectoguembelina cretacea* CUSHMAN and *Bifarina nodosaria* WHITE. *Bifarina* differs from *Rectobolivina* in having inflated chambers and a simple aperture, whereas *Rectobolivina* has an internal tube.]

Chiloguembelina LOEBLICH & TAPPAN, 1956, *1168, p. 340 [**Gümbelina midwayensis* CUSHMAN, 1940, *475, p. 65; OD]. Test free, flaring; inflated chambers biserially arranged, with tendency to become somewhat twisted; sutures distinct, depressed; wall calcareous, finely perforate, radial in structure, surface smooth to hispid; aperture a broad, low arch bordered by produced necklike

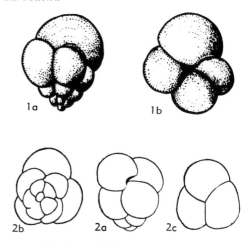

FIG. 524. Heterohelicidae (Guembelitriinae; *1,2, Gubkinella*) (p. C652).

extension of chamber, commonly forming more strongly developed flap at one side so that aperture appears to be directed toward one of flat sides of test. *Paleoc.-Oligo.,* cosmop.——FIG. 523,*8. C. crinita* (GLAESSNER), Paleoc.(Vincentown), USA (N.J.); *8a-c,* side, edge, and top views, ×143 (*1174).——FIG. 523,*9. *C. midwayensis* (CUSHMAN), Paleoc.(Midway), USA(Tex.); *9a,b,* side, edge views, ×97 (*1174).

[Differs from *Heterohelix* in the presence of an apertural necklike extension from the final chamber, in the tendency to develop a twisted test, and an asymmetrical aperture directed toward the flat side, instead of edge, of the test. Unlike true *Heterohelix,* it does not have an early coiled portion in the microspheric generation, all specimens being wholly biserial throughout.]

Gublerina KIKOÏNE, 1948, *1039, p. 26 [**G. cuvillieri* = *Ventilabrella ornatissima* CUSHMAN & CHURCH, 1929, *500, p. 512; OD] [=*Sigalia* REISS, 1957, *1528a, p. v (type, *Gümbelina (Gümbelina, Ventilabrella) deflaensis* SIGAL, 1952, *1746, p. 36)]. Test compressed, flabelliform, increasing rapidly in breadth; early stage planispiral, later biserial, with 2 series of chambers diverging widely, leaving broad nonseptate or incompletely divided central region, final stage may have chamber proliferation; sutures commonly thickened, nodose and elevated; wall calcareous, perforate, radial in structure, bilamellar, with double septa and septal peristomal canal; aperture arched. [*Sigalia* appears referable to *Gublerina,* the nonseptate central area being poorly defined because of the early proliferation of chambers. The characteristic ornamentation of *Gublerina* is present.] *U. Cret.,* Eu.-N.Am.-N.Afr.-W.Indies(Cuba).——FIG. 525,*1,2. *G. ornatissima* (CUSHMAN & CHURCH), Maastricht., S.Fr.; *1,* side view of specimen treated with acid to remove outer wall, showing position of septa and central nonseptate area, ×74 (*1303); *2,* side view, showing early coil, nodose early biserial chambers, later flaring biserial test with nonseptate central area although the surface grooves erroneously suggest septation, and final chamber proliferation, ×74 (*1303).

——FIG. 525,3. *G. deflaensis* (SIGAL), Turon., Algeria, ×50 (*1746).

Planoglobulina CUSHMAN, 1927, *428, p. 77 [*Güm-

FIG. 525. Heterohelicidae (Heterohelicinae; *1-3, Gublerina; 4, Planoglobulina; 5,6, Pseudoguembelina; 7, Pseudotextularia; 8, Racemiguembelina*) (p. C654-C656).

belina acervulinoides EGGER, 1899, *659, p. 36; OD] [=*Ventilabrella* CUSHMAN, 1928, *436, p. 2 (type, *V. eggeri*)]. Early stage coiled in microspheric form, later biserial, and finally with chamber proliferation in plane of biseriality, resulting in flabelliform test; exterior commonly ornamented with longitudinal striae; aperture multiple, on final series of chambers. U.Cret., Eu.-N.Am.——FIG. 525,4. **P. acervulinoides* (EGGER), Senon., Ger.; side view, ×116 (*2117).

[*Ventilabrella* was shown by MONTANARO GALLITELLI (1957, *1303) to be a synonym of *Planoglobulina*. The type-species of *Planoglobulina* is *Guembelina acervulinoides* EGGER, and CUSHMAN (1928, *436, p. 3) selected EGGER'S specimen in pl. 14, fig. 20 (*659) as the type-specimen (=lectotype). In 1946, CUSHMAN (*484, p. 111) placed the same specimen in his synonymy of *Ventilabrella eggeri* CUSHMAN, the type-species of *Ventilabrella*.]

Pseudoguembelina BRÖNNIMANN & BROWN, 1953, *234, p. 150 [**Gümbelina excolata* CUSHMAN, 1926, *425, p. 20; OD]. Test biserial in adult, rarely with microspheric coil in early stage; chambers subglobular; surface may have longitudinal striae or costae; aperture an interiomarginal arch, extending laterally, secondary sutural apertures may occur near zigzag suture between pairs of chambers. U.Cret., N.Am.-W.Indies(Cuba).——FIG. 525,5,6 **P. excolata* (CUSHMAN), Maastricht., USA(Tex.); *5*, hypotype showing early coil, ×155; *6*, hypotype showing biserial development throughout and well-developed secondary apertures, ×116 (*2117).

Pseudotextularia RZEHAK, 1891, *1604, p. 4 [**Cuneolina elegans* RZEHAK, 1891; OD (M)] [=*Pseudotextularia* RZEHAK, 1886, *1601, p. 6 (*nom. nud.*); *Bronnibrownia* MONTANARO GALLITELLI, 1955, *1300, p. 215, 220, 222 (*nom. nud.*); *Bronnimannella* MONTANARO GALLITELLI, 1956, *1302, p. 35 (type, *Gümbelina plummerae* LOETTERLE, 1937, *1188, p. 33)]. Early stage as in *Heterohelix*, later biserial chambers increasing rapidly in thickness and becoming laterally compressed, so that adult test has greater thickness than breadth, final chamber also may become nearly central in position; aperture a broad, low interiomarginal arch. U.Cret., Eu.-N.Am.-S.Am. ——FIG. 525,7. **P. elegans* (RZEHAK), Senon., USA(Tex.); *7a-c*, side, edge, and top views, ×100 (*1303).

[The *nomen nudum Pseudotextularia*, 1886, was originally used for a textularian form that was regarded as either a monstrosity or a new genus, but no description was given and no species included. In 1891, RZEHAK described *Cuneolina elegans*, stating that perhaps it represented a distinct genus, for which he had previously proposed the name *Pseudotextularia*. He thus validated the latter genus, whose type-species is *C. elegans*, by monotypy. The lectotype of *C. elegans* was designated by WHITE, 1929, *2055, p. 40, as RZEHAK, 1891, *1604, p. fig. 1a,b.]

Racemiguembelina MONTANARO GALLITELLI, 1957, *1303, p. 142 [**Gümbelina fructicosa* EGGER, 1899, *659, p. 35; OD]. Test subconical, early stage may be planispiral in microspheric forms, later biserial with globular chambers increasing regularly in size and with proliferation at crown perpendicular to previous axis of growth; surface may be ornamented by longitudinal striae or costae; aperture an interiomarginal arch on one or many of terminal chambers. U.Cret., Eu.-N.Am. ——FIG. 525,8. **R. fructicosa* (EGGER), Senon., USA(Tex.); *8a,b*, side, top views, ×116 (*1303).

Family PLANOMALINIDAE
Bolli, Loeblich, & Tappan, 1957

[*nom. transl.* SIGAL, 1958, p. 263 (*ex* subfamily Planomalininae BOLLI, LOEBLICH & TAPPAN, 1957, p. 21]

Coiling planispiral, primary aperture equatorial, or symmetrically paired, umbilical portions of successive apertures remaining as relict secondary apertures. *L.Cret. (Apt.)-Paleoc.(Dan.).*

Planomalina LOEBLICH & TAPPAN, 1946, *1154, p. 257 [**P. apsidostroba* LOEBLICH & TAPPAN, 1946, =*Planulina buxtorfi* GANDOLFI, 1942, *768, p. 103; OD]. Test free, planispiral, biumbilicate, involute to partially evolute, lobulate in outline; chambers angular-rhomboid; sutures radial, curved, elevated; wall calcareous, finely perforate, radial in structure, test ornamented by keel and thickened and nodose sutures; aperture an interiomarginal, equatorial arch, with opening extending back at either side to septum at base of chamber, lateral umbilical portions of successive apertures remaining open as supplementary relict apertures, each with remnant of bordering apertural lip. *L. Cret. (Alb.) - U. Cret. (Cenoman.)*, N. Am. - Eu.-Carib.-N.Afr.-Pak.——FIG. 526,1. **P. buxtorfi* (GANDOLFI), L.Cret.(Alb.), USA(Tex.); *1a,b*, side, edge views of holotype of *P. apsidostroba*, ×84 (*164).

Biglobigerinella LALICKER, 1948, *1081, p. 624 [**B. multispina*; OD]. Test free, planispiral, nearly or completely involute, biumbilicate, periphery rounded, peripheral margin lobulate; chambers globular, except for final 1 or 2 which may become broadly ovate, flattened, and finally replaced by 2 paired chambers, one on each side of plane of coiling, in some species tendency for chambers of final whorl to flare out in less involute coil is seen, with flange extending back on each side toward previous whorl, curving backward at umbilical margin, as in *Globigerinelloides*; sutures distinct, depressed, radial to curved or even sigmoid; wall calcareous, finely perforate, radial in structure, surface finely hispid to smooth or pitted; aperture an interiomarginal, equatorial, simple low arch in early stages, in later paired chambers one extraumbilical aperture present in each chamber of final pair. *L.Cret.(Apt.)-Paleoc. (Dan.)*, N.Am.-Carib.——FIG. 526,4,5. **B. multispina*, U.Cret.(Campan.); *4*, edge view of hypotype, USA(Tex.), ×119 (*164); *5a,b*, side, edge views of holotype, USA(Ark.), ×119 (*164).

Globigerinelloides CUSHMAN & TEN DAM, 1948, *501, p. 42 [**G. algeriana*; OD] [=*Biticinella* SIGAL, 1956, *1747, p. 35 (type, *Anomalina breggiensis* GANDOLFI, 1942, *768, p. 102)]. Test free, planispiral, biumbilicate, involute to partially evo-

Foraminiferida—Rotaliina—Globigerinacea

lute, lobulate in outline; chambers rounded to ovoid, may be somewhat elongated in specimens tending to become evolute; sutures depressed, radial, straight to curved or sigmoid; wall calcareous, finely perforate, radial in structure, surface smooth or roughened; aperture a broad, low,

FIG. 526. Planomalinidae; *1, Planomalina; 2,3, Hastigerinoides; 4,5, Biglobigerinella; 6,7, Globigerinelloides;* Schackoinidae; *8,9, Schackoina; 10, Leupoldina* (p. C656-C659).

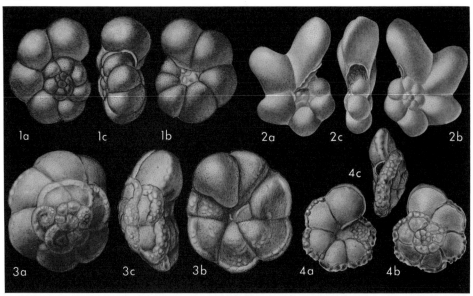

FIG. 527. Rotaliporidae (Hedbergellinae; *1, Hedbergella; 2, Clavihedbergella; 3,4, Praeglobotruncana*) (p. C659).

interiomarginal equatorial arch, with lateral umbilical portions of successive apertures remaining open as relict apertures. Cret., cosmop.——FIG. 526,6. **G. algeriana*, L.Cret.(Apt.), Algeria; *6a,b*, side, edge views, ×56 (*164).——FIG. 526,7. *G. eaglefordensis* (MOREMAN), L.Cret.(Alb.), Eng.; *7a,b*, side, edge views of holotype of *Planomalina caseyi*, ×135 (*164).

Hastigerinoides BRÖNNIMANN, 1952, *228, p. 52 [**Hastigerinella alexanderi* CUSHMAN, 1931, *450, p. 87; OD] [=*Eohastigerinella* MOROZOVA, 1957, *1316, p. 1112 (type, *Hastigerinella watersi* CUSHMAN, 1931, *450, p. 86)]. Test free, stellate in appearance, planispiral, biumbilicate, periphery rounded; early chambers globular, later chambers elongate-radial, much produced and tapering or clavate; sutures depressed, radial; wall calcareous, perforate, radial in structure, surface smooth, pitted or finely hispid; primary aperture interiomarginal, equatorial, a simple arch bordered above by protruding lip, with relict secondary apertures around umbilical region, representing umbilical portion of previous apertures, which may remain open or be closed. L.Cret.(Apt.)-U.Cret.(Turon.), N. Am.-Eu.-Carib.——FIG. 526,2. **H. alexanderi* (CUSHMAN), U.Cret.(Turon.), USA(Tex.); *2a,b*, side, edge views, ×70 (*164).——FIG. 526,3. *H. watersi* (CUSHMAN), U.Cret.(Turon.), USA(Tex.); *3a,b*, side, edge views, ×90 (*164).

Family SCHACKOINIDAE Pokorný, 1958

[Schackoinidae POKORNÝ, 1958, p. 348]

Test trochospiral to nearly planispiral, each chamber with one or rarely more hollow tubulospines; aperture equatorial, and may have broad spatulate lip. L.Cret.(Apt.)-U.Cret.(Maastricht.).

Schackoina THALMANN, 1932, *1894, p. 288 [**Siderolina cenomana* SCHACKO, 1897, *1635, p. 166; OD] [=*Hantkenina (Schackoina)* THALMANN, 1932, *1894, p. 288 (obj.)]. Test free, early portion may be more or less trochospiral, later becoming nearly planispiral; chambers radially elongate with one or more elongate, tapering, hollow tubulospines extending outward from mid-line of each chamber on periphery; sutures straight, radial, depressed; wall calcareous, finely perforate, surface smooth or very finely hispid; primary aperture an interiomarginal arch, extraumbilical and tending to become equatorial, may be bordered above by narrow lip. L.Cret.(Apt.)-U.Cret. (Campan.-?Maastricht.), cosmop.——FIG. 526, *8,9*. **S. cenomana* (SCHACKO), U.Cret.(Cenoman.), Eu.(Ger.) (*8*), N.Am. (*9*); *8a-c*, opposite sides and edge view, ×195 (*164); *9*, side view, USA (Kans.), ×158 (*1183).

Leupoldina BOLLI, 1958, *161, p. 275 [**L. protuberans*; OD]. Differs from *Schackoina* in having tubulospines which are bulbous at their extremities and in developing 2 interiomarginal apertures in final chamber of mature forms, one on each side of chamber, much as in *Biglobigerinella*. [BOLLI (*161) modified the generic description of *Schackoina* to include forms with bulbous tubulospines similar to those of *Leupoldina*. It seems probable that forms with such bulbous tubulospines are in reality immature specimens of *Leupoldina* in which the double aperture has not developed, as they occur in the same samples with *Leupoldina*.] L.Cret.(Apt.), W.Indies(Trinidad)-Eu.——FIG. 526,*10*. **L. protuberans*, Trinidad; *10a,b*, side, edge views of holotype, ×106 (*161).

Family ROTALIPORIDAE Sigal, 1958

[Rotaliporidae SIGAL, 1958, p. 264] [=Marginolamellidae HOFKER, 1951, p. 485 *(partim) (nom. nud.)*]

Coiling trochospiral; primary aperture extraumbilical-umbilical, with relatively prominent lip; may have secondary sutural apertures on umbilical side opening into posterior margin of chambers. *Cret.*

Subfamily HEDBERGELLINAE
Loeblich & Tappan, 1961

[Hedbergellinae LOEBLICH & TAPPAN, 1961, p. 309]

Primary aperture only, commonly with prominent apertural lip, those of previous chambers remaining as projections into umbilical region. *L.Cret.(Hauteriv.)-U.Cret. (Maastricht.).*

Hedbergella BRÖNNIMANN & BROWN, 1958, *236, p. 16 [**Anomalina lorneiana* D'ORBIGNY var. *trocoidea* GANDOLFI, 1942, *768, p. 98; OD] [=*Praeglobotruncana (Hedbergella)* BANNER & BLOW, 1959, *77, p. 18 (obj.); *Planogyrina* ZAKHAROVA-ATABEKYAN, 1961, *2098, p. 50 (type, *Globigerina gaultina* MOROZOVA, 1948, *1315, p. 41)]. Test free, trochospiral, biconvex, umbilicate, periphery rounded, with no indication of keel or poreless margin; chambers globular to ovate; sutures depressed, radial, straight to curved; wall calcareous, finely perforate, radial in structure, surface smooth to hispid or rugose; aperture an interiomarginal, extraumbilical-umbilical arch commonly bordered above by narrow lip or spatulate flap, and in forms with broad, open umbilicus, successive apertural flaps may remain visible to show serrate or scalloped border around umbilicus. [*Hedbergella* includes species which are otherwise similar to *Praeglobotruncana* but lacking a keel or poreless margin. The rounded chambers are reminiscent of *Globigerina* but the aperture is extraumbilical, rather than umbilical, and the umbilicus is commonly narrow.] L.Cret.(Hauteriv.)-U.Cret.(Maastricht.), cosmop.——FIG. 527, *1*. **H. trocoidea* (GANDOLFI), U.Cret.(Cenoman.), Blake Plateau, Atl.O.; *1a-c*, spiral, umbilical, and edge views, ×75 (*2117).

Clavihedbergella BANNER & BLOW, 1959, *77, p. 18 [**Hastigerinella subcretacea* TAPPAN, 1943, *1872, p. 513; OD] [=*Praeglobotruncana (Clavihedbergella)* BANNER & BLOW, 1959, *77, p. 8, 18 (obj.)]. Test free, low trochospiral, biconvex, broadly umbilicate, peripheral margin rounded, peripheral outline deeply lobulate, no keel or poreless margin; early chambers globular to ovate, later ones clavate to radial-elongate; sutures strongly constricted, radial, straight to curved; wall calcareous, finely perforate, radial in structure, surface smooth to hispid; aperture an interiomarginal, extraumbilical-umbilical arch, with narrowing bordering lip or spatulate flap (porticus). [Differs from *Hedbergella* in having radial-elongate chambers, and from *Hastigerinella* in having apertural flaps or portici. Although *Clavihedbergella* was described as ranging from upper Albian to Turonian, thus being more restricted than *Hedbergella* (*77, p. 17), we also have excellent examples of *Clavihedbergella* in Aptian strata of both hemispheres.] L.Cret.(Apt.)-U.Cret.(Turon.), cosmop. ——FIG. 527,*2*. **C. subcretacea* (TAPPAN), L.Cret.(Alb.), USA(Okla.); *2a-c*, umbilical, spiral, and edge views, ×78 (*2117).

Praeglobotruncana BERMÚDEZ, 1952, *127, p. 52 [**Globorotalia delrioensis* PLUMMER, 1931, *1463, p. 199; OD] [=*Rotundina* SUBBOTINA, 1953, *1847, p. 164 (type, *Globotruncana stephani* GANDOLFI, 1942, *768, p. 130)]. Test free, trochospiral, biconvex to spiroconvex, umbilicate, periphery rounded to subangular, with more or less well-developed peripheral keel, which is most prominent in earlier development; chambers ovate to subangular; sutures on spiral side radial or curved, depressed to elevated, commonly thickened or beaded, on umbilical side depressed and radial; wall calcareous, finely perforate, radial in structure, surface smooth to hispid; aperture an interiomarginal, extraumbilical-umbilical arch, bordered by apertural lip. [Regarded as containing both carinate and noncarinate species by BOLLI, LOEBLICH & TAPPAN (1957, *164), the genus is now restricted to include only species which have a peripheral keel or poreless margin. The noncarinate species are now placed in *Hedbergella*.] L.Cret.(U.Alb.)-U.Cret.(Cenoman.), cosmop. —— FIG. 527,*3*. *P. stephani* (GANDOLFI), Cenoman., Switz.; *3a-c*, opposite sides and edge view, ×75 (*1183).——FIG. 527,*4*. **P. delrioensis* (PLUMMER), Cenoman., USA(Tex.); *4a-c*, umbilical, spiral, and edge views of topotype, ×75 (*1183).

Subfamily ROTALIPORINAE Sigal, 1958

[*nom. transl.* BANNER & BLOW, 1959, p. 8 (*ex* family Rotaliporidae SIGAL, 1958)]

With primary aperture, and secondary sutural apertures on umbilical side. *L.Cret. (Alb.)-U.Cret.(Cenoman.-?Turon.).*

Rotalipora BROTZEN, 1942, *240, p. 32 [**R. turonica*, =*Globorotalia cushmani* MORROW, 1934, *1319, p. 199; OD] [=*Thalmanninella* SIGAL, 1948, *1743, p. 101 (type, *T. brotzeni*, =*Globorotalia greenhornensis* MORROW, 1934, *1319, p. 199)]. Test free, trochospiral, biconvex to planoconvex, umbilicate, periphery angular, with single keel; chambers angular-rhomboid; sutures curved on spiral side, depressed to elevated, and may be thickened or beaded, on umbilical side radial to slightly curved, flush to depressed; wall calcareous, perforate, radial in structure, surface smooth to nodose; primary aperture interiomarginal and extraumbilical-umbilical in position with bordering lip, single secondary sutural aperture per suture on umbilical side, or rarely 2 or more per suture, commonly also with bordering lip or thickened rim. [BOLLI, LOEBLICH & TAPPAN

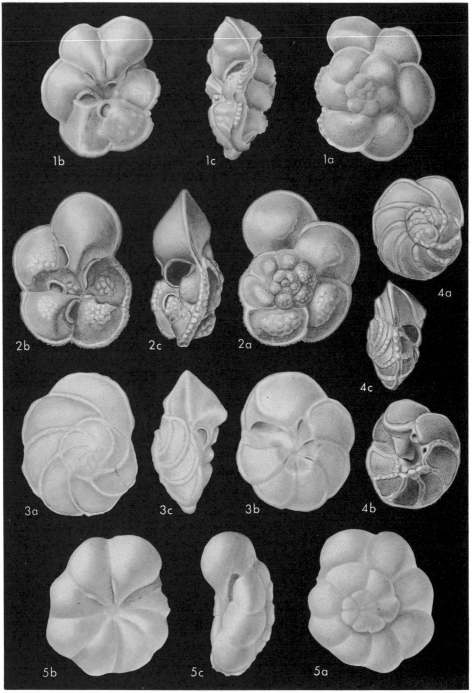

FIG. 528. Rotaliporidae (Rotaliporinae; *1-4, Rotalipora; 5, Ticinella*) (p. C659-C662).

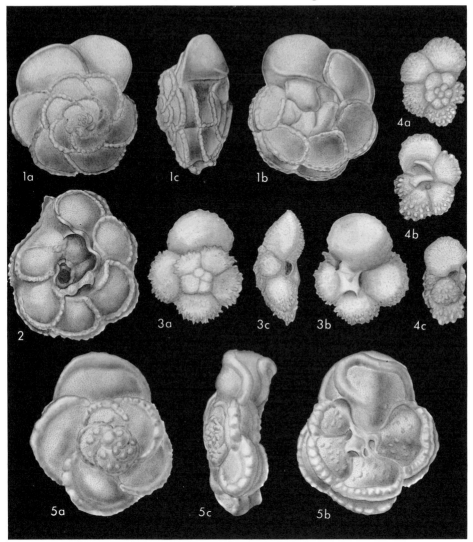

FIG. 529. Globotruncanidae; 1-4, Globotruncana; 5, Abathomphalus (p. C662-C663).

(1957, *164) included with this genus both the typically keeled species and the nonkeeled forms described as *Ticinella* by REICHEL (1950, *1522). The latter is here recognized as a distinct genus, characterized by the absence of a keel or poreless margin.] *U.Cret.(Cenoman.-?Turon.)*, cosmop. ——FIG. 528,*1,2*. *R. cushmani* (MORROW); *1a-c,* spiral, umbilical, and edge views, topotype of form described as *R. turonica* BROTZEN, Cenoman., ?Turon., Ger.-Pol.(Pomerania); *2a-c,* spiral, umbilical, and edge views of topotype of *R. cushmani* (MORROW), Cenoman., USA(Kans.), ×63 (*1183).——FIG. 528,*3,4*. *R. greenhornensis* (MORROW); *3a-c,* Cenoman., N.Afr.(Algeria); opposite sides and edge view of topotype of *Thalmanninella brotzeni* SIGAL, ×75; *4a-c,* Cenoman., USA(Kans.), opposite sides and edge view of topotype, ×60 (*1183).

Ticinella REICHEL, 1950, *1522, p. 600 [*Anomalina roberti* GANDOLFI, 1942, *768, p. 100; OD] [=*Globotruncana* (*Ticinella*) REICHEL, 1950, *1522, p. 600 (obj.)]. Test free, trochospiral, biconvex to plano-convex, umbilicate, periphery rounded, and lacking keel or poreless margin, chambers ovate; sutures on spiral side curved, depressed to elevated, on umbilical side flushed to depressed, radial or slightly curved; wall calcareous, perforate, radial in structure, surface smooth to spinose; primary aperture interiomarginal, extraumbilical-umbilical, and may be bordered above by lip, secondary sutural apertures on umbilical side, commonly one per suture, more rarely 2 or

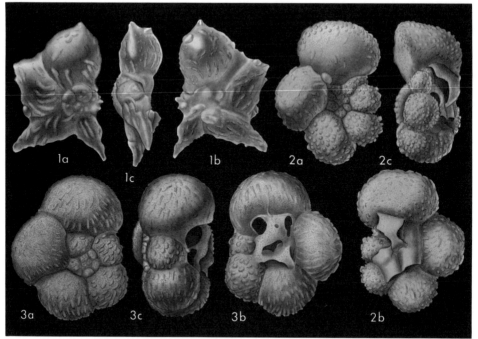

Fig. 530. Globotruncanidae; *1, Plummerita; 2, Trinitella; 3, Rugoglobigerina* (p. C663).

more, and each may be bordered by narrow lip, which in some specimens may be sufficiently large to give appearance of a cover plate, although not as extensive as umbilical tegilla of the Globotruncanidae. *L.Cret.(U.Alb.)-U.Cret.(Cenoman.)*, cosmop.——Fig. 528,5. **T. roberti* (GANDOLFI), U.Cret.(Cenoman.), Eu.(Switz.); *5a-c,* spiral, umbilical, and edge views of topotype, ×110 (*1183).

Family GLOBOTRUNCANIDAE
Brotzen, 1942

[*nom. transl.* MOROZOVA, 1957, p. 1111 (*ex* subfamily Globotruncaninae BROTZEN, 1942, p. 28)] [=Marginolamellidae HOFKER, 1951, p. 485 *(partim) (nom. nud.)*; =Rugoglobigerininae SUBBOTINA in RAUZER-CHERNOUSOVA & FURSENKO, 1959, p. 303]

Test trochospiral, chambers spherical to angular, commonly truncate or keeled; primary aperture umbilical, covered by spiral system of tegilla, wtih accessory intralaminal and infralaminal apertures. *U.Cret.(Turon.-Maastricht.).*

Globotruncana CUSHMAN, 1927, *431, p. 91 [*Pulvinulina arca CUSHMAN, 1926, *425, p. 23; OD] [—*Rosalinella* MARIE, 1941, *1215, p. 237, 256, 258 (type, *Rosalina linneiana* D'ORBIGNY, 1839, *1611, p. 101); *Marginotruncana* HOFKER, 1956, *947, p. 319 (type, *Rosalina marginata* REUSS, 1846, *1538, p. 36); *Rugotruncana* BRÖNNIMANN & BROWN, 1956, *235, p. 546 (type, *R. tilevi*); *Bucherina* BRÖNNIMANN & BROWN, 1956, *235, p. 557 (type, *B. sandidgei*); *Globotruncanella* REISS, 1957, *1529, p. 135 (type, *Globotruncana citae* BOLLI, 1951, *158, p. 197, =*Globotruncana havanensis* VOORWIJK, 1937, *2025, p. 195); *Globotruncanita* REISS, 1957, *1529, p. 136 (type, *Rosalina stuarti* DE LAPPARENT, 1918, *1096, p. 11); *Helvetoglobotruncana* REISS, 1957, *1529, p. 137 (type, *Globotruncana helvetica* BOLLI, 1945, *156, p. 226)]. Test free, trochospiral, biconvex, spiroconvex or umbilicoconvex, broadly umbilicate, periphery rounded with poreless margin, with single keel or truncate with double keel; chambers ovate, hemispherical, angular rhomboid or angular truncate; sutures on spiral side curved or radial, depressed to elevated, may be limbate and beaded, sutures on umbilical side curved or radial, depressed or more rarely elevated; wall calcareous, perforate, radial in structure, surface smooth, rugose or beaded; primary apertures interiomarginal, umbilical, in well-preserved specimens covered by tegilla, which are perforated by accessory infralaminal and intralaminal apertures that become sole openings to exterior; tegilla commonly partially or wholly broken out in fossilization or preserved only as scalloped fragments. *U.Cret.(Turon.-Maastricht.),* cosmop.——Fig. 529, *1,2.* **G. arca* (CUSHMAN), L.Maastricht, USA (Tex.); *1a-c,* spiral side, umbilical side with well-preserved tegilla covering entire umbilical region so as to obscure primary aperture, and edge view; *2,* umbilical view, tegilla broken out, exposing primary umbilical aperture; all ×70 (*164).——

Fig. 529,3. *G. havanensis* Voorwijk, Maastricht., W.Indies(Cuba); *3a-c,* opposite sides and edge view, ×79 (*2117).——Fig. 529,4. *G. tilevi* (Brönnimann & Brown), Maastricht., W.Indies (Cuba); *4a-c,* opposite sides and edge view, ×79 (*2117).

Abathomphalus Bolli, Loeblich & Tappan, 1957, *164, p. 43 [**Globotruncana mayaroensis* Bolli, 1951, *158, p. 198; OD]. Test free, trochospiral, biconvex to concavo-convex, almost nonumbilicate, periphery with single or double keel; sutures depressed, curved, and in some forms beaded on spiral side, depressed and radial on umbilical side; wall calcareous, perforate, radial in structure, commonly ornamented with fine nodes, peripheral keels and sutures may be beaded; primary aperture interiomarginal, extraumbilical, generally covered by continuous umbilical tegillum of irregular outline, with accessory infralaminal apertures situated at suture contacts with tegillum. *U.Cret. (Maastricht.),* W.Indies(Trinidad)-Mex.-Eu.-Afr. ——Fig. 529,5. **A. mayaroensis* (Bolli), Trinidad; *5a-c,* spiral, umbilical, and edge views, ×76 (*164).

[Differs from *Globotruncana* in lacking a wide and deep umbilicus with sharply angled rim and delicate tegilla extending from each chamber, and in having an interiomarginal, extraumbilical primary aperture. In *Abathomphalus* the umbilical area is not open, the final whorl of chambers all meeting ventrally, although their junction may be obscured by the single umbilical tegillum, which appears to be an extension from the final chamber. The accessory apertures are always infralaminal, not both infralaminal and intralaminal as in *Globotruncana.*]

Plummerita Brönnimann, 1952, *227, p. 146 [*pro Plummerella* Brönnimann, 1952 (*non* de Long, 1942)] [**Rugoglobigerina (Plummerella) hantkeninoides hantkeninoides* Brönnimann, 1952, *228, p. 37; OD] [=*Rugoglobigerina (Plummerella)* Brönnimann, 1952, *228, p. 37 (*non Plummerella* de Long, 1942) (obj.); *Rugoglobigerina (Plummerita)* Brönnimann, 1952, *227, p. 146 (obj.)]. Similar to *Rugoglobigerina* in form but with later chambers becoming radial-elongate; primary aperture interiomarginal, umbilical, with tegilla and infralaminal and intralaminal apertures. *U.Cret. (Maastricht.),* Carib.-USA.——Fig. 530,*1.* **P. hantkeninoides* (Brönnimann), W.Indies(Trinidad); *1a-c,* opposite sides and edge view of holotype, ×128 (*164).

Rugoglobigerina Brönnimann, 1952, *228, p. 16 [**Globigerina rugosa* Plummer, 1927, *1461, p. 38; OD] [=*Rugoglobigerina (Rugoglobigerina)* Brönnimann, 1952, *228, p. 17 (obj.); *Kuglerina* Brönnimann, 1956, *235, p. 557 (type, *Rugoglobigerina rugosa rotundata* Brönnimann, 1952, *228, p. 34)]. Test free, trochospiral, biconvex, umbilicate, periphery rounded; chambers rounded to spherical; sutures radial to slightly curved on spiral side, radial on umbilical side, depressed throughout; wall calcareous, perforate, radial in structure, surface typically rugose, with numerous large pustules which may coalesce into distinct ridges, radiating from mid-point of each chamber on periphery, more rarely smooth; primary apertures interiomarginal, umbilical, in well-preserved specimens covered by tegilla perforated by accessory infralaminal and intralaminal apertures which are only openings to exterior, tegilla tending to be partially or wholly broken out in preservation. *U.Cret.(Turon.-Maastricht.),* cosmop.—— Fig. 530,*3.* **R. rugosa* (Plummer), Maastricht., USA(Tex.); *3a-c,* opposite sides and edge view, ×90 (*164).

[*Rugoglobigerina* resembles *Globotruncana* in its apertural characters and presence of the umbilical tegilla, but differs in its prominent surface ornamentation and less angular chambers. *Rugoglobigerina* may be regarded as the form ancestral to *Globotruncana;* various species of the latter genus seem to have branched off from the main *Rugoglobigerina*-stem at different geologic times. *Rugoglobigerina* differs from *Globigerina* in having umbilical tegilla over the primary aperture, in having infralaminal and intralaminal accessory apertures, and commonly in displaying a characteristic rugose, highly ornamented surface.]

Trinitella Brönnimann, 1952, *228, p. 56 [**T. scotti;* OD]. Similar to *Rugoglobigerina* but with later chambers compressed and flattened on spiral side, producing subangular periphery; aperture umbilical in position, with tegilla and accessory apertures. *U.Cret.(Maastricht.),* Carib.-USA.—— Fig. 530,*2.* **T. scotti,* USA(Tex.); *2a-c,* opposite sides and edge view, ×98 (*164).

Family HANTKENINIDAE Cushman, 1927

[Hantkeninidae Cushman, 1927, p. 64]

Test planispiral or enrolled biserial; chambers spherical to elongate or clavate; primary aperture symmetrical and equatorial, single or multiple, and may have relict or areal secondary apertures. *Paleoc.-Rec.*

Subfamily HASTIGERININAE Bolli, Loeblich & Tappan, 1957

[Hastigerininae Bolli, Loeblich & Tappan, 1957, p. 29]
[=Hasterigerininae Loeblich & Tappan, 1961, p. 309 (*nom. null.*)]

Test planispiral; chambers spherical to clavate; primary aperture equatorial, without secondary apertures. *Paleoc.-Rec.*

Hastigerina Thomson *in* Murray, 1876, *1331, p. 534 [**H. murrayi* (=*Nonionina pelagica* d'Orbigny, 1839, *1393, p. 27; OD (M)] [=*Globigerinella* Cushman, 1927, *431, p. 87 (type, *Globigerina aequilateralis* Brady, 1879, *196b, p. 285, =*Globigerina siphonifera* d'Orbigny *in* de la Sagra, 1839, *1611, p. 83]. Test free, early stage may be slightly trochospiral, adult planispiral, ranging from involute to loosely coiled, biumbilicate, periphery broadly rounded; chambers spherical to ovate; sutures deeply depressed, radial; wall finely to coarsely perforate, radial in structure, surface smooth, hispid, or spinose; aperture interiomarginal, broad equatorial arch. *L.Mio.-Rec.,* cosmop.——Fig. 531,*1.* *H. siphonifera* (d'Orbigny), Rec., Pac.O.; *1a,b,* side, apert. views, ×54 (*164).——Fig. 531,2-4. *H. pelagica (d'Orbigny), Rec., S.Atl.O.; 2a,b, side, apert. views of hypotype (BMNH-ZF1563); 3, apert. view of lectotype of H. murrayi Thomson (=neotype of Nonionina pelagica d'Orbigny) (BMNH-ZF1562), specimen preserved in balsam,

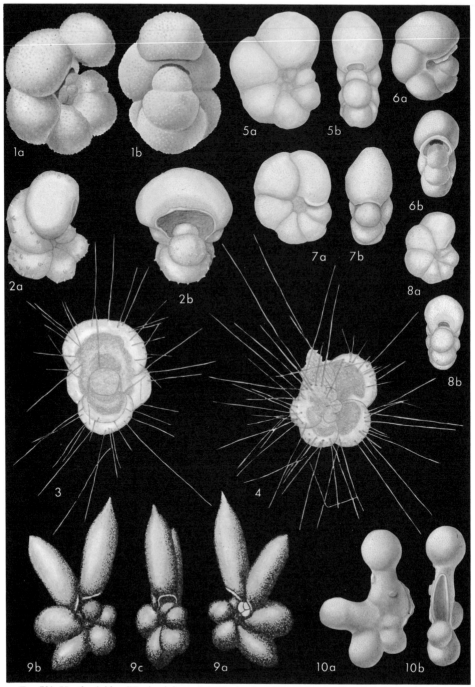

Fig. 531. Hantkeninidae (Hastigerininae; 1-4, Hastigerina; 5-8, Globanomalina; 9, Bolliella; 10, Clavigerinella) (p. C663-C666).

Fig. 532. Hantkeninidae (Hantkenininae; *1-3, Hantkenina;* 4, *Cribrohantkenina;* Cassigerinellinae; 5, *Cassigerinella*) (p. C666).

showing protoplasm preserved within test; *4*, side view of paratype; all ×36 (*164).

Bolliella BANNER & BLOW, 1959, *77, p. 12 [*Hastigerina (Bolliella) adamsi*; OD] [=*Hastigerina (Bolliella)* BANNER & BLOW, 1959, *77, p. 12 (obj.)]. Similar to *Hastigerina* but with radially elongate chambers in adult. Rec., Pac.O.——FIG. 531,9. *B. adamsi,* opposite sides and apert. view of holotype, ×38 (*77).

Clavigerinella BOLLI, LOEBLICH & TAPPAN, 1957, *164, p. 30 [*C. akersi;* OD]. Test free, planispiral, involute, radially lobulate in outline; early chambers spherical, later radially elongate or clavate; aperture an elongate interiomarginal, equatorial slit extending up apertural face and bordered laterally by wide flanges which narrow toward upper extremity of aperture where they join to form small lip. [*Clavigerinella* resembles *Hastigerinella* in the early globular chambers followed by later radial elongate and clavate chambers, but differs in being planispiral, with elevated equatorial aperture, instead of being trochospiral.] M.Eoc.-U.Eoc., W. Indies (Trinidad)-N. Am.——FIG. 531,10. *C. akersi,* M.Eoc., Trinidad; *10a,b,* side, apert. view of holotype, ×49 (*164).

Globanomalina HAQUE, 1956, *876, p. 147 [*G. ovalis;* OD] [=*Pseudohastigerina* BANNER & BLOW, 1959, *77, p. 19 (type, *Nonion micrus* COLE, 1927, *355, p. 22)]. Test free, planispiral to slightly asymmetrical, biumbilicate, chambers inflated, sutures curved and depressed; wall calcareous, finely perforate, radially built, and bilamellar, surface smooth; aperture an equatorial arch, with narrow lip, in some specimens with lip touching previous whorl at its periphery so as to form 2 lateral apertural openings. Paleoc.-Oligo., cosmop.——FIG. 531,5. *G. ovalis,* L.Eoc., Asia(Pak.); *5a,b,* side, apert. views, ×79 (*2117). ——FIG. 531,6-8. *G. micra* (COLE), M.Eoc., W. Indies(Trinidad) *(6),* Mex. *(7,8); 6a,b,* side, apert. views, ×109 (*160); *7a,b,* side and edge views of specimen with closely appressed final chamber closing aperture on periphery and leaving biglobigerinelloid double aperture; *8a,b,* side and edge of typical specimen, ×109 (*2117).

[*Globanomalina* was described originally as trochospiral and some species which have been assigned to this genus are trochospiral; they should be transferred to *Globorotaloides* as here redefined. The type-species of *Globanomalina* is involute on both sides, and although they are somewhat larger, topotype specimens are very similar to *Nonion micrus,* as figured by BANNER & BLOW (*77). The species are here regarded as congeneric, and *Pseudohastigerina* a synonym of *Globanomalina. Pseudohastigerina* was described as having an imperforate porticus (apertural lip) and to differ from *Globigerinelloides* in a reduced number of relict apertures. In topotypes of *Nonion micrus* a considerable degree of variation in the involution is found, some specimens being completely involute and a majority partially evolute. Although a distinct apertural lip is present, this is aparently perforate. In the rarer more involute specimens among middle Eocene topotypes and in Paleocene species, the apertural lip may attach to the previous whorl in the equatorial region, leaving the aperture open only laterally, as is common in *Biglobigerinella*. In a plate explanation given by BANNER & BLOW (*77, pl. 3, fig. 6) *Pseudohastigerina* was said to be monolamellar and to have imperforate septa. The apertural face (and hence septal face) is distinctly perforate, however. Furthermore, according to REISS (*1530, p. 68) the type-species of *Pseudohastigerina* is like the Hantkeninidae (bilamellar) in wall structure as are all of the Globigerinacea.]

Subfamily HANTKENININAE Cushman, 1927

[*nom. transl.* CHAPMAN & PARR, 1936, p. 145 (*ex* family Hantkeninidae CUSHMAN, 1927)]

Test planispiral; chambers globular, elongate or spinate; aperture equatorial or areal and multiple. *Eoc.*

Hantkenina CUSHMAN, 1924, *419, p. 1 [**H. alabamensis*; OD] [=*Hantkenia* CUSHMAN, 1924, *419, p. 1 (*nom. null.*) (*non* FISCHER, 1885; *nec* PREVER, 1902); *Sporohantkenina* BERMÚDEZ, 1937, *118, p. 151 (type, *Hantkenina brevispina* CUSHMAN, 1924, *419, p. 2); *Hantkenina (Aragonella)* THALMANN, 1942, *1901, p. 811 (type, *H. mexicana* var. *aragonensis* NUTTALL, 1930, *1371, p. 284); *Hantkenina (Applinella)* THALMANN, 1942, *1901, p. 812 (type, *H. dumblei* WEINZIERL & APPLIN, 1929, *2044, p. 402; *Hantkenina (Hantkeninella)* BRÖNNIMANN, 1950, *220, p. 399 (type, *H. alabamensis* var. *primitiva* CUSHMAN & JARVIS, 1929, *509, p. 16)]. Test free, planispiral, involute, biconvex, biumbilicate; chambers rounded, ovate or radial elongate, generally with single relatively long, heavy spine at forward margin of each chamber on periphery and in the plane of coiling, although spines rarely are absent on one or more chambers; sutures depressed, radial; wall calcareous, finely perforate, radial in structure, surface finely hispid, especially in area just beneath aperture on previous whorl; primary aperture interiomarginal, equatorial, triradiate, 2 "rays" forming slit across base of final chamber face, third ray arising from center of this slit and extending up face toward peripheral spine, flaring slightly to become rounded at its upper end, vertical slit bordered laterally by apertural flanges which join above as narrow lip. *Eoc.(Ypres.-Wemmel.),* N.Am.-Eu.-S.Am.-Afr.-M.East-Australia-N.Z.-E.Indies(Borneo).——FIG. 532,*1.* **H. alabamensis,* U.Eoc.(Jackson.), USA(Ala.); *1a,b,* side, edge views, ×27 (*164).——FIG. 532,*2. H. aragonensis* NUTTALL, Eoc., Mex.; *2a,b,* side, apert. view of paratype, ×45 (*164).——FIG. 532,*3. H. dumblei* WEINZIERL & APPLIN, Eoc., USA(Tex.); side view of lectotype, ×50 (*164).

Cribrohantkenina THALMANN, 1942, *1901, p. 812, 815, 819 [**Hantkenina (C.) bermudezi* (=*Hantkenina inflata* HOWE, 1928, *968A, p. 14); OD]. Test free, planispiral, biumbilicate; chambers subglobular, with prominent peripheral spine at forward margin of each chamber, succeeding chambers attached near base of spines, may partially or completely envelop spine of preceding chamber; sutures distinct, depressed, radial; wall calcareous, perforate, surface smooth, finely punctate, or finely spinose; primary aperture interiomarginal, equatorial, secondary multiple areal aperture consisting of small rounded or elongate openings above primary interiomarginal aperture, in well-developed specimens terminal portion of chamber may form a protruding "pore-plate," which lacks fine perforations in area between apertural pores and may cover primary interiomarginal aperture, attaching to peripheral margin of previous whorl, primary interiomarginal aperture and secondary areal apertures commonly bordered by distinct protruding lips, and multiple secondary openings may rarely be filled by later-formed shell growth. [Differs from *Hantkenina* in having the secondary multiple areal aperture in the region between the final spine and the primary interiomarginal aperture.] *U.Eoc.,* N. Am.-W. Indies(Cuba)-Afr.——FIG. 532,*4.* **C. inflata* (HOWE), Jackson, USA (Ala.); *4a,b,* side, apert. views, ×50 (*164).

Subfamily CASSIGERINELLINAE Bolli, Loeblich & Tappan, 1957

[Cassigerinellinae BOLLI, LOEBLICH & TAPPAN, 1957, p. 30]

Test planispiral in early stage, later enrolled biserial; chambers spherical to ovate; primary aperture equatorial in neanic stage, extraumbilical and alternating in adult. *Oligo.-Mio.*

Cassigerinella POKORNÝ, 1955, *1475, p. 136 [**C. boudecensis*; OD]. Test free, robust, early portion planispiral and similar to *Hastigerina,* later with biserially arranged chambers continuing to spiral in same plane, biumbilicate, periphery broadly rounded; chambers globular to ovate and only few pairs arranged as in *Cassidulina* to each whorl of test; sutures distinct, depressed, radial to curved; wall calcareous, perforate, radial in structure, surface smooth to pitted; aperture interiomarginal, an extraumbilical arch alternating in position from one side to next in successive chambers. [Differs from *Cassidulina* in having a perforate radial rather than granular wall structure and in having an early planispiral stage.] *Oligo.-Mio.,* Eu.-N. Am.-Carib.-S.Am.——FIG. 532,*5.* **C. boudecensis,* M.Oligo., Eu.(Czech.); *5a-c,* opposite sides and edge view showing biserial enrolled test and arched aperture, ×219 (*164).

Family GLOBOROTALIIDAE Cushman, 1927

[Globorotaliidae CUSHMAN, 1927, p. 91] [=Marginolamellidae HOFKER, 1951, p. 485 (*partim*) (*nom. nud.*)]

Test trochospiral; chambers ovate, spherical or angular; primary aperture interiomarginal, extraumbilical-umbilical, and secondary sutural apertures may occur on spiral side. *Paleoc.-Rec.*

Subfamily GLOBOROTALIINAE Cushman, 1927

[*nom. transl.* CHAPMAN & PARR, 1936, p. 145 (*ex* family Globorotaliidae CUSHMAN, 1927)]

Primary aperture only, on umbilical side. *Paleoc.-Rec.*

Foraminiferida—Rotaliina—Globigerinacea

Globorotalia CUSHMAN, 1927, *431, p. 91 [*Pulvinulina menardii* (D'ORBIGNY) var. *tumida* BRADY, 1877, *194, p. 535; OD] [=*G. (Truncorotalia)* CUSHMAN & BERMÚDEZ, 1949, *497, p. 35 (type, *Rotalina truncatulinoides* D'ORBIGNY in BARKER-WEBB & BERTHELOT, 1839, *86, p. 132); *Planoro-*

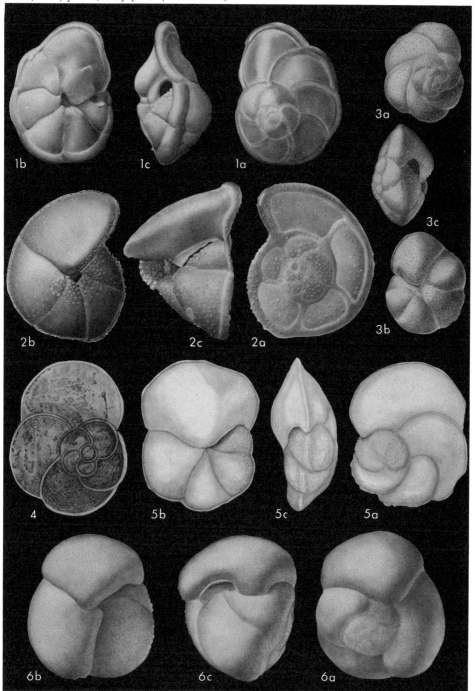

FIG. 533. Globorotaliidae (Globorotaliinae; *1-5, Globorotalia; 6, Turborotalia*) (p. C667-C668).

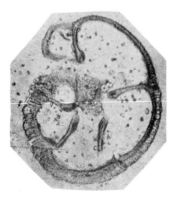

Fig. 534. Globorotaliidae (Globorotaliinae; *Globorotalia*) (p. C667-C668).

talia MOROZOVA, 1957, *1316, p. 1110 (type, *Planulina membranacea* EHRENBERG, 1854, *680, p. 25); *Planorotalites* MOROZOVA, 1957, *1316, p. 1112 (type, *Globorotalia pseudoscitula* GLAESSNER, 1937, *793, p. 32); *G. (Astrorotalia)* TURNOVSKY, 1958, *1956, p. 81 (type, *G. (A.) stellaria*)]. Test free, trochospiral, periphery carinate, chambers angular, rhomboid, or angular-conical; sutures may be thickened, depressed to elevated; wall calcareous, finely perforate, but with nonporous keel or peripheral band, surface smooth to cancellate or hispid; aperture interiomarginal, an extraumbilical-umbilical arch bordered by lip, varying from narrow rim to broad spatulate or triangular flap. *Paleoc.-Rec.*, cosmop.——FIG. 533,*1*. *G. tumida* (BRADY), Post-Tert., W.Pac.O.(New Ireland); *1a-c*, opposite sides and edge view, ×44 (*164). ——FIG. 533,*2*; 534. *G. truncatulinoides* (D'ORBIGNY), Rec., N.Atl.O.(Canary Is.), (533,2), N.Pac.O. (Bikini Atoll) (534); 533,*2a-c*, opposite sides and edge view of topotype, ×70 (*164); 534, equat. sec. showing radial bilamellar wall structure,

×75 (*1533).——FIG. 533,*3*. *G. pseudoscitula* GLAESSNER, Paleoc., USA(N.J.); *3a-c*, opposite sides and edge view, ×73 (*1174).——FIG. 533, *4,5*. *G. membranacea* (EHRENBERG), Plio., Eu. (Sicily); *4*, lectotype, here designated, enlarged (*680); *5a-c*, opposite sides and edge view of topotype, ×79 (*2117).

[Although the keeled and nonkeeled forms were previously regarded by us as congeneric (*164), the nonperforate peripheral band or keel is here considered to merit generic distinction; hence the nonkeeled genus *Turborotalia* is rcognized as valid. *Planorotalia* MOROZOVA, 1957, is based on the species *Planulina membranacea* EHRENBERG, to which a Paleocene species commonly has been referred. The species was recorded originally as occurring in the Weisser Kalkstein, Antilebanon, Syria (*680, pl. 25, fig. 41), and Cattolica, Sicily (*680, pl. 26, fig. 43). The Syrian specimen may not even be a planktonic species. The specimen of EHRENBERG'S fig. 43 (here reproduced) generally is regarded as more typical, hence is here designated as lectotype. A specimen from the type locality in Sicily, which is of Pliocene, rather than Paleocene age, is illustrated also to show the full character of this species. The Paleocene species is to be reterred to *G. pseudomenardii* BOLLI.]

Turborotalia CUSHMAN & BERMÚDEZ, 1949, *497, p. 42 [**Globorotalia centralis* CUSHMAN & BERMÚDEZ, 1937, *491, p. 26; OD] [=*Globorotalia (Turborotalia)* CUSHMAN & BERMÚDEZ, 1949, *497, p. 42 (obj.); *Acarinina* SUBBOTINA, 1953, *1847, p. 219 (type, *A. acarinata*)]. Test free, trochospiral, periphery noncarinate; chambers ovate or rounded; sutures commonly depressed; wall finely perforate, surface smooth to hispid; aperture interiomarginal, extraumbilical-umbilical, with bordering lip. [Differs from *Globorotalia* in lacking a keel or nonporous peripheral margin.] *Paleoc.-Rec.* cosmop.——FIG. 533,*6*. **T. centralis* (CUSHMAN & BERMÚDEZ), Eoc., W.Indies(Cuba); *6a-c*, opposite sides and edge view of holotype, ×84 (*164).

Subfamily TRUNCOROTALOIDINAE
Loeblich & Tappan, 1961

[Truncorotaloidinae LOEBLICH & TAPPAN, 1961, p. 309]

Primary aperture on umbilical side, and secondary sutural apertures on spiral side. *L.Eoc.-M.Eoc.*

FIG. 535. Globorotaliidae (Truncorotaloidinae; *1, Truncorotaloides*) (p. C669).

Foraminiferida—Rotaliina—Globigerinacea

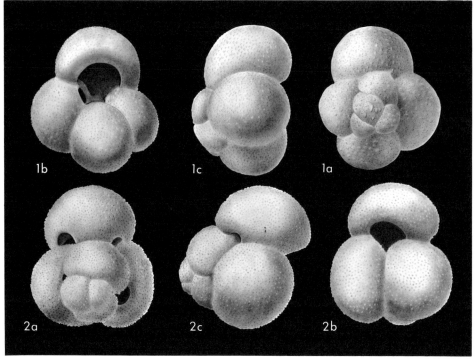

Fig. 536. Globigerinidae (Globigerininae; *1, Globigerina; 2, Globigerinoides*) (p. C669-C670).

Truncorotaloides Brönnimann & Bermúdez, 1953, *233, p. 817 [*T. rohri*; OD]. Test similar to *Globorotalia* but with secondary sutural apertures on spiral side. L.Eoc.-M.Eoc., W.Indies(Trinidad)-Mex.-USA-Eu.——Fig. 535,*1*. *T. rohri*, M.Eoc., Trinidad; *1a-c,* opposite sides and edge view of holotype, ×130 (*164).

Family GLOBIGERINIDAE
Carpenter, Parker & Jones, 1862

[*nom. correct.* Schulze, 1877, p. 29 (*pro* family Globigerinida Carpenter, Parker & Jones, 1862, p. 171) (*nom. conserv.* ICZN Opin. 552)]——[All names cited are of family rank; dagger(†) indicates *partim*]——[=Helicostèguest d'Orbigny, 1826, p. 268 (*nom. neg.*; *nom. nud.*); [=Uvellina† Ehrenberg, 1839, table opp. p. 120 (*nom. nud.*); =Turbinoidae d'Orbigny in de la Sagra, 1839, p. xxxviii, 71 (*nom. nud.*); =Uvellinida Schmarda, 1871, p. 164]——[=Orbulinida Schultze, 1854, p. 52; =Orbulinetta Haeckel, 1894, p. 164; =Orbulinidae Galloway, 1933, p. 326; =Globigerinidee Schwager, 1876, p. 479; =Globigerinidea Schwager, 1877, p. 20; =Globigerininae Bütschli in Bronn, 1880, p. 200; =Globigerínidos Gadea Buisán, 1947, p. 19 (*nom. neg.*)]

Test trochospiral, streptospiral or globular, chambers spherical, ovate or clavate; primary aperture umbilical or spiroumbilical, may have secondary sutural or areal apertures, bullae, and accessory infralaminal apertures. *U.Cret.(Maastricht.)-Rec.*

Subfamily GLOBIGERININAE
Carpenter, Parker & Jones, 1862

[*nom. correct.* Cushman, 1927, p. 87 (*pro* subfamily Globigerinae Carpenter, Parker & Jones, 1862, p. 181)]——[All names cited are of subfamily rank]——[=Globigerinina Jones in Griffith & Henfrey, 1875, p. 320; =Globigerinidae Schwager, 1877, p. 20; =Pulleniatininae Cushman, 1927, p. 89; =Globorotaloidinae Banner & Blow, 1959, p. 7]

Test trochospiral to streptospiral; primary aperture umbilical or spiroumbilical, and may have secondary sutural apertures. *U.Cret.(Maastricht.)-Rec.*

Globigerina d'Orbigny, 1826, *1391, p. 277 [*G. bulloides*; SD Parker, Jones, & Brady, 1865, *1419, p. 36] [=Globigenera Sowerby, 1842, *1819, p. 154 (*nom. null.*); Rhynchospira Ehrenberg, 1845, *675, p. 358 (type, *R. indica*); Pylodexia Ehrenberg, 1858, *683, p. 28 (type, *P. tetratrias*)]. Test free, trochospiral, chambers spherical to ovate; wall calcareous, perforate, radial in structure, surface may be smooth, pitted, cancellated, hispid or spinose; aperture interiomarginal, umbilical, with tendency in some species to extend to slightly extraumbilical position, previous apertures remaining open into umbilicus. *Paleoc.-Rec.,* cosmop.——Fig. 536,*1*. *G. bulloides,* Rec., Adriatic Sea (Porto Corsini, Italy); *1a-c,* opposite sides and edge view of hypotype, ×87 (*164).

Beella Banner & Blow, 1960, *78, p. 26 [*Globigerina digitata* Brady, 1879, *196b, p. 286; OD] [=*Globorotalia (B.)* Banner & Blow, 1960, *78, p. 26 (obj.)]. Test similar to *Globigerina,* but final chambers becoming radial-elongate, periphery non-carinate; aperture interiomarginal, extraumbilical-umbilical. [Because of the distinctive

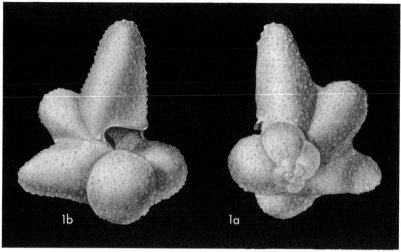

Fig. 537. Globigerinidae (Globigerininae; *1, Beella*) (p. C669-C670).

chamber form, *Beella* is elevated to generic status.] Mio.-Rec., S.Atl.O.-Carib.——Fig. 537,*1*. **B. digitata* (BRADY), Rec., Atl.O.; *1a,b*, opposite sides of hypotype, ×93 (*164).

Globigerinoides CUSHMAN, 1927, *431, p. 87 [**Globigerina rubra* D'ORBIGNY in DE LA SAGRA, 1839, *1611, p. 82; OD]. Test similar to *Globigerina* but with secondary sutural apertures on spiral side. L.Eoc.-Rec., cosmop.——Fig. 536,*2*. **G. ruber* (D'ORBIGNY), Rec., Carib.; *2a-c*, opposite sides and edge view of hypotype, ×73 (*164).

Globigerinopsis BOLLI, 1962, *163A, p. 281 [**G. aguasayensis*; OD]. Test free, trochospiral; chambers spherical to ovate; wall calcareous, perforate, radial in structure, surface smooth, punctate, cancellate, hispid or spinose; aperture in the early stage interiomarginal, umbilical, later becoming spiroumbilical. [Differs from *Globigerina* in the spiroumbilical aperture and from *Hastigerinella* in lacking the radially elongate or clavate chambers.] Mio., W.Indies(Dominican Republic)-S.Am. (E.Venez.).——Fig. 537A,*1*. **G. aguasayensis*, Mio., Venez.; *1a-c*, opposite sides and edge of holotype, showing extended aperture on the spiral side, ×43 (*163A).

Globoconusa KHALILOV, 1956, *1037, p. 249 [**G. conusa* (=*Globigerina daubjergensis* BRÖNNIMANN, 1953, *230, p. 340); OD]. Test small, trochospiral, similar to *Globigerina*, but commonly with strongly convex spiral side; chambers inflated and globular, increasing rapidly in size; wall characteristically spinose; aperture a small rounded umbilical opening, with one or more tiny secondary sutural openings on spiral side against early whorl. [The type-species, *G. conusa*, was described from the Danian of Azerbaidzhan, but is apparently conspecific with *Globigerina daubjergensis*, originally described from Denmark, but of world-wide occurrence in Danian strata. Although *Globoconusa* was described as high-spired, the type-species is quite variable as to height of spire.] Paleoc.(Dan.), Eu.-N.Am.-Carib.-USSR-S.Am.——Fig. 538,*1,2*. **G. daubjergensis* (BRÖNNIMANN), Sweden *(1)*, USA(Tex.) *(2)*; *1a-c*, opposite sides and edge of low-spired hypotype; *2a-c*, opposite sides and edge of high-spired hypotype, ×146 (*1174).

Globoquadrina FINLAY, 1947, *717e, p. 290 [**Globorotalia dehiscens* CHAPMAN, PARR, & COLLINS, 1934, *326, p. 569; OD]. Test free, trochospiral, umbilicate; aperture interiomarginal, umbilical, covered above by apertural flap which may vary from narrow rim to elongate toothlike projection, and in openly umbilicate forms earlier apertures remain open into umbilicus. [*Globoquadrina* differs from *Globigerina* in having prominent apertural flaps covering each aperture.] *U.*

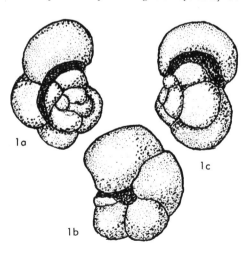

Fig. 537A. Globigerinidae (Globigerininae; *1, Globigerinopsis*) (p. C670).

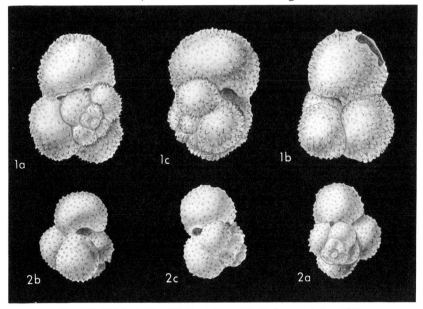

Fig. 538. Globigerinidae (Globigerininae; *1,2, Globoconusa*) (p. C670).

Eoc.-Mio., Australia-N.Z.-Carib.——Fig. 539,6. *G. dehiscens* (CHAPMAN, PARR, & COLLINS), Mio., Australia; *6a-c,* opposite sides and edge view of hypotype, ×107 (*164).——Fig. 539,5. *G. altispira* (CUSHMAN & JARVIS), Mio., Jamaica; umbilical view of holotype, showing prominent umbilical toothlike projections comprising apertural flaps of final whorl of chambers, ×54 (*164).

Globorotaloides BOLLI, 1957, *159, p. 117 [*G. variabilis*; OD] [=*Globigerina (Eoglobigerina)* MOROZOVA, 1959, *1317, p. 1115 (type, *G. (E.) eobulloides*)]. Test free, low trochospiral coil; chambers subglobular to spherical; sutures depressed; wall calcareous, finely perforate, surface smooth to pitted or hispid; aperture interiomarginal, extraumbilical to umbilical in position, and may have small lip. *U.Cret.(Maastricht.)-Rec.,* cosmop.——Fig. 540,*1,2.* *G. variabilis,* Mio., W.Indies(Trinidad); *1a-c,* opposite sides and edge view of holotype; *2a-c,* paratype, all×67 (*159). ——Fig. 540,*3.* *G. eobulloides* (MOROZOVA), Paleoc.(Dan.), USSR(Crimea); *3a-c,* opposite sides and edge view of holotype, ×100 (*1317).

[*Globorotaloides* was originally described as having an asymmetrical final chamber of bulla-like form. Gerontic specimens with atypical final chamber, of larger or smaller than normal size and commonly asymmetrical in position may occur in many species of planktonic and benthonic genera; hence, this feature is not regarded as of generic importance. The genus is here recognized as differing from *Globorotalia* and *Tuborotalia* in its globular chamber form and higher aperture of umbilical or nearly umbilical position. The type-species shows a relationship to both *Globigerina* and *Globoquadrina*, and *Globorotaloides* may have been ancestral to both. It first appears in the latest Maastrichtian.]

Hastigerinella CUSHMAN, 1927, *431, p. 87 [*Hastigerina digitata* RHUMBLER, 1911, *1572a, p. 202 (*non Globigerina digitata* BRADY, 1879) =*Hastigerinella rhumbleri* GALLOWAY, 1933, *762, p. 333; OD]. Test free, trochospiral, early portion with globular chambers, later chambers radially elongate, clavate or cylindrical, with elongate spines concentrated at outer ends of chambers, but commonly broken away in fossil or dead shells; aperture a broad interiomarginal, extraumbilical-umbilical arch, gradually increasing in extent to reach periphery or become spiroumbilical. [With the recognition of *Beella* as a valid genus and *Globigerina digitata* BRADY as its type, the species *Hastigerina digitata* RHUMBLER is no longer a homonym, as the 2 species were not originally placed in the same genus and are not now considered to be congeneric.] *Rec.,* Atl.O.-Pac.O.——Fig. 539,*4.* *H. digitata* (RHUMBLER), Atl.O.; *4a,b,* opposite sides, ×8.5 (*1572a).

Pulleniatina CUSHMAN, 1927, *431, p. 90 [*Pullenia obliqueloculata* PARKER & JONES, 1865, *1418, p. 368; OD]. Test free, globose, trochospiral to streptospiral, early portion as in *Globigerina,* with open umbilicus, later chambers completely enveloping entire umbilical side of previous trochospiral coil, and thus appearing involute; aperture interiomarginal, in young a broad umbilical arch, as in *Globigerina,* in adult a broad low extraumbilical arch at base of final enveloping chamber, bordered above by thickened lip but because of streptospiral plan of growth, not directly opening into earlier umbilicus. [*Pulleniatina* resembles *Globigerina* in early development but differs in its later streptospiral coiling with embracing final chamber and its characteristic extraumbilical peripheral aperture. It differs from *Globigerapsis* in having a single aperture, rather than multiple apertures in the final chamber against sutures of the early coil.] *Plio.-Rec.*, cosmop.——Fig. 539, *1,2. *P. obliqueloculata* (PARKER & JONES), Rec., S.Atl.O. *(1)*, Pac.O. *(2)*; *1a-c*, opposite sides and edge of paratype, ×82; *2*, dissected hypotype showing neanic *Globigerina* stage with typical

Fig. 539. Globigerinidae (Globigerininae; *1,2, Pulleniatina; 3, Subbotina; 4, Hastigerinella; 5,6, Globoquadrina*) (p. C670-C673).

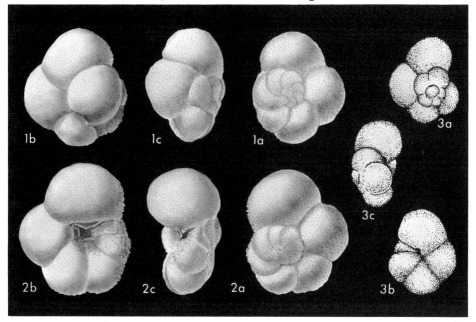

Fig. 540. Globigerinidae (Globigerininae; *1-3, Globorotaloides*) (p. C671).

umbilical aperture and change in plane of coiling with later development, ×57 (*164).

Subbotina BROTZEN & POZARYSKA, 1961, *243, p. 160 [*Globigerina triloculinoides PLUMMER, 1927, *1461, p. 134; OD]. Test trochospiral; chambers increasing rapidly in size and strongly inflated; sutures depressed; wall calcareous, perforate, radial in structure, surface reticulate or pitted, rather than spinose; aperture umbilical-extraumbilical, with distinct lip. [Originally defined solely on the basis of wall surface, the present genus apparently includes species which are closely similar in form and apertural character. The coarsely pitted surface is found in species with low and slightly extraumbilical aperture and distinctive lip, none of which are found in typical *Globigerina*.] *Paleoc.(Dan.)-Rec.*, cosmop.——Fig. 539, *3*. **S. triloculinoides* (PLUMMER), Midway., USA (Tex.); *3a-c*, opposite sides and edge view of topotype, ×73 (*1174).

Subfamily SPHAEROIDINELLINAE
Banner & Blow, 1959

[Sphaeroidinellinae BANNER & BLOW, 1959, p. 5]

Test trochospiral; chambers with flange-like margins; wall with secondary thickening and reduced perforations; primary aperture umbilical, may have secondary sutural apertures. *Mio.-Rec.*

Sphaeroidinella CUSHMAN, 1927, *431, p. 90 [*Sphaeroidina dehiscens PARKER & JONES, 1865, *1418, p. 369; OD]. Early portion trochospiral, with 2 or 3 much-embracing chambers of final whorl enveloping early whorl, chambers with marginal flanges extending out toward those of opposing chambers and partially obscuring arched apertures; wall calcareous, perforate, pores extremely large and closely arranged in early stage, giving an almost lattice-like appearance, area between pores raised and cancellated; in later chambers somewhat irregularly fimbriate or scalloped flange of clear shell material, relatively poreless, is formed around chamber base, tending to coalesce laterally and become much produced, exterior surface of final chambers becoming smooth and glassy due to external secondary deposit; primary aperture in young interiomarginal and umbilical, as in *Globigerina*, but later covered by embracing final chamber; one or more sutural secondary apertures may occur on opposite sides of final chamber, and may be partially obscured by overhanging chamber flanges which parallel sutures, or chambers may be distinctly separated, with wide open area between flanges of opposing chambers, with small arched bullae crossing the sutural slit and partially covering apertural regions, walls of bullae smoothly finished and with finer pores than in chambers, although similarly spaced. *U.Mio.-Rec.*, cosmop.——FIG. 541,*1-3*. **S. dehiscens* (PARKER & JONES), Rec., Pac.O. *(1,3)*, Atl.O. *(2)*; *1a,b*, side and edge views of paratype, showing well-developed sutural flanges; *3*, small paratype showing bulla over sutural aperture; *2*, dissected hypotype showing neanic *Globigerina* stage with large pores and umbilical aperture; all ×38 (*164).

Sphaeroidinellopsis BANNER & BLOW, 1959, *77, p. 15 [*Sphaeroidinella dehiscens subdehiscens BLOW,

1959, *149, p. 195; OD]. Test trochospiral, similar to *Globigerina*, with wall structure like that of *Sphaeroidinella*, primary wall covered by secondary layer reducing porosity; primary aperture umbilical, with bordering lip, no sutural secondary apertures. *L.Mio.-U.Mio.*, S.Am.-Carib.-Indon.-N.

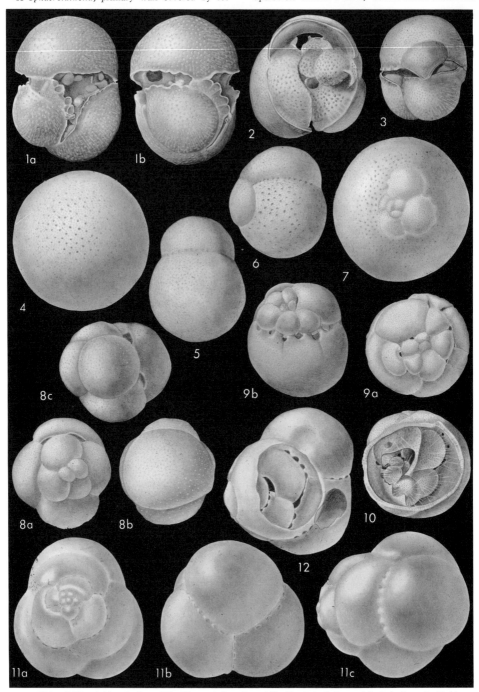

Fig. 541. Globigerinidae (Sphaeroidinellinae; *1-3, Sphaeroidinella;* Orbulininae; *4-7, Orbulina; 8, Globigerapsis; 9,10, Porticulasphaera; 11,12, Candeina*) (p. C673-C676).

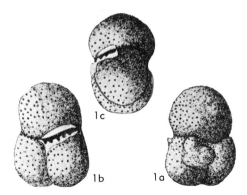

Fig. 542. Globigerinidae (Sphaeroidinellinae; *1*, *Sphaeroidinellopsis*) (p. C673-C675).

Z.-Eu.——Fig. 542,*1*. **S. subdehiscens* (Blow), Mio., Venez.; *1a-c,* opposite sides and edge of holotype, ×47 (*77).

Subfamily ORBULININAE Schultze, 1854

[*nom. transl.* Cushman, 1927, p. 89 (*ex* family Orbulinida Schultze, 1854)] [=Candeininae Cushman, 1927, p. 90]

Test trochospiral to streptospiral, later stage enveloping or globular; primary aperture not visible in adult, secondary apertures multiple and sutural or areal. *Eoc.-Rec.*

Orbulina d'Orbigny in de la Sagra, 1839, *1611, p. 2 [**O. universa*; OD (M)] [=*Coscinosphaera* Stuart, 1866, *1845, p. 328 (type, *C. ciliosa*); *Candorbulina* Jedlitschka, 1934, *986, p. 20 (type, *C. universa*); *Biorbulina* Blow, 1956, *148, p. 69 (type, *Globigerina bilobata* d'Orbigny, 1846, *1395, p. 164)]. Test free, adult generally spherical and composed of a single chamber, rarely 2- or 3-chambered, early chambers trochospiral in microspheric form, in adult the globigerine coil may remain visible at one side or may be completely enveloped by final spherical chamber, or test may consist of number of completely enveloping, concentric globular chambers (probably megalospheric form); primary aperture interiomarginal, umbilical in early globigerine stage where this is present, areal in adult, with numerous small openings scattered over one side or over much of test, small sutural secondary openings commonly found around early globigerine chambers of specimens where these are visible at surface; in sexual reproduction gametes formed within spherical test, accompanied by gradual resorption of wall of early globigerine chambers and in Recent forms by descending vertical migration in water column from surface to depth of about 300 m., gametes then escaping through large perforations in wall; gametes biflagellate, with homogeneous nucleus, no axostyle, and with large oily inclusion (*1105); cytoplasm with abundant areolated ectoplasm surrounding test; pseudopodia elongate, radiating, numerous and rigid. *L.Mio.-Rec.,* cosmop.——Fig. 541,*4-7*. **O. universa*, Rec., Atl.O. *(4-6)*, Mio., USA(Fla.) *(7)*; *4,* typical spherical microspheric specimen, with globigerine stage completely enclosed, ×40; *5,6,* 2- and 3-chambered (probably megalospheric forms), ×40; *7,* microspheric hypotype with globigerine stage visible at side of test, ×70 (*164).

[Surface specimens of *Orbulina* commonly contain embryonic globigerine chambers (*200, *1105), which are resorbed during gamete formation and accompanied by gradual descent of specimens in water column. Brady (*200, p. 609, 610) noted that specimens taken from the bottom had thicker walls, which were commonly laminated, "affording clear evidence that the increase in thickness has taken place . . . by the formation of successive layers . . . ." The enclosed spheres are loose and easily separated. Furthermore, 2-chambered shells were not infrequent in bottom-ooze, and rarely 3-chambered ones occur. Le Calvez (*1105) postulated that a benthonic stage might alternate with the planktonic one in *Orbulina* (as in *Tretomphalus*), but no direct evidence is available, the life cycle of the planktonic species being yet only partially known owing to the difficulties in culturing them. Possibly these bottom-specimens represent a megalospheric test, the so-called *Biorbulina* representing specimens with the proloculus visible at the side, just as the early coiled stage of the microspheric form can be seen at one side of some planktonic specimens, and specimens with concentric spherical chambers (*200, pl. 81, fig. 26) may represent completely involute megalospheric tests, just as some of the surface specimens completely enclose the small globigerine chambers of the microspheric test (*200, pl. 81, fig. 13). Further studies of living specimens of the planktonic genera are needed.]

Candeina d'Orbigny in de la Sagra, 1839, *1611, p. 107 [**C. nitida*; OD (M)]. Test free, trochospiral, relatively high-spired; chambers inflated; primary aperture in early stage interiomarginal, umbilical, later with tiny secondary sutural apertures on each side of primary aperture; no primary opening in adult tests, small rounded sutural secondary apertures almost completely surrounding later chambers. [In its development *Candeina* is similar to *Globigerina* and then to *Globigerinoides,* but differs in the absence of a primary aperture in the adult, and in the numerous small sutural secondary openings on both spiral and umbilical sides of the adult.] *Mio.-Rec.,* cosmop.——Fig. 541,*11,12*. **C. nitida,* Rec., Atl.O.; *11a-c,* opposite sides and edge of hypotype showing numerous sutural secondary apertures, ×82; *12,* dissected hypotype showing neanic *Globigerinoides* stage with primary umbilical aperture, ×77 (*164).

Globigerapsis Bolli, Loeblich & Tappan, 1957, *164, p. 33 [**G. kugleri*; OD]. Test free, subglobular; early portion trochospiral, with subglobular chambers, final chamber embracing and covering umbilical region of early coil; primary aperture interiomarginal, umbilical in young stage, covered in adult by enveloping final chamber, with 2 or more arched sutural secondary apertures at lower margin of final chamber, at contact with sutures of earlier whorl. *Eoc.,* Carib.-N.Am.-S.Am.-Eu.-N.Z.-Japan.——Fig. 541,*8*. **G. kugleri,* W. Indies (Trinidad); *8a-c,* opposite sides and edge view of holotype, showing early trochospiral stage and later embracing chamber with sutural openings, ×72 (*164).

[Differs from *Globigerinatheka* in lacking the small angular bullae covering the secondary apertures. It differs from *Globigerinoides* in the absence of an umbilical primary aperture in the adult and from *Globigerinoides* and *Porticulasphaera* in lacking multiple apertures in earlier chambers. SAITO (1962, *1620, p. 219, 220) erroneously regarded *G. kugleri* as a synonym of *Globigerapsis mexicana*, because of the poor original figures of the latter. As the holotypes of the 2 species are specifically distinct, as here recognized, *G. kugleri* is not a synonym of *G. mexicana*. SAITO also "emended" the generic description and erroneously "designated" *G. mexicana* as type-species for *Globigerapsis*, though it was already the type-species for *Porticulasphaera*. As comparison of the holotypes of *G. kugleri* and *P. mexicana* shows them to be both specifically and generically distinct, SAITO's generic emendations are invalid. New type designations are impossible under the Rules of Zoological Nomenclature as these have been previously designated and are irrevocably fixed; hence *Globigerina mexicana* cannot be regarded as the type-species of *Globigerapsis*.]

Porticulasphaera BOLLI, LOEBLICH & TAPPAN, 1957, *164, p. 34 [*Globigerina mexicana CUSHMAN, 1925, *421, p. 6 =Porticulasphaera beckmanni SAITO, 1962, *1620, p. 221; OD]. Test free, subglobular, early portion trochospiral, final chamber much inflated to almost spherical and strongly enveloping, covering umbilical region of early coil; primary aperture in early portion interiomarginal umbilical, covered by final enveloping chamber of adult, secondary sutural openings on spiral side. *M.Eoc.*, N.Am.-Eu.-Carib.-Japan.——FIG. 541,9,10. *P. mexicana* (CUSHMAN), W.Indies (Trinidad); *9a,b*, spiral side and edge view of hypotype, showing early trochospiral coil, final enveloping chamber and secondary sutural apertures; *10*, dissected hypotype showing neanic *Globigerina*-stage with umbilical aperture, coarse perforations, fine spines, and thick radially perforated final chamber wall, ×45 (*164).

[*Porticulasphaera* resembles *Orbulina* in its strongly embracing although less inflated final chamber but differs in having the early coil always visible and in having secondary sutural openings but no areal secondary apertures. SAITO (1962, *1620, p. 219-221) erroneously designated a new type-species (*P. beckmanni* SAITO, 1962) for *Porticulasphaera* in an "emendation" of the genus. The originally designated type-species' cannot be changed according to the International Rules. On the basis of the poor original figures of *G. mexicana* (the true type-species of *Porticulasphaera*), SAITO believed *Globigerapsis kugleri* to be a synonym. Examination of the holotype in the U.S. National Museum shows *G. mexicana* to be distinct from *G. kugleri* and similar to the better-preserved specimens here illustrated, even though not all the generic characters were well shown in the original figures. *Porticulasphaera beckmanni* is a junior synonym of *Globigerina mexicana*. Erroneous later identification as *G. mexicana* of other specimens has no bearing on the status of the species, which must include only forms conspecific with the holotype. SAITO's "emendations" of *Globigerapsis* and *Porticulasphaera* are therefore invalid, as are the "designations" of new type-species for these genera.]

Subfamily CATAPSYDRACINAE Bolli, Loeblich & Tappan, 1957

[Catapsydracinae BOLLI, LOEBLICH & TAPPAN, 1957, p. 36] [=Globigerinatellinae SIGAL, 1958, p. 263; =Globigerinitinae BERMÚDEZ, 1961, p. 1.261]

Test trochospiral to enveloping; chambers spherical to ovate; primary aperture umbilical, may have secondary sutural or areal apertures, one or more apertural bullae present in adult, with infralaminal accessory apertures. *M.Eoc.-Rec.*

Catapsydrax BOLLI, LOEBLICH & TAPPAN, 1957, *164, p. 36 [*Globigerina dissimilis CUSHMAN & BERMÚDEZ, 1937, *491, p. 25; OD]. Test free, similar to *Globigerina* in early development, with primary umbilical aperture; adult with single umbilical bulla over aperture, and with one or more accessory infralaminal apertures. *M.Eoc.-Mio.*, Carib.-N.Am.-Eu.——FIG. 543,1,2. *C. dissimilis* (CUSHMAN & BERMÚDEZ), Eoc., Cuba *(1)*, Oligo., W.Indies(Trinidad) *(2)*; *1a-c*, opposite sides and edge view of holotype; *2a,b*, edge and umbilical views of hypotype; all ×46 (*164).

Globigerinatella CUSHMAN & STAINFORTH, 1945, *525, p. 68 [*G. insueta*; OD]. Test free, subglobular, early portion trochospiral, final chamber embracing and obscuring interiomarginal, umbilical primary aperture, later chambers with secondary sutural and areal apertures which are surrounded by distinct lips and may be covered by small knobby pustule-like areal bullae and more irregular spreading sutural bullae, all bullae with infralaminal accessory apertures. *L.Mio.*, Carib.-N.Am.-Pac.O.——FIG. 543,3,4. *G. insueta*, L. Mio., W.Indies(Trinidad); *3a,b*, spiral and edge views of paratype showing early trochospiral stage, enveloping final chamber, and areal and sutural bullae; *4*, dissected topotype showing areal aperture exposed when bulla is partially removed and infralaminal accessory openings at margin of remaining part of the bulla, ×93 (*164).

Globigerinatheka BRÖNNIMANN, 1952, *226, p. 27 [*G. barri*; OD]. Test free, globular, early chambers trochospiral, as in *Globigerina*, later with large enveloping final chamber covering previous umbilical side, as in *Orbulina*; sutures depressed, radial; primary aperture interiomarginal, umbilical, but covered in adult by final enveloping chamber, secondary sutural apertures on spiral side, covered by small bullae, each with one or more infralaminal accessory apertures. [*Globigerinatheka* is similar to *Globigerapsis* but has bullae covering the sutural apertures.] *M.Eoc.-U.Eoc.*, W.Indies(Trinidad).——FIG. 543,5. *G. barri*, M.Eoc.; *5a-c*, opposite sides and edge of hypotype, ×72 (*164).

Globigerinita BRÖNNIMANN, 1951, *224, p. 18 [*G. naparimaensis*; OD] [=Turborotalita BLOW & BANNER in EAMES, BANNER, BLOW & CLARKE, 1962, *651, p. 122 (type, *Truncatulina humilis* BRADY, 1884, *200, p. 665)]. Test free, trochospiral, final chamber modified and extending across umbilical region; primary aperture interiomarginal and umbilical, but in adult covered by modified final chamber which extends across umbilical region, one or more small arched supplementary apertures present at umbilical margin of final chamber. *Mio.-Rec.*, cosmop.——FIG. 543, 8. *G. naparimaensis*, Mio., W.Indies(Trinidad); *8a-c*, opposite sides and edge view of holotype, with primary umbilical aperture visible on penultimate chamber through thin-walled modified last chamber which has 2 supplementary apertures,

×163 (*164).——Fig. 543,9. *G. parkerae* Loeblich & Tappan, Rec., Gulf Mex.; *9a,b*, umbilical and edge of holotype, ×140 (*1170).

[*Turborotalita* was based on species with the umbilical-extraumbilical aperture covered by a bulla that "may take the apparent form of a modified final chamber, which spreads ventrally partially or wholly to conceal the ventral

Fig. 543. Globigerinidae (Catapsydracinae; *1,2, Catapsydrax; 3,4, Globigerinatella; 5, Globigerinatheka; 6, Tinophodella; 7, Globigerinoita; 8,9, Globigerinita*) (p. C676-C678).

umbilicus." *Globigerinita parkerae* was also included in this genus by BLOW & BANNER. As can be seen in the illustration of the holotype of *G. naparimaensis* given here (Fig. 543,8), the primary aperture is extraumbilical and covered by a modified final chamber. Specimens included by various authors in *Globigerinita*, but which have true bullae rather than such a modified final chamber, are correctly referred to *Tinophodella*.]

Globigerinoita BRÖNNIMANN, 1952, *226, p. 26 [*G. morugaensis*; OD]. Test free, trochospiral; primary aperture umbilical in position, as in *Globigerina*, with one or more secondary sutural apertures on spiral side, as in *Globigerinoides*, with primary aperture covered by umbilical bulla, secondary apertures of spiral side may be covered by sutural bullae, with commonly 2 or 3 accessory infralaminal apertures at margins of each bulla. *U.Mio.*, W.Indies(Trinidad).——FIG. 543,7. *G. morugaensis*; 7a-c, opposite sides and edge view of holotype, ×130 (*164).

Tinophodella LOEBLICH & TAPPAN, 1957, *1170, p. 113 [*T. ambitacrena*; OD]. Test free, trochospiral, similar to *Globigerina*; primary aperture interiomarginal, umbilical, but in adult completely covered by irregular umbilical bulla expanding along earlier sutures, numerous accessory apertures along bulla margins at junction with sutures of earlier chambers and along contact with primary chambers. [Differs from *Globigerinita* in having a distinct umbilical bulla with numerous small accessory apertures opening beneath its margin, whereas *Globigerinita* has a modified final chamber with supplementary apertures.] *Mio.-Rec.*, Atl. O.-Carib.-Eu.——FIG. 543,6. *T. ambitacrena*, Rec., Atl.; 6a-c, opposite sides and edge view of holotype showing distinct umbilical bulla with marginal accessory apertures, ×73 (*1170).

Superfamily ORBITOIDACEA Schwager, 1876

[*nom. correct.* LOEBLICH & TAPPAN, 1961, p. 310 (*pro* superfamily Orbitoidicae BRÖNNIMANN, 1958, p. 167)]——[In synonymic citations superscript numbers indicate taxonomic rank assigned by authors (¹superfamily, ²family group, ³group); dagger(†) indicates *partim*]——[=³Fenestrifera GRAY, 1858, p. 270; =¹Orthoklinostegia† EIMER & FICKERT, 1899, p. 685 (*nom. nud.*); =²Flexostylidia† RHUMBLER in KÜKENTHAL & KRUMBACH, 1923, p. 87; =¹Bilamellidea† REISS, 1957, p. 127 (*nom. nud.*); =¹Discocyclinidea PURI, 1957, p. 139]

Test basically coiled, with radially laminated calcite walls, primarily formed double septa, walls of 2 layers, outer lamella covering all previously deposited parts of test as well as forming new chamber, inner lining confined to each chamber and wedging out at margins, present on distal face of chamber interior, on its roof and lateral walls. *Cret. Rec.*

Family EPONIDIDAE Hofker, 1951

[*nom. correct.* THALMANN, 1952, p. 984 (*pro* Eponidae HOFKER, 1951, p. 321)]——[Superscript numbers indicate taxonomic rank assigned by authors (¹family, ²subfamily; dagger(†) indicates *partim*] [=¹Radiolata† CROUCH, 1827, p. 41 (*nom. nud.*); =¹Radiolidida† BRODERIP, 1839, p. 321; =¹Cyclospiridae EIMER & FICKERT, 1899, p. 702 (*pro* Cyclospira EIMER & FICKERT, 1899, *non* HALL & CLARKE, 1894; =²Pulvinulininae SCHUBERT, 1921, p. 152; =¹Pulvinulinidae HOFKER, 1951, p. 448; =²Eponidinae SUBBOTINA in RAUZER-CHERNOUSOVA & FURSENKO, 1959, p. 269]

Test free, low trochospiral coil or may be uncoiled; aperture basal or areal, single or multiple, and may be covered by plate or spongy material. *Paleoc.-Rec.*

Eponides DE MONTFORT, 1808, *1305, p. 127 [*Nautilus repandus* FICHTEL & MOLL, 1798, *716, p. 35; OD] [=*Pulvinulus* LAMARCK, 1816, *1089, p. 14 (obj.); *Placentula* LAMARCK, 1822, *1090, p. 620 (type, *P. pulvinata*, =*Nautilus repandus* FICHTEL & MOLL, 1798) (*non Placentulae* SOLDANI, 1795, *1810, p. 237, pl. 161a-d); *Pulvulina* PARKER & JONES in CARPENTER, PARKER, & JONES, 1862, *281, p. 200, 210 (obj.); *Eponidopsis* REISS, 1960, *1533, p. 16 (type, *Eponides lornensis* FINLAY, 1939, *717a, p. 522)]. Test free, trochospiral, biconvex, periphery angled to distinctly carinate with narrow to broad depression in umbilical region (pseudoumbilicus), septa double, with intraseptal passages, sutures curved on spiral side, nearly radial to curved or sigmoid on umbilical side; wall calcareous, finely perforate, radial in structure, bilamellar, with septa primarily doubled, surface may have secondarily formed pustules or ridges formed on previous whorl below aperture; primary aperture an interiomarginal arch without internal tooth plate, intercameral foramen may be restricted and partly areal in position. *Eoc.-Rec.*, cosmop.——FIG. 544,1. *E. repandus* (FICHTEL & MOLL), Rec., Italy(Gulf Naples); 1a-c, opposite sides and edge view of neotype, ×39 (*1186).

[The validity of *Eponides* has recently been questioned because of poor original figures and illustrations and later erroneous references to it of dissimilar forms. For greater stability in nomenclature, a neotype was selected for *Nautilus repandus* from the type area (LOEBLICH & TAPPAN, 1962, *1186, p. 35, 36) and is here illustrated. The recently proposed *Eponidopsis* is a junior synonym of *Eponides*, *Pulvinulus*, *Placentula*, and *Pulvulina*, and all of the last 3 would have priority over *Eponidopsis* if *Eponides* were suppressed as a *nomen dubium*, as suggested by REISS (1960, *1533). Selection of a neotype places the genus *Eponides* on a firm basis, however.]

Cibicorbis HADLEY, 1934, *846, p. 26 [*C. herricki*; OD] [=*Sakhiella* HAQUE, 1956, *876, p. 155 (type, *S. nammalensis*)]. Test trochospiral, periphery angled and carinate, biconvex to planoconvex, spiral side evolute and flattened, umbilical side elevated and involute, later chambers tending to become inflated; chambers broad, low, arched, increasing more rapidly in breadth than in height, resulting in somewhat flaring outline; sutures curved, distinct, thickened and elevated; wall calcareous, coarsely perforate, radial in structure, viz., in *C. nammalensis* (HAQUE), unknown in *C. herricki*, lamellar character not described; aperture an interiomarginal slit on umbilical side, extending from near umbilicus to periphery and covered by large apertural flap which projects over umbilical region. *Paleoc.-Oligo.*, Carib.(Cuba)-

Asia(Pak.).——FIG. 544,2. *C. herricki, Oligo., Cuba; 2a-c, opposite sides and edge view of paratype, ×60 (*2117).——FIG. 544,3. C. nammalensis (HAQUE), Paleoc., Pak.; 3a-c, opposite sides and edge view of holotype, ×23 (*876).

Cincoriola HAQUE, 1958, *877, p. 103 [pro Punjabia

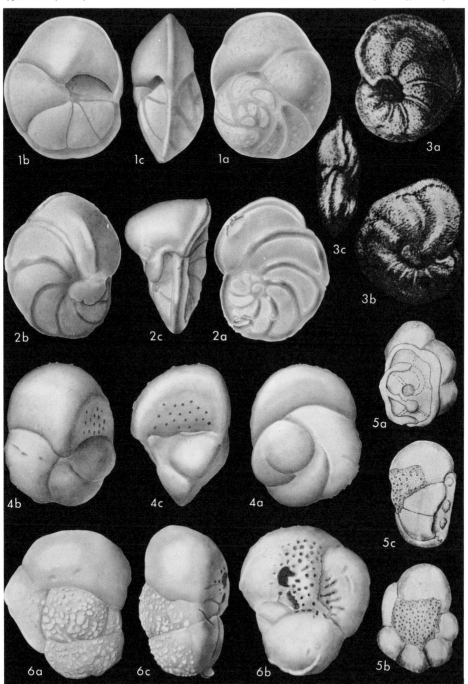

FIG. 544. Eponididae; *1, Eponides; 2,3, Cibicorbis; 4, Cribrogloborotalia; 5, Cincoriola; 6, Hofkerina* (p. C678-C680).

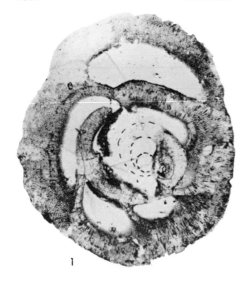

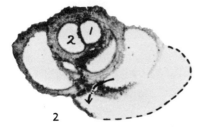

Fig. 545. Eponididae; *1,2, Hofkerina* (p. C680).

Haque, 1956, *876, p. 152 (obj.) (*non* Eames, 1952)] [*Punjabia ovoidea* Haque, 1956, *876, p. 153; OD]. Test trochospiral, spiral side truncate, flattened, evolute, opposite side umbilicate, involute, umbilical region covered with a perforate plate; wall calcareous, radially built, lamellar character unknown; aperture an interiomarginal slit near umbilicus or consisting of perforations in umbilical plate. *Paleoc.*, Asia(Pak.).——Fig. 544, 5. *C. ovoidea* (Haque); *5a-c,* opposite sides and edge of holotype, ×36 (*876).

Cribrogloborotalia Cushman & Bermúdez, 1936, *490, p. 63 [*C. marielina*; OD]. Test free, trochospiral, plano-convex, with flattened spiral side and elevated umbilical side, margin of apertural face sharply angled, resulting in subconical chambers; wall calcareous, finely perforate, wall structure and lamellar structure unknown; aperture consisting of numerous rounded areal pores scattered over sharply defined apertural face. *Eoc.*, Carib. (Cuba)-USA (Fla.).——Fig. 544,4. *C. marielina*, Cuba; *4a-c,* opposite sides and edge view of holotype, ×36 (*2117).

[Similar to *Eponides* in the sharply angled apertural face, differing in presence of the areal aperture and absence of an interiomarginal one. It differs from *Poroeponides* in the sharply angled apertural face, absence of an interiomarginal aperture, in the elevated, rather than depressed, umbilical side, and the closed umbilical area. It is not considered to be related to the planktonic *Globorotalia*.]

Hofkerina Chapman & Parr, 1931, *324, p. 237 [*Pulvinulina semiornata* Howchin, 1889, *966, p. 14; OD]. Test free, large, to 2.2 mm. diam., trochospirally coiled, chambers few and inflated, periphery broadly rounded, noncarinate, spiral side with "pillars" in wall, umbilical side inflated, may have slight umbilical depression; wall calcareous, finely perforate, radial in structure, thick and laminated, bilamellar, surface of spiral side ornamented with numerous irregular pustules; primary aperture a small, arched opening, interiomarginal and umbilical in position, additional areal openings and sutural pores occur in umbilical depression of final chamber, and primary opening may not be present in large specimens, openings from umbilical area into chambers of last whorl may form by resorption. [Originally assigned to the Victoriellidae, *Hofkerina* was placed in the Pegidiidae by Galloway (*762) and definitely excluded from the Victoriellidae by Glaessner & Wade (*797) because of the finely perforate wall and absence of axial spaces when seen in vertical section. It is here placed in the Eponididae, and considered closely related to *Sestronophora*, from which it differs in lacking a peripheral keel, in having a thick wall, and in the presence of "pillars."] *Mio.*, Australia(Vict.). ——Fig. 544,6; 545,*1,2*. *H. semiornata* (Howchin); *544,6a-c,* opposite sides and edge view, ×21 (*2117); *545,1,* horiz. sec. of paratype showing thick, bilamellar wall; *2,* axial sec., arrow showing position of intercameral foramen, ×48 (*797).

Neocribrella Cushman, 1928, *436, p. 6 [*Discorbina globigerinoides* Parker & Jones, 1865, *1418, p. 385, 421; OD]. Test free, trochospiral, umbilicus closed, chambers few, inflated, subglobular, rapidly enlarging; wall calcareous, perforate, radial in structure, lamellar character not known; umbilical region covered by platelike area with numerous large pores serving as an aperture. [A lectotype, here designated, was isolated by us from the original material of Parker & Jones. The lectotype (BMNH-P41661) and paratypes (P41660) are from the Middle Eocene (Lutetian) of Grignon, France.] *M.Eoc.(Lutet.),* Eu.——Fig. 546,*1*. *N. globigerinoides* (Parker & Jones), Fr.; *1a-c,* opposite sides and edge view of topotype, ×78 (*2117).

Neoeponides Reiss, 1960, *1533, p. 17 [*Rotalina schreibersii* d'Orbigny, 1846, *1395, p. 154; OD] [=*Cyclospira* Eimer & Fickert, 1899, *692, p. 702 (obj.) (*non* Hall & Clarke, 1894)]. Test free, trochospiral, plano-convex to inequally biconvex, periphery angled and carinate; sutures thickened, oblique and curved on elevated spiral side, radial on umbilicate opposite side, depressed and thickened near umbilical margin, septa pri-

Foraminiferida—Rotaliina—Orbitoidacea

marily double (bilamellar) with intraseptal passages; wall calcareous, coarsely perforate, radial in structure, secondary thickening of septa near umbilical margin may form an elevated ring; primary aperture an interiomarginal arch extending from periphery to umbilicus, bordered by im-

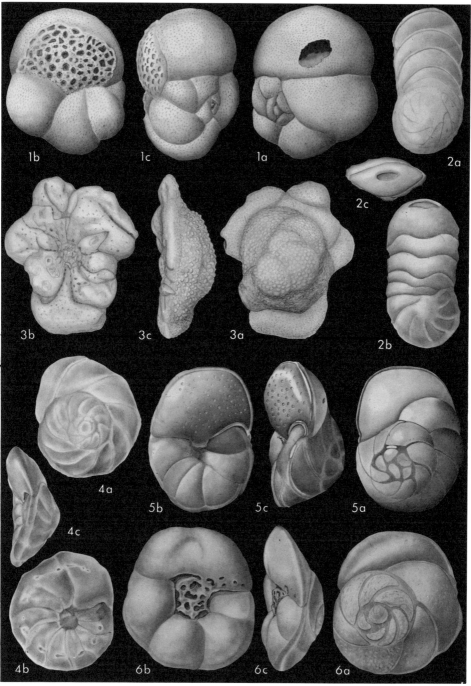

Fig. 546. Eponididae; *1, Neocribrella; 2, Rectoeponides; 3, Planopulvinulina; 4, Paumotua; 5, Poroeponides; 6, Sestronophora* (p. C680, C682-C684).

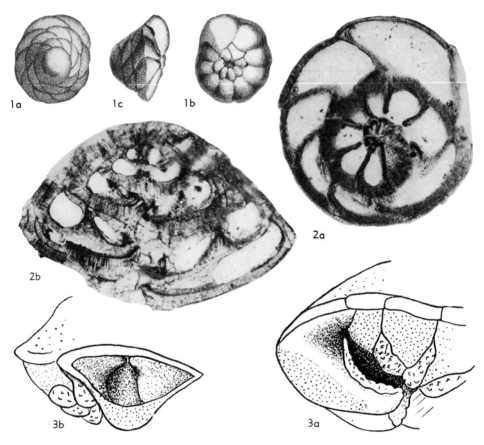

Fig. 547. Eponididae; *1-3, Neoeponides* (p. C680-C682).

perforate lip which may be pustulose or granular. [*Neoeponides* was described as a new genus by REISS but was preoccupied by the isogenotypic *Cyclospira* EIMER & FICKERT, 1899, which is a homonym of *Cyclospira* HALL & CLARKE, 1894.] *Paleoc. - Rec.,* Eu.-Asia (Israel)-Afr.-Pac. O.-Medit. Sea.-Red Sea-Atl.O.——FIG. 547,*1-3*. **N. schreibersii* (D'ORBIGNY), Mio., Aus. (*1*), Israel (*2a*), Morocco (*2b*); *1a-c,* opposite sides and edge view of holotype, ×17 (*700); *2a,* equat. sec. showing sulcus at inner edge of septa, ×55; *2b,* axial sec. with recurved edge of septa resulting in appearance of tooth plate, ×70; *3a,* apert. view, ×180; *3b,* chamber interior with recurved inner edge (inframarginal sulcus), ×180 (*1533).

Paumotua LOEBLICH, 1952, *1151, p. 192 [**Eponides terebra* CUSHMAN, 1933, *460, p. 89; OD]. Test free, trochoid, plano-convex, umbilical side flattened and umbilicate, spiral side convex, chambers numerous; wall calcareous hyaline; aperture a low interiomarginal arch between periphery and umbilicus, supplementary apertures in row paralleling periphery and in line with main aperture on umbilical side, consisting of one or more open pores or slits which increase in size and number as chambers increase in size. [Differs from *Eponides* in possessing supplementary apertures on the umbilical side and from *Poroeponides* in having fewer pores per chamber and in having these on earlier chambers rather than restricted to the final chamber.] *Rec.,* Pac.O.——FIG. 546,*4*. **P. terebra* (CUSHMAN), Paumotu Is.; *4a-c,* opposite sides and edge view of holotype, ×36 (*1151).

Planopulvinulina SCHUBERT, 1921, *1694, p. 153 [**Pulvinulina dispansa* BRADY, 1884, *200, p. 687; SD CUSHMAN, 1928, *439, p. 273]. Test attached, large, plano-convex, early chambers in irregular trochoid spire, later chambers more irregularly arranged, variable in size and outline, and spreading over attachment; spiral side strongly tuberculate, with numerous fine pores filling area between tubercles, early chambers with coarser and more closely spaced tubercles, peripheral and more spreading chambers having smaller and more

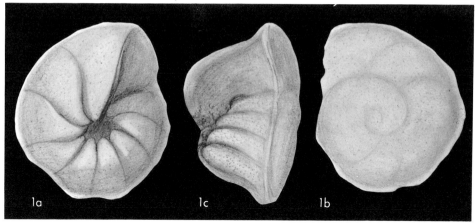

Fig. 548. Eponididae; *1, Pseudogloborotalia* (p. C683).

widely spaced tubercles and more numerous pores between them, umbilical surface flattened, rather smooth, outer margin of chambers with numerous fine pores like those of spiral surface; wall calcareous, hyaline, wall microstructure not known; aperture consisting of widely spaced large pores scattered over surface and in sutural rows on umbilical side. [The lectotype, here designated (BMNH-ZF3641, *ex* ZF2210) is one of the original syntypes of BRADY but not that originally figured, which, although larger, is incomplete. The remaining paratypes are BMNH-ZF2210.] *Late Tert.-Rec.*, Atl.O.——FIG. 546,*3*. **P. dispansa* (BRADY), Rec., off coast of Madeira Is.; *3a-c,* opposite sides and edge view of lectotype, ×12 (*2117).

Poroeponides CUSHMAN, 1944, *478, p. 34 [**Rosalina lateralis* TERQUEM, 1878, *1889, p. 25; OD]. Test free, trochospiral, plano-convex to biconvex, periphery angled and carinate, umbilical region excavated; chambers numerous, sutures oblique and curved on spiral side, radial on umbilical side; wall calcareous, perforate; primary aperture an interiomarginal arch extending from umbilicus to peripheral keel, with narrow bordering lip, small umbilical flap projects from mid-point of chamber into umbilical area, in addition rounded areal pores are scattered over face of final chamber on umbilical side, and some of those of earlier chambers may also remain open. [Differs from *Eponides* in having an areal multiple aperture and in the last chamber having a broad, flattened umbilical side, without the sharply defined apertural face defined by a distinct angle in the final chamber, as in *Eponides*. The umbilical region is depressed (pseudoumbilicus) in *Poroeponides*.] *Plio.-Rec.*, Medit. Sea-Is. Rhodes-Atl.O.-Pac.O.——FIG. 546,*5*. **P. lateralis* (TERQUEM), Rec., USA (R.I.); *5a-c,* opposite sides and edge view, ×44 (*2117).

Pseudogloborotalia HAQUE, 1956, *876, p. 184 [**P. ranikotensis*; OD]. Test trochospiral, plano-convex, periphery angular to keeled, spiral side flat, with sutures obscure, but oblique and curved, umbilical side strongly elevated, with incised straight and radial sutures; wall calcareous, perforate, radial in microstructure, surface smooth and unornamented; aperture a low interiomarginal arch, between periphery and umbilical shoulder. *Paleoc.*, Asia (Pak.).——FIG. 548,*1*. **P. ranikotensis*; *1a-c,* opposite sides and edge view of topotype, ×72 (*2117).

[Originally regarded as related to *Globorotalia*, this genus is here transferred to the Eponididae, as it does not appear to be a planktonic form. It is similar in general appearance to *Globorotalites* but differs in having a radially built wall. *Planulina membranacea* EHRENBERG, included in *Pseudogloborotalia* by HAQUE (1956, *876), had been selected as type-species of *Planorotalia* MOROZOVA, 1957, which on the basis of restudied topotypes is here regarded as synonymous with *Globorotalia*.]

Rectoeponides CUSHMAN & BERMÚDEZ, 1936, *489, p. 31 [**R. cubensis*; OD]. Test in early stage trochospiral, carinate, later uniserial, rectilinear and compressed; wall calcareous, finely perforate, microstructure unknown; aperture in adult terminal, an elongate slit slightly to one side of final chamber, on umbilical side of test. *Paleoc.-U.Eoc.*, Carib.(Cuba)-Eu.——FIG. 546,*2*. **R. cubensis*, U. Eoc., Cuba; *2a-c,* opposite sides and apert. view, ×44 (*2117).

Sestronophora LOEBLICH & TAPPAN, 1957, *1172, p. 229 [**S. arnoldi*; OD] [=*Sestranophora* RESIG, 1962, *1536, p. 55 *(nom. null.)*]. Test free, large, to 2 mm. diam., trochospiral, nearly plano-convex, periphery acute, carinate, spiral side strongly convex, with chambers of greater breadth than height, somewhat oblique and overlapping at periphery, opposite side flat, with broad umbilicus covered by series of plates arising from umbilical margin of each chamber and pierced by numerous very large openings which open into umbilical area beneath and which also connect laterally beneath plate into various chamber cavities; sutures distinct, somewhat thickened, gently curved and inclined back along periphery, depressed on spiral

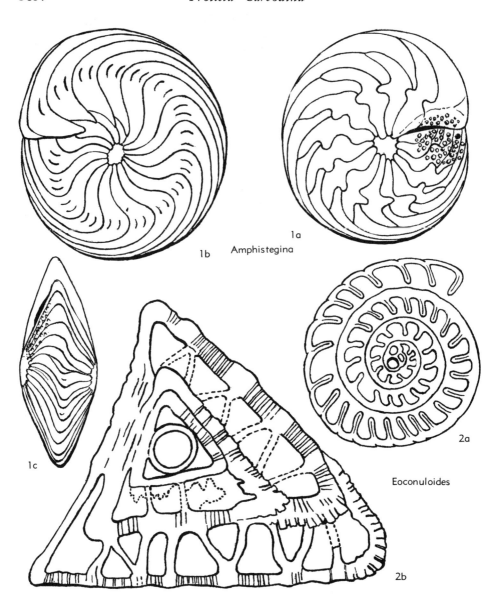

Fig. 549. Amphisteginidae; *1, Amphistegina; 2, Eoconuloides* (p. C685).

side, nearly radial and more strongly depressed on umbilical side; wall calcareous, finely perforate, surface smooth; aperture a low slitlike interiomarginal opening on umbilical side, with few small accessory pores in ventral face of final chamber. *Plio.-Pleist.*, N.Am.(USA)-Eu.(Eng.).——Fig. 546,6. **S. arnoldi*, Pleist., USA(Calif.); 6a-c, opposite sides and edge view of holotype, ×21 (*1172).

[Differs from *Eponides* in having umbilical perforated covering plates and supplementary areal openings on the umbilical side of the final chamber. It is similar to *Poroeponides* in having a few areal openings, but differs in possessing complex perforated umbilical plates. The type-species shows an ontogenetic development through *Eponides*- and *Poroeponides*-like juvenile stages, as illustrated by RESIG (1962, *1536). Such "biformed" ontogenetic stages are characteristic of many foraminiferal genera, but adult stages must be used in classification, and as the type-species of *Eponides* and *Poroeponides* do not have a *Sestronophora*-like adult, the three genera are regarded as distinct.]

AMPHISTEGINIDAE

By R. W. Barker

[Shell Development Company, Houston, Texas]

Family AMPHISTEGINIDAE Cushman, 1927

[Amphisteginidae Cushman, 1927, p. 79] [=Family Enthomostègues d'Orbigny, 1826, p. 304 *(partim)* *(nom. nud., nom. neg.)*; =family Helicotrochina Ehrenberg, 1839, opp. p. 120 *(partim)* *(nom. nud.)*; =*Amphistegininae* Chapman & Parr, 1936, p. 144]

Test free, calcareous, trochoid to asymmetrically lenticular; multichambered, chambers arranged in complex spiral, which (in some genera) splits up into chamberlets on ventral side or extends into peripheral flange; surface smooth, granulate or papillate; aperture consisting of narrow slit at inner margin of last chamber, usually with thin lip and generally surrounded by granulate area; no canal system (*431). [Warm, shallow water.] ?*Cret., Eoc.-Rec.*

Amphistegina d'Orbigny, 1826, *1391, p. 304 [*A. vulgaris*; SD Parker, Jones & Brady, 1865, *1419, p. 36] [=*Omphalophacus* Ehrenberg, 1840, *667, opp. p. 120 (type, *O. hemprichii*)]. Lenticular, generally unequally biconvex, with low turbinoid spire; multichambered, chambers equitant, with alar prolongations as in *Nummulites;* dorsal septa simple, radiate, falciform, and may undulate near umbo; ventral septa divided by deep, commonly imbricate constrictions forming secondary lobes that have appearance of secondary chamberlets in rosette around umbo; walls thick, laminated, and traversed by pores (*223, *1391, *1419). ?*U.Cret., Eoc.-Rec.,* cosmop.——Fig. 549, *1. A. gibbosa* d'Orbigny, Rec., W.Indies; *1a-c,* ext. views, ×30 (*2110).

Boreloides Cole & Bermúdez, 1947, *371, p. 197 [**B. cubensis*; OD]. Structurally similar to *Eoconuloides* but test subspherical to fusiform, with thick spiral wall resembling basal layer of alveolinellids but vitreocalcareous and perforate; apertural characters unknown (*371, *1519). [Originally placed in Alveolinellidae; transfer to Amphisteginidae suggested by Reichel.] *M.Eoc.,* Carib.(Cuba).——Fig. 550,*1.* **B. cubensis*; *1a,b,* transv. and axial secs., ×40 (*371).

Eoconuloides Cole & Bermúdez, 1944, *370, p. 340 [**E. wellsi*; OD]. Test conical, spiral, involute, final chambers subdivided on conical peripheral face into small chamberlets; embryonic apparatus bilocular, consisting of subspherical initial chamber and smaller second chamber; spiral wall thick initially, with irregularly developed pillars, but thinner in final stage and with less prominent pillars (*370). [*Eoconuloides* is readily distinguished from other amphistigenids by its conical form. Probably it developed from *Helicostegina* by axial elongation of the test.] *M.Eoc.,* Carib.(Cuba).——Fig. 549,*2.* **E. wellsi*; *2a,b,*

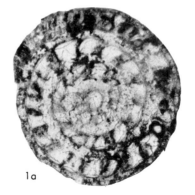

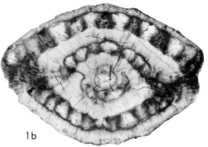

Fig. 550. Amphisteginidae; *1, Boreloides* (p. C685).

transv. and axial secs. of topotype, ×62, ×75 (*2110).

Tremastegina Brönnimann, 1951, *222, p. 256 [**Amphistegina senni* Cushman in Vaughan, 1945, *1995, p. 49; OD]. Similar to *Amphistegina,* differing in presence of parallel furrows and ridges on ventral surface near margin, absence of granulation near aperture and in form of apertures, which may consist of slitlike openings situated ventrally or ventromarginally in septa of dorsal chambers, with backward projecting lips (counterseptа) or otherwise comprise circular pores in ventral chambers where septa cross parallel furrows, pores near periphery communicating between ventral and dorsal chambers (*222). [This genus probably constitutes a link between *Amphistegina* and *Helicostegina.*] *Eoc.,* W.Indies-C.Am.——Fig. 551,*1.* **T. senni* (Cushman), M. Eoc., Barbados Is.; *1a,b,* side and edge ext. views; *1c,d,* transv. and axial secs.; all ×50 (*222).

Family CIBICIDIDAE Cushman, 1927

[*nom. transl.* Chapman, Parr & Collins, 1934, p. 556, 570 (*ex* subfamily Cibicidinae Cushman, 1927)]——[dagger(†) indicates *partim*]——[=Turbinacea† and Turbinacés† de Blainville, 1825, p. 390 *(nom. nud.)*; =Turbinoidae† d'Orbigny in de la Sagra, 1839, p. xxxviii, 71 *(nom. nud.)*; =Turbinoidea† Reuss, 1860, p. 151 *(nom. nud.)*; =Cibicidae Hofker, 1951, p. 332 *(nom. van.)*]

Test free or attached, trochospiral to nearly planispiral, or later spreading, irregular or cyclical; wall coarsely perforate, radial

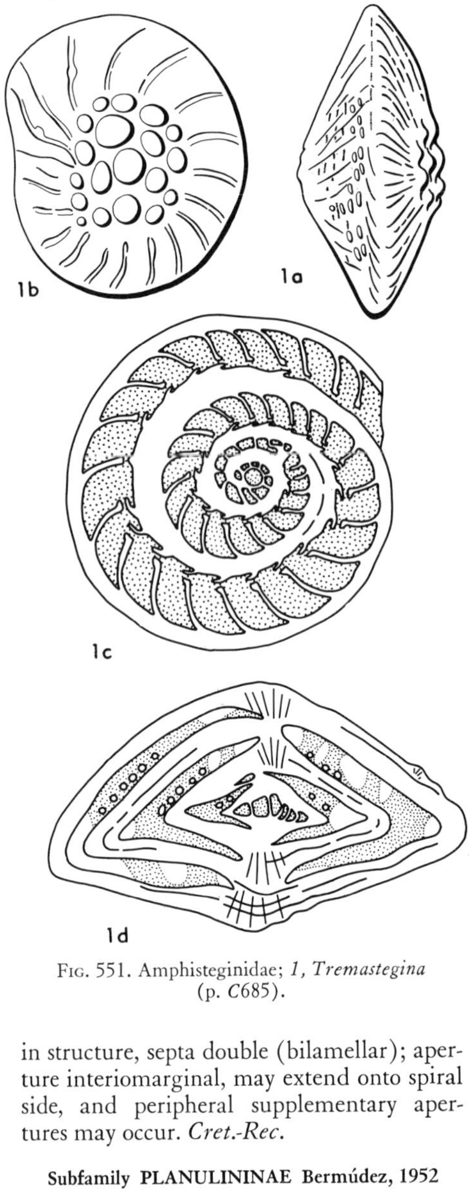

FIG. 551. Amphisteginidae; *1, Tremastegina* (p. C685).

in structure, septa double (bilamellar); aperture interiomarginal, may extend onto spiral side, and peripheral supplementary apertures may occur. *Cret.-Rec.*

Subfamily PLANULININAE Bermúdez, 1952

[Planulininae BERMÚDEZ, 1952, p. 91]

Test free, trochospiral to nearly planispiral; aperture single. *U.Cret.-Rec.*

Planulina D'ORBIGNY, 1826, *1391, p. 280 [*P. ariminensis*; SD GALLOWAY & WISSLER, 1927, *766, p. 66]. Test discoidal, compressed, low trochospiral, spiral side evolute, umbilical side partially evolute, periphery truncate, with thick marginal imperforate keel; sutures strongly arched, thickened, nonperforate, septa double (bilamellar); wall calcareous, radial in structure, finely perforate but with scattered large pores in addition, secondarily added lamellae covering umbilical region; aperture an equatorial, interiomarginal arch, with narrow bordering lip, extending somewhat onto less evolute umbilical side, beneath the flaplike chamber margin, both apertural lip and liplike margin of umbilical flaps imperforate. *U. Cret.-Rec.*, cosmop.——FIG. 552,1; 553. **P. ariminensis*, Rec., Italy (552,*1*), Plio., Italy (553); 552,*1a-c*, opposite sides and edge view of topotype, ×46 (*2117); 553, equat. sec. showing lamellar structure and bilamellar septa, ×80 (*1531).

Cibicidina BANDY, 1949, *70, p. 91 [*C. walli*; OD]. Test free, trochoid, plano-convex, periphery acutely angled and keeled; all chambers partially visible on flattened to concave spiral side although coil is partially involute, only chambers of final whorl visible from convex, umbilicate opposite side; sutures slightly depressed; wall calcareous, finely perforate, surface unornamented; aperture a low interiomarginal arch, against peripheral margin of previous whorl and extending very slightly onto spiral side. [Differs from *Cibicides* in being partially involute on the spiral side and in being more finely perforate. It was originally placed in the Rotaliidae, and transferred to the Anomalinidae by BERMÚDEZ (1952, *127, p. 88).] *Eoc.*, N.Am.——FIG. 552,*4*. **C. walli*, USA (Ala.); *4a-c*, opposite sides and edge view of topotype, ×78 (*2117).

Hyalinea HOFKER, 1951, *928c, p. 416, 508, 513 [**Nautilus balthicus* SCHRÖTER, 1783, *1677, p. 20; OD] [=*Hofkerinella* BERMÚDEZ, 1952, *127, p. 74 (*nom. subst. errore pro Hyalinea* HOFKER, 1951) (obj.)]. Test free, discoidal, slightly trochospiral to nearly planispiral, partially to nearly completely evolute on both sides, periphery angled, with broad imperforate keel; chambers numerous, about 10 to 12 in last of slowly enlarging whorls, thickened wall on all margins, including peripheral keel, septa, apertural face, and umbilical flaps at each side of chamber which form thickened ring of nodes along spiral suture; sutures slightly curved, thickened and elevated, nonperforate; wall calcareous, finely perforate, radial in structure, with septa and marginal keel nonperforate; aperture an equatorial interiomarginal arch with narrow bordering lip, and with low slits extending laterally beneath small umbilical chamber flaps along spiral suture on both sides of test, small rounded opening on each side beneath thickened umbilical flap communicating with chamber interior which remains open in earlier chambers until closed by lamellar thickening. *Pleist.-Rec.*, Eu.-Atl.O.-Pac.O.-Japan——FIG. 552,*2,3*. **H. balthica* (SCHRÖTER), Rec., N.Sea; *2a-c*, opposite sides and edge view showing nearly planispiral test with umbilical flaps and equatorial aperture; *3*, optical sec. showing umbilical chamber openings; all ×72 (*2117).

[*Hyalinea* is similar to *Planulina* in its flattened discoidal, partially evolute test, but differs in having lateral aper-

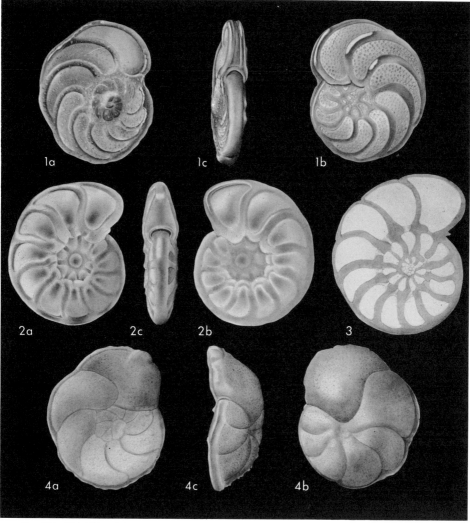

Fig. 552. Cibicididae (Planulininae; *1, Planulina;* *2,3, Hyalinea;* *4, Cibicidina*) (p. C686-C687).

tural extensions along the spiral sutures on both sides. The openings beneath the umbilical flaps were regarded by HOFKER (*928c) as a primitive canal system, suggesting a relationship to *Rotalia*. They are here interpreted as supplementary apertures rather than a true canal system. Although regrettably similar in spelling to *Hyalinia* AGASSIZ, 1837, and *Hyalina* SCHUMACHER, 1817 (*non* STUDER, 1820; *nec* ALBERS, 1850; *nec* JUNG, 1942), the difference in spelling of *Hyalinea* HOFKER is not among those regarded as constituting homonymy by the *International Code of Zoological Nomenclature*; hence according to Art. 56, "Even if the difference between two genus-group names is due to only one letter, these two names are not to be considered homonyms." Thus the replacement name *Hofkerinella* was unnecessary and invalid; the name is a junior objective synonym.]

Subfamily CIBICIDINAE Cushman, 1927

[Cibicidinae CUSHMAN, 1927, p. 93] [=Truncatulininae SCHUBERT, 1921, p. 151; =Orbitorotalininae HOFKER, 1933, p. 125 (*nom. nud.*)]

Test attached by spiral side; primary aperture equatorial, may extend onto spiral side, advanced forms may have multiple aperture. *Cret.-Rec.*

[NYHOLM (1961, *1380) has demonstrated a considerable variation in form of Recent *Cibicides lobatulus*. The so-called "monothalamous test resembling *Crithionina* or *Webbina*" described by NYHOLM is not a true test but an enclosure within which the young schizont develops. It is better referred to as a reproductive or growth cyst, similar to those reported in many other Recent foraminifers. Although attached forms obviously show great morphological variation, and some atypical specimens of *C. lobatulus* resemble "*Dyocibicides, Annulocibicides, Cyclocibicides, Stichocibicides,* or *Rectocibicides* according to the conditions of growth," random specimens may be found in many Recent and fossil species and genera which show characters of other genera. These may indicate a possible genetic relationship, but it will be necessary to restudy assemblages of each of the type-species of these other genera before they can be definitely regarded as synonymous. The supposed planorbulinoid stage reported by NYHOLM to be developed by *Cibicides* is somewhat doubtful, as it has been impossible to verify this in cultures. The mere association of planorbulinoid forms and *Cibicides* on the same ascidians is not definitive; since young forms found with planorbulinoid

Fig. 553. Cibicididae (Planulininae; *Planulina*) (p. C686).

adults invariably produced new planorbulinids, not young *Cibicides*, their assumed relationship to *Cibicides* seems doubtful. The detailed work on *Planorbulina mediterranensis* by LE CALVEZ (*1106) followed its life cycle completely and showed a regular alternation of generations of planorbulinoid forms, but no relationship to *Cibicides*. The various genera here included in the Cibicidinae are tentatively placed together until further studies are made of the many assemblages; they are not regarded as having close affinities with the Planorbulinidae.]

Cibicides DE MONTFORT, 1808, *1305, p. 122 [*C. refulgens*; OD] [=*Storilus* DE MONTFORT, 1808, *1305, p. 130 (type, *S. radiatus*); *Polyxenes* DE MONTFORT, 1808, *1305, p. 138 (type, *P. cribratus*); *Cymbicides* COSTA, 1839, *390, p. 186 (nom. null.?); *Truncatulina* D'ORBIGNY, 1826, *1391, p. 278 (obj.); *Lobatula* FLEMING, 1828, *722, p. 232 (type, *L. vulgaris*); *Soldanina* COSTA, 1856, *392, p. 246 (type, *S. exagona*); ?*Craterella* DONS, 1942, *609, p. 136 (type, *C. albescens*); (non *Craterella* SCHRAMMEN, 1901; nec KOFOID & CAMPBELL, 1929); ?*Crateriola* STRAND, 1943, *1844, p. 211 (type, *Craterella albescens* DONS, 1942, *609, p. 135)]. Test attached; plano-convex, trochospiral, spiral side flat to excavated, evolute, umbilical side strongly convex, involute, apertural face sharply angled, distinct from umbilical side, periphery angular, with nonporous keel; wall calcareous, radial in microstructure, bilamellar, coarsely perforate on spiral side, large pores of earlier chambers may be closed by lamellar thickening of wall, finely perforate on umbilical side, apertural face nonporous, aperture a low interiomarginal opening with narrow lip, may extend along spiral suture on spiral side. *Cret.-Rec.,* cosmop.——FIG. 554,*1*. *C. refulgens*, Rec., Atl.O.; *1a-c,* opposite sides and edge view, ×61 (*2117).

[NYHOLM (1961, *1380) showed the great variability in form of this attached genus, and described the agglutinated coniform reproductive cysts in which the young schizonts developed. *Craterella* DONS, 1942, and the substitute name *Crateriola* STRAND, 1943, are based on small, attached, conical or hemispherical agglutinated specimens, about 0.15 to 0.85 mm. diam., with an opening at the apex, occurring on the underside of rocks in tide pools near Trondheim Fjord, Norway. These are identical to the reproductive cysts described by NYHOLM, 1961, for *Cibicides* from the same general area; hence, *Crateriola* is here regarded as a probable synonym, based on an ontogenetic stage of *Cibicides*. As noted in connection with the subfamily description, it seems premature to regard *Dyocibicides, Cyclocibicides,* and other genera as synonyms of *Cibicides*, as was suggested by NYHOLM, until assemblage studies can be made of the type-species of each of these nominal genera. *Planorbulina* and *Gypsina* were also regarded as probable growth forms of *Cibicides* by NYHOLM, but their relationship to *Cibicides* seems still unproved, since cultures of these forms could not be maintained to prove their relationship; the earlier studies of the type-species of *Planorbulina* by LE CALVEZ (*1106) followed its life cycle completely without observing a *Cibicides*-like stage. WOOD (1949, *2073, p. 252) stated that *Cibicides refulgens* is granular in structure, but this was later corrected by WOOD & HAYNES (1957, *2076, p. 46). Some species previously referred to *Cibicides* have been noted by WOOD & HAYNES (*2076) and REISS (1959, *1531) to be granular, but these are referable to other genera. *Cibicides* is here restricted to coarsely perforate, plano-convex forms with radial microstructure of the wall.]

Annulocibicides CUSHMAN & PONTON, 1932, *520, p. 1 [*A. projectus*; OD]. Test similar to *Cyclocibicides* but lacking large sutural pores on spiral side, with all apertural openings peripheral and produced on slight necks. *Mio.,* USA(Fla.).——FIG. 554,*2*. *A. projectus*; *2a-c,* opposite sides and edge view of holotype, ×37 (*2117).

Caribeanella BERMÚDEZ, 1952, *127, p. 121 [*C. polystoma*; OD] [=*Oinomikadoina* MATSUNAGA, 1954, *1236, p. 163 (type, *O. ogiensis*); *Pseudocibicidoides* UJIIÉ, 1956, *1963, p. 263 (type, *P. katasensis*)]. Test free, plano-convex to biconvex, trochospiral, all whorls visible on flattened spiral side, only final whorl visible on umbilical side, periphery may be angular in early stages but commonly rounded; chambers increasing gradually in size; sutures arched backward at periphery on spiral side, nearly radial on umbilical side, may be thickened and imperforate; wall calcareous, later chambers very coarsely perforate, radial in structure; primary aperture a low arch on periphery and extending somewhat onto umbilical side, bordered above by prominent nonperforate lip, smaller secondary apertures, which also have distinct lip, at basal backward margin of each chamber on periphery, and additional series of supplementary apertures on spiral side consisting of low arches near inner margin of the later chambers against previous whorl, these also bordered by slight lips, final chamber thus possessing 3 openings, with 2 remaining open on each previous chamber of final whorl. *Plio.-Rec.,* Carib.-Atl.O.-Japan.——FIG. 555,*1*. *C. katasensis* (UJIIÉ), Rec., Japan; *1a-c,* opposite sides and edge view of holotype showing equatorial and umbilical openings, ×58 (*1963).——FIG. 555,*2*. *C. polystoma,* Rec., Atl.; *2a-c,* opposite sides and edge view of topotype, ×111 (*2117).——FIG. 555,*3,4*. *C. ogiensis* (MATSUNAGA), Plio., Japan; *3a-c,* opposite sides and edge view of topotype; *4,* spiral side of larger specimen showing 3 apert. openings in final chamber, ×57 (*2117).

Foraminiferida—Rotaliina—Orbitoidacea

[The original description of *Caribeanella* made no mention of secondary spiral apertures at the inner margin of the later chambers. These are less prominent in *C. polystoma* than in Recent specimens of *Oinomikadoina ogiensis*. As the two type-species are similar in all characters, *Oinomikadoina* is a junior synonym of *Caribeanella*. The original description of *Caribeanella* (*127) further stated that the peripheral apertures corresponded to earlier pri-

Fig. 554. Cibicididae (Cibicidinae; *1, Cibicides; 2, Annulocibicides; 3, Cibicidella; 4, Cyclocibicides; 5,6, Cycloloculina; 7, Rectocibicides*) (p. C688, C690, C692).

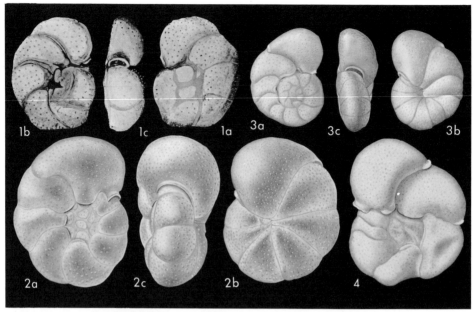

Fig. 555. Cibicididae (Cibicidinae; *1-4, Caribeanella*) (p. C688-C690).

mary apertures that remained open. Since these secondary apertures are on the rear portion of each chamber and open into the chamber in front of them, not into the one preceding, they cannot be relict openings. Furthermore, they are at the dorsal angle of the suture and periphery, whereas the primary apertures are interiomarginal and peripheral, against the preceding whorl. The third set of openings along the spiral suture were not mentioned in either original generic description, though they are present in both type-species. *Pseudocibicidoides* was described as having an umbilical opening on the umbilical side and a rounded equatorial aperture with prominent lip. Topotypes of the type-species show the umbilical opening along the spiral suture and the large equatorial aperture; also though less developed in this species, the peripheral openings at posterior margins of the chambers are observable in at least some of the better-preserved specimens. Thus *Pseudocibicidoides* is also regarded as a synonym of *Caribeanella*.]

Cibicidella CUSHMAN, 1927, *431, p. 93 [*Truncatulina variabilis* D'ORBIGNY, 1826, *1391, p. 279; OD]. Test attached, early stage trochospiral, as in *Cibicides*, later chambers added irregularly; wall calcareous, radial in structure, coarsely perforate; aperture in early stage as in *Cibicides*, more than one of irregularly arranged chambers in adult stage possessing arched apertural opening against attachment, each with narrow but distinct nonporous lip. *Mio.-Rec.*, Eu.——FIG. 554,3. *C. variabilis* (D'ORBIGNY), Rec., Medit.; *3a,b,* opposite sides, ×30 (*2117).

Cyclocibicides CUSHMAN, 1927, *431, p. 93 [*Planorbulina vermiculata* D'ORBIGNY, 1826, *1391, p. 280; OD]. Test attached, discoidal, almost flat in early stage, early chambers trochospirally arranged, attached by spiral side, later with irregular chambers and finally with annular chambers; sutures distinct; wall calcareous, radial in structure, coarsely perforate on unattached side; apertures consisting of large sutural pores on attached spiral side, and scattered peripheral pores, chambers connected internally by large pores through walls. *Rec.*, Medit. Sea.——FIG. 554,4. *C. vermiculata* (D'ORBIGNY), *4a-c,* opposite sides and edge view of lectotype, ×27 (*2117).

[*Cyclocibicides* differs from *Cibicidella* in the later annular chambers, which have a multiple aperture instead of a single rounded opening with surrounding lip. A lectotype was selected from the D'ORBIGNY collection in the Muséum National d'Histoire Naturelle, Paris (no. 12353) and here designated and refigured; it is from the Mediterranean.]

Cycloloculina HERON-ALLEN & EARLAND, 1908, *906, p. 533 [*C. annulata*; SD CUSHMAN, 1927, *433, p. 190] [=*Cycloloceilina* SHARP, 1910, *1722, p. 5 *(nom. null.)*]. Test discoidal, about 1 mm. diam., peripheral margin smooth, rounded; planispiral or slightly asymmetrical in early stage, later chambers uncoiled and enveloping, finally annular; wall calcareous, of radially built calcite, coarsely perforate, "pores" consisting only of regularly arranged large openings that serve as apertures, no fine perforations; aperture consisting of large perforations. *Paleoc.-Mio.*, Eu.-N.Am.-Asia(Pak.).——FIG. 554,5,6. *C. annulata*, Tert., Eng.; *5a,b,* side and edge views, ×71 (*2117); *6,* early stage in transmitted light, ×75 (*2075).

Dyocibicides CUSHMAN & VALENTINE, 1930, *532, p. 30 [*D. biserialis*; OD] [=*Rectocibicidella* McLEAN, 1956, *1201, p. 370 (type, *R. robertsi*)]. Test elongate, attached, early stage trochospirally coiled, attached by spiral side, later uncoiling and irregularly biserial or staggered uniserial, periphery carinate; wall calcareous, coarsely perforate; aperture terminal, elongate, with bordering lip. *Eoc.-Rec.*, N.Am.-S.Am.-Eu.-Japan.——FIG. 556, *1.* *D. biserialis*, Rec., USA(Calif.); *1a-c,* opposite sides and top view of holotype, ×74 (*2117).

——FIG. 556,2. *D. robertsi* (MCLEAN), Mio., USA (Va.); *2a-d,* opposite sides, edge, and top views of holotype, ×45 (*1201).

Falsocibicides POIGNANT, 1958, *1471, p. 117 [**F. aquitanicus*; OD]. Test attached, large, asymmetrical, trochospiral, plano-convex, spiral side flattened, peripheral outline lobulate, peripheral margin rounded, noncarinate; few chambers to whorl, increasing rapidly in size; internally thin plate divides aperture horizontally and extends back to previous foramen; wall calcareous, coarsely perforate; aperture large, rounded, equatorial and interiomarginal in position, extending somewhat onto spiral side with supplementary apertures at umbilical margin of chambers on umbilical side and more rarely secondary opening at opposite margin of final chamber on periphery, all apertures bordered by distinct nonperforate lips, spiral side also may have relict apertures or umbilical uncovered remnants of primary apertures of earlier chambers, relict apertures variable in occurrence within a species. *Oligo.(Stamp.)-Mio.(Burdigal.),* Fr.——FIG. 557,*1-3.* *F. aquitanicus,* Oligo.(Stamp.); *1a-c,* opposite sides and edge view of holotype showing large rounded equatorial aperture; *2,* sec. showing internal plate extending between foramina of final whorl; *3,* umbilical side of paratype showing umbilical supplementary apertures with bordering lips and border of additional peripheral aperture visible between last 2 chambers; all ×30 (*1471).

Planorbulinoides CUSHMAN, 1928, *436, p. 6 [**Planorbulina retinaculata* PARKER & JONES in CARPENTER, PARKER & JONES, 1862, *281, p. 209; PARKER & JONES, 1865, *1418, p. 380; OD]. Test attached, early stage trochospiral, later chambers added irregularly, as in *Cibicidella,* finally chambers irregularly and loosely arranged to form spreading network; wall calcareous, coarsely perforate, apertures at ends of short projecting necks, as in *Annulocibicides* and *Rectocibicides,* situated at chamber margins against attachment. [A search for the type-specimen in the British Museum (Natural History) was fruitless, and the species is apparently rare. It was stated to be "parasitic on shells, East and West Indies."] *Rec.,* E.Indies-W.Indies.——FIG. 558. *P. retinaculata* (PARKER & JONES), locality not stated, ×15 (*1418).

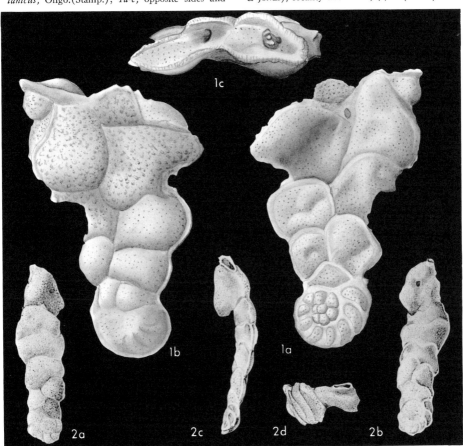

FIG. 556. Cibicididae (Cibicidinae; *1,2, Dyocibicides*) (p. C690-C691).

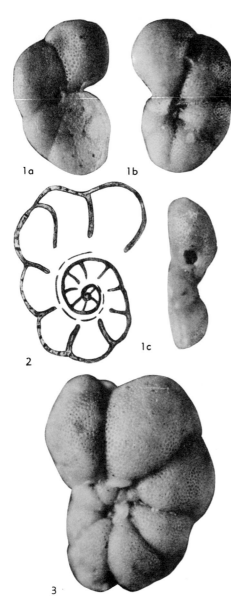

ing coarsely perforate and in having a multiple aperture.] *Mio.*, USA(Fla.).——FIG. 554,7. **R. miocenicus*; *7a-c,* opposite sides and apert. view of paratype, ×49 (*2117).

Stichocibicides CUSHMAN & BERMÚDEZ, 1936, *489, p. 33 [**S. cubensis*; OD]. Test attached, early portion in slight trochoid coil of one or more volutions, attached side showing earlier volutions, free convex side partially or completely involute, coil followed by uniserially arranged chambers, periphery with nonporous marginal keel; wall calcareous, coarsely perforate; aperture nearly terminal, rounded opening just above attachment. [Differs from *Karreria* in the angled, keeled periphery and coarsely perforate wall. *Dyocibicides* differs from *Stichocibicides* in having a biserial stage, and *Rectocibicides* has multiple terminal apertures.] *Eoc.*, Cuba-N.Am.-Haiti.——FIG. 559, *1-3.* **S. cubensis*, Cuba; *1a,b,* opposite sides of paratype; *2a,b,* opposite sides of paratype; *3a-c,* opposite sides and apertural view of holotype; all ×35 (*2117).

Family PLANORBULINIDAE
Schwager, 1877

[*nom. transl.* CUSHMAN, 1927, p. 95 (*ex* subfamily Planorbulinidae SCHWAGER, 1877, p. 20)]——[In synonymic citations superscript numbers refer to taxonomic rank assigned by authors ([1]family, [2]subfamily); dagger(†) indicates *partim*]——[=Hélicostégues† D'ORBIGNY, 1826, p. 268 *(nom. neg.; nom. nud.)*; =[1]Turbinoidae† D'ORBIGNY in DE LA SAGRA, 1839, p. xxxviii, 71 *(nom. nud.)*; =[2]Planorbulininae GALLOWAY, 1933, p. 297]

Test attached, early stage trochospiral, later with numerous chambers forming discoidal, cylindrical, conical, or subglobular

FIG. 557. Cibicididae (Cibicidinae; *1-3, Falsocibicides*) (p. C691).

Rectocibicides CUSHMAN & PONTON, 1932, *520, p. 2 [**R. miocenicus*; OD]. Test attached, early portion coiled, later with broad low chambers uniserially arranged; wall calcareous, coarsely perforate; aperture a series of ovate openings on slight projections from terminal face, each surrounded by lip. [Differs from *Dyocibicides* in being more regularly uniserial in later stages and in having a multiple aperture. Differs from *Karreria* in be-

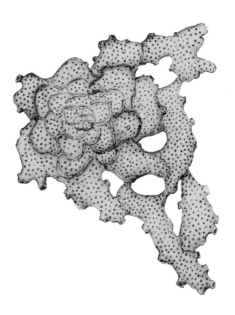

FIG. 558. Cibicididae (Cibicidinae; *Planorbulinoides*) (p. C691).

Foraminiferida—Rotaliina—Orbitoidacea

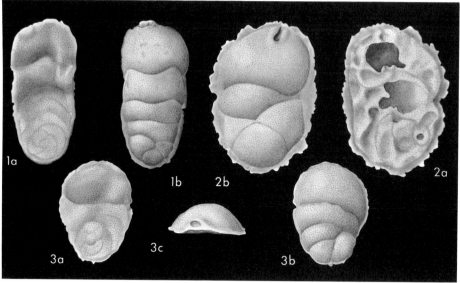

Fig. 559. Cibicididae (Cibicidinae; *1-3, Stichocibicides*) (p. C692).

test; aperture single or multiple, peripheral. Eoc.-Rec.

Planorbulina D'ORBIGNY, 1826, *1391, p. 280 [*P. mediterranensis*; SD CUSHMAN, 1915, *404e, p. 27] [=*Asterodiscus* EHRENBERG, 1840, *667, table opposite p. 120 (type, *A. forskalii*); *Spirobotrys* EHRENBERG, 1844, *674, p. 246, 247 (type, *S. aegaea*)]. Test discoidal, trochospiral, attached by spiral side, tests of both generations similar in size, proloculus of microspheric forms (about 4 per cent of specimens) 11-14µ diam., that of megalospheric forms (96 per cent) 23-56µ diam., coiling random, early portion spirally coiled, chambers each with single aperture, later 2 apertures developing on each chamber (Fig. 560,2), each giving rise to new biapertural chamber, thus making numerous spirals of chambers; wall calcareous, on pseudochitinous membrane, coarsely perforate, radiate in structure, early spire brownish due to thickness and pigmentation of pseudochitinous membrane, which is perforated only by apertures, not finer wall perforations; aperture multiple, peripheral, commonly 1 or 2 interiomarginal oval to semilunar openings on each chamber of final whorl, each with narrow bordering lip, smaller supplementary openings occur on both sides for extrusion of pseudopodia, appearing on third chamber of spiral side and on eighth chamber of ventral side in type-species; cytoplasm greenish-brown to salmon-rose, except during sexual reproduction when pigments are eliminated, central area of dense plasma with abundant fine refringent granules (microsomes), numerous fine vacuoles, a nucleus, and peripheral vegetative zone crowded with nutritive and excretive particles lacking microsomes and rich in ectoplasm; pseudopodia rectilinear, anastomosing slightly, about equal in length to test diameter, with slow circulation of granules; during vegetative reproduction (schizogony) some 60 to 100 embryos develop by division of parent nucleus and protoplasm, while protected by temporary encrusted "cyst," much of parent test becoming dissolved. *Eoc.-Rec.*, cosmop.——FIG. 560,*1,2*; 561. **P. mediterranensis*, Rec., USA(Fla.) (560,*1*), Medit. (560,*2*); 560,*1a-c*, opposite sides and edge view, ×44 (*2117); 560,*2*, central part of equat. sec. of microspheric specimen, ×100 (*1106); 561, biflagellate gamete, ×6,000 (*1103).

[Embryos at first have only pseudochitinous membrane; the calcareous test begins to be formed at about the 4-chamber stage by a progressive enrichment of the outer face of the pseudochitinous cover with calcium carbonate. After embryos attain 5 chambers, they gradually dislodge the sandy protective cyst and escape, move a short distance away, settle on the substratum by their flat spiral side, and begin to build additional chambers. Both uninucleate and plurinucleate megalospheric forms occur. These have been regarded by some as representing the A2 and A1 generations of trimorphism, but because the microspheric form in cultures invariably gives rise to uninucleate megalospheric forms only, trimorphism has not been proved (*1106). The youngest plurinucleate A1 specimen observed already had 18 chambers. The A1 generation always gave rise in turn to the microspheric B generation and after many rapid nuclear divisions, microspheric adults gave rise to many biflagellate gametes, which utilized all of the parent cytoplasm. The gametes have 2 unequal flagella inserted together at the base of the anterior part near a fatty inclusion. Commonly they escape from the parent test during the night.]

Eoannularia COLE & BERMÚDEZ, 1944, *370, p. 342 [*E. eocenica*; OD]. Test discoidal, flat to concavo-convex, may be umbonate on convex side, biloculine embryonic stage of megalospheric form consisting of proloculus, which is slightly or completely embraced by second chamber, later chambers in annular rings in single layer, chambers nearest center with arched outer walls, those of later annuli nearly rectangular and alternating in position with those of preceding and following

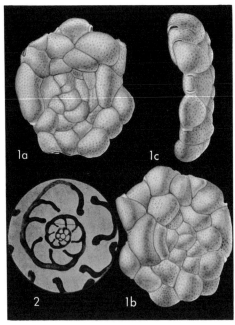

Fig. 560. Planorbulinidae; *1,2, Planorbulina* (p. C693).

annuli; wall calcareous, coarsely perforate, with some thickening of shell material in central position; numerous stoloniferous apertures. [The early stage is similar to *Linderina* but later chambers tend toward *Cycloclypeus* in form (*370).] M.Eoc., Carib.(Cuba).——Fig. 562,1-4. *E. eocenica; 1, ext., holotype and smaller paratypes, ×10; 2a,3, horiz. secs., ×41; 2b, central portion of 2a, ×163; 4, vert. sec., ×41 (*370).

Linderina Schlumberger, 1893, *1656, p. 120 [*L. brugesi; OD (M)]. Test discoidal, centrally thickened, 1-3.5 mm. diam.; megalospheric form with biloculine embryonic stage, later chambers arched, in concentric series in single plane, those of successive series alternating in position; wall calcareous, perforate, umbonal lamellar thickening pronounced, surface with numerous granules or pillars in central area; apertures at each side of base of chambers, as in position of stolons in orbitoidids. Eoc.-Mio., Eu.-Afr.-N.Am.-E.Indies (Indon.).——Fig. 562,5-10. *L. brugesi, U.Eoc., Fr.; 5, ext., ×40; 6, horiz. sec., central part, showing perforations of umbonal thickening, ×36; 7, equat. sec. showing stolon-like apertures, ×135 (*1352); 8, horiz. sec. of early portion of megalospheric test showing biloculine embryo and later arched chambers, ×75; 9,10, transv. secs. of microspheric and megalospheric tests showing equat. chambers and umbonal lamellar thickenings, ×73 (*1656).

Planorbulinella Cushman, 1927, *431, p. 96 [*Planorbulina vulgaris d'Orbigny var. larvata

Parker & Jones, 1865, *1418, p. 380; OD]. Test trochoid in early stage and may be attached, later chambers developing in annular series, those of outer row alternating with ones within, forming nearly bilaterally symmetrical test; wall calcareous, coarsely perforate, radial in structure, bilamellar; apertures 2 to each chamber, rarely one in median line on periphery, each with narrow bordering lip. Eoc.-Rec., Pac.O.-Australia-N.Z.-Cuba-Atl.O.-N.Am.——Fig. 563,1,2. *P. larvata (Parker & Jones), Rec., Australia; 1a-c, opposite sides and edge view, ×40 (*2117); 2, part of transv. sec., ×100 (*928a).

Family ACERVULINIDAE Schultze, 1854

[nom. correct. Eimer & Fickert, 1899, p. 702 (pro family Acervulinida Schultze, 1854, p. 53)] [=Gypsininae Silvestri, 1905, p. 5; Acervulininae Galloway, 1933, p. 308]

Test free or attached, early spiral stage followed by spreading chambers, in one or many layers; no canal system; no aperture except for mural pores. Eoc.-Rec.

Acervulina Schultze, 1854, *1695, p. 67 [*A. inhaerens; SD Galloway & Wissler, 1927, *766, p. 67] [=Aphrosina Carter, 1879, *295, p. 500 (type, A. informis)]. Test attached, early chambers coiled, later encrusting, with irregularly arranged inflated chambers; wall calcareous, coarsely perforate; no aperture other than coarse perforations. U.Tert.-Rec., Eu.-N.Am.-Pac.O.-Ind.O.——Fig. 564. *A. inhaerens, Rec., Italy; ext. of specimen attached to Corallina, ×72 (*1695).

Borodinia Hanzawa, 1940, *869, p. 790 [*B. septentrionalis; OD]. Test encrusting, with chambers of successive layers alternating in position, septa and walls approx. 12-25μ in thickness, outer wall 37-75μ in thickness and coarsely perforated; apertural stolons 37μ in diameter, at opposite ends of septum. Mio.(Aquitan.), Daito Is. (formerly Borodino Is.) [off E. China Sea E. of Okinawa].——Fig. 565,1. *B. septentrionalis; 1a, transv. sec. showing thickened outer wall of layer, ×13; 1b, tang. sec. through part of outer wall at upper right and part of chambered zone (scale not given by author) (*869).

Gypsina Carter, 1877, *292, p. 172 [*Polytrema

Fig. 561. Planorbulinidae; *Planorbulina* (p. C693).

planum CARTER, 1876, *291, p. 211, =*Gypsina melobesioides* CARTER, 1877, *292, p. 172; SD CARTER, 1880, *296, p. 445] [=*Discogypsina* A. SILVESTRI, 1937, *1787, p. 156 (type, *Tinoporus vesicularis* (PARKER & JONES) GÖES, 1882, *801, p. 104; *Hemigypsina* BERMÚDEZ, 1952, *127, p.

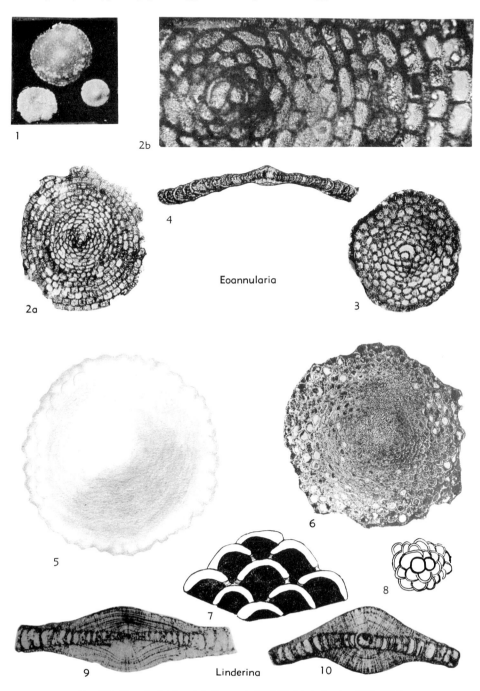

FIG. 562. Planorbulinidae; *1-4, Eoannularia; 5-10, Linderina* (p. C693-C694).

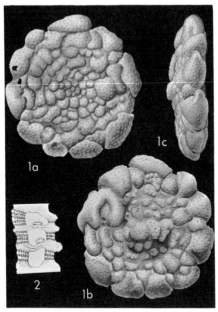

FIG. 563. Planorbulinidae; *1,2, Planorbulinella* (p. C694).

124 (type, *Gypsina mastelensis* BURSCH, 1947, *254, p. 37)]. Test relatively large, attached, encrusting or forming hemispherical mass; chambers roughly circular to rectangular or polygonal in outline and perforated by few large foramina, each about 5μ in diameter, with chambers of one layer alternating with those of row below, upper walls slightly convex outward, may have irregular

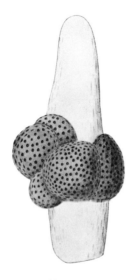

FIG. 564. Acervulinidae; *Acervulina* (p. C694).

knobby projections of groups of chambers, which are more polygonal in outline near center of these knobs; chamber walls of fibrous crystalline calcite, imperforate, embedded tetraxonid sponge spicules within it, but no other foreign matter, walls elevated at surface to form meshwork of clearly defined areolae about 120μ in diameter; no aperture other than large septal wall perforations visible on surface within meshwork of chamber walls. Eoc.-Rec., cosmop.——FIG. 566, *1-4.* *G. plana* (CARTER), Rec., Mauritius Is.; *1*, ext., ×1; *2*, sec. perpendicular to surface, ×75; *3*, diagram. sec. of specimen encrusting another shell, showing chambers, radially built walls, and perforated septa; *4*, sec. through center of a knob, showing less regular chambers and perforated septa, ×96 (*1139).——FIG. 566,5-8. *G. mastelensis* BURSCH, L.Oligo., E.Indies(Indon.); *5*, holo-

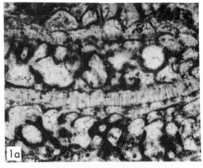

FIG. 565. Acervulinidae; *1, Borodinia* (p. C694).

type, axial sec., ×33; *6*, nearly equat. sec. of paratype, ×33; *7*, diagram. sec., early portion in equat. sec., ×200; *8*, schematic axial sec., enlarged (*254).——FIG. 567,1,2. *G. vesicularis* (PARKER & JONES), Rec., Australia; *1a-c*, side, base, top views of paratype, ×26; *1d*, portion of surface, ×88; *2a,b*, ext. edge view and vertically broken face of paratype, ×17 (*2117).

[*Orbitolina concava* var. *vesicularis* PARKER & JONES was designated the type-species of *Gypsina* by CUSHMAN (1915, *404e, p. 74), but the type had already been fixed by CARTER (1880, *296, p. 445). A lectotype was selected and isolated by us in the British Museum (Natural History), and is here designated for *O. concava vesicularis* PARKER &

Jones, 1860, BMNH-ZF3600 (*ex.* 94.4.3.1737) and paratypes BMNH-ZF3601 (*ex* 94.4.3.1737, 1738) from Jukes No. 2, 14 fathoms, north of Sir C. Hardy's inside reefs, northeast coast of Australia. Recent studies by Nyholm (*1381) suggest that *Gypsina* is a stage in the life cycle of *Cibicides*. The transformation from one "genus" to the other has not been followed in isolated specimens in cultures, and the mere association of the 2 forms in the same biotope

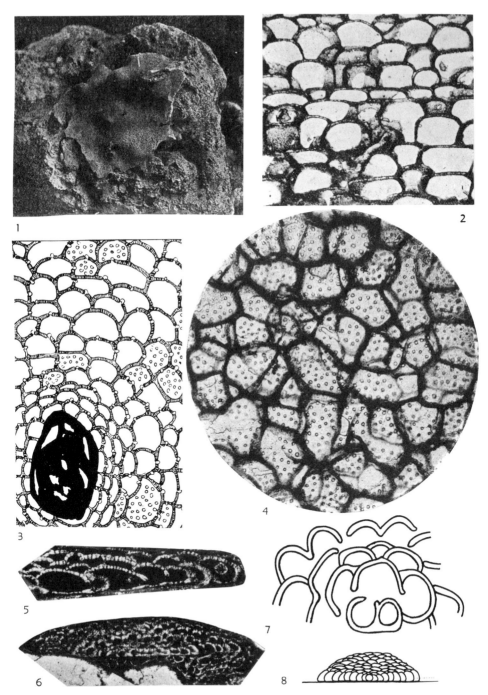

Fig. 566. Acervulinidae; *1-8, Gypsina* (p. C694-C698).

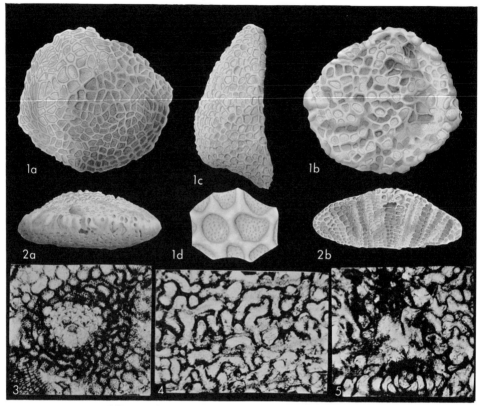

Fig. 567. Acervulinidae; *1,2, Gypsina; 3-5, Ladoronia* (p. C694-C698).

is not conclusive proof of their identity. Both are herein recognized as distinct.]

Ladoronia Hanzawa, 1957, *873, p. 68 [*Acervulina (Ladoronia) vermicularis*; OD] [=*Acervulina (Ladoronia)* Hanzawa, 1957, *873, p. 68 (obj.)]. Test attached, encrusting, early chambers in raspberry-like arrangement as in *Planorbulinella,* later chambers as in *Acervulina,* but elongate and irregularly sinuous as seen in horizontal section; intercameral stolons connecting chambers of same layer and fine pores connecting those of successive layers. *Mio.,* N.Pac.O.(Saipan Is.).——Fig. 567, *3-5.* *L. vermicularis*; *3,* horiz. sec. showing early stage in center of figure; *4,* horiz. sec. showing irregular chambers; *5,* vert. sec. showing early stage (juvenarium) at base of figure; all ×40 (*873).

Planogypsina Bermúdez, 1952, *127, p. 124 [*Gypsina vesicularis* var. *squamiformis* Chapman, 1901, *317, p. 200; OD]. Test large, discoidal, early stage with globular, planispirally arranged chambers, later chambers added irregularly and elongate to vermiform in outline; septal pores perforating walls; no aperture present other than pores. [A lectotype for the type-species was selected by us at the British Museum (Natural History) (BMNH-ZF3647, *ex* 03.2.5.14) and paratypes (BMNH 03.2.5.14), from Recent, Sample 8 of Chapman, 1901, at 26 fathoms, 4 miles from Mission Church, Funafuti Lagoon.] *U.Tert.-Rec.,* SW.Pac.O.——Fig. 568,*1.* *P. squamiformis* (Chapman), Rec., Funafuti Atoll; *1a-c,* top, base, and edge views of paratype, ×29 (*2117).

Sphaerogypsina Galloway, 1933, *762, p. 309 [*Ceriopora globulus* Reuss, 1848, *1539, p. 33; OD]. Test similar to *Gypsina,* but forming globular masses. *Eoc.-Rec.,* Eu.-Carib.-Pac.——Fig. 569, *1,2.* *S. globulus* (Reuss), Mio.(Torton.), Czech.; *1,* ext., enlarged (*1478); *2,* sec., ×20 (*873).

Family CYMBALOPORIDAE Cushman, 1927

[Cymbaloporidae Cushman, 1927, p. 81]——[In synonymic citations superscript numbers indicate taxonomic rank assigned by authors (¹family, ²subfamily)]——[=¹Cymbaloporettidae Cushman, 1928, p. 8; =²Cymbaloporinae Chapman & Parr, 1936, p. 143; =¹Halkyardiidae Kudo, 1931, p. 201]

Test trochospiral, later chambers in annular series in single flat or conical layer; apertures numerous, variously arranged circular pores. *U.Cret.-Rec.*

Cymbalopora von Hagenow, 1851, *859, p. 104 [*C. radiata*; OD (M)]. Test low conical, early

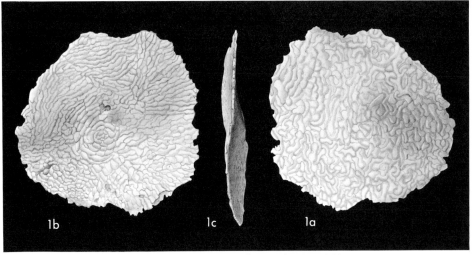

Fig. 568. Acervulinidae; *1, Planogypsina* (p. C698).

chambers trochospirally arranged, later in annular series, as in *Cymbaloporetta,* umbilicus open, commonly portions of chamber surfaces broken away near umbilicus; sutures completely obscured on spiral side, radial and depressed on umbilical side; wall calcareous, coarsely perforate, lamellar, lamellae obscuring sutures and chambers on spiral side where only large perforations can be seen; apertures at open umbilical ends of chambers. U.Cret., Eu.——Fig. 570,*1*; 571. **C. radiata,* Maastricht., Neth.; 570,*1a-c,* opposite sides and edge view, ×56 (*2117); 571, vert. sec. showing recrystallized wall, which has not completely obliterated lamellar structure, and coarse pores on spiral side, ×107 (*948).

[HOFKER (*928c, p. 477) regarded the wall as arenaceous and *Cymbalopora* as related to the Valvulinidae. Topotypes examined by us are distinctly calcareous, although the structure of the recrystallized wall is poorly preserved in this and many other associated calcareous species in the Maastrichtian chalk tuffs. Furthermore, HOFKER's figures (here reproduced) clearly show the lamellar development in *C. radiata.* Also, as noted by REISS (*1531, p. 355) lamellar structure is found only in the calcareous perforate foraminifers (suborder Rotaliina) and never in the agglutinated forms (suborder Textulariina).]

Archaecyclus A. SILVESTRI, 1908, *1771, p. 134 [**Planorbulina cenomaniana* SEGUENZA, 1882, *1714, p. 200; OD (M)]. Test large, discoidal, flat to concavo-convex, to 1.6 mm. diam., proloculus large, followed by coiled early portion of about 5 chambers to whorl, later in annular series with chambers of successive series alternating in position; sutures oblique; wall calcareous, perforate, bilamellar, with thin, dark, median layer; aperture in early stage interiomarginal, later with stolon-like pores at sides of each chamber. U.Cret.(Cenoman.), Eu.(Italy).——Fig. 572,*1,2.* **A. cenomaniana* (SEGUENZA); *1a,* ext., ×35; *1b,* portion of surface, ×100; *1c,* same in balsam to show internal structure, ×100; *2,* part of test with upper surface removed by HCl treatment, mounted in balsam, ×100 (*1714).

Cymbaloporella CUSHMAN, 1927, *431, p. 81 [**Cymbalopora tabellaeformis* BRADY, 1884, *200, p. 637; OD]. Test discoidal, early chambers trochospiral, later in annular series; all visible

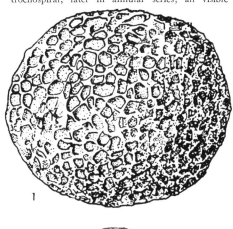

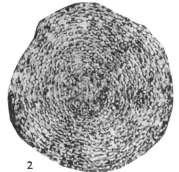

Fig. 569. Acervulinidae; *1,2, Sphaerogypsina* (p. C698).

from spiral side except where lamellar thickening obscures early portion, only final annulus visible from opposite side; wall calcareous, coarsely perforate, radial in structure, bilamellar; apertures in adult a series of openings at sides of chambers. *Eoc.-Rec.,* cosmop.——FIG. 570,2. **C. tabellae-*

FIG. 570. Cymbaloporidae; *1, Cymbalopora; 2, Cymbaloporella; 3, Cymbaloporetta* (p. C698-C701).

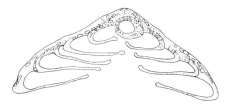

Fig. 571. Cymbaloporidae; *Cymbalopora* (p. *C*698-*C*699).

formis (BRADY), Rec., S.Pac.O.(Samoa Is.); *2a-c*, opposite sides and edge view, ×63 (*2117).

Cymbaloporetta CUSHMAN, 1928, *436, p. 7 [**Rosalina squammosa* D'ORBIGNY in DE LA SAGRA, 1839, *1611, p. 91; OD]. Test conical, early chambers trochospiral, later alternating in annular series, as in *Halkyardia*, with few chambers in each series, all chambers visible on highly convex spiral side, only few visible in last annulus on umbilical side; sutures oblique and flush on spiral side, deeply depressed and radial on umbilical side, with deep openings left between adjacent chambers, umbilicus small, open; wall calcareous, spiral side coarsely perforate, umbilical side nonporous, radial in structure, bilamellar; apertures consisting of one or more sutural openings at each side of chambers on umbilical side. *Mio.-Rec.,* cosmop.——FIG. 570,*3*; 573. **C. squammosa* (D'ORBIGNY), Rec., Bahama Is. (570,*3*), Carib. (573); *3a-c*, opposite sides and edge view, ×86 (*2117); 573, axial sec., ×160 (*951).

Eofabiania KÜPPER, 1955, *1070, p. 135 [**E. grahami*; OD]. Test conical, concavo-convex, early portion trochospiral, structure similar to *Fabiania* but without lateral chamberlets, exterior unknown. [HANZAWA (*874, p. 121) suggested that *Eofabiania* may be a synonym of *Fabiania*. Poorly known only from thin sections, the present genus is here tentatively recognized by the absence of lateral chamberlets, although these are not always well shown in axial sections of true *Fabiania*.] *M.Eoc.,* USA(Calif.).——FIG. 574,*1,2*. **E. grahami*; *1a,b*, axial sec. of paratype and sketch of same; *2*, sketch of axial sec. of holotype; all ×38 (*1070).

Fabiania A. SILVESTRI, 1924, *1778, p. 7 [**Patella (Cymbiola) cassis* OPPENHEIM, 1896, *1390, p. 55, 56; OD] [=*Eodictyoconus* COLE & BERMÚDEZ, 1944, *370, p. 336 (type, *Pseudorbitolina cubensis* CUSHMAN & BERMÚDEZ, 1936, *490, p. 59);

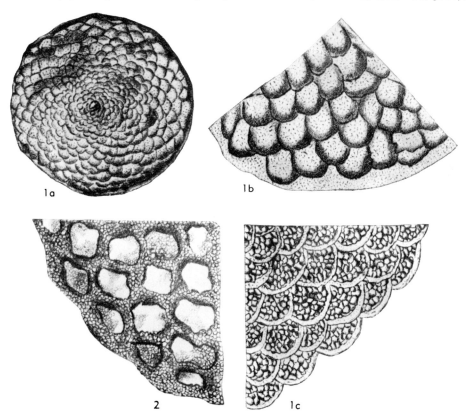

Fig. 572. Cymbaloporidae; *1,2, Archaecyclus* (p. *C*699).

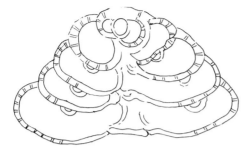

Fig. 573. Cymbaloporidae; *Cymbaloporetta* (p. C701).

Tschoppina KEIJZER, 1945, *1030, p. 213 (type, *Pseudorbitolina cubensis* CUSHMAN & BERMÚDEZ, 1936, *490, p. 59)]. Test of regular or flattened conical form with bluntly rounded apex; early stage of 3 simple globose chambers with basal aperture and thickened perforate wall, later chambers in cyclical series or tiers, area just beneath external wall subdivided by horizontal and vertical partitions forming coarse alveoli, which again are subdivided by thinner, shorter partitions into 2 or 3 smaller alveoli; sutures of chamber tiers visible externally but alveolar walls less distinct except on abraded specimens; wall calcareous, thick, bilamellar, outer wall coarsely perforate, wall of umbilical side and partitions imperforate. *Eoc.,* Eu.-Japan-Carib.(Cuba)-Pac. O.——FIG. 574,*3-5.* *F. cassis* (OPPENHEIM), Auvers., Italy *(3),* Lutet., Italy *(4,5); 3,* ext., ×13; *4,* horiz. sec., ×24 (*1781); *5,* tang. vert. sec., ×12 (*1781).——FIG. 574,*6,7. F. cubensis* (CUSHMAN & BERMÚDEZ), Cuba; *6,* horiz. sec. enlarged; *7,* axial sec., ×41 (*370).

Gunteria CUSHMAN & PONTON, 1933, *522, p. 25 [*G. floridana; OD]. Test compressed, flabelliform to reniform in outline, embryonic stage of large globular undivided chambers, later with concentric chambers divided by numerous transverse and radial partitions into chamberlets, as in *Fabiania;* sutures indistinct externally, except on abraded specimens; wall calcareous, perforate; aperture consisting of 2 rows of pores on terminal face. *M.Eoc.,* USA(Fla.)-Carib.(Cuba).——FIG. 575,*1,2;* 576,*1,2.* *G. floridana,* Fla. (575,*1,2*), Cuba (576,*1,2*); 575,*1a,b,* side and top views of paratype, ×12.5; 575,*2,* portion of apert. view of another paratype, ×33 (*2117); 576,*1,2,* axial and transv. secs., ×20(*372).

Halkyardia HERON-ALLEN & EARLAND in HALKYARD, 1918, *861, p. 107 [*Cymbalopora radiata* var. *minima* LIEBUS, 1911, *1135, p. 952; SD CUSHMAN, 1928, *439, p. 288]. Test small, plano-convex to lenticular with spiral side more strongly convex, periphery subacute, peripheral margin lobulate; early chambers in irregular or "raspberry" type of arrangement, later chambers small and numerous, alternating in annular series, thick wall lamellae obscuring chambers of early spire; umbilical area filled with horizontal lamellae and connecting hollow vertical pillars; sutures oblique, curved and flush on spiral side, radial and depressed on umbilical side; wall calcareous, distinctly perforate, radial in structure, inner walls nonporous; aperture consisting of small pores at periphery. [Although topotypes of *H. minima* were stated by WOOD (*2073, p. 250) to be radial in structure, the genus was regarded as microgranular by HOFKER (*951, p. 117).] *Eoc.,* Eu.-Pac.O.-N.Am.——FIG. 575,*5.* *H. minima* (LIEBUS), Eoc.(Barton.), Fr.; *5a-c,* opposite sides and edge view, ×130 (*2117).——FIG. 575,*6-8. H. bikiniensis* COLE, Eoc., Bikini Atoll; *6,* vert. sec. showing umbilical pillars and thickened spiral wall; *7,* transv. sec. nearer base cutting umbilical pillars and showing thickened outer wall; *8,* transv. sec. near apex, cutting embryonic chambers; all ×40 (*361).

Pyropilus CUSHMAN, 1934, *463, p. 100 [*P. rotundatus;* OD]. Early chambers trochospirally coiled, later added irregularly to form elongate mass, with all chambers visible on originally spiral side and only last series visible on opposite side around elongate or irregular umbilical depression; wall calcareous, coarsely perforate, with thin inner pseudochitinous layer; aperture consisting of numerous large pores along sutures and on apertural face on umbilical side of test, sutural openings remaining open on all chambers of final whorl. [Lamellar and microstructure of the wall have not been described, but general appearance suggests its relation to the Cymbaloporidae.] *Rec.,* Pac.O.——FIG. 575,*3,4.* *P. rotundatus,* S.Pac.O. (Rangiroa Atoll); *3a-c,* opposite sides and edge view of holotype; *4,* umbilical side of large paratype, ×49 (*2117).

Family HOMOTREMATIDAE Cushman, 1927

[*nom. correct.* LOEBLICH & TAPPAN, herein (*pro* Homotremidae CUSHMAN, 1927, p. 97)] [=Victoriellidae CHAPMAN & CRESPIN, 1930, p. 111; =Polytremidae CHAPMAN, PARR & COLLINS, 1934, p. 556, 573 (*recte* Polytrematidae); =Miniacinidae THALMANN, 1938, p. 208; =Eorupertiidae COLE, 1957, p. 337]

Test attached, early chambers irregularly trochospiral, later variously modified; wall coarsely perforate. *U.Cret.-Rec.*

Subfamily HOMOTREMATINAE Cushman, 1927

[*nom. transl.* CHAPMAN & PARR, 1936, p. 144; (*ex* Homotremidae CUSHMAN, 1927); Homotrematinae *nom. correct.* POKORNÝ, 1958, p. 333]

Test attached, early stage trochospiral, later growth irregular, extending upward from attachment, becoming branched; apertures large, and may be covered by perforated plate. *Eoc.-Rec.*

Homotrema HICKSON, 1911, *922, p. 445 [*Millepora rubra* LAMARCK, 1816, *1088, p. 202; OD

(M)]. Test attached, large, 2 to 8 mm. diam., variable in form, may be globose, hemispherical, encrusting with irregular swellings, with truncated conical projections or erect branches, possibly environmentally controlled; early chambers in spiral or "raspberry" arrangement, later cham-

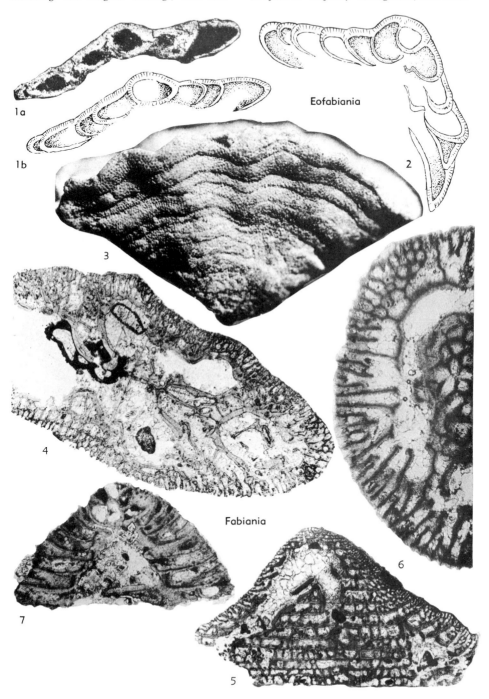

Fig. 574. Cymbaloporidae; *1,2, Eofabiania; 3-7, Fabiania* (p. C701-C702).

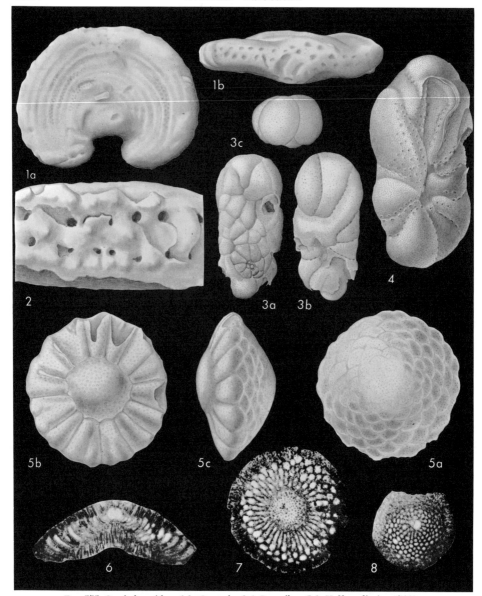

Fig. 575. Cymbaloporidae; *1,2, Gunteria; 3,4, Pyropilus; 5-8, Halkyardia* (p. C702).

bers in roughly concentric layers, outer surface with cribrate areolae surrounded by imperforate rims (chamber walls), beneath areolae containing large passages and irregular hollows, partially due to resorption; as additional layers of chambers are added, new cribrate plates appear above earlier ones, and at intervals continuous cribrate wall may cover imperforate areolae, later forming new areolae, earlier layer being covered simultaneously by nonperforate material; some large openings not covered by perforated plates may show protruding sponge spicules cemented by protoplasmic material; wall of early spiral portion pseudochitinous and insoluble, later portion calcareous, red; aperture consisting of large pores in areolae and large irregular openings with contained sponge spicules. *Rec.*, Atl.O.-Carib.-Ind.O.-Pac.O.——Fig. 577,*1-3.* **H. rubrum* (LAMARCK), Ind.O. *(1)*, Indon. *(2)*, Bermuda *(3)*; *1*, ext. of branching form, ×4.5 (*922); *2*, partially sectioned specimen showing perforated plates, imperforate areolae at surface, and inner walls with irregular openings, ×17 (*928a); *3*, continuous cribrate layer added over earlier areolae, ×170 (*702).

Miniacina GALLOWAY, 1933, *762, p. 305 [*pro* *Pustularia* GRAY, 1858, *812, p. 270, 271 (type, *P. rosea*) (*non* SWAINSON, 1840)] [**Millepora miniacea* PALLAS, 1766, *1407, p. 251; OD]. Test encrusting or branching, branches commonly more elongate and slender than in *Homotrema*, to 7 mm. in height; early stage with spiral or "raspberry" chamber arrangement, later with layers of perforated laminae, pores about 5µ diam., adjacent laminae connected by hollow pillars with imperforate double walls, which grow upward from foramina of previous lamina, central portion of branches with irregularly twisted elongate vertical tubes without perforate walls, which arise near base and extend to tips of branches; wall calcareous, red, pink, or white, surface with openings of 2 sizes, smaller wall perforations, and larger "pillar pores" or foramina 30-80µ diam. *Rec.*, Medit. Sea-Malay Arch.-Ind.O.-S.Pac.O.——— FIG. 577,*4-7*. **M. miniacea* (PALLAS), Indon.; *4a,b*, ext. of encrusting and branching types, ×4.5; *5*, diagram. transv. sec. of branch showing concentric layers of chambers, hollow "pillar pores" or foramina, and smaller perforations (*922); *6*, surface showing small perforations and larger foramina, ×50 (*922); *7*, part of transv. sec. showing pores, foramina, and double walls of "pillar pores," ×175 (*928a).

Sporadotrema HICKSON, 1911, *922, p. 447 [**Polytrema cylindricum* CARTER, 1880, *296, p. 441; OD]. Test attached, large, to 27 mm. in height, early juvenile stage coiled, later with large cylindrical branches; chambers large, at periphery of branches, communicating by large open passages, central portion of branches occupied by irregularly shaped tubes that spiral up trunk and branches and may open at tips of branches; inner septal walls nonperforate; wall calcareous, surface coarsely perforate, pores irregularly scattered, large ones at surface resulting from fusion within wall of numerous fine pores at inner surface of wall, lacking both areolae, found in *Homotrema*, and "pillar pores" of *Miniacina*, may incorporate siliceous sponge spicules in varying amounts; color, red, yellow, or orange. *Eoc.-Rec.*, Pac.O.-Ind.O. ———FIG. 578,*1-4*. **S. cylindricum* (CARTER), Rec., Ind.O. (*1,2,4*), Indon. (*3*); *1*, ext. showing branching form, ×2 (*922); *2*, portion of branch, enlarged (*922); *3*, diagram. long. sec. showing large peripheral chambers, internal long. stoloniferous tubes, finely porous inner chamber surface, pores fusing in wall to form fewer and larger openings at surface, enlarged (*928a); *4*, surface showing large pores formed by fusion of small inner pores, ×50 (*922).

Subfamily VICTORIELLINAE
Chapman & Crespin, 1930

[*nom. transl.* LOEBLICH & TAPPAN, herein (*ex* family Victoriellidae CHAPMAN & CRESPIN, 1930, p. 111)]

Test attached, early chambers trochospiral, later extending upward from base in

FIG. 576. Cymbaloporidae; *1,2, Gunteria* (p. C702).

loose spiral or becoming irregular rounded mass; wall calcareous, perforate, radial in structure, bilamellar; aperture interiomarginal. *U.Cret.-Rec.*

Victoriella CHAPMAN & CRESPIN, 1930, *323, p. 111, 112 [**Carpenteria proteiformis* var. *plecte* CHAPMAN, 1921, *321, p. 320, =*Carpenteria conoidea* RUTTEN, 1914, *1598, p. 47; OD (M)]. Test conical, commonly attached at apex, juvenile stage free, in low trochospiral coil of few chambers, when temporary or permanent attachment occurs direction of coiling may reverse and coiling is high-spired in adult, umbilicus depressed or forming axial hollow; chambers inflated, 3 or 4 to whorl, not embracing; sutures depressed but wall lamellae obscure early ones; septa of 3 layers, 2 layers of preceding chamber and inner lamella of following one; wall calcareous, thick, coarsely perforate, except for imperforate area surrounding aperture, radiate in structure, bilamellar, no canals between layers, but some interlocular spaces may occur in walls, numerous round to elliptical bosses interspaced between perforations, formed by pillar-like thickenings in wall which displace wall perforations; aperture umbilical in position, with thick lip on 3 sides. *U. Eoc.-Mio.*, Australia-N.Guinea-N.Z.-Eu.——— FIG. 579,*1-3*. **V. conoidea* (RUTTEN), Oligo., Australia; *1*, ext., ×25; *2a*, vert. sec. showing wall pillars, layering, and bilamellar structure, ×46; *2b*, portion enlarged to show 3-layered septa with single lamella of final chamber (at left) attached to bilamellar septal face of penultimate chamber, ×85; *3a*, diagram showing early chamber arrangement with proloculus and early whorl ob-

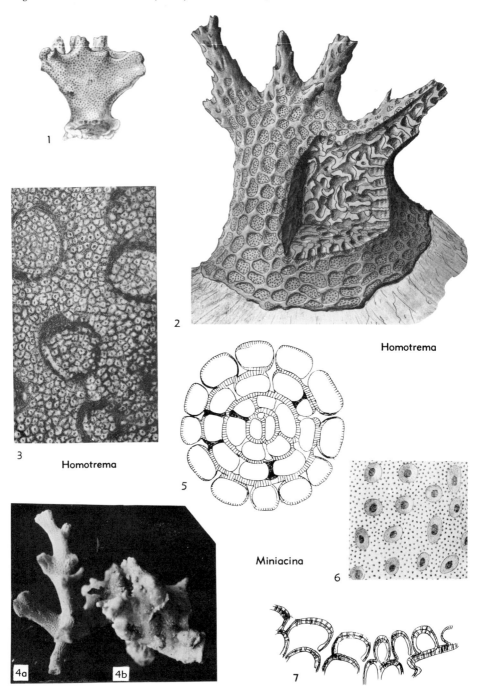

Fig. 577. Homotrematidae (Homotrematinae; *1-3, Homotrema; 4-7, Miniacina*) (p. C702-C705).

Foraminiferida—Rotaliina—Orbitoidacea

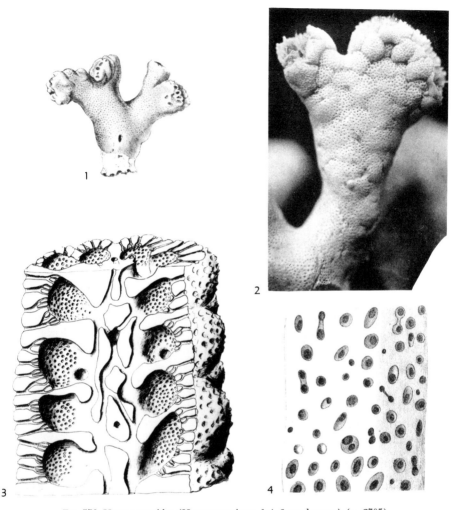

Fig. 578. Homotrematidae (Homotrematinae; *1-4, Sporadotrema*) (p. C705).

lique to later axis of coiling around axial hollow, attachment surface at side of third or fourth chamber, radial wall structure (shown only at right of figure), ×40; *3b*, portion of previous figure enlarged to show pillars and their displacement of pores, ×120 (*797).

Carpenteria GRAY, 1858, *812, p. 269, 270 [*C. balaniformis*; OD (M)] [=*Neocarpenteria* CUSHMAN & BERMÚDEZ, 1936, *489, p. 34 (type, *N. cubana*); *Carpenterella* BERMÚDEZ, 1949, *124, p. 313 (type, *C. truncata*) (*non Carpenterella* COLLENETTE, 1933; *nec* KRASHENINNIKOV, 1953); *Bermudezella* THALMANN, 1951, *1899d, p. 224 (*nom. subst. pro Carpenterella* BERMÚDEZ, 1949, *non* COLLENETTE, 1933); *Haerella* BELFORD, 1960, *110, p. 112 (type, *H. conica*)]. Test attached, plano-convex, trochospiral, all chambers visible from flat, attached spiral side, only those of last whorl visible on convex, centrally umbilicate opposite side, peripheral keel may spread slightly over attachment; wall calcareous, distinctly perforate over umbilical surface of chambers, radial in structure, only keel and small area around umbilical area and apertural margin being nonperforate, thickened shell material produced into pillar-like extensions around umbilicus on older specimens; aperture slitlike, extending from periphery along base of final chamber into open umbilicus. *U.Cret.-Rec.,* Pac.O.-Australia-W.Indies (Cuba-Carib.).——FIG. 580,*1*. *C. balaniformis*, Rec., Funafuti Atoll; *1a-c*, opposite sides and edge view, ×20 (*2117).——FIG. 580,*2*. *C. conica* (BELFORD), U.Cret.(Campan.), W.Australia; *2a-c*, opposite sides and edge view of holotype, ×37 (*110).——FIG. 580,*3*. *C. truncata* (BERMÚDEZ), M.Oligo., Haiti; *3a-c*, opposite sides and edge view of holotype, ×39 (*2117).——FIG. 580,*4*. *C. cubana* (CUSHMAN & BERMÚDEZ), Eoc., Cuba;

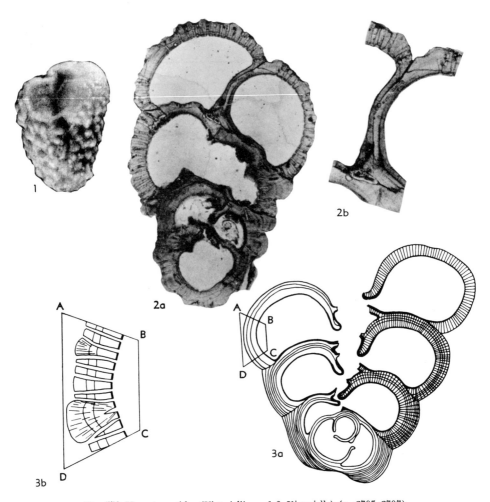

Fig. 579. Homotrematidae (Victoriellinae; 1-3, *Victoriella*) (p. C705-C707).

4a-c, opposite sides and edge view of paratype, ×76 (*2117).

[The original description of *Carpenteria* was somewhat generalized, and GRAY's types from the Philippine Islands are not preserved. The present redefinition is based on the type-species as shown by the specimen figured by CHAPMAN (1900, *314, p. 13, pl. 4, fig. 2) (BMNH Cat. No. 03.2.5.124, from off Funafuti at 115-200 fathoms, here redrawn). GRAY's original figures also show a low conical form which undoubtedly suggested the specific name of the type-species (barnacle-formed). Later workers have erroneously included much higher or uncoiled forms in *Carpenteria*. *Neocarpenteria* was proposed for a plano-convex trochoid form with semicircular ventral marginal aperture. This semicircular opening in the type-specimen (here redrawn) is merely an irregular remnant of the broken final chamber and the true aperture is not shown. CUSHMAN & BERMÚDEZ (*489, p. 34) stated, "There seems to be a tendency to grow upward slightly on the ventral side, suggesting the type of development seen in *Carpenteria.*" As the type-species of *Carpenteria* does not grow upward into a cylindrical form and the apertural distinction is nonexistent, the name *Neocarpenteria* is a junior synonym. *Carpenterella* was defined by BERMÚDEZ (1949, *124) to include similar plano-convex forms with a slitlike interiomarginal aperture. It was said to differ from *Carpenteria* in having a simple trochoid form and not becoming uniserial. A homonym of *Carpenterella* COLLENETTE, 1933, *Carpenterella* BERMÚDEZ was later renamed *Bermudezella*, but as the type-species of *Carpenteria* is also a low, rather than uniserial, form, both *Carpenterella* BERMÚDEZ and *Bermudezella* THALMANN are junior synonyms. The high cylindrical forms previously placed in *Carpenteria* should be referred to *Biarritzina*.]

Eorupertia YABE & HANZAWA, 1925, *2090, p. 77 [*pro Uhligina* YABE & HANZAWA, 1922, *2088, p. 71 (*non* SCHUBERT, 1899)] [*Uhligina boninensis* YABE & HANZAWA, 1922, *2088, p. 72; OD] [=*Gyroidinella* Y. LE CALVEZ, 1949, *1112, p. 27 (type, *G. magna*); *Neogyroidina* BERMÚDEZ, 1949, *124, p. 255 (type, *Gyroidina protea* CUSHMAN & BERMÚDEZ, 1937, *491, p. 22)]. Test trochospirally coiled, cylindrical or subconical in form, attached at spiral side of early stage, umbilicate, with chambers coiled about axial hollow, periphery angular to rounded; wall calcareous, radial in structure, perforate, except in apertural region, bilamellar, 2 laminae separated by dark layer, pillars developed in wall, septa 3-layered as in *Victoriella* and may enclose interseptal spaces; aperture umbilical, interiomarginal, slitlike, with

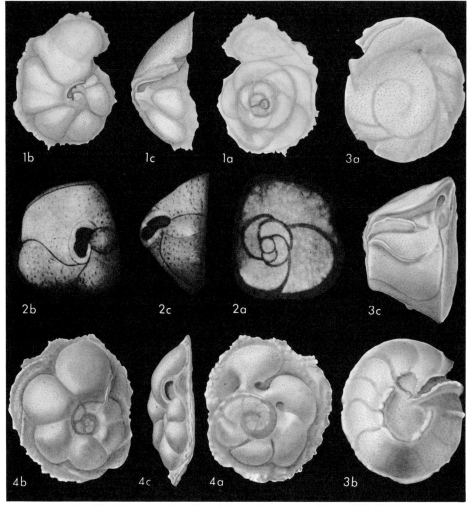

Fig. 580. Homotrematidae (Victoriellinae; *1-4, Carpenteria*) (p. C707-C708).

lip. [The synonymy of *Gyroidinella* with *Eorupertia* was demonstrated by REISS (*1528c, p. 6).] *M. Eoc.-U. Eoc.*, W.Pac.O.(Bonin Is.)-Eu.-S. Am.-M. East-Carib.-N. Am.-Japan.——FIG. 581,*1-6.* **E. boninensis* (YABE & HANZAWA), Eoc., Haha-jima, Japan; *1*, ext., ×10; *2,3,* long. secs. at side of axial hollow and nearly axial long. sec., ×20; *4,* transv. sec. showing chambers around axial hollow, ×20; *5,* diagram showing bilamellar wall and pillars, ×20; *6,* diagram of wall showing relation of pores and conical pillars (*2090).——FIG. 582,*1-3. E. magna* (Y. LE CALVEZ), M.Eoc. (Lutet.), Fr. *(1),* Eoc., Israel *(2,3); 1a-c,* opposite sides and edge view, ×23 (*2117); *2,* horiz. sec., ×14 (*1528c); *3,* axial section, ×37 (*1528c).

Maslinella GLAESSNER & WADE, 1959, *797, p. 203 [**M. chapmani*; OD]. Test large, early stage low, trochospiral, later pseudoplanispiral and semi-involute with axis of coiling perpendicular to that of early stage, but asymmetrical; chambers inflated, periphery subangular to rounded, increasing gradually in size; sutures straight to curved, radial, thickened and limbate on spiral side of test; wall calcareous, thick, coarsely perforate, radial in structure, bilamellar, 3-layered septa and apertural face nonperforate; aperture low interiomarginal equatorial opening with thickened lip. *U.Eoc.*, Australia.——FIG. 583,*1-4.* **M. chapmani; 1a-c,* opp. sides and edge view of holotype showing coarsely perforate wall and thickened sutures, ×25; *2,* edge view of paratype showing apert., ×25; *3,* part of median horiz. sec. showing 3-layered septa and thickened apert. lips, ×46; *4,* vert. sec. showing nepionic coil in plane of coiling and later coil at right angles to it, and thick wall with pillars and pores, ×95 (*797).

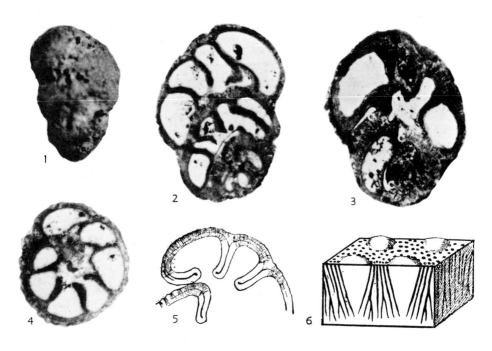

Fig. 581. Homotrematidae (Victoriellinae; *1-6, Eorupertia*) (p. C708-C709).

ORBITOIDIDAE
By W. Storrs Cole

Family ORBITOIDIDAE Schwager, 1876

[*nom. correct.* Eimer & Fickert, 1899, p. 688 (*pro* Orbitoidee Schwager, 1876, p. 481)] [=subfamily Orbitoidinae Prever, 1904, p. 111; =family Orbitoidinae Silvestri, 1907, p. 12 (*nom. van.*); =Orbitoidae Silvestri, 1937, p. 155; =Orbitoidida Copeland, 1956, p. 188; =Clypeorbinae Sigal in Piveteau, 1952, p. 259; =Lepidorbitoididae Pokorný, 1958, p. 388 (*ex* Lepidorbitoidinae Vaughan in Cushman, 1933, p. 285); =Pseudorbitellinae Hanzawa, 1962, p. 148]

Test biconcave to spherical, with embryonic chambers enclosed by thick perforate wall, or with thinner-walled bilocular embryonic chambers followed by several relatively large periembryonic chambers; equatorial and lateral chambers not differentiated, or equatorial chambers covered on each side with distinct zones of lateral chambers; equatorial chambers arcuate or short, spatulate; with stolons, but without canal system. *U.Cret.-Paleoc.*

An analysis of the initial chambers of microspheric specimens of *Orbitoides* led Küpper (*1068) to postulate that the orbitoidids were derived from a calcareous, biserial ancestor similar in structure to certain genera referred to the Guembelininae of the Heterohelicidae. Two genera, *Lepidorbitoides* (U.Cret.) and *Actinosiphon* (Paleoc.), which have been associated traditionally with the lepidocyclines, are here assigned provisionally to the Orbitoididae.

Although *Lepidorbitoides* and *Actinosiphon* resemble the lepidocyclines in form and structure, they cannot be related to them, inasmuch as the first true lepidocycline appeared in the middle Eocene. As the structures of certain species of *Orbitoides* are similar to those of *Lepidorbitoides*, it seems logical to postulate that *Lepidorbitoides* was derived from *Orbitoides* and in turn generated *Actinosiphon* as the final representative of this dominantly Upper Cretaceous family.

Orbitoides d'Orbigny in Lyell, 1848, *1192, p. 12 [*Lycophris faujasii* Defrance, 1823, *579b, p. 271; =*Orbitolites media* d'Archiac, 1837, *35, p. 178; SD Jones, Parker, & Brady, 1866, *1002, appendix I] [=*Hymenocyclus* Bronn, 1853, *214a, p. 94 (type, *Lycophris faujasii* Defrance, 1822, *579b, p. 271); *Simplorbites* deGregorio, 1882, *815, p. 10 (type, *Nummulites papyracea* Boubée, 1832, *176A, p. 115), *Silvestrina* Prever, 1904, *1482, p. 113, 122 (type, *Orbitoïdes apiculata* Schlumberger, 1901,*1661, p.465); *Schlumbergeria* A. Silvestri, 1910, *1771A, p. 118 (type, *Linderina? douvillei*); *Orbitella* Douvillé, 1915, *621, p. 666 (type, *Orbitolites media* d'Archiac, 1837, *35, p. 178); *Monolepidorbis* Astre, 1927, *54, p. 388 (type, *M. sanctae-pelagiae*); *Gallo-*

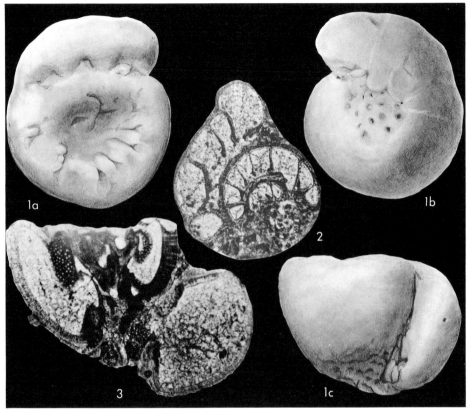

FIG. 582. Homotrematidae (Victoriellinae; *1-3, Eorupertia*) (p. C708-C709).

wayina ELLIS, 1932, *699, p. 1 (type, *G. browni*); *Hellenocyclina* REICHEL, 1949, *1521, p. 482 (type, *H. beotica*)]. Embryonic chambers surrounded by thick, perforated wall, bilocular to quadrilocular, or teratologically with more than 4 irregularly arranged chambers; equatorial chambers arcuate; lateral chambers reduced or well developed, slitlike. *U.Cret.*, Eu.-Asia(India)-N.Am.——FIG. 584, *3.* *O. faujasii* (DEFRANCE), Fr.; *3a,b,* equat. sec. vert. sec., ×40, ×20 (*2113c).

Actinosiphon VAUGHAN, 1929, *1990, p. 163, 166 [**A. semmesi*; OD] [=*Orbitosiphon* RAO, 1940, *1498, p. 414 (type, *Lepidocyclina (Polylepidina) punjabensis* DAVIES in DAVIES & PINFOLD, 1937, *563, p. 53)]. Embryonic chambers bilocular, large, completely surrounded by a ring of about 11 periembryonic chambers; equatorial chambers in rude radial rows with communication by large median stolon. *Paleoc.*, N.Am.(Mex.)-Indo-Pac. Reg.——FIG. 585,*1.* *A. semmesi*, Mex.; equat. sec., ×80 (*2122).

Lepidorbitoides A. SILVESTRI, 1907, *1766, p. 80 [**Orbitolites socialis* LEYMERIE, 1851, *1133, p. 191] [=*Clypeorbis* H. DOUVILLÉ, 1915, *621, p. 668, 669 (type, *Orbitoïdes mamillata* SCHLUMBERGER, 1902, *1662, p. 259); *Orbitocyclina* VAUGHAN, 1929, *1993, p. 291 (type, *Lepidorbitoides minima* DOUVILLÉ, 1927, *628A, p. 291); *Orbitocyclinoides* BRÖNNIMANN, 1944, *216, p. 5 (type, *Orbitocyclina (O.) schencki*); *Pseudorbitella* HANZAWA, 1962, *875, p. 148 (type, *P. americana*; OD)]. Embryonic chambers bilocular, small, with or without periembryonic chambers; equatorial chambers arcuate to hexagonal; lateral chambers well developed. *U.Cret.*, Eu.-N.Am.-Carib.-Asia.

L. (Lepidorbitoides). Test circular, *U.Cret.*, Eu.-Asia(India)-N.Am., trop.——FIG. 584,*1.* *L. (L.) socialis* (LEYMERIE), Fr.; *1a,b,* equat. sec., vert. sec., ×40, ×20 (*2113c).

L. (Asterorbis) VAUGHAN & COLE, 1932, *1996, p. 611 [**A. rooki*; OD] [=*Cryptasterorbis* M. G. RUTTEN, 1935, *1599, p. 533 (type, *?Asterorbis cubensis* PALMER, 1934, *1408, p. 249)]. Test stellate. *U.Cret.*, N.Am.-Carib., trop. zone.——FIG. 585,*2.* *L. (A.) rooki*, Cuba; *2a,b,* equat. and vert. secs., ×16 (*2113c).

Omphalocyclus BRONN, 1852, *214a, p. 95 [**Orbulites macroporus* LAMARCK, 1816, *1088, p. 197; OD]. Embryonic chambers of megalospheric generation similar to those of *Orbitoides*, but with lateral chambers of same kind and not differentiated from equatorial chambers. *U.Cret.*, Carib.-Asia.

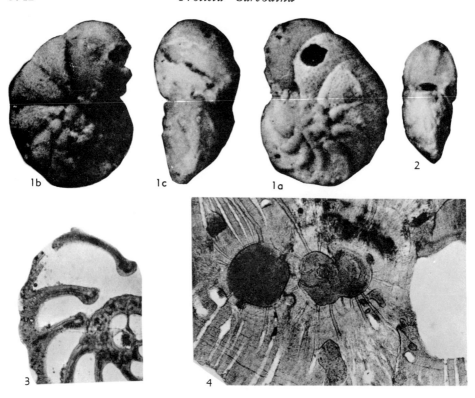

Fig. 583. Homotrematidae (Victoriellinae; *1-4, Maslinella*) (p. C709).

O. (Omphalocyclus). Test strongly biconvex. *U. Cret.*, Eu.-Asia (India)-Carib. (Cuba).——Fig. 584,2. **O. (O.) macroporus* (LAMARCK), Cuba; *2a,b*, equat. sec., vert. sec., ×40, ×20 (*2113c).

O. (Torreina) D. K. PALMER, 1934, *1408, p. 237 [**T. torrei*; OD]. Test nearly spherical. *U.Cret.*, Carib.——Fig. 584,4. **O. (T.) torrei*, Cuba; equat. sec., ×20 (*1408).

DISCOCYCLINIDAE

By W. STORRS COLE

Family DISCOCYCLINIDAE
Galloway, 1928

[*nom. transl.* VAUGHAN & COLE in CUSHMAN, 1940, p. 327 (*ex* Discocyclininae GALLOWAY, 1928, p. 55) [=Orthophragminidae WEDEKIND, 1937, p. 123, 124; =Orthophragmininae WEDEKIND, 1937, p. 125; =Asterocyclinidae BRÖNNIMANN, 1951, p. 208; =Orbitoclypeidae POKORNÝ, 1958, p. 393 (*ex* Orbitoclypeinae BRÖNNIMANN, 1946, p. 612)]

Test circular or stellate, thin or inflated, composed of equatorial layer with lateral chambers on each side; megalospheric generation with subspherical initial chamber partly or completely embraced by larger second chamber; microspheric generation with initial coil of small chambers; equatorial chambers rectangular to faintly hexagonal in plan; radial chamber walls, when present, arranged in annuli; equatorial chambers connected by annular and radial stolons with adjacent chambers in same annulus and with adjacent chambers in next inner and next outer annulus; intraseptal and intramural canal system present (Fig. 586,*1a*). *Paleoc.-Eoc.*

BRÖNNIMANN (*218) suggested that this family should be divided into two subfamilies; Discocyclininae, in which the equatorial layer is composed of chambers and chamberlets (Fig. 586,*1b*), and Orbitoclypeinae, in which the equatorial layer is composed only of chambers (Fig. 586,*2*). However, since the chambers of *Discocyclina (Discocyclina) anconensis* BARKER (Fig. 586,*1d*) are the same as those of typical representatives of the Orbitoclypeinae, it is doubtful if this family should be divided.

VAUGHAN (*1995) demonstrated that representatives of the Discocyclinidae possess interseptal canals (Fig. 586,*1c*), but BRÖNNIMANN (*221) interpreted these canals as a system of fissural interseptal spaces and,

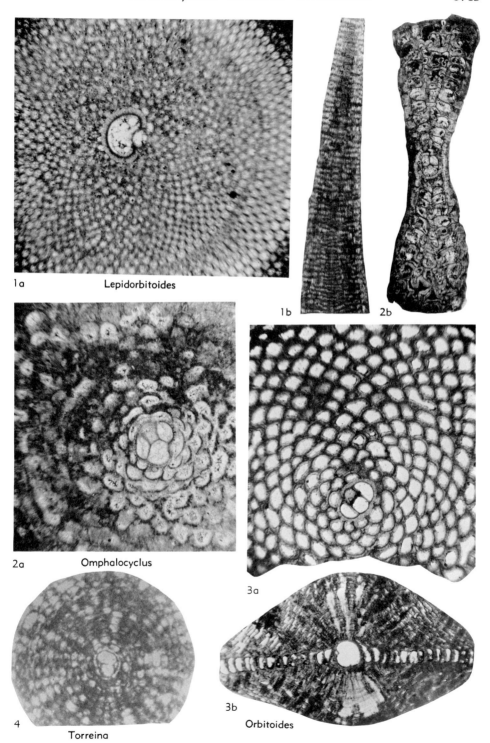

Fig. 584. Orbitoididae; *1,* Lepidorbitoides *(Lepitorbitoides); 2,* Omphalocyclus *(Omphalocyclus); 4, O. (Torreina); 3,* Orbitoides (p. *C710-C712).*

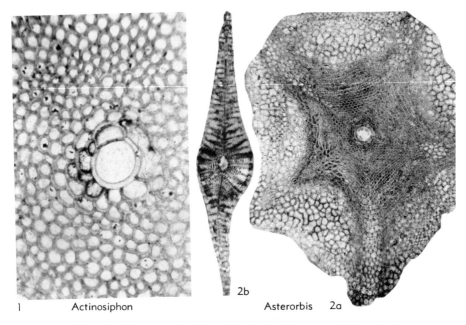

1 Actinosiphon   2b Asterorbis 2a

FIG. 585. Orbitoididae; *1, Actinosiphon; 2, Lepidorbitoides (Asterorbis)* (p. C711).

therefore, not true canals of the kind developed in the nummulitids. VAUGHAN (*1995) derived the Discocyclinidae from a *Nummulites*-like ancestor because of the presence of both intraseptal canals and annular canals which he assumed were "the morphological representation of the canals in the marginal plexus of the Camerinidae [Nummulitidae]" (*1995, p. 52). BRÖNNIMANN (*221, p. 211) questioned this origin for the Discocyclinidae.

Discocyclina GÜMBEL, 1870, *840, p. 687 [*Orbitulites pratti* MICHELIN, 1846, *1255, p. 278; SD GALLOWAY, 1928, *761, p. 56] [=*Rhipidocyclina* GÜMBEL, 1870, *840, p. 688 (type, *Orbitoïdes (R.) multiplicata* GÜMBEL, 1870); *Orthophragmina* MUNIER-CHALMAS, 1891, *1326, p. 17, 18, 19 (type, *Orbitulites pratti* MICHELIN, 1846, *1255, p. 278); *Orbitoclypeus* A. SILVESTRI, 1907, *1769, p. 106 (type, *O. himerensis*); *Exagonocyclina* CHECCHIA-RISPOLI, 1907, *330, p. 188 (type, *Orbitoïdes (E.) schopeni*, =*Orbitoclypeus himerensis* A. SILVESTRI); *Nodocyclina* HEIM, 1908, *893A, p. 271 (type, *Orthophragmina umbilicata* DEPRAT, 1905, *583, p. 497); *Eudiscodina* VAN DER WEIJDEN, 1940, *2042, p. 15 (type, *Orthophragmina archiaci* SCHLUMBERGER, 1903, *1663, p. 277); *Umbilicodiscodina* VAN DER WEIJDEN, 1940, *2042, p. 15 (type, *Orbitolites discus* RÜTIMEYER, 1850, *1594, p. 116); *Trybliodiscodina* VAN DER WEIJDEN, 1940, *2042, p. 15 (type, *Orthophragmina chudeaui* SCHLUMBERGER, 1903,

*1663, p. 282); *Hexagonocyclina* CAUDRI, 1944, *304, p. 362 (type, *Orbitoclypeus ?cristensis* VAUGHAN, 1924, *1988, p. 814); *Bontourina* CAUDRI, 1948, *305, p. 477 (type, *B. inflata*)]. Test circular in plan, discoidal or lenticular, with or without raised radiating ribs; annular stolon proximally situated; radial chamber walls of equatorial chambers in adjacent annuli usually alternating in position. *Paleo.-Eoc.*, Eu.-Indo.-Pac. Reg.-N.Am.-S.Am.

D. (Discocyclina). Test circular in plan; not stellate; without costae. *Paleoc.-Eoc.*, Eu.-Indo-Pac.-N.Am.-S.Am.——FIG. 587,1. *D. (D.) pratti* (MICHELIN), Eoc.(Auvers.), Eu.; *1a*, ext. view, ×5 (*2042); *1b*, equat. sec. with embryonic chambers, periembryonic chambers and equat. chambers, ×40 (*2042); *1c*, equat. chambers with proximally situated annular stolon, ×85 (*1994).

D. (Aktinocyclina) GÜMBEL, 1870, *840, p. 688 [*Orbitulites radians* D'ARCHIAC, 1848, *37A, p. 405; SD DOLLFUS, 1889, *607, p. 1226] [=*Actinocyclina* GÜMBEL, 1870, *840, p. 707 (nom. null.)]. With elevated rays formed by local increase in number of lateral chambers; rays not terminating in protuberant angles as in *Asterocyclina*. *Eoc.(Lutet.-Priabon.)*, Eu.——FIG. 587,2. *D. (A.) radians* (D'ARCHIAC), Priabon.; *2a*, ext. view, ×10 (*2119); *2b*, equat. sec. with embryonic chambers, periembryonic chambers and equat. chambers, ×50 (*217); *2c*, vert. sec., ×22 (*1995).

Asterocyclina GÜMBEL, 1870, *840, p. 689 [*nom. subst. pro* *Asterodiscus* SCHAFHÄUTL, 1863 (*non*

Foraminiferida—Rotaliina—Orbitoidacea

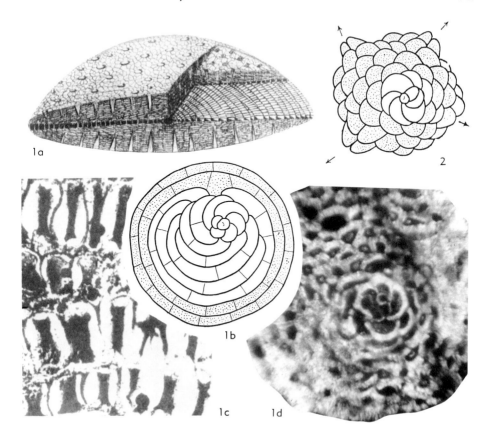

Fig. 586. Discocyclinidae: *1, Discocyclina,* structural features shown by oblique view of partly sectioned test *(1a)* and equatorial section *(1b)* (diagrammatic, not to scale), vertical section showing canals, ×400 *(1c)*, and equatorial section, ×180 *(1d)* (*1a*, *2121; *1b*, *217; *1c*, *1995; *1d*, *2113e); *2, Asterocyclina,* equatorial section (diagrammatic, not to scale) (*217).

EHRENBERG, 1840)] [*Asterodiscus pentagonalis SCHAFHÄUTL, 1863, *1638, p. 107, =*Calcarina? stellata D'ARCHIAC, 1846, *37, p. 199; OD (M)] [=?Asteriacites VON SCHLOTHEIM, 1822, *1649, p. 71 (type, *A. patellaris*); *Cisseis* GUPPY, 1886, *842, p. 584 (type, *C. astericus*) (non LAPORTE & GORY, 1839); *Asterodiscus* SCHAFHÄUTL, 1863, *1638, p. 107 (type, *A. pentagonalis*); *Asterodiscocyclina* BERRY, 1928, *130, p. 406 (type, *Orthophragmina (A.) stewarti*); *Orthocyclina* VAN DER VLERK, 1933, *2011, p. 93 (type, *O. soeroeanensis*); *Isodiscodina* VAN DER WEIJDEN, 1940, *2042, p. 15 (type, *Orthophragmina pentagonalis* DEPRAT, 1905, *583, p. 507)]. Test stellate, with radial zones of elongate equatorial chambers in equatorial plane. *M.Eoc.-U.Eoc.,* Eu.-Indo-Pac.Reg.-N.Am.-S.Am.——FIG. 587,*3a.* *A. stellata* (D'ARCHIAC), Lutet.-Auvers., Fr.; ext. view, ×4 (*2120).——FIG. 587,*3b,c.* *A. georgiana* (CUSHMAN), U.Eoc. (Ocala), USA; vert. and equat. secs., ×40 (*2113b). [See note, p. C796.]

Pseudophragmina DOUVILLÉ, 1923, *626, p. 106 [*Orthophragmina floridana* CUSHMAN, 1817, *408, p. 116; OD]. Test circular in plan, discoidal, or lenticular; annular stolon distally situated; radial chamber walls complete, incomplete, absent or indistinct, but when present, in alignment in adjacent annuli. *Paleoc.-Eoc.,* N.Am.-S.Am.-Asia(India).

P. (Pseudophragmina). Distal part of radial chamber walls degenerate, in places represented by rows of granules. *M.Eoc.-U.Eoc.,* N.Am.-S.Am.——FIG. 588,*1.* *P. (P.) floridana* (CUSHMAN), U.Eoc.(Ocala), USA(Fla.); *1a*, equat. sec. with embryonic chambers, periembryonic chambers and equat. chambers, ×40; *1b*, equat. sec. with radial chamber walls in alignment, incomplete at their distal ends, ×40; *1c*, vert. sec. with embryonic chambers, equat. layer and lateral chambers; ×20 (*2113c).

?**P. (Asterophragmina)** RAO, 1942 *1499, p. 9 [*P. (A.) pagoda*]. Possibly a defective specimen of *Asterocylina.* *U.Eoc.,* Asia(Burma).

P. (Athecocyclina) VAUGHAN & COLE in CUSHMAN,

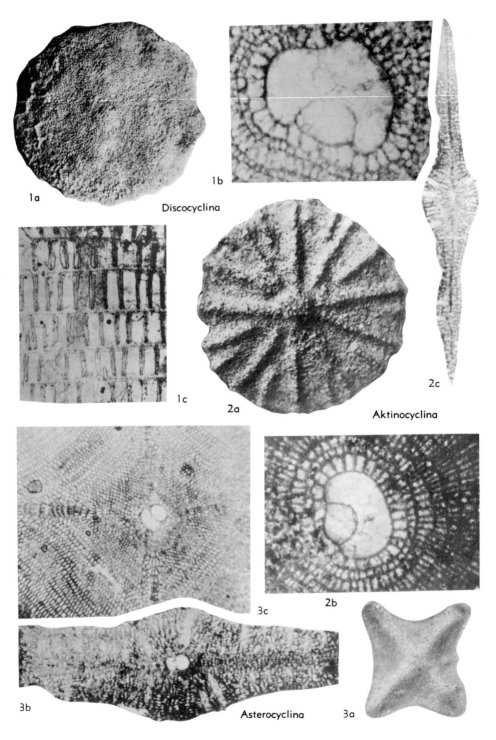

Fig. 587. Discocyclinidae; *1, Discocyclina (Discocyclina); 2, D. (Aktinocyclina); 3, Asterocyclina* (p. *C*714-*C*715).

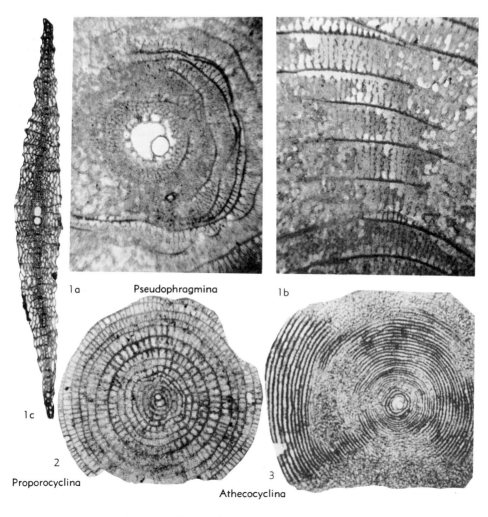

Fig. 588. Discocyclinidae; *1, Pseudophragmina (Pseudophragmina); 2, P. (Proporocyclina); 3, P. (Athecocyclina)* (p. C715-C717).

1940, *474, p. 330 [*Discocyclina cookei* VAUGHAN, 1936, *1994, p. 256]. Radial chamber walls absent or indistinct. *Paleoc.-M.Eoc.,* N.Am.-S.Am.——FIG. 588,*3.* **P. (A.) cookei* (VAUGHAN), L.Eoc.(Wilcox), USA(Ala.); equat. sec., ×15 (*1994).

P. (Proporocyclina) VAUGHAN & COLE in CUSHMAN, 1940, *474, p. 330. [*Discocyclina perpusilla* VAUGHAN, 1929, *1992, p. 9]. Radial chamber walls complete. *Paleoc.-Eoc.,* N.Am.-S.Am.——FIG. 588,*2.* **P. (P.) perpusilla* (VAUGHAN), M.Eoc.(Guayabal F.), Mex.; equat. sec., ×40 (*2113c).

LEPIDOCYCLINIDAE
By W. STORRS COLE

Family LEPIDOCYCLINIDAE
Scheffen, 1932

[Lepidocyclinidae SCHEFFEN, 1932, p. 251-252] [=Helicolepidinidae POKORNÝ, 1958, p. 395 (*nom. transl. ex* Helicolepidininae TAN, 1936)]

Test circular or radiate, compressed to inflated lenticular, composed of distinct equatorial layer overlain on each side by zones of lateral chambers or by laminated shell material with vacuoles; embryonic chambers bilocular, followed by distinct, long spiral of periembryonic chambers, or by short spiral of these chambers, or by reduced sequences of periembryonic chambers on periphery of embryonic chambers; equator-

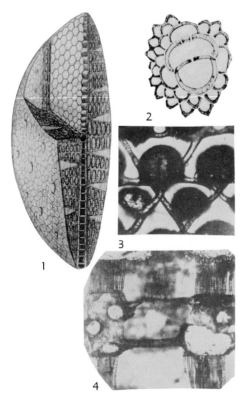

Fig. 589. Lepidocyclinidae; *1-4*, structural features.——*1*. Oblique view of sectioned *Lepidocyclina (Lepidocyclina)* test showing equatorial chambers and lateral regions with pillars (diagrammatic, not to scale) (*2121).——*2*. Embryonic apparatus of *L. (Nephrolepidina)* with 3 primary periembryonic chambers and 6 coils of additional periembryonic chambers (*2120A).——*3,4*. Decalcified Canada balsam preparations of *L. (L.) montgomeriensis* showing *(3)* diagonal and annular stolons of 6-stolonal system in part of equatorial section, and *(4)* fine tubules that perforate roofs and floors as seen in part of vertical section, annular stolons and 3 apertures for stolons visible at left, both ×140 (*1998).

ial chambers arcuate, ogival, rhombic, spatulate or hexagonal; chamber walls perforate with definite stolons, without canal system (Fig. 589,*1*). *M.Eoc.-M.Mio.*

The Helicolepidininae and Lepidocyclininae seemingly were derived from an *Amphistegina*-like ancestor. BARKER & GRIMSDALE (*84) have presented convincing evidence that such an ancestor could have generated two distinct lines, one developing into the Helicolepidininae, in which the equatorial layer is characterized by a well-developed sequence of chambers arranged in an open spire that persists beyond the initial periembryonic spire, and the other producing the Lepidocyclininae, in which the spiral and even the periembryonic chambers are reduced in importance so that the equatorial plane is composed only of the embryonic and equatorial chambers in advanced genera.

The evolutionary development postulated by BARKER & GRIMSDALE (*84) is accepted as the logical one, though certain disagreements concerning details should be noted. GRIMSDALE (*827), by analysis of the stolon systems, argued for a diphyletic origin of the lepidocyclines, dividing them into "lineage Y" with a "crossed stolon system" and "lineage X" with an "uncrossed stolon system." GRIMSDALE derived "lineage Y" from an *Amphistegina*-like ancestor, but did not identify the origin of "lineage X." COLE (*366) postulated that *Lepidocyclina (Polylepidina) antillea* CUSHMAN, the first true species of *Lepidocyclina*, which occurs in middle Eocene strata of the Caribbean region, was the original species from which the stratigraphically succeeding forms of *Lepidocyclina* were derived.

Although formerly *Lepidorbitoides* (Upper Cretaceous) and *Actinosiphon* (Paleocene) have been included in the Lepidocyclininae by many authors because the internal structure of their tests is similar to that of the lepidocyclines, these genera must be excluded from the Lepidocyclininae for stratigraphic reasons. The first representatives of the Lepidocyclinidae appear in the middle Eocene of the Caribbean region long after the disappearance of *Actinosiphon*.

The classification of the Lepidocyclinidae is based mainly on internal structures which are studied by means of equatorial and vertical thin sections. Although external shape and the sculpture of the surface of the test are important, internal structures normally reflect the surface features. For example, specimens with papillate surfaces have well-developed pillars, as the pillar heads project above the surface of the test so as to form the individual papillae, and stellate or rayed specimens have the equatorial chambers arranged so that this condition is shown in equatorial thin sections.

The major emphasis in classification of

genera and subgenera is based upon kind of megalospheric embryonic chambers observed, development of the periembryonic chambers, and characteristics of the equatorial chambers. These are shown best by equatorial thin sections, but vertical thin sections are helpful, particularly in determining whether the equatorial layer lies in a single plane or whether it becomes multiple or otherwise modified.

Generic determinations often can be made from vertical thin sections, but this kind of section is most useful for discrimination of species. The kind and arrangement of lateral chambers, the presence or absence of pillars, and relationships of the equatorial layer to the covering zones are features which assist in separating one species from another. However, the entire test must be studied in detail.

Because it is important that a correct correlation be made between equatorial and vertical thin sections in populations with several species present, matrix-free individuals should be ground to the equatorial plane. This plane can then be studied by reflected light. After several individuals are found with the same structures in the equatorial plane, some of them should be used for the making of vertical thin sections. Thus, equatorial thin sections may be correlated with vertical sections. Where individuals cannot be freed from the matrix, correlation may be made by means of the numerous tangential and oblique sections of specimens which normally show in thin sections made through the matrix and entombed specimens. These tangential and oblique sections often will show in a single rock sample structures both of the equatorial layer and the covering zones, although none of the zones will be exposed in its entirety.

In addition to the correlation between equatorial and vertical thin sections and in order to be absolutely certain that sections made from different individuals represent the same species, the association between megalospheric and microspheric specimens of the same species must be made also. If more than one species occurs in a given population difficulties may be encountered in recognizing which pairs represent a given species. Although microspheric individuals of a given species are larger than the megalospheric individuals, correspondence of all internal structures is found except for initial chambers of the equatorial layer. Thus, it is often possible to correlate specimens of the two generations by shape of the equatorial chambers, kind and arrangement of the lateral chambers, and degree of development of the pillars.

In megalospheric specimens the initial chambers (embryonic stage) are bilocular, consisting of an initial chamber (protoconch) followed by a second chamber (deuteroconch). These chambers have size relationships to each other varying from equality, as in *Lepidocyclina s.s.,* to a second chamber so large that it completely encloses the initial chamber except along the area of juncture of the two chambers, as in some species of *Lepidocyclina (Eulepidina)*.

However, this size relationship of the embryonic chambers is not an absolute criterion for generic or subgeneric designation, since individuals of a given species commonly exhibit variable relationships in size of the embryonic chambers. Moreover, abnormality of the initial chambers is a common occurrence. Specimens showing this commonly have an unusually large embryonic chamber with a sequence of smaller chambers lying around the margin. Although such specimens have been assigned to distinct genera, VAUGHAN & COLE (*1998) and COLE (*368) have attributed this development to one possible phase in the reproductive mechanism, inasmuch as some associated specimens have more than one set of otherwise typical embryonic chambers.

The initial (embryonic) chambers in most genera are surrounded partially or completely by periembryonic (nepionic) chambers before the equatorial (ephebic) chambers are developed. The first chamber (primary auxiliary) of any periembryonic sequence is connected to the second embryonic chamber by one or more stoloniferous passages, whereas the other periembryonic chambers (auxiliary) are not so connected. Four periembryonic sequences have been recognized: (1) **uniserial**, in which there is a single coil of periembryonic chambers which encircles the embryonic chambers in one direction; (2) **biserial**, in which two periembryonic coils originate from two distinct initial periembryonic chambers; (3) **quadriserial**, in which four periembryonic coils originate, though these are developed from only two initial periembryonic chambers; and (4) **multiserial**, in which more than four coils of periembryonic chambers and two initial periembryonic chambers are present (Fig 589,2).

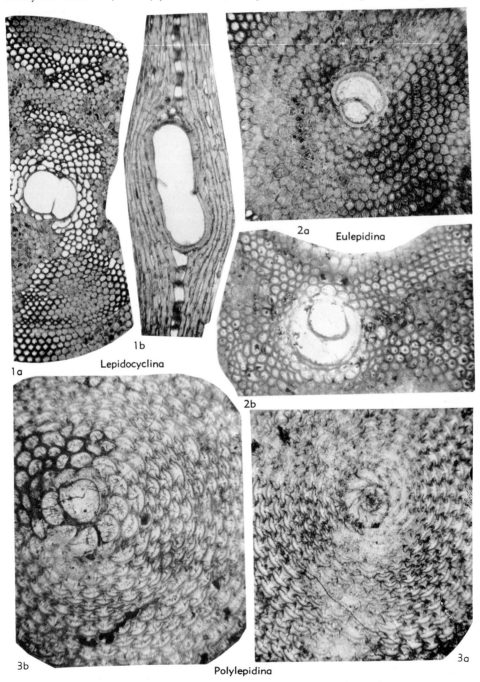

Fig. 590. Lepidocyclinidae (Lepidocyclininae; *1, Lepidocyclina (Lepidocyclina); 2, L. (Eulepidina); 3, L. (Polylepidina)*) (p. C721-C724).

The arrangement of the periembryonic chambers is useful both in specific and generic classification, but too much dependence cannot be placed on this characteristic alone, since it has been demonstrated that individuals of a given species may develop more than one of the periembryonic sequences.

The equatorial chambers are connected by stolon systems, of which five kinds, based on number and arrangement of the stolons, are recognized. Stolon systems are best studied in tests which have been infiltrated by colored matrix or in empty tests infiltrated artificially by Canada balsam or other material. In such tests it is possible to dissolve the substance of the test so that the infiltrating material outlines the stolon system (Fig. 589,3,4).

The lateral chambers lying above and below the equatorial layer may or may not be separated by conical masses of compact material known as **pillars**. Although the presence or absence of pillars and the degree of their development and characteristics have been much used in specific determinations, the value of these features may be questioned (Fig. 589,1). The development or lack of development of pillars may be a clue for specific determination, however. Conversely, the shape, arrangement, and configuration of the lateral chambers is extremely important. The chamber cavities may be open or slitlike. The floors and roofs of the chambers may be thin or thick, straight or rounded, and the chambers may be aligned in regular tiers or overlapping.

The equatorial chambers, viewed in vertical section, commonly are doubled in the peripheral zone and somewhat rarely are separated into two layers by a wedge of shell material. Although this doubling of the peripheral equatorial chambers may be significant for recognition of certain genera (e.g., *Pseudolepidina*), in others (e.g., *Lepidocyclina*), the doubled equatorial chambers are interrupted as structures of specific importance, or in some specimens of infraspecific occurrence, without being common to all specimens of the species.

Although numerous classifications of larger Foraminiferida have been attempted, based upon detailed analysis of a single internal structure (e.g., periembryonic chambers, stolon systems, lateral chambers), they have not been successful. Any natural classification must be based on a synoptic analysis of the whole test. The relationship and combination of all internal structures ultimately defines the genus. Unfortunately, many generic names have been based upon the relative development of single structures which characterize individual specimens only and are not even specific characters.

Subfamily LEPIDOCYCLININAE Scheffen, 1932

[nom. transl. TAN, 1936, p. 277 (ex Lepidocyclinidae SCHEFFEN, 1932] [=Lepidocyclinae VAUGHAN & COLE in CUSHMAN, 1940, p. 357 (nom. van.)]

Embryonic and periembryonic chambers with thin walls, periembryonic chambers present or lacking; lateral chambers normally numerous, well developed and distinctly separated from equatorial layer. *M.Eoc.-M.Mio.*

Lepidocyclina GÜMBEL, 1870, *840, p. 689 [*Nummulites mantelli MORTON, 1833, *1320, p. 291; SD H. DOUVILLÉ, 1898, *613, p. 594] [=Cyclosiphon EHRENBERG, 1855, *681, p. 288 (nom. reject., ICZN Op. 127); Astrolepidina A. SILVESTRI, 1931, *1785, p. 35 (type, Lepidocyclina asterodisca NUTTALL, 1932, *1371A, p. 34; SD COLE, herein)]. Embryonic chambers bilocular; equatorial chambers arcuate, rhombic, hexagonal or spatulate. *M.Eoc.-M.Mio.*, cosmop. [Trop.].

L. (Lepidocyclina) [=Isolepidina H. DOUVILLÉ, 1915, *622, p. 724 (type, Nummulites mantelli MORTON, 1833, *1320, p. 291)]. Embryonic chambers equal or subequal, separated by straight wall; equatorial chambers arcuate, hexagonal or spatulate, with 6- or 8-stolon system. *Oligo.-L.Mio.*, N.Am.-S.Am.——FIG. 590,1. *L. (L.) mantelli (MORTON), Oligo., USA(Fla.); 1a,b, equat. sec., vert. sec., ×20, ×40 (*2113c).

L. (Eulepidina) H. DOUVILLÉ, 1911, *620, p. 59, 68 [*Orbitoides dilatata MICHELOTTI, 1861, *1257, p. 17; SD YABE, 1919, *2085, p. 41] [=Nephrolepidina H. DOUVILLÉ, 1911, *620, p. 59, 70, 73 (type, Nummulites marginata MICHELOTTI, 1841, *1256, p. 297); Amphilepidina H. DOUVILLÉ, 1922, *625, p. 552 (type, Orbitoides sumatrensis BRADY, 1875, *192, p. 536); Trybliolepidina VAN DER VLERK, 1928, *2014, p. 10, 13 (type, Lepidocyclina ephippioides JONES & CHAPMAN in ANDREWS, 1900, *997, p. 251, 256); SD BERRY, 1929, *131, p. 37; Multilepidina HANZAWA, 1932, *865, p. 447 (type, Lepidocyclina (M.) irregularis); Cyclolepidina WHIPPLE, 1934, *2053, p. 143 (type, Lepidocyclina (C.) suvaensis)]. Embryonic chambers bilocular, smaller initial chamber slightly or completely surrounded by larger second chamber except along area of attachment, or teratologically with large chamber, on periphery of which are smaller chambers; equatorial chambers arcuate, rhombic, spatulate to hexagonal. *U.Eoc.-M.Mio.*, N.Am.-

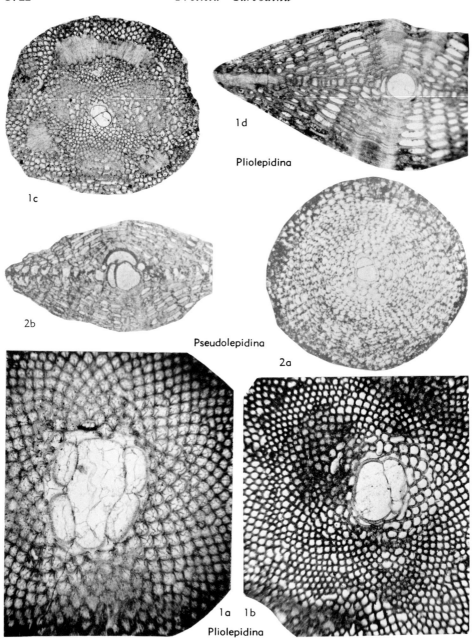

Fig. 591. Lepidocyclinidae (Lepidocyclininae; *1, Lepidocyclina (Pliolepidina); 2, Pseudolepidina*) (p. C722-C724).

S.Am.-Eu.-Afr.-C.Pac., trop.——Fig. 590,*2a*. *L. (E.) tournoueri* LEMOINE & R. DOUVILLÉ, Oligo., Mex.; equat. sec., ×40 (*366).——Fig. 590,*2b*. *L. (E.) ephippioides* JONES & CHAPMAN, L.Mio., Saipan Is.; equat. sec., ×40 (*366).

L. (Pliolepidina) H. DOUVILLÉ, 1915, *622, p. 727 [*L. (P.) tobleri* H. DOUVILLÉ, 1917, *623, p. 844 (=*Isolepidina pustulosa* H. DOUVILLÉ, 1917, *623, p. 843); SD (SM)] [=*Multicyclina* CUSHMAN, 1918, *410, p. 96 (type, *Lepidocyclina (M.) duplicata*); *Orbitoina* VAN DE GEYN & VAN DER VLERK, 1935, *786, p. 222, 227 *(nom. nud.)*;

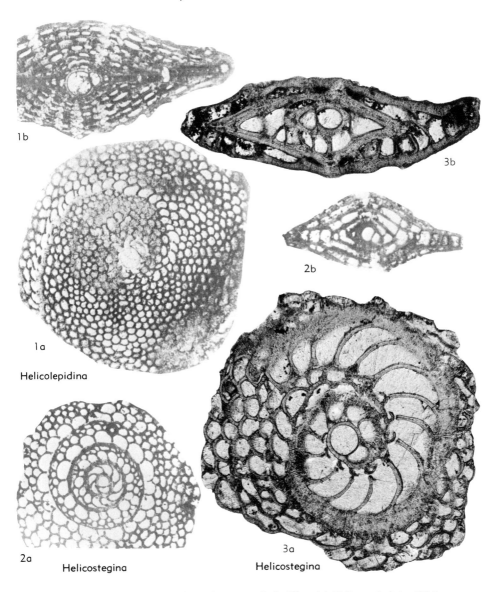

Fig. 592. Lepidocyclinidae (Helicolepidininae; *1, Helicolepidina; 2,3, Helicostegina*) (p. C724).

Isorbitoina VAN DE GEYN & VAN DER VLERK, 1935, *786, p.222, 227, 255 (nom. nud.); *Pliorbitoina* VAN DE GEYN & VAN DER VLERK, 1935, *786, p. 222, 227, 255 (type, *Lepidocyclina (Pliolepidina) tobleri* H. DOUVILLÉ); *Polyorbitoina* VAN DE GEYN & VAN DER VLERK, 1935, *786, p. 227 (type, *Lepidocyclina (Polylepidina) proteiformis* VAUGHAN, 1924, *1988, p. 810); *Multilepidina* A. SILVESTRI, 1937, *1787, p. 160 (non HANZAWA, 1932) (type, *Pliolepidina tobleri* H. DOUVILLÉ, 1917); *Isorbitoina* THALMANN, 1938, *1897c, p. 202 (type, *Lepidocyclina trinitatis* H. DOUVILLÉ, 1924, *627, p. 374); *Triplalepidina* VAUGHAN & COLE, 1938, *1997, p. 167 (type, *T. veracruziana*); *Neolepidina* BRÖNNIMANN, 1947, *219, p. 378 (type, *Isolepidina pustulosa* H. DOUVILLÉ, 1917, *623, p. 843)]. Embryonic chambers bilocular, initial chamber usually slightly larger than second chamber, with variable number of relatively large, distinct periembryonic chambers, or teratologically with one large chamber, on periphery of which are numerous smaller chambers, equatorial chambers rhombic to ogival with 4-stolon system. *M.Eoc.-U.Eoc.,* N.

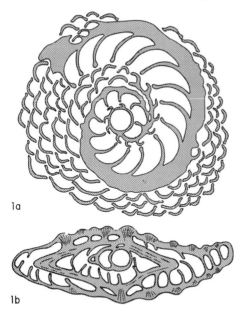

Fig. 592A. Lepidocyclinidae (Helicolepidininae; *1,* *Helicostegina*) (p. C724).

Am.-S.Am.-Afr.[Trop.].——Fig. 591,*1a,b*. **L. (P.) pustulosa* (H. Douvillé), U.Eoc., (*1a*, Panama; *1b*, Trinidad); *1a*, teratologic embryonic chambers; *1b*, normal embryonic chambers; both ×40 (*359).——Fig. 591,*1c,d*. *L. (P.) veracruziana* (Vaughan & Cole), U.Eoc., Mex.; equat. sec., vert. sec., ×20, ×40 (*2113c).

[H. Douvillé (1915, *622, p. 727) introduced the nominal subgenus named *Pliolepidina* without originally assigning species to it, but in 1917 (*623, p. 843) he described and illustrated a single species named *P. tobleri*, which thus was established as the type-species by subsequent monotypy. Also in 1917 Douvillé (*623, p. 844) described and figured a species named *Isolepidina pustulosa* and then in 1924 (*627, p. 43) expressed the opinion that *P. tobleri* "résulte seulement d'un accident tératologique." Vaughan & Cole (1941, *1998, p. 67) from a survey of extensive topotype materials concluded that Douvillé's surmise as to the teratological nature of his *P. tobleri* was undoubtedly correct. Therefore, they cited *Isolepidina pustulosa*, distinguished as the normal lepidocyclinid corresponding to teratological *P. tobleri*, as the correct designation of the type-species of *Pliolepidina*. This procedure accords with stipulations in zoological nomenclature, but in recognizing "*Lepidocyclina (Pliolepidina) pustulosa* forma *tobleri* (H. Douvillé) forma *teratologica*" (*1998, p. 66, pl. 24) they did not conform to the Zoological Code (1961), which specifies (Art. 1) that "names given to . . . teratological specimens . . . are excluded [from zoological nomenclature]." The name *Pliolepidina tobleri* is here rejected entirely, since it has the status of a *nomen nudum*.]

L. (Polylepidina) Vaughan, 1924, *1988, p. 794, 807 [**L. (P.) chiapasensis,* =**L. antillea* Cushman, 1919, *414, p. 63; OD] [=*Eulinderina* Barker & Grimsdale, 1936, *84, p. 237 (type, *Planorbulina (Planorbulinella) guayabalensis* Nuttall, 1930, *1371, p. 276); *Eolepidina* Tan, 1939 (type, *Eulinderina semiradiata* Barker & Grimsdale, 1936, *84, p. 238)]. Embryonic chambers bilocular, initial chamber usually slightly larger than second chamber, followed by partial but distinct coil of 4 to 9 large periembryonic chambers which gradually decrease in size; equatorial chambers arcuate, with 4-stolon system. M.Eoc., N.Am.-S.Am.——Fig. 590,*3*. **L. (P.) antillea* Cushman, W.Indies(St. Bartholomew); *3a,b*, equat. secs. of microspheric and megalospheric specimens, ×40 (*366).

Pseudolepidina Barker & Grimsdale, 1937, *85, p. 169 [**P. trimera*; OD]. Embryonic chambers bilocular in equatorial sections, trilocular in vertical sections; equatorial layer double in peripheral zone. M.Eoc., Carib.-N.Am.——Fig. 591,*2*. **P. trimera*, Mex.; *2a,b*, equat. sec., vert. sec., ×25, ×36 (*85).

Subfamily HELICOLEPIDININAE Tan, 1936

[Helicolepidininae Tan, 1936, p. 277] [=Helicolepidinae Vaughan & Cole in Cushman, 1940, p. 325 (*nom. van.*)]

Equatorial layer characterized by well-defined sequence of chambers arranged in open spiral which persists beyond initial periembryonic spire with chamberlets intercalated between whorls of chambers of spiral. M.Eoc.-U.Eoc.

Helicolepidina Tobler, 1922, *1937, p. 380 [**H. spiralis*; OD] [=*Helicocyclina* Tan, 1936, *1868, p. 995 (type, *Helicolepidina paucispira* Barker & Grimsdale, 1936, *84, p. 243)]. Megalospheric embryonic chambers bilocular, subequal, followed by open spiral of chambers which are bounded on their proximal side by perforate spiral band forming 1 or 2 volutions, and in some extending to periphery of test; equatorial chambers arcuate to rudely hexagonal; lateral chambers well developed. U.Eoc., N.Am.-S.Am.——Fig. 592,*1*. **H. spiralis*, W.Indies(Trinidad); *1a,b*, equat. and vert. secs., ×27 (*2122).

Helicostegina Barker & Grimsdale, 1936, *84, p. 233 [**H. dimorpha*; OD] [=*Helicolepidinoides* Tan, 1936, *1868, p. 992 (type, *Helicostegina gyralis* Barker & Grimsdale, 1936, *84, p. 236)]. Test lenticular, pustulose to papillose; earliest chambers coiled in involute trochoid spire, later ones subdividing ventrally into subsidiary chamberlets, in adult stage of some species forming distinct peripheral flange similar to other members of Lepidocyclinidae; aperture comprising narrow slit near inner margin of ventral face of last chamber, with backward projecting lip, as in *Tremastegina*; flange (if present) and chamberlets of ventral layer connected by paired foramina. M.Eoc.-U.Eoc., Carib.-N.Am.(Mex.).——Fig. 592,*2*. *H. polygyralis* (Barker), U.Eoc., W.Indies(Trinidad); *2a,b*, equat. and vert. secs., ×40 (*366).——Fig. 592, *3*; 592A,*1*. **H. dimorpha*, M.Eoc., Mex.; 592,*3a,b*, equat. and vert. secs., ×56 (*84); 592A, *1a,b*, sketches of equat. and vert. secs. showing structure, ×43 (*2110).

PSEUDORBITOIDIDAE

By W. STORRS COLE

Family PSEUDORBITOIDIDAE M. G. Rutten, 1935

[nom. transl. BRÖNNIMANN, 1958, p. 167 (ex Pseudoorbitoidinae M. G. RUTTEN, 1935, p. 544)]

Test lenticular, composed of equatorial layer covered on each side by zones of lateral chambers; embryonic chambers bilocular, followed by long or short rotaliid spire of nepionic chambers; equatorial layer beyond embryonic apparatus composed of radial vertical plates variously arranged except in microspheric specimens which have arcuate equatorial chambers in zone between embryonic apparatus and peripheral zone of radial plates; annular walls present in equatorial layer of some genera; protoplasmic communication by stolons and fine pores; canal system present. *U.Cret.*

The genera of this family are characterized by vertical radial plates which occur in the equatorial layer. The pseudorbitoidids are mutants of *Sulcoperculina* in which lateral chambers have developed and the radial vertical plates present in the sulcus of *Sulcoperculina* have become elongated.

Pseudorbitoides H. DOUVILLÉ, 1922, *624, p. 204 [*P. trechmanni; OD] [=Historbitoides BRÖNNIMANN, 1956, *231d, p. 61 (type, H. kozaryi); Aktinorbitoides BRÖNNIMANN, 1958, *232, p. 167 (type, A. browni)]. Embryonic chambers of microspheric generation forming distinct spire followed by arcuate equatorial chambers which are succeeded by radial plates; embryonic chambers of megalospheric form bilocular, with 2 or more periembryonic chambers forming irregular spire succeeded by radial plates which extend to periphery; lateral chambers well developed, resting directly on radial plates. *U.Cret.*, Carib.(Jamaica)-New Guinea-N.Am.——FIG. 593,*1a,b*. **P. trechmanni*, Jamaica; *1a,b,* equat. and vert. secs. of microspheric specimens, ×40 (*2123).——FIG. 593,*1c,d.* *P. israelskyi* VAUGHAN & COLE, USA (La.); *1c,d,* equat. and vert. secs. of megalospheric specimens, ×40 (*2123).

Sulcorbitoides BRÖNNIMANN, 1954, *231a, p. 55 [*S. pardoi; OD] [=Conorbitoides BRÖNNIMANN, 1958, *232, p. 173 (type, *C. cristalensis*)]. Nepionic coil long, rotaliid, followed by 2 alternating systems of vertical radial plates without annular walls; lateral chambers rest directly on radial rods of equatorial layer. *U.Cret.*, Carib.(Cuba)-USA (Texas).——FIG. 593,*2.* *S. pardoi,* Cuba; *2a,b,* equat. sec., vert. sec., ×28, ×40 (*231).

Vaughanina D. K. PALMER, 1934, *1408, p. 240 [*V. cubensis; OD] [=Rhabdorbitoides BRÖNNIMANN, 1955, *231c, p. 97 (type, *R. hedbergi*); *Ctenorbitoides* BRÖNNIMANN, 1958, *232, p. 171 (type, *C. cardwelli*)]. Nepionic coil short, followed by 2 alternating systems of radial plates; annular walls present; lateral chambers and radial plates separated by roof and floor of equatorial layer. *U.Cret.*, Carib.(Cuba)-Mex.-USA (Fla.).——FIG. 593,*3.* *V. cubensis,* Cuba; *3a,* ext. view, ×20; *3b,c,* equat. and vert. secs., ×37.5 (*2123).

Superfamily CASSIDULINACEA d'Orbigny, 1839

[nom. transl. LOEBLICH & TAPPAN, 1961, p. 313 (ex family Cassidulinidae d'ORBIGNY, 1839)]——[In synonymy citations superscript numbers indicate taxonomic rank assigned by authors ([1]superfamily, [2]family group); dagger(†) indicates *partim*]——[=Enclinostegia† EIMER & FICKERT, 1899, p. 682 (nom. nud.); =[1]Orthoklinostegia† EIMER & FICKERT, 1899, p. 685 (nom. nud.); =[2]Textulinidia† RHUMBLER in KÜKENTHAL & KRUMBACH, 1923, p. 88; =[2]Rotaliformes† BROTZEN, 1942, p. 9 (nom. neg.); =[1]Monolamellidea† REISS, 1957, p. 128 (nom. nud.); =[1]Bilamellidea† REISS, 1957, p. 127 (nom. nud.); =[1]Nonionidea SUBBOTINA in RAUZER-CHERNOUSOVA & FURSENKO, 1959, p. 282]

Test enrolled, planispiral, or low or high trochospiral; wall of perforate granular calcite; aperture slitlike, loop-shaped or multiple. *U.Trias.-Rec.*

Family PLEUROSTOMELLIDAE Reuss, 1860

[Pleurostomellidae REUSS, 1860, p. 151, 203] [=Pleurostomelleae GÜMBEL, 1870, p. 52; =Ellipsoidinidae A. SILVESTRI, 1923, p. 808; =Pleurostomellida COPELAND, 1956, p. 188 (nom. van.)]

Early stage triserial or biserial, later uniserial, or uniserial throughout; aperture a curved narrow slit, lateral or terminal, with internal siphon between those of adjacent chambers. *?Jur., L.Cret.-Rec.*

Subfamily PLEUROSTOMELLINAE Reuss, 1860

[nom. correct. LOEBLICH & TAPPAN, 1961, p. 315 (pro subfamily Pleurostomellidea REUSS, 1862, p. 368)]——[All names cited are of subfamily rank]——[=Cryptostegia BÜTSCHLI in BRONN, 1880, p. 203 (nom. nud.); =Ellipsonodosariinae A. SILVESTRI, 1901, p. 109; =Ellipsolageninae A. SILVESTRI, 1923, p. 265; =Ellipsodinininae PETTERS, 1954, p. 39]

Early stage biserial, later uniserial, or uniserial throughout. *?Jur., L.Cret.-Rec.*

Pleurostomella REUSS, 1860, *1548, p. 203 [*Dentalina subnodosa* REUSS, 1851, *1542, p. 24, =*Dentalina nodosa* d'ORBIGNY, REUSS, 1846, *1538, p. 28; SD CUSHMAN, 1911, *404b, p. 49] [=*Pleurostomellina* SCHUBERT, 1911, *1689b, p. 58 (type, *Pleurostomella barroisi* BERTHELIN, 1880, *133, p. 30); *Ellipsonodosaria* (*Ellipsodentalina*) FRANKE, 1928, *740, p. 54 (type, *Dentalina subnodosa* REUSS, 1851, *1542, p. 24; SD LOEBLICH & TAPPAN, herein) (obj.)]. Test small, elongate, chambers in early stage biserially arranged, or cuneate and alternating in position, later uniserial; sutures in early stage oblique, later becoming more nearly straight and horizontal, wall

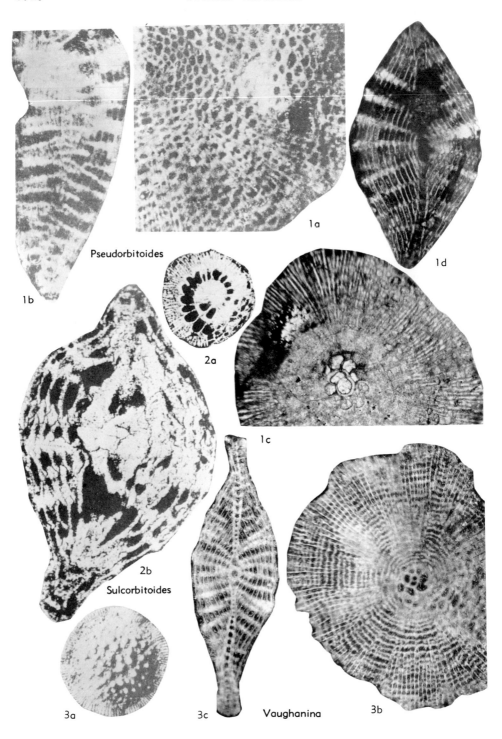

Fig. 593. Pseudorbitoididae; *1, Pseudorbitoides; 2, Sulcorbitoides; 3, Vaughanina* (p. C725).

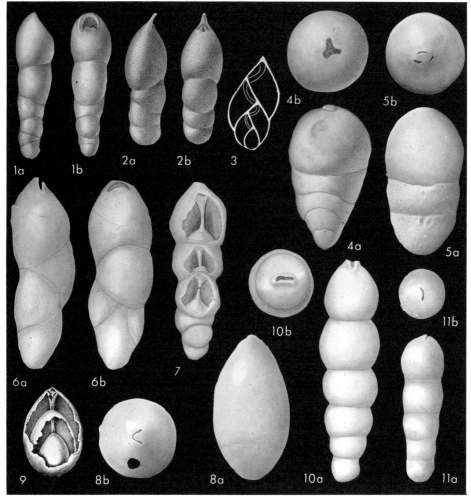

Fig. 594. Pleurostomellidae (Pleurostomellinae; *1-3, Pleurostomella; 4, Daucina; 5, Pinaria; 6,7, Ellipsoidella; 8,9, Ellipsoidina; 10,11, Nodosarella*) (p. C725-C728, C730).

calcareous, finely perforate, granular in structure; aperture terminal, with projecting hood at one side, 2 small teeth on opposite side, and internal tube. [*Pleurostomellina* was originally separated as being uniserial, but both type-species show considerable variation in the length and development of the biserial stage, or its indication by means of the alternating cuneate chambers.] L.Cret.-Rec., cosmop.——Fig. 594,*1. *P. subnodosa* (REUSS), U.Cret.(Campan.), Ger.(Bav.); *1a,b*, side and edge views, ×48 (*2117).——Fig. 594,*2. P. barroisi* BERTHELIN, L.Cret.(Alb.), Eng.; *2a,b*, side and edge views, ×74 (*2117).——Fig. 594,*3. P. brevis* SCHWAGER, Mio., Eu.(Italy); long. sec. showing internal tube extending between successive apertures, ×30 (*1757).

Daucina G. BORNEMANN in ERMAN, 1855, *710, p. 153 [*D. ermaniana*; OD (M)]. Test free, elongate, uniserial, with slowly enlarging and strongly overlapping chambers; wall calcareous, smooth, microstructure not described; aperture terminal, trilobate, apparent modification of bifid toothed aperture of *Pleurostomella*. Tert., S.Am.(Brazil).——Fig. 594,*4. *D. ermaniana*; *4a,b*, side and apert. view of lectotype, here designated (Cushman Coll. 14223), ×33 (*2117).

Ellipsobulimina A. SILVESTRI, 1903, *1757, p. 210 [*E. seguenzai*; OD (M)]. Test free, ovate or rounded with early biserial stage and later uniserial, each pair of biserial chambers completely overlapping all preceding ones, and uniserial chambers completely enveloping earlier test so that externally it resembles *Ellipsoidina*; wall calcareous; aperture terminal, semilunate, with internal tube connecting successive apertures. [Differs from *Ellipsoidina* in having an early biserial

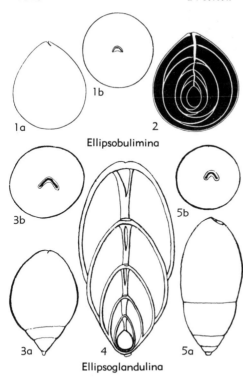

Fig. 595. Pleurostomellidae (Pleurostomellinae; *1,2, Ellipsobulimina; 3-5, Ellipsoglandulina*) (p. C727-C728).

stage.] Mio., Eu.——Fig. 595,*1,2*. **E. seguenzai,* Italy; *1a,b,* side, apert. views, ×33; *2,* long. sec. showing biserial early stage, enveloping chambers and internal tube, ×38 (*1758).

Ellipsodimorphina A. Silvestri, 1901, *1752, p. 16, 18 [**E. subcompacta* Liebus, 1922, *1136, p. 57; SD (SM) Liebus, 1922, *1136, p. 57]. Test elongate, biserial in early stage, later chambers cuneate and finally completely uniserial, rounded in section; sutures distinct, depressed; aperture an elongate arched silt. *U.Cret.-Eoc.,* Eu.——Fig. 596,*1*. **E. subcompacta* Liebus, Eoc., Czech. (Moravia); *1a,b,* side, apert. views, ×53 (*1136).

Ellipsoglandulina A. Silvestri, 1900, *1751, p. 12 [**E. laevigata;* OD (M)]. Test free, elongate, uniserial with strongly overlapping chambers and tapering base; wall calcareous; aperture terminal, semilunate, internally provided with entosolenian tube extending between successive apertures. [Differs from *Ellipsoidina* in not being completely involute.] *?Jur., ?Cret., Eoc.-Rec.,* Eu.-Carib.-N.Z.-N.Am.——Fig. 595,*3-5*. **E. laevigata,* Plio., Eu. (Sicily); *3a,b,* side, apert. views, megalospheric form, ×44; *4,* long. sec. showing tube, ×44; *5a,b,* side, apert. view of microspheric form, ×30 (*1751).

Ellipsoidella Heron-Allen & Earland, 1910, *907, p. 410, 414 [**E. pleurostomelloides;* OD (M)]. Test free, elongate, chambers cuneate and biserially arranged in early portion, later becoming less closely appressed and uniserial, but wedge-shaped and alternating, and may finally become completely rectilinear; sutures depressed, oblique; wall calcareous, perforate, granular in structure, surface smooth; aperture subterminal, an arched slit, with overhanging lip and internal tube extending downward from just beneath aperture where it is expanded, to attach to apertural region of preceding chamber. *U.Cret.,* Eu.——Fig. 594,*6,7*. **E. pleurostomelloides,* Brit.I.(Eng.); *6a,b,* side, edge views of lectotype, ×109; *7,* dissected paratype showing internal tube, ×79 (*2117).

[*Ellipsoidella* differs from *Nodosarella* in having an early biserial stage and later alternating chambers, whereas *Nodosarella* is uniserial throughout, with horizontal sutures. Cushman (1948, *486, p. 278) placed the biserial forms in *Nodosarella* and considered *Ellipsoidella* a synonym, but the type-species of *Nodosarella* is uniserial throughout, hence *Ellipsonodosaria* A. Silvestri, 1900, is a synonym of *Nodosarella,* and *Ellipsoidella* is a valid genus. A lectotype for *E. pleurostomelloides* is here designated and refigured (BMNH P41662, specimen figured by Heron-Allen & Earland, 1910, *907, pl. 10, fig. 4) and paratypes (BMNH-P41663, P41664), all from the Cretaceous chalk at Selsey Bill, Sussex, England.]

Ellipsoidina Seguenza, 1859, *1711, p. 12 [**E. ellipsoides;* SD Brady, 1868, *187, p. 338]. Test free, ovate, with completely enveloping uniserial chambers each attached to preceding ones at base of test; wall calcareous, finely perforate, granular in structure, white and opaque in appearance; aperture terminal, semilunate to chevron-shaped, provided with apertural tube which extends back internally to preceding aperture. *Eoc.-Plio.,* Eu.-Carib.——Fig. 594,*8,9*. **E. ellipsoides,* Mio., Eu. (Sicily); *8a,b,* side, apert. views, ×31 (*2117); *9,* partially dissected specimen, showing apertural entosolenian tube of final 2 chambers, ×12 (*187).

Ellipsolingulina A. Silvestri, 1907, *1765, p. 69 [**Lingulina impressa* Terquem, 1882, *1890, p.

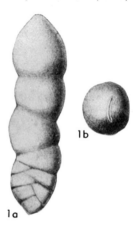

Fig. 596. Pleurostomellidae (Pleurostomellinae; *1, Ellipsodimorphina*) (p. C728).

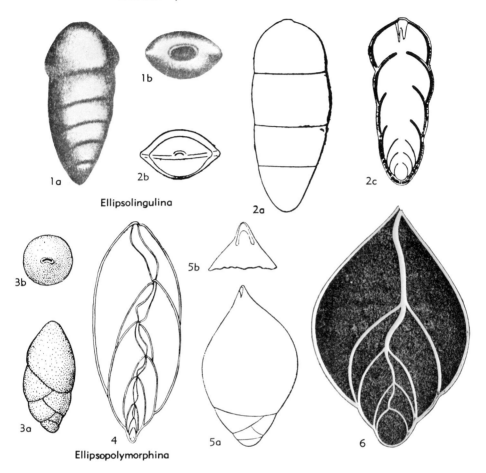

FIG. 597. Pleurostomellidae (Pleurostomellinae; *1,2, Ellipsolingulina; 3-6, Ellipsopolymorphina*) (p. C728-C730).

38; OD]. Test free, elongate, compressed, chambers uniserial and overlapping; sutures horizontal; wall calcareous, microstructure not known; aperture terminal, arcuate, with internal tube projecting inward from final chamber. *Eoc.-Oligo.*, Eu.——FIG. 597,*1*. **E. impressa* (TERQUEM), M. Eoc.(Lutet.), Fr.; *1a,b,* side, top views, ×140 (*700).——FIG. 597,*2*. *E. silvestrii* GALLOWAY, Oligo., Italy; *2a-c,* side, apert. view and long. sec., enlarged (*1765).

[The present genus is tentatively placed in the Pleurostomellinae because of its arcuate terminal aperture and internal tube. Whether the wall is granular or radial in microstructure and whether the internal tube is like that of the Pleurostomellinae or Glandulinidae requires study of topotypes. LE CALVEZ (1952, *1114, p. 35) stated that TERQUEM's type-specimen was not preserved in the Paris collections.]

Ellipsopolymorphina A. SILVESTRI, 1901, *1752, p. 14 [**Dimorphina deformis* (COSTA) FORNASINI, 1890, *730, p. 471 (*non Glandulina deformis* COSTA, 1853) =**Ellipsopolymorphina fornasinii* GALLOWAY, 1933, *762, p. 382; OD (M)]

[=*Ellipsopleurostomella* A. SILVESTRI, 1903, *1757, p. 209, 216 (type, *E. schlichti*)]. Test free, elongate, ovate, early stage biserial, later uniserial, with strongly overlapping chambers; sutures slightly depressed; wall calcareous; aperture terminal, semilunate or chevron-shaped slit, with internal tube connecting apertures of adjacent chambers. *Mio.-Plio.*, Eu.——FIG. 597,*3,4*. **E. fornasinii* GALLOWAY, Plio., Italy *(3),* Mio., Sicily *(4); 3a,b,* side, apert. views, approx. ×35 (*762); *4,* long. sec., ×65 (*1752).——FIG. 597,*5,6*. *E. schlichti* (SILVESTRI), Mio., Italy; *5a,* side view, ×50; *5b,* apert. end to show aperture, ×50; *6,* long. sec. showing connecting apertural tube, ×80 (*1758).

[*Ellipsopleurostomella* was proposed by A. SILVESTRI (1903, *1757) to include the forms previously placed in *Ellipsopolymorphina*, as he then regarded a number of earlier species as dimorphic variations of *Polymorphina labiata* SCHWAGER. The new name supposedly better indicated the relationship of the genus, and he stated that he then "repudiated" the earlier name, which obviously cannot be done under the Rules of Nomenclature. In 1903 SILVESTRI

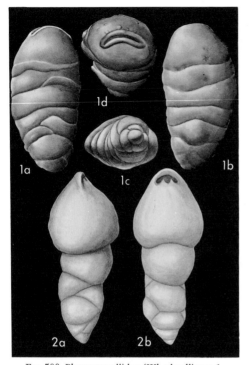

Fig. 598. Pleurostomellidae (Wheelerellinae; *1, Wheelerella*; *2, Bandyella*) (p. *C*730).

included two species in *Ellipsopleurostomella, E. labiata* (SCHWAGER) and *E. schlichti*, n.sp. The latter was selected as type by CUSHMAN (1933, *461) but was stated by ELLIS & MESSINA (*700) to be a *nomen nudum* in 1903. Although poorly described, the discussion of *E. schlichti* by SILVESTRI (1903, *1757, p. 216) appears sufficient to validate the species, and the type of *Ellipsopleurostomella* is here considered to be *E. schlichti*. The generic name is nevertheless a junior synonym of *Ellipsopolymorphina*.]

Nodosarella RZEHAK, 1895, *1605, p. 219 [*Lingulina tuberosa* GÜMBEL, 1870, *840, p. 629; SD CUSHMAN, 1928, *439, p. 261] [=*Ellipsonodosaria* A. SILVESTRI, 1900, *1751, p. 4 (type, *Lingulina rotundata* D'ORBIGNY, 1846, *1395, p. 61)]. Test free, uniserial; chambers inflated; sutures horizontal and constricted; wall calcareous, very finely perforate; aperture terminal, slitlike or faintly arcuate, bordered on each side by faint lip, or slightly overlapping hood on one side. *Paleoc.-Rec.,* cosmop.——FIG. 594,*10. N. rotundata* (D'ORBIGNY), Mio., Eu.(Aus.); *10a,b*, side, top views of lectotype, here designated (D'ORBIGNY Collection, MNHN, Paris), ×36 (*2117).——FIG. 594,*11. *N. tuberosa* (GÜMBEL), Up.M.Eoc. or Low.U. Eoc., Ger.(Bav.); *11a,b*, side, apert. views, ×25 (*2117).

[CUSHMAN, 1948, *486, p. 278) included biserial species in *Nodosarella* and uniserial ones in *Ellipsonodosaria*. As the type-species of *Nodosarella* is uniserial, with horizontal sutures, *Ellipsonodosaria* A. SILVESTRI, 1900, is thus a junior synonym, as was noted by STAINFORTH (1952, *1833, p. 7). Biserial forms are placed in *Ellipsoidella*.]

Pinaria BERMÚDEZ, 1937, *119, p. 242 [*P. heterosculpta*; OD]. Test free, robust, elongate, uniserial, sutures depressed, horizontal; wall calcareous, microstructure not known; aperture terminal, consisting of small slits, possibly due to fusion across opening of apertural teeth such as those of *Pleurostomella,* with internal tube. *Eoc.,* Carib.(Cuba).——FIG. 594,*5. *P. heterosculpta; 5a,b*, side, apert. view of holotype, ×18 (*2117).

Subfamily WHEELERELLINAE Petters, 1954

[Wheelerellinae PETTERS, 1954, p. 39]

Early stage triserial, later uniserial. U. Cret.

Wheelerella PETTERS, 1954, *1448, p. 38 [*W. magdalenaensis*; OD]. Test elongate, ovate in section, early portion with triserially arranged chambers, later uniserial, chambers low, broad, closely appressed, strongly overlapping; wall calcareous, finely perforate; aperture an elongate curved slit with bordering lip, which is slightly higher on outer curve, with internal siphon projecting inward from aperture. *U.Cret.(Coniac.),* S.Am.(Colom.).——FIG. 598,*1. *W. magdalenaensis; 1a-d*, opp. sides, basal, and apert. views of holotype, ×98 (*2117).

Bandyella LOEBLICH & TAPPAN, 1962, *1185, p. 111 [*Pleurostomella greatvalleyensis* TRUJILLO, 1960, *1954, p. 345; OD]. Test free, short, robust; chambers triserially arranged in early stage, later biserial, and final chambers cuneate, uniserial; wall calcareous, perforate-granular in structure; aperture subterminal, slightly eccentric, with a T-shaped opening consisting of crescentic slit just below hooded terminus, with short perpendicular slit extending down face. [Differs from *Wheelerella* in having a T-shaped eccentric or hooded aperture, instead of a straight terminal slitlike aperture. *Ellipsopolymorphina* resembles *Bandyella* in apertural form but has only a biserial early stage before the later uniserial development.] *U.Cret.(Coniac.-Campan.),* USA(Calif.).——FIG. 598,*2. *B. greatvalleyensis* (TRUJILLO), Campan.; *2a,b*, side, face views of holotype, ×79 (*2117).

Family ANNULOPATELLINIDAE Loeblich & Tappan, n.fam.

Test conical, proloculus followed by reniform second chamber, then uniserial, with annular chambers as seen from apex, overlapping on flattened side, chambers subdivided by many radial tubules opening as pores at surface; wall calcareous, perforate-granular in structure; no visible aperture other than surface pores. *Mio.-Rec.*

Annulopatellina PARR & COLLINS, 1930, *1430, p. 92 [*Orbitolina annularis* PARKER & JONES, 1860, *1417d, p. 31; OD (M)] [=*Anulopatellina* A. SILVESTRI, 1931, *1784, p. 65 *(nom. null.)*]. Test free, depressed uniserial, conical, concavo-convex, pairs of tests commonly found joined by their umbilical surfaces; proloculus followed by reni-

FIG. 599. Annulopatellinidae; *1, Annulopatellina* (p. C730-C731).

form second chamber, then by annular chambers, all visible from conical elevated side but completely overlapping previous chambers on concave umbilical side, chambers subdivided by many tiny radial tubules (which superficially resemble secondary septa of *Patellina*), being hollow and opening at surface as pores, curving and anastomosing to form area of many tiny vesicular pustules in center of umbilical side; wall calcareous, perforate granular in structure, not radial and composed of single crystal as Spirillinidae; no aperture visible. *Mio.-Rec.,* Australia-W.Indies(Trinidad).——FIG. 599,*1.* **A. annularis* (PARKER & JONES), Rec., S. Australia; *1a-c,* opposite sides and edge view, ×128 (*2117).

[Differs from *Patellina* in having a crescentic second chamber and in lacking an undivided spiraling chamber following the proloculus. It also differs in having uniserial depressed later chambers, instead of a biserial series, and in having the concave terminal face filled with vesicular tissue rather than the S-shaped columella typical of *Patellina* and *Patellinoides*. The test is composed of granular calcite, rather than formed of a single crystal as in *Patellina*. The types of *Orbitolina annularis* PARKER & JONES, 1860, the type-species of *Annulopatellina,* were isolated by us from the original material in the British Museum (Natural History). The lectotype, here designated (BMNH-ZF3597), and paratypes (BMNH-ZF3596) are from shore sand, Melbourne, Australia.]

Family CAUCASINIDAE N. K. Bykova, 1959

[*nom. transl.* LOEBLICH & TAPPAN, 1961, p. 314 (*ex* subfamily Caucasininae N. K. BYKOVA, 1959)] [=Virgulinidae HOFKER, 1951, p. 236; =Enallostègues D'ORBIGNY, 1826, p. 260 *(partim)* (*nom. neg., nom. nud.*); =?Silicotextulinidae SIGAL in PIVETEAU, 1952, p. 163]

Test elongate, early stage spiral about elongate axis, later may become uniserial; aperture loop-shaped, with internal tooth plate connecting those of adjacent chambers. *U.Cret.-Rec.*

Subfamily FURSENKOININAE Loeblich & Tappan, 1961

[Fursenkoininae LOEBLICH & TAPPAN, 1961, p. 314 *(nom. subst. pro* subfamily Virgulininae CUSHMAN, 1927, p. 68)]

Test basically biserial, but distinctly twisted, later may become uniserial; aperture loop-shaped in biserial stage, becoming terminal in uniserial stage. *U.Cret.-Rec.*

Fursenkoina LOEBLICH & TAPPAN, 1961, *1177, p. 314 [*pro Virgulina* D'ORBIGNY, 1826, *1391, p. 267 (*non* BORY DE ST. VINCENT, 1823)]. [**Virgulina squammosa* D'ORBIGNY, 1826, *1391, p. 267; OD]. Test free, narrow, elongate, rounded to ovate in section; chambers inflated, greater in height than breadth, early portion in highly twisted biserial arrangement, later becoming less sigmoid and more typically biserial, sutures distinct, depressed-oblique, wall calcareous, very finely perforate, granular in structure, surface smooth; aperture narrow, elongate, extending up face of final chamber, lower part may be closed, leaving only suture toward base of chamber, upper part open, resulting in comma-shaped opening, tooth plate attached to closed suture of aperture, with free folded part extending through apertural opening as slight denticulated tooth, opposite end of tooth plate attached to previous apertural foramen. [Numerous references have erroneously stated that *Virgulina* (=*Fursenkoina*) has a triserial base. Topotypes of the type-species, *V. squammosa* D'ORBIGNY, from the Pliocene of Italy, when examined from the base, show only the highly twisted biserial development of the test found in *Sigmavirgulina.*] *U.Cret.-Rec.*, cosmop. ——FIG. 600,*1-4.* **F. squammosa* (D'ORBIGNY), Plio., Italy *(1-3),* Rec., Indon. *(4); 1a-c,* opposite sides and edge view, ×44 (*2117); 2a,b, optical sec. of microspheric form showing tooth plates and edge view of aperture of same specimen, ×108 (*928c); 3, megalospheric specimen with last chamber broken away to show tooth plates, ×108 (*928c); 4, diagram. sketch of isolated tooth plate showing form and denticulate margin, enlarged (*928c).

Cassidella HOFKER, 1951, *928c, p. 264 [*Virgulina tegulata REUSS, 1846, *1538, p. 40; OD]

FIG. 600. Caucasinidae (Fursenkoininae; 1-4, Fursenkoina; 5-7, Cassidella; 8,9, Coryphostoma; 10, Suggrunda; 11-13, Virgulinella) (p. C731-C734).

[=*Praevirgulina* HOFKER, 1951, *935, p. 1 (*nom. nud.*)]. Test free, narrow, elongate, triserial in early stage, later biserial, very slightly twisted, chambers broad, low; sutures distinct, depressed; wall calcareous, finely perforate, granular in structure, surface smooth; aperture a long narrow slit, extending up face from base of final chamber, tooth plate simple, with folded or U-shaped section, arising at upper border of penultimate foramen, extending along basal wall of chamber to aperture where it becomes attached along lower apertural border. *U.Cret.-Rec.*, cosmop.——FIG. 600,*5-7*. *C. tegulata* (REUSS), U.Cret., USA (Ark.) *(5)*, Neth. *(6,7)*; *5a,b*, side, apert. views, ×93 (*2117); *6*, final chamber dissected to show tooth plate, ×103 (*928c); *7*, diagram showing tooth plate in relation to penultimate foramen below and at right, aperture in foreground, ×103 (*928c).

[The original type designation is somewhat ambiguous. HOFKER (*928c, p. 264) stated, "Genus *Cassidella*, nov. genus. Genotype, *Virgulina (Bolivina) tegulata* (Reuss)," and following the description, on p. 265 reported, "The type of the genus is *Cassidella oligocenica* Hofker." THALMANN (1952, *1897j, p. 971) in his bibliography and index cited *Virgulina tegulata* as type-species, recording the correct page number but incorrectly referring to it as published in a different paper by HOFKER, which did not describe the genus. As the first mentioned reference of HOFKER definitely stated "genotype, *V. tegulata*," this is regarded as original fixation of the type and therefore validation of the genus. Later workers have considered *Cassidella* a synonym of *Virgulina*, since *V. squammosa* was placed in *Cassidella* by HOFKER. As the Cretaceous type-species (*V. tegulata*) is a simple form, with less twisted test and simpler (nondenticulate) tooth plate than *V. squammosa* (=*Fursenkoina*), it is not here regarded as congeneric with the latter, and *Cassidella* is recognized as a valid and distinct genus. *Praevirgulina* was merely listed in combination with the specific name, as *P. tegulata*, but was not described. *Cassidella* differs from *Fursenkoina* in the less twisted test, broader and lower chambers, and simple, non-denticulate tooth plate with broader base and more U-shaped section.]

Coryphostoma LOEBLICH & TAPPAN, 1962, *1185, p. 111 [**Bolivina plaita* CARSEY, 1926, *282, p. 26; OD]. Test free, elongate, narrow, early chambers biserially arranged, later chambers becoming cuneiform with tendency to become uniserial; wall calcareous, finely perforate, granular in structure; aperture loop-shaped in early stage, extending from base of final chamber, becoming terminal in adult, with internal tooth plate. *U.Cret.(Campan.)-Rec.*, cosmop.——FIG. 600,*8,9*. *C. plaita* (CARSEY), U.Cret., USA(Tex.) *(8)*, Mex. *(9)*; *8a,b*, side, apert. views, ×64 (*2117); *9*, optical sec. showing internal tooth plates, ×104 (*948).

[Differs from *Loxostomum* in having an internal tooth plate, being rounded in section, and in the absence of sharply keeled margins. It differs from *Rectobolivina* in having a granular, rather than radially built, wall, and in the later chambers being cuneate, without an elongate uniserial and rectilinear stage. *Loxostomoides* differs in having a radially built wall and retral processes with reentrants and lobes or crenulations of the chamber margins along the sutures.]

Sigmavirgulina LOEBLICH & TAPPAN, 1957, *1172, p. 227 [**Bolivina tortuosa* BRADY, 1881, *196c, p. 57; OD]. Test free, biserial, with chambers added slightly more than 180° apart, with sigmoiline arrangement of 2 series of chambers that at first form tight low spire, later become high-spired

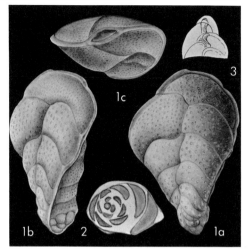

FIG. 601. Caucasinidae (Fursenkoininae; *1-3*, *Sigmavirgulina*) (p. C733).

and almost regularly biserial, though somewhat twisted throughout; periphery angled or with distinct keel, chambers numerous, increasing regularly in height as added, increasing more rapidly in breadth so that test flares; sutures distinct, thickened, depressed; wall calcareous, of calcite by X-ray determination, coarsely perforate, granular in structure, surface smooth or with short spines in early portion; aperture an elongate oval at inner margin of final chamber, surrounded by lip which passes gradually into peripheral keel, provided internally with simple flaring tooth plate which is also twisted; in some specimens aperture may tend to become terminal, and is situated a short distance above the base of the chamber. *Mio.-Rec.*, cosmop.——FIG. 601,*1-3*. *S. tortuosa* (BRADY), Rec., Fiji *(1,2)*, Indon. *(3)*; *1a-c*, side, edge, and apert. views, ×105 (*1172); *2*, basal view of partially etched specimen showing twisted biserial early chamber arrangement, ×105 (*1172); *3*, view in transmitted light showing twisted tooth plate, enlarged (*928c).

[Differs from *Bolivina* in having a granular wall structure, instead of radial, in the early sigmoiline type of development, the twisted adult test resulting from this process. *Sigmavirgulina* differs from *Fursenkoina* in having a compressed, rather than rounded, test, broad low chambers, rather than very high and elongate ones, and a coarsely perforate test.]

Suggrunda HOFFMEISTER & BERRY, 1937, *925, p. 29 [**S. porosa*; OD] [=?*Silicotextulina* DEFLANDRE, 1934, *574, p. 1447 (type, *S. diatomitarum*)]. Test small, tapering, biserial throughout, chambers broad and low, with lower margin commonly nodose or spinose; sutures nearly horizontal, straight, depressed; wall calcareous; finely perforate, granular in structure, may have larger pores near basal margin of chambers; aperture a hook-shaped opening in basal depression of final chamber, presence or absence of tooth plate not reported. *Mio.*, S.Am.(Venez.)-W.Indies(Trini-

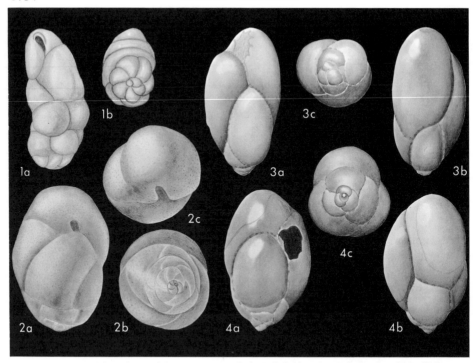

Fig. 602. Caucasinidae (Caucasininae; *1,2, Caucasina*); Delosinidae; *3,4, Delosina* (p. C734-C736).

dad)-USA(Calif.).——Fig. 600,*10*. **S. porosa*, M.Mio., Venez.; *10a-c*, side, edge, and apert. views of holotype, ×174 (*2117).

[The aperture of *Suggrunda* was originally described as a low basal arch, but the holotype of the type-species has a hook-shaped aperture, like that of *Grimsdaleinella* and *Gabonella*, though difficult to see because of its small size. *Silicotextulina* was described from isolated chambers and proculi which appeared siliceous or chitino-siliceous; they were found in Miocene diatomites of California. Apparently all forms referred to *Silicotextulina* consist either of internal casts or the silicified pseudochitinous inner membrane of an originally calcareous test. The small size, presence of pores near the basal margin of the chambers, and mode of occurrence strongly suggest that *Silicotextulina diatomitarum* DEFLANDRE might well be conspecific with *Suggrunda kleinpelli* BRAMLETTE, also described from California Miocene diatomites. Recent species referred to *Silicotextulina* appear to be internal casts of *Brizalina*.]

Virgulinella CUSHMAN, 1932, *453, p. 9 [**Virgulina pertusa* REUSS, 1861, *1550, p. 362; OD] [=*Virgulina (Virgulinella)* CUSHMAN, 1932, *453, p. 9 (obj.)]. Test free, elongate, rounded in section, early stage triserial, later biserial, chambers inflated, similar to *Fursenkoina* but with numerous small arched sutural openings, partially covered by bridges of basal chamber wall; wall calcareous, finely perforate, opaque, granular in structure; aperture an oblique loop-shaped opening in terminal face, with reduced tooth plate which begins near previous apertural foramen and attaches to lower part of chamber wall, then to lower border of aperture, supplementary sutural openings also present. [HOFKER (1956, *946, p. 98) regards *Candeina* as a descendant of *Virgulinella*, but *Candeina* has a radially built wall, and *Virgulinella*

a granular one.] Mio.-Plio., Eu.-N.Am.-Asia (Indon.).——Fig. 600,*11-13*. **V. pertusa* (REUSS), Mio., Ger. *(11)*, Neth. *(12,13)*; *11a-c*, opposite sides and apert. view, ×64 (*2117); *12a,b,13*, dissected specimen showing tooth plate and isolated tooth plate, enlarged (*946).

Subfamily CAUCASININAE N. K. Bykova, 1959

[Caucasininae N. K. BYKOVA in RAUZER-CHERNOUSOVA & FURSENKO, 1959, p. 328]

Early stage trochospiral, later biserial; aperture a loop in apertural face. *U.Cret.-Mio.*

[Although no information is available as to the microstructure of the wall of the type-species of *Caucasina*, the genus *Aeolostreptis* LOEBLICH and TAPPAN, 1957 (which on the basis of descriptions and illustrations appears to be a junior synonym of *Caucasina* KHALILOV, 1951), has a distinctly granular wall. Hence the subfamily and family are placed in the Cassidulinacea and removed from the Buliminacea.]

Caucasina KHALILOV, 1951, *1036, p. 58 [**C. oligocenica*; OD] [=*Aeolostreptis* LOEBLICH & TAPPAN, 1957, *1172, p. 227 (type, *Buliminella vitrea* CUSHMAN & PARKER, 1936, *515, p. 7)]. Test free, elongate, base bluntly rounded, early portion in low discorbine coil with up to 8 chambers per whorl, later whorls becoming high-spired and reduced in number of chambers to 3 per whorl, early chambers low, later about equal in breadth and height and may be inflated, but not extremely high and elongate; sutures distinct, depressed; wall calcareous, finely perforate, granular in structure, surface smooth; aperture an elongate loop at inner margin of final chamber, at right angles to sutures, with narrow lip at forward margin. *U.Cret.-Mio.,* Eu.-N.Am.-Asia.——FIG. 602,1. **C. schischkinskayae* (SAMOYLOVA), Oligo., USSR(Caucasus); *1a,b,* side and basal views, ×106 (*1509).——FIG. 602,2. *C. vitrea* (CUSHMAN & PARKER), U.Cret., USA(Miss.); *2a-c,* side, basal, and apert. views, ×200 (*1172). [**C. oligocenica* =*Bulimina schischkinskye* SAMOYLOVA, 1947, *1623, p. 82, 100 *(recte B. schischkinskayae).*]

[*Caucasina* was originally described from the Oligocene of the Caucasus, and later reported to occur from Eocene to Miocene. *Aeolostreptis* was defined for Upper Cretaceous species. As no morphologic distinction between these "genera" was observed and species referable to the genus have been described from the Upper Cretaceous, Paleocene, Eocene, Oligocene and Miocene, *Aeolostreptis* was later regarded by us as a synonym of *Caucasina* (*1177).]

Family DELOSINIDAE Parr, 1950

[Delosinidae PARR, 1950, p. 345]

Test triserial; no primary aperture, but large sutural pores open into subsutural canal. *Rec.*

Delosina WIESNER, 1931, *2063, p. 123 [**Polymorphina(?) complexa* SIDEBOTTOM, 1907, *1740, p. 16; OD (M)]. Test free, elongate, somewhat tapered at base, rounded in sections; chambers elongate, trochospirally arranged, with 3 strongly overlapping chambers per whorl, final whorl occupying much of entire length; sutures depressed; wall calcareous, finely perforate, granular in structure, surface smooth; distinct large sutural pores opening into subsutural canals which apparently terminate in spongy area of final chamber but do not open to exterior; apertural development consisting of perforations in somewhat restricted terminal area, in type-species forming somewhat loop-shaped arch upward from suture-contact with penultimate chamber at its apex. *Rec.,* Medit.-Antarctic-Pac.O.——FIG. 602, *3,4.* **D. complexa* (SIDEBOTTOM), Medit.(Delos

FIG. 603. Loxostomidae; *1-5, Loxostomum; 6,7, Trachelinella; 8, Aragonia* (p. C736).

ls.); *3a-c, 4a-c,* opposite sides and basal views of topotypes showing sutural pores opening into subsutural canals, ×87 (*2117).

[Originally regarded as *Polymorphina,* in the Polymorphinidae, this genus was later placed in the Buliminidae, Bulimininae, by EARLAND (1934, *653, p. 125), who stated that sections showed a loop-shaped aperture in one specimen of a megalospheric proloculus, suggesting a bulimine aperture. It was placed in the subfamily Uvigerininae by CUSHMAN (1948, *486) and in the newly proposed family Delosinidae by PARR (1950, *1429). The perforate-granular wall structure would eliminate this genus from the families Polymorphinidae, Buliminidae, and Uvigerinidae, and even from their superfamilies, showing that it belongs to the Cassidulinacea and suggesting its close relationship to the Fursenkoininae (the sutural pores of *Delosina* are reminiscent of those in *Virgulinella*). The internal tube of the Fursenkoininae has not yet been demonstrated in *Delosina,* however, and the absence of a distinct aperture allows retention of the monotypic family Delosinidae.]

Family LOXOSTOMIDAE Loeblich & Tappan, 1962

[Loxostomidae LOEBLICH & TAPPAN, 1962, p. 110]

Test free, biserial, or may become uniserial in later stage, commonly with flattened sides and carinate margins; wall calcareous, perforate-granular in structure; aperture interiomarginal in simpler forms, later may become terminal, no tooth plate or internal siphon. *U.Cret.(Senon.)-Eoc.*

Loxostomum EHRENBERG, 1854, *680, p. 22 [*L. subrostratum;* SD CUSHMAN, 1927, *434, p. 490] [=*Loxostoma* HOWE, 1930, *969, p. 329 *(nom. van.)* (non BIVONA-BERNARDI, 1838); *Bolivinitella* MARIE, 1941, *1215, p. 189 (type, *Bolivinita eleyi* CUSHMAN, 1927, *429, p. 91)]. Test elongate, compressed, quadrate in section, with flat or concave sides; chambers biserially arranged throughout, strongly overlapping and arched in adult with tendency to become uniserial; sutures limbate, arched, sutural thickening merging laterally into longitudinal carinae at 4 margins; wall calcareous, finely perforate; aperture terminal, slitlike to ovate, commonly with lip which may be very finely tuberculate but lacking any internal tooth plate. [The synonymy of *Loxostomum* and *Bolivinitella* was noted by HOFKER (1951, *928c, p. 44) and discussed by LOEBLICH & TAPPAN (1962, *1185, p. 110), who therefore emended the generic description.] *U.Cret.(Senon.)-Paleoc.,* Eu.-N.Am.——FIG. 603,*1-5.* *L. subrostratum,* U.Cret.(Senon.), Eu.(Fr.); *1,* side view of specimen mounted in balsam and viewed in transmitted light, copy of Ehrenberg's original figure (*472); *2a-c,* side, edge, and top views of holotype of *Bolivinita eleyi,* U.Cret., USA(Ark.), ×104 (*1303); *3,4,* side views of topotype of *B. eleyi,* ×100 (*1303); *5a,b,* side and edge views of specimen from U.Cret., Fr., figured originally as *Bolivinitella eleyi* (CUSHMAN) by MARIE; ×38 (*1215).

Aragonia FINLAY, 1939, *717c, p. 318 [*A. zelandica;* OD]. Test free, rhomboidal, compressed to fusiform in section, sides flat; chambers biserially arranged; sutures oblique, commonly limbate; wall calcareous, granular, not perforate, surface ornamented by limbate and elevated sutures and marginal keel, and may also have longitudinal and diagonally placed costae that form irregular network; aperture small, low opening at base of final chamber, no internal tooth plate. [Originally *Aragonia* was placed in the Heterohelicidae (*717c) as related to *Bolivinoides,* but because of the absence of a tooth plate and the character of the wall (lacking perforations, and resembling agglutinated calcareous grains) it was later regarded by REYMENT as related neither to the Heterohelicidae nor to the Bolivininae (*1558) but to be an agglutinated form.] *Paleoc.-Eoc.,* N.Z.-N.Am.-Carib.-Eu.——FIG. 603,*8.* *A. zelandica,* M.Eoc., N.Z.; *8a,b,* side, top views of paratype, ×192 (*2117).

Trachelinella MONTANARO GALLITELLI, 1956, *1302, p. 38 [*Bolivina watersi* CUSHMAN, 1927, *429, p. 88] [=*Trakelina* MONTANARO GALLITELLI, 1955, *1300, p. 215 *(nom. nud.)*]. Test elongate, somewhat compressed, flaring gradually, chamber arrangement biserial, test commonly twisted as much as 90° with growth, periphery subacute, commonly carinate; sutures arched, incised; wall calcareous, finely perforate, surface with prominent ribs aligned along major inflation of chambers and consequently strongly arched, commonly fusing laterally into marginal carinae; aperture terminal, rounded to ovate with short neck and lip, no apertural tooth observed. [Differs from *Bolivina* in lacking an internal tooth plate, and in having a well-developed terminal neck and terminal aperture. The oblique axis, short neck, biserial chamber arrangement and absence of a tooth plate suggest a relationship with *Loxostomum.*] *U.Cret.(Maastricht.),* USA(Tex.).——FIG. 603,*6,7.* *T. watersi* (CUSHMAN); *6a-c,* side, edge, and apert. views showing heavy ornamentation and terminal aperture; *7,* side view of holotype showing biserial chamber arrangement; all ×123 (*1302).

Family CASSIDULINIDAE d'Orbigny, 1839

[Cassidulinidae D'ORBIGNY in DE LA SAGRA, 1839, p. xxxix, 123]——[In synonymic citations superscript numbers indicate taxonomic rank assigned by authors (¹family, ²subfamily); dagger(†) indicates *partim*]——[=²Cassidulinida SCHULTZE, 1854, p. 52; =¹Cassidulinidea REUSS, 1862, p. 373; =²Cassidulinae BRADY, 1881, p. 44; =²Cassidulininae BRADY, 1884, p. 69; =¹Cassidulina LANKESTER, 1885, p. 847; =¹Cassiduline DELAGE & HÉROUARD, 1896, p. 140; =²Cassidulineae CALKINS, 1901, p. 108; =¹Cassidulinida COPELAND, 1956, p. 188 *(nom. van.)*]——[=¹Turbinoidat SCHULTZE, 1854, p. 52 *(nom. nud.);* =²Ehrenbergininae CUSHMAN, 1927, p. 84]

Test lenticular, subglobular or elongate; chambers biserially arranged, alternating chambers also planispirally enrolled at least in early stage, later may be uncoiled; aperture elongate, comma-shaped, slit extending from basal suture into apertural face. *Eoc.-Rec.*

Foraminiferida—Rotaliina—Cassidulinacea

Cassidulina D'ORBIGNY, 1826, *1391, p. 282 [*C. laevigata; OD (M)] [=Entrochus EHRENBERG, 1843, *672, p. 408 (type, E. septatus); Selenostomum EHRENBERG, 1858, *683, p. 12]. Test free, lenticular, commonly biumbonate, with clear central bosses; chambers biserially arranged in

FIG. 604. Cassidulinidae; *1,2, Cassidulina; 3,4, Favocassidulina; 5, Ehrenbergina; 6,7, Globocassidulina; 8, Burseolina; 9, Cassidulinella* (p. C737-C738).

coil, chambers alternating on each side of periphery, each reaching boss on one side and only extending part way to boss of opposite side; succeeding chamber extending to center on alternate sides; wall calcareous, hyaline, perforate, granular in structure, surface generally smooth; aperture an elongate slit, extending from base of final chamber upward in curve paralleling anterior margin of chamber with narrow bordering lip on lower margin but lacking internal tooth. [*Cassidulina*, as here recognized, excludes the radial-walled species, now placed in *Islandiella*, as well as those with globular, nonkeeled tests and tripartite aperture, now placed in *Globocassidulina*. *Cassidulina laevigata* was originally described from ballast sand of unknown provenance.] *Eoc.-Rec.*, cosmop.——FIG. 604,*1,2*. **C. laevigata*, Rec., Eu. (Italy) *(1)*, Atl.O. *(2)*; *1a-c*, opposite sides and edge view, ×78 (*2117); *2a,b*, final chamber showing apert. on exterior and partially dissected chamber showing apert. inside with inward bent apert. margin, ×40 (*1361).

Burseolina SEGUENZA, 1880, *1713, p. 138 [**B. calabra*; OD (M)]. Test free, subglobular, tiny, periphery broadly rounded; chambers biserially enrolled; wall calcareous, perforate, surface ornamented with striae, or with coarse ridges and reticulations which obscure sutures, as in *Favocassidulina*, apertural face smooth; aperture a narrow, elongate, arched slit, extending up face of final chamber, with narrow bordering lip. [*Burseolina* is similar to *Globocassidulina* in having a rounded, rather than angular to carinate, periphery, may resemble *Favocassidulina* in surface ornamentation, and has the apertural characters of *Cassidulina*, with elongate arched aperture extending up the face with narrow bordering lip.] *Mio.*, Eu.(Italy)-Carib.——FIG. 604,*8*. **B. calabra*, Torton., Italy; *8a-c*, side, dorsal, and face views, showing biserially arranged chambers, obscure ridges, and apert., ×111 (*2117).

Cassidulinella NATLAND, 1940, *1347, p. 568, 570 (non SUZIN in VOLOSHINOVA & DAIN, 1952) [**C. pliocenica*; OD]. Test free, flattened, chambers biserially enrolled as in *Cassidulina*, with later chambers much elongated and overlapping at periphery, tending to encircle much of peripheral margin, zigzag suture between biserially arranged chambers almost peripheral in position; wall calcareous, thin, finely perforate, microstructure not determined, as specimens available are pyritic casts; aperture a much elongated slit extending up face, near to and paralleling outer margin of final chamber. [Differs from *Cassidulina* in the encircling tendency of its later chambers. Whether it is to be finally placed with the Cassidulinidae or Islandiellidae depends on additional information as to wall structure and presence or absence of an internal tooth.] *U.Mio.-U.Plio.*, USA(Calif.).——FIG. 604,*9*. **C. pliocenica*, Plio.; *9a-d*, opposite sides, apert. and back peripheral views to show chamber alternation, holotype, ×56 (*2117).

Ehrenbergina REUSS, 1850, *1540, p. 377 [**E. serrata*; OD (M)]. Test flattened, compressed perpendicular to plane of coiling, periphery carinate; chambers broad, low, biserially arranged and enrolled, as in *Cassidulina*, but somewhat uncoiled; wall calcareous, finely perforate, granular in structure, surface smooth or with pustules or ridges; aperture an elongate curved slit, perpendicular to base of apertural face and paralleling peripheral keel. [The wall character and aperture of *Ehrenbergina* are similar to *Cassidulina*, but the test is uncoiled.] *Eoc.-Rec.*, cosmop.——FIG. 604,*5*. **E. serrata*, Mio., Eu.(Aus.); *5a-c*, opp. sides and edge view, ×78 (*2117).

Favocassidulina LOEBLICH & TAPPAN, 1957, *1172, p. 230 [**Pulvinulina favus* BRADY, 1877, *194, p. 535; OD]. Test free, lenticular, periphery acute; chambers biserially arranged and enrolled as in *Cassidulina*, each chamber extending to umbilicus on one side with only small triangular portion extending to opposite side; sutures not visible externally, obscured by coarse surface ornamentation; wall calcareous, finely perforate, granular in structure, ornamented by honeycomb-like secondary growth, with relatively wide hexagonal open areas separated by narrow, elevated ridges; aperture elongate, a slightly curved slit bordered by very narrow lip, and extending upward from base of final chamber, near to and paralleling anterior margin of chamber, opening toward side opposite that on which final chamber lies, each successive aperture appearing on alternate sides of test, region immediately surrounding aperture relatively smooth. *Rec.*, Pac.O.——FIG. 604,*3,4*. **F. favus* (BRADY), Chile *(3)*, Caroline Is. *(4)*; *3a,b*, side and edge views of topotype, ×44 (*1172); *4*, half-sectioned hypotype, ×48 (*1172).

Globocassidulina VOLOSHINOVA, 1960, *2020, p. 58 [**Cassidulina globosa* HANTKEN, 1875, *863, p. 64; OD] [=*Cassilongina* VOLOSHINOVA, 1960, *2020, p. 58 (type, *Cassidulina oblonga* REUSS, 1850, *1540, p. 376)]. Test free, subglobular, peripheral margin rounded; umbilicus closed; chambers biserially arranged and enrolled; wall calcareous, finely perforate, granular in structure, surface commonly smooth; aperture a narrow slit extending up face of final chamber, may have narrow infolded rim, but no apertural tooth plate. [*Cassilongina* was defined as having a tendency to elongate, but no true uncoiling occurs. *Cassilongina* was also stated to have a thin single-layered wall, and *Globocassidulina* a many-layered wall. Both are lamellar in character, and relative thickness of the wall varies in different species. *Cassilongina* is here regarded as synonymous with *Globocassidulina*.] *Eoc.-Rec.*, cosmop.——FIG. 604, *6*. **G. globosa* (HANTKEN), U.Eoc., USA(S.Car.); *6a,b*, side, edge views, ×90 (*467).——FIG. 604, *7*. *G. oblonga* (REUSS), Tert., Spain(Galicia); *7a,b*, side, edge views, approx. ×93 (*2022).

Family INVOLUTINIDAE Bütschli, 1880

[*nom. transl.* SIGAL in PIVETEAU, 1952, p. 159 (*ex* subfamily Involutinae BÜTSCHLI in BRONN, 1880, p. 209; Involutininae THALMANN, 1935, p. 715)]——[=Problematininae RHUMBLER, 1913, p. 389; =Arproblematoia RHUMBLER, 1913, p. 389 (*nom. van.*); =Ventrolaminidae WEYNSCHENK, 1950, p. 17; =Ventrolamininae LOEBLICH & TAPPAN, 1961, p. 292]

Test tubular and enrolled, with secondary

FIG. 605. Involutinidae; *1,2, Involutina; 3-5, Aulotortus; 6, Paalzowella* (p. C740-C741).

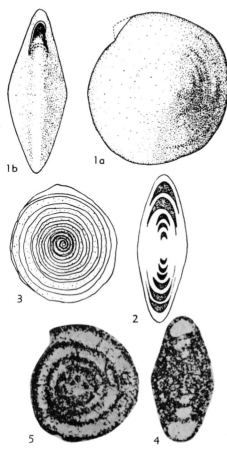

Fig. 606. Involutinidae; *1-5, Aulotortus* (p. *C*740-*C*741).

deposits in umbilical region on one or both sides; wall calcareous perforate, lamellar, microgranular. *U.Trias.-U.Cret.(Cenoman.-Turon.).*

Involutina TERQUEM, 1862, *1883, p. 450 [*I. jonesi* TERQUEM & PIETTE in TERQUEM, 1862, *1883, p. 461, =*Nummulites? liassicus* JONES in BRODIE, 1853, *208, p. 275; SD BORNEMANN, 1874, *174, p. 711] [=*Problematina* BORNEMANN, 1874, *174, p. 733 (type, *Involutina deslongchampsi* TERQUEM, 1864, *1885, p. 432); *Arinvolutoum* RHUMBLER, 1913, *1572b, p. 390 *(nom. van.);* *Arproblematoum* RHUMBLER, 1913, *1572b, p. 390 *(nom. van.)*]. Test free, lenticular, consisting of proloculus followed by planispirally coiled, nonseptate tubular second chamber, umbilical region on both sides filled with numerous secondarily deposited pillars or plugs; surface may be pitted; aperture at open end of tube. [The complex taxonomy and confusion as to the type-species has been recently reviewed in detail by LOEBLICH & TAPPAN (1961, *1176).] *U.Trias.-L.Jur.(Lias.),* Eu.——FIG. 605,*1,2*. **I. liassica* (JONES), Lias., Eng.*(1)*, Switz. *(2)*; *1a,b,* side, edge views of hypotype, ×35 (*2117); *2,* axial sec., ×27 (*1525).

Aulotortus WEYNSCHENK, 1956, *2052, p. 26 [*A. sinuosus;* OD] [=*Trocholina (Paratrocholina)* OBERHAUSER, 1957, *1383, p. 196 (type, *T. (P.) oscillens);* *Angulodiscus* KRISTAN, 1957, *1057, p. 278 (type, *A. communis);* *Arenovidalina* Ho, 1959, *923, p. 414 (type, *A. chialingchiangensis)*]. Test free, lenticular, compressed to nearly globular; small spherical proloculus followed by planispirally to slightly streptospirally enrolled and undivided tubular chamber, umbilical area of both sides of test filled with secondary deposit of crystalline calcite, so that only final whorl is visible at peripheral margin, earlier whorls and spiral suture obscured by secondary filling; wall calcareous, central area may be variously ornamented with irregular or radial ridges, ventral side in some species appearing granular and suggesting termination of umbilical pillars, which merge outward into radial ridges; aperture at open end of tubular chamber, slightly asymmetrical in position. *Trias.-U.Cret. (?Turon.),* Eu.-Asia(China-Turkey).——FIG. 605, *3.* **A. sinuosus,* M.Jur., Eu.(Aus.); ×37 (*2052).

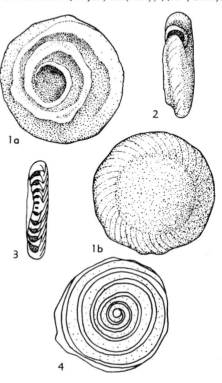

FIG. 607. Involutinidae; *1-4, Paalzowella* (p. *C*741).

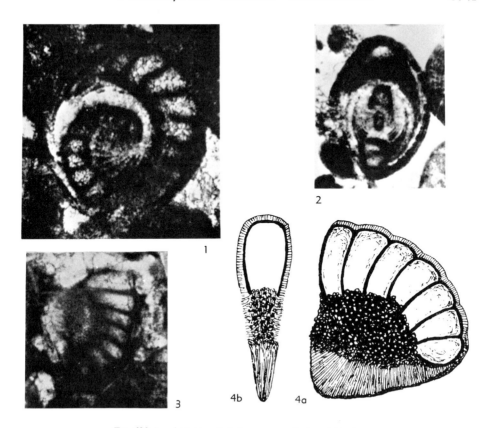

Fig. 608. Involutinidae; *1-4, Protopeneroplis* (p. C741-C742).

——Fig. 605,*4,5*. *A. oscillens* (OBERHAUSER), U. Cret.(?Turon.), Turkey; *4*, axial sec. of holotype, ×27 (*1383); *5*, equat. sec., ×27 (*1383).——Fig. 606,*1-3*. *A. communis* (KRISTAN), U.Trias. (Rhaet.), Eu.(Aus.); *1a,b,* side, apert. views of holotype; *2*, axial sec.; *3*, equat. sec.; all ×35 (*1057).——Fig. 606,*4,5*. *A. chialingchiangensis* (Ho), Trias., China; *4*, axial sec. of holotype, ×200 (*923); *5*, equat. sec., ×200 (*923).

[*Aulotortus* differs from *Involutina* in its slightly streptospiral coiling, and in its less well-differentiated umbilical pillars. It was originally described as calcareous and imperforate and was placed in the Ophthalmidiidae. A very nearly identical species was described by OBERHAUSER (1957, *1383) as *Paratrocholina*. Well-preserved specimens clearly show the perforate nature of the test. The synonymies of the type-species of both genera include reference to an earlier figure and description by WEYNSCHENK (1950, *2050) indicated as "Genus?, species?." In an appendix to his article, OBERHAUSER noted the probability that *Paratrocholina* and *Aulotortus* are synonymous, the apparently imperforate wall of *Aulotortus* possibly being due to later recrystallization. This is suggested also by the original figures of *Aulotortus*.]

Paalzowella CUSHMAN, 1933, *461, p. 234 [*Discorbina scalariformis* PAALZOW, 1917, *1403, p. 247; OD] [=*Coronella* KRISTAN, 1957, *1057, p. 280 (type, *C. austriaca*) (*non Coronella* LAURENTI, 1768; *nec* GOLDFUSS, 1828); *Coronipora* KRISTAN, 1958, *1058, p. 114 (*nom. subst. pro Coronella* KRISTAN, 1957) (obj.)]. Test free, conical, consisting of single tubular chamber, spirally enrolled, nearly completely involute on umbilical side, evolute on spiral side, periphery keeled; spiral suture may be thickened and elevated, showing remnant of earlier peripheral keel, radial markings giving ventral surface lobate appearance but do not reflect true septa; aperture at open end of tube. [Differs from *Trocholina* in being more completely involute on the umbilical side and lacking umbilical plugs and pillars, and in having radial ornamentation.] U.Trias.(Rhaet.)-U.Jur., Eu.——Fig. 605,*6*. **P. scalariformis* (PAALZOW), U.Jur., Ger.; *6a-c*, opposite sides and edge view of topotype, ×116 (*2117).——Fig. 607,*1-4*. *P. austriaca* (KRISTAN), U.Trias.(Rhaet.), Aus.; *1a,b*, opposite sides of holotype; *2*, edge view of paratype; *3*, axial sec.; *4*, equat. section; all ×35 (*1057).

Protopeneroplis WEYNSCHENK, 1950, *2050, p. 13 (*non* HOFKER, 1950, *933a, p. 393) [**P. striata*; OD] [=?*Ventrolamina* WEYNSCHENK, 1950, *2050, p. 17 (type, *V. cribrans*)]. Test planispirally enrolled, bilaterally symmetrical and involute, not close-coiled, successive whorls not touching in equatorial section; septa thickened,

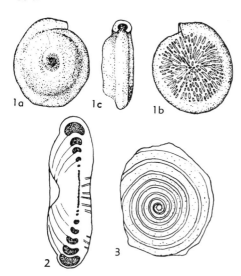

Fig. 609. Involutinidae; *1-3, Semiinvoluta* (p. C742).

slightly oblique; wall calcareous, granular, fibrous, finely perforate, lamellar, surface with regularly arranged spiraling costae; aperture areal. M.Jur.-U.Jur., Eu.(Aus.,Tirol)-Asia(Israel).——Fig. 608, *1,2*. *P. striata*, M.Jur., Aus.; *1*, equat. sec., ×110 (*2050); *2*, axial sec., ×130 (*2050).——Fig. 608,*3,4*. *P. cribrans* (WEYNSCHENK), U.Jur., Aus.; *3*, equat. or oblique sec., so-called sieve plate in lower part may be an oblique cut through fibrous wall, ×77 (*2050); *4a,b*, diagram. interpretation of equat. and axial secs., approx. ×120 (*2050).

[*Protopeneroplis* and *Ventrolamina* were both described from thin sections, hence are only partially known. It seems probable that the 2 nominal genera and possibly their type-species are synonymous, as they are from the same samples, show very similar chamber size, arrangement, and septal angle. *Ventrolamina* sections possibly represent oblique sections of *Protopeneroplis*. Additional study of free specimens and oriented sections is needful for better determining the relationships.]

Semiinvoluta KRISTAN, 1957, *1057, p. 276 [*S. clari*; OD]. Intermediate in character between *Trocholina* and *Involutina*; test similar to *Trocholina* in having umbilical pillars on one side only but with coiling nearly planispiral, as in *Involutina*, and with secondary thickening also on spiral side; aperture at open end of tube. U.Trias.(Rhaet.), Aus.——Fig. 609,*1-3*. *S. clari*; *1a-c*, opposite sides and edge views of holotype; *2*, axial sec.; *3*, equat. sec., all ×35 (*1057).

Trocholina PAALZOW, 1922, *1404, p. 10 [*Involutina conica* SCHLUMBERGER, 1898, *1659, p. 151; SD CUSHMAN, 1933, *461, p. 234] [=*Neotrocholina* REICHEL, 1956, *1525, p. 404 (type, *N. valdensis*); *Trocholina (Trochonella)* KRISTAN, 1957, *1057, p. 285 (type, *T. (T.) crassa*)]. Test free, conical, consisting of globular proloculus and spirally enrolled tubular second chamber, which is dorsally evolute with all whorls visible, ventral umbilical region completely filled with coarse calcite crystals, appearing as irregularly arranged pillars, nodes or beads on surface; wall calcareous, dorsally coarsely perforate, ventrally more finely perforate, granular in structure, surface smooth or with elevated spiral suture, ventral surface of final whorl may show somewhat curved faint growth striae, which end in pustules at umbilical margin; aperture at open end of tube. U.Trias.(Rhaet.)-U.Cret.(Cenoman.), Eu.-USA-Carib.-Afr.-M.East.——Fig. 610,*1,2*. *T. conica* (SCHLUMBERGER), M.Jur., Ger. *(1)*, M.Jur.(Bathon.), Fr. *(2)*; *1a-c*, spiral, umbilical, and edge views, ×66 (*2117); *2*, axial sec. of holotype, ×113 (*1525).——Fig. 610,*3,4*. *T. valdensis* (REICHEL), L.Cret.(Valangin.), Switz.; *3a,b*, umbilical and edge views, ×64 (*1525); *4*, axial sec. showing umbilical pillars, ×66 (*1525).——Fig. 610,*5-7*. *T. crassa* KRISTAN, U.Trias.(Rhaet.), Eu.(Aus.); *5*, umbilical view; *6*, edge view of different specimen; *7*, axial sec.; all ×24 (*1057).

[WICHER (1952, *2058, p. 275) showed a stratigraphic change in ornamentation in *Trocholina* with oldest forms possessing fewer and larger ventral pillars or nodes, and progressively younger species showing more numerous but smaller nodes. *Coscinoconus* LEUPOLD in LEUPOLD & BIGLER (1936, *1130, p. 618), regarded as a synonym of *Trocholina* by HENSON (1948, *900, p. 449) and WICHER (1952, *2058, p. 273) was shown by MASLOV (1958, *1232, p. 546) to belong to the Dasycladaceae (algae). HENSON and WICHER placed *Trocholina* in the subfamily Cornuspirinae (=Cyclogyrinae) and CUSHMAN (1948, *486, p. 284) placed it in the subfamily Turrispirillininae, family Rotaliidae. BERMÚDEZ assigned it to the Spirillininae, family Spirillinidae, though stating (*127, p. 29), "Probablemente el genero *Trocholina* estaria mejor situado en la subfamilia Rotaliinae de la familia Rotaliidae." *Trocholina* has a microgranular wall, and therefore cannot be placed with *Spirillina*, which has walls composed of a single crystal or several crystals of calcite, or with *Rotalia*, which has a radial and canaliculate wall structure.]

Family NONIONIDAE Schultze, 1854

[*nom. correct.* CUSHMAN, 1927, p. 49 (*pro* family Nonionida SCHMARDA, 1871, p. 165)]——[All names cited are of family rank; dagger(†) indicates *partim*]——[=Polythalama† LATREILLE, 1825, p. 161 (*nom. nud.*); =Hélicostèguest D'ORBIGNY, 1826, p. 268 (*nom. neg., nom. nud.*); =Nautiloida† SCHULTZE, 1854, p. 53 (*nom. nud.*); =Cryptostegia REUSS, 1862, p. 320, 372 (*nom. nud.*); =Nautiloidea† REUSS, 1860, p. 151 (*nom. nud.*)]——[=Nonioninidae REUSS, 1860, p. 151; =Nonioninideae REUSS, 1860, p. 221; =Nonionidea COPELAND, 1956, p. 187 (*nom. van.*)]——[=Chilostomellidae BRADY, 1881, p. 42, 44; =Chilostomellida HAECKEL, 1894, p. 185; =Quilostomélidos GADEA BUISÁN, 1947, p. 18 (*nom. neg.*)]

Test planispiral or trochospiral; finely perforate; aperture interiomarginal or areal. *Jur.-Rec.*

Subfamily CHILOSTOMELLINAE Brady, 1881

[*nom. transl.* A. SILVESTRI, 1906, p. 12 (23) (*ex* family Chilostomellidae BRADY, 1881)]——[=All names cited are of subfamily rank; dagger(†) indicates *partim*]——[=Cryptostegia BÜTSCHLI in BRONN, 1880, p. 203 (*nom. nud.*); =Allomorphininae CUSHMAN, 1927, p. 85; =Allomorphinellinae CUSHMAN, 1927, p. 86]

Test trochospiral, with few chambers to whorl, planispiral and involute; aperture interiomarginal on umbilical side. *Jur.-Rec.*

Chilostomella REUSS in CŽJŽEK, 1849, *546, p. 50 [*C. ovoidea* REUSS, 1850, *1540, p. 380; SD CUSHMAN, 1914, *404d, p. 2]. Test free, ovate,

FIG. 610. Involutinidae; *1-7, Trocholina* (p. C742).

planispiral and involute, with 2 chambers to whorl, chambers embracing; wall calcareous, perforate, granular in structure; aperture a narrow interiomarginal equatorial slit, which may have slight lip. U.Cret.-Rec., cosmop.——FIG. 611,*1*. **C. ovoidea*, Mio., Eu.(Aus.); *1a,b*, side, apert. views, ×63 (*2117).

Allomorphina REUSS in CŽJŽEK, 1849, *546, p. 50 [**A. trigona* REUSS, 1850, *1540, p. 380; SD (SM)]. Test trochospiral, commonly 3 chambers to whorl, involute, only final whorl visible externally; wall calcareous, perforate, granular in structure; aperture an elongate slit, paralleling suture and bordered with slight lip. [As here restricted, *Allomorphina* includes involutely coiled species and *Quadrimorphina* trochospiral species with early coil visible at one side of the test. The number of chambers to a whorl is regarded as a specific character only.] *Jur.-Rec.*, cosmop.—— FIG. 611,*3*. **A. trigona*, Mio., Eu.(Aus.); *3a-c*, side, edge, and apert. views, ×78 (*2117).

Allomorphinella CUSHMAN, 1927, *431, p. 86 [**Allomorphina contraria* REUSS, 1851, *1542, p. 43; OD]. Test free, planispiral, periphery rounded; chambers few to whorl, involute, increasing rapidly in size; wall calcareous, perforate, aperture an elongate, narrow interiomarginal equatorial slit. [Differs from *Allomorphina* in being planispiral, rather than trochospiral and involute.] *U. Cret.*, Eu.——FIG. 611,*2*. **A. contraria* (REUSS), U.Cret., Pol.; *2a-c*, side, edge, and top views, approx. ×60 (*700).

Chilostomelloides CUSHMAN, 1926, *427, p. 77 [**Lagena (Obliquina) oviformis* SHERBORN & CHAPMAN, 1886, *1732, p. 745; OD]. Test free, ovate in outline, adult with 2 chambers visible, latest formed almost completely embracing former; sutures oblique; wall calcareous, finely perforate; aperture offset from general contour of test, situated near suture line, circular, with slight bordering rim or lip. *Paleoc.-Mio.*, Eu.-N.Am.-Carib.-Afr.——FIG. 611,*4*. **C. oviformis* (SHERBORN & CHAPMAN), Eoc., Eng.; *4a,b*, side, edge views of lectotype, ×48 (*2117).

[Differs from *Chilostomella* in having a small, rounded, protruding aperture instead of a long, narrow slit. Wall structure and internal characters are unknown. A lectotype is here designated (BMNH-P41673) and paratypes (BMNH-P3648) from the London clay, in drainage works, Pic-

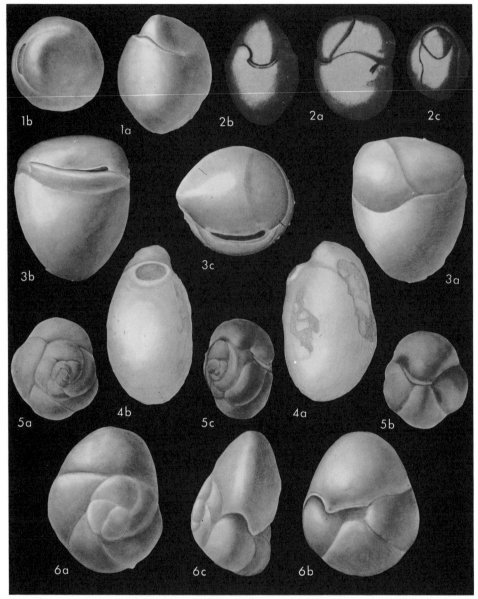

FIG. 611. Nonionidae (Chilostomellinae; *1, Chilostomella; 2, Allomorphinella; 3, Allomorphina; 4, Chilostomelloides; 5,6, Quadrimorphina*) (p. C742-C745).

cadilly, London, England. The lectoype is that originally figured by SHERBORN & CHAPMAN (*1732, pl. 14, fig. 19b).]

Quadrimorphina FINLAY, 1939, *717c, p. 325 [*Valvulina allomorphinoides REUSS, 1860, *1548, p. 223; OD] [=*Gyromorphina* MARIE, 1941, *1215, p. 230, 256 (obj.); *Pallaimorphina* TAPPAN, 1957, *1875, p. 220 (type, *P. ruckerae*)]. Test trochospiral, periphery rounded, all chambers visible on spiral side, commonly 3 or 4 chambers in final whorl; wall calcareous, finely perforate, granular in structure, surface smooth; aperture interiomarginal, umbilical or extraumbilical in position and partially covered by projecting umbilical flap. L.Cret.(Alb.)-Rec., Eu.-N.Am.-N.Z.——FIG. 611,5. *Q. ruckerae* (TAPPAN), L.Cret.(Alb.), Alaska; *5a-c,* opposite sides and edge view of paratype, ×75 (*1875).——FIG. 611,6. *Q. allomorphinoides* (REUSS), U.Cret., USA(Tenn.); *6a-c,* opposite sides and edge view, ×176 (*2117).

[*Pallaimorphina* was defined originally for a primitive species which does not show rapid chamber enlargement and had a narrower umbilical flap. As these features are

Foraminiferida—Rotaliina—Cassidulinacea

only of degree, the genera are regarded as synonymous. As here defined, *Quadrimorphina* also includes trochospiral species with evolute spiral side that had previously been placed in *Allomorphina*. *Allomorphina* is restricted to involute species, where only the final whorl of chambers is visible externally.]

Subfamily NONIONINAE Schultze, 1854

[*nom. correct.* CHAPMAN & PARR, 1936, p. 145 (*pro* subfamily Nonionida SCHULTZE, 1854, p. 53)]——[All names cited are of subfamily rank]——[=Pullenidae SCHWAGER, 1877, p. 18; =Pulleninae BÜTSCHLI in BRONN, 1880, p. 210; =Nonionininae A. SILVESTRI, 1950, p. 52; =Nonionellinae VOLOSHINOVA, 1958, p. 141]

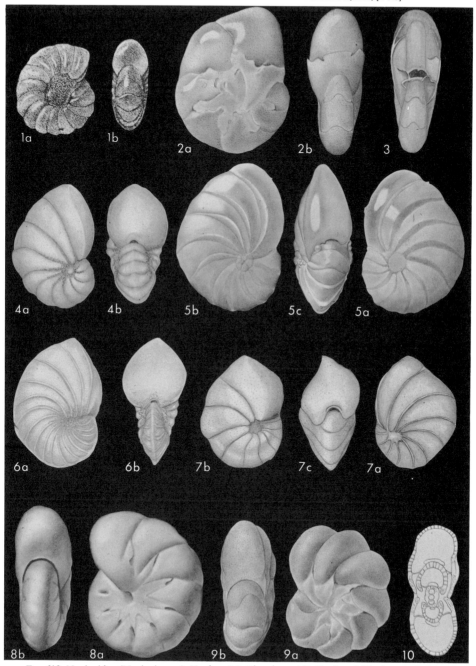

FIG. 612. Nonionidae (Nonioninae; *1, Nonion; 2,3, Bisaccium; 4-7, Florilus; 8-10, Astrononion*) (p. *C745-C748*).

Test planispiral and involute to slightly trochospiral; aperture interiomarginal and equatorial, or areal and multiple. *U.Cret.-Rec.*

Nonion DE MONTFORT, 1808, *1305, p. 210 [*Nautilus incrassatus* FICHTEL & MOLL, 1798, *716, p. 38; OD] [=*Nonionia* RISSO, 1826, *1579a, p. 22 (type, *Nautilus incrassatus* FICHTEL & MOLL, 1798, herein designated, obj.] Test free, planispiral and involute, slightly compressed, biumbonate, periphery rounded, peripheral outline lobulate; chambers numerous, increasing gradually in size as added; sutures distinct, depressed, radial, slightly curved; wall calcareous, finely perforate, granular in structure, surface smooth, umbonal region filled with secondarily deposited calcite, either as granules or solid boss; aperture an arched, equatorial, interiomarginal slit. [Differs from *Melonis* in having a filled, rather than open, umbilicus, thinner and more finely perforate and monolamellar walls (*1530). Many species previously included in *Nonion* should be referred to *Melonis* or *Florilus*.] ?*U.Cret., Paleoc.-Rec.*, cosmop.——FIG. 612,*1*. *N. incrassatus* (FICHTEL & MOLL), Mio., Albania; *1a,b,* side, edge views, ×100 (*2019).

Astrononion CUSHMAN & EDWARDS, 1937, *503, p. 30 [*Nonionina stelligera* D'ORBIGNY in BARKER-WEBB & BERTHELOT, 1839, *86, p. 128; OD]. Test free, planispiral and involute, umbilical region slightly excavated, peripheral margin rounded; chambers increasing gradually in size, each with backward-projecting, nonporous, umbilical flap which partially covers preceding suture and umbilical region but leaves small cavity open beneath it, giving appearance of secondary chamberlets; sutures radial, depressed, slightly curved; wall calcareous, finely perforate, granular in structure, monolamellar, surface smooth; aperture a low, interiomarginal, equatorial slit, with openings at outer edge of umbilical fillings along their sutural extension. [Differs from *Nonion* and *Florilus* in its more highly developed umbilical and sutural filling, with included cavities suggesting "chamberlets," although these nonporous fillings are not comparable to true chamberlets.] *Eoc.-Rec.*, cosmop.——FIG. 612,*8*. *A. stelligera* (D'ORBIGNY), Rec., E.Atl.O.(Canary Is.); *8a,b,* side, edge views, ×125 (*2117).——FIG. 612,*9*. *A. gallowayi* LOEBLICH & TAPPAN, Rec., N.Am. (Alaska); *9a,b,* side and edge views showing finely perforate walls and nonporous umbilical flaps, ×75 (*1162).——FIG. 612,*10*. *A. sidebottomi* CUSHMAN & EDWARDS, Rec., Medit.; axial sec. showing cavities below nonperforate chamber flaps, enlarged (*946).

Bisaccium ANDERSEN, 1951, *15, p. 32 [*B. imbricatum*; OD]. Test planispiral, bilaterally symmetrical, periphery rounded; chambers gradually enlarging, umbilical region and sutures of both sides covered by chamber extensions as in *Astrononion*, but more extensive, extending also across base of terminal chamber face to obscure aperture; sutures radial, slightly curved, depressed; wall calcareous, very thin, finely perforate, surface smooth and unornamented except for umbilical-sutural chamber extensions; apertural foramen an interiomarginal equatorial arch, but obscured in final chamber by secondary chamber flaps, communication of chamber cavities to exterior by means of openings along upper, lower, and peripheral sutural margins of secondary chamber flaps. *Rec.*, USA(La.).——FIG. 612,*2,3*. *B. imbricatum*; *2a*, side view of holotype, ×78 (*2117); *2b*, edge view, ×78 (*15); *3*, edge view of paratype with secondary covering removed to show equat. apert. foramen, ×78 (*15).

Chilostomellina CUSHMAN, 1926, *427, p. 78 [*C. fimbriata*; OD]. Test free, inflated to subglobular, planispiral throughout, and involute; chambers increasing rapidly in size, final one almost completely enveloping test and overlapping umbilical region on each side, with fimbriate margin at sides and base of apertural face; sutures slightly curved; wall calcareous, finely perforate, granular in structure; aperture a low, interiomarginal arch, with additional supplementary apertures at re-entrants between finger-like projections of final chamber. [Differs from *Chilostomella* in being planispiral throughout development, and in the final chamber having a fimbriate margin with supplementary apertures. It differs from *Nonion* in its multiple aperture and fimbriate final chamber margin, and in its inflated, subglobular test.] *Rec.*, Bering Sea-Pac.O.——FIG. 613,*11*. *C. fimbriata*, Pac.; *11a-d*, side, apert., peripheral and basal views, ×48 (*1162).

Cribropullenia THALMANN, 1937, *1898, p. 351 [*Nonion? marielensis* PALMER, 1936, *1409, p. 127; OD] [=*Antillesina* GALLOWAY & HEMINWAY, 1941, *764, p. 366 (obj.)]. Test free, planispiral and involute, close-coiled, periphery broadly rounded; chambers few to whorl, inflated; sutures radial, depressed; wall calcareous, finely perforate, wall structure and lamellar character not described, surface with low spiraling costae; aperture consisting of small openings near base of apertural face. [Although the surface ornamentation is not characteristic of the Nonionidae, the genus is retained here until information as to wall structure and lamellar character is available.] *Eoc.-Oligo.*, Carib.-Afr.(Egypt).——FIG. 613,*7*. *C. marielensis* (PALMER), Oligo., Cuba; *7a,b,* side, edge views of lectotype (here designated, USNM-498778), ×119 (*2117).

Florilus DE MONTFORT, 1808, *1305, p. 134 [*F. stellatus* (nom. subst. pro *Nautilus asterizans* FICHTEL & MOLL, 1798, *716, p. 37; OD] [=*Nonionina* D'ORBIGNY, 1826, *1391, p. 293 (type, *Nautilus asterizans* FICHTEL & MOLL, 1798, *716, p. 37, SD PARKER & JONES, 1863, *1417f,

p. 433); *Pseudononion* Asano, 1936, *47, p. 347 (type, *P. japonicum*); *Azera* Khalilov, 1958, *1038, p. 6 (type, *A. transversa*)]. Test free, planispiral, but may be asymmetrical, involute, but with broad, low chambers increasing rapidly in breadth and thickness resulting in flaring test,

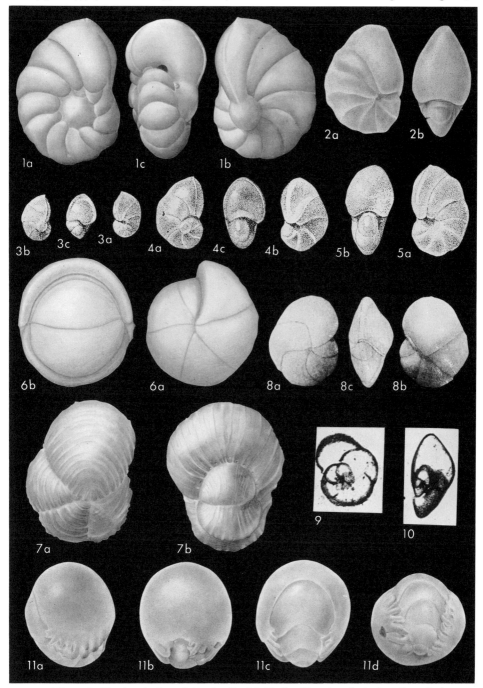

Fig. 613. Nonionidae (Nonioninae; *1, Nonionella; 2-5, Nonionellina; 6, Pullenia; 7, Cribropullenia; 8-10, Spirotecta; 11, Chilostomellina*) (p. C746, C748).

peripheral margin rounded to angled, umbilical region slightly depressed, filled with granular skeletal material which may extend slightly along sutures; wall calcareous, finely perforate, granular in structure, single-layered; aperture a narrow, interiomarginal, equatorial opening. [Differs from *Nonion* in the flaring test, due to the numerous broad low chambers, and from *Nonionella* in lacking the single umbilical chamber extension.] *Paleoc.-Rec.,* cosmop.——Fig. 612,4. **F. asterizans* (Fichtel & Moll), Rec., Eu.(Italy); *4a,b,* side, edge views, ×48 (*2117).——Fig. 612,5. *F. japonicus* (Asano), Plio., Japan; *5a-c,* opp. sides and edge view, ×71 (*2117).——Fig. 612,6. *F. costiferus* (Cushman), Mio., USA(Calif.); *6a,b,* side, edge views, ×48 (*2117).——Fig. 612,7. *F. transversus* (Khalilov), U.Eoc., USSR(Azerbaidzhan); *7a-c,* opp. sides and edge view of holotype, ×30 (*1038).

Nonionella Cushman, 1926, *426, p. 64 [**N. miocenica;* OD] [=*Nonionella* Rhumbler in Anonymous, 1949, *22, p. 40 (type, *N. aberrans*) *(nom. nud.)*]. Test free, trochospiral, slightly compressed, periphery rounded, spiral side partially evolute with umbonal boss, opposite side involute with final chamber overhanging umbilical region and may appear to form distinct umbilical flap; chambers relatively numerous, broad, low; wall calcareous, perforate, granular in structure; aperture interiomarginal, a low arch near periphery extending somewhat onto umbilical side. [Differs from *Nonion* in being asymmetrical and trochospiral, and in possessing an asymmetrically developed final chamber.] *U.Cret.-Rec.,* cosmop.—— Fig. 613,1. **N. miocenica,* Mio., USA(Calif.); *1a-c,* opp. sides and edge view, ×101 (*2117).

Nonionellina Voloshinova, 1958, *2019, p. 142 [**Nonionina labradorica* J. W. Dawson, 1860, *566, p. 191; OD]. Test free, trochospiral in early coiling, later becoming planispiral and involute; chambers enlarging rapidly around deep umbilicus; wall calcareous, finely perforate, granular in structure; aperture a low slit at base of apertural face. [*Nonionellina* is like *Nonionella* in the juvenile stages and like *Nonion* in the adult.] *Mio.-Rec.,* Eu.-N.Am.-Atl. O.-Japan-Pac. O.——Fig. 613,2-5. **N. labradorica* (J. W. Dawson), Rec., Alaska (2), Sea of Okhotsk (3-5); *2a,b,* side, edge views, ×36 (*1162); *3a-c,* opp. sides and edge view of juvenile specimen showing asymmetry; *4a-c,* somewhat older, more symmetrical specimen; *5a,b,* nearly adult specimen; *3-5,* ×50 (*2019).

Pullenia Parker & Jones in Carpenter, Parker & Jones, 1862, *281, p. 184 [**Nonionina bulloides* d'Orbigny, 1846, *1395, p. 107, =*Nonionina sphaeroides* d'Orbigny, 1826, *1391, p. 293 *(nom. nud.);* OD (M)]. Test free, spheroidal to compressed, planispiral and involute; chambers few, 3 to 6 in final whorl; sutures radial; wall calcareous, finely perforate, granular in structure; aperture a narrow crescentic interiomarginal slit extending nearly from umbilicus on one side to that opposite. *U.Cret.-Rec.,* cosmop.——Fig. 613, 6. **P. bulloides* (d'Orbigny), Mio., Eu.(Aus.); *6a,b,* side, edge views of lectotype, ×79 (*2117).

[Parker & Jones in Carpenter, Parker & Jones (1862, *281) described *Pullenia* as including the "form which has been represented by M. D'Orbigny (Modèles, No. 43) under the name of *Nonionina sphaeroides,* and has been subsequently described by him under the name of *N. bulloides.*" Both here and in later publications, Parker & Jones considered the two specific names synonymous. The plate in Carpenter, Parker & Jones (*281) cited the species under the name *Pullenia bulloides,* but nowhere in this original reference is *sphaeroides* cited in combination with *Pullenia.* This would therefore appear to be original designation of *bulloides* as type-species. Most later workers have also regarded *N. bulloides* as the type-species of *Pullenia* by original designation, although Cushman (1948, *486, p. 320) stated the type to be "*Nonionina sphaeroides* d'Orbigny, but, as that species is indeterminable, it seems better to use *Nonionina bulloides* d'Orbigny." Both specific names were cited by d'Orbigny in 1826 without description or illustration. *Nonionina sphaeroides* was included in the Modèles (No. 43) in 1826, but no description was ever given by d'Orbigny. The specimens came from ship ballast sand and the type locality and horizon are unknown. This species is not only unrecognizable but was a *nomen nudum;* hence *N. bulloides* must be the type by monotypy (only valid species originally included). *Nonionina bulloides* was well figured and described by d'Orbigny from the Vienna Basin (*1395). A lectotype from the Miocene of Nussdorf, Vienna Basin, is here designated and refigured; it is in the d'Orbigny collection in the Muséum National d'Histoire Naturelle, Paris.]

Spirotecta Belford, 1961, *111, p. 81 [**S. pellicula;* OD]. Test free, inequally biconvex, trochospiral but completely involute throughout, with closed umbilicus on both sides, periphery narrowly rounded; chambers few, increasing gradually in size; sutures curved on less convex side, nearly straight and radial on more convex side; wall calcareous, finely perforate, granular in structure, septa single, monolamellar; aperture an interiomarginal equatorial arch with lower extension to umbilicus of more convex side, bordered with thin lip. [Originally placed questionably in the Chilostomellidae, *Spirotecta* is here transferred to the Nonionidae.] *U.Cret.(Campan.-Maastricht.),* W.Australia.——Fig. 613,8-10. **S. pellicula,* Maastricht. *(8,9),* Campan. *(10); 8a-c,* opp. sides and edge view of holotype; *9,* horiz. sec. showing single septal walls; *10,* axial sec. showing completely involute but slightly trochospiral coiling; all ×48 (*111).

Family ALABAMINIDAE Hofker, 1951

[Alabaminidae Hofker, 1951, p. 389]——[In synonymic citations dagger(†) indicates *partim*]——[=Turbinoidae† d'Orbigny in de la Sagra, 1839, p. xxxviii, 71 *(nom. nud.)*]

Test lenticular, trochospiral; wall calcareous, perforate, granular, septa single-layered (monolamellar); aperture basal, or a slit extending up apertural face, or both. *U.Cret.-Rec.*

Alabamina Toulmin, 1941, *1944, p. 602 [**A. wilcoxensis;* OD] [=*Eponidoides* Brotzen, 1942, *240, p. 38 (type, *Eponides dorsoplana* Brotzen, 1940, *239, p. 31)]. Test free, lenticular, trochospiral, periphery subangular, with nonporous margin, all chambers visible on spiral side where

Foraminiferida—Rotaliina—Cassidulinacea C749

curved sutures are strongly oblique, only final whorl visible on opposite side where sutures are nearly radial around umbilical depression, chambers somewhat prolonged into projection at

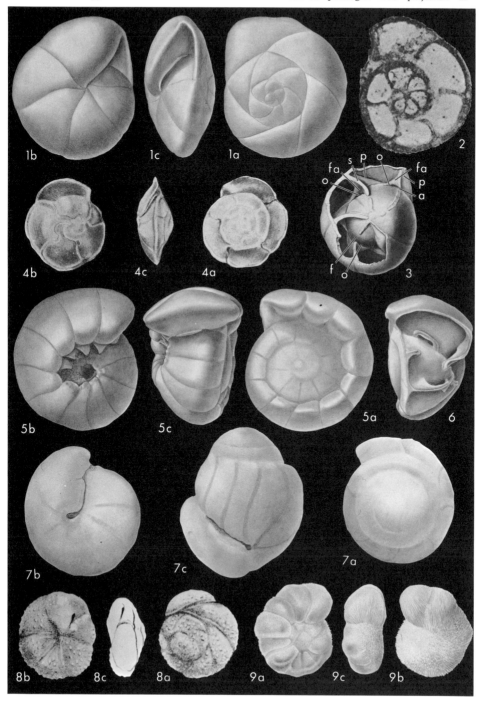

Fig. 614. Alabaminidae; *1-3, Alabamina; 4, Oridorsalis; 5,6, Gyroidina; 7, Rotaliatina; 8, Svratkina; 9, Trichohyalus* (p. C748-C751).

periphery on spiral side, apertural face sharply infolded below this projection; wall calcareous, finely perforate, granular in structure, with simple monolamellar septa; aperture an interiomarginal slit extending from near periphery almost to umbilicus, with narrow bordering lip. [HOFKER (1959, *951) regarded *Alabamina* as a synonym of *Eponides*, but *Alabamina* has a granular monolamellar wall, whereas that of *Eponides* is radial and bilamellar. The aperture of *Alabamina* is in an infolded area of the apertural face, unlike that of *Eponides*.] *U.Cret.(Santon.)-Rec.*, cosmop.——FIG. 614,*1,2*. **A. wilcoxensis*, L.Eoc., USA (Ala.); *1a-c*, opposite sides and edge view, ×140 (*2117); *2*, horiz. sec. showing monolamellar septa, ×100 (*1533).——FIG. 614,*3*. *A.* sp., Paleoc.(Dan.), Sweden; dissected specimen from umbilical side showing indentation of septal face (*fa*) and opening beneath it (*o*), peripheral projection of chambers (*p*), aperture (*a*), septum (*s*), and septal foramina (*f*), enlarged (*241).

Gyroidina D'ORBIGNY, 1826, *1391, p. 278 [**G. orbicularis*; SD CUSHMAN, 1927, *433, p. 190]. Test free, trochospiral, periphery rounded to subtruncate, spiral side flattened with all chambers visible, opposite side elevated and umbilicate with only chambers of final whorl visible; chambers rhomboidal in section, with angled umbilical shoulder; sutures radial to oblique, flush to depressed; wall calcareous, perforate, granular in structure, lamellar character unknown; primary aperture a low interiomarginal slit restricted to mid-portion of apertural face, bordered by narrow lip, small secondary apertures umbilical in position, against previous chamber wall with projecting umbilical flap extending backward over it, so that it is not evident except when test is viewed obliquely, or when final chamber is dissected so that secondary aperture may be seen. *Eoc.-Rec.*, cosmop.——FIG. 614,*5,6*. **G. orbicularis*, Rec., Eu.(Italy); *5a-c*, opp. sides and edge view; *6*, edge view of dissected specimen showing foramen; all ×74 (*2117).

[*Gyroidina* has never been completely described or well illustrated. The type-species was originally represented by one of D'ORBIGNY's models (figured later by other workers) but this model does not show details of the umbilical region. The umbilical flaps have been figured in some species of *Gyroidina* but not previously in the type-species. Their true character as lips over secondary umbilical apertures has not been noted. The presence of umbilical flaps was used by BROTZEN (1942, *240, p. 19) as a basis for distinguishing *Gyroidinoides* from *Gyroidina*, as the latter genus was erroneously said not to possess these structures. Although this basis is not valid, the genera are otherwise separable on apertural features. *Gyroidina* has a short, slitlike, primary interiomarginal aperture near the midline of the apertural face, and a secondary aperture opening from the chamber into the umbilicus, partially covered by an arched umbilical flap. *Gyroidinoides* has a single, more extensive interiomarginal aperture, extending from the periphery along the entire margin of the chamber to the umbilicus. It is partly covered by an umbilical chamber extension, but the umbilical flaps do not attach below, and the apertural opening is continuous beneath the flap.]

Oridorsalis ANDERSEN, 1961, *18, p. 107 [**O. westi*; OD]. Test free, lenticular, periphery carinate; chambers arranged in low trochospiral coil, chambers broad, low, all visible on spiral side but only those of final whorl visible on opposite side; sutures radial, slightly curved on spiral side, strongly sinuate on umbilical side; wall calcareous, very finely perforate, granular in structure; primary aperture interiomarginal, extending from periphery nearly to closed umbilicus of umbilical side; small secondary sutural openings on spiral side near junction of spiral and septal sutures, with similar small sutural openings at mid-point of sutures at sinuate curve on umbilical side. *Oligo.-Rec.*, N.Am.-Eu.-Japan-Carib.——FIG. 614,*4*. **O. westi*, Rec., USA(La.); *4a-c*, opp. sides and edge view of holotype, ×60 (*18).

[The genus was originally placed in the Discorbidae but is here transferred to the Alabaminidae because of its granular wall structure. Although the secondary sutural openings on the umbilical side were not reported in the original description, they are present in specimens obtained by us from the Miocene of Jamaica, and also appear to be indicated in the figures of the holotype. Some species previously placed in *Pseudoeponides* probably should be referred to *Oridorsalis*.]

Rotaliatina CUSHMAN, 1925, *421, p. 4 [**R. mexicana*; OD]. Test free, high trochospiral, with rounded periphery; all chambers visible on elevated spiral side, only those of final whorl visible around small, deep umbilicus on opposite side; sutures radial, nearly straight; wall calcareous, finely perforate, surface smooth, lamellar character and microstructure not described; aperture an elongate interiomarginal slit, extending from near periphery to open umbilicus, with narrow bordering lip. *Eoc.-Oligo.*, N.Am.——FIG. 614,*7*. **R. mexicana*, U.Eoc., Mex.; *7a-c*, opp. sides and edge view, ×65 (*2117).

Svratkina POKORNÝ, 1956, *1477, p. 257 [**Discorbis tuberculata* (BALKWILL & WRIGHT) var. *australiensis* CHAPMAN, PARR, & COLLINS, 1934, *326, p. 563; OD]. Test free, trochospiral, biconvex, periphery rounded, all chambers visible and sutures oblique and curved on spiral side, only final whorl visible and sutures radial on opposite side, umbilicus closed; wall calcareous, coarsely perforate, with large pores opening at ends of tubercles, lamellar character and wall structure not described; aperture an elongate opening extending from near umbilicus up face of chamber in slight depression, nearly to periphery. [Similar to *Alabamina* in apertural character, but characterized by large pores opening into tubercles at the surface.] ?*U.Cret., U.Eoc.-Rec.*, Australia-Eu.-N.Am.——FIG. 614,*8*. **S. australiensis* (CHAPMAN, PARR, & COLLINS), Oligo., Australia; *8a-c*, opp. sides and edge view, ×139 (*326).

Trichohyalus LOEBLICH & TAPPAN, 1953, *1162, p. 116 [**Discorbis bartletti* CUSHMAN, 1933, *457, p. 6; OD]. Test free, trochoid, plano-convex, all whorls visible on spiral side, umbilical side obscured by secondary growth of shell material, forming vesicular plate extending nearly to the periphery, perforations through this vesicular tissue opening into cavity beneath, exterior of plate variously ornamented; wall calcareous, coarsely perforate-granular in structure; no visible aperture

Foraminiferida—Rotaliina—Cassidulinacea C751

on final chamber, but interiomarginal intercameral openings occur on umbilical side near outer margin of chambers, which may be seen by dissection.

Rec., Arctic.——Fig. 614,9. *T. bartletti* (Cushman), Can.(Fox Basin); *9a-c,* opp. sides and edge view, ×17 (*1162).

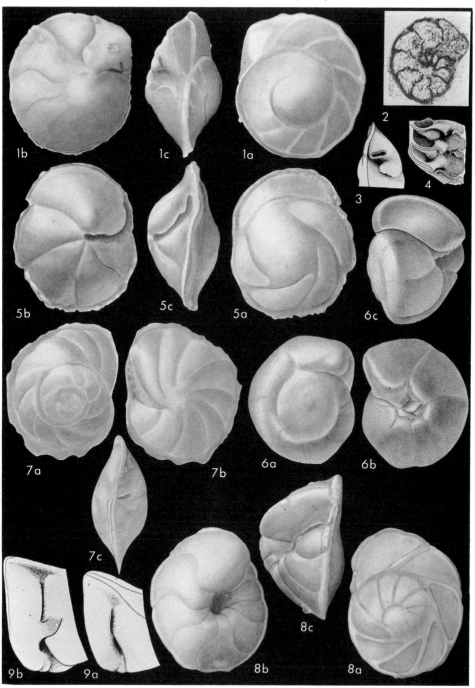

Fig. 615. Osangulariidae; *1-4, Osangularia; 5, Charltonina; 6, Gyroidinoides; 7, Cribroparrella; 8,9, Globorotalites* (p. C752-C753).

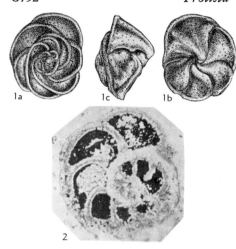

Fig. 616. Osangulariidae; *1,2, Conorotalites* (p. C752).

Family OSANGULARIIDAE
Loeblich & Tappan, n.fam.

Test trochospiral; wall calcareous, perforate granular in structure, bilamellar; aperture with interiomarginal portion and vertical or oblique portion extending up apertural face, 2 parts may be joined, or distinct, separate openings, and areal opening may be multiple. *L.Cret.-Rec.*

Osangularia BROTZEN, 1940, *239, p. 30 [*O. lens*; OD] [=*Parrella* FINLAY, 1939, *717a, p. 523 (type, *Anomalina bengalensis* SCHWAGER, 1866, *1703, p. 259) (non *Parrella* GINSBURG, 1938)]. Test free, trochospiral, lenticular, biumbonate, periphery carinate; all whorls visible on spiral side, only final whorl visible on opposite side, chambers increasing gradually in size, sutures curved and oblique on spiral side, radial and sinuate on umbilical side; wall calcareous, finely perforate, granular in structure, bilamellar; aperture a bent opening, lying along base of final chamber on umbilical side and bending at oblique angle up apertural face, or 2 angles may be separated openings, one interiomarginal and one areal. *L.Cret.-Rec.*, cosmop.——FIG. 615,*1,2*. *O. lens*, Paleoc.(Dan.), Sweden; *1a-c*, opp. sides and edge view, ×111 (*2117); *2*, horiz. sec., areal apert. openings visible in septa at lower left of figure, ×55 (*1530).——FIG. 615,*3,4*. *O. bengalensis* (SCHWAGER), Mio., Asia(Israel); *3*, apert. view showing areal opening; *4*, interior, from umbilical side with wall removed, showing bulging chamber ends, areal aperture and lip, and apert. face indentation extending to previous septum; all ×45 (*1533).

Charltonina BERMÚDEZ, 1952, *127, p. 69 [*Pseudoparrella madrugaensis* CUSHMAN & BERMÚDEZ, 1948, *496, p. 73; OD]. Test trochospiral, lenticular to inequally biconvex, periphery carinate; all chambers visible and sutures strongly oblique on spiral side, only final whorl visible and sutures radial on opposite side, umbilicus closed; wall calcareous, perforate, lamellar character and wall structure not described; aperture an elongate slit extending from umbilicus to periphery and bending up apertural face, parallel to peripheral keel. *U.Cret.-Paleoc.*, Carib.(Cuba).——FIG. 615,*5*. **P. madrugaensis* (CUSHMAN & BERMÚDEZ), Paleoc.; *5a-c*, opp. sides and edge view of holotype, ×108 (*2117).

Conorotalites KAEVER, 1958, *1007, p. 435 [*Globorotalites bartensteini aptiensis* BETTENSTAEDT, 1952, *137, p. 282; OD]. Test plano-convex or inequally biconvex with much elevated umbilical side, periphery acute and carinate; sutures distinct, may be limbate, curved and oblique on spiral side, nearly radial, curved to sinuate, flush or depressed on umbilical side around pseudoumbilicus, umbilical shoulder angular; wall calcareous, coarsely perforate, granular in structure, bilamellar; aperture similar to *Globorotalites* with narrow interiomarginal aperture, and deeply indented murus reflectus which gives appearance of second opening. [*Conorotalites* differs from *Globorotalites* in its coarsely perforate wall.] *L.Cret.(Barrem.-Alb.)*, Eu.——FIG. 616,*1,2*. **C. aptiensis* (BETTENSTAEDT), L.Apt., Ger.; *1a-c*, opp. sides and edge view of holotype, ×50 (*137); *2*, equat. sec., ×78 (*1533).

Cribroparrella TEN DAM, 1948, *556, p. 487 [*nom. imperf., nom. correct.* THALMANN, 1949, *1897h, p. 653] [**C. regadana*; OD] [=*Cribroparella, Dribroparella* TEN DAM, 1948,*556, p.486, pl. expl. (*nom. null.*)]. Test free, trochospiral, biconvex, periphery carinate; chambers numerous, broad, low, with oblique, curved septa and all chambers visible on spiral side, only final whorl with radial curved septa visible on umbilical side, umbilicus closed and umbonate; wall calcareous, finely perforate, granular in structure, bilamellar; aperture a narrow oblique slit near base of apertural face, with projecting lip and smaller supplementary circular areal openings occur over entire apertural face. *Mio.* N.Afr.(Algeria)-W.Indies(Jamaica).——FIG. 615,*7*. **C. regadana*, Mio., Algeria; *7a-c*, opposite sides and edge view of paratype, ×51 (*2117).

Globorotalites BROTZEN, 1942, *240, p. 31 [*Globorotalia multisepta* BROTZEN, 1936, *237, p. 161; OD]. Test free, trochospiral, plano-convex, spiral side flat or slightly concave or convex, umbilical side strongly convex, periphery carinate, with poreless keel; chambers increasing gradually in size, sutures oblique, thickened on spiral side, radial and curved or sinuate, depressed on umbilical side, which has broad pseudoumbilicus and angular umbilical shoulder, deep indentation of base of apertural face below aperture (murus reflectus) attached to previous septum and externally resembles aperture although it does not communicate with chamber interior; wall calcareous, finely perforate, granular in structure, bilamellar; aperture interiomarginal, on umbilical side, midway between umbilicus and periphery, deeply indented murus reflectus below aperture falsely appears to form second opening. *Cret.(Alb.-Maastricht.),* Eu.——Fig. 615,*8*. **G. multisepta* (Brotzen), U.Cret.(Coniac.), Sweden; *8a-c,* opp. sides and edge view of syntype, ×93 (*2117).——Fig. 615,*9*. *G. micheliniana* (d'Orbigny), U.Cret.(Campan.), Fr.; *9a,* apert. view showing aperture and indentation forming murus reflectus below it; *9b,* chamber interior showing broken murus reflectus; both ×90 (*1533).

Goupillaudina Marie, 1958, *1222b, p. 861 [**G. daguini;* OD] [=*Goupillaudina* Marie, 1957, *1222a, p. 247 *(nom. nud.)*]. Test free, lenticular to operculine, slightly trochospiral, early stage involute, later partially evolute on both sides, compressed, periphery acute; chambers numerous, broad, low, strongly arched; sutures strongly curved and oblique; wall calcareous, finely perforate, microstructure and lamellar character unknown; aperture interiomarginal, connecting with deep spiroumbilical suture and extending from umbilical region to periphery, then bending to extend up apertural face as in *Charltonina*. *U.Cret. (Coniac.-Maastricht.),* Eu.——Fig. 617,*1,2*. **G. daguini,* U.Campan., Fr.; *1a-c,* opp. sides and edge view of paratype, ×20; *2a-c,* opp. sides and edge view of holotype, ×20; *2d,* apert. detail, ×72 (*1222b).

Gyroidinoides Brotzen, 1942, *240, p. 19 [**Rotalina nitida* Reuss, 1844, *1537, p. 214; OD]. Test free, trochospiral, spiral side flattened, umbilical side elevated, periphery rounded; chambers rhomboidal in section, sutures radial to curved, flush to depressed; wall calcareous, perforate, bilamellar, granular in structure; aperture a continuous, low, interiomarginal slit extending from periphery to umbilicus, umbilical portion partially obscured by umbilical flap from each chamber. *Cret.-Rec.,* cosmop.——Fig. 615,*6*. **G. nitida* (Reuss), U.Cret.(Turon.), Sweden; *6a-c,* opp. sides and edge view, ×74 (*2117).

[Because of general misconception as to the characters of *Gyroidina,* forms with an open umbilicus and apertural lips were separated as *Gyroidinoides.* However, both of these morphological features occur in the type-species of *Gyroidina,* hence do not afford a valid distinction. The present genus was separated from *Gavelinella* as having a narrow umbilicus, high umbilical side, and reduced umbilical aperture. As here redefined on the basis of the type-species, *Gyroidinoides* differs from *Gyroidina* in having a single, continuous apertural opening from the periphery to the umbilicus, whereas *Gyroidina* has a restricted primary aperture at the mid-portion of the apertural face, and a secondary apertural opening into the umbilicus lying against the preceding chamber wall. The umbilical flap in *Gyroidinoides* is an extension of the chamber but does not divide the apertural opening. In *Gyroidina* it consists of an arched lip over the secondary aperture of each chamber. *Gyroidinoides* differs from *Pseudovalvulineria* in being plano-convex, rather than biconvex, and in having the final aperture as the only opening to the exterior, rather than having the umbilical portion of earlier apertures remaining open.]

Family ANOMALINIDAE Cushman, 1927

[Anomalinidae Cushman, 1927, p. 92]——[In synonymic

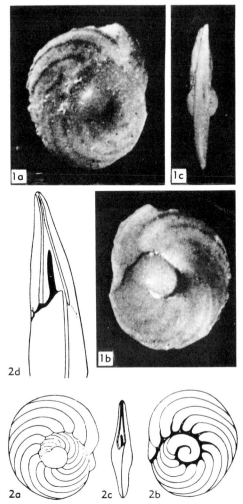

Fig. 617. Osangulariidae; *1,2, Goupillaudina* (p. *C753*).

citations dagger(†) indicates *partim*]——[=Hélicostèguest d'Orbigny, 1826, p. 268 *(nom. neg.; nom. nud.);* =Melonidae Chapman, Parr & Collins 1934, p. 556; =Parrelloididae Hofker, 1956, p. 936; =Gavelinellidae Hofker, 1956, p. 946]

Test trochospiral to nearly planispiral, evolute on one or both sides; chambers simple; wall calcareous, coarsely perforate, granular in structure, bilamellar; primary aperture interiomarginal equatorial or somewhat extending onto spiral or umbilical sides, and may also have additional peripheral apertures. *U.Trias.-Rec.*

Subfamily ANOMALININAE Cushman, 1927

[Anomalininae Cushman, 1927, p. 92] [=Praerotalininae Hofker, 1933, p. 125 *(partim) (nom. nud.);* =Melonisinae Voloshinova, 1958, p. 147; =Gavelinellinae Loeblich & Tappan, 1961, p. 316]

C754 *Protista—Sarcodina*

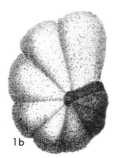

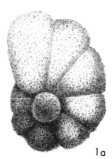

FIG. 618. Anomalinidae (Anomalininae; *1, Anomalina*) (p. C754-C755).

Single primary aperture, interiomarginal and equatorial or extending onto spiral or umbilical sides, may have apertural flaps on umbilical side beneath which aperture opens into chambers, and may also have secondary sutural openings on periphery. *U.Trias.-Rec.*

Anomalina D'ORBIGNY, 1826, *1391, p. 282 [*A. punctulata; SD CUSHMAN, 1915, *404e, p. 44] [=*Porospira* EHRENBERG, 1844, *673, p. 75 (type, *P. comes*)]. Test free, low trochospiral or nearly planispiral, spiral side with umbonal boss, opposite side with depressed umbilicus, periphery rounded; chambers few, sutures radiate; aperture an interiomarginal equatorial opening, extending slightly to umbilical side. [The status of *Anomalina* is somewhat in question, inasmuch as the type-species has not been recognized since its description. We searched for the original type in the D'ORBIGNY collection in Paris, but it is apparently not preserved. A search in the type locality, Recent, Mauritius Is.(Île de France) for this species would clarify the generic status, and determine whether or not *Anomalinoides* is distinct from *Anomalina*. Both are here tentatively recognized, *Anomalina* as based on the original figure and description.] *Rec.*, Ind.O.——FIG. 618, 1. **A. punctulata*, Rec., Mauritius Is.; *1a-c*, opp.

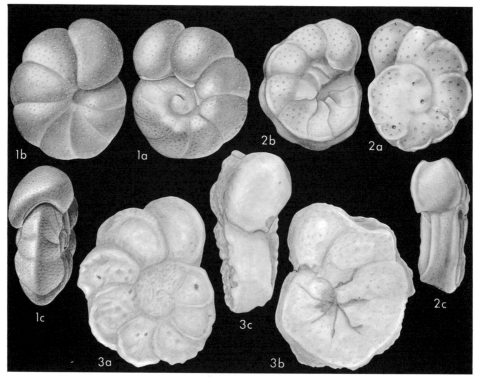

FIG. 619. Anomalinidae (Anomalininae; *1, Anomalinoides; 2,3, Boldia*) (p. C755-C757).

Foraminiferida—Rotaliina—Cassidulinacea

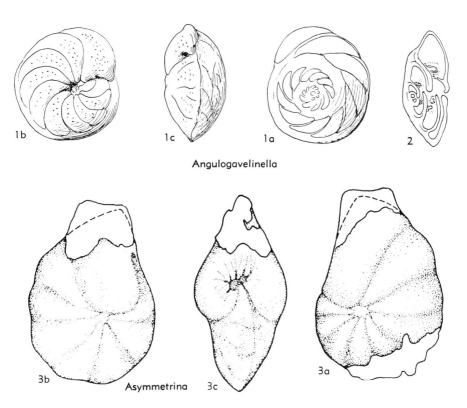

FIG. 620. Anomalinidae (Anomalininae; *1,2, Angulogavelinella; 3, Asymmetrina*) (p. C755).

sides and edge view of holotype, enlarged (*1391).

Anomalinoides BROTZEN, 1942, *240, p. 23 [*A. plummerae, =Anomalina pinguis JENNINGS, 1936, *989, p. 195; OD]. Test free, nearly planispiral, but asymmetrical, periphery broadly rounded, spiral side partially evolute with umbonal boss, opposite side involute and umbilicate; wall calcareous, coarsely perforate, granular in structure; aperture a low interiomarginal equatorial slit with narrow bordering lip, extending along spiral suture on evolute side under umbilical margin of later chambers. [*Anomalinoides* differs from *Anomalina* in that its aperture continues onto the spiral side instead of being entirely peripheral.] *U.Cret.-Rec.*, cosmop.——FIG. 619,1. *A. pinguis* (JENNINGS), U.Cret.(Maastricht.), USA(Tex.); *1a-c,* opp. sides and edge view, ×68 (*2117).

Angulogavelinella HOFKER, 1957, *948, p. 365 [*Discorbina gracilis* MARSSON, 1878, *1228, p. 166; OD]. Test trochospiral, lenticular, inequally biconvex, periphery with nonporous keel, small and deep umbilicus present; chambers numerous, low, arched; septa double (bilamellar), sutures curved, oblique; wall calcareous, coarsely perforate on umbilical side, nonperforate on spiral side, sutures and peripheral keel nonporous; aperture a somewhat oblique, high interiomarginal arch midway between periphery and umbilicus. *U.Cret.*, Eu.——FIG. 620,1,2. *A. gracilis* (MARSSON), Maastricht., Ger.; *1a-c,* opp. sides and edge view; *2,* axial sec. showing deep umbilicus, double septum (at right of figure), and apert. openings, ×60 (*948).

Asymmetrina KRISTAN-TOLLMANN, 1960, *1059, p. 74 [*A. biomphalica*; OD]. Test free, lenticular, planispiral, involute, but slightly asymmetrical, biumbilicate; wall calcareous, perforate, lamellar character and microstructure unknown; aperture an interiomarginal, equatorial arch with radiate margin. [The genus is known from a single specimen of the type-species and needs additional study for correct placement. It was originally included in the Anomalinidae.] *U.Trias.(Rhaet.)*, Eu. (Aus.).——FIG. 620,3. *A. biomphalica; 3a-c,* opp. sides and edge view of holotype, ×80 (*1059).

Boldia VAN BELLEN in VAN DEN BOLD, 1946, *155, p. 124, VAN BELLEN, 1946, *114, p. 122 [*nom. subst.* pro *Terquemia* VAN BELLEN (*non* TATE, 1868; *nec* VAN VEEN, 1932)] [*Rotalina lobata* TERQUEM, 1882, *1890, p. 63; OD] [=*Terquemia* VAN BELLEN, 1946, *113, p. 86 (obj.)]. Test free, trochospiral, plano-concave or biconcave, periphery broadly truncate; all chambers visible on slightly concave, nearly flat spiral side, only chambers of final whorl visible on concave, slightly umbilicate opposite side; sutures thickened and raised spirally, strongly incised on umbilical side; wall calcareous, perforate; aperture a low interiomarginal arch at umbilical edge of truncate periphery and extend-

Fig. 621. Anomalinidae (Anomalininae; *1,2, Cibicidoides; 3,4, Coleites; 5-7, Gavelinella*) (p. C757, C759).

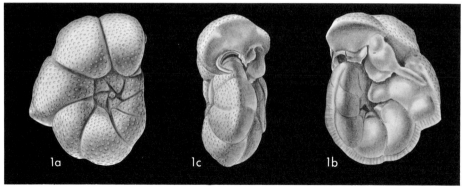

FIG. 622. Anomalinidae (Anomalininae; *1, Discanomalina*) (p. C757-C758).

ing onto umbilical side beneath flaplike margin of final chamber, earlier apertures also remaining open as sutural slits beneath imbricating flaps of previous chambers. Paleoc.-M.Eoc., Eu.-W.Indies (Cuba)-C.Am.(Guat.).——FIG. 619,2,3. *B. lobata* (TERQUEM), M.Eoc., Fr.; *2a-c,* opp. sides and edge view of hypotype; *3a-c,* opp. sides and edge view of holotype; all ×109 (*2117).

[In the original generic description, VAN BELLEN (*113, p. 86) cited *Rotalina lobata* TERQUEM as the type-species but on the plate explanation (pl. 13, figs. 13-15) he referred to *Terquemia lobata* (TERQUEM), nov. gen., nov. sp., stating that the illustrations are of the holotype. As he did not describe a new species, only a new genus, the holotype of *lobata* is the specimen of TERQUEM, which is in the collections of the Muséum National d'Histoire Naturelle, Paris (here refigured). Comparison of the illustrations suggests the possibility that VAN BELLEN's specimen belongs to a species distinct from TERQUEM's type, as it has an entire, rather than lobulate, periphery and numerous radiating grooves on the umbilical side, apparently covering the surface of the chambers. *Rotalina lobata* TERQUEM is the type-species of *Boldia* by original designation, regardless of the specific name eventually applied to the specimen of VAN BELLEN.——¶*Boldia* differs from *Anomalina* in its very truncate periphery and biconcave or planoconcave test with the aperture extending onto the umbilical side. It differs from *Pijpersia* in having deeply incised umbilical sutures, and in lacking the extremely inflated and angular chambers of that genus. The genus was originally described without definite family assignment, although in the chart arranged phylogenetically it is grouped with *Cibicides*. CUSHMAN (1948, *486, p. 333) apparently followed this in placing the genus in the Anomalinidae. Y. LE CALVEZ (1949, *1112, p. 8) stated that *Rotalina lobata* TERQUEM should be classified as *Anomalina lobata*, but apparently had not then noted VAN BELLEN's description of *Boldia*. BERMÚDEZ (1952, *127, p. 41) placed the genus in the Discorbisinae (=Discorbinae), considering that extension of the aperture onto the umbilical side was analogous to that of *Discorbis*.]

Cibicidoides THALMANN, 1939, *1897d, p. 448 [**Truncatulina mundula* BRADY, PARKER, & JONES, 1888, *203, p. 228; OD] [=*Cibicidoides* BROTZEN, 1936, *237, p. 186, 194 *(nom. nud.)*; *Parrelloides* HOFKER, 1956, *945, p. 936 (type, *Cibicides hyalinus* HOFKER, 1951, *928c, p. 359)]. Test free, trochospiral, biconvex and biumbonate, all chambers visible on spiral side, only those of final whorl visible on umbilical side; wall calcareous, hyaline, with series of coarse perforations on spiral side, appearing only near previous spiral suture in early portion of test, but covering large portion of spiral side of later chambers; aperture a low interiomarginal equatorial arch with slight projecting lip. [Although specimens were not available for determining the wall structure of the type-species, *C. proprius* is very similar to this species in other features and is of granular wall structure. The so-called radially built species listed by WOOD & HAYNES (1957, *2076) belong elsewhere (*Cibicidina*, etc.).] Rec., Ind.O.-Atl.O.——FIG. 621,1. *C. mundula* (BRADY, PARKER, & JONES); *1a-c,* opp. sides and edge view of lectotype (here designated and refigured), BMNH-ZF3585, from *Plumper* Station 4, 260 fathoms, lat. 22°54'S., long., 40°37'W., over Abroholos Bank, off coast of Brazil, S.Am., ×109 (*1166).——FIG. 621,2. *C. hyalinus* (HOFKER), Rec., Sumatra; *2a-c,* opp. sides and edge view; *2d,* axial sec., ×105 (*928c).

Coleites PLUMMER, 1934, *1466, p. 605 [**Pulvinulina reticulosa* PLUMMER, 1927, *1461, p. 152; OD]. Test with early stage trochospirally coiled, later uncoiling, periphery carinate; chambers low and broad; wall calcareous, hyaline, coarsely perforate, granular in structure, lamellar character not known, surface coarsely reticulate; aperture in early stage an irregular ovate areal opening near periphery on umbilical side, elongate and terminal in adult, with tooth on umbilical side of test, interior with solid column extending from inner margin of aperture to previous foramen. Paleoc.-L.Eoc., N.Am.-Eu.-C.Am.——FIG. 621,3,4. *C. reticulosa* (PLUMMER), Paleoc.(Midway.), USA (Ark.) *(3)*, Paleoc.(Dan.), Sweden *(4)*; *3a-c,* opp. sides and top view, ×57 (*2117); *4,* dissected specimen showing aperture *(a)*, intercameral column *(ic)*, outer wall *(ow)*, enlarged (*241).

Discanomalina ASANO, 1951, *52c, p. 13 [**D. japonica*; OD]. Test free, thick, planispiral, both sides excavated centrally, spiral side partially evolute, opposite side involute, periphery broadly rounded; chambers inflated, with backward-projecting flap on umbilical side, may have spinelike projections on periphery from one or more chambers; sutures radial; wall calcareous, granular in structure, coarsely perforate on spiral side, umbilical side and apertural face of clear, nonperforate shell material; aperture a low broad equatorial slit, interiomarginal, bordered by slight lip, slitlike supplementary openings may appear beneath umbilical chamber flaps. *Mio.-Rec.,* Japan-Pac.O.-Atl.O.-Carib.——Fig. 622,*1.* **D. japonica,* Rec., Pac.; *1a-c,* opp. sides and edge view, ×44 (*2117).

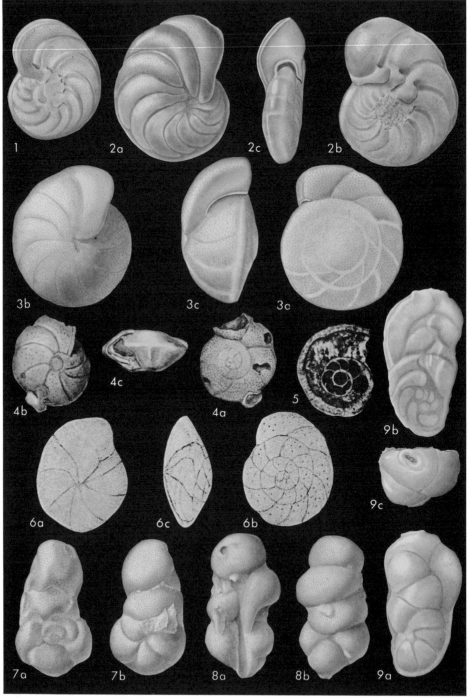

Fig. 623. Anomalinidae (Anomalininae; *1,2, Hanzawaia; 3-6, Heterolepa; 7-9, Karreria*) (p. C759-C761).

Gavelinella Brotzen, 1942, *240, p. 7 [*Discorbina pertusa* Marsson, 1878, *1228, p. 166; OD] [=*Pseudovalvulineria* Brotzen, 1942, *240, p. 20 (type, *Rosalina lorneiana* d'Orbigny, 1840, *1394, p. 36); *Anomalina (Brotzenella)* Vasilenko in N. K. Bykova et al., 1958, *265, p. 52 (type, *Anomalina monterelensis* Marie, 1941, *1215, p. 243)]. Test free, trochospiral, biconvex, sides flattened, periphery rounded; all whorls visible on spiral side, on opposite side only chambers of final whorl visible around umbilicus, which is partially closed by subtriangular flaps projecting from umbilical margins of each chamber; small umbilical boss may also be present; wall calcareous, perforate, granular in structure with double septal walls (bilamellar); aperture a low interiomarginal slit extending from near periphery to umbilicus, bordered above by narrow lip which broadens out into triangular flap at umbilical chamber margin, aperture continuous beneath flap with those of earlier chambers. *Cret.-Mio.*, Eu.-N. Am.-S. Am.-Australia-N. Z.——Fig. 621,5. *G. pertusa* (Marsson), U.Cret.(Maastricht.), Eu. (Denm.); *5a-c*, opp. sides and edge view, ×98 (*2117).——Fig. 621,6. *G. lorneiana* (d'Orbigny), U.Cret.(Senon.), Eu.(Fr.); *6a-c*, opp. sides and edge view of lectotype, ×61 (*2117).——Fig. 621,7. *G. monterelensis* (Marie), U.Cret. (Campan.), USSR; *7a-c*, opp. sides and edge view, ×38 (*265).

[*Pseudovalvulineria* was originally said to differ from *Gavelinella* in having a less open umbilicus, and an umbilical knob. The type-species of both genera lack an umbilical knob, and in other similar species this character is not constant, and the relative proportions of the umbilicus also vary considerably. Furthermore, the apertural features are identical; hence *Pseudovalvulineria* is regarded as a synonym as it was by Hofker (*948). A lectotype is here designated and refigured for *Rosalina lorneiana* d'Orbigny.]

Hanzawaia Asano, 1944, *50, p. 98 [*H. nipponica*; OD]. Test free, trochoid, plano-convex, periphery moderately angled with keel, flattened side partially involute with elevated flaps on lower margin of chamber partially or completely overlapping chambers of previous whorl and commonly coalescing over entire central area, opposite side involute but without open umbilicus, central area with clear boss; sutures strongly curved, thickened; wall calcareous, granular in microstructure, rather coarsely perforate except for clear area above aperture, central flaps of spiral side and thickened sutures and keel, all of which are of clear, apparently solid, calcite; aperture an arch on periphery, extending somewhat onto convex involute side but also laterally continuous with opening on flattened side, under central flap of final chamber, with supplementary openings under umbilical flaps, both on their outer and inner margins. *Mio.-Rec.*, cosmop.——Fig. 623,*1,2*. *H. nipponica*, Plio., Japan; *1*, evolute side of topotype; *2a-c*, opp. sides and edge view of hypotype, ×41 (*2117).

[*Hanzawaia* differs from *Cibicidina* in possessing central chamber lobes on the evolute side, and in being more

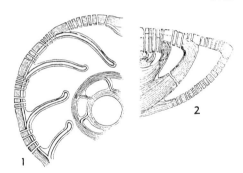

Fig. 624. Anomalinidae (Anomalininae; *1,2*, *Heterolepa*) (p. *C*759-*C*760).

coarsely perforate. *Cibicides* differs in lacking chamber flaps, having a radial wall structure, and a more elevated umbilical side. Reiss (1958, *1530, p. 65) mentioned *Hanzawaia* as belonging to the "Pulvinulinidae," and stated that all had radiate walls. *Hanzawaia* has a granular wall, however, and is not related to the other genera discussed by Reiss.]

Heterolepa Franzenau, 1884, *742, p. 214 [*H. simplex*=*Rotalina dutemplei* d'Orbigny, 1846, *1395, p. 157; SD Loeblich & Tappan, 1962, *1187, p. 72] [=*Pseudotruncatulina* Andreae, 1884, *19, p. 122 (type, *Rotalina dutemplei* d'Orbigny, 1846, *1395, p. 157); *Dendrina* Costa MS in Fornasini, 1898, *732, p. 206 (type, *D. succinea*) (non Quenstedt, 1848); *Pninaella* Brotzen, 1948, *241, p. 119 (type, *P. scanica*); *Cibicides (Gemellides)* Vasilenko, 1954, *1986, p. 186 (type, *C. (G.) orcinus*); *Hollandina* Haynes, 1956, *887, p. 94 (type, *H. pegwellensis*)]. Test free, trochospiral, inequally biconvex or plano-convex, periphery bluntly angled, may have nonperforate keel, flat to slightly convex evolute spiral side, with relatively numerous chambers in slowly enlarging whorls, more convex umbilical side involute, with radial sutures; wall calcareous, thick and lamellar, coarsely and regularly perforate, granular in structure, septa double (bilamellar); aperture slit-like, interiomarginal, extending about half of distance to umbilicus on umbilical side and extending across periphery on spiral side, may also extend for some distance along spiral suture. U. *Cret.(Maastricht.)-Rec.*, cosmop.——Fig. 623,3. *H. dutemplei* (d'Orbigny), Mio., Eu.(Aus.); *3a-c*, opp. sides and edge view, ×37 (*2117).——Fig. 624,*1*. *H. praecincta* Franzenau, Mio., Eu. (Hung.); horiz. sec. showing bilamellar wall character and coarse perforations, enlarged (*742).——Fig. 624,*2*. *H. bullata* Franzenau, Mio., Eu. (Hung.); vert. sec., enlarged (*742).——Fig. 623,*4,5*. *H. scanica* (Brotzen), Paleoc., Eu. (Sweden); *4a-c*, opp. sides and edge view of holotype; *5*, horiz. sec., with secondarily resorbed septa, probably due to preservation, ×38 (*241).——Fig. 623,*6*. *H. pegwellensis* (Haynes), Paleoc., Brit.I.(Eng.); *6a-c*, opp. sides and edge view of holotype, ×90 (*887).

[Franzenau originally included four species in *Heterolepa*, without designating a type-species, *H. simplex*, n. sp., *H.*

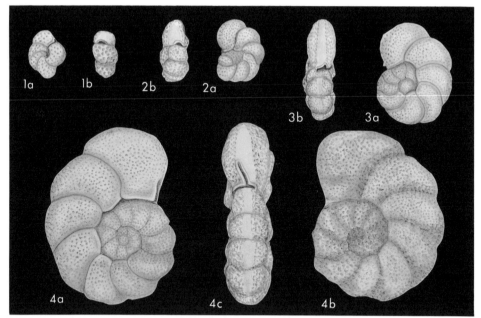

Fig. 625. Anomalinidae (Anomalininae; *1-4, Holmanella*) (p. C760).

costata, n. sp., *H. praecincta*, n. sp. and *H. bullata*, n. sp. ELLIS & MESSINA (*700) stated that FRANZENAU designated *Rotalina dutemplei* as the type in 1885, but this was not in the original list of species, hence was ineligible for selection as the type-species. In 1885 FRANZENAU (*743, p. 152) stated that *H. simplex* was a synonym of *Rotalina dutemplei* D'ORBIGNY. As the type must be one of the species originally included by FRANZENAU, we (*1187) so designated *H. simplex*.——¶During the same year (1884) *Pseudotruncatulina* was described on the basis of its bilamellar walls, also with *Rotalina dutemplei* as type-species. *Gemellides* (proposed as a subgenus of *Cibicides*) also originally included this species, but was separated on the basis of its apertural characters. Regardless of the basis for separation, both *Pseudotruncatulina* and *Gemellides*, including the same species, are junior synonyms of *Heterolepa*. *Pninaella* was regarded as having secondarily much enlarged foramina, but the figured section shows well-preserved septa in the early portion; hence it seems probable that the remaining septa were probably destroyed during preservation. *Pninaella scanica* seems otherwise much like *H. dutemplei* and certainly congeneric. The other species included by BROTZEN (*Pulvinulina nitidula*) is probably not congeneric, as it is a very thin-walled form. Although previously regarded as closely related to *Cibicides* (some species having been referred to it erroneously), *Heterolepa* has a granular wall structure and is free, not attached by the spiral side, thus related to the Anomalinidae, as here restricted, rather than to the Cibicididae.]

Holmanella LOEBLICH & TAPPAN, 1962, *1187, p. 72 [*Discorbinella valmonteensis* KLEINPELL, 1938, *1046, p. 350; OD]. Test free, large, compressed, enrolled, bievolute, nearly planispiral but somewhat asymmetrical, with nonporous, broadly rounded peripheral margin; chambers gradually enlarging; sutures distinct, depressed, curved backward at periphery; wall calcareous, thin, very coarsely perforate, granular in microstructure, bilamellar; aperture in young stage a low interiomarginal opening at one side of periphery, in later stages with low opening continuing along spiral suture to connect with previous apertures and with perpendicular slit extending obliquely up nonporous apertural face, all apertures bordered by narrow lip. *Mio.*, USA(Calif.).——FIG. 625,*1-4*. *H. valmonteensis* (KLEINPELL); *1a,b*, side and edge views of juvenile specimen showing slightly trochospiral development and low asymmetrical arched aperture; *2a,b*, side and edge views of somewhat older specimen with higher asymmetrical arch; *3a,b*, side and edge views of larger specimen with beginning of vertical slit shown as notch, imperforate area visible on periphery and along sutures of spiral side; *4a-c*, spiral, umbilical, and edge views, with well-developed oblique slit-like aperture extending up face and connecting with spiral suture on spiral side; all ×48 (*2117).

Involvina KRISTAN-TOLLMANN, 1960, *1059, p. 76 [*I. obliqua*; OD]. Test free, lenticular, trochospiral or with tendency to become planispiral, umbilical region closed or umbonate; wall calcareous perforate but coarse granular, with calcareous cement and some included sand grains; aperture a large oval equatorial opening that extends slightly to umbilical side, margin radiate. [The wall characters need clarification. The above description is from the original, and leaves doubt as to whether the wall is lamellar, hyaline perforate, and radial or granular in structure, or non-lamellar agglutinated calcareous, or granular. The genus was originally placed in the Anomalininae.] *U.Trias.(Rhaet.)*, Eu.(Aus.).——FIG. 626,*1*. *I. obliqua*; *1a-c*, opp. sides and edge view of holotype, ×125 (*1059).

Karreria RZEHAK, 1891, *1604, p. 4, 6 [*K. fallax*; OD] [=*Vagocibicides* FINLAY, 1939, *717c, p. 326 (type, *V. maoria*)]. Test attached, early portion trochospirally coiled with one or more volutions, attached by spiral side, free convex side

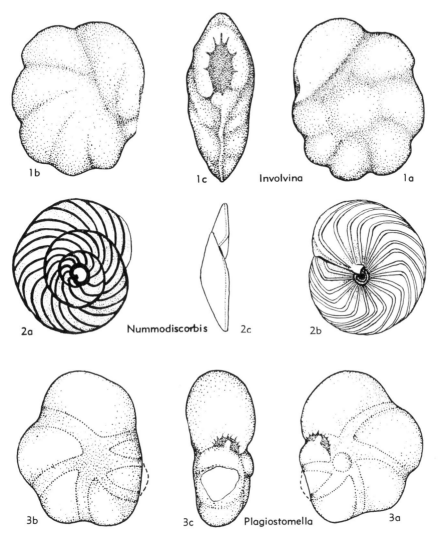

Fig. 626. Anomalinidae (Anomalininae; *1, Involvina; 2, Nummodiscorbis; 3, Plagiostomella*) (p. C760, C763).

involute, later portion uncoiling and rectilinear; sutures depressed, nearly straight; wall calcareous, thick, finely perforate, granular in structure, surface smooth; aperture terminal or subterminal, rounded. [*Stichocibicides* was regarded as a synonym by BROTZEN (*241), TEN DAM (*554), and BERMÚDEZ (*127), but it has a coarsely perforate wall and nonporous peripheral keel.] *L.Cret. (Alb.)-Rec.,* Eu.-N.Afr.-USA(Calif.)-N.Z.——FIG. 623,7,8. **K. fallax,* Paleoc., Eu.(Denm.); *7a,b, 8a,b,* opp. sides of two specimens, ×40 (*2117).
——FIG. 623,9. *K. maoria* (FINLAY), M.Oligo., N.Z.; *9a-c,* opp. sides and apert. view, ×51 (*2117).

Melonis DE MONTFORT, 1808, *1305, p. 66 [*M. etruscus=Nautilus pompilioides* FICHTEL & MOLL, 1798, *716, p. 31; OD] [=*Melossis* PALLAS in OKEN, 1815,*1385, p.333 (type, *Nautilus pompilioides* FICHTEL & MOLL, 1798, *716, p. 31; SD GALLOWAY, 1933, *762, p. 266); *Melonia* BRONN, 1849, *211, p. 720 (*non* LAMARCK, 1822; *nec* SCHINZ, 1825, *pro Melania* LAMARCK, 1799, *nom. van.*); *Gavelinonion* HOFKER, 1951, *935, p. 17 (*nom. nud.*); *Gavelinonion* THALMANN, 1953, *1897k, p. 876 (*nom. nud.*) (erroneously cited *Rotalia tuberculifera* REUSS, 1862, as type of *Gavelinonion* HOFKER, 1951); *Gavelinonion* HOFKER, 1956, *946, p. 116 (*nom. nud.*); *Gavelinonion* HOFKER, 1957, *948, p. 368 (type, *Nautilus umbilicatulus* WALKER & JACOB in KANMACHER, 1798, *1011, p. 641)]. Test free, early stage slightly trochospiral, adult planispiral, symmetrical and involute, deeply biumbilicate, with umbilicus commonly bordered by rim of nonperforate skeletal material, periphery broadly rounded; about 9 to 12 chambers per whorl; sutures flush to slightly depressed, radiate, straight to slightly curved, septa double, bilamellar (*946); wall calcareous, coarsely per-

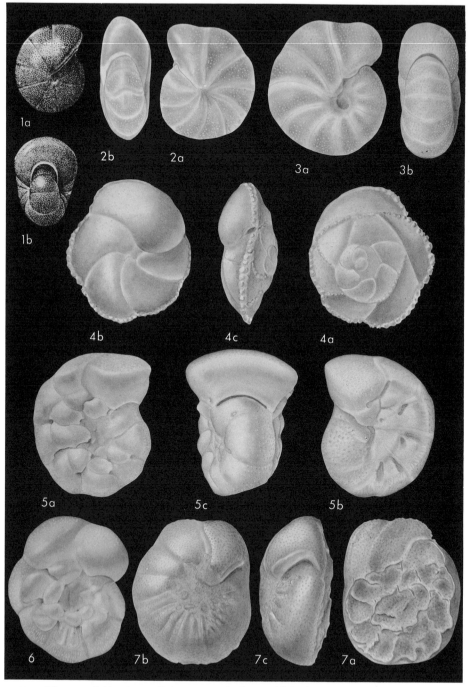

Fig. 627. Anomalinidae (Anomalininae; *1-3,* Melonis; *4,* Pulsiphonina; *5,6,* Paromalina; *7,* Stensioina) (p. C761-C763).

forate, granular in structure, apertural face, septa, and umbilical thickened rim imperforate, surface smooth; aperture an elongate interiomarginal, equatorial slit, extending laterally to umbilicus on both sides of test. *?U.Cret.(Maastricht.), Paleoc.-Rec.*, cosmop.——FIG. 627,1. **M. pompilioides* (FICHTEL & MOLL), Plio., Eu.(Albania); *1a,b*, side, edge views, ×100 (*2019).——FIG. 627,*2,3*. *M. zaandami* (VAN VOORTHUYSEN), Rec., Greenl.; *2a,b, 3a,b*, side and apert. views, ×75 (*1162).

Nummodiscorbis HORNIBROOK, 1961, *959, p. 106 [**N. novozealandica*; OD]. Test low and conical, plano-convex to concavo-convex, with angular periphery; low and numerous chambers trochospirally arranged, in numerous slowly enlarging whorls; sutures curved and oblique on evolute spiral side, sharply angled centrally on involute umbilicate opposite side; wall calcareous, finely perforate, lamellar character and microstructure not known; aperture an elongate interiomarginal slit, extending from near periphery to umbilicus, with small apertural flap projecting into umbilicus. *L.Mio.*, N.Z.——FIG. 626,2. **N. novozealandica*; *2a-c*, opp. sides and edge view of holotype, ×100 (*959).

[This genus was originally placed in the Discorbinae, as was *Gavelinella*, but no information was given (*959) as to whether or not *Nummodiscorbis* has the radially built monolamellar wall of this group. Since it appears closer in other characters to *Gavelinella*, it is here tentatively placed in the Anomalinidae, pending further study of its wall character.]

Paromalina LOEBLICH & TAPPAN, 1957, *1172, p. 230 [**P. bilateralis*; OD]. Test free, planispiral, biumbilicate, both sides somewhat excavated centrally, periphery truncate; chambers laterally inflated, with their umbilical margins extending backward in flap covering part of previous suture and chamber, flaps more rarely coalescing to obscure otherwise open umbilicus; sutures radial, depressed; wall calcareous, granular in structure, with clear imperforate wall on sides and apertural face, coarsely perforate truncate periphery; aperture a broad low slit on periphery, bordered above by narrow lip, at base of final chamber and against preceding whorl, with supplementary openings beneath umbilical chamber flaps on each side of test. [Differs from *Discanomalina* in having the clear imperforate-appearing shell wall on both sides of the test, and is coarsely perforate only on the truncate periphery.] *Rec.*, Atl.O.——FIG. 627, *5,6*. **P. bilateralis*; *5a-c*, opp. sides and edge view, ×53 (*1172); *6*, side view of paratype, ×48 (*1172).

Plagiostomella KRISTAN-TOLLMANN, 1960, *1059, p. 73 [**P. inflata*; OD]. Test biconvex, slightly trochospiral, tending to become planispiral, umbilicus closed, periphery rounded; wall calcareous, perforate, lamellar character and microstructure unknown; aperture an interiomarginal, equatorial arch, extending slightly onto umbilical side, upper apertural margin fimbriate, lower margin with tooth, or possibly double aperture. [This genus is imperfectly known, as it is represented by a single specimen of the type-species; hence its placement is questionable. It was originally placed in the Anomalininae.] *U.Trias.(Rhaet.)*, Eu.(Aus.).——FIG. 626,*3*. **P. inflata*; *3a-c*, opp. sides and edge view of holotype, ×125 (*1059).

Pulsiphonina BROTZEN, 1948, *241, p. 106 [**Siphonina prima* PLUMMER, 1927, *1461, p. 148; OD] [=*Siphonina (Pulsiphonina)* BROTZEN, 1948, *241, p. 106 (obj.)]. Test free trochospiral, biconvex, periphery angular and with carinate, limbate, or beaded margin; all whorls visible from spiral side, where chambers are broad, low, and semilunate in appearance, only final whorl visible on umbilical side, where sutures are curved but nearly radial; wall calcareous, coarsely perforate, granular in structure; aperture a low narrow opening at periphery on umbilical side, and lying against peripheral keel, with narrow bordering lip. [*Pulsiphonina* differs from the superficially similar *Siphonina* in having a granular, rather than radially, built wall and in lacking a distinct apertural neck]. *U.Cret.(Maastricht.)-L.Eoc.*, N.Am.-Eu.——FIG. 627,*4*. **P. prima* (PLUMMER), Paleoc. (Midway.), USA(Ark.); *4a-c*, opp. sides and edge view, ×185 (*2117).

Stensioina BROTZEN, 1936, *237, p. 315 [**Rotalia exsculpta* REUSS, 1860, *1548, p. 222; OD]. Test trochospiral, unequally biconvex to plano-convex, with flattened spiral side and elevated umbilical side; chambers enlarging gradually; sutures oblique and strongly elevated on spiral side, radial and depressed on umbilical side, septa double (bilamellar); wall calcareous, coarsely perforate, granular in structure, spiral side with characteristic ornamentation, with sutures forming elevated ridges resulting in irregularly reticulose pattern on spiral side, with chamber wall more finely reticulate and pitted; aperture a low interiomarginal opening between umbilicus and periphery. *U.Cret.*, cosmop. ——FIG. 627,7. **S. exsculpta* (REUSS), Eu.(Ger.); *7a-c*, opp. sides and edge view of topotype, ×74 (*2117).

Subfamily ALMAENINAE Myatlyuk, 1959

[Almaeninae MYATLYUK in RAUZER-CHERNOUSOVA & FURSENKO, 1959, p. 272]

Primary aperture interiomarginal, equatorial or slightly umbilical, with lip; secondary slitlike aperture at peripheral margin, in plane of coiling. *Eoc.-Rec.*

Almaena SAMOYLOVA, 1940, *1622, p. 377 [**A. taurica*; OD (M)] [=*Kelyphistoma* KEIJZER, 1945, *1030, p. 207 (type, *K. ampullolocultata*); *Planulinella* SIGAL, 1949, *1744, p. 158 (type, *P. escornebovensis*); *Almaena (Pseudoplanulinella)* SIGAL, 1950, *1745, p. 63, 68 (type, *A. (P.) hieroglyphica*)]. Test enrolled, compressed, planispiral, evolute on both sides, periphery carinate; wall calcareous, coarsely perforate, peripheral keel, apertural face and septa nonperforate; primary aperture ovate to slitlike, interiomarginal and equatorial to slightly asymmetrically equatorial in position, with distinct bordering lip, elongate slitlike secondary subperipheral aperture paralleling peripheral keel on one side of test, as in *Anomalinella*, those of earlier chambers secondarily filled. U.Eoc.-Mio., Eu.-Afr.-N.Am.——FIG. 628,1. *A. taurica*, U.Eoc., USSR(Crimea); *1a-c*, opp. sides and edge view of holotype, ×25 (*1332).——FIG. 628,2. *A. hieroglyphica* (SIGAL), Mio. (Aquitan.), Fr.; *2a,b*, side and edge views, ×71 (*2117).——FIG. 628,3. *A. escornebovensis* (SIGAL), Mio.(Aquitan.), Eu.(Fr.); *3a,b*, side and edge views, ×44 (*2117).

Anomalinella CUSHMAN, 1927, *431, p. 93 [*Truncatulina rostrata* BRADY, 1881, *196c, p. 65; OD]. Test free, lenticular, slightly trochoid to nearly planispiral in adult, involute, biumbonate; chambers relatively numerous, increasing gradually in size; sutures distinct, thickened, gently curved; wall calcareous, hyaline, coarsely perforate, granular in structure with clear and nonperforate peripheral keel and sutures; aperture consisting of low, rounded interiomarginal arch, against peripheral margin of previous whorl, bordered above by lip, supplementary aperture consisting of elongate slit just to one side of periphery, bordering and paralleling peripheral keel. Mio.-Rec., Pac.O.——FIG. 628,4. *A. rostrata* (BRADY), Rec., New Guinea (Papua); *4a,b*, side and edge views of lectotype, ×48 (*2117).

[Differs from *Almaena* in being completely involute, rather than partially evolute. It differs from *Queraltina* in being planispiral rather than trochoid. A lectotype is here designated and refigured for *Anomalinella rostrata* (BRADY) from the syntypes in the British Museum (Natural History), BMNH-ZF2549, from *Challenger* station 217A, Humboldt Bay, Papua (New Guinea), at a depth of 37 fathoms. HOFKER (1960, *953, p. 49) regarded *Anomalinella, Almaena, Planulinella, Pseudoplanulinella* and *Kelyphistoma* all as synonyms of *Planulina*. However, typical *Planulina* has a radially built wall, is perforate only on one side of the test, and does not have supplementary peripheral apertures (species not agreeing in all these characters should not be referred to *Planulina*), whereas *Anomalinella* is granular in wall structure, both sides of the test are perforate, and the supplementary peripheral apertures are characteristic. The genera do not even belong to the same superfamily.]

Ganella AUROUZE & BOULANGER, 1954, *57, p. 187 [*G. neumannae*; OD]. Test free, lenticular, trochospiral in the early stage, becoming nearly planispiral and evolute on both sides in adult, although slightly asymmetrical, periphery carinate; chambers gradually enlarging; sutures curved backward at periphery, slightly depressed, those of earlier chambers thickened and elevated; wall calcareous, coarsely perforate, granular in structure, with nonporous, beaded and elevated sutures, and peripheral keel; aperture an elongate vertical slit, in young forms extending up from base of somewhat obliquely situated, flat to concave apertural face and migrating up face to become areal opening in median position in face of adult test, bordered by elevated rim. Eoc. (Ypres.), Eu.(Fr.).——FIG. 628,5. *G. neumannae*; *5a-c*, opp. sides and edge view, ×72 (*2117).

Queraltina MARIE, 1950, *1218, p. 73 [*Q. epistominoides*; OD]. Test similar to *Almaena* but distinctly trochospiral, asymmetrical and inequally biconvex, with chambers distinctly inflated on umbilical side and nearly flat on spiral side; wall granular in structure; peripheral apertures on spiral side of test, paralleling peripheral keel. [*Queraltina* is probably ancestral to *Almaena*, with more pronounced trochospiral development, and inflated umbilical side, whereas *Almaena* is more nearly bilaterally symmetrical, and is strongly compressed.] Eoc.(U.Lutet.-Barton.), Eu.——FIG. 628,6,7. *Q. epistominoides*, Eoc.(Barton.), Fr.; *6a,b*, opp. sides of holotype, ×52 (*2117); *6c*, edge view, showing inflated chambers at left side, ×56 (*1218); *7a*, inflated umbilical side of paratype, ×52 (*2117); *7b*, edge view of more asymmetrical paratype, ×56 (*1218).

Superfamily CARTERINACEA Loeblich & Tappan, 1955

[nom. transl. LOEBLICH & TAPPAN, 1961, p. 317 (ex family Carterinidae LOEBLICH & TAPPAN, 1955)]

Test composed of secreted fusiform calcareous spicules, commonly oriented parallel to periphery and embedded in calcareous ground mass. *Rec.*

Family CARTERINIDAE Loeblich & Tappan, 1955

[Carterinidae LOEBLICH & TAPPAN, 1955, p. 27]

Test trochospiral, free or attached; later chambers subdivided by secondary septa. *Rec.*

Carterina BRADY, 1884, *200, p. 66, 345 [*Rotalia spiculotesta* CARTER, 1877, *294, p. 470; OD (M)]. Test free, trochospiral and umbilicate in early stages, attached and spreading irregularly in later stages, with wide, flangelike, undivided attachment spreading over surface of substratum; 3 to 5 crescentic chambers to whorl of approximately equal height throughout, becoming much more irregular and broader in later whorls, beginning in third whorl chambers subdivided by partial secondary septa projecting inward from lower and peripheral walls, in early stages only minor projections present, but in later ones almost complete partitions, true septa oblique and depressed on spiral side, secondary septa perpendicular to periphery, depressed on umbilical side but not visible on spiral side except when specimen is dampened, earlier chambers having only 2 or 3 of these secondary septa, but after third whorl increasing in number up to 15 to chamber as latter increase in relative length, leaving chamberlets all of approximately equal size; wall thin, composed of calcareous spicules (secreted by protoplasm) each forming single crystal with c-axis parallel to length of spicule, commonly aligned parallel to periphery of test, embedded in calcareous areolated ground mass; aperture not observed in attached specimens, ventral in free specimens. Rec., Philip. Is. - India (Ceylon) - Gulf Suez-

Foraminiferida—Rotaliina—Carterinacea

Medit.-Japan.——FIG. 629,1,2. *C. spiculotesta (CARTER), Philip. (1), Ceylon (2); 1a-c, opp. sides and edge view of free hypotype which may have broken from small attachment, ×60; 2a, complete, attached specimen with surrounding noncamerate flange also composed of secreted

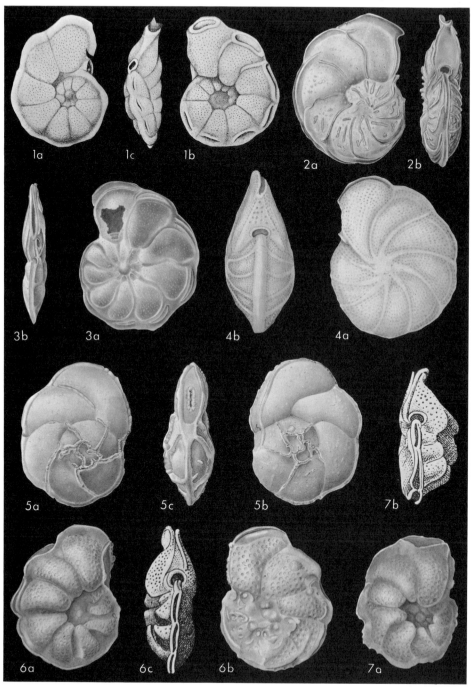

FIG. 628. Anomalinidae (Almaeninae; 1-3, Almaena; 4, Anomalinella; 5, Ganella; 6,7, Queraltina) (p. C763-C764).

Fig. 629. Carterinidae; *1,2, Carterina* (p. C764-C766).

spicules, ×21; *2b*, central area of same specimen dampened to show secondary septa in later whorls and undivided early chambers, ×79 (*1166).

Superfamily ROBERTINACEA Reuss, 1850

[*nom. transl.* LOEBLICH & TAPPAN, 1961, p. 317 (*ex* family Robertinidae REUSS, 1850)] [=superfamily Ceratobuliminidea MYATLYUK in RAUZER-CHERNOUSOVA & FURSENKO, 1959, p. 273]

Test trochospiral, chambers divided internally by partitions which become more important in advanced forms; wall perforate-radial in structure, of aragonite; aperture a low slit in chamber face, with secondary aperture in each septum above partition. ?*Trias., Jur.-Rec.*

Family CERATOBULIMINIDAE Cushman, 1927

[*nom. transl.* GLAESSNER, 1937, p. 27 (*ex* subfamily Ceratobulimininae CUSHMAN, 1927)] [=Epistominidae WEDEKIND, 1937, p. 115; =Conorbidae HOFKER, 1951, p. 414 (*pro Conorbis* HOFKER, 1951, *non* SWAINSON, 1840); =Conorboididae THALMANN, 1952, p. 984; =Ceratubuliminidae HOFKER, 1956, p. 103 (*nom. null.*)]

Test trochospiral; wall calcareous, perforate, of aragonite; primary aperture closed when new chambers added and new foramen opened by resorption above internal partition. ?*Trias., Jur.-Rec.*

Subfamily CERATOBULIMININAE Cushman, 1927

[Ceratobulimininae CUSHMAN, 1927, p. 84]

Primary aperture interiomarginal; coiling predominantly dextral. *Jur.-Rec.*

Ceratobulimina TOULA, 1915, *1943, p. 654 [*Rotalina contraria* REUSS, 1851, *1541, p. 76; OD (M)] [=*Fissistomella* CLODIUS, 1922, *350, p. 141 (type, *Rotalina contraria* REUSS, 1851; SD LOEBLICH & TAPPAN, herein); *Ceratobuliminoides* PARR, 1950, *1429, p. 358 (type, *C. bassensis*)]. Test trochospiral, deeply umbilicate, chambers enlarging rapidly, whorls few, coiling dextral; wall laminated, surface smooth, polished; aperture umbilical, consisting of elongate slit extending in groove up face of final chamber on umbilical side; internally incomplete, marginally serrate partition attached to posterior side of vertical apertural slit at interior of umbilical side, bends around aperture and extends across to be attached to spiral wall for short distance. [*Ceratobuliminoides* was said not to have internal septa but notches on the spiral side seem indicative of an internal partition similar to that of *Ceratobulimina*, and they are here regarded as synonymous.] *U.Cret.-Rec.*, cosmop.——FIG. 630,*1,2*. *C. contraria* (REUSS), M.Oligo., Ger. *(1)*, Denm. *(2)*; *1a-c*, opp. sides and edge view of topotype, ×90 (*2117); *2*, dissected specimen showing external aperture (*a*), septal foramen opening into penultimate chamber *(sf)*, and internal serrate partition *(p)*, ×92 (*1950).——FIG. 630,*3. C. bassensis* (PARR), Rec., Tasm.; *3a-c,* opp. sides and edge view, ×100 (*1429).

Cassidulinita SUZIN in VOLOSHINOVA & DAIN, 1952,

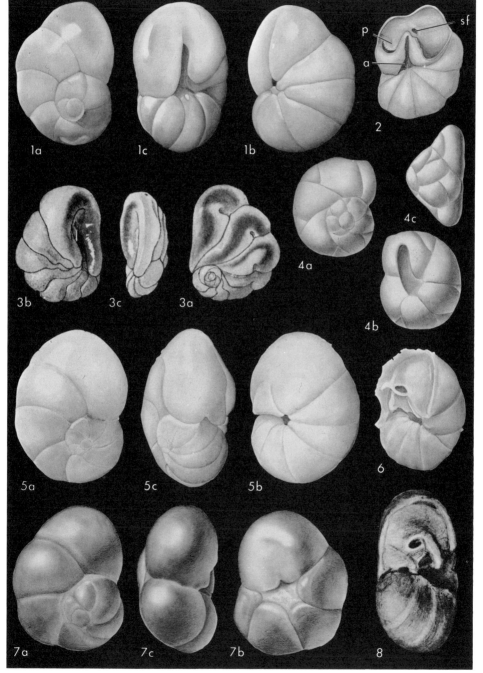

FIG. 630. Ceratobuliminidae (Ceratobulimininae; *1-3, Ceratobulimina; 4, Cassidulinita; 5,6, Ceratocancris; 7,8, Ceratolamarckina*) (p. C766-C769).

C768 *Protista—Sarcodina*

*2022, p. 102 [*C. prima*; OD] [=*Cassidulinella* Suzin, 1937, in Voloshinova & Dain, 1952, *2022, p. 102 (*non* Natland, 1940)]. Test free, extremely small, from 0.08-0.15 mm. diam., planoconvex; chambers biserially arranged and trochospirally enrolled, alternate chambers extending to

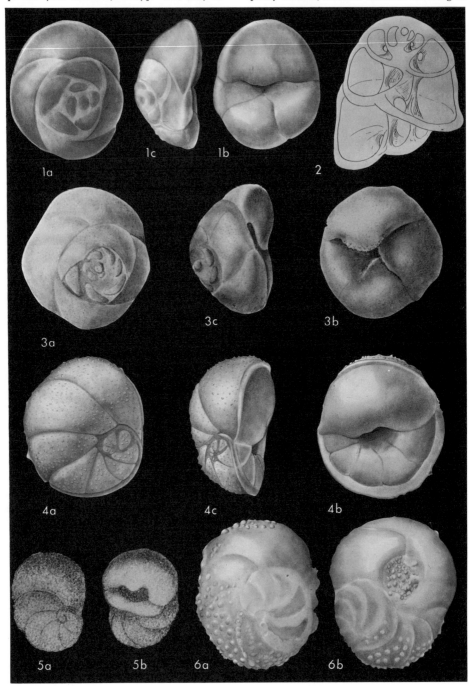

Fig. 631. Ceratobuliminidae (Ceratobulimininae; *1-3, Conorboides; 4,5, Lamarckina; 6, Roglicia*) (p. C769-C770).

umbilicus on flattened umbilical side, with only small triangular portions of other chambers visible as wedges between other chambers at peripheral margin, but these extend entirely to spiral suture on convex opposite side, with only triangular marginal portion of main series of umbilical chambers present on this side; wall calcareous, very finely perforate, smooth; aperture an elongate, crescentiform slit, in shallow depression paralleling outer margin of final chamber, but approximately at mid-line of chamber on umbilical side. *Plio.*, USSR(Caucasus).——Fig. 630, 4. **C. prima*; *4a-c*, opp. sides and edge view of holotype, ×260 (*2022).

[*Cassidulinita* resembles *Cassidulina* in having an enrolled biserial chamber arrangement but differs in being trochospiral. It resembles *Ceratobulimina* in apertural characters but differs in having a biserial chamber arrangement. Whether the wall is of calcite or aragonite is unknown, as is the microstructure (radial or granular). The large aperture and trochospiral coiling suggest placement of this genus with the Ceratobuliminidae, but assignment is tentative pending clarification of wall features. *Rubratella* may be synonymous, as noted under that genus.]

Ceratocancris FINLAY, 1939, *717b, p. 117 [**Ceratobulimina (Ceratocancris) clifdenensis*; OD] [=*Ceratobulimina (Ceratocancris)* FINLAY, 1939, *717b, p. 117 (obj.)]. Similar to *Ceratobulimina* but with low slitlike basal aperture, extraumbilical-umbilical in position, ending at small notch near periphery and at similar notch in umbilical margin; internal partition not attached to spiral wall, and with low accessory internal partition attached to surface of previous whorl just inside aperture. *Mio.*, Eu.-N.Z.——Fig. 630,5,6. **C. clifdenensis*, L.Mio., N.Z.; *5a-c*, opp. sides and edge view of topotype, showing low aperture with small notch at its umbilical and peripheral extremities; *6*, final chamber of topotype removed to show septal foramen, bordered below by main internal partition and small accessory partition just above primary apertural opening, ×69 (*2117).

Ceratolamarckina TROELSEN, 1954, *1950, p. 452 [**Ceratobulimina tuberculata* BROTZEN, 1948, *241, p. 124; OD] [=*Ceratobulimina (Ceratolamarckina)* TROELSEN, 1954, *1950, p. 452 (obj.)]. Test similar to *Ceratobulimina* but with short, wide umbilical aperture, with only small notch at posterior end, and internal partition not attached to interior of chamber on spiral side. *L.Cret.-Paleoc.*, Eu.-N.Am.——Fig. 630,7,8. **C. tuberculata* (BROTZEN), Paleoc., Eu.(Denm.); *7a-c*, opp. sides and edge view of topotype, ×115 (*2117); *8*, specimen figured as *C. perplexa*, Paleoc., Eu.(Sweden), final chamber removed, showing internal partition and intercameral foramen, ×100 (*241).

Conorboides HOFKER in THALMANN, 1952, *1903, p. 14 [*pro Conorbis* HOFKER, 1951, *936, p. 357 (type, *C. mitra*) (*non Conorbis* SWAINSON, 1840)] [**Conorbis mitra* HOFKER, 1951, *936, p. 357; OD] [=*Conorbis* HOFKER, 1950, *932, p. 68, 76 (*nom. nud.*); *Nanushukella* TAPPAN, 1957, *1875, p. 218 (type, *N. umiatensis*)]. Test free, low trochospiral, plano-convex, umbilicate, periphery subacute to rounded; few chambers to whorl; sutures oblique on spiral side, radiate on umbilical side; wall calcareous, of aragonite, by X-ray analysis; aperture a low interiomarginal umbilical arch with short, broad flap that may have fimbriate margin, apertures of earlier chambers of final whorl may remain open along suture beneath flaps, internal pillar extending from aperture parallel to axis of coiling to opposite chamber wall. [*Conorboides* differs from *Conorbina* in having an aragonitic wall (in the type-species of both *Conorboides* and *Nanushukella*), an open umbilicus, and a more extensive umbilical aperture.] *Jur.(Lias.)-L.Cret.(Alb.)*, Eu.-N.Am.——Fig. 631, 1,2. **C. mitra* (HOFKER), Alb., Neth.; *1a-c*, opp. sides and edge view, ×87 (*2117); *2*, vert. sec. showing internal pillars, ×120 (*928c).——Fig. 631,3. *C. umiatensis* (TAPPAN), Alb., Alaska; *3a-c*, opp. sides and edge view, ×112 (*1875).

Lamarckina BERTHELIN, 1881, *134, p. 555 [**Pulvinulina erinacea* KARRER, 1868, *1022, p. 187; OD] [=*Megalostomina* RZEHAK, 1891, *1604, p. 6 (type, *M. fuchsi* RZEHAK, 1895, *1605, p. 228, =*Discorbina fuchsii* RZEHAK, 1888, *1602, p. 228, *nom. nud.*)]. Test free, trochospiral, coiling dextral, plano-convex, spiral side may be pustulose, with chambers increasing rapidly in size, becoming relatively broad and low, periphery carinate, opposite side smooth and polished, deeply umbilicate, final chamber occupying nearly half of area; wall finely perforate, lamellar; aperture an umbilical interiomarginal arch, closed by a thin plate when new chamber is added, septal foramen not homologous to primary aperture; internal partition similar to that of *Ceratobulimina*. *U.Cret.-Rec.*, cosmop.——Fig. 631,4. **L. erinacea* (KARRER), Mio., Eu.(Hung.); *4a-c*, opp. sides and edge of topotype, ×70 (*2117).——Fig. 631, 5. *L. fuchsi* (RZEHAK), Eoc.(Barton.), Eu.(Aus.); *5a,b*, opp. sides, enlarged (*1605).

Praelamarckina KAPTARENKO-CHERNOUSOVA, 1956, *1016, p. 54; 1959, *1018, p. 86 [**P. humilis*; OD]. Test similar to *Lamarckina* but with closed umbilicus, may have umbonal boss on spiral side; aperture interiomarginal. [Nothing is known as to the presence or absence of an internal partition in this genus, nor has the wall composition been described. The original figures are reproduced, but are somewhat generalized.] *L.Jur.(Aalen.)-M.Jur.(Callov.)*, USSR.——Fig. 632,2. **P. humilis*, L.Aalen.; *2a-c*, opp. sides and edge view, ×108 (*1018).

Pseudolamarckina MYATLYUK in RAUZER-CHERNOUSOVA & FURSENKO, 1959, *1509, p. 278 [**Pulvinulina rjasanensis* UHLIG, 1883, *1962, p. 772; OD]. Test trochospiral, plano-convex, umbilicus closed; sutures oblique and thickened on spiral side, depressed and radial on umbilical side; wall thin, finely perforate; aperture interiomarginal, with extension up face of final chamber, internal partition parallel to plane of coiling. [*Pseudolamarckina*

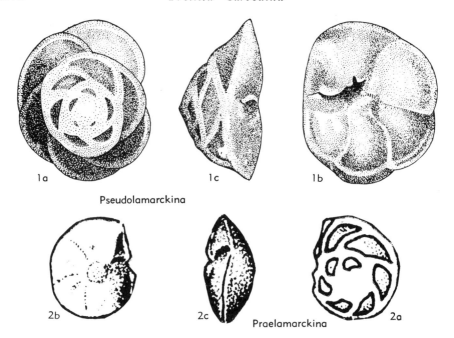

Fig. 632. Ceratobuliminidae (Ceratobulimininae; *1, Pseudolamarckina; 2, Praelamarckina*) (p. C769-C770).

is tentatively included in the Ceratobulimininae, although no information is available as to whether the wall composition is of aragonite or calcite.] M.Jur.-L.Cret., Eu.——Fig. 632,1. *P. rjasanensis (UHLIG), U.Jur.(Kimmeridg.), USSR; 1a-c, opp. sides and edge view, ×80 (*1332).

Roglicia VAN BELLEN, 1941, *112, p. 1000 [*R. sphaerica; OD (M)]. Test free, subglobular, trochospiral, final chambers large, somewhat embracing; surface of test with numerous short spines or pustules except in apertural region, which is smooth; aperture circular, umbilical in position, surrounded by thickened ring covered with thin plate. [*Roglicia* differs from *Ceratobulimina* in having an apertural plate.] Eoc., Eu.——Fig. 631, 6. *R. sphaerica, Yugo.; 6a,b, opp. sides of holotype, ×93 (*2117).

Rubratella GRELL, 1956, *819, p. 760 [*R. intermedia; OD]. Test free, small, trochospiral, to 7 chambers in type-species (4 to 7 in adult agamont, and 1 to 5 in somewhat smaller adult gamont), direction of coiling random, umbilicus closed, chambers broad and semilunate, with strongly curved sutures as seen from spiral side, wedgelike, with straight and radial sutures as seen from umbilical side, each chamber divided by radial internal partition into anterior and posterior half with small interconnecting foramen, only small portion of posterior half visible as triangular wedge near peripheral margin on spiral side, whereas anterior half of chamber occupies most of central portion of spiral side and only that of final chamber is visible on umbilical side, anterior half of previous chambers being covered by successive chambers, both halves formed simultaneously, not successively; lamellar structure not described, wall of anterior half of chamber nonperforate, that of posterior half distinctly perforate; aperture a large, open umbilical arch, occupying about half of diameter of final chamber; cytoplasm with numerous orange-red refringent inclusions (xanthosomes); agamont generation heterokaryotic, with one somatic or vegetative nucleus which disintegrates when reproduction occurs and 5 generative nuclei (in rare specimens, less than 10 per cent, total number of nuclei varying from usual 6 to 2-8), gamont generation with single nucleus, situated in proloculus; during reproduction inner chamber walls resorbed, sexual reproduction plastogamic, 2 individual gamonts joining by their umbilical surfaces to form amoeboid gametes and zygotes, in asexual reproduction entire protoplast escaping from test after nuclear division but before division of cytoplasm into individual young gamonts (*820b). Rec., Eu.(Fr.). ——Fig. 633,1-3. *R. intermedia; 1a,b, opp. sides, showing exterior; 2, decalcified protoplasmic body; 3, living specimen from spiral side, showing numerous xanthosomes, ×500 (*820b).

[Test morphology strongly suggests that *Rubratella* is a synonym of *Cassidulinita*, but the genera have not been compared by us and neither original description gives information as to wall structure (radial or granular), or lamellar character. The imperforate anterior half of the chamber described for *Rubratella* was not noted in *Cassidulinita*, which was merely stated to be finely perforate. Both type-species are extremely small, hence difficult to study in detail, and the 2 genera are therefore tentatively regarded as distinct. *Rubratella* was originally regarded

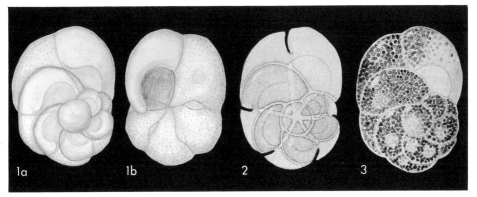

Fig. 633. Ceratobuliminidae (Ceratobulimininae; *1-3, Rubratella*) (p. *C*770).

as belonging to the Rotaliidae, subfamily Discorbinae. Because of its secondary partitions and umbilical aperture, *Rubratella* is tentatively placed in the Ceratobuliminidae.]

Subfamily EPISTOMININAE Wedekind, 1937

[*nom. transl.* LOEBLICH & TAPPAN, 1961, p. 317 (*ex* family Epistominidae WEDEKIND, 1937) [=Praerotalininae HOFKER 1933, p. 125 *(partim) (nom. nud.)*]

Coiling predominantly sinistral; primary aperture on peripheral margin of chambers; internal partition joined to dorsal lip of aperture. ?*Trias., Jur.-Rec.*

Epistomina TERQUEM, 1883, *1892, p. 37 [*E. regularis* TERQUEM, 1883, *1891, p. 379 (=*E. mosquensis* UHLIG, 1883, *1962, p. 766); SD GALLOWAY & WISSLER, 1927, *766, p. 60] [=*Brotzenia* HOFKER, 1954, *943, p. 169 (type, *Rotalia spinulifera* REUSS, 1863, *1554, p. 93); *Voorthuysenia* HOFKER, 1954, *943, p. 169 (type, *Epistomina tenuicostata* BARTENSTEIN & BRAND, 1951, *95, p. 327); *Sublamarckella* ANTONOVA, 1958, *24, p. 68 (type, *S. terquemi*)]. Test lenticular, trochospiral, periphery angular to carinate, umbilical area closed; internal partition crossing chamber cavity from outer margin of lateromarginal apertural opening parallel to periphery on umbilical side, extending nearly or completely to wall against previous whorl; sutures thickened, may be elevated; oblique areal oval aperture on umbilical side, later remaining as interseptal foramen, and additional lateromarginal opening paralleling periphery on umbilical side, in earlier chambers secondarily closed by shell material. ?*Trias., M.Jur.-L.Cret.*, Eu.-N.Am.-Afr.——FIG. 634,*1,2*. **E. regularis*, M.Jur.(Bajoc.), Fr. *(1)*, M.Jur.(Dogger γ), Eu.(Aus.) *(2)*; *1a-c*, opp. sides and edge of holotype, ×40 (*700); *2a-c*, opp. sides and edge of holotype of *E. mosquensis* UHLIG, ×72 (*1332).——FIG. 634,*3-5*. *E. spinulifera* (REUSS), L.Cret.(Alb.), Eu.(Neth.); *3a-c*, opp. sides and edge, ×50 (*555); *4*, horiz. sec. seen from umbilical side, showing internal partition, ×36; *5*, axial sec. through chamber to show septal foramen *(f)*, and internal partition *(p)*, ×195 (*943).——FIG. 634,*6*. *E. terquemi* (ANTONOVA), M.Jur.(Bajoc.), USSR(Caucasus); *6a-c*, opp. sides and edge view, ×80 (*24).——FIG. 634,*7*. *E. tenuicostata* BARTENSTEIN & BRAND, L.Cret.(Valangin.), N.Ger.; *7a-c*, opp. sides and edge view of holotype, ×65 (*95).

[*Epistomina* has been much divided recently and because of the poor illustrations and descriptions of the type-species, nearly all other species have been later assigned to one or another of these later genera. Many of TERQUEM's figures of other species are somewhat inaccurate, and the type of *E. regularis* was not located by us during an extensive search in the French museums in 1954, hence it is presumed to be lost. It is almost certainly identical to *E. mosquensis* UHLIG, 1883, however. Both species were described from equivalent strata, TERQUEM's species being from the Bajocian of Moselle, France, and that of UHLIG from the mid-Jurassic Dogger γ to basal upper Jurassic Malm α of Austria. UHLIG's species is commonly recognized and has been restudied both in Germany and the USSR. It was placed in *Brotzenia* by HOFKER (*943). *Brotzenia* is, therefore, here regarded as a synonym of *Epistomina*. *Voorthuysenia* was separated largely on relative size of the internal partition, which is here regarded as only of specific value. *Sublamarckella* was separated on the basis of the semicircular or reniform areas bordered by elevated ridges which lie near the umbilical region, and which are covered by thin shell material and regarded as representing apertures. Many species of *Epistomina* (including the type-species) show similar umbilical ornamentation, not here regarded as homologous to the lateromarginal aperture of *Epistomina*, which is directly related to the internal partition. *Sublamarckella* is also considered to be a junior synonym of *Epistomina*.]

Epistominita GRIGELIS, 1960, *825, p. 98 [**E. sudaviensis*; OD]. Test free, trochospiral, close-coiled, biconvex; chambers with internal secondary partition extending from spiral margin of peripheral aperture to attach obliquely to wall on umbilical side, as in *Epistominoides*, resulting in appearance of "supplementary chambers"; aperture a peripheral slit nearly in plane of coiling, with lip, apertures of earlier chambers closed by secondary skeletal material, but distinctly noticeable as peripheral grooves in these earlier chambers. [*Epistominita* has early apertural slits as in *Epistomina* and *Hoeglundina* and oblique supplementary sutures of the internal partition visible on the umbilical side of the test, as in *Epistominoides*.] U.Jur.(Oxford.), Eu.(Lith.).——FIG. 635, *1*. **E. sudaviensis*; *1a-c*, opp. sides and edge view of holotype, ×60 (*825).

Epistominoides PLUMMER, 1934, *1466, p. 602 [**Saracenaria wilcoxensis* CUSHMAN & PONTON, 1932, *521, p. 54; OD]. Test free, enrolled, slightly trochospiral, chambers triangular in section, enlarging rapidly, internally divided by partition which extends inward from aperture on spiral side of test across chamber cavity to attach at opposite wall where attachment forms supplementary suture; primary aperture a short peripheral slit at dorsal angle, lips merging gradually

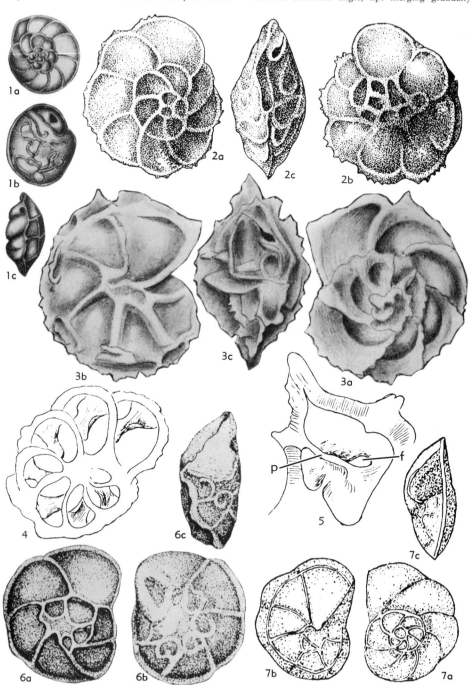

Fig. 634. Ceratobuliminidae (Epistomininae; *1-7, Epistomina*) (p. C771).

into peripheral keel, aperture closed when new chambers are added, and intercameral foramen formed by resorption about midway in septal face. *U.Jur.(Oxford.)-L.Eoc.*, N.Am.-Eu.——FIG. 636,1,2. *E. wilcoxensis* (CUSHMAN & PONTON), Paleoc.(Dan.), USA(Tex.); *1a*, side view showing primary and supplementary septa due to internal partition; *1b*, apert. view showing position of attachment of internal partition and external lateromarginal aperture; *2*, apert. view of specimen

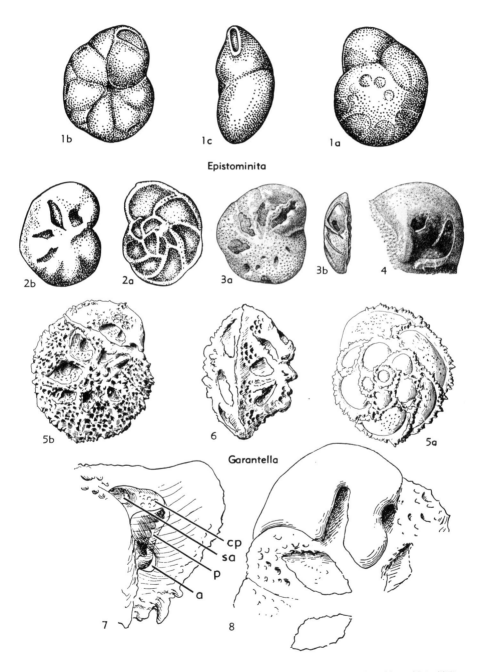

FIG. 635. Ceratobuliminidae (Epistomininae; *1, Epistominita; 2-8, Garantella*) (p. C771, C774-C775).

with last chamber removed showing areal intercameral foramen and remnant of internal partition; all ×93 (*2117).

Garantella KAPTARENKO-CHERNOUSOVA, 1956, *1016, p. 55; 1959, *1018, p. 102 [*G. rudia; OD]. Similar to Reinholdella but differs in umbilical-

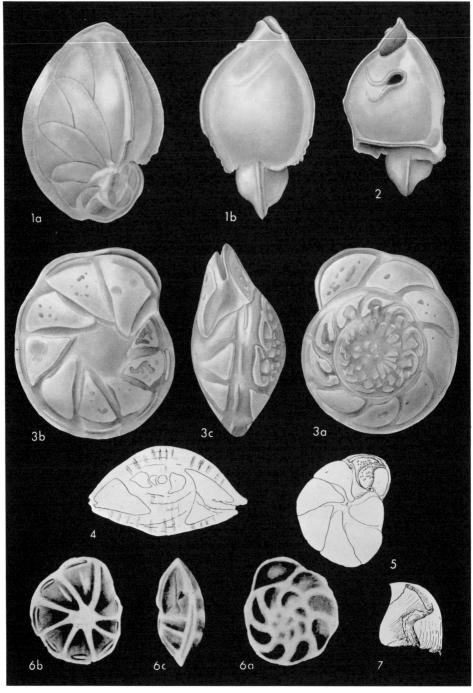

FIG. 636. Ceratobuliminidae (Epistomininae; 1,2, Epistominoides; 3-7, Hoeglundina) (p. C771-C776).

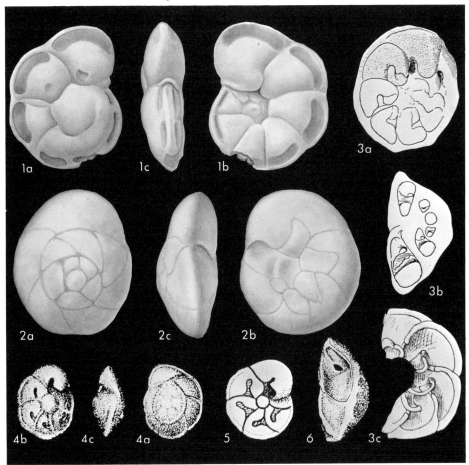

FIG. 637. Ceratobuliminidae (Epistomininae; *1, Mississippina; 2-6, Reinholdella*) (p. *C776-C777*).

sutural position of apertures and covering plates, which are thus parallel to sutures, instead of perpendicular to them. [In addition to Upper Bajocian species recorded from the Ukraine, *Reinholdella ornata* HOFKER, from strata of the same age in Germany, belongs to *Garantella*. *G. floscula* KAPTARENKO-CHERNOUSOVA is apparently a synonym of *R. ornata*, and the species *G. ornata* (HOFKER) thus ocurs in Germany and the Ukraine in Upper Bajocian strata (*Garantia garanti* zone, Dogger ε.] M.Jur.(U.Bajoc.), USSR(Ukraine)-Eu.(Ger.).——FIG. 635,2-4. *G. rudia*, Ukraine; 2a,b, spiral and umbilical sides of holotype, ×55; 3a,b, umbilical side and edge view of different specimens showing interseptal foramen, ×33; 4, last chamber from umbilical side, enlarged to show aperture and septal foramen before addition of cover plate (*1509).——FIG. 635,5-8. *G. ornata* (HOFKER), Ger.; 5a,b, spiral and umbilical sides showing ornamented test, umbilical-sutural supplementary apertures, and covering plates, ×80; 6, edge view, ×60; 7, optical sec. of final chamber (in clarifying oil) showing aperture (*a*), internal partition (*p*) with recurved margin, and porous cover plate (*cp*) over sutural aperture (*sa*); 8, ext. of final chamber with both apertures remaining open before addition of the cover plate, ×160 (*937).

Hoeglundina BROTZEN, 1948, *241, p. 92 [*Rotalia elegans* D'ORBIGNY, 1826, *1391, p. 272; OD] [=*Hiltermannia* HOFKER, 1954, *943, p. 169 (type, *Epistomina chapmani* TEN DAM, 1948, *555, p. 166)]. Test similar to *Epistomina* with more highly developed internal partition extending from posterior wall of chambers and always secondarily resorbed from earlier chambers; lateromarginal aperture nearly peripheral in position extending breadth of chambers; those of earlier chambers may remain open or be secondarily closed. [*Hiltermannia* was separated on the basis of a smaller internal partition, relative size being a feature here regarded as of only specific importance.] M.Jur.(Dogger)-Rec., cosmop.——FIG. 636,3-5. *H. elegans* (D'ORBIGNY), Rec., Carib.;

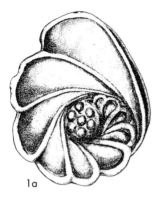

Fig. 638. Ceratobuliminidae (Epistomininae; *1, Pseudoepistominella*) (p. C776).

3a-c, opp. sides and edge view, ×31 (*2117); 4, axial sec. showing internal partition in final chamber only, ×20 (*943); 5, umbilical view of small specimen with final chamber clarified to show internal partition and position of areal and lateromarginal apertures, enlarged (*928c).——Fig. 636,6,7. *H. chapmani* (Ten Dam), L.Cret., Eu.(Neth.); 6a-c, opp. sides and edge view of holotype, ×57 (*555); 7, last chamber clarified to show internal partition, ×80 (*943).

Mississippina Howe, 1930, *969, p. 329 [*M. monsouri*; OD]. Test free, coiled, trochoid in early stage, later becoming nearly planispiral, spiral side umbonate, umbilical side somewhat more involute; sutures nearly radial; wall calcareous, perforate, monolamellar; aperture interiomarginal on periphery and extending somewhat to umbilical side beneath slight flap of final chamber, supplementary apertures near peripheral keel and paralleling it on both sides, filled with bands of clear shell material. [Differs from *Stomatorbina* in being nearly planispiral, in having a peripheral aperture, and supplementary apertural shell bands on both sides of the peripheral keel.] L.Oligo.-Rec., N.Am.-Pac.O.——Fig. 637,1. *M. monsouri*, L.Oligo., USA(Miss.); 1a-c, opp. sides and edge view of holotype, ×65 (*2117).

Pseudoepistominella Kuznetsova in N. K. Bykova et al., 1958, *265, p. 48 [*P. mirusa*; OD]. Test free, lenticular, biumbonate, umbilical region pustulose, with thickened knobs or pustules on both sides of test, early stage may be slightly trochospiral, later planispiral and bievolute, periphery carinate; chambers numerous, low, broad and curved, similar in form on both sides of test, with small internal diagonal partition; sutures curved, oblique, thickened; wall calcareous, smooth, probably perforate and aragonitic; aperture of 2 types, primary aperture interiomarginal, equatorial arch and additional oval areal aperture about one-third of distance from base of apertural face, both openings with thickened lip. [The wall of *Pseudoepistominella* was originally stated to be porcelaneous, smooth and dull, without statement as to presence or absence of pores, but as the genus was inferred to belong to the Epistominidae, it is here regarded to be perforate of aragonitic composition, as many of the other early Epistominidae have a porcelaneous appearance, but they are not imperforate like the Miliolacea.] L.Cret.(Barrem.), USSR(Caucasus).——Fig. 638,1. *P. mirusa*; 1a-c, opp. sides and edge view of holotype, showing 2 types of apertures and small transverse internal partition, near upper part of septal face, ×32 (*265).

Rectoepistominoides Grigelis, 1960, *825, p. 102 [*R. scientis*; OD]. Test similar in early development to *Epistominoides* but later stage uncoiling and becoming rectilinear; elongate slitlike aperture at dorsal angle of chamber, bordered by lip. U.Jur.(L.Oxford.), Eu.(Lith.).——Fig. 639,1. *R. scientis*; 1a-c, opp. sides and edge view of hypotype, ×60 (*825).

Reinholdella Brotzen, 1948, *241, p. 126 [*Discorbis dreheri* Bartenstein in Bartenstein & Brand, 1937, *92, p. 192; OD] [=*Lamarckella* Kaptarenko-Chernousova, 1956, *1016, p. 54, 1959, *1018, p. 91 (type, *L. media*)]. Test free, trochospiral, plano-convex to biconvex; supplementary cover plates surrounding umbilicus to cover sutural apertures, extending farthest toward periphery near mid-portion of primary chambers so that latter have saddle-shaped outline on umbilical side; sutures oblique dorsally, radiate ventrally; wall of aragonite (by X-ray powder diffraction film) finely perforate; aperture a low interiomarginal arch near periphery on umbilical side, with supplementary aperture in indentation at center of suture on umbilical side but secondarily closed in most specimens, internal pillar-like partition connected to this aperture extending from umbilical to spiral walls. L.Jur.(U.Lias.)-M.Jur. (L. Dogger), Eu.——Fig. 637,2,3. *R. dreheri* (Bartenstein), M.Jur.(Dogger), Ger.; 2a-c, opp. sides and edge view, ×185 (*2117); 3a-c, diagrams to show apert. characters; *3a*, last 2 chambers shaded showing final chamber with open aperture and secondary covering plate over this in earlier chambers; *3b,* transv. sec. showing pillarlike internal partitions and septal foramina; *3c,* horiz. sec. seen from spiral side, showing position and form of internal partitions and septal foramina, approx. ×70 (*937).——Fig. 637,*4-6*. *R. media* (KAPTARENKO-CHERNOUSOVA), M.Jur. (Bajoc.), USSR; *4a-c,* opp. sides and edge views; *5,* umbilical side; *6,* edge view showing intercameral foramen, ×54 (*1018).

[Originally the type-species was described as *Discorbis,* later placed in *Asterigerina* and finally made the type-species of the new genus *Reinholdella* on the basis of the "umbilical and interiomarginal aperture and an inner partition in the chambers as in *Lamarckina.*" BROTZEN considered that it approached closely "an ideal type of the primitive Ceratobuliminae," and that it was possibly ancestral to *Lamarckina, Ceratobulimina,* and *Asterigerina.* HOFKER (1952, *937, p. 20) described the 2 apertures in the type-species. The one near the periphery was designated by him as a deuteroforamen, formed by arching of the suture. The secondary aperture, in the chamber indentation, was the protoforamen of HOFKER and, as he stated (p. 22), "a well-developed tooth plate is connected with the proximal foramen, thus indicating that this foramen is a protoforamen." He also noted that the protoforamen is commonly "closed by a porous plate which forms the so-called supplementary chamber. This closing of the protoforamen leads to the forming of a small chamberlet, mainly formed by the lumen of the protoforamen itself. It is connected with the normal chamber by the opening of the nearly vertical toothplate." HOFKER stated that *Reinholdella* had been derived from *Conorboides,* and on the basis of a new species, *R. epistominoides,* he concluded that it was closely related to *Epistomina.* Our examination shows that the type-species of *Reinholdella* is composed of aragonite, as proved by X ray with powder diffraction film, thus upholding the suggested relationship of *Reinholdella* to other aragonitic genera (e.g., *Lamarckina, Ceratobulimina, Epistomina*) which was proposed originally on the basis of apertural and internal features. *Lamarckella* was said to be characterized by supplementary sutural apertures on the umbilical side, which also occur in the type-species of *Reinholdella,* and it was regarded as synonymous with *Reinholdella,* in the family Ceratobuliminidae, by MYATLYUK in RAUZER-CHERNOUSOVA & FURSENKO (1959, *1509, p. 277).]

Schlosserina HAGN, 1954, *860, p. 18 [**Rosalina asterites* GÜMBEL, 1870, *840, p. 658; OD]. Test free, trochoid, biconvex, with peripheral keel, ventrally umbilicate; all chambers visible on spiral side and sutures limbate, curved, only chambers of last whorl visible on umbilicate opposite side, where sutures are depressed and straight; wall calcitic (by X-ray powder diffraction film; see below), perforate; aperture multiple, of 4 types, all on umbilical side; primary aperture low slit at base of final chamber, supplementary sutural slits between later chambers, large areal pores scattered over final chamber face and wide-spiraling slits near periphery which are filled with secondary shell material. *Eoc.,* Eu.——Fig. 640,*1.* **S. asterites* (GÜMBEL); *1a-c,* opp. sides and edge view of neotype, ×35 (*2117).

[*Schlosserina* resembles *Stomatorbina* in being trochospiral with secondary spiraling slits only on the umbilical side but differs in possessing a multiple areal aperture. Since all of GÜMBEL's collection was destroyed during World War II, the types of *S. asterites* are lost. The specimen illustrated by HAGN (1954, *860, pl. 3, fig. 15) in describing the genus was referred to as a "genoholotype" (Coll. Munich Prot. 272) and is here designated as neotype of *Rosalina asterites.* It is from the "Stockletten" marls (Eocene) of the Rollgraben near Kressenberg, Bavaria, Germany. The X-ray diffraction film made for *Schlosserina* showed a dominantly calcite pattern, but portions of it

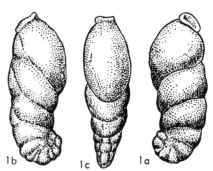

FIG. 639. Ceratobuliminidae (Epistomininae; *1, Rectoepistominoides*) (p. *C776*).

also showed that traces of aragonite were present. It is difficult to determine whether the aragonite traces represent the original wall or adherent material on the exterior of the shell, and whether the calcite represents the wall of the shell or a filling in the interior. It will be necessary to obtain clean and unfilled specimens in order to determine the exact wall composition more exactly. *Schlosserina* is tentatively placed with the morphologically similar *Mississippina* and *Stomatorbina* in the aragonite-walled Epistominidae.]

Stomatorbina DORREEN, 1948, *610, p. 295 [**Lamarckina torrei* CUSHMAN & BERMÚDEZ, 1937, *491, p. 21; OD]. Test free, trochoid, all chambers visible on convex spiral side where sutures are limbate and curved, only chambers of final whorl visible on umbilicate opposite side where sutures are radial; wall calcareous, of aragonite (by X-ray analysis), perforate; aperture consisting of interiomarginal slit on umbilical side, not reaching periphery, supplementary apertures represented by bands of clear shell material, paralleling peripheral keel only on umbilical side in adult. *Eoc.,* W.Indies(Cuba)-N.Z.——Fig. 640,*2.* **S. torrei* (CUSHMAN & BERMÚDEZ), Cuba; *2a-c,* opp. sides and edge view of holotype, ×41 (*2117).

[UCHIO (1952, *1958, p. 197) stated that young specimens of *Pulvulina concentrica* PARKER & JONES are planispiral and show bands of shell material on both sides of the test in young stages but become trochoid in later development, with loss of the supplementary apertures on the spiral side, thus showing a change from the characters of *Mississippina* in the juvenile forms to the adult characters of *Stomatorbina,* differing only in the aperture which remains peripheral while typical *Stomatorbina* has an aperture restricted to the umbilical side. A close relationship is shown between the genera, but the distinct adult characters are considered to be sufficient basis for their separation.]

Family ROBERTINIDAE Reuss, 1850

[Robertinidae REUSS, 1850, p. 375] [=Robertininae SIGAL in PIVETEAU, 1952, p. 220]

Test high, trochospiral, coiling predominantly dextral; septal foramen homologous with part of primary aperture, not a secondary feature as in the Ceratobuliminidae. *U.Cret.-Rec.*

Robertina D'ORBIGNY, 1846, *1395, p. 202 [**R. arctica*; OD (M)]. Test elongate, high, trochospiral, with several chambers in each whorl, chambers divided by double transverse partition

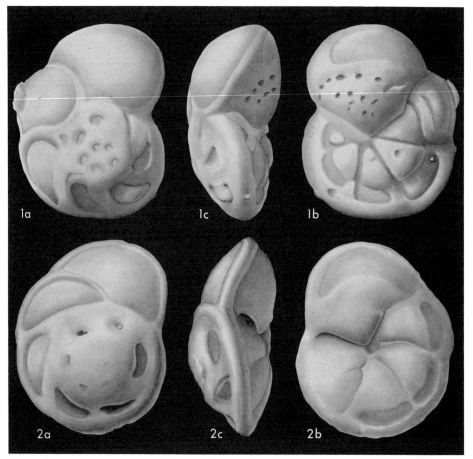

FIG. 640. Ceratobuliminidae (Epistomininae; *1, Schlosserina;* 2, *Stomatorbina*) (p. C777).

formed by infoldings of outer wall, chamber halves interconnected by low opening against previous chambers; primary aperture an elongate, loop-shaped opening extending up face of final chamber, with small supplementary triangular aperture on opposite side of test, where transverse internal partition meets preceding chamber, supplementary openings of earlier chambers secondarily closed as new chambers are added. L.Eoc.-Rec., Eu.-N.Am.-N.Z.-Tasm.-Atl.O.-Pac.O.-Arctic-Antàrctic.——FIG. 641,*1*. **R. arctica*, Rec., Spitz.; *1a-c*, opp. sides and edge view showing single primary loop-shaped aperture and supplementary aperture at suture junctions on opposite side of test, ×55 (*924); *1d*, detached chamber oriented as in *1a*, viewed from within, showing internal partition, primary aperture on far side, and small supplementary opening at the upper end of internal partition, ×105 (*924).

Alliatina TROELSEN, 1954, *1950, p. 464 [**Cushmanella excentrica* DI NAPOLI ALLIATA, 1952, *1346, p. 105; OD]. Test similar to *Cushmanella* but simpler and asymmetrically developed, internal partition consisting of oblique inverted V-shaped projection extending inward from oblique areal aperture to attach below septal foramen of penultimate chamber. Plio.-Rec., Eu.-N.Am.-Pac.O.-Malay Penin.-Kerimba Arch.——FIG. 641,*2,3*. **A. excentrica* (DI NAPOLI ALLIATA), Plio., Italy; *2a,b*, side, edge views of metatype, ×119 (*2117); *3a*, oblique view of dissected specimen showing areal septal foramen of penultimate chamber (*f*), primary aperture (*a*), and angular internal partition (*p*) surrounding aperture; *3b*, edge view, with final chamber partially dissected, showing angular asymmetrical internal partition, primary aperture at exterior lying within sharp upper angle, ×75 (*1950).

Alliatinella D. J. CARTER, 1957, *284, p. 82 [**A. gedgravensis*; OD]. Test similar to *Alliatina* but distinctly trochospiral, accessory chambers developed only on umbilical side; internal partition asymmetrical and chevron-shaped in section, extending obliquely across chamber; areal aperture asymmetrically placed somewhat to umbilical side of test and may be closed by thin plate and therefore nonfunctional until plate is resorbed to form septal foramen after addition of another chamber, basal, equatorial aperture always open. Plio., Brit. I.(Eng.)-Eu.(Italy).——FIG. 641,4,5. *A. gedgravensis, Eng.; 4a-c, holotype, opp. sides and edge view, ×92 (*284); 5, apert. view of dissected

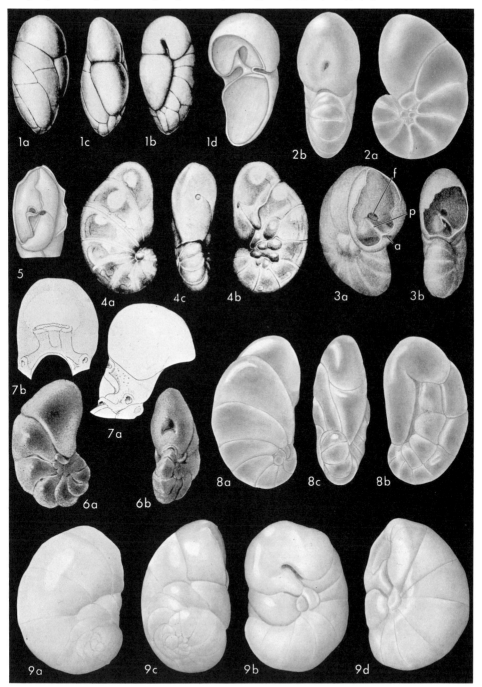

FIG. 641. Robertinidae; *1, Robertina; 2,3, Alliatina; 4,5, Alliatinella; 6,7, Cushmanella; 8, Cerobertina; 9, Pseudobulimina* (p. C777-C782).

paratype showing septal foramen with internal partition visible on left side of figure and sutural line showing position of partition on right side of chamber, ×92 (*284).

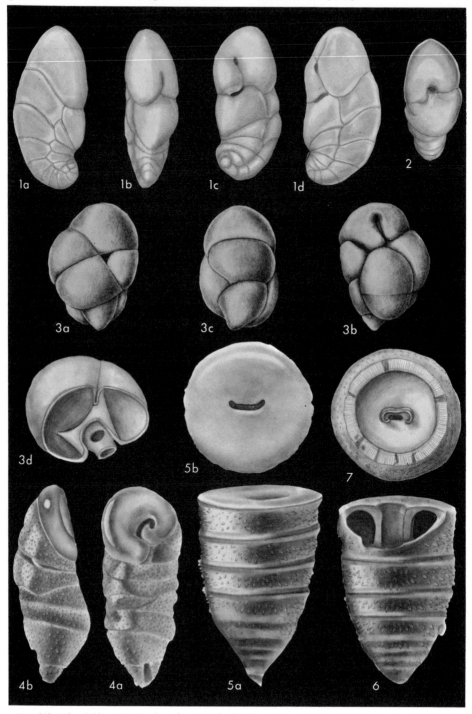

FIG. 642. Robertinidae; *1,2, Geminospira; 3, Robertinoides; 4, Ungulatella; 5-7, Colomia* (p. C781-C782).

Cerobertina FINLAY, 1939, *717b, p. 118 [*C. bartrumi; OD]. Test ovate to auriculate in outline, periphery rounded, chamber arrangement trochospiral in early stage, later uncoiling, internal secondary partition partially dividing chambers, sutures of these partitions visible on umbilical side where they appear to form supplementary chamberlets about equal in size to that of primary chamber on same side; wall of aragonite, perforate, surface smooth; aperture an interiomarginal slit, which extends in furrow-like depression of apertural face. *M.Eoc.-Rec.*, N.Z.-Antarctic-Malay Arch.——FIG. 641,8. *C. bartrumi*, L.Mio., N.Z.; *8a-c*, opp. sides and edge view of paratype, ×92 (*2117).

[*Cerobertina* differs from *Pseudobulimina* in having the smaller chamberlets and secondary partitions visible on the umbilical, rather than the spiral, side. In *Geminospira* the chamberlets can be seen from both sides and are peripheral in position. *Alliatinella* is similar to *Cerobertina*, but only the basal aperture is always open and the vertical slit is commonly closed, leaving only a small, rounded opening in the face.]

Colomia CUSHMAN & BERMÚDEZ, 1948, *495, p. 12 [*C. cretacea*; OD]. Test free, elongate, conical, early stage trochospiral, with 3 chambers in early microspheric whorl, followed by more or less well-developed biserial stage and finally uniserial, with low cylindrical chambers; sutures may be thickened and elevated; wall of aragonite, surface may be hispid or smooth; aperture a terminal crescentic slit, internal hemicylinder with thickened borders extending from inside of aperture to terminal wall of previous chamber, apertures of successive chambers and internal columella gradually changing in orientation at successive angles of about 80°. *U.Cret.-M.Eoc.(L.Lutet.)*, W. Indies(Cuba)-USA-Eu.——FIG. 642,5,6. *C. cretacea*, U.Cret., Cuba; *5a,b*, side, top views of holotype; *6*, paratype with final chamber dissected to show hemicylindrical columella; all ×133 (*2117).——FIG. 642,7. *C.* sp., U.Cret., USA (Calif.); interior of final chamber showing shape of hemicylindrical columella as seen in cross section, ×105 (*2117).

Cushmanella PALMER & BERMÚDEZ, 1936, *1411, p. 252 [*Nonionina brownii* D'ORBIGNY in DE LA SAGRA, 1839, *1611, p. 45; OD]. Test free, in nearly planispiral coil, involute, chambers increasing rapidly in size, with one or more small supplementary chambers at umbilical area of each chamber on both sides of test; wall calcareous, finely perforate except for equatorial oval area just above primary aperture on final septal face; arcuate, slitlike primary areal aperture, secondary interiomarginal equatorial aperture, and small accessory apertures at sides of test at supplementary chambers, tubelike internal partition attached only at upper and lateral inner margins of primary aperture, having free edges somewhat infolded at each side of lower margin of aperture; lateral tubular branches from partition opening at lateral accessory apertures. *Rec.*, Carib.——FIG. 641,6,7. *C. brownii* (D'ORBIGNY), Cuba; *6a,b*, side, edge views of topotype, ×30 (*117); *7a*, optical sec. of last chamber in side view, showing internal tubular partition extending back from primary areal aperture to attach at outer wall of previous chamber and extending laterally to small accessory apertures, secondary equatorial interiomarginal aperture also visible; *7b*, optical sec. of last chamber as seen from aperture, showing internal partition wtih free infolded basal margin, primary areal aperture, secondary basal aperture below and smaller accessory lateral apertures at ends of tubular extensions from partition, with oval nonperforate area of chamber wall above primary aperture; both ×120 (*946).

[On the original type slide of *Nonionina brownii* in the D'ORBIGNY collections in the Museum National d'Histoire Naturelle, Paris, examined by us in 1954, 3 specimens are mounted. One of these is crushed beyond recognition; the other 2 are conspecific but neither agrees with the original figures or descriptions given by D'ORBIGNY nor do they represent *Cushmanella* as generally understood. It is possible that the original illustration and description were based on the specimen which was later destroyed and that more than one species was originally erroneously regarded as identical. However, these specimens appear much closer in all respects to D'ORBIGNY's figures of *Valvulina inaequalis* (*1393, pl. 7, fig. 10-12), being distinctly trochospiral, with an umbilical flap of the last chamber covering the umbilical area and a simple interiomarginal aperture. This strongly suggests another possibility as to disposition of the types of *N. brownii*. The D'ORBIGNY types are mounted on tiny glass plates inserted in vials, which in turn are glued to boards bearing the printed labels. During our examination of this collection, it was noted that in some instances the glue attaching the vials to the labeled boards had been dried and cracked, and some of the vials had become detached. Some of them had obviously been later reglued to the boards, leaving open the possibility that the vial of *N. brownii* may have been so detached and perhaps erroneously later fastened to the wrong board.——¶As the above-mentioned specimens show none of the diagnostic features mentioned by D'ORBIGNY in his original description (nor even the same number of chambers per whorl), do not have supplementary chambers, are not planispiral, have no areal aperture, and thus do not resemble the original figures, it is probable that the original type has been lost, misplaced, or destroyed. In the interests of nomenclatural stability, we here recognize the species *N.brownii* (and *Cushmanella*, the genus based on it) as figured and described by D'ORBIGNY and all later workers, rather than as represented by the above-mentioned questionable specimens in the D'ORBIGNY collection.]

Geminospira MAKIYAMA & NAKAGAWA, 1941, *1206, p. 243 [*G. simaensis*; OD]. Test elongate, early chambers in trochospiral arrangement, later uncoiling and arcuate but somewhat asymmetrical, secondary series of smaller chamberlets at inner periphery and visible from both sides of test, somewhat more extensive on umbilical side; sutures radiate, curved, slightly depressed; wall of aragonite, finely perforate, surface smooth; aperture an interiomarginal slit at base of final chamber and nearly equatorial in position, with elongate groove extending up face of final chamber and broadening into ovate opening at upper end, which remains as intercameral foramen when next supplementary chamber is added. [*Geminospira* was regarded as a synonym of *Pseudobulimina* by ASANO (1950, *52a, p. 2), but *Geminospira* differs in having the secondary chambers visible from both sides of the test, owing to their equatorial position. The aperture was originally described as a vertical slit, and the interiomarginal opening was not previously noted.] *Plio.-Pleist.*, Japan.——FIG. 642,*1,2.* **G. simaensis*, Pleist.; *1a-d*, opp. sides, edge, and oblique views showing chamberlets and apertures; *2*, specimen with opened final chamber to show intercameral foramen, ×105 (*2117).

Pseudobulimina EARLAND, 1934, *653, p. 133 [**Bulimina chapmani* HERON-ALLEN & EARLAND, 1922, *911, p. 130; OD (M)]. Test ovate to elongate, with rapidly enlarging chambers in low trochospiral coil, chambers internally subdivided as in *Robertina*, suture formed by partial division visible on spiral side but not on umbilical side; aperture with 2 diverging slits as in *Robertinoides*, walls of vertical slit in face extending inward to connect with upper surface of preceding chamber, only sutural slit opening into larger chamber cavity, smaller chamber cavity without external opening, but connecting internally to larger chamber. *Eoc.-Rec.*, N.Am.-Eu.-Antarctic-Pac.O.——FIG. 641,*9.* **P. chapmani* (HERON-ALLEN & EARLAND), Rec., Antarctic; *9a-d*, spiral, umbilical, edge, and oblique umbilical views, ×56 (*2117).

Robertinoides HÖGLUND, 1947, *924, p. 222 [**Bulimina normani* GOËS, 1894, *804, p. 47; OD]. Test elongate, in high trochospiral coil, chambers divided by transverse partition formed by infolding of wall, as in *Robertina*, chamber halves connecting by low opening under this partition against earlier chambers; tubular chamber extension occurring between 2 divergent slits of primary aperture, opening into proximal half of chamber within, and connecting with exterior by means of oval opening into main aperture where its divergent branches originate, primary aperture double, with elongate slit or loop extending up face of final chamber in position of internal partition, and similar elongate slit along suture against previous whorl at distal margin of chamber; accessory aperture also present on opposite side of test where suture of internal partition meets spiral suture, as in *Robertina*. *Rec.*, Eu.(Sweden).——FIG. 642,*3.* **R. normani* (GOËS); *3a-c*, opp. sides and edge of neotype, ×50 (*924); *3d*, detached final chamber viewed from within, showing 2 diverging slits of primary aperture, internal septum, accessory aperture, and tubular chamber extension with large rounded opening between primary apertural slits, ×140 (*924).

Ungulatella CUSHMAN, 1931, *449, p. 81 [**U. pacifica*; OD]. Test elongate, trochospiral, with conical proloculus, later with sides nearly parallel, and ovate section, chambers as seen in clarifying oil (e.g., castor oil) each a full coil in length, suggesting uniserial appearance, but with one margin always overlapping other, and oblique sutures visible on side from which aperture may be seen; wall coarsely perforate, surface with tiny pustules but apertural face clear, smooth and polished, or may have faint radial striae; aperture umbilical, appearing as recurved loop extending up face from one side of final chamber. *Rec.*, Pac.——FIG. 642,*4.* **U. pacifica*; *4a,b*, side and edge views of paratype showing apert. character, ×192 (*2117).

[Originally included in the Buliminidae, *Ungulatella* was regarded as an uniserial derivative from *Buliminella* and *Buliminoides*. Later (*464, p. 101) it was stated to have a high-spired, undivided coil and to be related to *Conicospirillina*. It is here transferred to the Robertinidae and regarded as more closely related to *Robertina* and *Colomia* than to the above-mentioned Buliminidae.]

NOMINA INQUIRENDA

Diplostoma EBENSBERGER, 1962, *654A, p. 54 [**D. siamesia*; OD] [*non Diplostoma* RAFINESQUE, 1817, *nec* DE FROMENTEL, 1860]. Genus based on 16 specimens of type-species from *U.Cret. (Maastricht.)*, Ger.(Aachen). Placed in the Lagenidae (=Nodosariidae), it was said to occur as 1-, 2- or 4-chambered tests; chambers fusiform, with radiate apertures at both ends of proloculus; later chambers added simultaneously at each end, each with radiate terminal aperture. [As this form of growth is previously unknown in the Nodosariidae or Polymorphinidae, we believe it probable that these represent twinned specimens of a polymorphinid, perhaps one of the associated species of *Pyrulinoides* or *Pyrulina*. However, if additional evidence upholds the validity of this genus, it will have to be renamed, since the name here cited is a junior homonym.

Pseudonovella KIREEVA, 1949, *1040A bis?. *Pseudonovella* was stated by A. D. MIKLUKHO-MAKLAY, RAUZER-CHERNOUSOVA & ROZOVSKAYA in RAUZER-CHERNOUSOVA & FURSENKO (1959,*1509, p.208) to be a subgenus of *Novella* GROZDILOVA & LEBEDEVA. We have seen no citation for the original reference to the genus, but it is probably in the publication cited above, which we have been unable to locate in any U.S. library. The type-species and method of its fixation are also unknown to us. *Pseudonovella* differs from *Novella (Novella)* in the involute, rather than evolute, character of the final whorl.

GENERIC NOMINA NUDA APPLIED TO FORAMINIFERIDA

Acanthospira REINSCH, 1877, *1526, p. 177.
Amorphina PARKER in PARKER & JONES, 1857, *1416, p. 278.
Amphigramma REINSCH, 1877, *1526, p. 177.
Askopsis DE FOLIN, 1881, *724, p. 138.
Asterorbitoides A. SILVESTRI, 1907, *1768, p. 86 (*nom. nud.*, no species named). Seemingly proposed for radiate lepidocyclines.
Bigeneropolis MARIE, 1950, *1219, p. 50.
Calcidiscus GROZDILOVA, 1960, *830, p. 44.
Caspirella N. K. BYKOVA, 1960, *263, p. 324.
Caucasinella MYATLYUK, 1960, *1333, p. 208.
Chaetotrochus EHRENBERG, 1866, *686, p. 76, 81.
Cheirammina DE FOLIN, 1881, *724, p. 132.

Cheiropsis DE FOLIN, 1881, *724, p. 132.
Clavula DE FOLIN, 1881, *724, p. 132 (non WRIGHT, 1859).
Clyphogonium REINSCH, 1877, *1526, p. 177.
Cosinella EMBERGER, MAGNÉ, REYRE & SIGAL, 1955, *701, p. 113.
Cyclogypsinoides A. SILVESTRI, 1937, *1787, p. 201.
Cylindrospira DE FOLIN, 1883, *725, p. 318.
Dendropela DE FOLIN, 1883, *725, p. 328; 1887, *726a, p. 113.
Dillina MUNIER-CHALMAS & SCHLUMBERGER, 1883, *1329, p. 862.
Diplomasta DE FOLIN, 1881, *724, p. 136.
Discolita RAFINESQUE, 1815, *1496, p. 140.
Discorbitoides A. SILVESTRI, 1907, *1768, p. 86. [No type-species named. Seemingly proposed for non-radiate discocyclinids.]
Dyoxeia DE FOLIN, 1881, *724, p. 141.
Eilemammina DE FOLIN, 1881, *724, p. 132.
Eocyclammina BERMÚDEZ, 1950, *125, p. 225.
Eofrondicularia K. V. MIKLUKHO-MAKLAY, 1954, *1277, p. 42.
Eolituonella BERMÚDEZ, 1950, *125, p. 225.
Exseroammodiscus POYARKOV, 1957, *1480, p. 34, 36.
Glaesneria BROTZEN & BERMÚDEZ in BERMÚDEZ, 1950, *125, p. 341.
Glandulinaria DAIN, 1960, *549, p. 197.
Globalternina IVANOVA in SUBBOTINA, GLUSHKO & PISHVANOVA, 1955, *1848, p. 606.
Heterosteginella A. SILVESTRI, 1937, *1787, p. 117.
Ilyopegma DE FOLIN, 1881, *724, p. 139.
Ilyoperidia DE FOLIN, 1881, *724, p. 139.
Ilyosphaera DE FOLIN, 1883, *725, p. 328.
Ilyozotika DE FOLIN, 1881, *724, p. 139.
Julia DE FOLIN, 1881, *724, p. 141 (non GOULD, 1862).
Kikrammina DE FOLIN, 1881, *724, p. 132.
Limocaecum DE FOLIN, 1881, *724, p. 139.
Mallopela DE FOLIN, 1881, *724, p. 140; 1883, *725, p. 328.
Messina BROTZEN, 1960, *242, p. 13.
Neoarchaesphaera A. D. MIKLUKHO-MAKLAY, 1958, *1269, p. 131, fig. 1.
Nodulinella RHUMBLER in ANONYMOUS, 1949, *22, expl. pl. 8.
Nummularia WEDEKIND, 1937, *2041, p. 111 (non SOWERBY & SOWERBY, 1826).
Ophidionella DE FOLIN, 1881, *724, p. 140.
Ouladnailla EMBERGER, MAGNÉ, REYRE & SIGAL, 1955, *701, p. 113.
Ovulida DE FOLIN, 1887, *726a, p. 114.
Palaeocornuspira BOGDANOVICH, 1952, *152, p. 40, 41, 46, 57.
Pentasyderina NICOLUCCI, 1846, *1357, p. 205.
Praecosinella EMBERGER, MAGNÉ, REYRE & SIGAL, 1955, *701, p. 113.
Praerotalipora SALAJ & SAMUEL in SCHEIBNEROVA, 1962, *1643A, p. 215 [*Globotruncana ticinensis GANDOLFI, 1942, *768, p. 113].
Premnammina DE FOLIN, 1881, *724, p. 136.

Psammechinus DE FOLIN, 1881, *724, p. 136 (non AGASSIZ, 1864).
Psammolychna DE FOLIN, 1881, *724, p. 136.
Psammoperidia DE FOLIN, 1881, *724, p. 135.
Psammozotika DE FOLIN, 1881, *724, p. 138.
Pseudocoscinoconus SPECK, 1953, *1824.
Pseudolituola MARIE, 1941, *1215, p. 21, 256.
Pseudosigmoilina BOGDANOVICH, 1952, *152, p. 41, 42, 158.
Pseudospiroloculina BOGDANOVICH, 1952, *152, p. 41, 42, 152.
Ptyka DE FOLIN, 1881, *724, p. 139.
Rectotrochamminoides FISCHER, 1954, *719, p. 9.
Rhizopela DE FOLIN, 1881, *724, p. 140.
Ropalozotika DE FOLIN, 1881, *724, p. 141.
Scarificatina MARIE, 1950, *1219, p. 50.
Sphaerophthalmidium POKORNÝ, 1954, *1474, p. 59.
Stephanopela DE FOLIN, 1881, *724, p. 140.
Toxinopsis DE FOLIN, 1881, *724, p. 138.

UNRECOGNIZABLE GENERIC NAMES APPLIED TO FORAMINIFERIDA

Adherentina SPANDEL, 1909, *1823, p. 212 [*A. rhenana]. Placed by some authors as a synonym of Cibicides, but it possesses a terminal aperture and original description stated that it lacked pores. Possibly similar to Karreria, but impossible to determine from the inadequate figures and description.

Aeolides DE MONTFORT, 1808, *1305, p. 143 [*A. squammatus].

Annulina TERQUEM, 1862, *1883, p. 432 [*A. metensis]. Siliceous discs with thickened rim, from L.Jur.(Lias.) of France and Germany, have been referred to echinoderms (*1348a) and regarded as spicules (*1890), or may possibly belong to Radiolaria.

Apiopterina ZBORZEWSKI, 1834, *2101, p. 311 [*A. orbignyi]. A polymorphinid but unrecognizable generically from description and figures; regarded as synonym of Pyrulina D'ORBIGNY, 1839 (*762, p. 258). Should investigation prove this correct, it would take precedence over Pyrulina.

Arethusa DE MONTFORT, 1808, *1305, p. 303 [*A. corymbosa]. Perhaps a member of the Polymorphinidae.

Aristeropora EHRENBERG, 1858, *683, p. 11. A turbinate rotaliid form.

Arthrocena MODEER, 1791, *1291, p. 91.

Aspidodexia EHRENBERG, 1872, *687, p. 276 [*A. lineolata].

Aspidospira EHRENBERG, 1844, *673, p. 75.

Auriculina COSTA, 1856, *392, p. 259 [*A. crenata] (non Auriculina GRATELOUP, 1838; nec GRAY, 1847).

Buliminopsis RZEHAK, 1895, *1605, p. 217 [*B. conulus] (non Buliminopsis HEUDE, 1890).

Calatharia ZALESSKY, 1926, *2099, p. 87 [*C. perforata]. Unrecognizable form in thin section.

Cameroconus MEUNIER, 1888, *1254, p. 234 [*C. marmoris]. Apparently axial section of an enrolled foraminifer, unrecognizable.

Canopus DE MONTFORT, 1808, *1305, p. 291 [*C. fabeolatus] (non Canopus FABRICIUS, 1803; nec RAFINESQUE, 1840; nec WALKER, 1855; nec FELDER, 1861; nec WOLLASTON, 1864).

Cantharus DE MONTFORT, 1808, *1305, p. 295 [*C. calceolatus] (non Cantharus BOLTEN, 1798; nec CUVIER, 1817; nec SCUDDER, 1882).

Canthropes DE MONTFORT, 1808, *1305, p. 47 [*Canthrope galet (nom. neg., =Canthropes sp.)] [=Canthropus PALLAS in OKEN, 1815, *1385, p. 335 (nom. van.); Cantharipes AGASSIZ, 1846, *6, p. 64 (nom. van.)].

Cepinula SCHAFHÄUTL, 1851, *1637, p. 49.

Cerataria ZALESSKY, 1926, *2099, p. 92 [*C. pulchella].

Chelibs DE MONTFORT, 1808, *1305, p. 307 [*C. gradatus] [=Celibs SHERBORN, 1893, *1731a, p. 38 (nom. van.)].

Cidarollus DE MONTFORT, 1808, *1305, p. 111 [*C. plicatus].

Cimelidium EHRENBERG, 1858, *683, p. 22 [*Guttulina? homeri EHRENBERG, 1858].

Clypeocyclina A. SILVESTRI, 1908, *1771, p. 154 [type, no recognizable species named.] "An invalid genus, defined theoretically, resembling Linderina and Cycloclypeus," *762, p. 456.

Colpopleura EHRENBERG, 1844, *673, p. 74 [*Rotalia ocellata EHRENBERG, 1838].

Cortalus DE MONTFORT, 1808, *1305, p. 115 [*C. pagodus].

Craterularia RHUMBLER, 1911, *1572a, p. 90, 100, 136. No species named in original paper and no valid species yet described; probably represents a Trochammina with boring organism.

Crustula ALLIX in LECOINTRE & ALLIX, 1913, *1117, p. 46 [*C. complanata]. Type in Lecointre collection, BRGG, Paris, mounted in balsam, cracked and unrecognizable.

Cucurbitina COSTA, 1856, *392, p. 363 [*C. cruciata] (non Cucurbitina ALEXANDER, 1833).

Cyclopavonina SILVESTRI, 1937, *1787, p. 93 [*C. cyclica].

Cylindria DE GREGORIO, 1930, *817, p. 48 [*C. minuta] (non Cylindria ZETTERSTEDT, 1849, err. pro Cylidria DESVOIDY, 1830).

Dexiopora EHRENBERG, 1861, *685, p. 304 [*D.? megapora].

Dorbignyaea DESHAYES, 1830, *590, p. 231.

Dujardinia GRAY, 1858, *812, p. 270 [*D. mediterranea] (non QUATREFAGES, 1844; nec. GEDOELST, 1916). Stated to be calcareous, with pores, and intermediate between Rhizopoda and Porifera.

Elliptina HARTING, 1852, *883, p. 116. Included E. inflata and E. truncata.

Epistominites ZALESSKY, 1926, *2099, p. 92 [*E. formosulus].

Fusulinella (Ozawaina) LEE, 1927, *1119, p. 13 [*Nummulina antiquior ROUILLIER & VOSINSKY, 1849, *1588A, p. 337; SD GALLOWAY, 1933, *762, p. 396].

Glandiolus DE MONTFORT, 1808, *1305, p. 315 [*G. gradatus].

Grammobotrys EHRENBERG, 1844, *673, p. 95 [*Polymorphina? aculeata EHRENBERG, 1844]. CUSHMAN, 1944, *480, stated the type from Loandra, South Africa, belongs to Virgulina, but he used Virgulina in a more inclusive sense than at present, and no information is available as to wall structure and other diagnostic features. Could be Cassidella, Fursenkoina, Brizalina, or Bolivina.

Gyrammina EIMER & FICKERT, 1899, *692, p. 669 [*Trochammina annularis BRADY, 1876]. Unrecognizable as based on the types in the BRADY collection in the British Museum (Natural History).

Hedbergina BRÖNNIMANN & BROWN, 1956, *235, p. 529 [*Globigerina seminolensis HARLTON, 1927, *879, p. 24]. Probably a Cretaceous form but described from Pennsylvanian; unrecognizable (*164, p. 39, 40).

Hemistegina KAUFMANN, 1867, *1026, p. 150 [*H. rotula].

Hemisterea EHRENBERG, 1872, *687, p. 276 [*H. nautilus].

Hemisticta EHRENBERG, 1872, *687, p. 276 [*H. amplificata].

Heterostomum EHRENBERG, 1854, *680, p. 22 [*H. cyclostomum] [non DIESING, 1850 (pro Heterostoma FILIPPI, 1837)].

Lagenopsis DE GREGORIO, 1930, *817, p. 48 [*L. maliarda].

Lekithiammina DE FOLIN, 1887, *727Aa, p. 128 [*L. aculeata] [=Lekithiammina DE FOLIN, 1881, *724, p. 136 (nom. nud.)].

Lepista ZALESSKY, 1926, *2099, p. 90 [*L. ornata] (non Lepista WALLENGREN, 1863).

Lobularia COSTA, 1839, *390, p. 186 [*L. vesiculosa] (non Lobularia LAMARCK, 1816).

Lyrina ZBORZEWSKY, 1834, *2101, p. 311 [*L. fischeri].

Mesopora EHRENBERG, 1854, *679, p. 377 [*M. chloris] (non Mesopora WESMAEL, 1852).

Metarotaliella GRELL, 1962, *822, p. 214 [*M. parva]. Incompletely described (only reproductive characters), not illustrated and test character not mentioned, stated to be a small heterokaryotic rotaliid with asexual development as in Rotaliella (with 3 generative nuclei and one vegetative nucleus), the agamont commonly giving rise to 12 gamonts; sexual reproduction with association in pairs as in Rubratella, and resulting in a variable number of amoeboid gametes. Apparently it is to be more completely described later in the Archiv für Protistenkunde. Rec., Fr.

Mirfa DE GREGORIO, 1890, *816, p. 260 [*M. subtetraedra].

Mirga DE GREGORIO, 1930, *817, p. 49 [*Orbulina (M.) permiana].

Misilus DE MONTFORT, 1808, *1305, p. 295 [*M. aquatifer].

Molnaria ZALESSKY, 1926, *2099, p. 89 [*M. spinulata].

Monocystis EHRENBERG, 1854, *680, p. 22 [*Miliola (Monocystis) arcella] (non Monocystis STEIN, 1848).

Nummulitella DORREN, 1948, *610, p. 291 [*N. polystylata; OD]. Assigned by author to Nummulitidae; probably a rotaliid. U.Eoc., N.Z.

Oncobotrys EHRENBERG, 1856, *682, p. 172 [*O. buccinum].

Orobias EICHWALD, 1860, *691, p. 22, fig. 16 [*O. aequalis].

Orthocerina D'ORBIGNY in DE LA SAGRA, 1839, *1611, p. 17 [*Nodosaria (O.) quadrilatera; OD (M)] [=Nodosaria (Orthocérine) D'ORBIGNY, 1826, *1391, p. 255 (nom. neg.)].

[The type cannot be Nodosaria clavulus LAMARCK, as stated by GALLOWAY, *762, as in 1826 D'ORBIGNY (*1391, p. 255) did not use a Latin name for the subgenus, only the French vernacular; hence it was invalid. In 1839 when the Latin designation was used by D'ORBIGNY (*1611, p. 17) only O. quadrilatera was mentioned by name, hence is the type-species by monotypy, although D'ORBIGNY stated that there were 2 species known to him, the other being fossil from the Tertiary of the Paris area. PARKER & JONES (*1417f, p. 433) stated that Nodosaria (Orthocerina) clavulus did not belong to the genus, restricting it to include only O. quadrilatera and 4 later described species by REUSS. They erroneously stated O. murchisoni (REUSS) to be the type, however. HERON-ALLEN & EARLAND (1930, *915, p. 172) correctly regarded O. quadrilatera as the type-species. Other forms included by PARKER & JONES have agglutinated tests, some of which are now placed in Triplasia. HERON-ALLEN & EARLAND included calcareous species now regarded as Tristix. Other species placed in Orthocerina by various authors are to be placed in Nodosaria, Pseudonodosaria, Geinitzina, and Amphimorphina, representing 4 or 5 families. The type-species is poorly known from the original brief description only, which does not state whether it is calcareous or agglutinated, and it has not since been recognized in the type area (Cuba, Jamaica), where it was stated to be rare. Thus it is regarded as unrecognizable.]

Otostomum EHRENBERG, 1872, *687, p. 276 [*O. strophoconus].

Ovolina TERQUEM, 1864, *1884, p. 285 [*Ovolina fusiformis, =Oolina fusiformis TERQUEM, 1863].

Paronia PREVER in CHELUSSI, 1903, *330A, p. 74 [non DIAMARE, 1900] [*Nummulites complanata LAMARCK, 1804].

Pectinaria ZALESSKY, 1926, *2099, p. 94 [*P. costata] (non Pectinaria LAMARCK, 1818).

Phanerostomum EHRENBERG, 1843, *672, p. 409 [*P. integerrimum].

Physomphalus EHRENBERG, 1856, *682, p. 172 [*P. porosus].

Platyoecus EHRENBERG, 1854, *680, p. 23 [*P.? squama].

Pleurites EHRENBERG, 1854, *680, p. 23 [*P. cretae].

Pleurostomina A. COSTA, 1862, *389, p. 94 [*P. bimucronata].

Pleurotrema EHRENBERG, 1840, *667, chart opposite p. 120 [*P. calcarina].

Pseudastrorhizula WETZEL, 1940, *2047, p. 122 [*P. eisenacki]. Internal cast or "steinkern" of a foraminifer from an Upper Cretaceous glacial pebble in Denmark.

Pteroptyx EHRENBERG, 1873, *689, p. 151, 152 (non OLIVIER, 1902) [*P. vespertilio].

Ptygostomum EHRENBERG, 1843, *672, p. 409 [*P. oligoporum].

Raphanulina ZBORZEWSKI, 1834, *2101, p. 311 [*R. humboldtii]. A polymorphinid regarded as equivalent to Globulina D'ORBIGNY (*762, p. 259) but unrecognizable generically from the description and figures.

Renulina BLAKE, 1876, *144, p. 262 (non LAMARCK, 1805; nec DE BLAINVILLE, 1825) [*R. sorbyana].

Rhabdella D'ARCHIAC & HAIME, 1853, *38, p. 351 [*R. malcolmi].

Rhaphidodendron MÖBIUS, 1876, *1292, p. 115 [*R. album].

Rhynchoplecta EHRENBERG, 1854, *679, p. 405 [*R. punctata].

Rotalites LAMARCK, 1801, *1084, p. 401 [*R. tuberculosa]. Grignon, Fr.

Semseya FRANZENAU, 1893, *745, p. 358 [*S. lamellata].

Septammina MEUNIER, 1888, *1254, p. 235 [*S. renaulti].

Siderospira EHRENBERG, 1845, *675, p. 376 [*Siderolina? indica EHRENBERG, 1845].

Spiroplectina SCHUBERT, 1902, *1682, p. 84 (non CUSHMAN, 1927). No species named. Early stage as in Spiroplecta, later as in Frondicularia, but not stated whether calcareous or agglutinated.

Spiropleurites EHRENBERG, 1854, *678, p. 237.

Strophoconus EHRENBERG, 1843, *670, p. 166 [*S. cribosus]. CUSHMAN (1927, *434) stated that one of the species, S. auricula, was a young Virgulina, but no information is available about the type-species.

Synspira EHRENBERG, 1854, *680, p. 24 [*S. triquetra].

Tinoporus DE MONTFORT, 1808, *1305, p. 147 [*T. baculatus]. See discussion by LOEBLICH & TAPPAN (1962, *1186).

Trioxeia DE FOLIN, 1888, *727, p. 110 [*T. edwardsi] [=Trioxeia DE FOLIN, 1881, *724, p. 141 (nom. nud.)].

Upsonella W. L. MOORE, 1959, *1308A, p. 995 [*U. typus; OD]. "Unilocular, subspherical, spinose foraminifer characterized by a distinctive furrow or attachment scar which is developed along the base of the test and which has a narrow flap or rim around its periphery. The multiple apertures of this form are probably associated with the spines (*1308A)." L.Penn., USA(Tex.).

[Whether the genus is agglutinated or granular calcareous in wall character it is not stated, hence it is uncertain whether it is close to Parathurammina, Astrorhiza, Thurammina, or Archaeochitinia. Although nomenclatorially validated, the above genus is thus unrecognizable, without further published description and illustration. Although more complete description may be found in the unpublished dissertation, of which the above reference is an abstract, the dissertation is not a publication. Neither the sale of microfilm nor Xerox reproduction from the microfilm consists of publication, and (ICZN, 1961, Art. 8) "a work when first issued must (1) be reproduced in ink on paper by some method that assures numerous identical copies; and (4) not . . . reproduced or distributed by a forbidden method." According to Art. 9 (1) "distribution of microfilms, or microcards, or matter reproduced by similar methods" does not constitute publication.]

Volutaria ZALESSKY, 1926, *2099, p. 95 [*V. potoniei].

Volvotextularia G. Termier & H. Termier, 1950, *1882, p. 33, 39 [*V. polymorpha] [=Volvotextularia G. Termier & H. Termier, 1947, *1881, table p. 146, 147, 271 (nom. nud.)].

GENERIC NAMES ERRONEOUSLY APPLIED TO FORAMINIFERIDA

Aguayoina Bermúdez, 1938, *120, p. 386 [*A. asterostomata]. Anthozoan.

Ammosphaeroides Cushman, 1910, *404a, p. 51 [*A. distoma] [=Arammosphaerium Rhumbler, 1913, *1572b, p. 348 (nom. van.) (obj.)]. Inorganic, mineral coating on a sand grain.

Archaelagena Howchin, 1888, *965, p. 539 [*Lagena howchiniana Brady, 1876] [=Archealagena Harlton, 1927, *879, p. 24 (nom. null.) (obj.)]. A plant.

Balanulina Rzehak, 1888, *1603, p. 265 [*B. kittlii]. May be a barnacle, bryozoan, or coral, or a foraminifer. Unrecognizable.

Birrimarnoldia Hovasse & Couture, 1961, *964, p. 1054 [pro Arnoldia Hovasse, 1956, *963, p. 2584 (non Mayer, 1887, non Kieffer, 1895, non Wlassenko, 1931)] [*Arnoldia antiqua Hovasse, 1956; OD]. Minute siliceous and iron oxide globules from Precambrian of Africa, probably inorganic.

Cadosina Wanner, 1941, *2038, p. 79 [*C. fusca]. Member of family Cadosinidae of Tintinnina.

Cadosinella Vogler, 1941, *2015, p. 282 [*C. gracillimoides]. Member of family Cadosinidae of Tintinnina.

Capsulina Seguenza, 1880, *1713, p. 375 [*C. loculicida]. Originally described as a foraminifer, probably echinoderm pedicellaria.

Cayeuxina Galloway, 1933, *762, p. 156 [*C. precambrica]. Probably inorganic.

Cellulina Zborzewski, 1834, *2101, p. 308. Alga.

Cercidina Vogler, 1941, *2015, p. 290 [*C. supracretacea]. Probably member of Tintinnina.

Cheilosporites Wähner, 1903, *2029, p. 98 [*C. tirolensis]. Problematica, described from non-oriented limestone sections, originally and here regarded as algal in nature, later variously referred to sponges and foraminifers. Consists of large branching colonies (to 5 cm.), of uniserial chambers up to 4 mm. diam., with axial siphon; wall of calcite grains. The chambers show very little increase in size as added and some apparent branches have series of chambers approximately half of normal size, without a gradual change in size as is common in foraminiferal ontogeny. Made the monotypic basis for the family Cheilosporitidae A. G. Fischer (1962, *718, p. 123). U.Trias., Bavaria.

Chuaria Walcott, 1899, *2032, p. 234 [*C. circularis]. Algonkian, Chuar terrane, USA(Ariz.).

Cochleatina E. V. Bykova, 1956, *258, p. 12 [*C. plavinensis]. A bryozoan, probably Corynotrypa Bassler, 1911.

Coelotrochium Schlüter, 1879, *1670, p. 668. Alga.

Coscinoconus Leupold in Leupold & Bigler, 1936, *1130, p. 618 [*C. alpinus]. According to Maslov (1958, *1232) this is an alga.

Cysteodictyina Carter, 1880, *296, p. 448 [*C. compressa]. Placed by Carter in a new group Testamoebiformia; probably calcareous alga.

Dexiospira Ehrenberg, 1859, *684, p. 309 (non Dexiospira Caullery & Mesnil, 1897). Inorganic concretionary masses.

Discoidina Terquem & Berthelin, 1875, *1893, p. 15 [*D. liasica] (non Discoidina Stein, 1850). Incertae sedis; not a foraminifer.

Girvanella Nicholson & Etheridge, 1878, *1356, p. 23 [*G. problematica] [=Argirvanellum Rhumbler, 1913, *1572b, p. 386 (nom. van.)]. Alga.

Goniolina d'Orbigny, 1850, *1397b, p. 41 [*G. hexagona]. A plant fossil.

Holocladina Carter, 1880, *296, p. 447 [*H. pustulifera]. Placed by Carter in a new group Testamoebiformia; probably calcareous alga.

Keramosphaerina Stache, 1913, *1829, p. 659, 666 [*Bradya tergestina Stache, 1889] [=Bradya Stache, 1889, *1828, p. 35, 89 (obj.) (non Bradya Boeck, 1873; nec Bradya Carter, 1877)]. Probably a hydrocoralline.

Ladinosphaera Oberhauser, 1960, *1384, p. 44 [*L. geometrica]. Questionably organic, probably small limonitic "concretions" in geometric arrangement.

Matthewina Galloway, 1933, *762, p. 157 [*Globigerina cambrica Matthew, 1895]. Probably inorganic.

Millarella Carter, 1888, *298, p. 178 [*M. cantabrigiensis]. Not a foraminifer.

Nodoplanulis Hussey, 1943, *975, p. 166 [*N. elongata]. An isopod appendage (*1303, p. 151, 152).

Polytrema Risso, 1826, *1579b, p. 340 (non Rafinesque, 1819, non Férussac, 1822, non d'Orbigny, 1850). A bryozoan.

Protocyclina Paalzow, 1922, *1404, p. 35 [*P. liasina]. Not a foraminifer, but echinoderm ossicle.

Psammosiphon Vine, 1882, *2006, p. 390 [*P. amplexus] (non Psammosiphon Rhumbler, 1913). Not a foraminifer; possibly Annelida.

Pseudogypsina Trauth, 1918, *1948, p. 243 [*P. multiformis]. Probably a calcareous alga.

Rhaphidoscene Jennings, 1896, *988, p. 320 [*R. conica] [=Arrhaphoscenum Rhumbler, 1913, *1572b, p. 346 (obj.) (nom. van.)]. Represents young of the sponge Tentorium.

Siphonema Bornemann, 1886, *175, p. 17. Alga.

Spirocerium Ehrenberg, 1858, *684, p. 310 [*S. priscum]. Inorganic; globular mass of "glauconite."

Spongina de Gregorio, 1930, *817, p. 8, 48 [*Globigerina (S.) permica]. Described as a subgenus of Globigerina; not a foraminifer.

Stoliczkiella CARTER, 1888, *298, p. 173 [*S. theobaldi*]. Probably an echinoid.

Stomiosphaera WANNER, 1940, *2038, p. 76 [*S. moluccana*]. Similar to *Cadosina*, but with perforate walls, probably related to Tintinnina.

Terquemina GALLOWAY, 1933, *762, p. 157 [*T. devonica*]. Not a foraminifer.

Wetheredella WOOD, 1948, *2072, p. 20 [*W. silurica*]. Composed of subcircular, radially layered calcite tubes, encrusting and irregularly branching in habit; doubtfully a foraminifer, probably algal.

Order REITLINGERELLIDA Vologdin, 1958

[Order Reitlingerellida VOLOGDIN, 1958, p. 405]

Shell free, consisting of narrow tubular chamber of constant diameter (0.016-0.017 mm.) coiled in expanding spire, cylindrical helical spire, or with early glomerate coil. [Genera here included were originally regarded as foraminifers, and some have since been considered as algae (e.g., *Obruchevella, Cavifera, Glomovertella, Syniella*) (ELIAS, 1954, *697, p. 52). Although their systematic position is doubtful, they are here listed and figured. No attempt is made to evaluate the validity of these similar-appearing forms.] *L.Cam.-U.Ord.*

Family REITLINGERELLIDAE Loeblich & Tappan, n.fam.

Characters of order. *L.Cam.-U.Ord.*

Reitlingerella VOLOGDIN, 1958, *2018, p. 408 [*R. densa*; OD]. Test with system of tubular chambers, closely appressed, with curved loops of differing form and orientation. *L.Cam.*, USSR(Tuva).——FIG. 643,*1*. *R. densa*, ×210 (*2018).

Bostrychosaria VOLOGDIN, 1958, *2018, p. 406 [*B. bistorta*; OD]. Closely spiraling tube, 0.017 mm. diam., with elongate axis of spiraling, entire specimen being of equal diameter throughout and approximately cylindrical. *L.Cam.*, USSR(Tuva).——FIG. 643,*2*. *B. bistorta*, holotype, ×210 (*2018).

Cavifera REYTLINGER, 1948, *1559, p. 80 [*C. concinna*; OD]. Tube coiling in single whorl, approx. 0.08-0.09 mm. diam., leaving broad central cavity; wall calcareous, microgranular; end of tube open. *Cam.*, USSR(Yakutiya).——FIG. 644,*1*. *C. concinna*, ×215 (*1559).

Chabakovia VOLOGDIN, 1939, *2017, p. 221, 255 [*C. ramosa*, OD]. Small dendritic branches formed by series of bulbous chambers, with partitions convex in direction of growth; wall calcareous, nearly opaque. [Originally described as an alga, *Chabakovia* was regarded by ELIAS (*696) as a foraminifer belonging to the Ptychocladiinae. The complex, chambered branching form is of a more advanced nature than would be expected in early Paleozoic foraminifers, and *Chabakovia* is tentatively here placed with the Reitlingerellida, although it may possibly belong to the algae.] *M.Cam.*, USSR(Ural Mts.).——FIG. 645,*1*. *C. ramosa*; *1a-c*, typical specimens, ×40 (*2017).

Flexurella VOLOGDIN, 1958, *2018, p. 407 [*F. obvoluta*; OD]. Shell tubular, flattened, more or

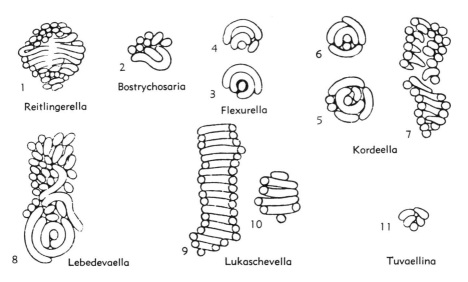

FIG. 643. Reitlingerellidae; *1, Reitlingerella; 2, Bostrychosaria; 3,4, Flexurella; 5-7, Kordeella; 8, Lebedevaella; 9,10, Lukaschevella; 11, Tuvaellina* (p. C787-C789).

Protista—Sarcodina

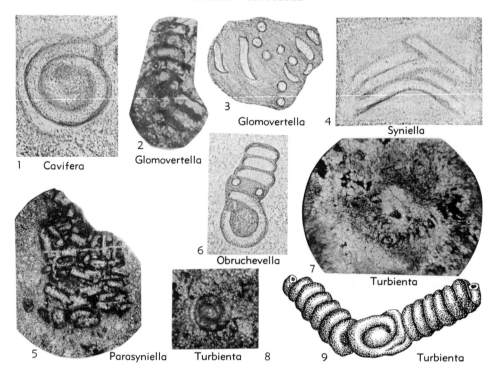

FIG. 644. Reitlingerellidae; *1, Cavifera; 2,3, Glomovertella; 4, Syniella; 5, Parasyniella; 6, Obruchevella; 7-9, Turbienta* (p. C787-C789).

less discoidal, in early stage coiled in 2 or 3 whorls, later stage with arcuate loops. *L.Cam.*, USSR (Tuva).——FIG. 643,*3,4*. **F. obvoluta*; ×210 (*2018).

Glomovertella REYTLINGER, 1948, *1559, p. 80 [**G. firma*; OD]. Test free or attached, with globular initial chamber followed by tubular chamber of 1 or 2 whorls and with later loops in changing planes of coiling; wall calcareous, finely granular; aperture at open end of tube. *Cam.*, USSR.——FIG. 644,*2,3*. **G. firma*; *2*, paratype, ×240 (*1565); *3*, holotype, ×244 (*1559).

Kordeella VOLOGDIN, 1958, *2018, p. 407 [**K. campylodroma*; OD]. Test consisting of compact closely looped tube of 0.017 mm. diam., somewhat produced in direction of growth. *L.Cam.*, USSR(Tuva).——FIG. 643,*5-7*. **K. campylodroma*; ×210 (*2018).

Lebedevaella VOLOGDIN, 1958, *2018, p. 408 [**L. involventis*; OD]. Test to 0.28 mm. in length and 0.14 mm. in breadth, consisting of narrow tubular chamber 0.020 mm. diam., forming interwoven loops transverse to flat axis. *L.Cam.*, USSR(Tuva).——FIG. 643,*8*. **L. involventis*, ×210 (*2018).

Lukaschevella VOLOGDIN, 1958, *2018, p. 408 [**L. spiralis*; OD]. Tubular chamber coiled in high spire, with slight variation in dimensions. *L.Cam.*, USSR(Tuva).——FIG. 643,*9,10*. **L. spiralis*; ×210 (*2018).

Obruchevella REYTLINGER, 1948, *1559, p. 78 [**O. delicata*; OD]. Elongate cylindrical form, consisting of elongate tube of equal diameter coiled in tightly closed spire, not around central cavity; wall calcareous, finely granular; communication with exterior at open end of tube. *L.Cam.*, USSR (Yakutiya).——FIG. 644,*6*. **O. delicata*; holotype, ×244 (*1559).

Parasyniella E. V. BYKOVA, 1961, *260, p. 67 [**P. geniculosa*; OD]. Test free, globular, consisting of numerous chambers of rectangular section or irregularly arranged tubes without visible orderly arrangement; wall calcareous, dark in thin section, fine-grained; aperture not observed, chambers interconnected by openings in walls or by open tubular branches. *U.Ord.(Caradoc.)*, USSR (N.Kazakh.).——FIG. 644,*5*. **P. geniculosa*; holotype, ×47 (*2112).

Rectangulina ANTROPOV, 1959, *25A, p. 30 [**Syniella tortuosa* ANTROPOV, 1950, *25, p. 31; OD]. Test irregularly angular in form, consisting of groups of closely arranged, regular, prismatic, quadrate chambers in parallel rows, groups of parallel chambers variously oriented relative to each other and to test exterior; aperture unknown. [*Syniella* was regarded originally as of uncertain systematic position, and later placed in the Order Reitlingerellida, Subclass Foraminifera by VOLOGDIN, 1958 (*2018). ANTROPOV regarded *Rec-*

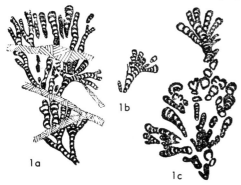

FIG. 645. Reitlingerellidae; *1, Chabakovia* (p. C787).

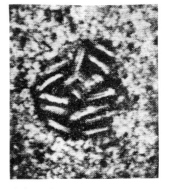

FIG. 645A. Reitlingerellidae; *Rectangulina* (p. C788-C789).

tangulina as of uncertain position, possibly algal. It undoubtedly is related correctly to the Reitlingerellida, although the true systematic position of the entire group is uncertain.] *U.Dev.(L. Frasn.)*, USSR(Tatar - Bashkir - Kuybyshevsk - Udmurt).——FIG. 645A. *R. tortuosa* (ANTROPOV), N.Russian platform(Shugurian region); sec. of holotype, ×67 (*25).

Syniella REYTLINGER, 1948, *1559, p. 81 [*S. invenusta*; OD]. Test appears to be elongate irregularly bending tube which may be bent double; wall calcareous, finely granular; aperture not observed. *L.Cam.*, USSR(Yakutsk).——FIG. 644,4. *S. invenusta*; long. sec. of holotype, ×244 (*1559).

Turbienta E. V. BYKOVA, 1961, *260, p. 65 [*T. bifida*; OD]. Globular proloculus followed by tubelike undivided second chamber which at first coils in 1 or 2 flat whorls, then with 2 coils in opposite directions in high cylindrical open spires; wall calcareous, fine-grained; aperture at open end of tube. *U.Ord.(Caradoc.)*, USSR(N.Kazakh.). ——FIG. 644,7-9. *T. bifida*; 7, holotype, ×107; 8, paratype, ×220; 9, diagram, enlarged (*2112).

Tuvaellina VOLOGDIN, 1958, *2018, p. 406 [*T. prima*; OD]. Low spiraling tube of 0.016 mm. diam., with slight connections of one whorl to another. *L.Cam.*, USSR(Tuva).——FIG. 643,11. *T. prima*; ×260 (*2018).

Order XENOPHYOPHORIDA Schulze, 1904

[*nom. correct.* LOEBLICH & TAPPAN, 1961, p. 318 (*pro* order Xenophyophora SCHULZE, 1912, p. 41, *nom. transl. ex* group Xenophyophora SCHULZE, 1904, p. 1387)]——[All synonymic citations refer to order status unless otherwise stated; dagger(†) indicates *partim*]——[=Domatocoelat† HAECKEL, 1889, p. 8; =Xenophyophoren SCHULZE, 1905, p. 6 (*nom. neg.*); =suborder Arxenophyria RHUMBLER, 1913, p. 339 (*nom. van.*); =Xenophiophorae CHATTON, 1925, p. 76; =suborder Xenophyophora JIROVEC, 1953, p. 335 (*nom. transl.*)]——[=Myxozoa† SCHEPOTIEFF, 1912, p. 267; =Mycetozoida† SCHEPOTIEFF, 1912, p. 267)]——[=Psamminidea POCHE, 1913, p. 202]

Multinucleate plasmodium containing numerous clear solid bodies (granellae), and forming pseudopodial network enclosed in system of hollow tubes (granellarium), some tubes (stercomarium) also containing dark bodies (stercomata), probably of fecal nature, and may contain xanthosomes, tiny red or yellow highly refractive spherical bodies; tube system composed of hyaline organic substance resembling spongin, and interspaces containing pseudoskeleton of foreign matter (xenophya), including sand grains, sponge spicules, tests of foraminifers, radiolarians or diatoms; reproduction probably by swarm spores. [Deep-water forms.] *Rec.*

These organisms have been described as sponges (*851), agglutinated foraminifers (*803), Labyrinthulida or Mycetozoa (*1647). There is no trace of cell differentiation, tissue or organ formation, such as is found in sponges. As living organisms have not been studied (only preserved material was used) the pseudopodial character is unknown. The presence of stercomata relates these to many rhizopods, as such bodies have been reported in the orders Gromida and Foraminiferida (*1701). The plasmodium in the granellarium, containing nuclei, may disintegrate into single isolated mononucleate cells. The plasma lumps at the ends of the granellarium branches are comparable to such lumps formed in other rhizopods, the pseudopodial complex of which retracts under unfavorable conditions.

The Xenophyophorida differ from the Labyrinthulida in having a skeleton of foreign particles, in the character of the tube

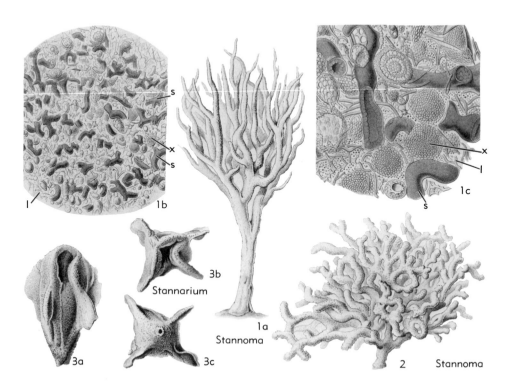

Fig. 646. Stannomidae; *1,2, Stannoma; 3, Stannarium* (p. C790-C792).

systems, granellae and stercomarium, and the linellae of the Stannomidae. The linellae are similar in form and perhaps correspond to the capillitium of the Mycetozoida, but the capillitium is formed within the plasmodium fruiting body and the linellae lie outside of this. The Xenophyophorida differ from Foraminiferida in having the loose internal skeleton of xenophya in which the protoplasma-filled tubes are freely suspended, the test of agglutinated Foraminiferida enclosing the protoplasm. The linellae of the family Stannomidae are completely different from anything found in the Foraminiferida. The Xenophyophorida are regarded as belonging to a separate order intermediate between the Foraminiferida and Labyrinthulida (*1700).

Family STANNOMIDAE Haeckel, 1889

[Stannomidae HAECKEL, 1889, p. 7, 8, 54] [=subfamily Stannomida LANKESTER, 1909, p. 286 *(nom. transl.)*; =†Xenophyophoridae LANKESTER, 1909, p. 286 *(nom. nud.)*]
——[=Neusininae CUSHMAN, 1910, p. 129; =Neusinidae CUSHMAN, 1927, p. 29 *(nom. transl.)*]

Expanded flabelliform or branching body, flexible in life, with xenophya (foreign bodies) held together by smooth, strongly refractive spongin-like threads, rounded in section, up to several mm. in length and 1 to 12 microns in diameter (linellae), expanding in size where they attach to the xenophya. *Rec.*

Stannoma HAECKEL, 1889, *851, p. 72 [*S. dendroides*; SD LOEBLICH & TAPPAN, herein] [=*Stannoplegma* HAECKEL, 1889, *851, p. 74 (type, *Stannoma coralloides* HAECKEL, 1889)]. Arborescent body (height to 8 cm.), with numerous free or anastomosing cylindrical branches, originating from nearly cylindrical pedicle (length 1-3 cm., diam. 2-5 cm.) terminating basally in a soft, finely fibrous mass; internal structure with abundant linellae (av. diam. 4 microns). [Originally described with two included species but no type designated, although *S. coralloides* was stated possibly to represent a distinct genus.] *Rec.,* C.Pac. (2,400-2,600 fathoms).——FIG. 646,*1*. **S. dendroides*, trop. Pac.; *1a*, exterior, ×1.3; *1b*, fragment of section, showing stercomarium *(s)*, xenophya *(x)*, linellae *(l)*, ×26; *1c*, same, ×100 (*851).——FIG. 646,*2*. *S. coralloides* HAECKEL, trop.Pac.; ×2 (*851).

Stannarium HAECKEL, 1889, *851, p. 69 [*S. concretum* HAECKEL, 1889, p. 71; SD LOEBLICH &

Xenophyophorida

Tappan, herein]. Branched lamellar body, with 2 primary vertical leaves, which are either free or grown together, and secondary leaves budding from these; xenophya consisting of Radiolaria or *Globigerina* tests, linellae regular, thin (diam. 2-8μ) and elongate. *Rec.*, C.Pac. (2,600-2,900

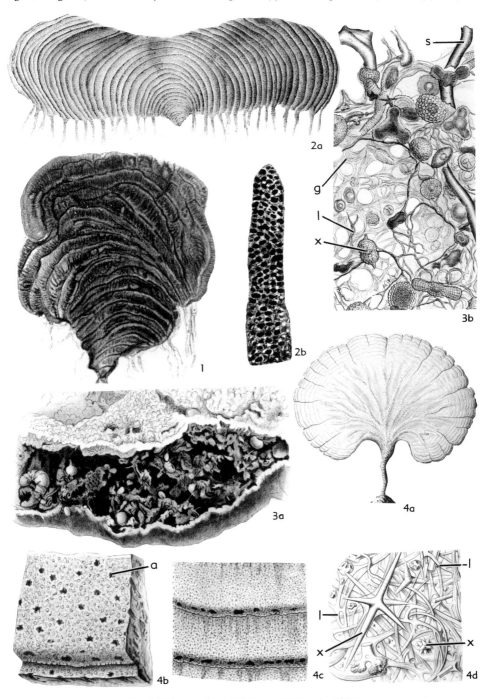

Fig. 647. Stannomidae; *1-4, Stannophyllum* (p. *C*792).

fathoms).——Fig. 646,3. *S. concretum; 3a-c, side, top, and base, ×1 (*851).

Stannophyllum Haeckel, 1889, *851, p. 60 [*S. zonarium Haeckel, 1889, p. 62; SD Loeblich & Tappan, herein (=S. flabellum Haeckel, 1889, =Neusina agassizi Goës, 1892)] [=Psammophyllum Haeckel, 1889, p. 49 (type, P. flustraceum Haeckel, 1889, p. 51; SD Loeblich & Tappan, herein); Neusina Goës, 1892, p. 195 (type, N. agassizi Goës, 1892, obj.)]. Thin foliaceous or flabelliform erect body (diam. 4-24 cm.) arising from simple short pedicle expanding basally; surface may be marked by concentric furrows, may have loosely bound free linellae (length to 2 cm.) near margins, linellae forming dense network on both surfaces, with numerous embedded xenophya; stercomarium dendritic, occupying considerable portion of body and containing numerous xanthosomes; granellarium filled with uniform plasma, containing granellae and evenly distributed nuclei (diam. 4μ) with some larger nuclei (diam. 6-8μ) that have a distinct nuclear membrane, a dense network of chromatin, and 1 or 2 spherical homogeneous nucleoli. *Rec.*, E.Pac. (1,740-2,200 fathoms), trop. Pac.-N.Pac.-W.Ind.O. (2,100-2,900 fathoms).——Fig. 647,1-3. *S. zonarium; 1, trop. Pac., ×1 (*1700); 2a, E.Pac., ×0.5; 2b, transv. sec. (transmitted light), ×11 (*803); 3a, margin, ×30; 3b, sec. showing granellarium (g), stercomarium (s), linellae (l), and xenophya (x), ×70 (*851).——Fig. 647,4. S. flustraceum (Haeckel), N.Pac.; 4a, ×0.5; 4b, distal surface, showing apertural openings (a), ×12; 4c, same, ×4; 4d, section showing linellae (l) and xenophya (x), ×150 (*851).

Family PSAMMINIDAE Haeckel, 1889

[Psamminidae Haeckel, 1889, p. 7, 8, 32] [=Psamminae Lendenfeld, 1886, p. 589 (nom. nud., pro Psammella Lendenfeld, ms.); =subfamily Psamminidae Lankester, 1909, p. 286 (nom. transl.)]——[=†Xenophyophoridae Lankester, 1909, p. 286 (nom. nud.)]

Body discoidal or an irregular lump or crust; with xenophya cemented together and enclosed by transparent maltha; no linellae. *Rec.*

Psammina Haeckel, 1889, *851, p. 34 [*P. globigerina Haeckel, 1889; SD Loeblich & Tappan, herein] [=Psammoplakina Haeckel, 1889, *851, p. 35 (type, P. discoidea Haeckel, 1889 =Psammina plakina Haeckel, 1889)]. Body discoidal (diam. 20-30 mm., thickness 1.5-3.5 mm.), with thin flat plates of cemented xenophya, commonly foraminiferal, rarely radiolarian tests; oriented arborescent stercomarium, branches (diam. 0.3-0.5 mm.), and dichotomously branching granellarium, with jelly-like mass predominant and granellae scattered; distinct and large pores on peripheral margin or upper surface. [*Psammina* originally included 3 species without type citation (*851). The description of *P. plakina* stated that it differed sufficiently from the 2 typical species to be the type of a new genus. *Psammoplakina discoidea* is thus an objective synonym of *Psammina plakina*. *P. plakina* and *P. globigerina* are congeneric (*1700).] *Rec.*, S.Atl.-trop.Pac. (1,100 to 2,750 fathoms).——Fig. 648,1-3. *P. globigerina, trop. Pac.; 1a,b, top and edge, ×10; 2, section showing radiating stercomarium (s), anastomosing granellarium (g), ×10; 3, decalcified fragment, as above, with few xenophya (x) remaining consisting of radiolaria, ×100.——Fig. 648,4,5. P. plakina Haeckel, S.Atl.; 4a,b, top and edge, ×5; 5, vert. sec. showing platelike upper and lower layers of xenophya (x) and apertural pores (a), ×35 (*851).

Cerelasma Haeckel, 1889, *851, p. 45 [*C. gyrosphaera Haeckel, 1889; SD Loeblich & Tappan, herein]. Globular or tuberose body; differing from *Psammina* in rich secretion of spongin-like organic matter (maltha) forming a thin lamellar framework for entire body, and also enclosing xenophya (usually Radiolaria) in small sacculi; numerous anastomosing tubes of stercomarium containing plasmodia, with some dark-colored grains or stercomata; granellarium containing nuclei and granellae. *Rec.*, trop.Pac. (2,000-2,425 fathoms). ——Fig. 648,6. *C. gyrosphaera; 6a, exterior, ×0.5; 6b, part of transv. sec., showing maltha (m), surrounding xenophya (x), and anastomosing stercomarium (s), ×50; 6c, same, without xenophya, ×150 (*851).

Holopsamma Carter, 1885, *297, p. 211 [*H. laevis Carter, 1885, p. 212; SD Loeblich & Tappan, herein]. Body massively tuberose or lumpy, with groups of apertural pores at crest of prominent ridges or projecting lobes; differing internally from *Cerelasma* in absence of sacculi around xenophya, and from *Psammopemma* in restriction of apertures to ridges or lobes; internal structure similar to *Psammetta*, but with addition of dark clublike masses near granellarium (latter may form network enclosed in clear sheath, and may contain small bodies as does stercomarium). [Original description (*297) included 5 species, type not cited, of which 2 were removed to *Psammopemma* (*851), remaining species not since recognized, and never figured. *H. argillaceum* Haeckel probably=*H. laevis* Carter, as both are described as lobose, other species hemispherical, globular, massive.] *Rec.*, N.Atl.-S.Pac.-S.Australia (1,675-2,270 fathoms).——Fig. 649,1. H. argillaceum Haeckel, S.Pac.; 1a, exterior, ×2; 1b, vert. sec., ×2.5 (*851).

Psammetta Schulze, 1905, *1700, p. 6 [*P. erythrocytomorpha Schulze, 1905]. Biconcave circular discs (diam. 2-3 cm.), periphery rounded, thickness constant (5-12 mm. depending on test size); surface roughened, feltlike texture, olive or brownish green; xenophya consisting largely of dense network of radially oriented siliceous sponge spicules and less abundant foraminiferal tests; granellarium of dichotomously branched but not oriented, light yellow tubes, open at ends but with some short branches having viscous spherical terminations, filled with granellae (diam. 1-3μ) and with definite cell nuclei (diam. 3μ) at intervals of 10μ within the tubes; dark brown, nearly straight strands of arborescent stercomarium (diam. 0.1-0.2 mm.), arising at center of disc and enlarging and branching outward with irregu-

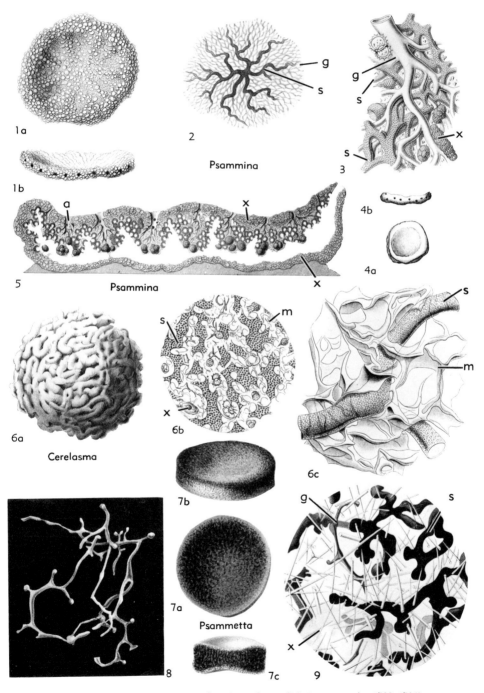

Fig. 648. Psamminidae; *1-5, Psammina; 6, Cerelasma; 7-9, Psammetta* (p. C792-C794).

Protista—Sarcodina

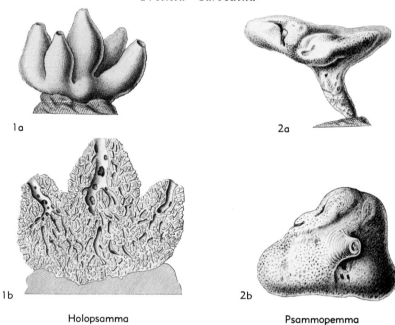

Fig. 649. Psamminidae; *1, Holopsamma*; *2, Psammopemma* (p. C792, C794).

larly spaced knotlike thickenings and containing yellowish to dark greenish brown globular stercomata (diam. 10-40μ) which are acid- and dye-resistant; all structures surrounded by thin membranous spongin-like sheath or binding material, with expansions as putty-like mass at points of junction, sheath thin over xenophya, solid and firm over stercomarium, and dense over granellarium, except for spherical bulbous ends of branches; smooth, spherical xanthosomes (diam. 1-10μ) occurring inside stercomarium and free between strands, are highly refractive and yellow-red (garnet) in color. *Rec.*, Ind.O.-E.Afr. (depth 1,668 m.).——Fig. 648,7-9. **P. erythrocytomorpha*, Ind.O.; *7a-c*, top, edge and vert. sec., ×1; *8*, single complete portion of granellarium, ×10; *9*, sec. showing stercomarium *(s)*, granellarium *(g)* and xenophya *(x)* of sponge spicules, ×65 (*1700).

Psammopemma MARSHALL, 1881, *1227, p. 113 [**P. densum* MARSHALL, 1881, OD (M)]. Irregular massive or lumpy body, entire surface with numerous small pores, no large openings as in *Psammina* and *Holopsamma*; xenophya of foraminifers or radiolarians, not enclosed in sacculi like those of *Cerelasma*; narrow branched tubes of granellarium interwoven with anastomosing tubes of stercomarium. *Rec.*, trop.Atl.-trop.Pac. (2,400-2,600 fathoms).——Fig. 649,2. *P. radiolarium* HAECKEL, trop.Pac.; *2a,b*, side and basal views, ×2 (*851).

Order LABYRINTHULIDA Lankester, 1877

[In synonymic citations superscript numbers indicate taxonomic rank assigned by authors ([1]series, [2]section, [3]suborder, [4]order, [5]subclass, [6]class) and a dagger(†) indicates *partim*]
——[=[4]Labyrinthulida LANKESTER, 1877, p. 442; =[6]Labyrinthulidea LANKESTER, 1885, p. 838; =[3]Labyrinthuleae ZOPF, 1892, p. 46; =[4]Labyrinthulés DELAGE & HÉROUARD, 1896, p. 79 *(nom. neg.)*; =[4]Labyrinthuleae OLIVE, 1902, p. 453; =[4]Labyrinthulida POCHE, 1913, p. 194; =[4]Labyrinthuloidea VALKANOV, 1940, p. 245; =[4]Labyrinthulales MARTIN in AINSWORTH and BISBY, 1950, p. 411]——[=Monadinen (Monadineae)† ZOPF, 1885, p. 98; =Monadineae azoosporeae ZOPF, 1885, p. 99]——[=[6]Proteomyxa LANKESTER, 1885, p. 839; =[5]Proteomyxés *(nom. neg.)* and =[5]Proteomyxiae DELAGE & HÉROUARD, 1896, p. 66; =[3]Protomyxidea DOFLEIN, 1901, p. 40; =[5]Proteomyxa CALKINS, 1909, p. 38; =[4]Proteomyxae CHATTON, 1925, p. 76; =[4]Proteomyxa KUDO, 1931, p. 177; =[4]Proteomyxida T. L. JAHN & F. F. JAHN, 1949, p. 108; =Protomyxidea, Protomyxées *(nom. neg.)* TRÉGOUBOFF in GRASSÉ, 1953, p. 466]——[=[2]Filosa†, [2]Proteana LANKESTER, 1885, p. 838; =[4]Filoplasmodiés *(nom. neg.)*, [4]Filoplasmodida DELAGE & HÉROUARD, 1896, p. 79; =Filoplasmodida CALKINS, 1909, p. 38; =[3]Filoplasmodinos *(nom. neg.)*, [3]Filoplasmodinae FERNÁNDEZ GALIANO, 1921, p. 40]——[=Zoosporeae† BERLESE in SACCARDO, 1888, p. 453; =[4]Zoosporés *(nom. neg.)*, [4]Zoosporida DELAGE & HÉROUARD, 1896, p. 72]——[=Azoosporeae† BERLESE in SACCARDO, 1888, p. 453; =[4]Azoosporés *(nom. neg.)*, [4]Azoosporida DELAGE & HÉROUARD, 1896, p. 67]——[=[4]Acystosporést *(nom. neg.)*, [4]Acystosporidia† DELAGE & HÉROUARD, 1896, p. 66]——[=[4]Athalamia† SCHMARDA, 1871, p. 160; =[4]Oomycétes† VAN TIEGHEM, 1898, p. 22 *(nom. neg.)*; =[4]Vampyrellida WEST, 1901, p. 308, 333; =[4]Vampyrellidea POCHE, 1913, p. 182; =[4]Myxoidea HARTOG in HARMER & SHIPLEY, 1906, p. 89; =[3]Reticulosa (Proteomyxa) MINCHEN, 1912, p. 217; =[4]Myxozoa†, Mycetozoa†, Mycetozoida† SCHEPOTIEFF, 1912, p. 267 *(non*=Mycetozoa DE BARY, 1859; *non* =Mycetozoida CALKINS, 1901]——[=[1]Hydromyxales† E. JAHN in ENGLER & PRANTL, 1928, p. 311; =Myxothallophyta† FITZPATRICK, 1930, p. 5; =Pseudo-Heliozoaires TRÉGOUBOFF in GRASSÉ, 1953, p. 466 *(nom. neg.)*]

Branching and anastomosing radiating filopodia or rhizopodia; no test or shell; majority parasitic on algae or higher plants in fresh or marine water; flagellate swarmers and encystment occur in life cycle. No hard parts. *Rec.*

Family LABYRINTHULIDAE Cienkowski, 1867

[*nom. correct.* DOFLEIN, 1901, p. 47 (*pro* Labyrinthuleae CIENKOWSKI, 1867, p. 274)]——=Labyrinthuleen ZOPF, 1892, p. 46 (*nom. neg.*); =Laberintúlidos GADEA BUISÁN, 1947, p. 28 (*nom. neg.*); =Labyrinthulida COPELAND, 1956, p. 201, 203 (*nom. van.*); =Filoplasmodieae HARTOG in HARMER and SHIPLEY, 1906, p. x, 90 (*nom. nud.*)]

Small fusiform bodies grouped in a network of filopodia, or pseudoplasmodium, individuals encyst independently, may have flagellate stage in life cycle. *Rec.*

Family PSEUDOSPORIDAE Berlese, 1888

[*nom. correct.* POCHE, 1913, p. 197 (*pro* Pseudosporeae BERLESE in SACCARDO, 1888, p. 453)]——[=Monadineae Zoosporeae CIENKOWSKI, 1865, p. 213 (*nom. nud.*); =Zoosporeae HARTOG in HARMER and SHIPLEY, 1909, p. x, 89 (*nom. nud.*); =Pseudosporeen ZOPF, 1885, p. 115 (*nom. neg.*); =Pseudosporinae DELAGE and HÉROUARD, 1896, p. 74; =Pseudosporea COPELAND, 1956, p. 191 (*nom. van.*); =Ectobiellidae POCHE, 1913, p. 199]

Solitary and heliozoan-like, with flagellate swarmers. *Rec.*

Family VAMPYRELLIDAE Zopf, 1885

[*nom. correct.* KLEBS, 1892, p. 428 (*pro* Vampyrellaceae ZOPF, 1885, p. 99]——[=Vampyrellae BERLESE in SACCARDO, 1888, p. 453; =Vampyrellacées VAN TIEGHEM, 1898, p. 22 (*nom. neg.*); =Vampyrellida CASH & HOPKINSON, 1905, p. 36; =Vampyrellacea COPELAND, 1956, p. 191 (*nom. van.*)]——=Monadinae Tetraplastae CIENKOWSKI, 1865, p. 218 (*nom. nud.*); =Hydromyxaceae KLEIN, 1882, p. 254 (*nom. nud.*); =Bursullineen ZOPF, 1885, p. 111 (*nom. neg.*); =Bursullineae BERLESE in SACCARDO, 1888, p. 453; =Bursullidae POCHE, 1913, p. 183——[=Azoosporidae DOFLEIN, 1901, p. 40 (*nom. nud.*); =Azoosporeae† HARTOG in HARMER & SHIPLEY, 1906, p. x, 89 (*nom. nud.*); =Azoosporida VALKANOV, 1940, p. 240 (*nom. nud.*)]——[=Monobidiidae POCHE, 1913, p. 183; =Pseudo-Heliozoa SANDON, 1927, p. 146 (*nom. nud.*); =Plakopodaceae E. JAHN in ENGLER & PRANTL, 1928, p. 313]

Solitary and heliozoan-like, multinucleate; without flagellate swarmers. *Rec.*

ADDENDUM

The following genera were published after families to which they belong were submitted to the Editor.

Accordiella FARINACCI, 1962, *711A, p. 7, 9 [*A. conica*; OD]. Test large, 1.2 mm. in height, 0.6 to 1.0 mm. in diameter, conical, circular in section; chambers large, in high trochospiral coil of 3 or more chambers per whorl and 8 to 12 volutions, exterior region simple and undivided, axial part of test with numerous horizontal plates and vertical pillars, resulting in labyrinthic appearance, chambers communicating with inner labyrinthic region by means of evenly aligned perforations at inner edge of chamber roof; wall calcareous, imperforate, microgranular, with rare agglutinated grains, inner layer darker, possibly originally pseudochitinous, outer layer with hyaline calcite crystals; aperture cribrate, consisting of perforations between pillars, over convex terminal face. *U. Cret. (Coniac.-Santon.)*, Eu.(Italy-Fr.-Spain).——FIG. 651,*1*. **A. conica*, Santon., Italy(S. Lazio); *1a*, nearly axial sec. of holotype, *1b*, transv. sec. of paratype, *1c*, subtang. sec. of paratype showing undivided outer region of chambers, all ×35 (*711A).

[Originally placed in the family Verneuilinidae, subfamily Eggerellinae, this genus is here transferred to the Pfenderininae (family Pavonitinidae) because of wall character and composition, trochospiral coiling, and complex interior. It is similar to *Pfenderina*, differing in having fewer chambers per whorl, and broadly conical test.] (See p. C292.)

Dainella BRAZHNIKOVA, 1962, *204A, p. 22 [*Endothyra(?) chomatica* DAIN in BRAZHNIKOVA, 1962, *204A, p. 23; OD]. Subglobular test, slightly appressed along axis; numerous chambers per whorl, increasing slowly and regularly in size; streptospirally enrolled, involute; wall calcareous, single homogeneous layer, with massive chomata; aperture simple, basal. [Differs from *Endothyra* in the homogeneous wall, numerous chambers per whorl and massive chomata, and from *Quasiendothyra* in the involute coiling and inflated test, simple wall, and strongly streptospiral coiling throughout.] *L.Carb.(L.Visean)*, USSR(Don Basin-Ukraine).——FIG. 650,*1*. **D. chomatica* (DAIN); *1a*, axial sec. of holotype, ×100; *1b,c*, horiz. and axial secs. of paratypes, ×90, ×75 (*204A). (See p. C346.)

Goatapitigba NARCHI, 1962, *1346A, p. 277 [*G. jurara*; OD]. Test attached; globular proloculus followed by few somewhat inflated pyriform cham-

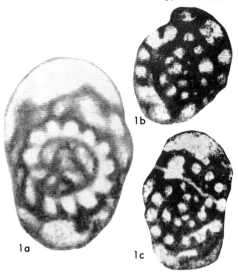

FIG. 650. Endothyridae (Endothyrinae; *1, Dainella*) (p. C795).

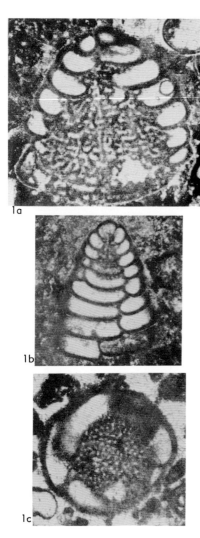

Fig. 651. Pavonitinidae (Pfenderininae; *1, Accordiella*) (p. C795).

bers; wall agglutinated, with considerable cement and pseudochitinous inner layer; aperture terminal, against attachment. *Rec.,* S.Am.(Brazil, Cabo Frio).——FIG. 652,1. **G. jurara*; *1a,* holotype, ×14; *1b,c,* top and edge views of paratype, ×14 (*1346A).

[The chamber form is reminiscent of the porcelaneous Nubeculariinae, some of which may also have surficial agglutinated material. As no information is available as to the amount or character of the cement, the present genus is retained tentatively in the Saccamminidae and because of its attached nature is placed in the Hemisphaerammininae. It differs from *Saccamminis* in the regular and pyriform chambers, and finely agglutinated wall.] (See p. C204.)

Petchorina REYTLINGER in VARSANOFEVA & REYT-

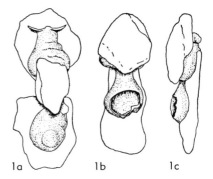

FIG. 652. Saccamminidae (Hemisphaerammininae; *1, Goatapitigba*) (p. C795-C796).

LINGER, 1962, *1980A, p. 56 [**P. schezhimovensis*; OD]. Test oval to angularly irregular in outline, 0.31 to 0.45 mm. in diameter; interior partially subdivided by short pseudosepta into 2 or 3 successively larger pseudochambers; wall calcareous, microgranular, about 36μ in thickness; aperture not observed. [Originally placed in the Parathuramminidae, and stated to differ from *Bisphaera* by the presence of pseudosepta, and from *Baituganella* by the test form and minutely granular wall. Tentatively recognized herein, it is also similar in outline to *Uslonia,* which occurs in correlative strata in Kazan, but differs in the presence of partial chamber subdivisions.] *Dev.* (Frasn.), USSR (Pechora distr.).——FIG. 653,1. **P. schezhimovensis*; *1a,* holotype; *1b,* paratype, ×70 (*1980A). (See p. C315.)

Asterocyclina GÜMBEL, 1870 (see p. C714).

[The name *Asteriatites* was published by VON SCHLOTHEIM in 1813 (Taschenb. Mineralogie, v. 7, p. 68, 109) as designation of an asterozoan (p. 68) and of a foraminifer (p. 109). In 1820 the same author (Die Petrefactenkunde, p. 324) referred to the echinoderm as *Asteriacites,* interpreted by NEAVE as a *lapsus* (pro *Asteriatites* SCHLOTHEIM, 1813, p. 68). Then, in 1822 SCHLOTHEIM again published *Asteriacites* (Petrefactenkunde Nachtrag 1, p. 71) as the name for his (1813, p. 109) foraminifer. The ambiguous application of *Asteriatites* in 1813 is resolved by fixing it herein as the name for an ophiuroid (p. 68), with typespecies by subsequent monotypy as *Asteriacites ophiurus* SCHLOTHEIM, 1820, p. 324. *Asteriacites* SCHLOTHEIM, 1822 (type, *A. patellaris*) is a junior homonym, cited properly as (non *Asteriacites* SCHLOTHEIM, 1820).1

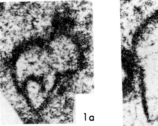

FIG. 653. Parathuramminidae; *1, Petchorina* (p. C796).

REFERENCES

(1) **Abich, H.**, 1859, *Ueber das Steinsalz und seine geologische Stellung im russischen Armenien:* Acad. Impér. Sci., St. Pétersbourg, Cl. Sci., Math. & Phys., Mém., ser. 6, v. 7, p. 61-150, pl. 1-10.
——(1A) 1859, *Vergleichende Grundzüge der Geologie des Kaukasus wie der armenischen und nordpersischen Gebirge:* Acad. Impér. Sci. St. Pétersbourg, Cl. Sci., Math. & Phys., Mém., ser. 6, v. 7, p. 359-534, pl. 1-8.

(2) **Abrard, R.**, 1926, *Un foraminifère nouveau du Campanien de la Charente-Inférieure:* Soc. géol. France, Comptes Rendus, no. 4, p. 31-32.
——(2A) 1956, *Une Operculine cordelée de l'Eocene inférieur de la Côte-d' Ivoire Operculina (Nummulitoides) tessieri n. subgen., n. sp.:* Soc. géol. France, Bull., ser. 6, v. 5 (1955), p. 489-493, pl. 23.

(3) **Acosta, J. T.**, 1940, *Algunos foraminíferos nuevos de las costas Cubanas:* Torreia, no. 5, p. 1-6, pl. 1.

(3A) **Agalarova, D. A.**, 1960, *Stratigrafiya i mikrofauna yurskikh otlozheniy severo-zapadnogo Turkmenistana:* Voprosy Geologii, Bureniya i Dobychi Nefti, Trudy, Azerbaydzhanskiy Nauchno-Issledovatelskiy Institut po Dobyche Nefti (AzNIIDN), no. 10, p. 56-87, pl. 1-9. [*Stratigraphy and microfauna of the Jurassic deposits of northwestern Turkmenistan.*]

(4) **Agardh, C. A.**, 1827, *Aufzählung einiger in den österreichischen Ländern gefundenen neuen Gattungen und Arten von Algen:* Flora oder allgemeine (Botanische Zeitschr.), v. 10, pt. 2, p. 625-646 (Regensburg).

(5) **Agassiz, Louis**, 1844, *Nomina systematica generum polyporum (Anthozoorum et bryozoorum cum polythalamiis) tam viventium quam fossilium:* Nomenclator Zoologicus, pt. 5, Polypi, p. 1-28 (Soloduri).——(6) 1846, *Nomenclatoris zoologici index universalis:* vii+393 p. (Soloduri). [Title page, 1846; cover of fascicle, 1847.]

(7) **AGIP Mineraria**, 1957, *Foraminiferi Padani (Terziario e Quaternario):* AGIP Mineraria, Milano, pl. 1-52.

(8) **Ainsworth, G. C., & Bisby, G. R.**, 1950, *A dictionary of the fungi:* ed. 3, 447 p. (Kew).

(9) **Akimets, V. S.**, 1958, *O novom rode i vide Foraminifer iz Verkhnemelovykh otlozheniy Belorussii:* Akad. Nauk Belorusskoy SSR, Doklady, v. 2, no. 1, p. 35-36. [*Concerning a new genus and species of Foraminifera from Upper Cretaceous deposits of Belorussia.*]——(10) 1961, *Stratigrafiya i Foraminifery Verkhnemelovykh otlozheniy Belorussii:* Akad. Nauk Belorusskoy SSR, Inst. Geol. Nauk, Paleont. i Strat. BSSR, v. 3, p. 3-245, pl. 1-19, text-fig. 1-8, tables. [*Stratigraphy and Foraminifera of the Upper Cretaceous deposits of Belorussia.*]

(11) **Alexander, C. J., & Smith, J. P.**, 1932, *Foraminifera of the genera Flabellammina and Frankeina from the Cretaceous of Texas:* Jour. Paleontology, v. 6, p. 299-311, pl. 45-47, text-fig. 1-2.

(12) **Almela, Antonio**, 1946, *Una nueva especie de "Dictyoconus" del Cenomaniense valenciano:* Inst. Geol. & Minero España, Notas & Comun., Bull., no. 16, p. 151-156, 1 pl.

(13) **Alth, A.**, 1850, *Geognostisch-paläontologische Beschreibung der nächsten Umgebung von Lemberg:* Naturwiss. Wien, Abhandl., v. 3, p. 171-284, pl. 9-13.

(14) **Altpeter, Otto**, 1913, *Beiträge zur Anatomie und Physiologie von Alveolina:* Neues Jahrb. Mineral. Geol. & Paläont., Beil.-Bd. 36, p. 82-112, pl. 6-8, fig. ser. A-D.

(15) **Andersen, H. V.**, 1951, *Two new genera of Foraminifera from Recent deposits of Louisiana:* Jour. Paleontology, v. 25, p. 31-34, text-fig. 1-2.——(16) 1951, *An addenda to Arenoparrella and Arenoparrella mexicana (Kornfeld):* Cushman Found. Foram. Research, Contrib., v. 2, p. 96-97, pl. 11.——(17) 1952, *Buccella, a new genus of the rotalid Foraminifera:* Washington Acad. Sci., Jour., v. 42, no. 5, p. 143-151, text-fig. 1-13.——(18) 1961, *Genesis and paleontology of the Mississippi River mudlumps, Pt. II, Foraminifera of the mudlumps, lower Mississippi River delta:* Louisiana Dept. Conserv., Geol. Bull. 35, p. 1-208, pl. 1-29.

(19) **Andreae, A.**, 1884, *Beitrag zur Kenntniss des Elsässer Tertiars; Theil II, Die Oligocän-Schichten:* Geol. Spezialkarte Elsass-Loth., Abhandl., v. 2, no. 3, p. 1-239, pl. 1-12.——(20) 1895, *Eine merkwürdige Nodosariidenform aus dem Septarienthon von Lobsann im Unter-Elsass:* Mittheil. geol. Landesanst. Elsass-Lothringen, v. 4, no. 4, p. 171-173, text-fig. 1-2.——(21) 1898, *Die Foraminiferen des Mitteloligocäns der Umgegend von Lobsann und Pechelbronn im Unter-Elsass und Resultate der neueren Bohrungen in dortiger Gegend:* Mitt. geol. Landesanst. von Elsass-Lothringen, v. 4, no. 4, p. 287-303, 9 text-fig.

(22) **Anonymous**, 1949, *Plate explanations of Rhumbler's "Plankton-Expedition":* Micropaleontologist, v. 3, p. 33-40.

(23) **Antonova, Z. A.**, 1958, *K voprosy ob evolyutsii nekotorykh predstaviteley Oftalmidiid na primere razvitiya ikh v yurskoe vremya v basseyne r. laby:* Akad. Nauk SSSR, Doklady, v. 122, no. 5, p. 913-916, 2 tables, text-figs. [*On the question of the evolution of certain representatives of the Ophthalmidiidae as an example of the development in Jurassic time in the basin of the Laby River.*]——(24), 1958, *Foraminifery sredney Yury Basseyna r. Laby:* Vsesoy. Neft. Nauchno-Issledov. Institut (VNII),

Trudy, no. 17, p. 41-79, pl. 1-5. [*Foraminifera of the middle Jurassic of the basin of the Laby River.*]

(25) **Antropov, I. A.**, 1950, *Novye vidy Foraminifer Verkhnego Devona nekotorykh rayonov vostoka Russkoy Platformy:* Akad. Nauk SSSR, Geol. Inst. Kazan, Izvestiya Kazanskogo Filiala, v. 1, p. 21-33, pl. 1-3. [*New species of Foraminifera from the Upper Devonian of certain areas of the eastern Russian Platform.*]

(25A) 1959, *Foraminifery Devona Tatarii:* Same, no. 7, p. 11-34, pl. 1, tables 1-7. [*Foraminifera from the Devonian of Tatar.*]

(26) **Applin, E. R.,** & **Jordan, Louise**, 1945, *Diagnostic Foraminifera from subsurface formations in Florida:* Jour. Paleontology, v. 19, p. 129-148, pl. 18-21, 2 text-fig.

(27) ———, **Loeblich, A. R., Jr.,** & **Tappan, Helen**, 1950, *Two new Lower Cretaceous lituolid Foraminifera:* Washington Acad. Sci., Jour., v. 40, no. 3, p. 75-79, text-fig. 1-6.

(27A) **Arapova, N. D.**, 1961, *K sistematike semeistva Ammodiscidae:* Trudy Tashkent Gosudarstvennogo Universiteta im V. I. LENINA, Geol., no. 180 (1960), p. 151-154, pl. 1. [*On systematics of the family Ammodiscidae.*]

(28) **Archer, William**, 1867, *Proceedings Dublin Microscopical Club, Session November 16, 1866:* Quart. Jour. Micro. Sci., new ser., v. 7, p. 173-175.———(29) 1869, *On some freshwater Rhizopoda, new or little-known:* Same, v. 9, p. 250-271, pl. 16-17.———(30) 1869, *Proceedings Dublin Microscopical Society, Session 18th March, 1869:* Same, v. 9, p. 322.———(31) 1869, *On some freshwater Rhizopoda, new or little-known:* Same, v. 9, p. 386-397, pl. 16, 20.———(32) 1876, *Proceedings of the Dublin Microscopical Club, 23rd March, 1876:* Same, v. 16, p. 343-344.———(33) 1877, *Proceedings Dublin Microscopical Club, session July 13, 1876:* Same, v. 17, p. 102-104.———(34) 1877, *Résumé of recent contributions to our knowledge of "Freshwater Rhizopoda," Pt. IV:* Same, v. 17; (a) p. 107-124, pl. 8; (b) p. 330-353, pl. 21.

(35) **Archiac, Adolphe d'**, 1837, *Mémoire sur la formation crétacé du sud-ouest de la France:* Soc. géol. France, Mém., v. 2, no. 7, p. 157-192, pl. 11-13.———(36) 1843, *Description géologique du département de l'Aisne:* Same, mém. 3, v. 5, p. 129-420, pl. 21-31, 4 tab. (also numbered p. 1-292, pl. A-K, 4 tab.).———(37) 1846, *Description des fossiles recueillis par M. Thorent, dans les couches à nummulines des environs de Bayonne:* Same, ser. 2, v. 2, pt. 1, p. 189-217, pl. 7.———(37A) 1848, *Description des fossiles du groupe Nummulitique recueillis par M.S.P. Pratt et M.J. Delbos aux environs de Bayonne et de Dax:* Same, ser. 2, v. 3, p. 397-456, pl. 8-13.

(38) ———, & **Haime, Jules**, 1853-1854, *Description des animaux fossiles du groupe nummulitique de l'Inde, précédé d'un résumé géologique et d'une monograph des Nummulites:* v. 1(1853); v. 2(1854); 373 p., 36 pl., Gide & J. Baudry (Paris).———(39) 1854, *Coupe géologique des environs de Baines de Rennes (Aude) suivie de la description de quelques fossiles de cette localité:* Soc. géol. France, Bull., ser. 2, v. 11 (1853-54), p. 205-206, pl. 2.

(40) **Arnold, Z. M.**, 1948, *A new foraminiferan belonging to the genus Allogromia:* Am. Micro. Soc., Trans., v. 67, no. 3, p. 231-235.———(41) 1952, *Structure and paleontological significance of the oral apparatus of the foraminiferoid Gromia oviformis Dujardin:* Jour. Paleontology, v. 26, p. 829-831, 1 text-fig.———(42) 1954, *Discorinopsis aguayoi (Bermúdez) and Discorinopsis vadescens Cushman and Brönnimann: A study of variation in cultures of living Foraminifera:* Cushman Found. Foram. Research, Contrib., v. 5, pt. 1, p. 4-13, pl. 1-2. ———(43) 1954, *A note on foraminiferan sieveplates:* Same, v. 5, pt. 2, p. 77.———(44) 1955, *An unusual feature of miliolid reproduction:* Same, v. 6, pt. 3, p. 94-96.———(45) 1955, *Life history and cytology of the foraminiferan Allogromia laticollaris:* Univ. Calif. Publ. Zool., v. 61, no. 4, p. 167-252, pl. 27-35, 3 text-fig.

(46) **Asano, Kiyoshi**, 1936, *New Foraminifera from the Kakegawa district, Tôtômi, Japan (Studies on the fossil Foraminifera from the Neogene of Japan, Pt. 4):* Japan. Jour. Geol. & Geog., v. 13, no. 3-4, p. 325-331, pl. 36-37. ———(47) 1936, *Pseudononion, a new genus of Foraminifera found in Muracks-mura, Kamakura-gori, Kanagawa Prefecture:* Geol. Soc. Japan, Jour., v. 43, p. 347-348.———(48) 1936, *Rotalidium, a new genus of Foraminifera from the Pacific:* Imper. Acad. Tokyo, Proc., v. 12, no. 10, p. 350-351, text-fig. 1-3.———(49) 1938, *Japanese fossil Nodosariidae, with notes on the Frondiculariidae:* Tôhoku Imper. Univ., Sci. Repts., ser. 2 (Geol.), v. 19, no. 2 (1937-38), p. 179-200, pl. 24-31, 4 text-fig.——— (50) 1944, *Hanzawaia, a new genus of Foraminifera from the Pliocene of Japan:* Geol. Soc. Japan, Jour., v. 51, no. 606, p. 97-98, pl. 4. ———(51) 1950, *Cretaceous Foraminifera from Teshio, Hokkaido:* Tôhoku Univ., Inst. Geol. Paleont., Short Papers, no. 2, p. 13-22, pl. 3. ———(52) 1950-51, *Illustrated catalogue of Japanese Tertiary smaller Foraminifera:* Petrol. Branch, Natural Resources Sec., General Headquarters, Supreme Commander for Allied Powers (Tokyo); (a) *Pt. 2, Buliminidae*, p. 1-19 (1950); (b) *Pt. 7, Cassidulinidae*, p. 1-7 (1951); (c) *Pt. 13, Anomalinidae*, p. 12-19 (1951).

———(53) 1952, *Paleogene Foraminifera from*

the Ishikari and Kushiro Coal-Fields, Hokkaido: Tôhoku Univ., Institute Geol. & Paleont., Short Papers, no. 4, p. 23-46, pl. 3-5, 1 text-fig.

(54) **Astre, Gaston,** 1927, *Sur Monolepidorbis foraminifère voisin des Lindérines et des Orbitoides:* Soc. géol. France, Bull., ser. 4, v. 27, pt. 6-9, p. 387-394, pl. 20, text-fig.

(55) **Auerbach, Leopold,** 1856, *Ueber die Eingelligkeit der Amoeben:* Zeitschr. Wiss. Zool., v. 7, p. 365-430, pl. 19-22.

(56) **Aurouze, Germaine, & Bizon, J. J.** 1958, *Rapports et différences des deux genres de fora minifères, Kilianina (Pfender) et Meyendorffina n. gen.:* Revue Micropaléont., v. 1, no. 2, p. 67-74, pl. 1-3.

(57) ———, **& Boulanger, D.,** 1954, *Ganella n. gen., nouveau genre de foraminifères de l'Ypresien de Gan (Basses-Pyrénées:)* Soc. géol. France, Comptes Rendus, Somm., no. 9-10, p. 186-188, text-fig. 1-3.

(58) **Averintsev [Awerinzew], S.,** 1903, *Über die Struktur der Kalkschalen mariner Rhizopoden:* Zeitsch. Wiss. Zool., v. 74, p. 478-490, pl. 24.———(59) 1906, *Rhizopoda priesnykh vod. Vyp. 2. Sistematika Rhizopoda testacea:* Imper. St. Petersbourg Obshch. Estestvoisp., Trudy, v. 36, no. 2, p. 121-346, pl. 5.———(60) 1906, *Über die Süsswasser-protozoen der Insel Waigatsch:* Zool. Anzeiger, v. 31, p. 306-312, text-fig. 1-5.———(61) 1907, *Die Struktur und die chemische Zusammensetzung der Gehaüse bei den Süsswasserrhizopoden:* Archiv Protistenkunde, v. 8, p. 91-111.———(62) 1911, *Zur Foraminiferen-Fauna des Sibirischen Eismeeres:* Acad. Impér. Sci., St. Pétersbourg, Cl. Phys.-Math., Mém., ser. 8, v. 29, no. 3, p. 1-28, 1 pl.

(63) **Avnimelech, Moshè,** 1952, *Revision of the tubular Monothalamia:* Cushman Found. Foram. Research, Contrib., v. 3, p. 60-68, text-fig. 1-17.

(64) ———, **Parness, A., & Reiss, Zeev,** 1954, *Mollusca and Foraminifera from the Lower Albian of the Negev (southern Israel):* Jour. Paleontology, v. 28, p. 835-839, text-fig. 1-9.

(65) **Bailey, J. W.,** 1851, *Microscopical examination of soundings made by the U.S. Coast Survey off the Atlantic coast of the U.S.:* Smithsonian Contr. Know., v. 2, art. 3, p. 1-15, 1 pl. ———(66) 1853, *Observations on a newly discovered animalcule:* Am. Jour. Sci. & Arts, ser. 2, v. 15, p. 341-347, text-fig. 1-40.———(67) 1856, *Notice of microscopic forms found in the soundings of the Sea of Kamtschatka—with a plate:* Same, ser. 2, v. 22, p. 1-6.

(68) **Bakx, L. A. J.,** 1932, *De genera Fasciolites en Neoalveolina in het Indo-Pacifische gebied:* Geol.-Mijnb. Genoot. Nederland Kolon., Verhandl., Geol. Ser., v. 9, p. 205-266, pl. 1-4.

(69) **Bandy, O. L.,** 1944, *Eocene Foraminifera from Cape Blanco, Oregon:* Jour. Paleontology, v. 18, p. 366-377, pl. 60-62.———(70) 1949, *Eocene and Oligocene Foraminifera from Little Stave Creek, Clarke County, Alabama:* Bull. Am. Paleontology, v. 32, no. 131, 210 p., pl. 1-27.———(71) 1949, *Textularia vs. Spiroplectammina:* Micropaleontologist, v. 3, no. 1, p. 22.———(72) 1952, *The genotype of Siphogenerina:* Cushman Found. Foram. Research, Contrib., v. 3, pt. 1, p. 17-18.———(73) 1954, *Aragonite tests among the Foraminifera:* Jour. Sed. Petrology, v. 24, no. 1, p. 60-61.——— (74) 1960, *General correlation of foraminiferal structure with environment:* Internatl. Geol. Cong., Session 21 Norden, Copenhagen, pt. 22, p. 7-19, fig. 1-9.———(75) 1960, *The geologic significance of coiling ratios in the foraminifer Globigerina pachyderma (Ehrenberg):* Jour. Paleontology, v. 34, p. 671-681, 7 text-fig.

(76) ———, **& Burnside, R. J.,** 1951, *The genus Siphogenerina Schlumberger:* Cushman Found. Foram. Research, Contrib., v. 2, pt. 1, p. 13-15.

(77) **Banner, F. T., & Blow, W. H.,** 1959, *The classification and stratigraphical distribution of the Globigerinaceae:* Palaeontology, v. 2, pt. 1, p. 1-27, pl. 1-3.———(78) 1960, *The taxonomy, morphology and affinities of the genera included in the subfamily Hastigerininae:* Micropaleontology, v. 6, no. 1, p. 19-31, text-fig. 1-11. ———(79) 1960, *Some primary types of species belonging to the superfamily Globigerinacea:* Cushman Found. Foram. Research Contrib., v. 11, pt. 1, p. 1-41, pl. 1-8.

(80) **Bargoni, E.,** 1894, *Di un foraminifero parassita nelle Salpe e considerazioni sui corpuscoli amilacei dei protozoi superiori:* Rome, R. Univ., Biol. Lab., Ricerche v. 4 (1894-95), pt. 1-2, p. 43-64, pl. 3-4.

(81) **Barker, John,** 1868, *Proceedings of the Dublin Microscopical Club, December 19, 1867:* Quart. Jour. Micro. Sci., new ser., v. 8, p. 122-124.

(82) **Barker, R. W.,** 1939, *Species of the foraminiferal family Camerinidae in the Tertiary and Cretaceous of Mexico:* U.S. Natl. Museum, Proc., v. 86, no. 3052, p. 305-330, pl. 11-22.——— (83) 1944, *Some larger Foraminifera from the Lower Cretaceous of Texas:* Jour. Paleontology, v. 18, p. 204-209, pl. 35.

(84) ———, **& Grimsdale, T. F.,** 1936, *A contribution to the phylogeny of the orbitoidal Foraminifera, with descriptions of new forms from the Eocene of Mexico:* Jour. Paleontology, v. 10, p. 231-247, pl. 30-38, 4 fig.———(85) 1937, *Studies of Mexican fossil Foraminifera:* Ann. & Mag. Nat. History, ser. 10, v. 19, p. 161-178, 5 pl., 2 fig.

(86) **Barker-Webb, P., & Berthelot, S.,** 1839, *Foraminifères:* in Histoire Naturelle des Îles Canaries, v. 2, pt. 2, Zool., p. 119-146, pl. 1-3.

(87) **Barnard, Tom,** 1958, *Some Mesozoic adherent Foraminifera:* Palaeontology, v. 1, pt. 2, p. 116-124, pl. 22-25.

(88) ———, & **Banner, F. T.,** 1953, *Arenaceous Foraminifera from the Upper Cretaceous of England:* Geol. Soc. London, Quart. Jour., v. 109, p. 173-216, pl. 7-9.

(89) **Barrier, J.,** & **Neumann, Madeleine,** 1959, *Contribution a l'étude de Nonionina cretacea Schlumberger:* Revue Micropaléont. v. 1, no. 4, p. 223-229, pl. 1, 2.

(90) **Bartenstein, Helmut,** 1948, *Taxonomische Abgrenzung der Foraminiferen-Gattungen Palmula Lea, Flabellina Orbigny und Falsopalmula n.g., gleichzeitig eine Revision der Jura-Arten von "Flabellina":* Senckenbergiana, v. 28, no. 4/6, p. 119-137, pl. 1-2, text-fig. 1-5.———
(91) 1952, *Taxonomische Bemerkungen zu den Ammobaculites, Haplophragmium, Lituola und verwandten Gattungen.* (For.): Same, v. 33, p. 313-342.

(92) ———, & **Brand, Erich,** 1937, *Mikropaläontologische Untersuchungen zur Stratigraphie des nordwest-deutschen Lias und Doggers:* Senckenberg. naturforsch. Gesell. Abhandl., no. 439, p. 1-224, pl. 1-20.———(93) 1938, *Die Foraminiferen-Fauna des Jade-Gebietes. 1. Jadammina polystoma n.g. n.sp. aus dem Jade-Gebiet* (For.): Senckenbergiana, v. 20, no. 5, p. 381-385, text-fig. 1-3.———(94) 1949, *New genera of Foraminifera from the Lower Cretaceous of Germany and England:* Jour. Paleontology, v. 23, p. 669-672, text-fig. 1-10.———(95) 1951, *Mikropaläontologische Untersuchungen zur Stratigraphie des nordwestdeutschen Valendis:* Senckenberg naturforsch. Gesell., Abhandl., no. 485, p. 239-336, pl. 1-25.

(96) **Bartoš, E.,** 1938, *Eine neue moosbewohnende Nebella-Art, Nebella pulchra m.n.sp.:* Archiv Protistenkunde, v. 90, no. 2, p. 346-347, 1 text-fig.

(97) **Bary, Anton de,** 1859, *Die Mycetozoan, Ein Beitrag zur Kenntniss der niedersten Thiere:* Zeitschr. Wiss. Zool., v. 10, p. 88-175, pl. 6-10.———(98) 1864, *Die Mycetozoen:* ed. 2, revised, xii+132 p., 5 pl. (Leipzig).———(99) 1884, *Vergleichende Morphologie und Biologie der Pilze Mycetozoen und Bacterien:* xvi+558 p., 198 fig. (Leipzig).———(100) 1887, *Comparative morphology and biology of the fungi Mycetozoa and Bacteria,* authorized English translation by H. E. F. Garnsey, revised by I. B. Balfour: xviii+525 p., 198 fig., Clarendon Press (Oxford).

(101) **Basset, Charles,** 1885, *Foraminifères de la Société des Sciences naturelles de la Charente-Inférieure:* Soc. Sci. Nat. Charente-Inférieur (1884), no. 21, p. 153-174, pls.

(102) **Batsch, A. I. G. C.,** 1791, *Sechs Kupfertafeln mit Conchylien des Seesandes, gezeichnet und gestochen von A. J. G. K. Batsch:* 6 pl. (Jena).

(103) **Beckmann, Heinz,** 1950, *Rhenothyra, eine neue Foraminiferengattung aus dem rheinischen Mitteldevon:* Neues Jahrb. Geol. & Paläont., Monatshefte (1950), no. 6, p. 183-187, 5 text-fig.———(104) 1953, *Palachemonella torleyi n.gen. et n.sp., eine neue Foraminifere aus den Schleddenhofer Schichten (Mitteldevon):* Geol. Jahrb., v. 67, p. 259-272, pl. A-B, 6 text-fig.

(105) **Beede, J. W.,** & **Kniker, H. T.,** 1924, *Species of the genus Schwagerina and their stratigraphic significance:* Univ. Texas, Bull. 2433, 96 p., 9 pl.

(106) **Beissel, Ignaz,** 1891, *Die Foraminiferen der Aachener Kreide:* K. Preuss. Geol. Landesanst., Abhandl., new ser., no. 3, p. 1-78; atlas, pl. 1-16.

(107) **Bělař, Karl,** 1921, *Untersuchungen über Thecamöben der Chlamydophrys-Gruppe:* Archiv Protistenkunde, v. 43, p. 287-354, pl. 3-10, 24 text-fig.

(108) **Belford, D. J.,** 1958, *The genera Nuttallides Finlay, 1939, and Nuttallina, n. gen.:* Cushman Found. Foram. Research, Contrib., v. 9, pt. 4, p. 93-98, pl. 18-19, text-fig. 1-4.———(109) 1959, *Nuttallinella, new name for Nuttallina Belford, 1958 (non Nuttallina Dall, 1871):* Same, v. 10, pt. 1, p. 20.———(110) 1960, *Upper Cretaceous Foraminifera from the Toolonga calcilutite and Gingin chalk, western Australia:* Australia Bur. Mineral. Res., Geol. & Geophys., Bull. 57, p. 1-198, pl. 1-35.———
(111) 1961, *Spirotecta pellicula, n.gen., n.sp., from the Upper Cretaceous and Giraliarella triloba, n.sp., from the Permian of Western Australia:* Cushman Found. Foram. Research, Contrib., v. 12, pt. 3, p. 81-82, pl. 3.

(112) **Bellen, R. C. van,** 1941, *Some Eocene Foraminifera from the Neighbourhood of Ričice near Imotski, E. Dalmatia, Yugoslavia:* Nederland. Akad. Wetensch., Proc., v. 44, no. 8, p. 996-1005, 1 pl.———(113) 1946, *Foraminifera from the middle Eocene in the southern part of the Netherlands Province of Limburg:* Meded. Geol. Stichting, ser. C, v. 5, no. 4, p. 1-144, pl. 1-13.———(114) 1946, *Some homonyms in "Foraminifera from the middle Eocene in the southern part of the Netherlands province of Limburg":* Cushman Lab. Foram. Research, Contrib., v. 22, pt. 4, p. 120-123, text-fig. 1.

(115) **Bennett, A. W.,** & **Murray, G. A.,** 1889, *A handbook of cryptogamic botany:* viii+473 p., 382 fig., Longmans, Green & Co. (London & New York).

(116) **Bermúdez, P. J.,** 1934, *Un genero y especie nueva de Foraminíferos viventes de Cuba:* Soc. Cubana Historia Nat., Mem., v. 8, no. 2, p.

83-86, fig. 1-3.──(117) 1935, *Foraminíferos de la Costa Norte de Cuba:* Same, v. 9, no. 3, p. 129-224, pl. 10-17, text-fig. 1-3.──(118) 1937, *Notas sobre Hantkenina brevispina Cushman:* Same, v. 11, no. 3, p. 151-152.── (119) 1937, *Nuevas especies de Foraminíferos del Eoceno de las cercanías de Guanajay, Provincia Pinar del Rio:* Same, v. 11, p. 237-247, pl. 20, 21.──(120) 1938, *Resultados de la primera expedición en las Antillas del ketch Atlantis bajo los auspicios de las Universidades de Harvard y Habana, Aguayoina asterostomata, un Foraminífero nuevo del Mar Caribe:* Same, v. 12, no. 5, p. 385-388, pl. 29.──(121) 1939, *Resultados de la primera expedición en las Antillas del Ketch Atlantis bajo los auspicios de las Universidades de Harvard y Habana:* Same; (a) v. 13, no. 1, p. 9-12, pl. 1-2; (b) v. 13, no. 4, p. 247-251, pl. 33, text-fig. 1.── (122) 1940, *Barbourinella, nuevo nombre par Barbourina, Foraminífero:* Same, v. 14, p. 410. ──(123) 1949, *Pavoninoides, a new genus of the Miliolidae from Panama:* Cushman Lab. Foram. Research. Contrib., v. 25, pt. 3, p. 58, text-fig.──(124) 1949, *Tertiary smaller Foraminifera of the Dominican Republic:* Same, Spec. Publ. 25, p. 1-322, pl. 1-26.──(125) 1950,*Contribución al estudio del Cenozoico Cubano:* Soc. Cubana Historia Nat., Mem., v. 19, no. 3, p. 205-375.──(126) 1951, *Heminwayina, un genero nuevo de los Foraminíferos rotaliformes y sus especies:* Soc. Ciencias Nat. La Salle, Mem., v. 11, no. 30, p. 325-329, 1 pl.──(127) 1952, *Estudio sistemático de los Foraminíferos rotaliformes:* Venezuela Minist. Minas & Hidrocarb., Bull. Geol., v. 2, no. 4, p. 1-230, pl. 1-35.── (128) 1961, *Contribución al estudio de las Globigerinidea de la region Caribe-Antillana (Paleoceno-Reciente):* 3rd Congr. Geol. Venezolano, Bol. Geol., Mem., v. 3, spec. publ. 3 (1960), p. 1.119-1.393, pl. 1-20.

(129) ──, & Key, C. E., 1952, *Tres generos nuevos de Foraminíferos de las familias Reophacidae y Valvulinidae:* Soc. Ciencias Nat. La Salle, Mem., v. 12, no. 31, p. 71-76, 1 pl.

(130) **Berry, E. W.,** 1928, *Asterodiscocyclina, a new subgenus of Orthophragmina:* Eclogae geol. Helv., v. 21, p. 405-407, pl. 33.──(131) 1929, *Larger Foraminifera of the Verdun Formation of northwestern Peru:* Johns Hopkins Univ., Studies Geol., no. 9, p. 9-166, pl. 1-22, text-fig. 1-3.

(132) **Berthelin, Georges,** 1879, *Foraminifères du Lias moyen de la Vendée:* Revue Mag. Zool., Paris, ser. 3, v. 7, p. 24-41, pl. 1.──(133) 1880, *Mémoire sur les Foraminifères fossiles de l'Etage Albien de Moncley (Doubs):* Soc. géol. France, Mém., ser. 3, v. 1, no. 5, p. 1-84, pl. 24-27.──(134) 1881, *Coup d'oeil sur la faune rhizopodique du Calcaire Grossier inférieur de la Marne:* Assoc. Franç. Avanc. Sci., Comptes Rendus, Sess. 9 (Reims, 1880), p. 553-559.──(135) 1893, *Sur l'Orbicula elliptica d'Archiac, du Bathonien supérieur de l'Aisne et des Ardennes:* Soc. géol. France, Comptes Rendus, Somm., p. lxxiii.

(136) **Bessels, Emil,** 1875, *Haeckelina gigantea. Ein Protist aus der Gruppe der Monothalamien:* Jenaische Zeitschr. Naturwiss., v. 9, p. 265-271, pl. 14.

(137) **Bettenstaedt, F.,** 1952, *Stratigraphisch wichtige Foraminiferen-Arten aus dem Barrême vorwiegend Nordwest-Deutschlands:* Senckenbergiana, v. 33, no. 4-6, p. 263-295, pl. 1-4.

(137A) **Bieda, F.,** 1950, *Sur quelques foraminifères nouveaux ou peu connus du flysch des Karpates Polonaises:* Rocznika Polskiego Towarzystwa Geologicznego z Roku, v. 18 (1948), p. 167-179, pl. 3-4.

(138) **Bignot, G., & Neumann, Madeleine,** 1962, *La structure des tests des Foraminifères analyse bibliographique:* Revue Micropaléont., v. 4, no. 4, p. 237-248, pl. 1-2, text-fig. 1-2.

(139) **Birina, L. M.,** 1948, *Novye vidy izvestkovykh vodorosley i foraminifer pogranichnykh sloev Devona i Karbona:* Sovetskaya Geologiya, Sbornik 28, Minist. Geol. Soyuzo SSR, p. 154-159, pl. 1-2. [*New species of calcareous algae and Foraminifera of the boundary strata of the Devonian and Carboniferous.*]

(140) **Blackwelder, R. E.,** 1959, *The functions and limitations of classification:* Systematic Zoology, v. 8, no. 4, p. 202-211, 1 fig.

(141) **Blainville, H. M. Ducrotay de,** 1824-30, *Dictionnaire des Sciences Naturelles:* (a) 1824, v. 32, p. 1-567; (b) 1824, Zoologie, Conchyliologie et Malacologie, atlas v. 31, pl. 1-33; (c) 1826, v. 41, p. 1-558; (d) 1830, v. 60, p. 1-631; F. G. Levrault (Paris).──(142) 1825, *Manuel de malacologie et de conchyliologie:* 664 p., 87 pl. (1827), F. G. Levrault (Paris). ──(143) 1828, in VIEILLOT, L. J. P., *Faune française, ou histoire naturelle, géenerale et particulière des animaux qui se trouvent en France* (1821-28, v. 1-29): v. 18, p. 66 (Paris).

(144) **Blake, J. F.,** 1876, *On Renulina sorbyana:* Monthly Micro. Jour., v. 15, p. 262-264.

(145) **Blanc, Henri,** 1886, *Un nouveau Foraminifère de la faune profonde du Lac:* Biblio. Universelle (Archives Sci., Phys., & Nat.), ser. 3, v. 16, p. 362-366.

(146) **Blanckenhorn, Max,** 1900, *Neues zur Geologie und Paläontologie Aegyptens:* Deutsche geol. Gesell., Zeitschr., v. 52, p. 403-479.

(147) **Blochmann, Friedrich,** 1895, *Die Mikroskopische Thierwelt des Süsswassers Abth. I, Protozoa* in *Die Mikroskopische Pflanzen und Thierwelt des Süsswassers:* pt. 2, p. 1-134, pl. 1-8, Lucas Gräfe & Sillem (Hamburg).

(148) **Blow, W. H.**, 1956, *Origin and evolution of the foraminiferal genus Orbulina d'Orbigny:* Micropaleontology, v. 2, no. 1, p. 57-70, fig. 1-4.
———(149) 1959, *Age, correlation, and biostratigraphy of the upper Tocuyo (San Lorenzo) and Pozón formations, Eastern Falcón, Venezuela:* Bull. Am. Paleontology, v. 39, no. 178, p. 67-251, pl. 6-191.

(150) **Blumenbach, J. F.**, 1799-1805, *Abbildungen naturhistorischer Gegenständ:* (a) 1799, no. 4 (40), p. 1-2, 1 pl.; (b) 1805, no. 8 (80), fig. 80; H. Dieterich (Göttingen).

(151) **Bogdanovich, A. K.**, 1935, *O novoy Foraminifere Meandroloculina bogatschovi nov. gen. et sp. iz Miotsenovykh otlozheniy Zakavkazya:* Akad. Nauk SSSR, Izvestia, ser. 7, no. 5, p. 691-696. [*On a new foraminifer Meandroloculina bogatschovi nov. gen. and sp. from Miocene deposits of the Caucasus.*]———(152) 1952, *Miliolidy i Peneroplidy, Iskopaemye Foraminifery SSSR:* VNIGRI, Trudy, new ser., no. 64, p. 1-338, 39 pl., 70 text-fig. [*Miliolidae and Peneroplidae, fossil Foraminifera of the USSR.*]
———(153) 1960, *O novom predstavitele Miliolid s probodennoy stenkoy:* Akad. Nauk SSSR, Voprosy Mikropaleontologii, no. 3, Otdel. Geol.-Geogr. Nauk, Geol. Inst. Akad. Nauk, p. 17-21, 1 pl. [*On a new representative of the Miliolidae with a foraminate wall.*]

(154) ———, & **Voloshinova, N. A.**, 1949, *O novom predstavitele semeystva Miliolidae—Dogielina sarmatica gen. et sp. n. iz Srednesarmatskikh otlozheniy Krymsko-Kavkazskoy oblasti:* VNIGRI, Mikrofauna Neftyanykh Mestorozhdeniy SSSR, Sbornik 2, Trudy, new ser., no. 34, p. 183-186, pl. 1. [*On a new representative of the family Miliolidae—Dogielina sarmatica gen. and sp. n. from middle Sarmatian deposits of the Crimea-Caucasus district.*]

(155) **Bold, W. A. van den**, 1946, *Contribution to the study of Ostracoda with special reference to the Tertiary and Cretaceous microfauna of the Caribbean region:* Dissertation, Rijks-Univ. Utrecht, 167 p., 18 pl., 8 text-fig.

(156) **Bolli, H. M.**, 1945, *Zur Stratigraphie der oberen Kreide in den höheren helvetischen Decken:* Eclogae geol. Helv., v. 37, no. 2 (1944), p. 217-328, pl. 9, 6 text-fig.———
(157) 1950, *The direction of coiling in the evolution of some Globorotaliidae:* Cushman Found. Foram. Research, Contrib., v. 1, pts. 3-4, p. 82-89, pl. 15, text-fig. 1-5.———(158) 1951, *The genus Globotruncana in Trinidad, B. W. I.:* Jour. Paleontology, v. 25, p. 187-199, pl. 34-35, 1 text-fig.———(159) 1957, *Planktonic Foraminifera from the Oligocene-Miocene Cipero and Lengua formations of Trinidad, B. W. I.:* U.S. Natl. Museum, Bull. 215, p. 97-121, pl. 22-29, text-fig. 17-21.———(160) 1957, *Planktonic Foraminifera from the Eocene Navet and San Fernando formations of Trinidad, B. W. I.:* Same, Bull. 215, p. 155-172, pl. 35-39.———
(161) 1958, *The foraminiferal genera Schackoina Thalmann, emended, and Leupoldina, n. gen., in the Cretaceous of Trinidad, B. W. I.:* Eclogae geol. Helv., v. 50, no. 2 (1958), p. 271-278, pl. 1-2.———(162) 1959, *Grimsdaleinella, a new genus of the foraminiferal family Heterohelicidae:* Same, v. 52, no. 1, p. 1-4, pl. 1.———(163) 1961, *Bireophax, a new genus of the foraminiferal family Reophacidae:* Same, v. 53, no. 2 (1960), p. 493-496, pl. 1.
———(163A) 1962, *Globigerinopsis, a new genus of the foraminiferal family Globigerinidae:* Same, v. 55, no. 1, p. 281-284, pl. 1.

(164) ———, **Loeblich, A. R., Jr.**, & **Tappan, Helen**, 1957, *Planktonic foraminiferal families Hantkeninidae, Orbulinidae, Globorotaliidae, and Globotruncanidae:* U.S. Natl. Museum, Bull. 215, p. 3-50, pl. 1-11, text-fig. 1-9.

(165) **Boltovskoy, Esteban**, 1956, *Application of chemical ecology in the study of the Foraminifera:* Micropaleontology, v. 2, p. 321-325.———
(166) 1961, *Algunos Foraminíferos nuevos de las aguas Brasileñas:* Neotropica (notas zoologicas sudamericanas), v. 7, no. 24, p. 73-79, fig. 1-10.

(167) **Bonner, J. T.**, 1959, *The cellular slime molds:* 150 p., Princeton Univ. Press.

(168) **Bonnet, L.**, 1959, *Dékystement, phase trophique et enkystement chez Plagiopyxis minuta Bonnet (Thécamoebiens), Incidences systématiques:* Acad. Sci. Paris, Comptes Rendus, v. 249, p. 2617-2619.———(169) 1959, *Nouveaux Thécamoebiens du sol:* Soc. Histoire Nat. Toulouse, Bull., v. 94, p. 177-188, pl. 1-2.
———(170) 1960, *Nouveaux Thécamoebiens du sol (III):* Same, Bull., v. 95, p. 1-3, text-fig. 1-7.

(171) ———, & **Thomas, R.**, 1955, *Étude sur les Thécamoebiens du sol:* Soc. Histoire Nat. Toulouse, Bull., v. 90, pt. 3-4, p. 411-428, text-fig. 1-41.

(172) **Bonte, Antoine**, 1944, *Orbitammina elliptica d'Arch. sp., Foraminifère de grande taille du Bathonien supérieur de l'Aisne et des Ardennes:* Soc. géol. France, Bull., ser. 5, v. 12, p. 329-350, pl. 9.

(173) **Boomgaart, Lubbartus**, 1949, *Smaller foraminifera from Bodjonegoro (Java):* Dissertation, Univ. Utrecht, 175 p., 14 pl., chart.

(174) **Bornemann, L. G.**, 1874, *Ueber die Foraminiferengattung Involutina:* Deutsche geol. Gesell., Zeitschr., v. 26, p. 702-749, pl. 18-19.

(175) **Bornemann, J. G.**, 1886 (1887), *Die Versteinerung des cambrischen Schichtensystem der Insel Sardinien: Erste Abteilung:* K. Leop.-Carol. Deutsch. Akad. Naturf. (Nova Acta), v. 51 (1886), no. 1, p. 1-148, pl. 2.

(176) Bosc, L. A. G., 1816, *Nouvelle Dictionnaire d'Histoire Naturelle:* ed. 2, v. 5, p. 491 (fide Neave, 1939, p. 625).

(176A) Boubée, Nérée, 1832, *Présentation à la Société de deux nouvelles espèces de Nummulites:* Soc. géol. France, Bull., ser. 1, v. 2 (1831-32), p. 444-445.

(177) Bourdon, M., & Lys, M., 1955, *Foraminifères du Stampien de la carrière de la Souys-Floirac (Gironde):* Soc. géol. France, Comptes Rendus, no. 16, p. 336-338, 2 text-fig.

(178) Boussac, Jean, 1906, *Développement et morphologie de quelques Foraminifères de Priabona:* Soc. géol. France, Bull., ser. 4, v. 6, pt. 2-3, p. 89-97, pl. 1-3.———(179) 1911, *Etudes paléontologiques sur le nummulitique alpin:* Mém. Carte Géol. France, 437 p. 22 pl.

(180) Bovee, E. C., 1957, *Protozoa of Amazonian and Andean waters of Colombia, South America:* Jour. Protozoology, v. 4, p. 63-66.———(181) 1960, *Protozoa of the Mountain Lake region, Giles County, Virginia:* Same, v. 7, p. 352-361, 1 fig.

(182) Bowen, R. N. C., 1955, *Observations on the foraminiferal genus Gaudryina d'Orbigny, 1839:* Micropaleontology, v. 1, p. 359-364, text-fig. 1-6.

(183) Bowerbank, J. S., 1862, *On the anatomy and physiology of the Spongiadae—Part 3:* Royal Soc. London, Philos. Trans., v. 152, p. 1087-1135, pl. 72-74.

(184) Bradshaw, J. S., 1957, *Laboratory studies on the rate of growth of the foraminifer "Streblus beccarii (Linné) var. tepida Cushman":* Jour. Paleontology, v. 31, p. 1138-1147, 5 text-fig.———(185) 1959, *Ecology of living planktonic Foraminifera in the north and equatorial Pacific Ocean:* Cushman Found. Foram. Research, Contrib., v. 10, pt. 2, p. 25-64, pl. 6-8, text-fig. 1-43.

(186) Brady, H. B., 1864, *Contributions to the knowledge of the Foraminifera.—On the rhizopodal fauna of the Shetlands:* Linnean Soc. London, Trans., v. 24, p. 463-476, pl. 48.———(187) 1868, *On Ellipsoidina, a new genus of Foraminifera. By Giuseppe Seguenza, Professor of Natural History in the Royal Lyceum, Messina:* Ann. & Mag. Nat. History, ser. 4, v. 1, p. 333-343, pl. 13.———(187A) 1870, *Notes on the Foraminifera of mineral veins and the adjacent strata:* British Assoc. Advanc. Sci., London, Rept. 1870, 39th meeting (1869), p. 381-382.———(188) 1871, *On Saccammina carteri, a new foraminifer from the Carboniferous limestone of Northumberland:* Ann. & Mag. Nat. History, ser. 4, v. 7, p. 177-184, pl. 12.———(189) 1873, *On Archaediscus Karreri, a new type of Carboniferous Foraminifera:* Same, ser. 4, v. 12, p. 286-290, pl. 11.———(190) 1873, *Explanation of sheet 23, Lanarkshire, central districts:* in Geol. Survey Scotland, Mem., p. 94-96 (Edinburgh).———(191) 1874, *On a true Carboniferous nummulite:* Ann. & Mag. Nat. History, ser. 4, v. 13, p. 222-231, pl. 12.———(192) 1875, *On some fossil Foraminifera from the West-Coast district, Sumatra:* Geol. Mag., new ser., decade 2, v. 2, p. 532-539, pl. 13-14.———(193) 1876, *A monograph of the Carboniferous and Permian Foraminifera (the genus Fusulina excepted):* Palaeontograph. Soc. London, p. 1-166, pl. 1-12.———(194) 1877, *Supplementary note on the Foraminifera of the Chalk (?) of the new Britain group:* Geol. Mag., new ser., decade 2, v. 4, p. 534-536.———(195) 1878, *On the reticularian and radiolarian Rhizopoda (Foraminifera and Polycystina) of the North Polar Expedition of 1875-76:* Ann. & Mag. Nat. History, ser. 5, v. 1, p. 425-440, pl. 20-21.———(196) 1879-81, *Notes on some of the reticularian Rhizopoda of the Challenger Expedition:* Quart. Jour. Micro. Sci., new ser., v. 19; (a) Part 1, *On new or little-known arenaceous types,* p. 20-63, pl. 3-5 (1879); (b) Part 2, *Additions to the knowledge of porcellanous and hyaline types,* p. 261-299, pl. 8 (1879); (c) Part 3, *1—Classification, 2—Further notes on new species, 3—Note on Biloculina mud,* v. 21, p. 31-71.———(197) 1881, *Ueber einige arktische Tiefsee-Foraminiferen gesammelt während der oesterreichisch-ungarischen Nordpol-Expedition in den Jahren 1872-74:* K. Akad. Wiss. Wien, Denkschr., v. 43, p. 9-110, pls.———(198) 1882, *Notes on Keramosphaera, a new type of porcellanous Foraminifera:* Ann. & Mag. Nat. History, ser. 5, v. 10, p. 242-245, pl. 13.———(199) 1883, *Note on Syringammina, a new type of arenaceous Rhizopoda:* Royal Soc. London, Proc., v. 35, p. 155-161, pl. 2-3.———(200) 1884, *Report on the Foraminifera dredged by HMS Challenger, during the years 1873-1876:* Rept. Scientific Results Explor. Voyage HMS Challenger, Zoology, v. 9, p. 1-814, pl. 1-115.———
(201) 1889, *On a new type of Astrorhizidae from the Bay of Bengal:* Ann. & Mag. Nat. History, ser. 6, v. 3, p. 293-296, text-fig. 1-2.———
(202) 1890, *Note on a new type of Foraminifera of the family Chilostomellidae:* Royal Micro. Soc. London, Jour., p. 567-571, text-fig.

(203) ———, Parker, W. K., & Jones, T. R., 1888[1890], *On some Foraminifera from the Abrohlos bank:* Zool. Soc. London, Trans., v. 12, pt. 7, p. 211-239, pl. 40-47.

(204) Bray, D. M., 1944, *The determination of calcite and aragonite in invertebrate shells:* Royal Soc. New S. Wales, Jour., v. 78, p. 113-117.

(204A) Brazhnikova, N. E., 1962, *Quasiendothyra i blizkie k nim formy iz Nizhnego Karbona Donetzkogo Basseina i drugikh rayonov Ukrainy:* Akad. Nauk URSR, Inst. Geol. Nauk,

Trudy, ser. strat. paleont., no. 44, p. 3-48, pl. 1-14. [*Quasiendothyra and related forms from the Lower Carboniferous of the Donetz basin and neighboring regions of the Ukraine.*]
(205) ———, & **Yartseva, M. V.**, 1956, *K voprosu ob evolyutsii roda Monotaxis*: Akad. Nauk SSSR, Voprosy Mikropaleontologii, v. 1, p. 62-68, pl. 1, fig. 1. [*On the evolution of the genus Monotaxis.*]
(206) **Breyn, J. P.**, 1732, *Dissertatio physica de Polythalamiis, nova Testaceorum classe, etc.*: Gedani, plates.
(207) **Broderip, W. J.**, 1839, [in] *The Penny Cyclopaedia of the Society for the diffusion of useful knowledge*: v. 14 (Limonia-Massachusetts), p. 1-486, Charles Knight & Co. (London).
(208) **Brodie, P. B.**, 1853, *Remarks on the Lias at Fretherne near Newnham, and Purton near Sharpness; with an account of some new Foraminifera discovered there; and on certain Pleistocene deposits in the Vale of Gloucester*: Ann. & Mag. Nat. History, ser. 2, v. 12, p. 272-277.
(209) **Bronn, H. G.**, 1825, *System der urweltlichen Pflanzenthiere*: iv+47 p., J. C. B. Mohr (Heidelberg).———(210) 1838, *Lethaea Geognostica*: v. 2, p. 545-1346, pl. 1-47, E. Schweizerbart (Stuttgart).———(211) 1849, *Index Palaeontologicus oder Übersicht der bis jetzt bekannten Fossilen Organismen*: lxxxiii+980 p., E. Schweizerbart (Stuttgart).———(212) 1859, *Die Klassen und Ordnungen des Thier-Reichs, wissenschaftlich dargestellt in Wort und Bild*: v. 1, p. 1-142, C. F. Winter (Leipzig & Heidelberg).———(213) 1880, *Klassen und Ordnungen des Thier-Reichs*: v. 1, Protozoa, pt. 1, Sarkodina und Sporozoa, p. 1-1097, pl. 1-55, C. F. Winter (Leipzig & Heidelberg).
(214) ———, & **Roemer, C. F.**, 1851-1856, *Lethaea Geognostica; Vierte Periode; Kreide-Gebirge*: ed. 3; (a) v. 2, pt. 5 (1851-1852), p. 81-96, pl. 29, 29^1, 29^2, 33 (1853); (b) v. 3, pt. 6 (1853-1856), p. 1-1130, pl. 1-63 (1854), E. Schweizerbart (Stuttgart).
(215) **Brönnimann, Paul**, 1940, *Über die tertiären Orbitoididen und die Miogypsiniden von Nordwest-Marokko*: Schweiz. Palaeont. Gesell. Zurich, Abhandl., v. 63, p. 1-113, pl. 1-11.———(216) 1944, *Ein neues Subgenus von Orbitocyclina aus Iran nebst Bemerkungen über Helicolepidina Tobler und verwandte Formen*: Same, Abhandl., v. 64, p. 1-42, pl. 1-3, 15 text-fig.———(217) 1945, *Zur Morphologie von Aktinocyclina Gümbel 1868*: Eclogae geol. Helv., v. 38, p. 560-578, pl. 20.———(218) 1946, *Zur Frage der verwandtschaftlichen Beziehungen zwischen Discocylina ss. und Asterocyclina*: Same, v. 38 (1945), p. 579-615, pl. 21-22, 23 text-fig.———(219) 1947, *Zur neu-Definition von Pliolepidina H. Douvillé, 1915*: Same, v. 39, no. 2 (1946), p. 373-379.———(220) 1950, *The genus Hantkenina Cushman in Trinidad and Barbados, B. W. I.*: Jour. Paleontology, v. 24, p. 397-420, pl. 55-56.———(221) 1951, *A model of the internal structure of Discocyclina s.s.*: Same, v. 25, p. 208-211, 1 fig.———(222) 1951, *Tremastegina, ein neues Genus der Familie Asterigerinidae d'Orbigny*: Eclogae geol. Helv., v. 43, no. 2 (1950), p. 255-265, text-fig. 1-7.———(223) 1951, *Bemerkungen über den Bau von Amphistegina d'Orbigny*: Same, v. 43, no. 2, p. 251-254, text-fig. 1-6.———(224) 1951, *Globigerinita naparimaensis, n.gen., n.sp., from the Miocene of Trinidad, B. W. I.*: Cushman Found. Foram. Research, Contrib., v. 2, pt. 1, p. 16-18, fig. 1-14.———(225) 1951, *Guppyella, Alveovalvulina and Discamminoides, new genera of arenaceous Foraminifera from the Miocene of Trinidad, B. W. I.*: Same, v. 2, pt. 3, p. 97-105, pl. 11, text-fig. 1-12.———(226) 1952, *Globigerinoita and Globigerinatheka, new genera from the Tertiary of Trinidad, B. W. I.*: Same, v. 3, pt. 1, p. 25-28, fig. 1-2.———(227) 1952, *Plummerita, new name for Plummerella Brönnimann, 1952 (not Plummerella De Long, 1942)*: Same, v. 3, pt. 3-4, p. 146.———(228) 1952, *Globigerinidae from the Upper Cretaceous (Cenomanian-Maestrichtian) of Trinidad, B. W. I.*: Bull. Am. Paleontology, v. 34, no. 140, p. 1-61, pl. 1-4.———(229) 1953, *Arenaceous Foraminifera from the Oligo-Miocene of Trinidad*: Cushman Found. Foram. Research, Contrib., v. 4, p. 87-100, pl. 15, text-fig. 1-15.———(230) 1953, *Note on Planktonic Foraminifera from Danian localities of Jutland, Denmark*: Eclogae geol. Helv., v. 45, no. 2 (1952), p. 339-341, fig. 1.———(231) 1954-56, *Upper Cretaceous orbitoidal Foraminifera from Cuba*: Cushman Found. Foram. Research, Contrib.; (a) Pt. I. *Sulcorbitoides* n. gen., v. 5, pt. 2, p. 55-61, pl. 9-10, text-fig. 1-5 (1954); (b) Pt. III. *Pseudorbitoides H. Douvillé, 1922*, v. 6, pt. 2, p. 57-76, pl. 9-12, 17 text-fig. (1955); (c) Pt. IV. *Rhabdorbitoides*, n. gen., v. 6, pt. 3, p. 97-104, pl. 15, text-fig. 1-5 (1955); (d) Pt. V. *Historbitoides*, n. gen., v. 7, pt. 2, p. 60-66, pl. 13, text-fig. 1-7 (1956).———(232) 1958, *New Pseudorbitoididae from the Upper Cretaceous of Cuba, with remarks on encrusting Foraminifera*: Micropaleontology, v. 4, p. 165-185, pl. 1-7, fig. 1-11.

(233) ———, & **Bermúdez, P. J.**, 1953, *Truncorotaloides, a new foraminiferal genus from the Eocene of Trinidad, B. W. I.*: Jour. Paleontology, v. 27, p. 817-820, pl. 87.

(234) ———, & **Brown, N. K., Jr.**, 1953, *Observations on some planktonic Heterohelicidae from the Upper Cretaceous of Cuba*: Cushman Found. Foram. Research, Contrib., v. 4, pt. 4, p. 150-156, text-fig. 1-14.———(235) 1956, *Taxonomy of the Globotruncanidae*: Eclogae geol.

Helv., v. 48 (1955), p. 503-561, pl. 20-24, text-fig. 1-24.———(236) 1958, *Hedbergella, a new name for a Cretaceous planktonic foraminiferal genus:* Washington Acad. Sci., Jour., v. 48, no. 1, p. 15-17, text-fig. 1.

(237) **Brotzen, Fritz,** 1936, *Foraminifera aus dem schwedischen untersten senon von Eriksdal in Schonen:* Sver. Geol. Undersök., v. 30, no. 3, ser. C, no. 396, p. 1-206, pl. 1-14.———(238) 1937, *Die Foraminiferen in Sven Nilssons Petrificata Suecana 1827:* Geol. Fören. Stockholm, Förhandl., v. 59, no. 1, p. 59-76, pl. 2, text-fig. 1-6.———(239) 1940, *Flintrännans och trindelrännans Geologi:* Sver. Geol. Undersök., v. 34, no. 5, ser. C, no. 435, p. 1-33, 8 fig., 1 pl.———(240) 1942, *Die Foraminiferengattung Gavelinella nov. gen. und die Systematik der Rotaliiformes:* Same, v. 36, no. 8, ser. C, no. 451, p. 1-60, pl. 1, text-fig. 1-18.———(241) 1948, *The Swedish Paleocene and its foraminiferal fauna:* Same, v. 42, no. 2, ser. C, no. 493, p. 1-140, pl. 1-19, 41 text-fig.———(242) 1960, *On Tylocidaris species (Echinoidea) and the stratigraphy of the Danian of Sweden with a bibliography of the Danian and the Paleocene:* Same, v. 54, no. 2, ser. C, no. 571, 1959, p. 1-81, pl. 1-3.

(243) ———, & **Pozaryska, K.** 1961, *Foraminifères du Paléocène et de l'Éocène inférieur en Pologne septentrionale remarques paleogéographiques:* Revue Micropaléont., v. 4, no. 3, p. 155-166, pl. 1-4.

(244) **Brown, Thomas,** 1827, *Illustrations of the conchology of Great Britain and Ireland:* p. i-v, 52 pl., W. H. & D. Lizar (Edinburgh).———(245) 1843, *The elements of fossil conchology; according to the arrangement of Lamarck; with the newly established genera of other authors:* 133 p., 12 pl., Houlston & Stoneman (London).———(246) 1844, *Illustrations of the Recent Conchology of Great Britain and Ireland, with descriptions and localities of all the species:* ed. 2, 145 p., 59 pl. (London).

(247) **Bruguière, J. G.,** 1792, *Encyclopédia méthodique. Histoire naturelle des Vers:* v. 1, A-Cone, Panckoucke (Paris).

(248) **Brünnich, M. T.,** 1772, *M. T. Brünnich Zoologiae fundamenta:* 253 p., Grunde i Dyreloeren (Hafniae et Lipsiae).

(249) **Buchanan, J. B.,** & **Hedley, R. H.,** 1960, *A contribution to the biology of Astrorhiza limicola (Foraminifera):* Jour. Marine Biol. Assoc. United Kingdom, v. 39, p. 549-560, text-fig. 1-5.

(250) **Buchner, P.,** 1942, *Die Lingulinen des Golfes von Neapel und der marinen Ablagerungen auf Ischia:* K. Leop.-Carol. Deutsch. Akad. Naturf., Nova Acta Leopoldina. Abhandl., new ser., v. 11, no. 75, p. 103-145, text-fig. 1-18.

(251) **Buck, Emil,** 1878, *Einige Rhizopodenstudien:* Zeitschr. Wiss. Zool., v. 30, p. 1-49, pl. 1-2.

(252) **Bukalova, G. V.,** 1957, *O novom rode Foraminifer iz Albskikh otlozheniy severo-zapadnogo Kavkaza:* Akad. Nauk. SSSR, Doklady, v. 114, no. 1, p. 185-188, text-fig. 1-2. [*On a new genus of Foraminifera from Albian deposits of the northwestern Caucasus.*]

(253) **Burbach, O.,** 1886, *Beiträge zur Kenntniss der Foraminiferen des mittleren Lias von grossen Seeberg bei Gotha:* Zeitschr. Naturwiss., v. 59 (Ser. 4, v. 5); (a) *I-Die Gattung Frondicularia Defr.,* p. 30-53, pl. 1-2; (b) *II-Die Milioliden,* p. 493-502, pl. 5 (Halle).

(254) **Bursch, J. G.,** 1947, *Mikropaläontologische Untersuchungen des Tertiärs von Gross Kei (Molukken):* Schweiz. Palaeont. Gesell. Zurich., Abhandl., v. 65, p. 1-69, pl. 1-5.———(255) 1952, *Praeammoastuta, new foraminiferal genus of the Venezuelan Tertiary, with an emendation of Ammoastuta Cushman and Brönnimann:* Jour. Paleontology, v. 26, p. 915-923, pl. 132, 4 text-fig.

(256) **Bütschli, Otto,** 1908, *Untersuchung über organische Kalkgebilde nebst Bemerkungen über organische Kieselgebilde:* Gesell. Wiss. Göttingen, Math.-Phys. Kl., Abhandl., new ser., v. 6, no. 3, p. 1-177.

(257) **Bykova, E. V.,** 1952, *Foraminifery Devona Russkoy Platformy i Priural'ya:* VNIGRI, Trudy, no. 60, Mikrofauna SSSR, new ser., v. 5, p. 5-64, pl. 1-14. [*Foraminifera of the Devonian of the Russian Platform and Pre-Urals.*]———(258) 1956, *Foraminifery Ordovika i Silura Sovetskoy Pribaltika:* Same, new ser., no. 98, Mikrofauna SSSR, v. 8, p. 6-27, pl. 1-5. [*Foraminifera of the Ordovician and Silurian of the Soviet Pre-Baltic.*]———(259) 1958, *O Nakhodke khitinoidnykh foraminifer v otlozheniyakh ordovika severnogo kazakhstana:* Akad. Nauk SSSR, Doklady, v. 120, no. 4, p. 879-881, 1 pl. [*On an occurrence of a chitinoid Foraminifera in deposits of the Ordovician of northern Kazakhstan.*]———(260) 1961, *Foraminifery Karadoka vostochnogo Kazakhstana:* Akad. Nauk Kazakhskoy SSR, Inst. Geol. Nauk, p. 1-119, pl. 1-25, text-figs. 1-32. [*Foraminifera of the Caradocian of eastern Kazakhstan.*]———

(261) ———, & **Polenova, E. N.,** 1955, *Foraminifery, radiolyarii i ostrakody Devona Volgo-Ural'skoi oblasti:* VNIGRI, Trudy, new ser., no. 87, p. 5-190, pl. 1-25. [*Foraminifera, Radiolaria and Ostracoda of the Devonian of the Volga-Ural district.*]

(262) **Bykova, N. K.,** 1947, *Materialy k izucheniyu fauny foraminifer Senomana Bykharskoy oblasti:* VNIGRI, Mikrofauna Neftyanykh Mestorozhdeniy Kavkaza, Emby i Sredney Azii, p. 222-238, pl. 1. [*Material for study of the*

foraminiferal fauna of the Cenomanian of the Bykharskoy district.]——(263) 1960, *K voprosu o tsiklichnosti filogeneticheskogo razvitiya u foraminifer:* VNIGRI, Trudy, no. 163, Geol. Sbornik v. 5, p. 309-327, pl. 1-5. [*On the question of cyclic recurrence in phylogenetical development of Foraminifera.*] —— (264) 1962, *Dymia N. K. Bykova, new name for Candela N. K. Bykova, 1958 not Herrmannsen, 1846:* Cushman Found. Foram. Research, Contrib., v. 13, pt. 1, p. 22.

(265) ——, Balakhmatova, V. T., Vasilenko, V. P., Voloshinova, N. A., Grigelis, A., Dain, L. G., Ivanova, L. V., Kuzina, V. I., Kuznetsova, Z. V., Kozyreva, V. F., Morozova, V. G., Myatlyuk, E. V., & Subbotina, N. N., 1958, *Novye Rody i Vidy Foraminifer:* VNIGRI, Trudy, no. 115, Mikrofauna SSSR, v. 9, p. 4-81, pl. 1-12. [*New genera and species of Foraminifera.*]

(266) Calkins, G. N., 1901, *The Protozoa:* p. 1-347, Columbia Press (New York).—— (267) 1909, *Protozoology:* 349 p., 4 pl., 125 fig., Lea & Febiger (New York & Philadelphia).——(268) 1926, *The biology of the Protozoa:* 623 p., 238 text-fig., Lea & Febiger (Philadelphia).——(269) 1933, *The biology of the Protozoa:* ed. 2, 607 p., Lea & Febiger (Philadelphia).

(270) Carman, K. W., 1933, *Dentostomina, a new genus of the Miliolidae:* Cushman Lab. Foram. Research, Contrib., v. 9, pt. 2, p. 31-32, pl. 3.

(271) Carpenter, W. B., 1856-83, *Researches in the Foraminifera:* Royal Soc. London, Philos. Trans.; (a) *Pt. II. On the genera Orbiculina, Alveolina, Cycloclypeus and Heterostegina*, v. 146, p. 547-569, pl. 28-31 (1856); (b) *Pt. III. On the genera Peneroplis, Operculina, and Amphistegina*, v. 149 (1859), p. 1-41, pl. 1-6 (1860); (c) *Supplemental memoir. On an abyssal type of the genus Orbitolites; a study in the theory of descent*, v. 174, p. 551-573, pl. 37-38 (1883).——(272) 1861, *On the systematic arrangement of the Rhizopoda:* Nat. History Review, v. 1, no. 4, p. 456-472.——(273) 1868, *Preliminary report of dredging operations in the seas to the north of the British Islands, carried on in Her Majesty's steam-vessel Lightning by Dr. Carpenter and Dr. Wyville Thomson:* Royal Soc. London, Proc., v. 17, p. 168-197.——(274) 1869, *On the rhizopodal fauna of the deep sea:* Same, v. 18 (1868), no. 114, p. 59-62.——(275) 1870, *Descriptive catalogue of objects from deep-sea dredgings, exhibited at the soirée of the Royal Microscopical Society, King's College, April 20, 1870:* p. 1-11 (London).——(276) 1875, *The microscope and its revelations:* ed. 5, xxxii+848 p., 25 pl., 449 fig., J. & A. Churchill (London).——
(277) 1879, *Foraminifera:* in Encyclopaedia Britannica, ed. 9, v. 9, p. 371-387, text-fig. 1-37, Charles Scribner's Sons (New York).

(278) ——, & Brady, H. B., 1870, *Description of Parkeria and Loftusia, two gigantic types of arenaceous Foraminifera:* Royal Soc. London, Philos. Trans. (1869), v. 159, p. 721-754, pl. 72-80, text-fig. A-C.

(279) ——, & Jeffreys, J. G., 1870, *Report on deep-sea researches carried on during the months of July, August, and September, 1870, in H. M. Surveying-ship Porcupine:* Royal Soc. London, Proc., v. 19, p. 146-221 (1870-1871).

(280) ——, ——, & Thomson, W., 1870, *Preliminary report of the scientific exploration of the deep sea in H. M. Surveying-vessel "Porcupine" during the summer of 1869:* Royal Soc. London, Proc., v. 18 (1869-70), no. 121, p. 397-453, pl. 4-6.

(281) ——, Parker, W. K., & Jones, T. R., 1862, *Introduction to the study of the Foraminifera:* Ray Soc. Publs., p. 1-319, pl. 1-22.

(282) Carsey, D. O., 1926, *Foraminifera of the Cretaceous of central Texas:* Texas Univ., Bull., no. 2612, p. 1-56.

(283) Carter, D. J., 1951, *Indigenous and exotic Foraminifera in the Coralline Crag of Sutton, Suffolk:* Geol. Mag., v. 88, p. 236-248, text-fig. 1-4.——(284) 1957, *The distribution of the foraminifer Alliatina excentrica (di Napoli Alliata) and the new genus Alliatinella:* Palaeontology, v. 1, pt. 1, p. 76-86, pl. 14, text-fig. 1-2.

(285) Carter, H. J., 1854, *On the true position of the canaliferous structure in the shells of fossil Alveolina (d'Orbigny):* Ann. & Mag. Nat. History, ser. 2, v. 14, p. 99-101.——(286) 1856, *Notes on the fresh-water Infusoria of the Island of Bombay, No. 1. Organization:* Same, ser. 2, v. 18, p. 221-249.——(287) 1861, *Further observations on the structure of Foraminifera, and on the larger fossilized forms of Scinde, etc., including a new genus and species:* (a) Same, ser. 3, v. 8, p. 446-470, pl. 15-17; (b) Royal Asiatic Soc., Bombay Branch, Jour., v. 6, p. 31-96.——(288) 1864, *On fresh-water Rhizopoda of England and India; with illustration:* Ann. & Mag. Nat. History, ser. 3, v. 13, p. 18-39, pl. 1-2.——(289) 1865, *On the fresh- and salt-water Rhizopoda of England and India:* Same, ser. 3, v. 15, p. 277-293, pl. 12.
——(290) 1870, *On two new species of the foraminiferous genus Squamulina; and on a new species of Difflugia:* Same, ser. 4, v. 5, p. 309-326, pl. 4-5.——(291) 1876, *On the Polytremata (Foraminifera), especially with reference to their mythical hybrid nature:* Same, ser. 4, v. 17, p. 185-214.——(292) 1877, *On a Melobesian form of Foraminifera (Gypsina melobesioides, mihi); and further observations on Carpenteria monticularis:* Same, ser. 4, v. 20, p. 172-176.——(293) 1877, *Description of*

Bdelloidina aggregata a new genus and species of arenaceous Foraminifera, in which their so-called "Imperforation" is questioned: Same, ser. 4, v. 19, p. 201-209, pl. 13.———(294) 1877, *Description of a new species of Foraminifera (Rotalia spiculotesta):* Same, ser. 4, v. 20, p. 470-473, pl. 16.———(295) 1879, *On a new genus of Foraminifera (Aphrosina informis), and spiculation of an unknown sponge:* Royal Micro. Soc., Jour., v. 2, p. 500-502, pl. 17a. ———(296) 1880, *Report on specimens dredged up from the Gulf of Manaar, and presented to the Liverpool Free Museum by Capt. W. H. Cawne Warren:* Ann. & Mag. Nat. History, ser. 5, v. 5, p. 437-457.———(297) 1885, *Descriptions of sponges from the neighborhood of Port Phillip Heads, South Australia, continued:* Same, ser. 5, v. 15, p. 196-222.— (298) 1888, *On two new genera allied to Loftusia from the Karakoram Pass and the Cambridge Greensand respectively:* Same, ser. 6, v. 1, p. 172-184, pl. 13.

(299) **Cash, James,** 1904, *On some new and little-known British fresh-water Rhizopoda:* Linnean Soc. Zool., Jour., v. 29, p. 218-225, pl. 26.

(300) ———, & **Hopkinson, John,** 1905, *The British fresh-water Rhizopoda and Heliozoa, vol. 1, Rhizopoda, Pt. 1:* Ray Soc. Publ. 85, p. 1-148, pl. 1-16.———(301) 1909, *The British fresh-water Rhizopoda and Heliozoa, vol. 2, Rhizopoda, Pt. 2:* Ray Soc. Publ. 89, p. 1-166, pl. 17-32.

(302) ———, **Wailes, G. H.,** & **Hopkinson, John,** 1915-19, *The British fresh-water Rhizopoda and Heliozoa, by G. H. Wailes:* Ray Soc. Publs.; (a) v. 3, pt. 3, publ. 98, p. 1-156, pl. 33-57 (1915); (b) v. 4, *Supplement to the Rhizopoda,* publ. 103, p. 1-130, pl. 58-63 (1919).

(303) **Cati, F.,** 1959, *Nuovo Lituolide nei calcari grigi liassici del vincento:* Giornale Geol., Ann. Mus. Geol. Bologna, ser. 2, v. 27, p. 1-10, pl. 1.

(304) **Caudri, C. M. Bramine,** 1944, *The larger Foraminifera from San Juan de los Morros, State of Guarico, Venezuela:* Bull. Am. Paleontology, v. 28, no. 114, p. 351-404, 5 pl.— (305) 1948, *Note on the stratigraphic distribution of Lepidorbitoides:* Jour. Paleontology, v. 22, p. 473-481, pl. 73-74.

(306) **Cecioni, Giovanni,** 1955, *Noticias preliminares sobre et hallazgo del Paleozoico Superior en el Archipiélago Patagónico:* Univ. Chile, Inst. Geol., publ. 6, p. 257-258, 1 fig., Editorial Universitaria, S.A.

(307) **Certes, A.,** 1891, *Protozoaires:* Mission Scientifique du Cap Horn 1882-83, v. 6, Zool., pt. 3, p. L1-L43, pl. 1-6 (Paris).

(308) **Chapman, Frederick,** 1891, *The Foraminifera of the Gault of Folkestone; Part I:* Royal Micro. Soc. London, Jour., p. 565-575, pl. 9. ———(309) 1892, *Some new forms of hyaline Foraminifera from the Gault:* Geol. Mag., new ser., decade 3, v. 9, p. 52-54, pl. 2.———(310) 1894, *Bargate beds of Surrey and their microscopic contents:* Quart. Jour. Geol. Soc. London, v. 50, p. 677-730, pl. 33-34.———(311) 1894, *The Foraminifera of the Gault of Folkestone, V:* Royal Micro. Soc., Jour., p. 153-163, pl. 3-4. ———(312) 1895, *On Rhaetic Foraminifera from Wedmore, in Somerset:* Ann. & Mag. Nat. History, ser. 6, v. 16, p. 305-329, pl. 11-12. ———(313) 1898, *On Haddonia, a new genus of the Foraminifera, from Torres Straits:* Linnean Soc. London, Jour., Zool., v. 26, p. 452-456, pl. 28, text-fig. 1.———(314) 1900, *On some new and interesting Foraminifera from the Funafuti Atoll, Ellice Islands:* Same, v. 28, p. 1-27, pl. 1-4.———(315) 1900, *On some Foraminifera of Tithonian age from the Stramberg limestone of Nesselsdorf:* Same, v. 28, p. 28-32, pl. 5.———(316) 1900, *On a Patellina-limestone and another foraminiferal limestone from Egypt:* Geol. Mag., new ser., v. 7, p. 3-17, pl. 2.———(317) 1901, *Foraminifera from the lagoon at Funafuti:* Linnean Soc., Jour., v. 28, p. 161-210, pl. 19, 20.———(318) 1904, *On the mineralogical structure of the porcellanous Foraminifera:* Ann. & Mag. Nat. History, ser. 7, v. 14, p. 310.———(319) 1906, *On some Foraminifera and Ostracoda obtained off Great Barrier Island, New Zealand:* New Zealand Inst., Trans. & Proc., v. 38 (new ser. 21), p. 77-107, pl. 3.———(320) 1916, *Report on the Foraminifera and Ostracoda out of marine muds from soundings in the Ross Sea:* British Antarctic Exped. 1907-1909, Repts. Sci. Investigations, Geol., v. 2, pt. 3, p. 53-80, pl. 1-6.———(321) 1921, *Report of an examination of material obtained from a bore at Torquay:* Victoria Geol. Survey, Records, v. 4, p. 315-324, pl. 51.— (322) 1922, *Sherbornina: a new genus of the Foraminifera from Table Cape, Tasmania:* Linnean Soc. London, Jour., Zool., v. 34 (1918-22), no. 230, p. 501-503, p. 32.

(323) ———, & **Crespin, Irene,** 1930, *Rare Foraminifera from deep borings in the Victorian Tertiaries—Victoriella, gen. nov., Cycloclypeus communis Martin, and Lepidocyclina borneënsis Provale:* Royal Soc. Victoria, Proc., new ser., v. 42, p. 110-115, pl. 7-8.

(324) ———, & **Parr, W. J.,** 1931, *Notes on new and aberrant types of Foraminifera:* Royal Soc. Victoria, Proc., new ser., v. 43, pt. 2, p. 236-240, pl. 9, text-fig. 1.———(325) 1936, *A classification of the Foraminifera:* Same, new ser., v. 49, pt. 1, p. 139-151.

(326) ———, ———, & **Collins, A. C.,** 1934, *Tertiary Foraminifera of Victoria, Australia—The Balcombian deposits of Port Phillip, Part III:* Linnean Soc., Jour., Zool., v. 38, no. 262, p. 553-577, pl. 8-11.

(327) **Chatton, Édouard,** 1925, *Pansporella per-*

plexa. Réflexions sur la Biologie et la phylogénie des Protozoaires: Ann. Sci. Nat. Zool., ser. 10, v. 8, p. 5-84, 1 pl.

(328) **Chave, K. E.**, 1954, *Aspects of the biogeochemistry of magnesium; I. Calcareous marine organisms:* Jour. Geol., v. 62, no. 3, p. 266-283, fig. 1-16.

(329) **Checchia-Rispoli, Giuseppe**, 1905, *Sopra alcune Alveoline eoceniche della Sicilia:* Palaeont. Italica, Mem. Paleont., v. 11, p. 147-167, pl. 12-13.——(330) 1907, *Nota preventiva sulla serie Nummulitica dei dintorni in provincia di Palermo:* Giornale Sci., Nat. & Econ. Palermo, v. 26, p. 156-188.

(330A) **Chelussi, Italo**, 1903, *Sulla geologia della Conca Aquilana:* Soc. Ital. Sci. Nat., Milano, Atti, v. 42, p. 58-87.

(331) **Chen, S.**, 1934, *A new species of Fusulinidae from the Meitien Limestone:* Geol. Soc. China, Bull., v. 13, no. 2, p. 237-242, pl. 1.——(332) 1934, *Fusulinidae of South China, Part I:* Geol. Surv. China, Palaeont. Sinica, ser. B, v. 4, pt. 2, 185 p., 16 pl.——(333) 1937, *Permian Fusulinidae of Texas* in DUNBAR, C. O., & SKINNER, J. W., *The Geology of Texas, Pt. 2:* Univ. Texas, Bull. 3701, v. 3, p. 517-825, pl. 42-81, fig. 89-97.

(334) **Chernysheva, N. E.**, 1940, *K stratigrafii nizhnego Karbona Makarovskogo rayona yuzhnogo Urala po faune foraminifer:* Moskov. Obschch. Ispyt. Prirody, Otdel Geol., Bull., v. 18 (no. 5-6), p. 113-135, pl. 1, 2. [*On the stratigraphy of the Lower Carboniferous of the Makarovskoy district of the Southern Urals, on the basis of the foraminiferal fauna.*]—— (335) 1941, *A new genus of Foraminifera from the Tournaisian deposits of the Urals:* Akad. Nauk SSSR, Doklady (Acad. Sci. URSS, Comptes Rendus), v. 32, no. 1, p. 69-70.—— (336) 1948, *Ob Archaediscus i blizkikh k nemu formakh iz nizhnego karbona SSSR:* Akad. Nauk SSSR, Inst. Geol. Nauk, Trudy, no. 62 (Geol. ser., no. 19), p. 150-158, pl. 2. [*About Archaediscus and similar forms from the Lower Carboniferous of the USSR.*]

(337) **Children, J. G.**, 1823, *Lamarck's genera of shells, translated from the French by J. G. Children with plates from original drawings by Miss Anna Children:* 177 p., 10 pl., The Author (London).

(337A) **Chodat, Robert**, 1920, *Algues de la région du Grand St-Bernard:* Soc. Bot. Genève, Bull., ser. 2, v. 12, p. 293-305, text-fig. 1-10.

(337B) **Choffat, Paul**, 1885, *Quelques points importants de la geologie du Portugal:* Travaux Soc. Helv. Sci. Nat., Comptes Rendus, Sess. 68 (Aug. 11-13), p. 22-26.

(338) **Christiansen, B.**, 1958, *The foraminifer fauna in the Dröbak Sound in the Oslo Fjord (Norway):* Nytt Magasin Zool., v. 6, p. 5-91.

(339) **Cienkowski, Leon**, 1865, *Beiträge zur Kenntniss der Monaden:* Archiv. Mikro. Anat., v. 1, p. 203-232, pl. 12-14.——(340) 1867, *Ueber den Bau und die Entwicklung der Labyrinthuleen:* Same, v. 3, p. 274-310, pl. 15-17.——(341) 1876, *Ueber einige Rhizopoden und verwandte Organismen:* Same, v. 12, p. 15-50, pl. 4-8.

(342) **Ciry, Raymond**, 1948, *Un nouveau fusulinidé Permien Dunbarula mathieui:* Bull. Sci. Bourgogne, v. 11, p. 103-110, pl. 1, fig. 1. Imprimerie Veuve Paul Berthier (Dijon).

(343) ——, & **Rat, Pierre**, 1951, *Un foraminifère nouveau du Crétacé supérieur de la Navarre Espagnole:* Bull. Sci. Bourgogne, v. 13, p. 75-86, pl. 2, text-fig. 1, 2.——(344) 1953, *Description d'un nouveau genre de foraminifère: Simplorbitolina manasi nov. gen., nov. sp.:* Same, v. 14, p. 85-100, 1 pl., 5 text-fig.

(344A) **Cita, M. B.**, & **Scipolo, C.**, 1951, *Chapmanina gassinensis (Silvestri) dans l'Oligocène du Monte Baldo (Italie):* Revue Micropaléont., v. 4, no. 3, p. 121-134, pl. 1-3, text-fig. 1-6.

(345) **Claparède, Édouard**, & **Lachmann, Johannes**, 1859, *Étude sur les Infusoires et les Rhizopodes, vol. 1, livraison 2:* l'Inst. Genèvois, Mém., v. 6, p. 261-482 (Genève).

(346) **Clarke, F. W.**, & **Wheeler, W. C.**, 1922, *The inorganic constituents of marine invertebrates:* U.S. Geol. Survey, Prof. Paper 124, 62 p.

(347) **Claus, Carl**, 1872, *Grundzüge der Zoologie:* 1170 p. (Marburg in Leipzig).——(348) 1905, *Lehrbuch der Zoologie, revised by Karl Grobben:* 955 p. (Marburg in Hessen).

(349) **Clements, F. E.**, & **Shear, C. L.**, 1931, *The genera of Fungi:* 496 p., Wilson Co. (New York).

(350) **Clodius, G.**, 1922, *Die Foraminiferen des obermiozänen Glimmertons in Norddeutschland mit besonderer Berüchsichtigung der Aufschlüsse in Mecklenburg:* Vereins. Freunde Naturg. Mecklenburg, Archiv, v. 75, p. 76-145, pl. 1.

(351) **Cockerell, T. D. A.**, 1909, *New names for two genera of Protozoa:* Zool. Anzeiger, v. 34, p. 565.——(352) 1911, *The nomenclature of the Rhizopoda:* Same, v. 38, p. 136-137.——(353) 1930, *Siliceous shells of Protozoa:* Nature, v. 125, p. 975.

(354) **Colani, Madeleine**, 1924, *Nouvelle contribution à l'étude des fusulinidés de l'extrême-Orient:* Service Geol. Indochine, Mém., v. 11, pt. 1, 191 p., 29 pl., 28 graph.

(355) **Cole, W. S.**, 1927, *A foraminiferal fauna from the Guayabal Formation in Mexico:* Bull. Am. Paleontology, v. 14, p. 1-46, pl. 1-5.——(356) 1938, *Stratigraphy and micropaleontology*

of two deep wells in Florida: Florida Geol. Survey, Bull. 16, 73 p., 12 pl.——(357) 1941, *Stratigraphic and paleontologic studies of wells in Florida:* Same, Bull. 19, vi+91 p., 18 pl., 4 text-fig.——(358) 1947, *Internal structure of some Floridian Foraminifera:* Bull. Am. Paleontology, v. 31, no. 126, p. 227-254, pl. 21-25.
——(359) 1952 [1953], *Eocene and Oligocene larger Foraminifera from the Panama Canal Zone and vicinity:* U.S. Geol. Survey, Prof. Paper 244, 41 p., 28 pl., 2 text-fig.——(360) 1953, *Criteria for the recognition of certain assumed camerinid genera:* Bull. Am. Paleontology, v. 35, no. 147, p. 1-22, 3 pl.——(361) 1954, *Larger Foraminifera and smaller diagnostic Foraminifera from Bikini drill holes:* U.S. Geol. Survey, Prof. Paper 260-O, p. 569-608, pl. 204-222.——(362) 1956, *The genera Miscellanea and Pellatispirella:* Bull. Am. Paleontology, v. 36, no. 159, p. 239-254, pl. 32-34.
——(363) 1957, *Late Oligocene larger Foraminifera from Barro Colorado Island, Panama Canal Zone:* Same, v. 37, no. 163, p. 313-330, pl. 24-30.——(364) 1957, *Larger Foraminifera,* in *Geology of Saipan Mariana Islands Pt. 3, Paleontology:* U.S. Geol. Survey, Prof. Paper 280-I, p. 321-360, pl. 94-118.——(365) 1958, *Names of and variation in certain American larger Foraminifera, particularly the camerinids:* Bull. Am. Paleontology, v. 38, p. 261-284, pl. 32-34.——(366) 1960, *Revision of Helicostegina, Helicolepidina and Lepidocyclina (Polylepidina):* Cushman Found. Foram. Research, Contrib., v. 11, pt. 2, p. 57-63, pl. 10-13.——(367) 1960, *The genus Camerina:* Bull. Am. Paleontology, v. 41, no. 190, p. 189-205, pl. 23-26.——(368) 1960, *Variability in embryonic chambers of Lepidocyclina:* Micropaleontology, v. 6, no. 2, p. 133-140, pl. 1-4.
——(369) 1961, *Names of and variation in certain Indo-Pacific camerinids, No. 2. A reply:* Bull. Am. Paleontology, v. 43, no. 195, p. 111-128, pl. 14-16.

(370) ——, & Bermúdez, P. J., 1944, *New foraminiferal genera from the Cuban middle Eocene:* Bull. Am. Paleontology, v. 28, no. 113, p. 333-334, pl. 27-29.——(371) 1947, *Eocene Discocyclinidae and other Foraminifera from Cuba:* Same, v. 31, no. 125, p. 191-224, pl. 14-20.

(372) ——, & Gravell, D. W., 1952, *Middle Eocene Foraminifera from Peñon Seep, Matanzas Province, Cuba:* Jour. Paleontology, v. 26, p. 708-727, pl. 90-103.

(373) **Collin, Bernard,** 1912, *Sur un amibe à coque, pourvu de tentacules: Chlamydamoeba tentaculifera, n.g., n.sp.:* Arch. Zool. Expér. & Générale, Notes & Revue, ser. 5, v. 10, p. lxxxviii-xcv, text-fig. 1-2.——(374) 1914, *Notes Protistologiques:* Arch. Zool. Expér. & Générale, v. 54, p. 85-97, text-fig. 1-5.

(375) **Collins, A. C.,** 1958[1960], *Foraminifera:* in Great Barrier Reef Expedition 1928-29, Sci. Repts., v. 6, no. 6, British Mus. Nat. History, p. 335-437, pl. 1-5. [Issued Sept. 16, 1960.]

(376) **Colom, Guillermo,** 1956, *Los Foraminíferos del Burdigaliense de Mallorca:* Real Acad. Cien. & Art. Barcelona, Mem., v. 32, no. 5 (tercera época, no. 653), p. 7-140, pl. 1-25.——(377) 1959, *Gymnesina glomerosa, n.gen., n.sp. (Fam. Ophthalmidiidae) from the Mediterranean:* Cushman Found. Foram. Research, Contrib., v. 10, pt. 1, p. 16-19.

(378) **Conklin, J. E.,** 1954, *Hyperammina kentuckyensis n. sp. from the Mississippian of Kentucky, and discussion of Hyperammina and Hyperamminoides:* Cushman Found. Foram. Research, Contrib., v. 5, pt. 4, p. 165-169, pl. 31.

(379) **Conrad, T. A.,** 1846, *Description of new species of organic remains from the upper Eocene limestone of Tampa Bay:* Am. Jour. Sci., ser. 2, v. 2, p. 399-400.——(380) 1865, *Catalogue of the Eocene Annulata, Foraminifera, Echinodermata and Cirripedia of the United States:* Acad. Nat. Sci. Philadelphia, Proc., v. 17, p. 73-75.

(381) **Coogan, A. H.,** 1960, *Stratigraphy and paleontology of the Permian Nosoni and Dekkas Formations (Bollibokka Group):* Univ. California Publs., Geol. Sci., v. 36, no. 5, p. 243-315, pl. 22-27, fig. 1-22.

(382) **Cook, W. R. I.,** 1933, *A monograph of the Plasmodiophorales:* Archiv Protistunkunde, v. 80, p. 179-254, pl. 5-11.

(383) **Cooke, W. B.,** 1951, *Some Myxomycetes from south central Washington:* Northwest Sci., v. 25, no. 4, p. 171-175.

(384) **Copeland, H. F.,** 1956, *The classification of lower organisms:* 302 p., Pacific Books (Palo Alto, Calif.).

(385) **Corliss, J. O.,** 1960, *Comments on the systematics and phylogeny of the Protozoa:* Systematic Zoology, v. 8, no. 4, p. 169-190 (1959).
——(386) 1962, *Taxonomic procedures in classification of Protozoa:* Symposia Soc. for General Microbiology, no. 12, Microbial Classification, p. 37-67 (Great Britain).

(387) **Cornish, Vaughan, & Kendall, P. F.,** 1888, *On the mineralogical constitution of calcareous organisms:* Geol. Mag., decade 3, v. 5, p. 66-73.

(388) **Cosijn, A. J.,** 1938, *Statistical studies on the phylogeny of some Foraminifera. Cycloclypeus and Lepidocyclina from Spain, Globorotalia from the East-Indies:* Leidsche Geol. Meded., v. 10, pt. 1, p. 1-61, pl. 1-5.

(389) **Costa, Achille,** 1862, *Di un novello genere di foraminiferi:* Univ. Napoli, Mus. Zool., v. 1, p. 94-95.

(390) **Costa, O. G.,** 1839, *Descrizione di alcune specie nuove di testacei freschi e fossili del regno*

delle due Sicilie: R. Accad. Sci. Napoli, Cl. Fis. Storia Nat., Atti, v. 4, p. 175-192.———(391) 1855, *Foraminiferi fossili della marna blù del Vaticano:* R. Accad. Sci. Napoli, v. 2 (1855-57), p. 113-126, pl. 1.———(392) 1856, *Paleontologia del regno di Napoli, Parte II:* Accad. Pont. Napoli, Atti, v. 7, pt. 2, p. 113-378, pl. 9-27.———(393) 1861, *Microdoride Mediterranea:* v. 1, p. i-xviii, 41-55, Stamperia dell' Iride (Napoli).

(394) **Crespin, Irene,** 1958, *Permian Foraminifera of Australia:* Australia Bur. Mineral. Res., Geol. & Geophys., Bull. 48, 207 p., 33 pl.———
(394A) 1962, *Lacazinella, a new genus of trematophore Foraminifera:* Micropaleontology, v. 8, no. 3, p. 337-342, pl. 1-2.

(395) ———, & **Belford, D. J.,** 1957, *New genera and species of Foraminifera from the Lower Permian of Western Australia:* Cushman Found. Foram. Res., Contrib., v. 8, pt. 2, p. 73-76, 80-81, pl. 11-12.

(396) ———, & **Parr, W. J.,** 1941, *Arenaceous Foraminifera from the Permian rocks of New South Wales:* Royal Soc. New S. Wales, Jour. & Proc., v. 74, p. 300-311, pl. 12-13.

(397) **Crouch, E. A.,** 1827, *An illustrated introduction to Lamarck's conchology:* 47 p., 22 pl., Longman, Rees, Orme, Brown & Green & J. Mawe (London).

(398) **Cummings, R. H.,** 1955, *New genera of Foraminifera from the British Lower Carboniferous:* Washington Acad. Sci., Jour., v. 45, no. 1, p. 1-8, text-fig. 1-5.———(399) 1955, *Stacheoides, a new foraminiferal genus from the British Upper Paleozoic:* Same, Jour., v. 45, no. 11, p. 342-346, text-fig. 1-8.———(400) 1955, *Nodosinella Brady, 1876, and associated upper Paleozoic genera:* Micropaleontology, v. 1, no. 3, p. 221-238, pl. 1, text-fig. 1-10.———
(401) 1956, *Revision of the upper Palaeozoic textulariid Foraminifera:* Same, v. 2, no. 3, p. 201-242, pl. 1, text-fig. 1-24.

(402) **Cushman, J. A.,** 1909, *Ammodiscoides, a new genus of arenaceous Foraminifera:* U.S. Natl. Museum, Proc., v. 36, no. 1676, p. 423-424, pl. 33.———(403) 1910, *New arenaceous Foraminifera from the Philippines:* Same, Proc., v. 38, p. 437-442, fig. 1-19.———(404) 1910-17, *A monograph of the Foraminifera of the North Pacific Ocean:* Same, Bull. 71; (a) *Pt. 1. Astrorhizidae and Lituolidae* (1910), 134 p., 203 text-fig.; (b) *Pt. 2. Textulariidae* (1911), 108 p., 156 text-fig.; (c) *Pt. 3. Lagenidae* (1913), 125 p., 47 pl.; (d) *Pt. 4. Chilostomellidae, Globigerinidae, Nummulitidae* (1914), 46 p., 19 pl.; (e) *Pt. 5. Rotaliidae* (1915), 81 p., 31 pl., 62 text-fig.; (f) *Pt. 6. Miliolidae* (1917), 108 p., 39 pl., 52 text-fig.———(405) 1912, *New arenaceous Foraminifera from the Philippine Islands and contiguous waters:* Same, Proc., v. 42, p. 227-230, pl. 28.———(406) 1913, *New Textulariidae and other arenaceous Foraminifera from the Philippine Islands and contiguous waters:* Same, Proc., v. 44, no. 1973, p. 633-638, pl. 78-80.———(407) 1917, *New species and varieties of Foraminifera from the Philippines and adjacent waters:* Same, Proc., v. 51, no. 2172, p. 651-662.———(408) 1917, *Orbitoid Foraminifera of the genus Orthophragmina from Georgia and Florida:* U.S. Geol. Survey, Prof. Paper 108-G. p. 115-118, pl. 40-44.———(409) 1918, *The smaller fossil Foraminifera of the Panama Canal Zone:* U.S. Natl. Museum, Bull. 103, p. 45-87, pl. 19-33.
———(410) 1918, *The larger fossil Foraminifera of the Panama Canal Zone:* Same, Bull. 103, p. 89-102, pl. 34-45.———(411) 1918-23, *The Foraminifera of the Atlantic Ocean:* Same, Bull. 104; (a) *Pt. 1. Astrorhizidae* (1918), 111 p., 39 pl.; (b) *Pt. 2. Lituolidae* (1920), 111 p., 18 pl.; (c) *Pt. 3. Textulariidae* (1922), 143 p., 26 pl.; (d) *Pt. 4. Lagenidae* (1923), 228 p., 42 pl.———(412) 1919, *The relationships of the genera Calcarina, Tinoporus, and Baculogypsina as indicated by Recent Philippine material:* Same, Bull. 100, v. 1, pt. 6, p. 363-368, pl. 44-45.———(413) 1919, *Recent Foraminifera from off New Zealand:* Same, Proc., v. 56, p. 593-640, pl. 74-75.———(414) 1919, *Fossil Foraminifera from the West Indies:* in Vaughan, T. W., Contributions to the geology and paleontology of the West Indies: Carnegie Inst. Washington, publ. 291, p. 23-71.———(415) 1921, *Foraminifera of the Philippine and adjacent seas:* U.S. Natl. Museum, Bull. 100, v. 4, p. 1-608, pl. 1-100, text-fig. 1-52.———(416) 1922, *Shallow-water Foraminifera of the Tortugas region:* Carnegie Inst. Washington, Publ. 311 (Dept. Marine Biol., Papers, v. 17), p. 1-85, pl. 1-14.———(417) 1922, *The Foraminifera of the Mint Spring calcareous marl member of the Marianna Limestone:* U.S. Geol. Survey, Prof. Paper 129-F, p. 123-143, pl. 29-35.
———(418) 1924, *Samoan Foraminifera:* Carnegie Inst. Washington, Publ. 342 (Dept. Marine Biol. Papers, v. 21), p. 1-75, pl. 1-25.
———(419) 1924, *A new genus of Eocene Foraminifera:* U.S. Natl. Museum, Proc., v. 66, art. 30, p. 1-4, pl. 1-2.———(420) 1925, *An introduction to the morphology and classification of the Foraminifera:* Smithsonian Misc. Coll., v. 77, no. 4, p. 1-77, pl. 1-16.———
(421) 1925, *New Foraminifera from the upper Eocene of Mexico:* Cushman Lab. Foram. Research, Contrib., v. 1, pt. 1, p. 4-8, pl. 1.
———(422) 1926, *Foraminifera of the genera Siphogenerina and Pavonina:* U.S. Natl. Museum, Proc., v. 67, art. 25, p. 1-24, pl. 1-6.
———(423) 1926, *The Foraminifera of the Velasco shale of the Tampico embayment:* Am. Assoc. Petroleum Geologists, Bull., v. 10, 581-612, pl. 15-21.——— (424) 1926,

Eouvigerina a new genus from the Cretaceous: Cushman Lab. Foram. Research, Contrib., v. 2, pt. 1, p. 3-6, pl. 1.──(425) 1926, *Some Foraminifera from the Mendez shale of eastern Mexico:* Same, Contrib., v. 2, pt. 1, p. 16-26, pl. 2-3.──(426) 1926, *Foraminifera of the typical Monterey of California:* Same, Contrib., v. 2, pt. 3, p. 53-69, pl. 7-9. ──(427) 1926, *The genus Chilostomella and related genera:* Same, Contrib., v. 1, pt. 4, p. 73-80, pl. 11.──(428) 1927, *Some new genera of the Foraminifera:* Same, Contrib., v. 2, pt. 4, p. 77-81, pl. 11.──(429) 1927, *American Upper Cretaceous species of Bolivina and related species:* Same, Contrib., v. 2, pt. 4, p. 85-91, pl. 12.──(430) 1927, *Sporadogenerina a degenerate foraminiferal genus:* Same, Contrib., v. 2, pt. 4, p. 94-95, pl. 11.── (431) 1927, *An outline of a re-classification of the Foraminifera:* Same, Contrib., v. 3, pt. 1, p. 1-105, pl. 1-21.──(432) 1927, *Some notes on the early foraminiferal genera erected before 1808:* Same, Contrib., v. 3, pt. 2, p. 122-126, pl. 24.──(433) 1927, *The designation of some genotypes in the Foraminifera:* Same, Contrib., v. 3, pt. 4, p. 188-190.──(434) 1927, *Notes on Foraminifera in the collection of Ehrenberg:* Washington Acad. Sci., Jour., v. 17, p. 487-491.──(435) 1927, *Recent Foraminifera from off the West Coast of America:* Univ. California Scripps Inst. Oceanog., Bull., tech. ser., v. 1, p. 119-188, pl. 1-6.── (436) 1928, *Additional genera of the Foraminifera:* Cushman Lab. Foram. Research, Contrib., v. 4, pt. 1, p. 1-8, pl. 1, 3.──(437) 1928, *On Rotalia beccarii (Linné):* Same, Contrib., v. 4, pt. 4, p. 103-107, pl. 15.── (438) 1928, *Fistulose species of Gaudryina and Heterostomella:* Same, Contrib., v. 4, pt. 4, p. 107-112, pl. 16.──(439) 1928, *Foraminifera their classification and economic use:* Same, Spec. Publ. 1, p. 1-401, pl. 1-59.──(440) 1929, *Kyphopyxa, a new genus from the Cretaceous of Texas:* Same, Contrib., v. 5, pt. 1, p. 1-4, pl. 1.──(441) 1929, *The genus Bolivinella and its species:* Same, Contrib., v. 5, pt. 2, p. 28-34, pl. 5.──(442) 1929, *A late Tertiary fauna of Venezuela and other related regions:* Same, Contrib., v. 5, p. 77-101, pl. 12-14.──(443) 1929, *The genus Trimosina and its relationships to other genera of the foraminifera:* Washington Acad. Sci., Jour., v. 19, no. 8, p. 155-159, text-fig. 1-3.──(444) 1930, *The Foraminifera of the Atlantic Ocean, Pt. 7. Nonionidae, Camerinidae, Peneroplidae and Alveolinellidae:* U.S. Natl. Museum, Bull. 104, pt. 7, vi+79 p., 18 pl.──(445) 1930, *The Foraminifera of the Choctawhatchee Formation of Florida:* Florida State Geol. Survey, Bull. 4, 63 p., 12 pl.──(446) 1930, *Note sur quelques Foraminifères jurassiques d'Auberville (Calvados):* Soc. Linnéenne de Normandie, Bull., ser. 8, v. 2, p. 132-135, pl. 4.── (447) 1930, *A resumé of new genera of the Foraminifera erected since early 1928:* Cushman Lab. Foram. Research, Contrib., v. 6, pt. 4, p. 73-94, pl. 10-12.──(448) 1931, *Parrina, a new generic name:* Same, Contrib., v. 7, pt. 1, p. 20.──(449) 1931, *Two new foraminiferal genera from the South Pacific:* Same, Contrib., v. 7, pt. 4, p. 78-82, pl. 10.──(450) 1931, *Hastigerinella and other interesting Foraminifera from the Upper Cretaceous of Texas:* Same, Contrib., v. 7, pt. 4, p. 83-90, pl. 11.── (451) 1931, *The Foraminifera of the Atlantic Ocean, Pt. 8. Rotaliidae, Amphisteginidae, Calcarinidae, Cymbaloporettidae, Globorotaliidae, Anomalinidae, Planorbulinidae, Rupertiidae and Homotremidae:* U.S. Natl. Museum, Bull. 104, pt. 8, ix+179 p., 26 pl.──(452) 1932, *Rectogümbelina, a new genus from the Cretaceous:* Cushman Lab. Foram. Research, Contrib., v. 8, pt. 1, p. 4-7, pl. 1.──(453) 1932, *Notes on the genus Virgulina:* Same, Contrib., v. 8, pt. 1, p. 7-23, pl. 2-3.──(454) 1932, *The relationships of Textulariella and description of a new species:* Same, Contrib., v. 8, pt. 4, p. 97-98.──(455) 1932, *The Foraminifera of the tropical Pacific collections of the "Albatross", 1899-1900, Pt. 1. Astrorhizidae to Trochamminidae:* U.S. Natl. Museum, Bull. 161, pt. 1, 88 p., 17 pl.──(456) 1933, *Two new genera, Pernerina and Hagenowella, and their relationships to genera of the Valvulinidae:* Am. Jour. Sci., ser. 5, v. 26, p. 19-26, pl. 1-2.──(457) 1933, *New Arctic Foraminifera collected by Capt. R. A. Bartlett from Fox Basin and off the northeast coast of Greenland:* Smithsonian Misc. Coll., v. 89, no. 9, p. 1-8, pl. 1, 2.──(458) 1933, *Some new foraminiferal genera:* Cushman Lab. Foram. Research, Contrib., v. 9, pt. 2, p. 32-38, pl. 3, 4.──(459) 1933, *New American Cretaceous Foraminifera:* Same, Contrib., v. 9, pt. 3, p. 49-64, pl. 5-6.──(460) 1933, *Some new Recent Foraminifera from the tropical Pacific:* Same, Contrib., v. 9, pt. 4, p. 77-95, pl. 8-10.──(461) 1933, *Foraminifera their classification and economic use:* Same, Spec. Publ. 4, p. 1-349, pl. 1-40.──(462) 1933, *The Foraminifera of the tropical Pacific Collections of the "Albatross", 1899-1900, Pt. 2. Lagenidae to Alveolinellidae:* U.S. Natl. Museum, Bull. 161, pt. 2, vi+79 p., 19 pl.──(463) 1934, *Notes on the genus Tretomphalus, with descriptions of some new species and a new genus, Pyropilus:* Cushman Lab. Foram. Research, Contrib., v. 10, pt. 4, p. 79-101, pl. 11-13.──(464) 1934, *The relationships of Ungulatella, with descriptions of additional species:* Same, Contrib., v. 10, pt. 4, p. 101-104, pl. 13.──(465) 1935, *Some new Foraminifera from the late Tertiary of Georges Bank:* Same, Contrib., v. 11, pt. 4,

p. 77-83, pl. 12.———(466) 1935, *Fourteen new species of Foraminifera*: Smithsonian Misc. Coll., v. 91, no. 21 (publ. 3327), p. 1-9, pl. 1-3.———(467) 1935, *Upper Eocene Foraminifera of the southeastern United States*: U.S. Geol. Survey, Prof. Paper 181, 88 p., 23 pl.
———(468) 1936, *New genera and species of the families Verneuilinidae and Valvulinidae and of the subfamily Virgulininae*: Cushman Lab. Foram. Research, Spec. Publ. 6, 71 p., 8 pl.
———(469) 1936, *Some new species of Elphidium and related genera*: Same, Contrib., v. 12, pt. 4, p. 78-91, pl. 14, 15.———(470) 1937, *A monograph of the foraminiferal family Verneuilinidae*: Same, Spec. Publ. 7, 157 p., 20 pl.———(471) 1937, *A monograph of the foraminiferal family Valvulinidae*: Same, Spec. Publ. 8, 210 p., 24 pl.———(472) 1937, *A monograph of the subfamily Virgulininae of the foraminiferal family Buliminidae*: Same, Spec. Publ. 9, xv+228 p., 24 pl.———(473) 1939, *A monograph of the foraminiferal family Nonionidae*: U.S. Geol. Survey, Prof. Paper 191, 100 p., 20 pl.———(474) 1940, *Foraminifera their classification and economic use*: ed. 3, 535 p., 48 pl., Harvard Univ. Press (Cambridge, Mass.).———(475) 1940, *Midway Foraminifera from Alabama*: Cushman Lab. Foram. Research, Contrib., v. 16, pt. 3, p. 51-73, pl. 9-12.
———(476) 1942, *The Foraminifera of the tropical Pacific collections of the "Albatross", 1899-1900, Part 3. Heterohelicidae and Buliminidae*: U.S. Natl. Museum, Bull. 161, pt. 3, 67 p., 15 pl.———(477) 1943, *A new genus of the Trochamminidae*: Cushman Lab. Foram. Research, Contrib., v. 19, pt. 4, p. 95-96, pl. 16.
———(478) 1944, *Foraminifera from the shallow water of the New England coast*: Same, Spec. Publ. 12, p. 1-37, pl. 1-4.———(479) 1944, *Poroarticulina, a new genus of Foraminifera*: Same, Contrib., v. 20, pt. 2, p. 52, pl. 8
———(480) 1944, *Additional notes on Foraminifera in the collection of Ehrenberg*: Washington Acad. Sci., Jour., v. 34, p. 157-158.
———(481) 1945, *The species of Foraminifera recorded by d'Orbigny in 1826 from the Pliocene of Castel Arquato, Italy*: Cushman Lab. Foram. Research, Spec. Publ. 13, 27 p., 6 pl.
———(482) 1945, *A foraminiferal fauna from the Twiggs clay of Georgia*: Same, Contrib., v. 21, pt. 1, p. 1-11, pl. 1, 2.———(483) 1946, *Polysegmentina, a new genus of the Ophthalmidiidae*: Same, Contrib., v. 22, pt. 1, p. 1, pl. 1.———(484) 1946, *Upper Cretaceous Foraminifera of the Gulf Coastal region of the United States and adjacent areas*: U.S. Geol. Survey Prof. Paper 206, 241 p., 66 pl.———(485) 1947, *A supplement to the monograph of the foraminiferal family Valvulinidae*: Cushman Lab. Foram. Research, Spec. Publ. 8A, 69 p., 8 pl.———(486) 1948, *Foraminifera their classification and economic use*: ed. 4, 605 p., 55 pl., Harvard Univ. Press (Cambridge, Mass.).

(487) ———, & **Alexander, C. I.**, 1929, *Frankeina, a new genus of arenaceous Foraminifera*: Cushman Lab. Foram. Research, Contrib., v. 5, pl. 61-62, pl. 10.———(488) 1930, *Some Vaginulinas and other Foraminifera from the Lower Cretaceous of Texas*: Same, Contrib., v. 6, pt. 1, p. 1-10, pl. 1-2.

(489) ———, & **Bermúdez, P. J.**, 1936, *New genera and species of Foraminifera from the Eocene of Cuba*: Cushman Lab. Foram. Research, Contrib., v. 12, pt. 2, p. 27-38, pl. 5-6.
———(490) 1936, *Additional new species of Foraminifera and a new genus from the Eocene of Cuba*: Same, Contrib., v. 12, pt. 3, p. 55-63, pl. 10, 11.———(491) 1937, *Further new species of Foraminifera from the Eocene of Cuba*: Same, Contrib., v. 13, pt. 1, p. 1-29, pl. 1-2.
———(492) 1941, *Cuneolinella, a new genus from the Miocene*: Same, Contrib., v. 17, pt. 4, p. 101-102, pl. 24.———(493) 1946, *A new genus, Cribropyrgo, and a new species of Rotalia*: Same, Contrib., v. 22, pt. 4, p. 119-120, pl. 19.———(494) 1947, *Some Cuban Foraminifera of the genus Rotalia*: Same, Contrib., v. 23, pt. 2, p. 23-29, pl. 5-10.———(495) 1948, *Colomia, a new genus from the upper Cretaceous of Cuba*: Same, Contrib., v. 24, pt. 1, p. 12, pl. 2.———(496) 1948, *Some Paleocene Foraminifera from the Madruga formation of Cuba*: Same, Contrib., v. 24, pt. 3, p. 68-75, pl. 11-12.———(497) 1949, *Some Cuban species of Globorotalia*: Same, Contrib., v. 25, pt. 2, p. 26-45, pl. 5-8.

(498) ———, & **Brönnimann, Paul**, 1948, *Some new genera and species of Foraminifera from brackish water of Trinidad*: Cushman Lab. Foram. Research, Contrib., v. 24, pt. 1, p. 15-21, pl. 3, 4.———(498A) 1948, *Additional new species of arenaceous Foraminifera from shallow waters of Trinidad*: Same, Contrib., v. 24, pt. 2, p. 37-42, pl. 7, 8.

(499) ———, & **Campbell, A. S.**, 1936, *A new Siphogenerinoides from California*: Cushman Lab. Foram. Research, Contrib., v. 12, pt. 4, p. 91-92, pl. 13.

(500) ———, & **Church, C. C.**, 1929, *Some Upper Cretaceous Foraminifera from near Coalinga, California*: Calif. Acad. Sci., Proc., ser. 4, v. 18, no. 16, p. 497-530, pl. 36-41.

(501) ———, & **Dam, Abraham ten**, 1948, *Globigerinelloides, a new genus of the Globigerinidae*: Cushman Lab. Foram. Research, Contrib., v. 24, p. 42-43, pl. 8.———(502) 1948, *Pseudoparrella, a new generic name, and a new species of Parrella*: Same, Contrib., v. 24, pt. 3, p. 49-50, pl. 9.

(503) ———, & **Edwards, P. G.**, 1937, *Astro-*

nonion, a new genus of the Foraminifera, and its species: Cushman Lab. Foram. Research, Contrib., v. 13, pt. 1, p. 29-36, pl. 3.

(504) ——, & Hanzawa, Shoshiro, 1936, *New genera and species of Foraminifera of the late Tertiary of the Pacific:* Cushman Lab. Foram. Research, Contrib., v. 12, pt. 2, p. 45-48, pl. 8.

——(505) 1937, *Notes on some of the species referred to Vertebralina and Articulina, and a new genus Nodobaculariella:* Same, Contrib., v. 13, pt. 2, p. 41-46, pl. 5.

(506) ——, & Hedberg, H. D., 1935, *A new genus of Foraminifera from the Miocene of Venezuela:* Cushman Lab. Foram. Research, Contrib., v. 11, pt. 1, p. 13-16, pl. 3.—— (507) 1941, *Upper Cretaceous Foraminifera from Santander del Norte, Colombia, S.A.:* Same, Contrib., v. 17, pt. 4, p. 79-102, pl. 21-24.

(508) ——, & Hughes, D. D., 1925, *Some later Tertiary Cassidulinas of California:* Cushman Lab. Foram. Research, Contrib., v. 1, pt. 1, p. 11-17, pl. 2.

(509) ——, & Jarvis, P. W., 1929, *New Foraminifera from Trinidad:* Cushman Lab. Foram. Research, Contrib., v. 5, p. 6-17, pl. 2-3.

(510) ——, & LeRoy, L. W., 1939, *Cribrolinoides, a new genus of the Foraminifera, its development and relationships:* Cushman Lab. Foram. Research. Contrib., v. 15, pt. 1, p. 15-19, pl. 3-4.

(511) ——, & McCulloch, I., 1939, *A report on some arenaceous Foraminifera:* Allan Hancock Pacific Exped., v. 6, no. 1, p. 1-113, pl. 1-12.

(512) ——, & Martin, L. T., 1935, *A new genus of Foraminifera, Discorbinella, from Monterey Bay, California:* Cushman Lab. Foram. Research, Contrib., v. 11, pt. 4, p. 89-90, pl. 14.

(513) ——, & Ozawa, Yoshiaki, 1928, *An outline of a revision of the Polymorphinidae:* Cushman Lab. Foram. Research, Contrib., v. 4, pt. 1, p. 13-21, pl. 2.——(514) 1930, *A monograph of the foraminiferal family Polymorphinidae, Recent and fossil:* U.S. Natl. Museum, Proc., v. 77, p. 1-195, pl. 1-40.

(515) ——, & Parker, F. L., 1936, *Notes on some Cretaceous species of Buliminella and Neobulimina:* Cushman Lab. Foram. Research, Contrib., v. 12, pt. 1, p. 5-10, pl. 2.—— (516) 1937, *Notes on some European species of Bulimina:* Same, Contrib., v. 13, pt. 2, p. 46-54, pl. 6-7.——(517) 1938, *The Recent species of Bulimina named by d'Orbigny in 1826:* Same, Contrib., v. 14, pt. 4, p. 90-94, pl. 16.——(518) 1940, *The species of the genus Bulimina having Recent types:* Same, Contrib., v. 16, pt. 1, p. 7-23, pl. 2-3.—— (519) 1947, *Bulimina and related foraminiferal genera:* U.S. Geol. Survey, Prof. Paper 210-D, p. 55-176, pl. 15-30.

(520) ——, & Ponton, G. M., 1932, *Some interesting new Foraminifera from the Miocene of Florida:* Cushman Lab. Foram. Research, Contrib., v. 8, pt. 1, p. 1-4, pl. 1.——(521) 1932, *An Eocene foraminiferal fauna of Wilcox age from Alabama:* Same, Contrib., v. 8, pt. 3, p. 51-72, pl. 7-9.——(522) 1933, *A new genus of the Foraminifera, Gunteria, from the middle Eocene of Florida:* Same, Contrib., v. 9, pt. 2, p. 25-30, pl. 3.

(523) ——, J. A., & Renz, H. H., 1941, *New Oligocene-Miocene Foraminifera from Venezuela:* Cushman Lab. Foram. Research, Contrib., v. 17, no. 1, p. 1-27, pl. 1-4.

(524) ——, & Stainbrook, M. A., 1943, *Some Foraminifera from the Devonian of Iowa:* Cushman Lab. Foram. Research, Contrib., v. 19, no. 4, p. 73-79, pl. 13.

(525) ——, & Stainforth, R. M., 1945, *The Foraminifera of the Cipero marl formation of Trinidad, British West Indies:* Cushman Lab. Foram. Research, Spec. Publ., no. 14, p. 1-74, pl. 1-16.——(526) 1947, *A new genus and some new species of Foraminifera from the upper Eocene of Ecuador:* Same, Contrib., v. 23, pt. 4, p. 77-80, pl. 17.

(527) ——, & Todd, Ruth, 1941, *The structure and development of Laticarinina pauperata (Parker and Jones):* Cushman Lab. Foram. Research, Contrib., v. 17, no. 4, p. 103-105, pl. 24.——(528) 1942, *The genus Cancris and its species:* Same, Contrib., v. 18, pt. 4, p. 72-94, pl. 17-24.——(529) 1943, *The genus Pullenia and its species:* Same, v. 19, pt. 1, p. 1-23, pl. 1-4.——(530) 1949, *The genus Sphaeroidina and its species:* Same, Contrib., v. 25, pt. 1, p. 11-21, pl. 3, 4.

(531) ——, ——, & Post, R. J., 1954, *Recent Foraminifera of the Marshall Islands:* U.S. Geol. Survey, Prof. Paper 260-H, p. 319-384, pl. 82-93.

(532) ——, & Valentine, W. W., 1930, *Shallow-water Foraminifera from the Channel Islands of southern California:* Stanford Univ., Contrib. Dept. Geol., v. 1, no. 1, p. 5-51, pl. 1-10.

(533) ——, & Warner, W. C., 1940, *A preliminary study of the structure of the test in the so-called porcellanous Foraminifera:* Cushman Lab. Foram. Research, Contrib., v. 16, p. 24-26, pl. 4.

(534) ——, & Waters, J. A., 1927, *Arenaceous Paleozoic Foraminifera from Texas:* Cushman Lab. Foram. Research, Contrib., v. 3, pt. 3, p. 146-153, pl. 26-27.——(535) 1928, *Some Foraminifera from the Pennsylvanian and Permian of Texas:* Same, Contrib., v. 4, pt. 2, p.

31-55, pl. 4-7.———(536) 1928, *Additional Cisco Foraminifera from Texas:* Same, Contrib., v. 4, pt. 3, p. 62-67, pl. 8.———(537) 1928, *Hyperamminoides, a new name for Hyperamminella Cushman and Waters:* Same, Contrib., v. 4, pt. 4, p. 112.———(538) 1928, *The development of Climacammina and its allies in the Pennsylvanian of Texas:* Jour. Paleontology, v. 2, p. 119-130, pl. 17-20.———(539) 1928, *Upper Paleozoic Foraminifera from Sutton County, Texas:* Same, v. 2, p. 358-371, pl. 47-49.———(539A) 1930, *Foraminifera of the Cisco Group of Texas:* Univ. Texas, Bull. 3019, p. 22-81, pl. 2-12.

(540) ———, & White, E. M., 1936, *Pyrgoella, a new genus of the Miliolidae:* Cushman Lab. Foram. Research, Contrib., v. 12, pt. 4, p. 90-91.

(541) ———, & Wickenden, R. T. D., 1928, *A new foraminiferal genus from the Upper Cretaceous:* Cushman Lab. Foram. Research, Contrib., v. 4, pt. 1, p. 12-13, pl. 1.

(542) **Cuvier, Georges A.**, 1817, *Le Règne Animal distribué d'apres son organisation, pour servir de base à l'histoire naturelle des animaux et d'introduction à l'anatomie comparée:* (a) v. 2, p. 359-378; (b) v. 4, Zoophytes, p. 1-255, Deterville (Paris).———(543) 1851, *The animal kingdom, arranged according to its organization, forming a natural history of animals and an introduction to comparative anatomy. With additions by W. B. Carpenter and J. O. Westwood:* 708 p. (London).

(544) **Cuvillier, Jean, & Szakall, V.**, 1949, *Foraminifères d'Aquitaine, Pt. 1. Reophacidae à Nonionidae:* Soc. Nat. Pétroles d'Aquitaine, 112 p., 32 pl. (Paris).

(545) **Czjžek, Johann**, 1848, *Beitrag zur Kenntniss der fossilen Foraminiferen des Wiener Beckens:* Haidinger's Naturwiss. Abhandl., v. 2, pt. 1, p. 137-150, pl. 12-13.———(546) 1849, *Über zwei neue Arten von Foraminiferen aus dem Tegel von Baden und Möllersdorf:* Freunde Naturwiss. Wien, Ber. Mitteil., v. 5 (1848-49), no. 6, p. 50-51.

(547) **Dabagyan, N. V., Myatlyuk, E. V., & Pishvanova, L. S.**, 1956, *Novye dannye po stratigrafii Tretichnykh otlozheniy Zakarpatya na osnovanii izucheniya fauny foraminifer:* Geol. Sbornik L'vovskogo Geol. ob.-va., no. 2-3, p. 220-236, pl. 1, 2. [*New data on the stratigraphy of the Tertiary deposits of the Carpathians on the basis of the study of the foraminiferal fauna.*]

(548) **Daday, Jenö**, 1883, *Adatok a Devai vizek faunájának ismeretéhez:* Orvos termesz. Értesitö, Kolozsvárt., v. 8, p. 197-228, pl. 5.

(549) **Dain, L. G.**, 1960, *Kratkiy obzor literatury po foraminiferam Yury za poslednie 15 let:* VNIGRI, Trudy pervogo seminara po mikrofaune, Leningrad, p. 188-206. [*Brief survey of literature on Jurassic Foraminifera of the last 15 years.*]

(550) ———, & Grozdilova, L., 1953, *Iskopaemye Foraminifery SSSR: Turneyellidy i Archedistsidy:* VNIGRI, Trudy, new ser., no. 74, 115 p., 11 pl. [*Fossil Foraminifera of the USSR: Tournayellidae and Archaediscidae.*]

(551) **Dalmatskaya, I. I.**, 1951, *Novyi rod fuzulinid iz nizhnei chasti Srednekamennougolnykh otlozhenii Russkoi Platformy:* Moskov. Obshch. Ispyt. Prirody, Trudy, Otdel Geol., v. 1, p. 194-196, pl. 1. [*New genus of fusulinid from the lower part of Middle Carboniferous deposits of the Russian Platform.*]

(552) **Dam, Abraham ten**, 1946, *Les espèces du genre de Foraminifères Quadratina, genre nouveau de la famille des Lagenidae:* Soc. géol. France, Bull. ser. 5, v. 16, p. 65-69.———(553) 1947, *Structure of Asterigerina and a new species:* Jour. Paleontology, v. 21, p. 584-586, text-fig. 1-6.———(554) 1948, *Observations sur le genre de Foraminifères Karreria Rzehak, 1891:* Soc. géol. France, Bull., ser. 5, v. 18, p. 285-288, pl. 13.———(555) 1948, *Les espèces du genre Epistomina Terquem, 1883:* Revue Inst. Français Pétrole & Ann. Comb. liquides, v. 3, no. 6, p. 161-170, pl. 1-2.———(556) 1948, *Cribroparella, a new genus of Foraminifera from the upper Miocene of Algeria:* Jour. Paleontology, v. 22, no. 4, p. 486-487, pl. 76.

(557) ———, & Reinhold, Th., 1942, *Some Foraminifera from the Lower Liassic and the Lower Oolitic of the eastern Netherlands:* Geologie Mijnbouw, v. 4, no. 1, p. 8-11, fig. 1-2.

(558) ———, & Schijfsma, Ernest, 1945, *Sur un genre nouveau de la famille des Lagenidae:* Soc. géol. France, Comptes Rendus, no. 16, p. 233-234.

(559) **Davies, L. M.**, 1927, *The Ranikot beds at Thal (North-West Frontier Provinces of India):* Geol. Soc. London, Quart. Jour., v. 83, p. 260-290, pl. 18-22.———(560) 1930, *The genus Dictyoconus and its allies: A review of the group, together with a description of three new species from the Lower Eocene beds of northern Baluchistan:* Royal Soc. Edinburgh, Trans., v. 56, p. 2, no. 20, p. 485-505, 2 pl., 9 text-fig.———(561) 1932, *The genera Dictyoconoides Nuttall, Lockhartia nov., and Rotalia Lamarck: Their type species, generic differences, and fundamental distinction from the Dictyoconus group of forms:* Same, v. 57, pt. 2, p. 397-428, pl. 1-4, text-fig. 1-10.———(562) 1939, *An early Dictyoconus, and the genus Orbitolina: their contemporaneity, structural distinction, and respective natural allies:* Same, v. 59, p. 773-790, pl. 1-2.

(563) ———, & Pinfold, E. S., 1937, *The Eocene beds of the Punjab Salt Range:* India Geol. Survey, Mem., Paleont. Indica, new ser., v. 24, p. 1-79, pl. 1-7, text-fig. 1-4.

(564) **Davis, A. G.**, 1951, *Howchinia bradyana (Howchin) and its distribution in the Lower Carboniferous of England:* Geologists' Assoc., Proc., v. 62, pt. 4, p. 248-253, pl. 10-11.

(565) **Dawson, G. M.**, 1870, *On Foraminifera from the Gulf and River St. Lawrence:* Canadian Nat., new ser., v. 5, p. 172-177.

(566) **Dawson, J. W.**, 1860, *Notice of Tertiary fossils from Labrador, Maine, etc., and remarks on the climate of Canada in the Newer Pliocene or Pleistocene period:* Canadian Nat., v. 5, p. 188-200, text-fig. 1-5.

(567) **Debourle, André**, 1955, *Cuvillierina eocenica, nouveau genre et nouvelle espèce de foraminifère de l'Yprésien d'Aquitaine:* (a) Soc. géol. France, Comptes Rendus Somm., no. 2, p. 19; [Also (b) Soc. géol. France, Bull., ser. 6, v. 5, p. 55-57, pl. 2.]

(568) **Deecke, W.**, 1884, *Die Foraminiferenfauna der zone des Stephanoceras humphriesianum im Unter-Elsass:* Geol. Spezialk., Elsass-Lothringen, Abhandl., v. 4, no. 1, p. 3-68, pl. 1-2.

(569) **Deflandre, Georges**, 1928, *Le genre Arcella Ehrenberg. Morphologie-Biologie, Essai phylogénétique et systématique:* Archiv Protistenkunde, v. 64, p. 152-287.———(570) 1928, *Deux genres nouveaux de Rhizopodes testacés:* Ann. Protistologie, v. 1, p. 37-43, text-fig. 1-13.———(571) 1928, *A propos du genre "Arcella" Ehr.:* Same, v. 1, p. 198.——— (572) 1929, *Le genre Centropyxis Stein:* Archiv Protistenkunde, v. 67, p. 322-375, text-fig. 1-176.——
(573) 1932, *Paraquadrula nov. gen. irregularis (Archer) conjugaison et enkystement:* Société de Biologie, Comptes Rendus Hebdomadaires des Séances & Mém., v. 109, p. 1346-1347.———(574) 1934, *Sur un foraminifère siliceux fossile des diatomites miocènes de Californie: Silicotextulina diatomitarum n. g. n. sp.:* Acad. Sci. Paris, Comptes Rendus, v. 198, p. 1446-1448.———(575) 1936, *Remarques sur le comportement des pseudopodes chez quelques Thécamoebiens:* Ann. Protistologie, v. 5, p. 65-71, text-fig. 1-34.———(576) 1936, *Étude monographique sur le genre Nebela Leidy (Rhizopoda-Testacea):* Same, v. 5, p. 201-286, pl. 10-27, text-fig. 1-161.

(577) ———, & **Deflandre-Rigaud, Marthe**, 1959, *Difflugia? marina Bailey, une espèce oubliée synonyme de Quadrulella symmetrica (Wallich), Rhizopode testacé d'eau douce:* Hydrobiologia, v. 12, p. 299-307, fig. 1-9.

(578) **Deflandre-Rigaud, Marthe**, 1958, *Annexe IV. Index systématique* in DEFLANDRE, G., *Eugène Penard (1855-1954) Correspondance et souvenirs. Bibliographie et bilan systématique de son oeuvre:* Hydrobiologia, v. 10, p. 20-37.

(579) **Defrance, M. J. L.**, 1820-28, *Dictionnaire des Sciences Naturelles:* (a) v. 16, p. 1-567 (1820); (b) v. 24 (1822); (c) v. 25, p. 1-485 (1822); (d) v. 26, p. 1-555 (1823); (e) v. 32, p. 1-567 (1824); (f) v. 35, p. 1-534 (1825); (g) v. 53, p. 1-508 (1828). F. G. Levrault (Paris).

(580) **Delage, Yves**, & **Hérouard, Edgard**, 1896, *Traité de Zoologie Concrète. Tome I. La Cellule et les Protozoaires:* 584 p., 868 text-fig. (Paris).

(580A) **Deleau, P.**, & **Marie, Pierre**, 1959[1961], *Les Fusulinidés du Westphalien C du Bassin d'Abadla et quelques autres Foraminifères du Carbonifère algérien (Région de Colomb-Béchar):* Travaux des Collaborateurs, Publications du Service de la Carte Géologique de l'Algérie, new ser., Bull. 25, p. 43-160, pl. 1-12, for 1958 (Alger).

(581) **Delmas, M.**, & **Deloffre, R.**, 1961, *Découverte d'un nouveau genre d'Orbitolinidae dans la base de l'Albien en Aquitaine:* Revue Micropaléont., v. 4, no. 3, p. 167-172, pl. 1.

(582) **Deloffre, R.**, 1961, *Sur la découverte d'un nouveau Lituolidé du Crétacé inférieur des Basses-Pyrénées: Pseudochoffatella cuvillieri n. gen., n.sp.:* Revue Micropaléont., v. 4, no. 2, p. 105-107, pl. 1.

(583) **Deprat, J.**, 1905, *Les dépots Éocène néo-Calédoniens:* Soc. géol. France, Bull., ser. 4, v. 5, pt. 5, p. 485-516, pl. 16-19, text-fig. A-G.———(584) 1912, *Étude géologique du Yun-Nan Oriental, Pt. 3. Etude des Fusulinidés de Chine et d'Indochine et classification des calcaires à fusulines:* Service géol. Indochine, Mém., v. 1, pt. 3, 76 p., 9 pl., 30 fig.———(585) 1912, *Sur deux genres nouveaux de Fusulinidés de l'Asie orientale, intéressants au point de vue phylogénique:* Acad. Sci. Paris, Comptes Rendus, v. 154, p. 1548-1550.———(586) 1913, *Étude des Fusulinidés de Chine et d'Indochine. Les Fusulinidés des calcaires carbonifériens et permiens du Tonkin, du Laos et du Nord-Annam:* Service géol. Indochine, Mém., v. 2, pt. 1, 74 p., 10 pl., 25 fig.———(587) 1914, *Étude des Fusulinidés du Japon, de Chine et d'Indochine. Etude comparative des Fusulinidés d'Akasaka (Japon) et des Fusulinidés de Chine et d'Indochine:* Same, v. 3, pt. 1, Mém. 3, 45 p., 8 pl., 8 fig.———(587A) 1915, *Étude des Fusulinidés de Chine et d'Indochine et classification des calcaires à fusulines (IVe Mémoire). Les Fusulinidés des calcaires carbonifériens et permiens du Tonkin, du Laos et du Nord-Annam:* Same, v. 4, pt. 1, p. 1-30, pl. 1-3.

(588) **Dervieux, Ermanno**, 1894, *Osservazioni sopra le Tinoporinae e descrizione del nuovo genere Flabelliporus:* R. Accad. Sci. Torino, Atti, v. 29, p. 57-61, pl. 1.

(589) **Derville, Henry**, 1950, *Contribution à l'ètude des calcisphères du calcaire de Bachant:* Soc. géol. Nord, Ann., v. 70, p. 273-283.

(590) **Deshayes, G. P.**, 1830, *Encyclopédie métho-*

dique. Histoire naturelle des Vers: v. 2 (with suppl.), 594 p., Mme. V. Agasse (Paris).

(591) **Desjardins, Felix,** 1835, *Observations nouvelles sur les Céphalopodes microscopiques:* Ann. Sci. Nat., ser. 2, v. 3, p. 108-109.

(592) **Dick, A. B.,** 1928, *On needles of rutile in the test of Bathysiphon argenteus:* Edinburgh Geol. Soc., Trans., v. 12, p. 19-21, pl. 4.

(593) **Didkovskiy, V. Ya.,** 1957, *O novom predstavitele semeystva Miliolidae-Tortonella bondartschuki gen. et sp. nov. iz Tortonskikh otlozheniy USSR:* Akad. Nauk SSSR, Doklady, v. 113, no. 5, p. 1137-1139, text-fig. 1-3. [*On a new representative of the family Miliolidae-Tortonella bondartschuki, gen. et sp. nov., from Tortonian deposits of the Ukraine.*]———(594) 1958, *Noviy predstavnik peneroplid Neoperoplis sarmaticus gen. et sp. nov. z Serednosarmats'kikh vidkladiv Ukrayni ta Moldaviy:* Akad. Nauk Ukrain RSR, Kiev, Dopovidi, v. 11, p. 1251-1254, pl. [*A new representative of the Peneroplidae, Neoperoplis sarmaticus gen. et sp. nov., from middle Sarmatian deposits of the Ukraine and Moldavia.*]———(595) 1960, *Pro novogo predstavnika rodini Miliolidae-Flintinella volhynica gen. et sp. n. z Seredn'osarmatskikh vidkladiv URSR:* Same, no. 10, p. 1432-1435, text-fig. 1-4. [*On a new representative genus of the Miliolidae-Flintinella volhynica gen. et sp. nov. from middle Sarmatian deposits of the Ukraine.*]

(596) **Diesing, C. M.,** 1848, *Systematische Uebersicht der Foraminiferen monostegia und Bryozoa anopisthia:* K. Akad. Wiss. Wien, Sitzungsber., v. 1, p. 494-527.

(597) **Dietrich, W. O.,** 1935, *Zur Stratigraphie der kolumbianischen Ostkordillere:* Zentralbl. Mineral., Geol. & Paläont., Jahrgang 1935, pt. B, p. 74-82, text-fig. 1-8.

(598) **Döderlein, L.,** 1892, in *"Demonstrationen":* Deutsch. Zool. Gesell., Verhandl., v. 2, p. 143-146.

(599) **Doflein, Franz,** 1901, *Die Protozoen als Parasiten und Krankheitserrager nach biologischen Gesichtspunkten dargestellt:* 274 p., 220 text-fig. (Jena).———(600) 1902, *Das System der Protozoen:* Archiv Protistenkunde, v. 1, p. 169-192.———(601) 1909, *Lehrbuch der Protozoenkunde eine Darstellung der Naturgeschichte der Protozoen, mit besonderer Berücksichtigung der parasitischen und pathogenen Formen:* ed. 2, 914 p., 825 fig., G. Fischer (Jena).———(602) 1911, *Lehrbuch der Protozoenkunde:* ed. 3, xii+1043 p., 951 fig. (Jena).———(603) 1916, *Lehrbuch der Protozoenkunde, Eine Darstellung der Naturgeschichte der Protozoen mit besonderer Berücksichtigung der parasitischen und pathogenen Formen:* ed. 4, 1190 p., 1198 fig.

(604) ———, & **Reichenow, Eduard,** 1929, *Lehrbuch der Protozoenkunde:* ed. 5, 1262 p., 1201 fig. (Jena).———(605) 1952, *Lehrbuch der Protozoenkunde. ed. 6, Pt. 2 Spezielle Naturgeschichte der Protozoen. Hälfte 1: Mastigophoren und Rhizopoden:* p. i-iv, 411-776, 393 fig. (Jena).

(606) **Dogel, V. A.,** 1951, *Obshchaya Protistologiya:* Gosudarstvennoe izdatelstvo Sovetskaya Nauka, p. 1-603, text-fig. 1-322 (Moscow). [General Protistology.]

(607) **Dollfus, G. F.,** 1889, *Foraminifères:* Annuaire Geol. Universal, ann. 1888, v. 5, p. 1217-1231.

(608) **Donceiux, Louis,** 1905, *Catalogue descriptif des fossiles nummulitiques de l'Aude et de L'Hérault; Première partie—Montagne Noire et Minervois:* Lyon Univ., Ann., new ser. 1 (Sci.-Méd.), pt. 17, p. 1-128, pl. 1-5, text-fig. 1-3.

(609) **Dons, Carl,** 1942, *Craterella albescens, n. gen., n.sp., ein neuer Foraminifer:* K. Norske Vidensk. Selsk., Forhandl., v. 14 (1941), no. 36, p. 136.

(610) **Dorreen, J. M.,** 1948, *A foraminiferal fauna from the Kaiatan Stage (upper Eocene) of New Zealand:* Jour. Paleontology, v. 22, p. 281-300, pl. 36-41.

(611) **Douglass, R. C.,** 1960, *Revision of the family Orbitolinidae:* Micropaleontology, v. 6, p. 249-270, pl. 1-6, fig. 1-3.———(612) 1960, *The foraminiferal genus Orbitolina in North America:* U.S. Geol. Survey, Prof. Paper 333, 52 p., 17 pl., 32 text-fig.

(613) **Douvillé, Henri,** 1898, *Sur l'âge des couches traversées par le canal de Panama:* Soc. géol. France, Bull., ser. 3, v. 26, pt. 6, p. 587-600.———(614) 1902, *Essai d'une revision des Orbitolites:* Same, ser. 4, v. 2, p. 289-306, pl. 9-10.———(615) 1905, *Les Foraminifères dans le Tertiaire de Bornéo:* Same, ser. 4, v. 5, pt. 4, p. 435-464, pl. 14, text-fig. 1-2.———(616) 1906, *Sur la structure du test dans les Fusulinés:* Acad. Sci. Paris, Comptes Rendus, v. 143, p. 258-261.———(617) 1906[1907], *Les calcaires à fusulines de l'Indo-Chine:* Soc. géol. France, Bull., ser. 4, v. 6 (1906), pt. 7, p. 576-587, pl. 17-18, fig. 1-10.———(618) 1906[1907], *Evolution et enchaînements des Foraminifères:* Same, ser. 4, v. 6 (1906), pt. 7, p. 588-602, pl. 18, fig. 11-13.———(619) 1910, *La Craie et le Tertiaire des environs de Royan:* Same, ser. 4, v. 10, p. 51-61.———(620) 1911, *Les Foraminifères dans le Tertiaire des Philippines:* Philippine Jour. Sci., v. 6, p. 53-80, pl. A-D, text-fig. 1-9.———(621) 1915, *Les Orbitoïdes: développement et phase embryonnaire; leur évolution pendant le Crétacé:* Acad. Sci. Paris, Comptes Rendus, v. 161, p. 664-670, text-fig.———(622) 1915, *Les Orbitoïdes du Danien et de tertiaire: Orthophragmina et Lepidocyclina:* Same, v. 161, p. 721-728, text-fig.———(623)

1917, *Les Orbitoïdes de l'île de la Trinité:* Same, v. 164, p. 841-847, text-fig.——(**624**) 1922, *Orbitoïdes de la Jamaïque. Pseudorbitoides Trechmanni, nov.gen., nov. sp.:* Soc. géol. France, Comptes Rendus, Somm., no. 17, p. 203-204, text-fig. 1.——(**625**) 1922, *Les Lépidocyclines et leur évolution: un genre nouveau "Amphilepidina":* Acad. Sci. Paris, Comptes Rendus, v. 175, p. 550-555.——(**626**) 1923, *Les Orbitoïdes en Amérique:* Soc. géol. France, Comptes Rendus, Somm., no. 10, p. 106-107.——(**627**) 1924, *Les Orbitoïdes et leur évolution en Amérique:* Soc. géol. France, Bull., ser. 4, v. 23 (1923), pt. 7-8, p. 369-376, pl. 13.——(**628**) 1924-1925, *Revision des lépidocyclines:* Soc. géol. France, Mém. 2, new ser., v. 1, p. 1-50, pl. 1-2, 47 text-fig.; v. 2, p. 51-115, pl. 3-7, 25 text-fig.——(**628A**) 1927, *Les Orbitoïdes de la région pétrolifère du Mexique:* Soc. géol. France, Comptes Rendus, Somm., no. 4, p. 34-35.——(**629**) 1930, *Une Miliolideé géante du Sénonien du Maroc Lacazopsis termieri:* Soc. géol. France, Bull., ser. 4, v. 29, no. 3-5 (1929), p. 245-250, pl. 21, text-fig. 7-8.

(**630**) Drooger, C. W., 1952, *Study of American Miogypsinidae:* 80 p., 18 text-fig., 1 table, Vonk & Co. (Zeist).——(**631**) 1960, *Some early rotaliid Foraminifera;* (a) I, K. Nederland. Akad. Wetensch., Proc., ser. B, no. 3, p. 287-301, pl. 1-2; (b) II, p. 302-318, pl. 3-5.

(**632**) Dujardin, Félix, 1835, *Observations sur les Rhizopodes et les Infusoires:* Acad. Sci. Paris, Comptes Rendus, v. 1, p. 338-340.——(**633**) 1835, *Observations nouvelles sur les prétendus Céphalopodes microscopiques:* Ann. Sci. Nat., ser. 2, v. 3, p. 312-314.——(**634**) 1835-36, *Recherches sur les organismes inférieurs:* Same, ser. 2, v. 4; (a) p. 343-376, pl. 9-11 (1835); (b) p. 193-205, pl. 9 (1836).——(**635**) 1840, *Mémoire sur une classification des Infusoires en rapport avec leur organization:* Acad. Sci. Paris, Comptes Rendus, v. 11, p. 281-286.——(**636**) 1841, *Histoire naturelle des Zoophytes-Infusoires:* 648 p., atlas, 22 pl. (Paris).——(**637**) 1852, *Note sur les Infusoires vivant dans les mousses et dans les Jungermannes humides:* Ann. Sci. Nat., Zool., ser. 3, v. 18, p. 240-242.

(**637A**) Dunbar, C. O., 1933, *Fusulinidae:* in CUSHMAN, J. A., *Foraminifera, their classification and economic use:* ed. 2, p. 126-140, Cushman Lab. Foram. Research, spec. publ. 4 (Sharon, Mass.).——(**637B**) 1940, *Fusulinidae:* in CUSHMAN, J. A., *Foraminifera, their classification and economic use,* ed. 3, p. 132-156, Harvard Univ. Press (Cambridge, Mass.).——
(**638**) 1944, *Pt. 2, Permian and Pennsylvanian (?) fusulines* in KING, R. E., DUNBAR, C. O., CLOUD, P. E., JR., & MILLER, A. K., Geology and paleontology of the Permian area northwest of Las Delicias, southwestern Coahuila, Mexico; Geol. Soc. America, Spec. Paper 52, p. 35-48, p. 9-16.——(**639**) 1946, *Parafusulina from the Permian of Alaska:* Am. Museum Nat. History, Novitates, no. 1325, 4 p., 9 fig.

(**639A**) ——, Baker, A. A·, et al., 1960, *Correlation of the Permian formations of North America:* Geol. Soc. America, Bull., v. 71, p. 1763-1806, 1 pl., 2 fig.

(**640**) ——, & Condra, G. E., 1927[1928], *The Fusulinidae of the Pennsylvanian System in Nebraska:* Nebraska Geol. Surv., Bull. 2, (1927), ser. 2, 135 p., 15 pl., 13 fig.

(**641**) ——, & Henbest, L. G., 1930, *The fusulinid genera Fusulina, Fusulinella and Wedekindella:* Am. Jour. Sci., ser. 5, v. 20, p. 357-364, text-fig. 1.——(**642**) 1931, *Wedekindia, a new fusulinid name:* Same, ser. 5, v. 21, p. 458.——(**643**) 1942, *Pennsylvanian Fusulinidae of Illinois:* Illinois Geol. Surv., Bull. 67, 218 p., 23 pl., 13 fig.

(**644**) ——, & Skinner, J. W., 1931, *New fusulinid genera from the Permian of West Texas:* Am. Jour. Sci., ser. 5, v. 22, p. 252-268, pl. 1-3.——(**645**) 1936, *Schwagerina versus Pseudoschwagerina and Paraschwagerina:* Jour. Paleontology, v. 10, p. 83-91, pl. 10-11.——(**646**) 1937, *Pt. 2, Permian Fusulinidae of Texas:* Univ. Texas, Bull. 3701, v. 3, p. 517-825, pl. 42-81, fig. 89-97.

(**647**) ——, & King, R. E., 1936, *Dimorphism in Permian fusulines:* Univ. Texas, Bull. 3501, p. 173-190, pl. 1-3, fig. 30.

(**647A**) ——, Troelsen, John, Ross, C. A., Ross, J. P., & Norford, Brian, 1962, *Faunas and correlation of the late Paleozoic rocks of northeast Greenland. Part I. General discussion and summary. Part II.* (C. A. Ross & DUNBAR) *Fusulinidae:* Meddel. Grønland, Kommissionen Videnskabelige Undersøgelser I Grønland, pt. 1, v. 167, no. 4, 16 p., 4 fig.; pt. 2, v. 167, no. 5, 55 p., 7 pl.

(**648**) Dunn, P. H., 1942, *Silurian Foraminifera of the Mississippi Basin:* Jour. Paleontology, v. 16, p. 317-342, pl. 42-44.

(**649**) Durkina, A. V., 1959, *Foraminifery Nizhnekamennougolnykh otlozheniy Timano-Pechorskoy probintsii:* VNIGRI, Trudy, no. 136, Mikrofauna SSSR, Sbornik 10, p. 132-335, pl. 1-27, text-fig. 1-7. [*Foraminifera of Lower Carboniferous deposits of the Timano-Pechorsky province.*]

(**650**) Dutkevich, G. A., 1934, *Permskaya fauna fusulinid naydennaya v razrezakh Kara-Su i Kubergandy na vostochnom Pamire* in DUTKEVICH, G. A. & KHABAKOV, A. V., *Permskie otlozheniya vostochnogo Pamira i paleogeografiya verkhnego paleozoya Tsentralnoy Azii:* Akad. Nauk SSSR, Trudy, Tadzh. Kompleks eksped. 1932, Geol., no. 8, p. 53-112, pl. 1-3. [*Permian fusulinid fauna found in sections of Kara-Su* *and Kubergandy in eastern Pamir: in* Permian deposits of eastern Pamir and paleogeography of upper Paleozoic of Central Asia.]

(650A) **Dyhrenfurth, Günter,** 1909, *Die asiatischer Fusulinen,* in SCHELLWIEN, E., *Monographie der Fusulinen,* Teil II: Palaeontographica, v. 56, p. 137-176, pl. 13-16, text-fig. 1-10.

(651) **Eames, F. E., Banner, F. T., Blow, W. H., & Clark, W. J.,** 1962, *Fundamentals of Mid-Tertiary stratigraphical correlation:* viii+163 p., 17 pl., 20 fig., University Press (Cambridge).

(652) **Earland, Arthur,** 1933, *Foraminifera, Part II. South Georgia:* Discovery Repts., v. 7, p. 27-138, pl. 1-7.——(653) 1934, *Foraminifera, Part III. The Falklands sector of the Antarctic (excluding South Georgia):* Same, v. 10, p. 1-208, pl. 1-10.

(654) **Easton, W. H.,** 1960, *Invertebrate paleontology:* 701 p., illus., Harper & Brothers (New York).

(654A) **Ebensberger, Hans,** 1962, *Stratigraphische und mikropaläontologische Untersuchungen in der Aachener Oberkreide, besonders der Maastricht-Stufe:* Palaeontographica, v. 120, pt. A, no. 1-3, p. 1-20, text-fig. 1-19, pl. 1-12.

(655) **Edmondson, C. H.,** 1906, *The Protozoa of Iowa:* Davenport Acad. Sci., Proc., v. 11, p. 1-124.

(656) **Edmondson, W. T.,** 1959, in WARD & WHIPPLE, Freshwater Biology: ed. 2, 1248 p., illus.

(657) **Egger, J. G.,** 1857, *Die Foraminiferen der Miocän-Schichten bei Ortenburg in Nieder-Bayern:* Neues Jahrb. Mineral. Geogn. Geol. Petref., p. 266-311, pl. 5-15.——(658) 1893, *Foraminiferen aus Meeresgrundproben, gelothet von 1874 bis 1876 von S. M. Sch. Gazelle:* K. Bayer, Akad. Wiss., München, Math.-Phys. Cl., Abhandl., v. 18, pt. 2, p. 193-458, pl. 1-21.——(659) 1899-1902, *Foraminiferen und Ostrakoden aus den Kreidemergeln der Oberbayerischen Alpen:* Same, Abhandl., v. 21, pt. 1, p. 1-230, pl. 1-27 (1902).——(660) 1902, *Der Bau der Orbitolinen und verwandter Formen:* Same, Abhandl., v. 21, pt. 3, p. 575-600, pl. 1-6.——(661) 1902, *Ergänzungen zum Studium der Foraminiferenfamilie der Orbitoliniden:* Same, Abhandl., v. 21, no. 3, p. 673-682, pl. A-B.——(662) 1909, *Foraminiferen der Seewener Kreideschichten:* K. Bayer. Akad. Wiss. München, Math.-Phys. Kl., Sitzber., p. 3-52, pl. 1-6.

(663) **Ehrenberg, C. G.,** 1830, *Organization, Systematik und Geographische Verhältniss der Infusions-thierchen:* Folio (Berlin).——(664) 1832, *Beiträge zur Kenntniss der Organisation des Infusorien und ihrer geographischen Verbreitung, besonders in Sibirien:* K. Preuss. Akad. Wiss. Berlin, Abhandl. (1830), p. 1-88, pl. 1-8.——(665) 1834, *Dritter Beitrag zur Erkenntniss grosser Organisation in der Richtung des kleinsten Raumes:* Same, Abhandl. (1883), p. 145-336, pl. 1-11.——(666) 1837, *Zusätze zur Erkenntniss grosser organischer Ausbildung in den kleinsten Thierischen Organismen:* Same, Abhandl. (1835), p. 151-180, pl. 1.—— (666A) 1838, *Über dem blossen Auge unsichtbare Kalkthierchen und Kieselthierchen als Hauptbestandtheile der Kreidegebirge:* Same, Ber., Jahrg. 1838, v. 3, p 192-200.—— (667) 1839[1840], *Über die Bildung der Kreidefelsen und des Kreidemergels durch unsichtbare Organismen:* Same, Abhandl. (1838), p. 59-147, pl. 1-4, 2 tables.——(668) 1839, *Die Infusionthierchen als vollkommene Organismen:* 547 p., atlas, 64 pl., L. Voss (Leipzig).—— (669) 1840, *Das grössere Infusorienwerke:* K. Preuss Akad. Wiss. Berlin, Ber. (1840), p. 198-219.——(669A) 1842, *Der Bergkalk am Onega-See aus Polythalamien bestehend:* Same, p. 273-275.——(670) 1843, *Über den sichtlichen Einfluss der mikroskopischen Meeres-Organismen auf den Boden des Elbbettes bis oberhalb Hamburg:* Same, p. 160-167. (671) 1843, *Beobachtungen über die Verbreitung des jetzt wirkenden kleinsten organischen Lebens in Asien, Australien und Afrika und über die vorherrschende Bildung auch des Oolithkalkes der Juraformation aus kleinen polythalamischen Thieren:* K. Preuss. Akad. Wiss. Berlin, Verhandl., Ber., p. 101-106.—— (672) 1843, *Verbreitung und Einfluss des Mikroskopischen Lebens in Süd- und Nord-Amerika:* K. Preuss. Akad. Wiss. Berlin, Abhandl. (1841), pt. 1, p. 291-446, pl. 1-4. ——(673) 1844, *Eine Mittheilung über 2 neue Lager von Gebirgsmassen aus Infusorien als Meeres-Absatz in Nord-Amerika und eine Vergleichung derselben mit den organischen Kreide-Gebilden in Europa und Afrika:* K. Preuss. Akad. Wiss. Berlin, Ber., p. 57-98.—— (674) 1844, *Ueber Spirobotrys, eine neue physiologisch merkwürdige Gattung von Polythalamien:* Same, p. 245-248.——(675) 1845, *Ueber das kleinste organische Leben an mehreren bisher nicht untersuchten Erdpunkten, mikroskopische Lebensformen von Portugal und Spanien, Sud-Afrika, Hinter-Indien, Japan und Kurdistan:* Same, Ber., p. 357-381.——(676) 1948, *Über eigenthümliche auf den Bäumen des Urwaldes in Süd-Amerika zahlreich lebende mikroskopische oft kieselschalige Organismen:* K. Preuss. Akad. Wiss. Berlin, Monatsber. p. 213-220.——(677) 1948, *Fortgesetzte Beobachtungen über jetzt herrschende atmosphärische mikroskopische Verhältnisse:* K. Preuss. Akad. Wiss. Berlin, Ber., Monatsber., p. 370-381.——(678) 1854, *Das organische Leben des Meeresgrundes; Weitere Mittheilung über die aus grossen Meerestiefen gehobenen Grund-Massen; Charakteristik der neuen mikroskopischen Organismen des tiefen atlantischen Oceans:*

K. Preuss. Akad. Wiss. Berlin, Ber., p. 235-251.
———(679) 1854, *Beitrag zur Kentniss der Natur und Entstehung des Grünsandes; Weitere Mittheilungen über die Natur und Entstehung des Grünsandes:* Same, p. 374-377, 384-410.
———(680) 1854, *Mikrogeologie:* 374 p., 40 pl., L. Voss (Leipzig).———(681) 1855, *Über neue Erkenntniss immer grösser Organisation der Polythalamien durch deren urweltliche Steinkerne:* K. Preuss Akad. Wiss., Berlin, Ber., p. 272-290.———(682) 1856, *Über den Grünsand und seine Erläuterung des organischen Lebens:* K. Akad. Wiss. Berlin, Physik. Abhandl. (1855), p. 85-176, pl. 1-7.———(683) 1858, *Kurze Characteristik der 9 neuen Genera und der 105 neuen species des ägäischen Meeres und des Tiefgrundes des Mittel-Meeres:* K. Preuss. Akad. Wiss. Berlin, Monatsber., p. 10-40.———(684) 1858, *Fortschreitende Erkenntniss massenhafter mikroskopischer Lebensformen in den untersten silurischen Thonschichten bei Petersburg; weitere Mittheilungen über andere massenhafte mikroskopische Lebensformen der ältesten silurischen Grauwachen-Thone bei Petersburg:* K. Preuss. Akad. Wiss. Berlin, Phys.-Math. Kl., Monatsber., p. 295-311, 324-337, pl. 1.———(685) 1861, *Elemente des tiefen Meeresgrundes im Mexikanischen Golfstrome bei Florida; über die Tiefgrund-Verhältnisse des Oceans am Eingange der Davisstrasse und bei Island:* K. Preuss Akad. Wiss. Berlin, Monatsber., p. 275-315.———(686) 1866, *Die mikroskopischen Lebensformen auf der Insel St. Paul:* Novara Exped. 1857-59, v. 2, geol. pt., p. 71-82.———(687) 1872, *Mikrogeologische Studien als zusammenfassung seiner Beobachtungen des kleinsten Lebens der Meeres-Tiefgründe aller zonen und dessen geologischen Einfluss:* K. Preuss. Akad. Wiss., Monatsber., p. 265-322.———(688) 1872, *Nachtrag zur übersicht der Organischen Atmosphärillien. Systematische und Geographische Studien über die Arcellinen:* K. Akad. Wiss. Berlin, Abhandl. (1871), p. 233-275, pl. 3.———(689) 1873, *Mikrogeologische Studien über das kleinste Leben der Meeres-Tiefgründe aller Zonen und dessen geologischen Einfluss:* Same, Abhandl. (1872), p. 131-397, pl. 1-12.

(690) **Eicher, D. L.**, 1960, *Stratigraphy and micropaleontology of the Thermopolis shale:* Peabody Museum Nat. History, Yale Univ., Bull. 15, 126 p., 6 pl.

(691) **Eichwald, Eduard von**, 1860, *Lethaea Rossica ou Paléontologie de la Russie, Prémier section de l'ancienne période:* v. 1, xix+681 p., atlas, 59 pl. (1859), E. Schweizerbart (Stuttgart).

(692) **Eimer, G. H. T., & Fickert, C.**, 1899, *Die Artbildung und Verwandtschaft bei den Foraminiferen, Entwurf einer natürlichen Eintheilung derselben:* Zeitschr. Wiss. Zool., v. 65, no. 4, p. 527-636, text-fig. 1-45.

(693) **Eisenack, Alfred**, 1932-38, *Neue Mikrofossilien des baltischen Silurs;* Paläont. Zeitschr., (a) II, v. 14, p. 257-277 (1932); (b) IV, v. 19, no. 3-4, p. 233-243, pl. 15-16, text-fig. 8-22 (1938).———(694) 1954, *Foraminiferen aus dem baltischen Silur.:* Senckenbergiana Lethaea, v. 35, no. 1-2, p. 51-72, pl. 1-5.———(695) 1959, *Chitinöse Hüllen aus Silur und Jura des Baltikums als Foraminiferen:* Paläont. Zeitschr., v. 33, no. 1-2, p. 90-95, pl. 9, text-fig. 1.

(696) **Elias, M. K.**, 1950, *Paleozoic Ptychocladia and related Foraminifera:* Jour. Paleontology, v. 24, p. 287-306, pl. 43-45, 2 text-fig.———(697) 1954, *Cambroporella and Coeloclema, Lower Cambrian and Ordovician bryozoans:* Same, v. 28, p. 52-58, pl. 9-10.

(698) **Elliott, G. F.**, 1958, *Fossil microproblematica from the Middle East:* Micropaleontology, v. 4, no. 4, p. 419-428, pl. 1-3.

(699) **Ellis, B. F.**, 1932, *Gallowayina browni, a new genus and species of orbitoid from Cuba, with notes on the American occurrence of Omphalocyclus macropora:* Am. Museum Nat. History, Novitates, no. 568, p. 1-8, illus.

(700) ———, **& Messina, Angelina**, 1940, *Catalogue of Foraminifera:* Am. Museum Nat. History (supplements, post-1940).

(701) **Emberger, Jacques; Magné, Jean; Reyre, Dominique; & Sigal, Jacques**, 1955, *Note préliminaire sur quelques Foraminifères nouveaux ou peu connus dans le Crétacé superieur de faciès sub-récifal d'Algérie:* Soc. Géol. France, Comptes Rendus, somm. séances, p. 110-114.

(702) **Emiliani, Cesare**, 1951, *On the species Homotrema rubrum (Lamarck):* Cushman Found. Foram. Research, Contrib., v. 2, pt. 4, p. 143-147, pl. 15-16.———(703) 1954, *Depth habitats of some species of pelagic Foraminifera as indicated by oxygen isotope ratios:* Am. Jour. Sci., v. 252, p. 149-158, text-fig. 1-4, tables 1-6.———(704) 1955, *Mineralogical and chemical composition of the tests of certain pelagic Foraminifera:* Micropaleontology, v. 1, p. 377-380, text-fig. 1-3, table 1-4.

(705) **Emory, W. H.**, 1857, *Report on the United States and Mexican boundary survey, made under the direction of the Secretary of the Interior:* U.S. 34th Congr. Sess. 1, Senate Exec. Doc. 108 & House Exec. Doc. 135, v. 1, pt. 2, p. 141-174, pl. 1-21.

(706) **Engler, Adolf, & Prantl, K.**, 1928, *Myxomycetes* in *Die natürlichen Pflanzenfamilien:* ed. 2, v. 2, p. 304-339.

(707) **Epsteyn, G. V.**, 1926, *Testamoeba hominis n.g., n.sp., novye dannye k poznaniiu kishechnykh Prosteyshikh cheloveka:* Russkiy Arkhiv Protistologiy, v. 5, no. 3-4, p. 181-204, pl. 12-

16, text-fig. 1-3. [German summary, p. 204-209.] [*Testamoeba hominis* n.g., n.sp., new data on knowledge of human intestinal Protozoa.]

(708) **Ericson, D. B., Wollin, Goesta, & Wollin, Janet,** 1954, *Coiling direction of Globorotalia truncatulinoides in deep-sea cores:* Deep-Sea Research, v. 2, p. 152-158, pl. 1, fig. 2-4.

(709) **Erk, A. S.,** 1941[1942], *Sur la présence du genre Codonofusiella Dunb. et Skin. dans le Permien de Bursa (Turquie):* Eclogae geol. Helv., v. 34 (1941), p. 243-253, pl. 12-14.

(710) **Erman, Adolph,** 1855, *Einige palaeographische und zoologische Beobachtungen während der Reise von Kamtschatka nach Europa, II. Ueber einige bisher nicht beachtete Tertiär-Gesteine aus der Umgegend von Rio de Janeiro:* Erman's Archiv Wiss. Russland, v. 14, p. 144-161, pl. 1 (Berlin).

(711) **Etheridge, Robert, Jr.,** 1873, *Notes on certain genera and species mentioned in the foregoing lists:* Scotland Geol. Survey, Mem., Explan. Sheet 23, appendix 2, p. 93-107.

(711A) **Farinacci, Anna,** 1962, *Nuovo genere di Verneuilinidae (Foraminifera) marker di zona del Senoniano:* Geol. Romana, v. 1, p. 5-10, pl. 1-5.

(712) **Faujas de Saint-Fond, Barthélemy,** 1799, *Histoire naturelle de la montagne de Saint-Pierre de Maestricht:* 263 p., 54 pl. (Paris).

(713) **Fauré-Fremiet, E.,** 1911, *La constitution du test chez les Foraminifères arenacés:* Bull. Inst. Océanog. Monaco, no. 216, p. 1-7.

(714) **Feray, D. E.,** 1941, *Siphonides, a new genus of Foraminifera:* Jour. Paleontology, v. 15, p. 174-175, text-fig. 1-4.

(715) **Fernández Galiano, Emilio,** 1921, *Morfología y biologia de los protozoos:* 266 p., 152 fig. (Madrid).

(716) **Fichtel, Leopold von, & Moll, J. P. C., von,** 1798,*Testacea microscopica, aliaque minuta ex generibus Argonauta et Nautilus, ad naturam picta et descripta (Microspische und andere Klein Schalthiere aus den Geschlechtern Argonaute und Schiffer):* vii+123 p., 24 pl., Camesina (Wien). [Reprinted 1803.]

(717) **Finlay, H. J.,** 1939-47, *New Zealand Foraminifera, Key species in stratigraphy:* (a-d) Royal Soc. New Zealand, Trans.; (e) New Zealand Jour. Sci. Tech; (a) v. 68, p. 504-543, pl. 68-69 (1939); (b) v. 69, pt. 1, p. 89-128, pl. 11-14 (1939); (c) v. 69, pt. 3, p. 309-329, pl. 24-29 (1939); (d) v. 69, pt. 4, p. 448-472, pl. 62-67 (1940); (e) v. 28, no. 5, sec. B, p. 259-292, pl. 1-9 (1947).

(718) **Fischer, A. G.,** 1962, *Fossilien aus Riffkomplexen der alpinen Trias: Cheilosporites Wähner, eine Foraminifere?:* Paläont. Zeitschr., v. 36, no. 1-2, p. 118-124, pl. 13-14.

(719) **Fischer, W. A.,** 1954, *The Foraminifera and stratigraphy of the Colorado group in central and eastern Colorado:* Univ. Colo. Studies, general ser., v. 29, no. 3, p. 9.

(720) **Fischer de Waldheim, Gotthelf,** 1817, *Adversaria Zoologica:* Soc. Impér. Nat. Moscou, Mém., v. 5, p. 357-471, pl. 13.———(720A) 1829, *Les céphalopodes fossiles de Moscou et de ses environs, en montrant ces objets en nature:* Soc. Impér. Nat. Moscou, Bull., v. 1, p. 300-362.———(720B) 1837, *Oryctographie du gouvernement de Moscou:* Soc. Impér. Nat. Moscou, p. 1-202, pl. 1-51.

(721) **Fitzpatrick, H. M.,** 1930, *The lower fungi Phycomycetes:* xi+331 p., 112 fig., McGraw-Hill Book Co. (New York).

(722) **Fleming, John,** 1828, *A history of British animals, exhibiting the descriptive characters and systematic arrangement of the genera and species of quadrupeds, birds, fishes, mollusca and radiata of the United Kingdom* (Edinburgh).

(723) **Flint, J. M.,** 1899, *Recent Foraminifera, A descriptive catalogue of specimens dredged by the U.S. Fish Commission Steamer Albatross:* U.S. Natl. Museum, Rept. (1897), p. 249-349, pl. 1-80.

(724) **Folin, L. A. G. de,** 1881, *Exploration de l'aviso à vapeur "Le Travailleur" dans de Golfe de Gascogne, en Juillet 1880:* Soc. d'Histoire Nat. Toulouse, Bull., v. 15, p. 130-141.———(725) 1883, *Recherches sur quelques Foraminifères à l'effet d'obtenir des preuves à l'appui de la classification de certaines organismes vaseux:* Congr. Sci. Dax, Sess. 1 (1882), p. 297-329. ———(726) 1887, *Les Rhizopodes réticulaires:* Naturaliste, Paris, ser. 2, v. 9; (a) p. 102-103, 113-115, text-fig. 1-11; (b) *Tribus des Arénacés et des Globigerinacés,* p. 127-128, text-fig. 12-15.———(727) 1888, *Considérations physiologiques sur les Rhizopodes réticulaires:* Naturaliste, ser. 2, v. 10, p. 109-111.

(727A) ———, **& Périer, L.,** 1875-87, *Les Fonds de la Mer, étude internationale sur les particuliarités nouvelles des régions sous-marines, commencée et dirigée par M.M.L. de Folin et L. Périer:* (a) v. 2, pt. 1, chapter 8 (1875); (b) v. 4, pt. 2, chapter 12, pt. 3, chapter 3 (1887) (Paris).

(728) **Fomina, E. V.,** 1958, *K voprosu o stroenii stenok rakovin nekotorykh Viseyskikh foraminifer Podmoskovnogo basseyna:* Akad. Nauk SSSR, Otdel Geol.-Geog. Nauk, Geol. Inst., Voprosy Mikropaleont., no. 2, p. 121-123, text-fig. 1-2. [*On the question of wall structure of the test of certain Visean Foraminifera of the lower Moscow Basin.*]

(729) **Fornasini, Carlo,** 1889, *Minute forme di rizopodi reticolari nella marna pliocenica del*

Ponticello di Savena presso Bologna: Tipografia Fava e Garagnani (Bologna).——**(730)** 1890, *Primo contributo alla conoscenza della microfauna terziaria italiana:* R. Accad. Sci. Ist. Bologna, Mem. Sci. Nat., ser. 4, v. 10 (1889), p. 463-472, pl.——**(731)** 1894, *Quinto contributo alla conoscenza della microfauna terziaria italiana:* Same, ser. 5, v. 4, p. 201-230, pl. 1-3.——**(732)** 1898, *Contributo alla conoscenza della microfauna terziaria italiana. Foraminiferi del Pliocene superiore di San Pietro in Lama presso Lecce:* Same, ser. 5, v. 7, p. 205-212, 1 pl.——**(733)** 1904, *Illustrazione di specie orbignyane di Foraminiferi istituite nel 1826:* Same, ser. 6, v. 1, p. 1-17, pl. 1-4.——**(734)** 1905, *Illustrazione de specie orbignyane di Miliolidi Institute nel 1826:* Same, ser. 6, v. 2, p. 1-14, pl. 1-4.

(735) Fortis, Alberto, 1801, *Sur quelques nouvelles espèces de Discolithes (Camerines, Lenticulaires, Helicites, Numismales, etc.):* Jour. Phys. Chimie & Histoire Nat. Arts, v. 52, p. 106-115, pl. 2.——**(735A)** 1802, *Mémoires pour servir a l'histoire naturelle et principalement a l'oryctographie de l'Italie, et des pays adjacens:* v. 2, p. 5-137, pl. 1-4, J. J. Fuchs (Paris).

(736) Føyn, Bjørn, 1936, *Ueber die Kernverhältnisse der Foraminifere Myxotheca arenilega Schaudinn:* Archiv Protistenkunde, v. 87, p. 272-295.

(737) Francé, R. H., 1913, *Das Edaphon, Untersuchungen zur Oekologie der bodenbewohnenden Mikroorganismen:* Arbeiten aus dem Biolog. Institut München, no. 2, pl. 1-99, text-fig. 1-35.

(738) Franke, Adolf, 1912, *Die Foraminiferen der Kreideformation des Münsterchen Beckens:* Verein. Preuss. Rheinlande Westfalens, Verhandl., v. 69, p. 255-285, pl. 2.——**(739)** 1912, *Die Foraminiferen der Tiefbohrung Th. XVI auf Blatt Allermöbe bei Hamburg:* Hamburg. Wiss. Anst., v. 29 (1911), p. 29-33.——**(740)** 1928, *Die Foraminiferen der Oberen Kreide Nord-und Mitteldeutschlands:* Preuss. Geol. Landesanst., Abhandl., new ser., no. 111, p. 1-207, pl. 1-18, text-fig. 1.——**(741)** 1936, *Die Foraminiferen des deutschen Lias:* Same, new ser., no. 169, p. 1-138, pl. 1-11.

(742) Franzenau, Agoston, 1884, *Heterolepa, egy uj genus a Foraminiferák rendjében:* Természetrajzi Füzetek, v. 8, pt. 3, p. 181-184, 214-217, pl. 5.——**(743)** 1885, *Adalék nehány foraminifera héjszerkezetének ismeretéhez (Beitrag zur Kenntniss der Schalenstruktur einiger Foraminiferen):* Magyar Nemz. Múz., Termész. Füzetek-Budapest, v. 9, no. 2, p. 92-94 (also p. 151-153), pl. 7.——**(744)** 1888, *Pleiona, n.gen. a foraminiferák rendjében és a Chilostomella eximia n.sp.-röl.:* Same, v. 11, p. 146-147, 203-204, text-fig.——**(745)** 1893, *Semseya, eine neue Gattung aus der Ordnung der Foraminiferen:* Math. & Naturwiss. Berichten aus Ungarn., v. 11, p. 358-361, pl. 25.

(746) Frenguelli, Joaquin, 1953, *Analisis microscopico de una segunda serie de muestras de la turbera del Rio de la Mision, Rio Grande, Tierra del Fuego:* Suomal. Tiedeakat. Toimituksia Ann. Acad. Scient. Fennicae, ser. A, III, v. 34, p. 1-52, text-fig. 1-7.

(747) Frentzen, K., 1944, *Die agglutinierenden Foraminiferen der Birmensdorferschichten (Transversarius-Horizont in Schwammfazies) des Gebietes von Blumberg in Baden:* Paläont. Zeitschr., v. 23, p. 317-343, 2 pl.

(748) Fries, E. M., 1821-29, *Systema mycologicum, sistens Fungorum ordines, genera et species, huc usque cognitas, quas ad normam methodi naturalis determinavit, disposuit atque descripsit:* (a) v. 1, lxii+520 p. (1821); (b) v. 2, 621 p. (1823); (c) v. 3, 259 p. (1829).

(749) Fries, R. E., 1903, *Myxomyceten von Argentinien und Bolivia:* Arkiv Botanik, v. 1, p. 57-70.

(750) Frizzell, D. L., 1943, *Upper Cretaceous Foraminifera from northwestern Peru:* Jour. Paleontology, v. 17, p. 331-353, pl. 55-57.——**(751)** 1949, *Rotaliid Foraminifera of the Chapmanininae: their natural distinction and parallelism to the Dictyoconus lineage:* Same, v. 23, p. 481-495, text-fig. 1-20.

(752) ——, & Keen, A. M., 1949, *On the nomenclature and generic position of Nautilus beccarii Linné (Foraminifera, "Rotaliidae"):* Jour. Paleontology, v. 23, p. 106-108.

(753) ——, & Schwartz, Ely, 1950, *A new lituolid foraminiferal genus from the Cretaceous with an emendation of Cribrostomoides Cushman:* Univ. Missouri, Bull., tech. ser. no. 76, p. 1-12, pl. 1.

(754) Fujimoto, Haruyoshi, & Igo, Hisayoshi, 1955, *Hidaella, a new genus of the Pennsylvanian fusulinids from the Fukuji District, eastern part of the Hida Mountainland, Central Japan:* Palaeont. Soc. Japan, Trans. & Proc., new ser., no. 18, p. 45-48, pl. 7.

(755) ——, & Kanuma, Mosaburo, 1953, *Minojapanella, a new genus of Permian fusulinids:* Jour. Paleontology, v. 27, p. 150-152, pl. 19.

(756) ——, & Kawada, Shigema, 1953, *Hayasakaina, a new genus of fusulinids from the Omi-Limestone, Niigata Prefecture, Japan:* Tokyo Bunrika Daigaku, Sci. Repts., sec. C, v. 2, no. 13, p. 207-209, pl. 1.

(757) Furrer, M. A., 1961, *Siphogenerita, new genus, and a revision of California Cretaceous "Siphogenerinoides" (Foraminiferida):* Biol. Soc. Washington, Proc., v. 74, p. 267-274, text-fig. A, 1-3.

(758) Fursenko, A. V., 1958, *Osnovnye etapy razvitiya faun foraminifer v geologicheskom* *proshlom:* Akad. Nauk Belorusskoi SSR, Inst. Geol. Nauk, Trudy, no. 1, p. 10-29. [*Fundamental state of development of foraminiferal faunas in the geologic past.*]

(759) **Gabriel, B.**, 1876, *Untersuchungen über Morphologie, Zeugung und Entwickelung der Protozoen:* Gegenbaurs Morphologisches Jahrbuch, v. 1, p. 535-572, pl. 20.

(760) **Gadea Buisán, Enrique**, 1947, *Clasificación de los protozoos clave para la determinación hasta familias:* Consejo Superior Invest. Cien., Publ. Inst. Biol. Aplicada, Serie Taxonómica 1, p. 1-84, 175 fig.

(761) **Galloway, J. J.**, 1928, *A revision of the family Orbitoididae:* Jour. Paleontology, v. 2, p. 45-69, 4 fig.———(762) 1933, *A manual of Foraminifera:* James Furman Kemp Memorial Ser., publ. 1, xii+483 p., 42 pl., Principia Press (Bloomington, Indiana).

(763) ———, & **Harlton, B. H.**, 1928, *Some Pennsylvanian Foraminifera of Oklahoma with special reference to the genus Orobias:* Jour. Paleontology, v. 2, p. 338-357, pl. 45-56.

(764) ———, & **Heminway, C. E.**, 1941, *The Tertiary Foraminifera of Porto Rico:* N.Y. Acad. Sci., Scientific Survey of Porto Rico & Virgin Islands, v. 3, pt. 4, p. 275-491, pl. 1-36.

(765) ———, & **Ryniker, Charles**, 1930, *Foraminifera from the Atoka Formation of Oklahoma:* Oklahoma Geol. Survey, Circ. 21, 36 p., 5 pl.

(766) ———, & **Wissler, S. G.**, 1927, *Pleistocene Foraminifera from the Lomita Quarry, Palos Verdes Hills, California:* Jour. Paleontology, v. 1, p. 35-87, pl. 7-12.———(767) 1927, *Correction of names of Foraminifera:* Same, v. 1, p. 193.

(768) **Gandolfi, Rolando**, 1942, *Ricerche micropaleontologiche e stratigrafiche sulla Scaglia e sul flysch Cretacici dei Dintorni di Balerna (Canton Ticino):* Rivista Italiana Paleont., v. 48, mem. 4, p. 1-160, pl. 1-14, text-fig. 1-49.

(769) **Ganelina, R. A.**, 1956, *Foraminifery vizeyskikh otlozheniy Severo-Zapadnykh rayonov Podmoskovnoy Kotloviny:* VNIGRI, Trudy, new ser., no. 98, Mikrofauna SSSR, Sbornik 8, p. 61-159, pl. 1-12. [*Foraminifera of Visean deposits of the northwestern area of the lower Moscow Valley.*]

(770) **Gaümann, Ernst**, 1926, *Vergleichende Morphologie der Pilze:* 626 p., 398 fig. (Jena).

(771) **Gaümann, E. A.**, & **Wynd, F. L.**, 1952, *The fungi, a description of their morphological features and evolutionary development:* 420 p., 440 fig., Hafner Publishing Co. (New York).

(772) **Gauthier-Lièvre, L.**, 1935, *Sur une des singularités de l'Oued Rhir: Des Foraminifères thalassoïdes vivant dans des eaux Sahariennes:* Soc. Histoire Nat. Afrique Nord, Bull., v. 26, p. 142-147, text-fig. 1-2 A-C.———(773) 1954, *Les genres Nebela, Paraquadrula et Pseudonebela (Rhizopodes testacés) en Afrique:* Same, v. 44, no. 7-8 (1953), p. 324-366, text-fig. 1-20.

(774) ———, & **Thomas, Raymond**, 1958, *Les genres Difflugia, Pentagonia, Maghrebia et Hoogenraadia (Rhizopodes testacés) en Afrique:* Archiv Protistenkunde, v. 103, p. 241-370, pl. 8-14, text-fig. 1-57.———(775) 1960, *Le genre Cucurbitella Penard:* Same, v. 104, no. 4, p. 569-602, pl. 39-43, text-fig. 1-13.

(776) **Geinitz, H. B.**, & **Gutbier, A. von**, 1848, *Die Versteinerungen des Zechsteingebirges und Rothliegenden:* no. 1, 26 p., 8 pl., Arnold (Dresden).

(776A) ———, & **Marck, W. von der**, 1876, *Zur Geologie von Sumatra:* Palaeontographica, v. 22, p. 399-404.

(777) **Gerke, A. A.**, 1952, *Mikrofauna Permskikh otlozheniy Nordvikskogo rayona i ee stratigraficheskoe znachenie:* NIIGA, Trudy, v. 28. [*Microfauna of Permian deposits of the Nordvik district and its stratigraphic indications..*]———
(778) 1957, *Nekotorye novye predstaviteli Foraminifer iz Verkhnetriasovykh i Nizhneyurskikh otlozheniy Arktiki:* Statey po Paleontologii i Biostratigrafii, NIIGA, Minist. Geol. i Okhrany Nedr SSSR, v. 3, p. 31-52, pl. 1-3. [*Certain new representatives of the Foraminifera from the Upper Triassic and Lower Jurassic deposits of the Arctic.*]———(779) 1959, *O novom rode Permskikh Nodozarievidnykh Foraminifer i ..utochnenii kharakteristiki roda Nodosaria:* Same, Minist. Geol. i Okhrany Nedr SSSR, v. 17, p. 41-59, pl. 1-3. [*On a new genus of Permian Nodosarian-like Foraminifera and the limiting characteristics of the genus Nodosaria.*]———
(780) 1960, *Lingulinelly i Linguliny (Foraminifera) iz Permskikh i Nizhnemezozoyskikh otlozheniy severa tsentral'noy Sibiri:* Same, Minist. Geol. i Okhrany Nedr SSSR, v. 21, p. 29-70, pl. 1-4. [*Lingulinella and Lingulina (Foraminifera) from Permian and lower Mesozoic deposits of north central Siberia.*]———
(781) 1960, *Ob odnom iz spornykh voprosov sistematiki i nomenklatury foraminifer (K revizii rodov Ammodiscus i Involutina):* Same, Minist. Geol. i Okhrany Nedr SSSR, v. 19, p. 5-18. [*One of the disputable questions in the systematics and nomenclature of the Foraminifera (with revision of the genera Ammodiscus and Involutina).*]———(782) 1961, *Foraminifery Permskikh, Triasovykh i Leyasovykh otlozheniy neftenosnykh rayonov severa tsentral'noy Sibiri:* NIIGA, Trudy, v. 120, p. 1-518, pl. 1-122. [*Foraminifera of the Triassic and Liassic deposits of the petriferous region of north central Siberia.*]

(783) **Geroch, Stanisław**, 1955, *Saccamminoides, nowa otwornica z Eocenu Karpat Fliszowych:*

Polskiego Towarzystwa Geologicznego, Rocznik, v. 23 (1953), p. 53-63, pl. 5, text-fig. 1a-b.——(784) 1957, *Uvigerinammina jankói Majzon (Foraminifera) in the Carpathian Flysch:* Same, (Ann. Soc. Géol. Pologne), v. 25, pt. 3, p. 231-244, pl. 14-15.——(785) 1961, *Pseudoreophax nowy rodzaj otwornic z dolnej Kredy Karpat fliszowych:* Same, v. 31, pt. 1, p. 159-165, pl. 17, text-fig. 1-2.

(786) **Geyn, W. A. E. van de, & Vlerk, I. M. van der,** 1935, *A monograph on the Orbitoididae, occurring in the Tertiary of America compiled in connection with an examination of a collection of larger Foraminifera from Trinidad:* Leidsche Geol. Meded., v. 7, p. 221-272, pl.

(787) **Giard, A.,** 1900, *Sur un protozoaire nouveau de la famille des Gromidae (Amoebogromia cinnabarina Gd):* Soc. Biol. Paris, Comptes Rendus, v. 52, p. 377-378.

(788) **Gignoux, Maurice, & Moret, Léon,** 1920, *Le genre Orbitopsella Mun.-Chalm. et ses relations avec Orbitolina:* Soc. géol. France, Bull., ser. 4, v. 20, pt. 4-6, p. 129-140, pl. 6.

(789) **Girty, G. H.,** 1904, *Triticites, a new genus of Carboniferous foraminifers:* Am. Jour. Sci., ser. 4, v. 17 (whole ser., v. 167), art. 21, p. 234-240, fig. 1-5.——(790) 1911, *On some new genera and species of Pennsylvanian fossils from the Wewoka formation of Oklahoma:* N.Y. Acad. Sci., Ann., v. 21, p. 119-156.——(791) 1915, *Fauna of the Wewoka Formation of Oklahoma:* U.S. Geol. Survey, Bull. 544, 353 p., 35 pl.

(792) **Glaessner, M. F.,** 1936, *Die Foraminiferengattungen Pseudotextularia und Amphimorphina:* Moscow Univ., Prob. Paleont., Lab Paleont., v. 1, p. 95-134, pl. 1-2.——(793) 1937, *Planktonforaminiferen aus der Kreide und dem Eozän und ihre stratigraphische Bedeutung:* Moscow Univ., Lab. Paleont., Studies Micropaleont., v. 1, pt. 1, p. 27-46, pl. 1-2.—— (794) 1937, *On a new family of Foraminifera:* Same, v. 1, pt. 3, p. 19-29, 2 pl.——(795) 1937, *Die Entfaltung der Foraminiferenfamilie Buliminidae:* Moscow Univ., Prob. Paleont., Lab. Paleont., v. 2-3, p. 411-422, text-fig. 1-2.—— (796) 1945, *Principles of micropaleontology:* 296 p., 14 pl., 64 text-fig., 7 tables, Melbourne Univ. Press.

(797) ——, **& Wade, Mary,** 1959, *Revision of the foraminiferal family Victoriellidae:* Micropaleontology, v. 5, no. 2, p. 193-212, pl. 1-3, text-fig. 1-6.

(798) **Gmelin, J. F.,** 1791, *Systema naturae Linnaei:* ed. 13, v. 1, pt. 6, Vermes, G. E. Beer (Lipsiae, Germania).

(799) **Goddard, E. J., & Jensen, H. I.,** 1907, *Contributions to a knowledge of Australian Foraminifera, Pt. 2:* Linnean Soc. New S. Wales, v. 32, pt. 2, no. 126, p. 291-318, pl. 6.

(800) **Goës, Axel,** 1881, *Om ett oceaniskt Rhizopodum reticulatum, Lituolina scorpiura Montf., funnet i Östersjön:* K. Svenska Vetenskapakad. Förhandl., Öfvers., v. 38, no. 8, p. 33-35.—— (801) 1882, *On the reticularian Rhizopoda of the Caribbean Sea:* Same, Handl., v. 19, no. 4, p. 1-151, pl. 1-12.——(802) 1889, *Om den sa Kallade "Verkliga" dimorfismen hos Rhizopoda reticulata:* Same, Handl., Bihang., v. 15, pt. 4, no. 2, p. 1-14, pl. 2.——(803) 1892, *On a peculiar type of arenaceous foraminifer from the American tropical Pacific, Neusina agassizi:* Harvard Univ., Museum Comp. Zool., Bull., v. 23, no. 5, p. 195-198, fig. 1-9.—— (804) 1894, *A synopsis of the Arctic and Scandinavian Recent marine Foraminifera hitherto discovered:* K. Svenska Vetenskapsakad., Handl., v. 25, no. 9, pl. 1-127, pl. 1-25.——
(805) 1896, *The Foraminifera,* in Reports on *the dredging operations off the West Coast of Central America to the Galapagos, to the West Coast of Mexico, and in the Gulf of California, in charge of Alexander Agassiz, carried on by the U.S. Fish Commission Steamer "Albatross," during 1891, Lieut. Commander Z. L. Tanner U.S.N., Commanding:* Harvard Univ., Museum Comp. Zool., Bull., v. 29, no. 1, p. 1-103, pl. 1-10.

(806) **Goldschmidt, R.,** 1907, *Lebensgeschichte der Mastigamöben Mastigella vitrea n.sp. u. Mastigina setosa n.sp.:* Archiv Protistenkunde, suppl. I, p. 83-168, pl. 5-9.

(807) **Golev, B. T.,** 1961, *O rode Operculinoides Hanzawa:* Akad. Nauk SSSR, Otdel. Geol. & Geog. Nauk, Geol. Inst., Voprosy Mikropaleont., no. 5, p. 112-120, pl. 1-2, text-fig. 1. [*The genus Operculinoides Hanzawa.*]

(808) **Gorbenko, V. F.,** 1957, *Pseudospiroplectinata—Novyy rod Foraminifer iz Verkhnemelovykh otlozheniy severo-zapadnogo Donbassa:* Akad. Nauk SSSR, Doklady, v. 117, no. 5, p. 879-880. [*Pseudospiroplectinata—a new genus of Foraminifera from Upper Cretaceous deposits of the northwestern Don Basin.*]——(808A) 1960, *Novye vidy Foraminifer iz otlozheniy verkhnego Mela severo-zapadnoy okrainy Donetskogo Basseyna:* Isvestiya Vysshikh Uchevnykh Zavedeniy, Geol. i Razved. (1960), no. 1, p. 67-76. [*New species of Foraminifera from deposits of the Upper Cretaceous of northwestern Ukraine Donets Basin.*]

(809) **Gorsky, I. I., et al.,** 1939, *Atlas rukovodiashchikh form iskopaemykh faun SSSR, v. 5, Srednii i verkhnii otdely Kamennougolnoy sistemy:* GONTI-Gosgeolizdat, 180 p., 37 fig., 36 pl. [*Atlas of leading forms of the fossil faunas of USSR, v. 5, Middle and upper strata of the Carboniferous System.*]

(809A) **Grabau, A. W.,** 1936, *Early Permian fossils of China, Pt. II, Fauna of the Maping limestone of Kwangsi and Kweichow:* Geol. Surv.

China, Paleont. Sinica, ser. B, v. 8, 327 p., 31 pl.
(810) **Grassé, P.-P.**, 1953, *Traité de Zoologie. Protozoaires:* v. 1, pt. 2, 1160 p., 833 text-fig.
(811) **Gray, J. E.**, 1840, *Synopsis of the contents of the British Museum:* ed. 42, iv+370 p.————(812) 1858, *On Carpenteria and Dujardinia, two genera of a new form of Protozoa with attached multilocular shells filled with sponge, apparently intermediate between Rhizopoda and Porifera:* Zool. Soc. London, Proc., v. 26, p. 266-271, text-fig. 1-4.
(813) **Greeff, Richard**, 1866, *Ueber einige in der Erde lebende Amöben und andere Rhizopoden:* Archiv Mikro. Anat., v. 2, p. 299-331, pl. 17-18.————(814) 1888, *Land-Protozoen:* Gesellschaft zur Beförderung der gesammten Naturwissenschaften zu Marburg, Sitzungsber. (1888), p. 90-158.
(815) **Gregorio, Antonio de**, 1882, *Fossili dei dintorni di Pachino:* Il Tempo, p. 3-23, pl. 1-6 (Palermo).————(816) 1890, *Monographie de la faune éocénique de l'Alabama et surtout de celle de Claiborne de l'étage Parisien (horizon à Venericardia planicosta Lamk.):* Ann. Géol. & Paléont. (Palermo), v. 7-8, p. 1-316, pl. 1-46.
————(816A) 1894, *Description des faunes Tertiares de la Vénétie fossiles des environs de Bassano, surtout du Tertiaire inferieur de l'horizon à Conus diversiformis Desh. et Serpula spirulaea Lamk. (Recueilles par M. Andrea Balestra):* Same, v. 13, p. 1-40, pl. 1-5.————
(817) 1930, *Sul Permiano di Sicilia:* Same, v. 52, p. 1-70, pl. 1-21.
(818) **Grell, K. G.**, 1954, *Die Generationswechsel der polythalamen Foraminifere Rotaliella heterocaryotica:* Archiv Protistenkunde, v. 100, no. 2, p. 268-286, text-fig. 1.————(819) 1956, *Über die Elimination somatischer Kern bei heterokaryotischen Foraminiferen:* Zeitschr. Naturforschung, v. 11B, p. 759-761.————(820) 1957-59, *Untersuchungen über die Fortpflanzung und Sexualität der Foraminiferen:* Archiv Protistenkunde; (a) *I. Rotaliella roscoffensis*, v. 102, no. 2, p. 147-164, pl. 1-11, text-fig. 1-2 (1957); (b) *II. Rubratella intermedia*, v. 102, no. 3-4, p. 291-308, pl. 22-23, text-fig. 1-3 (1958); (c), *III. Glabratella sulcata*, v. 102, no. 3-4, p. 449-472, pl. 34-40, text-fig. 1-6 (1958); (d) *IV. Patellina corrugata*, v. 104, no. 2, pl. 211-234, text-fig. 1-8, pl. 8-21 (1959)————(821) 1958, *Studien zum Differenzierungsproblem an Foraminiferen:* Naturwissenschaften, v. 45, no. 2, p. 3-32, text-fig. 1-12.————(822) 1962, *Entwicklung und Geschlechtsdifferenzierung einer neuen Foraminifere:* Same, v. 49, no. 9, p. 214.
(823) **Grice, C. R.**, 1948, *Manorella, a new genus of Foraminifera from the Austin chalk of Texas:* Jour. Paleontology, v. 22, p. 222-224, 5 text-fig.

(824) **Griffith, J. W.**, & **Henfrey, Arthur**, 1875, *The micrographic dictionary:* ed. 3, v. 1, p. 316-320, van Voorst (London).
(825) **Grigelis, A. A.**, 1960, *O predpolagaemom filogenetischeskom ryade semeystva Epistominidae iz yurskikh otlozheniy litvy*, in *Dochetvertichnaya mikropaleontologiya: Mezhdunarodniy geologicheskiy congress sessiya 21*, Doklady Sovetskikh geologov, Problema 6, p. 98-104, text-fig. 1-5. [*On assumed phylogenetic lines in the family Epistominidae in the Jurassic deposits of Lithuania*, in Pre-Quaternary micropaleontology.]
(826) **Grimsdale, T. F.**, 1952, *Cretaceous and Tertiary Foraminifera from the Middle East:* British Museum (Nat. History), Bull., Geol., v. 1, no. 8, p. 221-248, pl. 20-25, 3 text-fig.————(827) 1959, *Evolution in the American Lepidocyclinidae (Cainozoic Foraminifera)—an interim view, Pt. I-II:* K. Nederland. Akad. Wetensch. Amsterdam, Proc., ser. B, v. 62, no. 1, p. 8-33, text-fig. 1a-b.
(828) **Gronovius, L. T.**, 1781, *Zoophylacii Gronoviani:* pt. 3, p. 241-380, pl. 18-20, T. Haak & Soc. (Leyden).
(829) **Grospietsch, T.**, 1958, *Wechseltierchen (Rhizopoden):* Kosmos. Gesell. Naturfreunde Frankh'sche Verlagshandlung, 80 p., 4 pl. (Stuttgart).
(830) **Grozdilova, L. J.**, 1960, *Metodika izucheniya Paleozoyskikh foraminifer:* VNIGRI, Trudy Pervogo seminara po mikrofaune, p. 22-47, text-fig. 1-16. [*Methods of study of Paleozoic Foraminifera.*]
(830A) ————, & **Lebedeva, N. S.**, 1950, *Nekotorye vidy shtaffell srednekamennougolnykh otlozheniy zapadnogo sklona Urala:* VNIGRI, Trudy, new ser., no. 50, Microfauna Neftyanykh mestorozhdeniy SSSR, Sbornik 3, p. 5-46, pl. 1-5. [*Certain species of Staffella from Middle Carboniferous deposits of the western slope of the Urals.*]————(831) 1954, *Foraminifery nizhnego karbona i bashkirskogo yarusa srednego karbona Kolvo-Visherskogo kraya:* Same, (new ser., no. 81), Mikrofauna SSSR, Sbornik 7, p. 4-203, 15 pl. [*Foraminifera of the lower Carboniferous and Baskir strata of the middle Carboniferous of the Kolvo-Vishersky border.*]
(832) **Grubbs, D. M.**, 1939, *Fauna of the Niagaran nodules of the Chicago area:* Jour. Paleontology, v. 13, p. 543-560, pl. 61-62.
(833) **Gruber, Auguste**, 1884, *Die Protozoen des Hafens von Genua:* K. Acad. Leop.-Carol., Deutsch. Akad. Naturf. (Nova Acta), Halle, v. 46, p. 475-539, pl. 7-11.————(834) 1888, *Ueber einige Rhizopoden aus dem Genueser Hafen:* Naturforsch. Gesell. Freiburg, Ber., Freiburg i. B., v. 4, p. 1-12, pl. 1.
(835) **Grzybowski, J.**, 1896, *Otwornice czerwonych iłow z Wadowic:* Akad. Umiej. Kra-

kówie, Wydz. Mat.-Przyr., Rozpravy, ser. 2, v. 10, p. 261-308, pl. 8-11.——(836) 1897, *Otvornice pokładów naftonosnych, okolicy krosna:* Same, v. 33, p. 257-305, pl. 10-12.

(837) **Gubler, Jean,** 1934, *Structure et sécrétion du test des fusulinidés:* Ann. Protistologie, v. 4, 24 p., 15 fig.——(838) 1935, *Les Fusulinidés du Permien de l'Indochine:* Soc. géol. France, Mém. 26, new ser., v. 11, p. 1-173, pl. 1-8, 54 fig.

(839) **Gümbel, C. W.,** 1861, *Geognostische Beschreibung des bayerischen Alpengebirges und seines Vorlandes:* v. 1, 950 p., 42 pl., T. Perthas (Gotha).——(840) 1868[1870], *Beiträge zur Foraminiferenfauna der nordalpinen Eocängebilde:* K. Bayer. Akad. Wiss., Abhandl., Cl. II, v. 10, pt. 2 (1868), p. 581-730 (also p. 1-152), pl. 1-4.——(841) 1872, *Ueber zwei jurassische Vorläufer des Foraminiferen-Geschlechtes Nummulina und Orbitulites:* Neues Jahrb. Mineral., p. 241-260, pl. 6-7.

(842) **Guppy, R. J. L.,** 1866, *On the relations of the Tertiary formations of the West Indies:* Geol. Soc. London, Quart. Jour., v. 22, p. 570-590, pl. 26, text-fig. 1-3.——(843) 1894, *On some Foraminifera from the Microzoic deposits of Trinidad, West Indies:* Zool. Soc. London, Proc., p. 647-652, pl. 41.——(844) 1904, *Observations on some of the Foraminifera of the oceanic rocks of Trinidad:* Victoria Inst. Trinidad, Proc., v. 2, pt. 1, p. 7-16, pl. 1-2.

(844A) **Gutschick, R. C.,** 1962, *Arenaceous Foraminifera from oncolites in the Mississippian Sappington Formation of Montana:* Jour. Paleontology, v. 36, p. 1291-1304, pl. 174-176, 6 text-fig.

(845) **Haan, Guilielmo de,** 1825, *Monographiae Ammoniteorum et Goniatiteorum specimen:* 168 p., Lugdun Balavorum.

(846) **Hadley, W. H., Jr.,** 1934, *Some Tertiary Foraminifera from the north coast of Cuba:* Bull. Am. Paleontology, v. 20, no. 70A, p. 1-40, pl. 1-5.

(847) **Haeckel, Ernst,** 1862, *Die Radiolarien (Rhizopoda Radiaria):* Pt. 1, p. 1-572 (Berlin).——(848) 1870, *Biologische Studien, Heft 1. Studien über Moneren und andere Protisten:* 184 p., 4 pl. (Leipzig).——(849) 1877, *Die Physemarien (Haliphysema und Gastrophysema), Gastraeaden der Gegenwart:* Jenaische Zeitschr. Naturwiss., v. 11, p. 1-54, pl. 1-6.——(850) 1877, *Biologische Studien, Heft 2. Studien zur Gastraea-Theorie:* 270 p., 14 pl. (Jena).——(851) 1889, *Report on the deep-sea Keratosa collected by H. M. S. Challenger during the years 1873-1876:* Rept. Sci. Results Explor. Voyage H. M. S. Challenger, Zool., v. 32, p. 1-92, pl. 1-8.——(852) 1894, *Systematische Phylogenie. Entwurf eines natürlichen Systems der Organismen auf Grund ihrer Stammesgeschichte. Theil 1, Systematische Phylogenie der Protisten und Pflanzen:* xv+400 p., Georg Reimer (Berlin).

(853) **Haeusler, Rudolf,** 1883, *Ueber die neue Foraminiferengattung Thuramminopsis:* Neues Jahrb. Mineral., v. 2, p. 68-72, pl. 4——(854) 1890, *Monographie der Foraminiferen-Fauna der schweizerischen Transversarius-Zone:* Schweiz. Palaeont. Gesell., Abhandl., v. 17, p. 1-134, pl. 1-15.

(855) **Hagelstein, R.,** 1932, *Revision of the Myxomycetes:* N.Y. Acad. Sci., Scientific Survey Porto Rico & Virgin Islands, v. 8, pt. 2, p. 241-248.——(856) 1942, *A new genus of the Mycetozoa:* Mycologia, v. 34, no. 5, p. 593-594.——(857) 1944, *The Mycetozoa of North America:* p. 1-306, pl. 1-16, The author (Mineola, N.Y.).

(858) **Hagenow, Friedrich von,** 1842, *Monographie der Rügen'schen Kreide-Versteinerungen, Abt. III-Mollusken:* Neues Jahrb. Mineral., Geog. & Geol. Petrefactenkunde, p. 528-575, pl. 9.——(859) 1851, *Die Bryozoen der Maastrichter Kreidebildung:* 111 p., 12 pl., T. Fischer (Cassel).

(860) **Hagn, Herbert,** 1954, *Some Eocene Foraminifera from the Bavarian Alps and adjacent areas:* Cushman Found. Foram. Research, Contrib., v. 5, pt. 1, p. 14-20, pl. 3-4.

(861) **Halkyard, Edward,** 1918, *The fossil Foraminifera of the Blue Marl of the Côte des Basques, Biarritz:* Manchester Lit. & Philos. Soc., Mem. & Proc., v. 62, pt. 2, p. 1-145, pl. 1-9.

(862) **Hall, R. P.,** 1953, *Protozoology:* 682 p., illus., Prentice-Hall Publ. Co. (New York).

(863) **Hantkin, Miksa von,** 1875[1876], *A Clavulina Szabói rétegek Faunája, I. Foraminiferak:* Magyar Kir. földt. int. evkönyve, v. 4 (1875), p. 1-82, pl. 1-16.

(864) **Hanzawa, Shôshirô,** 1932, *Foraminifera:* Iwanami Lectures, Geol. & Paleont., p. 1-134, text-fig. 1-124. [In Japanese.]——(865) 1932, *A new type of Lepidocyclina with a multilocular nucleoconch from the Taitô Mountains, Taiwan (Formosa):* Imper. Acad. Japan, Proc., v. 8, p. 446-449.——(866) 1935, *Some fossil Operculina and Miogypsina from Japan and their stratigraphical significance:* Tohoku Imper. Univ., Sci. Repts., ser. 2 (Geol.), v. 18, no. 1, p. 1-29, pl. 1-3.——(867) 1937, *Notes on some interesting Cretaceous and Tertiary Foraminifera from the West Indies:* Jour. Paleontology, v. 11, p. 110-117, pl. 20-21.——(868) 1938, *An aberrant type of the Fusulinidae from the Kitakami Mountainland, northeastern Japan:* Imper. Acad. Tokyo, Proc., v. 14, no. 7, p. 255-259, fig. 1-16.——(869) 1940, *Micropalaeontological studies of drill cores from a deep well in Kita-Daitô-Zima (North Borodino*

Island): Jubilee Publication in Commemoration of Prof. H. Yabe's 60th Birthday, p. 775-802, pl. 39-42.——(870) 1947, *Reinstatement of the genus Heterosteginoides, and the classification of the Miogypsinidae:* Jour. Paleontology, v. 21, p. 260-263, pl. 41.——(871) 1949, *A new type of the fusulinid Foraminifera from central Japan:* Same, v. 23, p. 205-209, pl. 43, fig. 1-3.——(872) 1952, *Notes on the Recent and fossil Baculogypsinoides spinosus Yabe and Hanzawa from the Ryukyu Islands and Taiwan (Formosa), with remarks on some spinose Foraminifera:* Tohoku Univ., Inst. Geol. & Paleont., Short Papers, no. 4, p. 1-22, pl. 1-2.——(873) 1957, *Cenozoic Foraminifera of Micronesia:* Geol. Soc. America, Mem. 66, 163 p., 38 pl., maps.——(874) 1959, *The foraminiferal species Fabiania cassis (Oppenheim), in Japan:* Cushman Found. Foram. Research, Contrib., v. 10, pt. 4, p. 119-122, pl. 9.——(875) 1962, *Upper Cretaceous and Tertiary three-layered larger Foraminifera and their allied forms:* Micropaleontology, v. 8, no. 2, p. 129-186, pl. 1-8.

(876) **Haque, A. F. M. Mohsenul,** 1956, *The Foraminifera of the Ranikot and the Laki of the Nammal Gorge, Salt Range:* Geol. Survey Pakistan, Palaeont. Pakistanica, v. 1, p. 1-300, pl. 1-34.——(877) 1958, *Cincoriola, a new generic name for Punjabia Haque, 1956:* Cushman Found. Foram. Research, Contrib., v. 9, pt. 4, p. 103.

(878) **Harker, Peter, & Thorsteinsson, Raymond,** 1960, *Permian rocks and faunas of Grinnell Peninsula, Arctic Archipelago:* Geol. Survey Canada, Dept. Mines & Tech. Surveys, Mem. 309, 89 p., 25 pl., 9 fig.

(879) **Harlton, B. H.,** 1927, *Some Pennsylvanian Foraminifera of the Glenn Formation of southern Oklahoma:* Jour. Paleontology, v. 1, p. 15-27, pl. 1-5.——(880) 1928, *Pennsylvanian Foraminifera of Oklahoma and Texas:* Same, v. 1, p. 305-310, pl. 52-53.

(881) **Harmer, S. F., & Shipley, A. E.,** eds., 1906, *Cambridge Natural History, v. 1 (Protozoa by Marcus Hartog, Porifera, Coelenterata, Ctenophora, Echinodermata):* p. 3-162, Macmillan & Co., Ltd. (London).

(882) **Harris, R. W., & Sutherland, B. W.,** 1954, *A new foraminiferal genus and species from the Midway Formation of southwest Arkansas:* Oklahoma Acad. Sci., Proc., v. 33 (1952), p. 207-208, fig. 1-2.

(883) **Harting, Pieter,** 1852, *De bodem onder Amsterdam ondersocht en beschreven:* K. Nederland Inst. Wetenschap. Let. Schoone Kunst., Kl. 1, Verhandl., ser. 3, pt. 5, p. 73-232, pl. 1-4 (Amsterdam).

(884) **Hartmann, Max,** 1907, *Das System der Protozoen:* Archiv Protistenkunde, v. 10, p. 139-158.

(885) **Hayden, H. H.,** 1909, *Fusulinidae from Afghanistan:* Geol. Survey, India, Records, v. 38, pt. 3, p. 230-256, pl. 17-22, fig. 1.

(886) **Haynes, John,** 1954, *Taxonomic position of some British Palaeocene Buliminidae:* Cushman Found. Foram. Research, Contrib., v. 5, pt. 4, p. 185-191, pl. 35, text-fig. 1-20.——(887) 1956, *Certain smaller British Paleocene Foraminifera, Pt. I. Nonionidae, Chilostomellidae, Epistominidae, Discorbidae, Amphisteginidae, Globigerinidae, Globorotaliidae and Gümbelinidae:* Same, v. 7, pt. 3, p. 79-101, pl. 16-18.

(888) **Hedley, R. H.,** 1957, *Microradiography applied to the study of Foraminifera:* Micropaleontology, v. 3, no. 1, p. 19-28, pl. 1-4, text-fig. 1.——(889) 1958, *A contribution to the biology and cytology of Haliphysema (Foraminifera):* Zool. Soc. London, Proc., v. 130, pt. 4, p. 569-576, pl. 1-3.——(890) 1960, *New observations on Pelosphaera cornuta:* Cushman Found. Foram. Research, Contrib., v. 11, pt. 2, p. 54-56, pl. 9.——(891) 1960, *The iron-containing shell of Gromia oviformis (Rhizopoda):* Quart. Jour. Micro. Sci., v. 101, pt. 3, p. 279-293, text-fig. 1-6.

(892) ——, & **Bertaud, W. S.,** 1962, *Electron-microscope observations of Gromia oviformis (Sarcodina):* Jour. Protozoology, v. 9, no. 1, p. 79-87, fig. 1-15.

(893) **Heilprin, Angelo,** 1883, *On the occurrence of nummulitic deposits in Florida, and the association of Nummulites with a fresh-water fauna:* Acad. Nat. Sci. Philadelphia, Proc., pt. 2 (1882), p. 189-193.

(893A) **Heim, Arnold,** 1908, *Die Nummuliten- und Flyschbildungen der Schweizeralpen:* Schweiz. Paläont. Gesell., Abhandl. (Soc. Paläont. Suisse, Mém.), v. 35, art. 4, p. 1-301, pl. 1-8.

(893B) **Henbest, L. G.,** 1928, *Fusulinellas from the Stonefort Limestone Member of the Tradewater Formation:* Jour. Paleontology, v. 2, p. 70-85, pl. 8-10.——(894) 1931, *The species Endothyra baileyi (Hall):* Cushman Lab. Foram. Research, Contrib., v. 7, pt. 4, p. 90-93, pl. 11-12.——(895) 1935, *Nanicella, a new genus of Devonian Foraminifera:* Washington Acad. Sci., Jour., v. 25, no. 1, p. 34-35.——(896) 1937, *Keriothecal wall-structure in Fusulina and its influence on fusuline classification:* Jour. Paleontology, v. 11, p. 212-230, pl. 34-35.——(897) 1953, *The name and dimorphism of Endothyra bowmani Phillips, 1846:* Cushman Found. Foram. Research, Contrib., v. 4, pt. 2, p. 63-65, text-fig. 1-2.——(898) 1960, *Paleontologic significance of shell composition and diagenesis of certain late Paleozoic sedentary Foraminifera:* U.S. Geol. Survey, Prof. Paper 400-B, p. B386-B387.

(899) **Henrici, Heinz,** 1934, *Foraminiferen aus dem Eozän und Altmiozän von Timor:* Palaeon-

tographica, Suppl. v. 4, p. 1-56, pl. 1-4, text-fig. 1-26.
(900) **Henson, F. R. S.,** 1948, *Foraminifera of the genus Trocholina in the Middle East:* Ann. & Mag. Nat. History, ser. 11, v. 14 (1947), p. 445-459, pl. 11-13.———(901) 1948, *New Trochamminidae and Verneuilinidae from the Middle East:* Same, ser. 11, v. 14 (1947), p. 605-630, pl. 14-18.———(902) 1948, *Larger imperforate Foraminifera of southwestern Asia, Families Lituolidae, Orbitolinidae and Meandropsinidae:* British Museum (Nat. History), Mon., p. 1-127, pl. 1-16, fig. 1-16.———(903) 1950, *Middle eastern Tertiary Peneroplidae (Foraminifera), with remarks on the phylogeny and taxonomy of the family:* 70 p., 10 pl., 3 text-fig., West Yorkshire Printing Co. (Wakefield, England).
(904) **Heron-Allen, Edward,** 1915, *Contributions to the study of bionomics and reproductive processes in the Foraminifera:* Royal Soc., London, Philos. Trans., ser. B, v. 206, p. 227-279.
(905) ———, & **Barnard, J. E.,** 1918, *Application of X-rays to determine the interior structure of microscopic fossils:* Geol. Mag., new ser., decade 6, v. 5, p. 90-92.
(906) ———, & **Earland, Arthur,** 1908, *On Cycloloculina, a new generic type of the Foraminifera, with a preliminary study of the foraminiferous deposits and shore-sands of Selsey Bill:* Royal Micro. Soc. London, Jour., p. 529-543, pl. 12.———(907) 1910, *On the Recent and fossil Foraminifera of the shore-sands of Selsey Bill, Sussex, Part V. The Cretaceous Foraminifera:* Same, p. 401-426, pl. 6-11.———(908) 1913, *On some Foraminifera from the North Sea, etc., dredged by the Fisheries Cruiser "Goldseeker" (International North Sea Investigations Scotland), III. On Cornuspira diffusa, a new type from the North Sea:* Royal Micro. Soc. London, Jour., Trans. & Proc., p. 272-276, pl. 12.———(909) 1913, *Clare Island Survey. Foraminifera:* Royal Irish Acad., Proc., v. 31, pt. 64, p. 1-188, pl. 1-13.———(910) 1914-15, *The Foraminifera of the Kerimba Archipelago (Portugese East Africa):* Zool. Soc. London, Trans., v. 20 (1912-1915); (a) *Part 1,* pt. 12, p. 363-390, pl. 35-37 (1914); (b) *Part 2,* pt. 17, p. 543-794, pl. 40-53, text-fig. 42-44 (1915).———(911) 1922, *Protozoa, Part II. Foraminifera:* British Antarctic ("Terra Nova") Exped., 1910, Zool., v. 6, no. 2, p. 25-268, pl. 1-8.———(912) 1924, *The Foraminifera of Lord Howe Island, South Pacific:* Linnean Soc. London, Jour., Zool., v. 35, p. 599-647, pl. 35-37.———(913) 1928, *On the Pegididae, a new family of Foraminifera:* Royal Micro. Soc. London, Jour., v. 48, p. 283-299, pl. 1-3.———(914) 1929-32, *Some new Foraminifera from the South Atlantic:* Same, ser. 3; (a) *Pt. I,* v. 49, p. 102-108, pl. 1-3 (1929); (b) *Pt. II.* v. 49, pt. 4, art. 27, p. 324-334, pl. 1-4 (1929); (c) *Pt. III. Miliammina, a new siliceous genus,* v. 50, p. 38-45, pl. 1 (1930); (d), *Pt. IV. Four new genera from South Georgia,* v. 52, p. 253-261, pl. 1, 2 (1932).———(915) 1930, *The Foraminifera of the Plymouth District, II:* Same, ser. 3, v. 50, pt. 2, p. 161-199, pl. 4-5.———(916) 1932, *Foraminifera, Part 1. The ice-free area of the Falkland Islands and adjacent seas:* Discovery Repts., v. 4, p. 291-460, pl. 6-17.
(917) **Hertwig, Richard,** 1874, *Ueber Mikrogromia socialis, eine Colonie bildende Monothalamie des süssen Wassers:* Archiv Mikro. Anat., v. 10, suppl., p. 1-34, pl. 1.———(918) 1876, *Bemerkungen zur Organisation und Systematischen Stellung der Foraminiferen:* Jenaische Zeitschr. Naturwiss. (new ser., v. 3), v. 10, p. 41-55, pl. 2.———(919) 1893, *Lehrbuch der Zoologie:* rev. ed. 2, 576 p., 568 text-fig. (Jena).———(920) 1919, *Lehrbuch der Zoologie:* ed. 12, 686 p., 588 text-fig. (Jena).
(921) ———, & **Lesser, E.,** 1874, *Ueber Rhizopoden und denselben nahestehende Organismen:* Archiv. Mikro. Anat., v. 10, suppl., p. 35-243, pl. 2-5.
(922) **Hickson, S. J.,** 1911, *On Polytrema and some allied genera. A study of some sedentary Foraminifera based mainly on a collection made by Prof. Stanley Gardiner in the Indian Ocean:* Linnean Soc. London, Trans., Zool., ser. 2, v. 14, p. 443-462, pl. 30-32.
(923) **Ho, Yen,** 1959, *Triassic Foraminifera from the Chialingkiang Limestone of south Szechuan:* Acta Palaeont. Sinica, v. 7, p. 387-405 (Chinese); p. 405-418 (English); pl. 1-8.
(924) **Höglund, Hans,** 1947, *Foraminifera in the Gullmar Fjord and the Skagerak:* Zoologiska Bidrag Uppsala, v. 26, p. 1-328, pl. 1-32, 312 text-fig., 2 maps, 7 tables.
(925) **Hoffmeister, W. S.,** & **Berry, C. T.,** 1937, *A new genus of Foraminifera from the Miocene of Venezuela and Trinidad:* Jour. Paleontology, v. 11, p. 29-30, pl. 5.
(926) **Hofker, Jan,** 1925, *On heterogamy in Foraminifera:* Tijdschr. Nederland. Dierk Vereen., ser. 2, v. 19, p. 68-70.———(927) 1927, *Die Foraminiferen aus dem Senon Limburgens, VII:* Natuurhist. Maandblad Maastricht., v. 16, p. 173-176, fig. 1-8.———(928) 1927-51, *The Foraminifera of the Siboga Expedition:* Siboga Expeditie, Mon. IV; (a) *Pt. I. Tinoporidae, Rotaliidae, Nummulitidae, Amphisteginidae,* p. 1-78, pl. 1-38 (1927); (b) *Pt. 2. Families Astrorhizidae, Rhizamminidae, Reophacidae, Anomalinidae, Peneroplidae,* p. 79-170, pl. 39-64 (1930); (c) *Pt. 3,* p. 1-513, 348 fig.; E. J. Brill (Leiden) (1951).———(929) 1928, *On Faujasina d'Orbigny:* Cushman Lab. Foram. Research, Contrib., v. 4, pt. 3, p.

80-83, pl. 11.——(930) 1933, *Papers from Dr. Th. Mortensen's Pacific Expedition 1914-16, LXII. Foraminifera of the Malay Archipelago:* Vidensk. Medd. Dansk naturhist. Foren., v. 93, p. 71-167, pl. 2-6.——(931) 1949, *On Foraminifera from the upper Senonian of south Limburg (Maestrichtian):* Inst. Royal Sci. Nat. Belgique, Mém. 112, p. 3-69, text-fig. 1-23.——(932) 1950, *Wonderful animals of the sea, Foraminifera:* Amsterdam Naturalist, v. 1, no. 3, p. 60-79, text-fig. 1-42.——(933) 1950-52, *Recent Peneroplidae:* Royal Micro. Soc. London, Jour.; (a) v. 70, p. 388-396 (1950); (b) ser. 3, v. 71, p. 223-239, text-fig. 2-18 (1951); (c) ser. 3, v. 71, p. 450-463, text-fig. 36-51 (1951) [1952].——(934) 1951, *Pores of Foraminifera:* Micropaleontologist, v. 5, p. 38.——(935) 1951, *On Foraminifera from the Dutch Cretaceous:* Natuurhist. Genoot. Limburg, ser. 4, p. 1-40, text-fig. 1-47.——(936) 1951, *The toothplate-Foraminifera:* Arch. Néerlandaises Zool., v. 8, pt. 4, p. 353-372, fig. 1-30.——(937) 1952, *The Jurassic genus Reinholdella Brotzen (1948) (Foram.):* Paläont. Zeitschr., v. 26, no. 1-2, p. 15-29, text-fig. 1-17.——(938) 1953, *The genus Epistomaria Galloway, 1933, and the genus Epistomaroides Uchio, 1952:* Same, v. 27, no. 3/4, p. 129-142, 14 text-fig.——(939) 1953, *Types of genera described in Part III of the "Siboga Foraminifera":* Micropaleontologist, v. 7, p. 26-28.——(940) 1953, *Arenaceous tests in Foraminifera—chalk or silica:* Same, v. 7, p. 65-66.——(941) 1954, *Chamber arrangement in Foraminifera:* Same, v. 8, p. 30-32.——(942) 1954, *Notes on the generic names of some rotaliform Foraminifera:* Same, v. 8, p. 34-35.——(943) 1954, *Über die Familie Epistomariidae (Foram):* Palaeontographica, v. 105, pt. A, p. 166-206, 56 fig.——(944) 1955, *Foraminifera from the Cretaceous of southern Limburg, Netherlands, IX: Dictyoconus mosae nov. spec.:* Natuurhist. Maandblad Limburg, v. 44, no. 11-12, p. 115-117, 2 text-fig. ——(945) 1956, *Tertiary Foraminifera of coastal Ecuador, Part II. Additional notes on the Eocene species:* Jour. Paleontology, v. 30, p. 891-958, 101 text-fig.——(946) 1956, *Foraminifera Dentata-Foraminifera of Santa Cruz and Thatch-Island Virginia-Archipelago West-Indies:* Spolia Zool. Musei Hauniensis XV, p. 1-237, pl. 1-35.——(947) 1956, *Die Globotruncanen von Nordwest-Deutschland und Holland:* Neues Jahrb. Geol. & Paläont., Abhandl., v. 103, no. 3, p. 312-340, text-fig. 1-26.——(948) 1957, *Foraminiferen der Oberkreide von Nordwestdeutschland und Holland:* Beihefte Geol. Jahrbuch, no. 27, p. 1-464, text-fig. 1-495.——(949) 1958, *The taxonomic status of Palmerinella palmerae Bermúdez:* Cushman Found. Foram. Research, Contrib., v. 9, pt. 2, p. 32-33, text-fig. A-E.

——(950) 1958, *The taxonomic position of the genus Pseudoeponides Uchio, 1950:* Same, v. 9, pt. 2, p. 46-48, text-fig. 1-2.——(951) 1959, *The genera Eponides, Lacosteina, Nuttallides, Planorbulina, and Halkyardia:* Same, v. 10, pt. 4, p. 111-118, text-fig. 1-27.——(952) 1959, *Les Foraminifères du Crétacé supérieur du Cotentin:* Congrès Soc. savantes savois., 84th Sess., p. 369-397, fig. 1-68.——(953) 1960, *The taxonomic positions of the genera Boldia van Bellen, 1946, and Anomalinella Cushman, 1927:* Cushman Found. Foram. Research, Contrib., v. 11, pt. 2, p. 47-52, text-fig. 1-11.

(954) **Honjo, Susumu,** 1959, *Neoschwagerinids from the Akasaka Limestone (A palaeontological study of the Akasaka Limestone, 1st report):* Hokkaido Univ., Jour. Faculty Sci., ser. 4, v. 10, no. 1, p. 111-161, pl. 1-12, fig. 1-8.

(955) **Hoogenraad, H. R.,** 1933, *Einige Beobachtungen an Bullinula indica Penard:* Archiv. Protistenkunde, v. 79, p. 119-130, pl. 12, text-fig. 1.——(956) 1935, *Studien über die sphagnicolen Rhizopoden der niederländischen Fauna:* Same, v. 84, no. 1, p. 1-100, pl. 1-2, text-fig. 1-48.

(957) ——, **& De Groot, A. A.,** 1940, *Zootwaterrhizopoden en Heliozoen:* Fauna van Nederland, Afl. 9, 303 p.

(958) **Hornibrook, N. de B.,** 1951, *Permian fusulinid Foraminifera from the North Auckland Peninsula, New Zealand:* Royal Soc. New Zealand, Trans., v. 79, pt. 2, p. 319-321, pl. 50. ——(959) 1961, *Tertiary Foraminifera from Oamaru District (N.Z.), Pt. 1. Systematics and distribution:* New Zealand Geol. Survey, Paleont. Bull. 34 (1), 192 p., 28 pl.

(960) ——, **& Vella, Paul,** 1954, *Notes on the generic names of some rotaliform Foraminifera:* Micropaleontologist, v. 8, p. 24-28.

(961) **Hottinger, Lukas,** 1960, *Ueber paleocaene und eocaene Alveolinen:* Eclogae geol. Helv., v. 53, no. 1, p. 265-283, pl. 1-21, fig. 1-3, tab. 1.——(962) 1960[1962], *Recherches sur les Alvéolines du Paléocène et de l'Eocène:* Schweiz. Palaeont., Abhandl. (Soc. Paléont. Suisse, Mém.), v. 75-76 (1960), p. 1-243, pl. 1-18, text-fig. 1-117, 1 table.

(963) **Hovasse, Raymond,** 1956, *Arnoldia antiqua, gen. nov., sp. nov., Foraminifère probable du Pré-Cambrien de la Côte-d'Ivoire:* Acad. Sci. Paris, Comptes Rendus, v. 242, p. 2582-2584, text-fig. 1-5.

(964) ——, **& Couture, R.,** 1961, *Nouvelle découverte dans l'Antécambrian de la Côte-d'Ivoire, de Birrimarnoldia antiqua (gen. nov.) =Arnoldia antiqua Hovasse, 1956:* Acad. Sci. Paris, Comptes Rendus, v. 252, no. 7, p. 1054-1056, text-fig. 1-2.

(965) **Howchin, Walter,** 1888, *Additions to the*

knowledge of the Carboniferous Foraminifera: Royal Micro. Soc. London, Jour., pt. 2, p. 533-545.———(966) 1889, *The Foraminifera of the Older Tertiary of Australia (No. 1, Muddy Creek, Victoria):* Royal Soc. S. Australia, Trans. & Proc., v. 12 (1888-1889), p. 1-20, pl. 1.——— (967) 1895, *Carboniferous Foraminifera of western Australia, with descriptions of new species:* Same, v. 19, p. 194-198, pl. 10.

(968) ———, & Parr, W. J., 1938, *Notes on the geological features and foraminiferal fauna of the Metropolitan Abattoirs bore, Adelaide:* Royal Soc. S. Australia, Trans., v. 62, no. 2, p. 287-317, pl. 15-19.

(968A) Howe, H. V., 1928, *An observation on the range of the genus Hantkenina:* Jour. Paleontology, v. 2, p. 13-14, text-fig. 1-2.——— (969) 1930, *Distinctive new species of Foraminifera from the Oligocene of Mississippi:* Same, v. 4, p. 327-331, pl. 27.———(970) 1934, *Bitubulogenerina, a Tertiary new genus of Foraminifera:* Same, v. 8, p. 417-421, pl. 51. ———(971) 1939, *Louisiana Cook Mountain Eocene Foraminifera:* Louisiana Geol. Survey, Geol. Bull. 14, xi+122 p., 15 pl., table 1.

(972) ———, & Wallace, W. E., 1932, *Foraminifera of the Jackson Eocene at Danville Landing on the Ouachita, Catahoula Parish, Louisiana:* Louisiana Dept. Conserv., Geol. Bull. 2, p. 1-118, pl. 1-15.

(973) Hsu, Y. C., 1942, *On the type species of Chusenella:* Geol. Soc. China, Bull., v. 22, no. 3-4, p. 175-176, fig. 1-3.

(974) Husezima, Reiko, & Maruhasi, Masaho, 1944, *A new genus and thirteen new species of Foraminifera from the core-sample of Kasiwazaki oil-field, Nigata-ken:* Sigenkagaku Kenkyusyo, Jour. (Research Inst. for Nat. Resources, Japan), v. 1, no. 3, p. 391-400, pl. 34.

(975) Hussey, K. M., 1943, *Distinctive new species of Foraminifera from the Cane River Eocene of Louisiana:* Jour. Paleontology, v. 17, p. 160-167, pl. 26-27.

(976) Ireland, H. A., 1939, *Devonian and Silurian Foraminifera from Oklahoma:* Jour. Paleontology, v. 13, p. 190-202, 75 text-fig.——— (977) 1956, *Upper Pennsylvanian arenaceous Foraminifera from Kansas:* Same, v. 30, p. 831-864, 7 text-fig.———(978) 1960, *Emendations to Upper Pennsylvanian arenaceous Foraminifera from Kansas:* Same, v. 34, p. 1217-1218.

(978A) Ishii, Ken-ichi, & Nogami, Yasuo, 1961, *On the new genus Metadoliolina:* Palaeont. Soc. Japan, Trans. & Proc., new ser., no. 44, p. 161-166, pl. 25.

(979) Israelsky, M. C., 1949, *Oscillation chart:* Am. Assoc. Petroleum Geologists, Bull., v. 33, p. 92-98, 3 text-fig., 1 chart.———(980) 1951, *Foraminifera of the Lodo Formation central California, General introduction and Part 1,* *Arenaceous Foraminifera:* U.S. Geol. Survey, Prof. Paper 240-A, p. 1-29, pl. 1-11.

(981) Jahn, Brigitte, 1953, *Elektronenmikroskopische Untersuchungen an Foraminiferenschalen:* Zeitschr. Wiss. Mikroskopie & Microskopische Technik, v. 61, no. 5, p. 294-297, 9 text-fig.

(982) Jahn, Eduard, 1928, *Myxomycetenstudien 12. Das System der Myxomyceten:* Deutsch. Botan. Gesell., Ber., v. 46, p. 8-17, pl. 1.

(983) Jahn, T. L., & Jahn, F. F., 1949, *How to know the Protozoa:* 234 p., W. C. Brown (Dubuque, Iowa).

(984) ———, & Rinaldi, R. A., 1959, *Protoplasmic movement in the foraminiferan, Allogromia laticollaris, and a theory of its mechanism:* Biol. Bull., v. 117, p. 100-118.

(984A) James, E., 1823, *Account of an expedition from Pittsburg to the Rocky Mountains in the years 1819-1820 under the command of Maj. S. H. Long:* v. 1, note 24, p. 323-329, Longman, Hurst, Rees, Orme and Brown (London).

(985) Jedlitschka, Heinrich, 1931, *Neue Beobachtungen über Dentalina Verneuilli (d'Orb.) und Nodosaria abyssorum (Brady):* Firgenwald, v. 4, p. 121-127 [Reichenberg (Liberec), Czech.].———(986) 1934, *Über Candorbulina, eine neue foraminiferen-Gattung und zwei neue Candeina-Arten:* Naturforsch. Ver. Brünn, Verhandl., v. 65 (1933), p. 17-26.

(987) Jeffreys, J. G., 1876, *On the Crustacea, tunicate Polyzoa, Echinodermata, Actinozoa, Foraminifera, Polycystina, and Spongida* in Preliminary reports of the biological results of a cruise in H. M. S. "Valorous" to Davis Straits in 1875: Royal Soc. London, Proc., v. 25, p. 212-215, pl. 2-3.

(988) Jennings, A. V., 1896, *On a new genus of Foraminifera of the family Astrorhizidae:* Linnean Soc. London, Jour., Zool., v. 25, p. 320-321.

(989) Jennings, P. H., 1936, *A microfauna from the Monmouth and basal Rancocas Groups of New Jersey:* Bull. Am. Paleontology, v. 23, no. 78, p. 161-232, pl. 23-34.

(990) Jepps, M. W., 1926, *Contribution to the study of Gromia oviformis Duj.:* Quart. Jour. Micro. Sci., new ser., v. 70, p. 701-719.——— (991) 1934, *On Kibisidytes marinus, n.gen., n.sp., and some other Rhizopod Protozoa found on surface films:* Same, new ser., v. 77, p. 121-127, pl. 5-6.———(992) 1942, *Studies on Polystomella Lamarck:* Jour. Marine Biol. Assoc., v. 25, p. 607-666.———(993) 1956, *The Protozoa, Sarcodina:* 183 p., 80 text-fig., Oliver & Boyd (Edinburgh & London).

(994) Jírovec, O., 1953, *Protozoologie:* Naklad Českoslov. Acad. Věd, p. 1-643, pl., 238 fig. (Praha).

(995) **Jollos, Victor**, 1917, *Untersuchungen zur Morphologie der Amöbenteilung:* Archiv Protistenkunde, v. 37, no. 3, p. 229-275, pl. 13-16 (publ. May 14, 1917).

(996) **Jones, T. R.**, 1895, *A monograph of the Foraminifera of the Crag, Pt. 2:* Palaeont. Soc. London, p. i-vii+73-210, pl. 5-7.

(997) ———, & **Chapman, Frederick**, 1900, *On the Foraminifera of the orbitoidal limestones and reef rocks of Christmas Island* in ANDREWS, C. W., A monograph of Christmas Island (Indian Ocean): British Museum (Nat. History), p. 226-264, pl. 20-21.

(998) ———, & **Parker, W. K.**, 1860, *On the rhizopodal fauna of the Mediterranean, compared with that of the Italian and some other Tertiary deposits:* Geol. Soc. London, Quart. Jour., v. 16, p. 292-307.———(999) 1860, *On some fossil Foraminifera from Chellaston, near Derby:* Same, v. 16, p. 452-458, pl. 19-20.

———(1000) 1863, *Notes on some fossil and Recent Foraminifera collected in Jamaica by the late Lucas Barrett, F. G. S.:* Rept. Brit. Assoc. (Newcastle-on-Tyne meeting), Trans. Secs., p. 80.———(1001) 1876, *Notice sur les Foraminifères vivants et fossiles de Jamaique: —suivie de la description d'une espèce nouvelle* [*Tinoporus pilaris*] *des Couches Miocènes de la Jamaique, par H. B. Brady:* Soc. Malacol. Belg., Ann., v. 11, Mém., p. 91-103, fig.

(1002) ———, ———, & **Brady, H. B.**, 1866, *A monograph of the Foraminifera of the Crag, Pt. 1:* Palaeont. Soc. London, v. 19 (1865), p. 1-72, pl. 1-4.

(1003) **Jordan, Louise,** & **Applin, E. R.**, 1952, *Choffatella in the Gulf Coastal regions of the United States and description of Anchispirocyclina n.gen.:* Cushman Found. Foram. Research, Contrib., v. 3, pt. 1, p. 1-5, pl. 1-2.

(1004) **Joukowsky, Étienne,** & **Favre, J.**, 1913, *Monographie géologique et paléontologique du Salève (Haute-Savoie, France):* Soc. Phys. & Histoire Nat. Genève, Mém., v. 37, pt. 4, p. 295-523.

(1005) **Jung, Wilhelm**, 1942, *Südchilenische Thekamöben (aus dem südchilenischen Küstengebiet, Beitrag 10):* Archiv Protistenkunde, v. 95, p. 253-356, text-fig. 1-79.———(1006) 1942, *Illustrierte Thekamöben-Bestimmungstabellen: I. Die Systematik der Nebelinen:* Same, v. 95, p. 357-390.

(1007) **Kaever, Mathias**, 1958, *Über Globorotalites Brotzen, 1942 und Conorotalites nov. gen.:* Geol. Jahrb., v. 75, p. 433-436, text-fig. 1-2.

(1008) **Kahler, Franz**, 1946, *Die Foraminiferengattung Nummulostegina Schubert, 1907:* Geol. Bundesanst. Wien, Verhandl., no. 7-9, p. 102-107, 1 fig.

(1009) ———, & **Kahler, Gustava**, 1937, *Beiträge zur Kenntnis der Fusuliniden der Ostalpen. Die Pseudoschwagerinen der Grenzlandbänke und des oberen Schwagerinenkalkes:* Palaeontographica, v. 87, pt. A, pl. 1-42, pl. 1-3, fig. 1-2.

———(1010) 1946, *Zur Nomenklatur und Entwicklung der Fusuliniden:* K. K., geol. Reichsanst. (Bundesanst.), Wien, Verhandl., no. 10-12, p. 167-172.

(1011) **Kanmacher, Frederick**, 1798, *Adam's Essays on the microscope; the second edition, with considerable additions and improvements:* Dillon & Keating (London).

(1012) **Kanmera, Kametoshi**, 1954, *Fusulinids from the Upper Permian Kuma Formation, southern Kyushu, Japan—with special reference to the fusulinid zone in the Upper Permian of Japan:* Kyushu Univ., Faculty Sci., Mem., ser. D, Geol., v. 4, no. 1, 38 p., 6 pl., 2 fig.———
(1013) 1956, *Toriyamaia, a new Permian fusulinid genus from the Kuma Massif, Kyushu, Japan:* Palaeont. Soc. Japan, Trans. & Proc., new ser., no. 24, p. 251-257, pl. 36.

(1014) **Kanuma, Mosaburo,** & **Sakagami, Sumio**, 1957, *Mesoschubertella, a new Permian fusulinid genus from Japan:* Paleont. Soc. Japan, Trans. & Proc., new ser., no. 26, p. 41-46, pl. 8, 1 fig.

(1015) **Kaptarenko-Chernousova, O. K.**, 1956, *Pro novi rodi Foraminifer z rodini Epistominid:* Dopovidi Akad. Nauk Ukraïn. RSR, no. 2, p. 157-161, text-fig. 1-5. [*About a new genus of Foraminifera of the family Epistominidae.* In Ukrainian.]———(1016) 1956, *K voprosu o vidoobrazovanii i sistematike yurskikh Epistominid:* Voprosy Mikropaleontologii, v. 1, Akad. Nauk SSSR, p. 49-61, pl. 1, text-fig. 1. [*On the question of erection of species and systematics of Jurassic Epistominidae.*] [In Russian.]———(1017) 1956, *Foraminiferi küvskogo yarusu Dniprovsko-Donetskoï zapadini ta pivnichno-zakhidnikh okraïn Donets'kogo baseynu:* Akad. Nauk Ukraïn. RSR, Trudi, Inst. Geol. Nauk., Ser. strat. & paleont. no. 8, p. 1-64, pl. 1-11. [*Foraminifera of the Kiev strata of the Dnieper-Donets depression and northwest periphery of the Donets Basin.*] [In Ukrainian.]
———(1018) 1959, *Foraminiferi Yurskikh vidkladiv Dniprovsko-Donetskoï zapadini:* Same, no. 15, p. 1-121, pl. 1-18. [*Foraminifera of Jurassic sediments of the Dnieper-Donets depression.*] [In Ukrainian.]

(1019) **Karling, J. S.**, 1942, *The Plasmodiophorales:* 144 p., The Author (New York).

(1020) **Karrer, Felix**, 1865, *Die Foraminiferen-Fauna des tertiären Grünsandsteines der Orakei-Bay bei Auckland:* Novara Exped. 1857-59, v. 1, Geol. Theil., p. 69-86, pl. 16.———(1021) 1866, *Ueber das Auftreten von Foraminiferen in den älteren Schichten des Wiener Sandsteins:*

K. Akad. Wiss. Wien, Sitzungsber., v. 52, pt. 1 (1865), p. 492-497, pl. 1.————(1022) 1868, *Die miocene Foraminiferenfauna von Kostej im Banat:* K. Akad. Wiss. Wien. Math.-Naturwiss. Cl., Sitzungber., v. 58, pt. 1, p. 121-193, pl. 1-5.
————(1023) 1877, *Geologie der Kaiser Franz-Josefs Hochquellen-Wasserleitung, Eine Studie in den Tertiär-Bildungen am Westrande des alpinen Theiles der Niederung von Wien:* K.K. geol. Reichsanst., Abhandl., v. 9, p. 1-420, pl. 1-20.

(1024) ————, & Sinzow, Johann, 1877, *Über das Auftreten des Foraminiferen Genus Nubecularia im sarmatischen Sande von Kischenew:* K. Akad. Wiss. Wien, Math.-Naturwiss. Cl., Sitzungber., v. 74, pt. 1 (1876), no. 7, p. 272-284, pl. 1.

(1025)   **Karsten, Hermann**, 1858, *Über die geognostischen Verhältnisse des westlichen Columbien. Der heutigen Republiken Neu-Granada und Equador:* Deutsch. Naturforsch. Ärzte Wien, Amtl. Ber., v. 32 (1856), p. 80-117, pl. 1-6.

(1026)   **Kaufmann, F. J.**, 1856, *Der Pilatus, geologisch untersucht und beschrieben:* Beitrage Geol. Karte Schweiz, Lief. 5, p. 1-166, pl. 1-10 (Bern).

(1027)   **Kawai, K., Uchio, T., Ueno, M., & Hozuki, M.**, 1950, *Natural gas in the vicinity of Otaki, Chiba-ken:* Assoc. Petrol. Technology Jour., v. 15, no. 4, p. 151-219, text-fig. 1-25.

(1028)   **Keijzer [Keyzer], F. G.**, 1941, *Eine neue eozäne Foraminiferengattung aus Dalmatien:* Nederland. Akad. Wetensch., Proc., v. 44, no. 8, p. 1006-1007, text-fig. 1-4.————(1029) 1942, *On a new genus of arenaceous Foraminifera from the Cretaceous of Texas:* Same, v. 45, p. 1016-1017, text-fig. a-j.————(1030) 1945, *Outline of the geology of the eastern part of the province of Oriente, Cuba (E. of 76°WL) with notes on the geology of other parts of the island:* Dissertation, Univ. Utrecht, p. 1-239, pl. 1-11, De Vliegende Hollander (Utrecht).
————(1031) 1953, *Reconsideration of the so-called Oligocene fauna in the asphaltic deposits of Buton (Malay Archipelago), 2. Young-Neogene Foraminifera and calcareous algae:* Leidse Geol. Meded., pt. 17, p. 259-293, pl. 1-4.————
(1032) 1955, *Lamarckinita, new name, replacing Ruttenella Keyzer, 1953 (non Ruttenella van den Bold, 1946):* Cushman Found. Foram. Research, Contrib., v. 6, pt. 3, p. 119.

(1033)   **Keller, B. M.**, 1946, *Foraminifery verkhnemelovykh otlozheniy Sochinskogo rayona:* Moskov. Obshch. Ispyt. Prirody, Byull., v. 51, Otdel geol., v. 21, no. 3, p. 83-108, pl. 1-3, 2 tables. [*Foraminifera of Upper Cretaceous deposits of the Sochinsky district.*]

(1034)   **Kent, W. S.**, 1878, *The foraminiferal nature of Haliphysema tumanowiczii, Bow.* *(Squamulina scopula Carter),* demonstrated: Ann. & Mag. Nat. History, ser. 5, v. 2, p. 68-78, pl. 4-5.————(1035) 1880, *A manual of the Infusoria; including a description of all known flagellate, ciliate, and tentaculiferous Protozoa, British and foreign, and an account of the organization and affinities of the sponges:* v. 1, 472 p., D. Bogue (London).

Keyzer, F. G. [*see* Keijzer, F. G.]

(1036)   **Khalilov, D. M.**, 1951, *O faune foraminifer i raschlenenii oligotsenovykh otlozheniy severo-vostochnogo predgorya Malogo Kavkaza:* Akad. Nauk Azerbaidzhanskoi SSR, Izvestya, no. 3, p. 43-61, pl. 1-4. [*On a foraminiferal fauna and isolated Oligocene deposits of the northeast foothills of the lesser Caucasus.*]————
(1037) 1956, *O Pelagicheskoy faune foraminifer paleogenovykh otlozheniy Azerbaydzhana:* Akad. Nauk Azerbaidzhanskoi SSR, Inst. Geol., Trudy, v. 17, p. 234-255, pl. 1-5. [*On a pelagic foraminiferal fauna of Paleogene deposits of Azerbaidzhan.*]————(1038) 1958, *Novye predstaviteli foraminifer paleogenovykh otlozheniy Azerbaydzhana:* Akad. Nauk Azerbaidzhanskoi SSR, Izvestiya, Ser. Geol. & Geog. Nauk, no. 2, p. 3-14, pl. 1-2. [*New representatives of Foraminifera of Paleogene deposits of Aberbaidzhan.*]

(1039)   **Kikoïne, J.**, 1948, *Les Heterohelicidae du Crétacé superieur pyrénéen:* Soc. géol. France, ser. 5, v. 18, pt. 1-3, p. 15-35, pl. 1-2.

(1039A)   **King, William**, 1850, *Monograph of the Permian fossils of England:* Palaeontograph. Soc. London, v. 3, xxxvii+250 p., 28 pl. [Foraminifera by T. R. JONES, p. 15-20, pl. 6] (London).

(1040)   **Kiparisova, L. D., Markovsky, B. P., & Radchenko, G. P.**, 1956, *Materialy po paleontologii, novye semeystva i rody:* Vses. Nauchno-Issledov. Geol. Inst. (VSEGEI), Minist. Geol. & Okhrany Nedr. SSSR, p. 1-354, pl. 1-43. [*Material on paleontology, new families and genera.*]

(1040A)   **Kireeva, G. D.**, 1949, *Pseudofuzuliny tastubskogo i sterlitamakskogo gorizontov pogrebennykh massivov Bashkirii:* Akad. Nauk SSSR, Inst. Geol., Trudy, no. 105 (geol. ser. 35), p. 171-191, pl. 1-6. [*Pseudofusulinas of the Tastubsky and Sterlitamaksky horizons of the buried Bashkir massif.*]————(1040A bis) 1949, *Nekotorye novye vidy fuzulinid iz Kamennougol'nykh izvestnyakov tsentral'nogo raiona Donbassa:* Glavnoe Upravlenie Razvedkam Uglya, Geol. Issled. Byuro, Trudy, no. 6 (Moscow), *non vidi*. [*Some new species of the Fusulinidae from the Carboniferous limestone of the Donbass region.*]————(1040B) 1950, *Novye vidy fuzulinid iz izvestnyakov svit C_3^1 i C_3^2 Donetskogo basseyna:* Geol.-Issled. Raboty, Glavnoe Upravlenie Razvedkam Uglya, p. 193-212. [*New*

species of Fusulinidae from the well-known formations C_3^1 and C_3^2 of the Donets Basin.]
———(1040C) 1953, O nizhney granitse verkhnego karbona v Donetskom basseyne: Akad. Nauk SSSR, Doklady, v. 88, no. 1, p. 117-119. [The lower boundary of the Upper Carboniferous in the Donets Basin.]

(1041) Klasz, Ivan de, 1953, Quadratobuliminella n.gen., eine neue Foraminiferengattung von der Wende Kreide-Tertiär: Neues Jahrb. Paläont., Monatshefte, v. 10, p. 434-436, text-fig. 1-2.

(1042) ———, Marie, Pierre, & Meijer, M., 1960, Gabonella, nov. gen., un nouveau genre de Foraminifères du Crétacé supérieur et du Tertiaire basal de l'Afrique Occidentale: Revue Micropaléont., v. 3, no. 3, pl. 167-182, pl. 1-2.

(1043) ———, & Rérat, Daniel, 1962, Quelques nouveaux Foraminifères du Crétacé et du Tertiaire du Gabon (Afrique Equatoriale): Revue Micropaléont., v. 4, no. 4, p. 175-189, pl. 1-3.

(1044) Klebs, Georg, 1892, Flagellatenstudien, Theil II: Zeitschr. Wiss. Zool., v. 55, p. 353-445, pl. 17-18.

(1045) Klein, J., 1882, Vampyrella Cnk., ihre Entwicklung und systematische Stellung: Bot. Centralbl., v. 11, no. 7, p. 247-264, pl. 1-4.

(1046) Kleinpell, R. M., 1938, Miocene stratigraphy of California: Am. Assoc. Petroleum Geologists, 450 p., 22 pl.

(1047) Kobayashi, Manabu, 1957, Paleontological study of the Ibukiyama Limestone, Shiga Prefecture, Central Japan: Tokyo Kyoiku Daigaku, Sci. Rept., sec. C, v. 5, no. 47-48, p. 247-311, pl. 1-10, fig. 1-2.

(1047A) Kochansky-Devidé, V., & Ramovš, A., 1955, Neoschwagerinski skladi in njih fuzulinida favna pri Bohinjski Beli in Bledu: Slovenska Akad. Znanosti Umetnosti, Razred Prirodoslovne Vede, Classis 4 (Hist. Nat.), Razprave, p. 361-424, pl. 1-8, fig. 1-3. [Neoschwagerine beds and their fusulinid fauna in Bohinjska Bela and Bled.] [In Serbian & German.]

(1048) Kornfeld, M. M., 1931, Recent littoral Foraminifera from Texas and Louisiana: Stanford Univ., Dept. Geol., Contrib., v. 1, no. 3, p. 77-101, pl. 13-16.

(1049) Köváry, J., 1956, Thékamöbák (Testaceák) a magyarországi alsópannóniai korú üled ékekből: Földtani Közlöny, v. 86, no. 3, p. 266-273, pl. 35-39.

(1050) Krasheninnikov, V. A., 1953, K morfologii i sistematike foraminifer sem. Nonionidae: Moskov. Obshch. Ispyt. Prirody., new ser., v. 8 (58), Otdel. Geol. Bull., v. 28, no. 3, p. 88-89. [On the morphology and systematics of the foraminiferal family Nonionidae.]———(1051) 1958, Rotaliidy i Anomalinidy Miotsenovykh otlozheniy podolii: VNIGNI, Trudy, no. 9, Paleont. Sbornik, p. 212-250, pl. 1-9. [Rotaliidae and Anomalinidae of the Miocene deposits of Podolia.]———(1052) 1960, Mikrostruktura stenki u Miotsenovykh diskorbid i rotaliid: Voprosy Mikropaleontologii, no. 3, Akad. Nauk SSSR, Otdel. Geol. & Geog. Nauk, Geol. Inst., p. 41-49, pl. 1, 2. [Microstructure of the wall in Miocene discorbids and rotaliids.]———(1053) 1960, Izmenenie kompleksov foraminifer v ritmakh sedimentatsii Miotsenovykh otlozheniy yugo-zapada russkoy platformy: Mezhdunarodnyy Geol. Kongress Moskva, Sess. 21, Doklady Sov. Geol., Prob. 6, p. 78-84. [Variation in the foraminiferal assemblage in rhythmic sedimentation of Miocene deposits of the southwest Russian Platform.]———(1054) 1960, Elfidiidy Miotsenovykh otlozheniy Podolii: Akad. Nauk SSSR, Trudy, Geol. Inst., no. 21, p. 1-141, pl. 1-11. [Elphidiidae of the Miocene deposits of Podolia.]

(1055) Krestovnikov, V. N., & Teodorovich, G. I., 1936, Novyy vid roda Archaediscus iz karbona yuzhnogo Urala: Moskov. Obshch. Ispyt. Prirody, v. 44, Otdel. Geol. Byull., v. 14(1), p. 86-89. [New species of the genus Archaediscus from the Carboniferous of the southern Urals.]

(1056) Krinsley, David, 1960, Trace elements in the tests of planktonic Foraminifera: Micropaleontology, v. 6, p. 297-300, tables 1-2.

(1057) Kristan [-Tollmann], Edith, 1957, Ophthalmidiidae und Tetrataxinae (Foraminifera) aus dem Rhät der Hohen Wand in Nieder-Österreich: Geol. Bundesanstalt, Jahrb., v. 100, no. 2, p. 269-298, pl. 22-27.———(1058) 1958, Neue Namen für zwei Foraminiferengattungen aus dem Rhät: Geol. Bundesanstalt, Verhandl., no. 1, p. 114.

(1059) Kristan-Tollmann, Edith, 1960, Rotaliidea (Foraminifera) aus der Trias der Ostalpen: Geol. Bundesanstalt, Jahrb., spec. vol. 5, p. 47-78, pl. 7-21, 2 text-fig.———(1059A) 1962, Stratigraphisch wertvolle Foraminiferen aus Obertrias- und Liaskalken der voralpinen Fazies bei Wien: Erdoel-Zeitschrift, no. 4, p. 228-233, pl. 1-2.

(1060) Kübler, J., & Zwingli, Heinrich, 1866, Mikroskopische Bilder aus der Urwelt der Schweiz; Heft II: Winterthur, Bürgersbibl., Neujahrsbl., p. 1-28, pl. 1-3.———(1061), Die Foraminiferen des Schweizerischen Jura: p. 5-49, pl. 1-4, Steiner (Winterthur).

(1062) Kudo, R. R., 1931, Handbook of Protozoology: 451 p., 175 fig., C. C. Thomas (Baltimore).———(1063) 1939, Protozoology: ed. 2, 689 p., 291 fig., C. C. Thomas (Springfield & Baltimore).———(1064) 1954, Protozoology: ed. 4, 966 p., 376 text-fig., C. C. Thomas (Springfield).

(1065) Kufferath, Hubert, 1932, Rhizopodes du Congo: Revue Zool. Bot. Africaines, v. 23, pt. 1, p. 52-60, pl. 3-4.

(1066) **Kühn, Alfred,** 1926, *Morphologie der Tiere in Bildern. Heft 2, Protozoen; 2. Teil: Rhizopoden:* i-iv+107-272 p., fig. 202-407, Gebrüder Borntraeger (Berlin).

(1067) **Kükenthal, W. G., & Krumbach, Thilo,** 1923, *Handbuch der Zoologie:* v. 1, p. 51-112 (1923-1925), W. de Gruyter & Co. (Berlin).

(1068) **Küpper, Klaus,** 1954, *Notes on Upper Cretaceous larger Foraminifera. II, Genera of the subfamily Orbitoidinae with remarks on the microspheric generation of Orbitoides and Omphalocyclus:* Cushman Lab. Foram. Research, Contrib., v. 5, pt. 4, p. 179-184, pl. 33-34, 3 text-fig.——(1069) 1954, *Note on Schlumbergerella Hanzawa and related genera:* Same, v. 6, pt. 1, p. 26-30, text-fig. 1-4.——(1070) 1955, *Eocene larger Foraminifera near Guadalupe, Santa Clara County, California:* Same, v. 6, pt. 4, pl. 133-139, pl. 19, text-fig. 1-7.

(1071) **Kuwano, Y.,** 1950, *New species of Foraminifera from the Pliocene of Tama Hills in the vicinity of Tokyo:* Geol. Soc. Japan, Jour., v. 56, p. 311-321, text-fig. 1-13.

(1072) **Lacroix, Eugene,** 1923, *Texture chitineuse fondamentale de la coquille des Foraminifères porcelanés:* Acad. Sci. Paris, Comptes Rendus, v. 176, p. 1673.——(1073) 1926, *Du choix des coccolithes par les Foraminifères arénacés pour l'édification de leurs tests:* Assoc. Franç. Avanc. Sci. Lyon, p. 418-421.——(1074) 1929, *Textularia sagittula ou Spiroplecta wrightii?:* Inst. Océanog. Monaco, Bull., no. 532, p. 1-12.——(1075) 1931, *Microtexture du test des Textulariidae:* Same, no. 582, p. 1-18, text-fig. 1-10.——(1076) 1932, *Discammina, nouveau genre méditerranéen de Foraminifères arénacés:* Same, no. 600, p. 1-4, text-fig. a-e.——(1077) 1935, *Discammina fallax et Haplophragmium emaciatum:* Same, no. 667, p. 1-16.——(1078) 1938, *Sur une texture méconnue de la coquille de diverse Massilines des mers tropicales:* Same, no. 750, p. 1-8, text-fig. 1-4.——(1079) 1938, *Révision du genre Massilina:* Same, no. 754, p. 1-11, text-fig. 1-9.

(1080) **Lalicker, C. G.,** 1948, *Dwarfed protozoan faunas:* Jour. Sed. Petrology, v. 18, p. 51-55, pl. 1.——(1081) 1948, *A new genus of Foraminifera from the Upper Cretaceous:* Jour. Paleontology, v. 22, p. 624, pl. 92.——(1082) 1950, *Foraminifera of the Ellis group, Jurassic, at the type locality:* Univ. Kansas Paleont. Contrib., Protozoa, Art. 2, p. 3-20, pl. 1-4, fig. 1-5.

(1083) **Lamarck, J. B.,** 1799, *Prodrôme d'une nouvelle classification des coquilles, comprenant une rédaction appropriée des caractères génériques, et l'établissement d'un grand nombre de genres nouveaux:* Soc. Histoire Nat. Paris, Mém., p. 63-91.——(1084) 1801, *Système des animaux sans vertèbres:* 432 p., The Author (Paris).——(1085) 1804, *Suite des mémoires sur les fossiles des environs de Paris:* Museum Natl. Histoire Nat. Paris, Ann., v. 5; (a) p. 179-188, pl. 62; (b) p. 237-245, pl. 62; (c) p. 349-357, pl. 17.——(1086) 1809, *Philosophie zoologique, ou exposition des considérations relatives à l'histoire naturelle des Animaux, etc.:* v. 1, xxv+428 p., v. 2, 475 p., Dentu (Paris).——(1087) 1812, *Extrait du cours de zoologie du Muséum d'Histoire Naturelle sur les animaux invertébres:* 127 p. (Paris).——(1088) 1816, *Histoire naturelle des animaux sans vertèbres:* v. 2, 568 p., Verdière (Paris).——(1089) 1816, *Tableau encyclopédie et méthodique de trois règnes de la nature, Partie 23. Mollusques et Polypes divers:* p. 1-16, pl. 391-488, Mme. V. Agasse (Paris).——(1090) 1822, *Histoire naturelle des animaux sans vertèbres:* v. 7, 711 p., The Author (Paris).

(1091) **Lange, Erich,** 1925, *Eine Mittelpermische Fauna von Guguk Bulat (Padanger Oberland, Sumatra):* Geol.-Mijnb. Genoot. Nederland. Kolon., Verhandl. Geol. Ser., v. 7, p. 213-295, pl. 1-5, 10 text-fig.

(1092) **Lankester, E. R.,** 1877, *Notes on the embryology and classification of the animal kingdom: comprising a revision of speculations relative to the origin and significance of the germ-layers:* Quart. Jour. Micro. Sci., new ser., v. 17, p. 399-454, pl. 25.——(1093) 1885, *Protozoa,* in The Encyclopaedia Britannica: ed. 9, v. 19, p. 830-866.——(1094) 1903, *Introduction and Protozoa,* in A Treatise on Zoology, Pt. 1, fasc. 2, p. 47-149, Adam and Charles Black (London).——(1095) 1909, *Introduction and Protozoa,* in A Treatise on Zoology: Pt. 1, p. 68-93, Adam and Charles Black (London).

(1096) **Lapparent, Jacques de,** 1918, *Étude lithologique des terrains crétacés de la région d'Hendaye:* Serv. Carte géol. France, Mém., p. 1-155, pl. 1-10, text-fig. 1-27.

(1097) **Latreille, P. A.,** 1825, *Familles naturelles du Règne Animal, exposées succinctement et dans un ordre analytique, avec l'indication de leurs genres:* 570 p., J. B. Baillière (Paris.)——(1097A) 1827, *Natürlichen Familien des Thierreichs:* Translated from French, with annotations and additions by A. A. BERTHOLD (Weimar).

(1098) **Lauterborn, Robert,** 1895, *Protozoenstudien, II. Paulinella chromatophora nov. gen. nov. spec., ein beschalter Rhizopode des Süsswassers mit blaugrünen chromatophorenartigen Einschlüssen:* Zeitschr. Wiss. Zool., v. 59, p. 537-544, pl. 30.

(1099) **Lea, Isaac,** 1833, *Contributions to Geology:* 227 p., 6 pl., Carey, Lea & Blanchard (Philadelphia).

(1100) **Lebedeva, N. S.**, 1954, *Foraminifery nizhnego karbona Kuznetskogo basseyna:* Mikrofauna SSSR, Sbornik 7, VNIGRI, new ser., Trudy, no. 81, p. 237-295, pl. 1-12, text-fig., tables. [*Lower Carboniferous Foraminifera of the Kuznets Basin.*]———(1101) 1956, *Foraminifery etrenskikh otlozheniy Tengizskoy vpadiny:* Same, new ser., no. 98, Mikrofauna SSSR Sbornik 8, p. 39-53, pl. 1-3. [*Foraminifera of the Etroeungtian deposits of the Tengizsky Basin.*]

(1102) **Le Calvez, Jean**, 1935, *Sur quelques Foraminifères de Villefranche et de Banyuls:* Archives Zool. Expér. & Générale, v. 77 (Notes & Revue), no. 2, p. 79-98, text-fig. 1-11.———
(1103) 1935, *Les gamètes de quelques Foraminifères:* Acad. Sci. Paris, Comptes Rendus, v. 201, p. 1505-1507.———(1104) 1936, *Observations sur le genre Iridia:* Archives Zool. Expér. & Générale, v. 78, pt. 3, p. 115-131, text-fig. 1-7, pl. 1.———(1105) 1936, *Modifications du test des Foraminifères pélagiques en rapport avec la reproduction: Orbulina universa d'Orb. et Tretomphalus bulloides d'Orb.:* Ann. Protistologie, v. 5, p. 125-133, text-fig. 1-8.
———(1106) 1938, *Recherches sur les Foraminifères—I. Developpement et reproduction:* Archives Zool. Expér. & Générale, v. 80, pt. 3, p. 163-333, pl. 2-7, text-fig. 1-26.———(1107) 1938, *Un Foraminifère Géant Bathysiphon filiformis G. O. Sars:* Same, v. 79, no. 2, p. 82-88.———(1108) 1947, *Les perforations du test de Discorbis erecta (Foraminifère):* Lab. maritime Dinard, Bull., v. 29, p. 1-4.———(1109) 1950, *Recherches sur les Foraminifères. 2. Place de la méiose et sexualité:* Archives Zool. Expér. & Générale, v. 87, pt. 4, p. 211-243, 1 pl., 4 text-fig.———(1110) 1952, *Le couple Discorbis patelliformis (Brady)—erecta (Sidebottom) et les Discorbis plastogamiques:* Same, v. 89, p. 56-62.

(1111) ———, & **Le Calvez, Yolande**, 1951, *Contribution a l'étude des Foraminifères des eaux saumatres. I. Etangs de Canet et de Salses:* Vie et Milieu, v. 2, pt. 2, p. 237-254, text-fig. 1-5.

(1112) **Le Calvez, Yolande**, 1949, *Révision des Foraminifères Lutétiens du Bassin de Paris. II. Rotaliidae et familles affines:* Carte Géol. Détaillée France, Mém., 54 p., 6 pl.———(1113) 1950, *Révision des Foraminifères Lutétiens du Bassin de Paris, III. Polymorphinidae, Buliminidae, Nonionidae:* Same, 64 p., 4 pl.———
(1114) 1952, *Révision des Foraminifères Lutétiens du Bassin de Paris, IV. Valvulinidae, Peneroplidae, Ophthalmidiidae, Lagenidae:* Same, p. 1-64, pl. 1-4.———(1115) 1959, *Étude de quelques Foraminifères nouveaux du Cuisien Franco-Belge:* Revue Micropaléont., v. 2, no. 1, p. 88-94, pl. 1.

(1116) **Leclerc, L.**, 1816, *Note sur la Difflugie, nouveaux genres de polype amorphe:* Museum Histoire Nat., Mém., v. 2, p. 474-478, pl. 17, Sept. 1816 (Paris).

(1117) **Lecointre, Georges**, & **Allix, Henri**, 1913, *Les formes diverses de la vie dans les Faluns de Touraine; Treizième suite—Les Foraminifères:* Feuille Jeunes Nat., v. 43 (ser. 5, v. 3), p. 6-8, 29-35, 41-47, text-fig. 1-10.

(1118) **Lee, J. S.**, 1924, *Grabauina, a transitional form between Fusulinella and Fusulina:* Geol. Soc. China, Bull., v. 3, p. 51-54, fig. 1-2.
———(1119) 1927, *Fusulinidae of North China:* Geol. Surv. China, Palaeont. Sinica, ser. B, v. 4, pt. 1, 172 p., 24 pl., 21 fig.———
(1119A) 1931, *Distribution of the dominant types of the fusulinoid Foraminifera in the Chinese seas:* Geol. Soc. China, Bull., v. 10, p. 273-290, pl. 1.———(1120) 1933[1934], *Taxonomic criteria of Fusulinidae with notes on seven new Permian genera:* Natl. Research Inst. Geol., Mem., no. 14, p. 1-32, pl. 1-5, fig. 1-9.
———(1120A) 1942, *Note on a new fusulinid genus Chusenella:* Geol. Soc. China, Bull., v. 22, no. 3, p. 171-173.

(1121) ———, **Chen, S.**, & **Chu, S.**, 1930, *The Huanglung Limestone and its fauna:* Natl. Research Inst. Geol., Mem., no. 9, p. 85-143, pl. 2-15.

(1122) **Lehmann, R.**, 1962, *Strukturanalyse einiger Gattungen der Subfamilie Orbitolitinae:* Eclogae geol. Helvet., v. 54, no. 2, p. 597-667, pl. 1-14.

(1123) **Leidy, Joseph**, 1874, *Notice of some new fresh-water rhizopods:* Acad. Nat. Sci. Philadelphia, Proc., p. 77-79.———(1124) 1875, *On a curious rhizopod:* Same, p. 124-125.———
(1125) 1875, *Notice of some rhizopods:* Same, (1874), pt. 3, p. 155-157.———(1126) 1877, *Remarks on rhizopods, and notice of a new form:* Same, pt. 3, p. 293-294.———(1127) 1879, *Freshwater rhizopods of North America:* U.S. Geol. Survey Terr., v. 12, p. 1-324, pl. 1-48.

(1128) **Leischner, W.**, 1961, *Zur Kenntnis der Mikrofauna und -flora der Salzburger Kalkalpen:* Neues Jahrb. Geol. & Paläont., Abhandl., v. 112, no. 1, p. 1-47, pl. 1-14.

(1129) **Lendenfeld, R. von**, 1886, *On the systematic position and classification of sponges:* Zool. Soc. London, Proc., p. 558-662.

(1130) **Leupold, Wolfgang**, & **Bigler, H.**, 1936, *Coscinoconus eine neue Foraminiferenform aus Tithon-Unterkreide-Gesteinen der helvetischen Zone der Alpen:* Eclogae geol. Helvet., v. 28 (1935), no. 2, p. 606-624, pl. 18.

(1131) ———, & **Maync, Wolf**, 1935, *Das Auftreten von Choffatella, Pseudocyclammina, Lovćenipora (Cladocoropsis) und Clypeina im alpinen Faziesgebiet:* Eclogae geol. Helvet., v. 28, p. 129-139.

(1132) **Levine, N. D.,** 1962, *Protozoology today:* Jour. Protozoology, v. 9, no. 1, p. 1-6, text-fig. 1-3, table 1-6.

(1132A) **Leymerie, A. F. G. A.,** 1846, *Mémoire sur le terrain à Nummulites (épicrétacé) des Corbières et de la Montagne Noire:* Soc. géol. France, Mém., ser. 2, v. 1, pt. 2, p. 337-373, p. 13.———(1133) 1851, *Mémoire sur un nouveau type pyrénéen:* Same, ser. 2, v. 4, pt. 1, p. 177-202, pl. A-C (9-11).

(1134) **Liebus, Adalbert,** 1902, *Ergebnisse einer mikroskopischen Untersuchung der organischen Einschlusse der oberbayerischen Molasse:* K.K. geol. Reichsanst., Jahrb. (1902), v. 52, no. 1, p. 71-104, pl. 5, text-fig. 1-7.———(1135) 1911, *Die Foraminiferenfauna der Mitteleocänen Mergel von Norddalmatien:* K. Akad. Wiss. Wien, Math.-Naturwiss., Kl., Sitzungsber., v. 120, pt. 1, p. 865-956, pl. 1-3.———(1136) 1922, *Zur Altersfrage der Flyschbildungen im nordöstlichen Mähren:* Naturwiss. Zeitschr. Lotos, v. 70, p. 23-66, pl. 2.

(1137) **Likharev [Licharew], B. K.,** 1926, *Palaeofusulina nana sp. nova iz antrakolitovykh otlozheny sev. Kavkaza:* Izvestiya Geol. Komiteta, Izdanie Geol. Kom., v. 45, no. 2, p. 59-66, pl. 2, 1 fig. [*Palaeofusulina nana sp. nov. from anthracolithic deposits of the northern Caucasus.*]

(1138) ———, et al., 1939, *Atlas rukovodyashchikh form iskopayemykh faun SSSR, VI Permskaya Sistema:* Tsentralnyi Nauchno-issledov. Geologo-razved. Institut SSSR, 269 p., 56 pl., 113 fig. [*The atlas of the leading forms of the fossil fauna USSR, VI. Permian System.*]

(1139) **Lindsey, Marjorie,** 1913, *On Gypsina plana Carter, and the relations of the genus:* Linnean Soc. London, Trans., ser. 2, Zool., v. 16, pt. 1, p. 45-51, text-fig. 1-6.

(1140) **Linné, Caroli,** 1758, *Systema naturae per regna tria naturae, secundum classes, ordines, genera, species, cum characteribus, differentiis, synoymis, locis:* ed. 10, v. 1, p. 1-824, G. Engelman (Lipsiae).

(1141) **Lipina, O. A.,** 1948, *Foraminifery Chernyshinskoy svity turneyskogo yarusa Podmoskovnogo nizhnego karbona:* Akad. Nauk SSSR, Inst. Geol. Nauk, Trudy, no. 62, Geol. ser. no. 19, p. 251-259, pl. 19, 20. [*Foraminifera of the Chernyshinsky formation of the Tournaisian Stage of the Lower Moscovian, Lower Carboniferous.*]———(1142) 1950, *Foraminifery verkhnego devona Russkoy platformy:* Same, no. 119, Geol. ser., no. 43, p. 110-133, pl. 1-3. [*Foraminifera of the upper Devonian of the Russian Platform.*]———(1143) 1955, *Foraminifery turneyskogo yarusa i verkhney chasti devona Volgo-Ural'skoy oblasti i zapadnogo sklona Srednego Urala:* Same, v. 163, p. 1-96, pl. 1-13, text-fig. 1-7. [*Foraminifera of the Tournaisian Stage and upper part of the Devonian of the Volgo-Ural district and western slope of the middle Urals.*]———(1144) 1959, *Nakhodka foraminifer v Silure i Ordovike Sibiri:* Akad. Nauk SSSR, Doklady, v. 128, no. 4, p. 823-826, fig. 1-25. [*Discovery of Foraminifera in the Silurian and Ordovician of Siberia.*]——— (1145) 1960, *Foraminifery turneyskikh otlozheniy Russkoy platformy i Urala:* Mezhdunarodnyy Geol. Kongress, Sess. 21, 1960, Doklady Sovetskikh Geol., Prob. 6, Akad. Nauk Soyuza SSR Moscow, p. 48-55. [*Foraminifera of the Tournaisian deposits of the Russian Platform and Urals.*]

(1146) **Lister, Arthur,** 1894, *A monograph of the Mycetozoa, being a descriptive catalog of the species in the herbarium of the British Museum:* British Museum (Nat. History), p. 1-224, pl. 1-78, 51 fig.

(1147) ———, & **Lister, Gulielma,** 1925, *A monograph of the Mycetozoa, a descriptive catalogue of the species in the herbarium of the British Museum:* Ed. 3, British Museum (Nat. History), xxxii+296 p., 222 pl., 60 text-fig.

(1148) **Lister, Gulielma,** 1918, *The Mycetozoa: A short history of their study in Britain; an account of their habitats generally; and a list of species recorded from Essex:* Essex Field Club, Spec. Mem., v. 6, p. 1-54, 1 pl.

(1149) **Lister, J. J.,** 1895, *Contributions to the life history of the Foraminifera:* Royal Soc. London, Philos. Trans., ser. B, v. 186, p. 401-453.

(1150) **Loeblich, A. R., Jr.,** 1951, *Coiling in the Heterohelicidae:* Cushman Found. Foram. Research, Contrib., v. 2, pt. 3, p. 106-111, pl. 12.———(1151) 1952, *New Recent foraminiferal genera from the tropical Pacific:* Washington Acad. Sci., Jour., v. 42, no. 6, p. 189-193, fig. 1-5.———(1152) 1952, *Ammopemphix, new name for the Recent foraminiferal genus Urnula Wiesner:* Same, v. 43, no. 3, p. 82.——— (1153) 1958, *The Foraminiferal genus Halyphysema and two new tropical Pacific species:* U.S. Natl. Museum, Proc., v. 107, no. 3385, p. 123-126, 1 pl.

(1154) ———, & **Tappan, Helen,** 1946, *New Washita Foraminifera:* Jour. Paleontology, v. 20, p. 238-258, pl. 35-37, 4 text-fig.——— (1155) 1949, *New Kansas Lower Cretaceous Foraminifera:* Washington Acad. Sci., Jour., v. 39, no. 3, p. 90-92.———(1156) 1949, *Foraminifera from the Walnut Formation (Lower Cretaceous) of northern Texas and Southern Oklahoma:* Jour. Paleontology, v. 23, no. 3, p. 245-266, pl. 46-51.———(1157) 1950, *North American Jurassic Foraminifera II: characteristic western interior Callovian species:* Washington Acad. Sci., Jour., v. 40, no. 1, p. 5-19, pl. 1. ———(1158), 1952, *Cribrotextularia, a new* foraminiferal genus from the Eocene of Florida: Same, v. 42, no. 3, p. 79-81, fig. 1-5.——— (1159) 1952, Adercotryma, a new Recent foraminiferal genus from the Arctic: Same, v. 42, no. 5, p. 141-142, text-fig. 1-4.———(1160) 1952, Poritextularia, a new Recent foraminiferal genus: Same, v. 42, no. 8, p. 264-266, text-fig. 1-3.———(1161) 1952, The foraminiferal genus Triplasia Reuss, 1854: Smithsonian Misc. Coll., v. 117, no. 15, p. 1-61, pl. 1-8.——— (1162) 1953, Studies of Arctic Foraminifera: Same, v. 121, no. 7, p. 1-150, pl. 1-24.——— (1163) 1953, Olssonina Bermúdez, 1949 for Cribrotextularia Loeblich and Tappan, 1952: Micropaleontologist, v. 7, no. 2, p. 44-45.——— (1164) 1954, The type species of Bulbophragmium Maync, 1952: Same, v. 8, no. 4, p. 32-33. ———(1165) 1954, Emendation of the foraminiferal genera Ammodiscus Reuss, 1862, and Involutina Terquem, 1862: Washington Acad. Sci., Jour., v. 44, no. 10, p. 306-310, text-fig. 1-2.———(1166) 1955, Revision of some Recent foraminiferal genera: Smithsonian Misc. Coll., v. 128, no. 5 (Publ. 4214), p. 1-37, pl. 1-4.———(1167) 1955, A revision of some glanduline Nodosariidae (Foraminifera): Same, v. 126, no. 3, p. 1-9, pl. 1.———(1168) 1956, Chiloguembelina, a new Tertiary genus of the Heterohelicidae (Foraminifera): Washington Acad. Sci., Jour., v. 46, no. 11, p. 340.——— (1169) 1957, Woodringina, a new foraminiferal genus (Heterohelicidae) from the Paleocene of Alabama: Same, v. 47, no. 2, p. 39-40, text-fig. 1.———(1170) 1957, The new planktonic foraminiferal genus Tinophodella, and an emendation of Globigerinita Brönnimann: Same, v. 47, no. 4, p. 112-116, fig. 1-3.———(1171) 1957, Morphology and taxonomy of the foraminiferal genus Pararotalia Le Calvez, 1949: Smithsonian Misc. Coll., v. 135, no. 2, p. 1-24, pl. 1-5.———(1172) 1957, Eleven new genera of Foraminifera: U.S. Natl. Museum, Bull. 215, p. 223-232, pl. 72-73.———(1173) 1957, The Foraminiferal genus Cruciloculina d'Orbigny, 1839: Same, Bull. 215, p. 233-235, pl. 74. ———(1174) 1957, Planktonic Foraminifera of Paleocene and early Eocene age from the Gulf and Atlantic Coastal Plains: Same, Bull. 215, p. 173-198, pl. 40-64, fig. 27-28.———(1175) 1960, Saedeleeria, new genus of the family Allogromiidae (Foraminifera): Biol. Soc. Washington, Proc., v. 73, p. 195-196.———(1176) 1961, The status and type species of the foraminiferal genera Ammodiscus, 1862, and Involutina Terquem, 1862: Micropaleontology, v. 7, no. 2, p. 189-192.———(1177) 1961, Suprageneric classification of the Rhizopodea: Jour. Paleontology, v. 35, p. 245-330.———(1178) 1961, The genera Microaulopora Kuntz, 1895, and Guembelina Kuntz, 1895, and the status of Guembelina Egger, 1899: Same, v. 35, p. 625-627, 1 text-fig.———(1179) 1961, The type species of Marginulina d'Orbigny, 1826: Cushman Found. Foram. Research, Contrib., v. 12, pt. 3, p. 77-78.——— (1180) 1961, The type species of the foraminiferan genus Saccammina Carpenter, 1869: Same, v. 12, pt. 3, p. 79-80.———(1181) 1961, Remarks on the systematics of the Sarkodina (Protozoa), renamed homonyms and new and validated genera: Biol. Soc. Washington, Proc., v. 74, p. 213-234.———(1182) 1961, The status of Hagenowella Cushman, 1933 and a new genus Hagenowina: Same, v. 74, p. 241-244.———(1183) 1961, Cretaceous planktonic Foraminifera: Part 1—Cenomanian: Micropaleontology, v. 7, no. 3, p. 257-304, pl. 1-8.——— (1184) 1962, Quinqueloculina d'Orbigny, 1826 (Foraminifera); Proposed validation under the plenary powers and designation of a neotype for Serpula seminulum Linnaeus, 1758. Z.N.(S.) 1494: Bull. Zool. Nomenclature, v. 19, pt. 2, p. 118-124.———(1185) 1962, Six new generic names in the Mycetozoida (Trichiidae) and Foraminiferida (Fischerinidae, Buliminidae, Caucasinidae and Pleurostomellidae), and a redescription of Loxostomum (Loxostomidae, new family): Biol. Soc. Washington, Proc., v. 75, p. 107-113.———(1186) 1962, The status and type species of Calcarina, Tinoporus and Eponides: Cushman Found. Foram. Research, Contrib., v. 13, pt. 2, p. 33-38, text-fig. 1a-c. ———(1187) 1962, The foraminiferal genera Cibicides, Heterolepa, Planulina and Holmanella, new genus: Same, v. 13, pt. 3, p. 71-73. ———(1187A) 1963, Discolithus Fortis, 1802 (Foraminiferida), and its type species: Jour. Paleontology, v. 37, p. 488-490.

(1188) **Loetterle, G. J.,** 1937, *The micropaleontology of the Niobrara Formation in Kansas, Nebraska, and South Dakota:* Nebraska Geol. Survey, Bull. 12, ser. 2, 73 p., 11 pl.

(1189) **Logue, L. L., & Haas, M. W.,** 1943, *Paranonion, a new genus of Foraminifera from the Miocene of Venezuela:* Jour. Paleontology, v. 17, p. 177-178, pl. 30.

(1190) **Luerssen, Christian,** 1879, *Handbuch der systematischen Botanik, Band I:* xii+657 p., 181 fig., H. Haessel (Leipzig).

(1191) **Lütken, C. F.,** 1876, *Protozoa:* Zool. Record, v. 11 (1874), p. 531-545.

(1192) **Lyell, Charles,** 1848, *On the relative age and position of the so-called Nummulite limestone of Alabama:* Geol. Soc. London, Quart. Jour., v. 4, p. 10-16.

(1193) **MacBride, T. H.,** 1892, *The Myxomycetes of eastern Iowa:* State Univ. Iowa, Lab. Nat. History, Bull., v. 2, no. 2, p. 99-162, pl. 1-10. ———(1194) 1899, *North American slime moulds:* xvii+231 p., 18 pl., Macmillan Co. (New York).———(1195) 1922, *The North American slime-moulds, new and revised edi-*

tion: xvii+299 p., 23 pl., Macmillan Co. (New York).

(1196) **M'Coy, Frederick**, 1849, *On some new genera and species of Paleozoic corals and Foraminifera:* Ann. & Mag. Nat. History, ser. 2, v. 3, p. 119-136.

(1197) **Macfadyen, W. A.**, 1933, *A note on the foraminiferal genus Bolivinopsis Yakovlev:* Royal Micro. Soc. London, Jour., ser. 3, v. 53, p. 139-141, text-fig.———(1198) 1936, *D'Orbigny's Lias Foraminifera:* Same, v. 56, p. 147-153, pl. 1.———(1199) 1939, *On Ophthalmidium, and two new names for Recent Foraminifera of the family Ophthalmidiidae:* Same, ser. 3, v. 59, p. 162-169, text-fig. 1-3.———(1200) 1941, *Foraminifera from the Green Ammonite beds, Lower Lias, of Dorset:* Royal Soc. London, Philos. Trans., ser. B, no. 576, v. 231, p. 1-73, pl. 1-4.

(1201) **McLean, J. D., Jr.**, 1956, *The Foraminifera of the Yorktown Formation in the York-James Peninsula of Virginia, with notes on the associated mollusks:* Bull. Am. Paleontology, v. 36, no. 160, p. 261-394, pl. 35-53.

(1202) **Maitland, R. T.**, 1851, *Descriptio systematica animalium Belgii septentrionalis, etc., Pt. 1. Rhizopodes:* 234 p., Lugduni-Batavorum, C. C. van der Hoek (Leiden).

(1203) **Majzon, László**, 1943, *Adatok Egyes Kárpátaljai flis-rétegekhez, tekintellel a Globotruncanákra:* Évkönyve, Magyar Kiralyi Földtani Intézet, v. 37, no. 1, p. 1-170, pl. 1-2.——— (1204) 1948, *Centenarina, n.gen., and Cassidulina vitalisi, n.sp., from the lower Rupelian strata at Budai:* Földtani Közlöny, v. 78, p. 22-25.———(1205) 1954, *Contributions to the stratigraphy of the Dachstein limestone:* Acad. Sci. Hungary, Acta Geologica, v. 2, fasc. 3-4, p. 243-249, pl. 1-3.

(1206) **Makiyama, J., & Nakagawa, T.**, 1941, *Pleistocene Foraminifera from Simi, Mie Prefecture:* Geol. Soc. Japan, Jour., v. 48, p. 239-242 (p. 242-243, English résumé).

(1207) **Malakhova, N. P.**, 1954, *Foraminifery kizelovskogo izvestnyaka zapadnogo sklona Urala:* Moskov. Obshch. Ispyt. Prirody, Otdel Geol. Bull., v. 29, no. 1, p. 49-60, pl. 1-2. [*Foraminifera of the Kizelovsky limestone of the western slopes of the Urals.*]———(1208) 1956, *Foraminifery verkhnego turne zapadnogo sklona severnogo i srednego Urala:* Akad. Nauk SSSR, Uralskiy Filial, Trudy Gorno-Geol. Inst., no. 24, p. 72-155, pl. 1-15. [*Foraminifera of the upper Tournaisian of the western slopes of the northern and middle Urals.*]

(1209) **Małecki, Jerzy**, 1954, *Flabellamminopsis, nowy rodzaj otwornic aglutynujacych z doggeru okolic Częstochowy:* Soc. géol. Pologne, Ann., v. 22 (1952), p. 101-122, pl. 3-5, text-fig. 1-3. ———(1210) 1954, *O Nowych rodzajach otwornic aglutynujacych z Polskiego Miocenu:* Osobne Odbicie z Rocznika Pol Towarzystwa Geologicznego, v. 22, no. 4, p. 497-513, text-fig. 1-5, pl. 12-13.

(1211) (*see* **1787A.**)

(1212) **Mangin, J. P.**, 1954, *Description d'un nouveau genre de Foraminifère: Fallotella alavensis:* Bull. Sci. Bourgogne, v. 14, p. 209-219, pl. 1, 3 fig.

(1213) **Mantell, G. A.**, 1850, *A pictorial atlas of fossil remains consisting of illustrations selected from Parkinson's "Organic remains of a former world" and Artis' "Antediluvian phytology":* 207 p., 74 pl., text-fig., H. G. Bohn (London).

(1214) **Marie, Pierre**, 1938, *Sur quelques Foraminifères nouveaux ou peu connus du Crétacé du Bassin de Paris:* Soc. géol. France, Bull., ser. 5, v. 8, p. 91-104, pl. 7-8.———(1215) 1941, *Les Foraminifères de la Craie à Belemnitella mucronata du Bassin de Paris:* Museum Natl. Histoire Nat., Mém., new ser., v. 12, pt. 1, p. 1-296, pl. 1-37.———(1216) 1945, *Sur un Foraminifère nouveaux du Crétacé Supérieur Marocain: Lacosteina gouskovi nov. gen. et nov. sp.:* Soc. géol. France, Bull., ser. 5, v. 13 (1943), p. 295-298, text-fig. 1-6.———(1217) 1946, *Sur Laffitteina bibensis et Laffitteina monodi nouveau genre et nouvelles espèces de Foraminifères du Montien:* Same, ser. 5, v. 15 (1945), p. 419-434, pl. 5.———(1218) 1950, *Queraltina, noveau genre de Foraminifères de l'Eocène pyrénéen:* Same, ser. 5, v. 20, p. 73-80, text-fig. 1-9.———(1219) 1950, *Sur l'evolution de la faune de Foraminifères des couches de passage du Crétacé au Tertiaire:* Internatl. Geol. Congr., 18th Sess. (1948), Great Britain, Rept., Pt. 15 (Internatl. Paleont. Union), p. 50.——— (1220) 1955, *Quelques genres nouveaux de Foraminifères du Crétacé à faciès récifal:* Internatl. Geol. Congr., 19th Sess. (1952), Alger, Proc., sec. 13, pt. 15, p. 117-124, 5 text-fig.———(1221) 1956, *Sur quelques Foraminifères nouveaux du Crétacé supérieur belge:* Soc. géol. Belg., Ann., v. 80, p. B235-237, 3 pl. ———(1222) 1957-58, *Goupillaudina, nouveau genre de Foraminifère du Crétacé supérieur:* (a) 1957, Soc. géol. France, Comptes Rendus, no. 12, p. 247-248; (b) 1958, Same, Bull., ser. 6, v. 7 (1957), p. 861-876, pl. 43, text-fig. 1-3. ———(1223) 1958, *Peneroplidae du Crétacé supérieur a faciès récifal, 1. A propos des genres Broekina et Praesorites et sur le nouveau genre Vandenbroekia:* Revue Micropaléont., v. 1, no. 3, p. 125-139, pl. 1.———(1224) 1960, *Sur les faciès à Foraminifères du Coniacien subrécifal de la région de Foissac (Gard) et sur le nouveau genre Sornayina:* Soc. géol. France, Bull., ser. 7, v. 1 (1959), no. 3, p. 320-326, pl. 19b, fig. 1.

(1225) **Marks, Peter**, 1951, *Arenonionella, a new arenaceous genus of Foraminifera from the Miocene of Algeria:* K. Nederland. Akad. Wetensch.

Proc., ser. B, v. 54, no. 4, p. 375-378, text-fig. 1-4.

(1226) **Marriott, W. K.**, 1878, *The classification of the Foraminifera:* Hardwicke's Science-Gossip, v. 14, p. 30-31.

(1227) **Marshall, W.**, 1881, *Untersuchungen über Dysideiden und Phoriospongien:* Zeitschr. Wiss. Zool., v. 35, p. 88-129, pl. 6-8.

(1228) **Marsson, Theodor**, 1878, *Die Foraminiferen der weissen Schreibkreide der Inseln Rügen:* Mitt. nat. ver. Neu-Vorpommern und Rügen, v. 10, p. 115-196, pl. 1-5.

(1229) **Martin, Karl**, 1880, *Die Tertiärschichten auf Java:* Lief. 3, Paläont. Theil (1879-1880), p. 150-164, I-VI, Tab. 1-28, E. J. Brill (Leiden).
———(1230) 1890, *Untersuchungen über den Bau von Orbitolina (Patellina auct.) von Borneo:* Geol. Reichs-Museum Leiden, Samml., ser. 1, v. 4 (1884-1889), p. 209-229, pl. 24-25.

(1231) **Maslov, V. P.**, 1935, *Novye dannye o Foraminiferakh Donbassa i ikh rod', kak markiruyushchikh organizmov:* Geologiya na fronte industrializatsiy "Azchergeogidro-geodeziya," no. 4, p. 9-16. [*New data on Foraminifera of the Don Basin and their genera, as index organisms.*]———(1232) 1958, *Nakhodka v yure Kryma roda Coscinoconus Leupold i ego istinnaya priroda:* Akad. Nauk SSSR, Doklady, v. 121, no. 3, p. 545-548, text-fig. 1-3. [*Occurrence in the Jurassic of the Crimea of the genus Coscinoconus Leupold and its true nature.*]

(1233) **Massee, G. E.**, 1892, *A monograph of the Myxogastres:* 359 p., 12 pl., Methuen & Co. (London).

(1234) **Mathews, R. D.**, 1945, *Rectuvigerina, a new genus of Foraminifera from a restudy of Siphogenerina:* Jour. Paleontology, v. 19, p. 588-606, pl. 81-83.

(1235) **Matouschek, Franz**, 1895, *Beiträge zur Paläontologie des böhmischen Mittelgebirges; II. Mikroskopische Fauna des Baculitenmergels von Tetschen:* Naturwiss. Zeitschr., Lotos, new ser., v. 15, p. 117-163, pl. 1.

(1236) **Matsunaga, Takashi**, 1954, *Oinomikadoina ogiensis, n.gen., n.sp. from the Pliocene of Niigata, Japan:* Paleont. Soc. Japan, Trans. & Proc., new ser., no. 15, p. 163-164, text-fig. 1-3.
———(1237) 1955, *Spirosigmoilinella, a new foraminiferal genus from the Miocene of Japan:* Same, new ser., no. 18, p. 49-50, 2 text-fig.

(1238) **Mayer, F. K.**, 1932, *Ueber die Modifikation des Kalzium Karbonats in Schalen und Skeletten rezenter und fossiler Organismen:* Chemie der Erde, v. 7, no. 2, p. 346-350, text-fig. 1-4.

(1239) **Maync, Wolf**, 1950, *The foraminiferal genus Choffatella in the Lower Cretaceous (Urgonian) of the Caribbean Region (Venezuela, Cuba, Mexico, and Florida):* Eclogae geol. Helv., v. 42, no. 2 (1949), p. 529-547, pl. 11-12, 1 fig.———(1240) 1952, *Critical taxonomic study and nomenclatural revision of the Lituolidae based upon the prototype of the family, Lituola nautiloidea Lamarck, 1804:* Cushman Found. Foram. Research, Contrib., v. 3, pt. 2, p. 35-56, pl. 9-12.———(1241) 1952, *Alveolophragmium venezuelanum n.sp. from the Oligo-Miocene of Venezuela:* Same, v. 3, pt. 3-4, p. 141-144, pl. 26.———(1242) 1953, *Hemicyclammina sigali, n.gen., n.sp., from the Cenomanian of Algeria:* Same, v. 4, pt. 4, p. 149-150, text-fig. 1.———(1243) 1954, *The genus Navarella Ciry and Rat, 1951, in the Maestrichtian of Switzerland:* Same, v. 5, pt. 3, p. 138-144, pl. 25-27.———(1244) 1955, *Reticulophragmium, n. gen., a new name for Alveolophragmium Stschedrina, 1936 (Pars):* Jour. Paleontology, v. 29, p. 557-558.———(1245) 1958, *Feurtillia frequens, n.gen., n.sp., a new genus of lituolid Foraminifera:* Cushman Found. Foram. Research, Contrib., v. 9, pt. 1, p. 1-3, pl. 1-2.———(1246) 1958, *Ammocycloloculina, n.gen., an unknown foraminiferal genus:* Same, v. 9, pt. 3, p. 53-57, pl. 13-14.———(1247) 1959, *Deux nouvelles espèces Crétacées du genre Pseudocyclammina (Foraminifères):* Revue Micropaléont., v. 1, no. 4, p. 179-189, pl. 1-4.———(1248) 1959, *Martiguesia cylamminiformis n.gen., n.sp., un nouveau genre de Lituolidés à structure complexe:* Same, v. 2, no. 1, p. 21-26, pl. 1-3.———(1249) 1959, *The foraminiferal genera Spirocyclina and Iberina:* Micropaleontology, v. 5, no. 1, p. 33-68, pl. 1-8, text-fig. 1-3.———(1250) 1960, *Torinosuella, n.gen., eine mesozoische Gattung der lituoliden Foraminiferen:* Eclogae geol. Helv., v. 52, no. 1, p. 5-14, pl. 1.———(1251) 1961, *Remarks on the foraminiferal genus Sornayina:* Same, v. 53 (1960), no. 2, p. 497-500, pl. 1-2.

(1251A) **Meek, F. B.**, 1864, *Carboniferous and Jurassic fossils; Sect. 1. Description of the Carboniferous fossils:* Geol. Survey California, Paleont., v. 1, p. 1-4, pl. 2.

(1251B) ———, & **Hayden, F. V.**, 1859, *Remarks on the Lower Cretaceous beds of Kansas and Nebraska, together with descriptions of some new species of Carboniferous fossils from the valley of Kansas River:* Acad. Nat. Sci. Philadelphia, Proc., v. 10 (1858), p. 256-264.———(1252) 1865, *Paleontology of the upper Missouri; invertebrates:* Smithsonian Contrib. Knowledge, v. 14, art. 5 (172), p. 1-135.

(1253) **Melville, R. V.**, 1959, *Proposed use of the plenary powers to suppress the generic names Orthoceros Brünnich, 1771, and Orthocera Modeer, 1789, so as to stabilize the generic name Orthoceras Bruguière, 1789 (Class Cephalopoda, order Nautiloidea)* Z.N. (S). 44: Bull. Zool. Nomenclature, v. 17, p. 9-24.

(1254) **Meunier, Stanislas**, 1888, *Examen paléontologique du calcaire à Saccamina de Cussy-en-*

Morvan: Soc. Histoire Nat. Autun, no. 1, p. 232-236, pl. 7.

(1255) **Michelin, Hardouin,** 1846, *Inconographie zoophytologique:* livr. 21-26, p. 222-320; atlas, pl. 61-76, P. Bertrand (Paris).

(1256) **Michelotti, Giovanni,** 1841, *Saggio storico dei Rizopodi carratteristici dei terreni sopracretacei:* Soc. Ital. Sci., Mem. Fis., v. 22, p. 253-302, pl. 1-3.———(1257) 1861, *Études sur le Miocène inferieur de l'Italie septentrionale:* Natuurk. Verhandl. Holland, Maatsch. Wetensch., v. 2, pt. 15, p. 1-183, pl. 1-16.

(1258) **Migula, Walter,** 1910, *Kryptogamen-Flora von Deutschland, Deutsch-Österreich und der Schweiz, v. 3, Pilze:* pt. 1, p. 1-510.

(1259) **Mikhaylov, A. V.,** 1935, *K voprosu o filogenii kamennougolnykh foraminifer:* Izvestia Leningrad Geol. -gidro-geodez, tresta, no. 2-3 (7-8), p. 33-42, 1 pl. [*About the question of the phylogeny of Carboniferous foraminifera.*]———(1260) 1939, *K kharakteristike rodov nizhnekamennougol'nykh foraminifer territorii SSSR:* Leningrad. Geol. Upravl., no. 3, p. 47-62, pl. 1-4. [*On characteristics of the genera of Lower Carboniferous Foraminifers in the territory of the U.S.S.R.*]

(1261) **Miklukho-Maklay, A. D.,** 1949, *Verkhnepaleozoyskie fusulinidy Sredney Azii, Fergana, Darvaz i Pamir:* Izd. Leningrad Gos. Univ., p. 1-111, 14 pl. [*Upper Paleozoic fusulinids of Central Asia—Ferghana, Darvaz and Pamir.*]———(1261A) 1950, *Triticites ferganensis sp. n. iz verkhnekamennougolnykh otlozheniy khrebta Kara-Chatyr (Yuzhnaya Fergana):* Uchenye Zapiski Leningrad. Gosud. Univ., no. 102, ser. Geol. Nauk, no. 2, p. 59-70, 1 pl. [*Triticites ferganensis sp.n. from Upper Carboniferous deposits of the Kara-Chatyr range (southern Ferghana).*]———(1262) 1953, *K sistematike semeystva Archaediscidae:* Ezhegodnik Vses. Paleont. Obshch., v. 14 (1948-53), Otdel Ottisk, p. 127-131, pl. 6. [*On the systematics of the family Archaediscidae.*]———(1263) 1953, *K sistematike semeystva Fusulinidae Moeller:* Uchenye Zapiski Leningrad Univ., no. 159, ser. Geol. Nauk. no. 3, p. 12-24. [*On systematics of the family Fusulinidae Möller.*]———(1264) 1955, *Novye dannye o permskikh fuzulinidakh yuzhnykh rayonov SSSR:* Akad. Nauk SSSR, Doklady, v. 105, no. 3, p. 573-576, 1 fig. [*New data on Permian fusulinids in the southern regions of the USSR.*]———(1265) 1956, *Biostratigraficheskoe razdelenie verkhnego paleozoya khr. Kara-Chatyr, Yuzhnaya Fergana:* Same, v. 108, p. 1152-1155. [*Contribution to the biostratigraphic subdivision of the upper Paleozoic in the Kara-Chatyr Mountain Ridge, South Ferghana.*]———(1266) 1957, *Novye dannye po sistematike i filogenii Arkhedistsid:* Vestnik Leningrad. Univ., no. 24, ser. Geol. & Geogr., no. 4, p. 34-46, 4 text-fig. [*New data on the systematics and phylogeny of the Archaediscidae.*]———(1267) 1957, *Nekotorye fuzulinidy permi Kryma:* Uchenye Zapiski Leningrad. Univ., no. 225, ser Geol. Nauk, no. 9, p. 93-159, pl. 1-14. [*Some fusulinids from the Permian in Crimea.*]———(1268) 1958, *Sistematika vysshikh Fuzulinid:* Vestnik Leningrad. Univ., no. 12, ser. Geol. & Geogr., no. 2, p. 5-14. [*Systematics of the higher Fusulinidae.*]———(1269) 1958, *Novoe semeystvo foraminifer-Tuberitinidae M.-Maclay fam. nov.:* Voprosy Mikropaleontologii, v. 2, Akad. Nauk. SSSR, Otdel Geol. & Geogr. Nauk, p. 130-135, 1 text-fig., 1 table. [*A new foraminiferal family, Tuberitinidae M.-Maclay, fam. nov.*]———(1270) 1959, *O stratigraficheskom znachenii, sistematike i filogenii Shtaffelloobraznykh Foraminifer:* Akad. Nauk SSSR, Doklady, v. 125, no. 3, p. 628-631. [*On the stratigraphic significance, systematics and phylogeny of Staffellaformed Foraminifera.*]———(1270A) 1959, *Sistematika i filogeniya fuzulinid—rod Triticites i blizkie k nemu rody:* Vestnik Leningrad. Univ., no. 6, ser. Geol. & Geog., no. 1, p. 5-23, 1 fig. [*Systematics and phylogeny of the Fusulinidae (genus Triticites and related genera).*]———(1271) 1959, *Znachenie gomeomorfii dlya sistematiki fuzulinid:* Uchenye Zapiski Leningrad. Gosud. Univ., no. 268, ser. Geol. Nauk, no. 10, p. 155-172, pl. 1-2. [*The significance of homeomorphy for the systematics of fusulinids.*]———(1272) 1960, *Korrelyatsiya verkhnepaleozoiskikh otlozhenii srednei Azii, Kavkaza i dalnego Vostoka po dannym izucheniya foraminifer:* Mezhdunarodnyi Geol. Congress, Sess. 21, 1960, Doklady Sovetskikh Geologov, p. 69-77. [*Correlation of upper Paleozoic deposits of central Asia, Caucasus to the Far East by means of data from studied Foraminifera.*]———(1273) 1960, *Novye Rannekamennougolnye Endotiridy in Novye vidy drevnikh rasteniy i bespozvonochnykh SSSR, pt. 1:* Vses. Nauchno-Issledov. Geol. Inst. (VSEGEI), Minist. Geol. i Okhrany Nedr SSSR, p. 140-143, pl. 25. [*New Early Carboniferous Endothyridae: in New species of older plants and invertebrates of the USSR.*]———(1274) 1960, *Novye rannekamennougolnye Arkhedistsidy* in novye vidy drevnikh rasteniy i bespozvonochnykh SSSR, pt. 1: Same, p. 149-151, pl. 25. [*New Early Carboniferous Archaediscidae:* in *New species of the older plants and invertebrates of the USSR.*]

(1275) ———, **Rauzer-Chernousova, D. M.,** & **Rozovskaya, S. E.,** 1958, *Sistematika i filogeniya fusulinidey:* Voprosy Mikropaleontologii, v. 2, Akad. Nauk SSSR, Otdel, Geol. & Geogr. Nauk, p. 5-21, 2 text-fig. [*Systematics and phylogeny of the fusulinids.*]

(1276) **Miklukho-Maklay, K. V.,** 1952, *Novye*

dannye o verkhnepaleozoyskikh fuzulinidakh severnogo kavkaza, sredney azii i dalnego vostoka: Akad. Nauk SSSR, Doklady, v. 82, no. 6, p. 989-992. [*New data on the upper Paleozoic fusulinids of the northern Caucasus, central Asia and the Far East.*]———(1277) 1954, *Foraminifery verkhnepermskikh otlozheniy Severnogo Kavkaza:* Vses. Nauchno-Issledov. Geol. Inst. (VSEGEI), Minist. Geol. i Okhrany Nedr, Moscow, p. 1-162, pl. 1-19, 3 tables. [*Foraminifera of the Upper Permian deposits of the northern Caucasus.*]———(1278) 1958, *O filogenii i stratigraficheskom znachenii Paleozoyskikh Lagenid:* Akad. Nauk SSSR, Doklady, v. 122, no. 3, p. 481-484, text-fig. 1. [*On the phylogeny and stratigraphical significance of Paleozoic Lagenidae.*]———(1279) 1960, *Novye kazanskie Lagenidy Russkoy platformy* in Novye vidy drevnikh rasteniy i bespozvonochnykh SSSR, pt. 1: Vses. Nauchno-Issledov. Geol. Inst. (VSEGEI), Minist. Geol. i Okhrany Nedr SSSR, p. 153-161, pl. 27. [*New Kazanian Lagenidae of the Russian Platform:* in *New species of the older plants and invertebrates of the USSR.*]

(1280) **Miller, A. K.,** 1933, *Age of the Permian limestones of Sicily:* Am. Jour. Sci., ser. 5, v. 26, p. 409-427.

(1281) ———, & **Carmer, A. M.,** 1933, *Devonian Foraminifera from Iowa:* Jour. Paleontology, v. 7, p. 423-431, pl. 50.

(1282) **Miller, D. N.,** 1953, *Ecological study of the Foraminifera of Mason Inlet, North Carolina:* Cushman Found. Foram. Research, Contrib., v. 4, pt. 2, p. 41-63, pl. 7-10, text-fig. 1-4, tables 1-3.

(1283) **Miller, S. A.,** 1889, *North American geology and paleontology for the use of amateurs, students and scientists:* 664 p., 1194 fig., Western Methodist Book Concern (Cincinnati).

(1284) **Millett, F. W.,** 1898-1904, *Report on the Recent Foraminifera of the Malay Archipelago collected by Mr. A. Durrand, F.R.M.S.:* Royal Micro. Soc., Jour.; (a) p. 258-269, pl. 5-6 (1898); (b) Pt. 3, p. 607-614, pl. 13 (1898); (c) Pt. 4, p. 249-255, pl. 4 (1899); (d) Pt. 8, p. 273-281, pl. 2 (1900); (e) Pt. 9, p. 539-549, pl. 4 (1900); (f) Pt. 17, p. 597-609 (1904).

(1285) **Milne-Edward, Alphonse,** 1881, *Compte rendu sommaire d'une exploration zoologique, faite dans la Méditerranée, à bord du navire de l'Etat "le Travailleur":* Acad. Sci. Paris, Comptes Rendus, v. 93, p. 876-882.———(1286) 1882, *Rapport sur les travaux de la Commission chargée d'étudier la faune sous-marine dans le grandes profondeurs de la Méditerranée et de l'océan Atlantique:* Missions Sci. Litteraires, Paris, Arch., ser. 3, v. 9, p. 1-59.

(1287) **Minato, Masao, & Honjo, Susumu,** 1958, *Shell structure of Metaschwagerina n.g. from Akasaka Limestone:* Earth Science, no. 38, frontispiece (Tokyo).———(1288) 1959, *The axial septula of some Japanese Neoschwagerininae with special remarks on the phylogeny of the subfamily Neoschwagerininae Dunbar and Condra, 1928:* Hokkaido Univ., Jour. Faculty Sci., ser. 4, v. 10, no. 2, p. 305-336, pl. 1-6, fig. 1-2.

(1289) **Minchen, E. A.,** 1912, *Introduction to the study of the Protozoa:* 517 p., E. Arnold (London).

(1290) **Mityanina, I. V.,** 1957, *O foraminiferakh yurskikh otlozheniy yugo-zapada Belorussii:* Akad. Nauk Belorusskoi SSR, Inst. Geol. Nauk, Paleont. & Strat. BSSR, v. 2, p. 210-239, pl. 1-2. [*On Foraminifera of Jurassic deposits of south-western Belorussia.*]

(1291) **Modeer, Adolf,** 1791, *Illustrationes quaedam in R. D. Ambrosii Soldani opus egregium Saggio Orittografico dictum:* Nova Acta Acad. Caes. Leop.-Carol., v. 8, Appendix, p. 85-94.

(1292) **Möbius, K. A.,** 1876, *Neue Rhizopoden:* Gesell. Deutsch. Naturforsch. Ärzte, Tagebl. Versamml. 49, p. 115 (Hamburg).———(1293) 1880, *Foraminifera von Mauritius,* in K. MÖBIUS, F. RICHTER, & E. VON MARTENS, Beiträge zur Meeresfauna der Insel Mauritius und der Seychellen: p. 65-112, pl. 1-14, Gutman (Berlin).

(1294) **Möller, Valerian von,** 1877, *Ueber Fusulinen und ähnliche Foraminiferen-Formen des russischen Kohlenkalks:* Neues Jahrb. Mineral., Geol. & Paläont., v. 1877, p. 139-146, 1 fig.———(1295) 1878, *Die spiral-gewundenen Foraminiferen des russischen Kohlenkalks:* Acad. Imper. Sci. St.-Pétersbourg, Mém., ser. 7, v. 25, no. 9, 147 p., 15 pl., 6 fig.———(1296) 1879, *Die Foraminiferen des russischen Kohlenkalks:* Same, ser. 7, v. 27, no. 5, p. 1-131, pl. 1-7, text-fig. 1-30.

(1297) **Mohler, Willi,** 1938, *Mikropaläontologische Untersuchungen in der nordschweizerischen Juraformation:* Schweiz. Palaeont. Gesell., Abhandl., v. 60, p. 1-53, pl. 1-4, text-fig. 1-10.

(1298) **Montagu, George,** 1803, *Testacea Britannica, or natural history of British shells, marine, land, and fresh-water, including the most minute:* 606 p., 16 pl., J. S. Hollis (Romsey, England).———(1299) 1808, *Testacea Britannica, Supplement:* 183 p., 30 pl., S. Woolmer (Exeter, England).

(1299A) **Montanaro Gallitelli, Eugenia,** 1947, *Per la geologia delle argille ofiolitifere appenniniche. Nota III. Foraminiferi dell'argilla scagliosa di Castelvecchio (Modena):* Atti Soc. Toscana Sci. Nat., Mem., v. 54, p. 174-195, text-fig. 1, 2.———(1300) 1955, *Una revisione della famiglia Heterohelicidae Cushman:* Accad. Sci. Lettere & Arti Modena, Atti., Mem., ser. 5, v. 13, p.

213-223.——(1301) 1955, *Foraminiferi cretacei delle marne a fucoidi di Serramazzoni (Appennino modenese)*: Accad. Sci. Lettere & Arti Modena, ser. 5, v. 13, p. 175-204.——(1302) 1956, *Bronnimannella, Tappanina and Trachelinella, three new foraminiferal genera from the Upper Cretaceous*: Cushman Found. Foram. Research, Contrib., v. 7, pt. 2, p. 35-39, pl. 7.——(1303) 1957, *A revision of the foraminiferal family Heterohelicidae*: U.S. Natl. Museum, Bull. 215, p. 133-154, pl. 31-34.
——(1304) 1958, *Specie nuove e note di Foraminiferi del Cretaceo superiore di Serramazzoni (Modena)*: Accad. Sci. Lettere & Arti Modena, Atti & Mem., ser. 5, v. 16, p. 3-28, pl. 1-4.

(1305) **Montfort, Denys de,** 1808, *Conchyliologie systématique et classification methodique des coquilles*: v. 1, lxxxvii+409 p.

(1306) **Moore, Charles,** 1870, *Report on mineral veins in Carboniferous limestone, and their organic content*: Rept. British Association, 39th Meeting (Exeter, 1869), p. 360-388.

(1307) **Moore, R. C.,** 1936, *Stratigraphic classification of the Pennsylvanian rocks of Kansas*: Kansas State Geol. Survey, Bull. 22, 256 p., 12 fig.

(1308) ——, et al., 1934, *Pennsylvanian and Permian rocks of Kansas. Composite section along Kansas River and in west-central Missouri (chart)*: Kansas State Geol. Survey.

(1038A) **Moore, W. L.,** 1959, *Pennsylvanian Foraminifera from the Big Saline formation of the Llano Uplift of Texas*: Dissertation Abstracts, v. 20, no. 3, p. 995-996.

(1309) **Moreman, W. L.,** 1930, *Arenaceous Foraminifera from Ordovician and Silurian limestones of Oklahoma*: Jour. Paleontology, v. 4, p. 42-59, pl. 5-7.——(1310) 1933, *Arenaceous Foraminifera from the lower Paleozoic rocks of Oklahoma*: Same, v. 7, p. 393-397, pl. 47.

(1311) **Morgan, A. P.,** 1893, *The Myxomycetes of the Miami Valley, Ohio*: Cincinnati Soc. Nat. History, Jour., v. 15, p. 1-17, pl. 3.——(1312) 1900, *The Myxomycetes of the Miami Valley, Ohio*: Same, v. 22, p. 111-130.

(1312A) **Morikawa, Rokuro,** 1952, *On a new genus Fujimotoella*: Saitama Univ., Sci. Rept., ser. B, v. 1, no. 1, p. 35-38, pl. 1.

(1313) ——, & **Isomi, Hiroshi,** 1960, *A new genus Biwaella, Schwagerina-like Schubertella*: Saitama Univ., Sci. Rept., ser. B, v. 3, no. 3, p. 301-305, pl. 54.

(1314) **Morishima, Masao,** 1948, *The accumulation of foraminiferal tests in inlets of Wakasa Bay on the Inland Sea of Japan*: Natl. Research Council, Rept. of Committee on Treatise on Marine Ecology & Paleoecology, no. 7, 1946-1947, p. 89-91 (Washington).

(1315) **Morozova, V. G.,** 1948, *Foraminifery nizhnemelovykh otlozheniy rayona g. Sochi (yugo-zapadnyy Kavkaz)*: Moskov. Obschch. Ispyt., Prirody, Otdel. Geol., Byull., v. 23(3), p. 23-43, pl. 1-2. [*Foraminifera of the Lower Cretaceous deposits in the region of the Sochi Mountains, southwest Caucasus*.]——(1316) 1957, *Nadsemeystvo foraminifer Globigerinidea superfam. nova i nekotorye ego predstaviteli*: Akad. Nauk SSSR, Doklady, v. 114, no. 5, p. 1109-1112, text-fig. 1. [*Foraminiferal superfamily Globigerinidea, superfam. nov., and certain of its representatives.*]——(1317) 1959, *Stratigrafiya datsko-montskikh otlozheniy kryma po foraminiferam*: Same, v. 124, no. 5, p. 1113-1116, text-fig. 1. [*Stratigraphy of the Danian-Montian deposits of Crimea, by means of Foraminifera.*]

(1318) ——, & **Moskalenko, T. A.,** 1961, *Planktonnye foraminifery pogranichnykh otlozheniy bayosskogo i batskogo yarusov tsentralnogo Dagestana (severo-vostochnyy Kavkaz)*: Voprosy Mikropaleontologii no. 5, Akad. Nauk SSSR, Otdel Geol.-Geog. Nauk, Geol. Inst., p. 3-30, pl. 1-2, text-fig. 1-9. [*Planktonic Foraminifera of the boundary deposits of the Bajocian and Bathonian stages of central Dagestan (northeast Caucasus).*]

(1319) **Morrow, A. L.,** 1934, *Foraminifera and Ostracoda from the Upper Cretaceous of Kansas*: Jour. Paleontology, v. 8, p. 186-205, pl. 29-31.

(1320) **Morton, S. G.,** 1833, *Supplement to the "Synopsis of the organic remains of the ferruginous sand formation of the United States," contained in Vols. XVII and XVIII of this journal*: Am. Jour. Sci. & Arts, v. 23, p. 288-294, pl. 5, 8-9.

(1321) **Mound, M. C.,** 1961, *Arenaceous Foraminifera from the Brassfield Limestone (Albion) of southeastern Indiana*: Indiana Geol. Survey, Bull. 23, p. 1-38, pl. 1-3, text-fig. 1-5.

(1322) **Munier-Chalmas, E.,** 1882, *La structure des Triloculines et des Quinqueloculines, Caractéres de Miliolidae*: Soc. géol. France, Bull., ser. 3, v. 10 (1881-82), pt. 6, p. 424-425.——(1323) 1882, *La connaissance des phases successives par lesquelles passent les Foraminifères*: Same, ser. 3, v. 10 (1881-82), p. 470-471.——(1324) 1882, *Un genre nouveau de Foraminifères sénoniens*: Same, ser. 3, v. 10 (1881-82), p. 471-472.——(1325) 1887, *Sur la Cyclolina et trois nouveaux genres de Foraminifères de couches à Rudistes: Cyclopsina, Dicyclina et Spirocyclina*: Soc. géol. France, Comptes Rendus, Somm., no. 7, p. xxx-xxxi.——(1326) 1891, *Étude du Tithonique, du Crétacé et du Tertiaire du Vicentin*: Thèses Faculté Sci. Paris, p. 1-182 (Paris).——(1327) 1902, *Sur les Foraminifères ayant un réseau de mailles polygonales*: Soc. géol. France, Bull., ser. 4, v. 2, p. 349-351.

———(1328) 1902, *Sur les Foraminifères rapportés au groupe des Orbitolites:* Same, ser. 4, v. 2, p. 351-353.

(1329) ———, & **Schlumberger, Charles,** 1883, *Nouvelles observations sur le dimorphisme des Foraminifères:* Acad. Sci. Paris, Comptes Rendus, v. 96, p. 862-866.———(1330) 1885, *Note sur les Miliolidées trématophorées:* Soc. géol. France, Bull., ser. 3, v. 13 (1884-85), pt. 4, p. 273-323, pl. 13-14, text-fig. 1-44.

(1331) **Murray, John,** 1876, *Preliminary reports to Professor Wyville Thomson, F.R.S., director of the civilian scientific staff, on work done on board the "Challenger":* Royal Soc. London, Proc., v. 24, p. 471-544, pl. 20-24.

(1332) **Myatlyuk, E. V.,** 1953, *Spirillinidy, Rotaliidy, Epistominidy i Asterigerinidy:* Iskopaemye Foraminifery SSSR, VNIGRI, Trudy, new ser., no. 71, p. 1-273, 39 pl. [*Spirillinidae, Rotaliidae, Epistominidae and Asterigerinidae:* in Fossil Foraminifera of the USSR.]———(1333) 1960, *Novye dannye po issledovaniyu foraminifer verkhneoligotsenovykh i nizhnemiotsenovykh otlozheniy:* VNIGRI, Trudy Pervogo Seminara po Mikrofaune, p. 207-227. [*New data on foraminiferal research in upper Oligocene and lower Miocene deposits.*]

(1334) **Myers, E. H.,** 1933, *Multiple tests in the Foraminifera:* Natl. Acad. Sci. Washington, Proc., v. 19, no. 10, p. 893-899.———(1335) 1935, *Morphogenesis of the test and the biological significance of dimorphism in the foraminifer Patellina corrugata Williamson:* Univ. Calif., Scripps Inst. Oceanog., Bull., tech. ser., v. 3, p. 393-404, 1 fig.———(1336) 1935, *The life history of Patellina corrugata Williamson a foraminifer:* Same, v. 3, p. 355-392, pl. 10-16, 1 fig.———(1337) 1936, *The life-cycle of Spirillina vivipara Ehrenberg, with notes on the morphogenesis, systematics, and distribution of the Foraminifera:* Royal Micro. Soc. London, Jour., v. 56, p. 120-146, pl. 1-3.———(1338) 1938, *The present state of our knowledge concerning the life cycle of the Foraminifera:* Natl. Acad. Sci., Washington, Proc., v. 24, no. 1, p. 10-17.———(1339) 1940, *Observations on the origin and fate of flagellated gametes in multiple tests of Discorbis (Foraminifera):* Marine Biol. Assoc. United Kingdom, Jour., v. 24, p. 201-226, pl. 1-3.———(1340) 1943, *Life activities of Foraminifera in relation to marine ecology:* Am. Philos. Soc., Proc., v. 86, no. 3, p. 439-458, text-fig. 1-7, pl. 1.———(1341) 1943, *Biology, ecology, and morphogenesis of a pelagic foraminifer:* Stanford Univ. Publ., Biol. Sci., v. 9, no. 1, p. 5-30, pl. 1-4.———(1342) 1945, *Recent studies of sediments in the Java Sea and their significance in relation to stratigraphic and petroleum geology,* in Science and Scientists in the Netherlands Indies: p. 265-269, fig. 74, Board for the Netherlands Indies, Surinam & Curaçao (New York).

(1343) ———, & **Cole, W. S.,** 1957, *Foraminifera,* in Treatise on marine ecology and paleoecology, v. 1, Ecology, HEDGPETH, J. W., Ed.; Geol. Soc. America, Mem. 67, p. 1075-1081.

(1344) **Nagappa, Yedatore,** 1957, *Direction of coiling in Globorotalia as an aid in correlation:* Micropaleontology, v. 3, p. 393-398, pl. 1, text-fig. 1-8.

(1345) **Nakkady, S. E.,** 1955, *The stratigraphy and geology of the district between the northern and southern Galala Plateaus (Gulf of Suez Coast, Egypt):* Inst. Egypte, Bull., v. 36, p. 253-268, 1 pl., 1 text-fig.

(1346) **Napoli Alliata, Enrico di,** 1952, *Nuove specie di Foraminiferi nel Pliocene e nel Pleistocene della zona di Castel-l'Arquato (Piacenza):* Rivista Italiana, Paleont. & Strat., v. 58, no. 3, p. 95-109, pl. 5.

(1346A) **Narchi, Walter,** 1962, *A new genus of Foraminifera from South Atlantic:* Acad. Brasileira de Ciencias, Anais, v. 34, no. 2, p. 277-279, text-fig. 1-12.

(1347) **Natland, M. L.,** 1940, *New genus of Foraminifera from the later Tertiary of California:* Jour. Paleontology, v. 14, p. 568-571, pl. 69, text-fig. 1-2.

(1348) **Neave, S. A.,** 1939-40, *Nomenclator Zoologicus:* Richard Clay & Co. (London); (a) v. 1, A-C, xiv+947 p. (1939); (b) v. 2, D-L, 1025 p. (1939); (c) v. 3, M-P, 1065 p. (1940); (d) v. 4, Q-Z, 758 p. (1940); (e) Suppl., v. 5, 308 p. (1950).

(1349) **Neugeboren, J. L.,** 1850, *Foraminiferen von Felsö-Lapugy; zweiter Artikel:* Siebenb. Vereins Naturwiss. Hermannstadt, Verhandl. Mitt., Jahrg. 1, p. 118-127, pl. 3,4.———(1350) 1852, *Foraminiferen von Ober-Lapugy; vierter Artikel (Schluss):* Vereins Naturwiss. Hermannstadt, Verhandl. Mitt., Jahrg. 3, no. 4, p. 50-59, pl. 1.———(1351) 1856, *Die Foraminiferen aus der Ordnung der Stichostegier von Ober-Lapugy in Siebenbürgen:* K. Akad. Wiss. Wien, math.-naturwiss. Cl., Denkschr., v. 12, pt. 2, p. 65-108, pl. 1-5.

(1352) **Neumann, Madeleine,** 1954, *Le genre Linderina et quelques autres Foraminifères l'accompagnant dans le Nummulitique d'Aquitaine:* Soc. géol. France, Bull., ser. 6, v. 4, p. 55-59, pl. 4, 5, text-fig. 1.

(1353) ———, & **Damotte, Renée,** 1960, *Abrardia, nouveau genre du Crétacé superieur d'Aquitaine:* Revue Micropaléont., v. 3, no. 1, p. 60-64, pl. 1, text-fig. 1-3.

(1354) **Neumayr, Melchior,** 1887, *Die natürlichen Verwandtschaftsverhältnisse der schalentragenden Foraminiferen:* K. Akad. Wiss. Wien, math.-naturwiss. Cl., Sitzungsber., v. 95, pt. 1, p.

156-186.———(1355) 1899, *Die Stämme des Thierreiches; wirbellose thiere:* v. 1, 603 p., text-fig. 1-192, F. Tempsky (Wien).

(1356) **Nicholson, H. A., & Etheridge, Robert, Jr.,** 1878, *A monograph of the Silurian fossils of the Girvan district in Ayrshire, with especial reference to those contained in the "Gray Collection":* v. 1, 341 p., 24 pl., pt. 1(1878); pt. 2-3(1880), William Blackwood & Sons (London).

(1357) **Nicolucii, Gustiniano,** 1846, *Politalami fossili della Italia meridionale:* Nuovi Ann. Sci. Nat. Bologna, ser. 2, v. 6, p. 161-216.

(1358) **Nilsson, Sven,** 1826, *Om de mångrummiga snäckor som förekomma i kritformationen i sverige:* K. Vetenskaps. Acad. Stockholm, Handl., v. 1825, p. 329-343.

(1359) **Nørvang, Aksel,** 1945, *The zoology of Iceland, Foraminifera:* v. 2, pt. 2, 79 p., 14 text-fig., Ejnar Munksgaard (Copenhagen & Reykjavík).———(1360) 1957, *The Foraminifera of the Lias series in Jutland, Denmark:* Meddel. Dansk Geol. Foren., v. 13, pt. 5, p. 1-135, pl. with fig. 1-182, text-fig. 1-5.———(1361) 1959, *Islandiella n.g. and Cassidulina d'Orbigny:* Vidensk. Medd. Dansk naturhist. Foren., v. 120 (1958), p. 25-41, pl. 6-9.———(1362) 1961, *Schizamminidae, a new family of Foraminifera:* Atlantide Rept. No. 6 (Sci. results Danish Exped. coasts tropical West Africa, 1945-1946), p. 169-201, pl. 6-9 (Copenhagen).

(1363) **Norman, A. M.,** 1878, *On the genus Haliphysema with a description of several forms apparently allied to it:* Ann. & Mag. Nat. History, ser. 5, v. 1, p. 265-284, pl. 16.———(1364) 1892, *Museum Normanianum:* pt. 7-8, p. 14-21, The Author (Durham).

(1365) **Noth, Rudolf,** 1952, *Plectorecurvoides eine neue Foraminiferengattung:* Verhandl. Geol. Bundesanst. 1952, no. 3, p. 117-119, text-fig. 1-2.

(1366) **Nusslin, O.,** 1884, *Ueber einige Urthiere aus dem Herrenwieser See im badischen Schwarzwalde:* Zeitschr. Wiss. Zool., v. 40, p. 697-724, pl. 35, 36.

(1367) **Nuttall, W. L. F.,** 1925, *Two species of Eocene Foraminifera from India; Alveolina elliptica and Dictyoconoides cooki:* Ann. & Mag. Nat. History, ser. 9, v. 16, p. 378-388, pl. 20-21.———(1368) 1925, *The stratigraphy of the Laki Series (Lower Eocene) of parts of Sind and Baluchistan (India); with a description of the larger Foraminifera contained in those beds:* Geol. Soc. London, Quart. Jour., v. 81, pt. 3, p. 417-453, pl. 23-27, text-fig. 1-5, 1 table.———(1369) 1926, *The zonal distribution of the larger Foraminifera of the Eocene of Western India:* Geol. Mag., v. 63, p. 495-504, table 1-4.———(1370) 1928, *Notes on the Tertiary Foraminifera of southern Mexico:* Jour. Paleontology, v. 2, p. 372-376, pl. 50.———(1371) 1930, *Eocene Foraminifera from Mexico:* Same, v. 4, p. 271-293, pl. 23-25.———(1371A) 1932, *Lower Oligocene Foraminifera from Mexico:* Same, v. 6, p. 3-35, pl. 1-9.———(1372) 1933, *Two species of Miogypsina from the Oligocene of Mexico:* Same, v. 7, p. 175-177, pl. 24.

(1373) **Nyholm, K.-G.,** 1951, *A monothalamous foraminifer, Marenda nematoides, n.gen., n.sp.:* Cushman Found. Foram. Research, Contrib., v. 2, p. 91-95, text-fig. 1-14.———(1374) 1952, *Studies on Recent Allogromiidae: 1. Micrometula hyalostriata, n.gen., n.sp., from the Gullmar Fjord, Sweden:* Same, v. 3, pt. 1, p. 14-17, pl. 4, text-fig.———(1375) 1953, *Studies on Recent Allogromiidae (2): Nemogullmia longevariabilis, n.g., n.sp., from the Gullmar Fjord:* Same, v. 4, p. 105-106, text-fig. 1-5, pl. 18.———(1376) 1954, *Studies on Recent Allogromiidae (3): Tinogullmia hyalina, n.gen., n.sp., from the Gullmar Fjord, Sweden:* Same, v. 5, pt. 1, p. 36, pl. 7.———(1377) 1955, *Studies on Recent Allogromiidae (4), Phainogullmia aurata, n.gen., n.sp.:* Zool. Bidrag Uppsala, v. 30, p. 465-474, pl. 1-5, text-fig. 1-18.———(1378) 1956, *On the life cycle and cytology of the foraminiferan Nemogullmia longivariabilis:* Same, v. 31, p. 483-495, pl. 1-3, text-fig. 1-9.———(1379) 1957, *Orientation and binding power of Recent monothalamous Foraminifera in soft sediments:* Micropaleontology, v. 3, p. 75-76, text-fig. 1.——— (1380) 1961, *Morphogenesis and biology of the foraminifer Cibicides lobatulus:* Zool. Bidrag Uppsala, v. 33, p. 157-196, pl. 1-5.———(1381) 1962, *A study of the foraminifer Gypsina:* Same, v. 33, p. 201-206, pl. 1-2.

(1382) **Nyirö, M. R.,** 1954, *Új oligocén foraminiferák a Budapest-környéki katti rétegekböl— Nouveaux Foraminifères oligocènes des couches chattiennes des environs de Budapest:* Földtani Közlöny, v. 84, no. 1-2, p. 67-74.

(1383) **Oberhauser, Rudolf,** 1957, *Neue mesozoiche Foraminiferen aus der Türkei:* R. V. Klebelsberg-Festschrift, Geol. Gesell. Wien, v. 48, p. 193-200, pl. 1.———(1384) 1960, *Foraminiferen und Mikrofossilien "incertae sedis" der ladinischen und karnischen Stufe der Trias aus den Ostalpen und aus Persien:* Geol. Bundesanst., Wien, Jahrb., spec. v. 5, p. 5-46, pl. 1-6.

(1385) **Oken, Lorenz,** 1815, *Oken's Lehrbuch der Naturgeschichte:* Pt. 3, Zoologie, no. 1, Fleischlose Thiere, p. 1-842, C. H. Reclam (Leipzig).

(1386) **Okimura, Yuji,** 1958, *Biostratigraphical and paleontological studies on the endothyroid Foraminifera from the Atetsu Limestone Plateau, Okayama Prefecture, Japan:* Hiroshima Univ., Jour. Sci., ser. C, v. 2, no. 3, p. 235-264, pl. 32-36.

(1387) Olive, E. W., 1901, *A preliminary enumeration of the Sorophoreae:* Am. Acad. Arts & Sci., Proc., v. 37, 1901-1902, p. 333-344.——— (1388) 1902, *Monograph of the Acrasieae:* Boston Soc. Nat. History, Proc., v. 30, p. 451-513, pl. 5-8.

(1389) Omara, S. M., 1956, *New Foraminifera from the Cenomanian of Sinai, Egypt:* Jour. Paleontology, v. 30, p. 883-890, pl. 101-102, 6 text-fig.

(1390) Oppenheim, Paul von, 1896, *Das Alttertiär der Colli Berici in Venetien, die Stellung der Schichten von Priabona, und die Oligocäne Transgression in alpinen Europa:* Zeitschr. deutsch. geol. Gesell., v. 48, p. 27-152, pl. 2-5.

(1391) Orbigny, Alcide Dessalines d', 1826, *Tableau méthodique de la classe des Céphalopodes:* Ann. Sci. Nat. Paris, ser. 1, v. 7, p. 245-314; atlas, pl. 10-17, Crochard (Paris).———
(1392) 1839, *Foraminifères* in Sagra, Ramon de la (=1611), Histoire physique, politique et naturelle de l'île de Cuba: xlviii+224 p., atlas, 12 pl.———(1393) 1839, *Voyage dans l'Amérique Méridionale-Foraminifères:* v. 5, pt. 5, 86 p., 9 pl., Pitois-Levrault et Cᵉ (Paris), V. Levrault (Strasbourg).———(1394) 1840, *Mémoire sur les Foraminifères de la craie blanche du bassin de Paris:* Soc. géol. France, Mém., v. 4, pt. 1, p. 1-51, pl. 1-4.———(1395) 1846, *Foraminifères fossiles du Bassin Tertiaire de Vienne (Autriche):* 312 p., 21 pl., Gide et Compᵉ (Paris).———(1396) 1849, *Foraminifères:* in Dictionnaire universel d'histoire naturelle, v. 5, p. 662-671, Renard, Martinet & Cie. (Paris).
———(1397) 1849[1850], *Prodrôme de paléontologie stratigraphique universelle des animaux mollusques & rayonnés:* V. Masson (Paris); (a) v. 1, ix+392 p. (1849); (b) v. 2, 427 p. (1850). [1850, fide Ellis & Messina, (*700, Bibliog., p. 181).]——— (1398) 1851, *Cours élémentaire de paléontologie et de géologie stratigraphique:* v. 2, pt. 1, p. 189-207, V. Masson (Paris).

(1399) Orlova, I. N., 1955, *Novyy rod semeystva Archaediscidae E. Tchern.:* Akad. Nauk SSSR, Doklady, v. 102, no. 3, p. 621-622, text-fig. 1. [*New genera of the family Archaediscidae E. Tchern.*]

(1400) Osimo, Giuseppina, 1909, *Studio critico sul genere Alveolina d'Orb.:* Paleont. Italica, Mem., Paleont., v. 15, p. 70-100, pl. 4(1)-6(4).

(1401) Ozawa, Yoshiaki, 1925, *On the classification of Fusulinidae:* Tokyo Imper. Univ., College Sci., Jour., v. 45, art. 4, 26 p., 4 pl., 3 fig.———(1401A) 1925, *Paleontological and stratigraphical studies on the Permo-Carboniferous limestone of Nagato, Part 2. Paleontology:* Same, v. 45, art. 6, p. 1-90, pl. 1-14.———
(1401B) 1927, *Stratigraphical studies of the Fusulina limestone of Akasaka, Province of Mino:* Tokyo Imper. Univ., Faculty Sci., Jour., Sec. 2 (Geol.), v. 2, pt. 3, p. 121-164, pl. 34-46.———(1401C) 1928, *Fusulinidae;* in Cushman, J. A., Foraminifera, their classification and economic use: p. 131-139, Cushman Lab. Foram. Research (Sharon, Mass.).———(1402) 1928, *A new genus, Depratella, and its relation to Endothyra:* Cushman Lab. Foram. Research, Contrib., v. 4, pt. 1, p. 9-10, pl. 1.

(1403) Paalzow, Richard, 1917, *Beiträge zur Kenntnis der Foraminiferenfauna der Schwammergel des Unteren Weissen Jura in Süddeutschland:* Naturhist. Gesell. Nürnberg, Abhandl., v. 19, p. 203-248, pl. 41-47.———(1404) 1922, *Die Foraminiferen der Parkinsoni-Mergel von Heidenheim am Hahnenkamm:* Same, v. 22, p. 1-35, pl. 1-4.———(1405) 1932, *Die Foraminiferen aus den Transversarius-Schichten und Impressa-Tonen der nordöstlichen schwäbischen Alb:* Jahresh. Verein. Vaterländ. Naturk. Württemberg, v. 88, p. 81-142, pl. 4-11.———
(1406) 1935, *Die Foraminiferen im Zechstein des östlichen Thüringen:* Preuss. Geol. Landesanst. Jahrb. 1935, v. 56, p. 26-45, pl. 3-5.

(1407) Pallas, P. S., 1766, *Elenchus Zoophytorum sistens generum adumbrationes generaliores et specierum cognitarum succinctas descriptiones cum selectis auctorum synonymis:* 451 p., P. van Cleef (Hagae).

(1408) Palmer, D. K., 1934, *Some large fossil Foraminifera from Cuba:* Soc. Cubana Historia Nat., Mem., v. 8, p. 235-264, pl. 12-16, 19 text-fig.———(1409) 1936, *New genera and species of Cuban Oligocene Foraminifera:* Same, v. 10, no. 2, p. 123-128, pl. 5, text-fig. 1-3.
———(1410) 1941, *Foraminifera of the upper Oligocene Cojimar Formation of Cuba:* Same, (a) Pt. 4, v. 15, no. 2, p. 181-200, pl. 15-17; (b) Pt. 5, v. 15, p. 281-306, pl. 28-31, 1 text-fig.

(1411) ———, & Bermúdez, P. J., 1936, *Late Tertiary Foraminifera from the Matanzas Bay region, Cuba:* Soc. Cubana Historia Nat., Mem., v. 9, p. 237-257, pl. 20-22.———(1412) 1936, *An Oligocene foraminiferal fauna from Cuba:* Same, v. 10, no. 4, p. 227-271, pl. 13-20.

(1413) Papp, Adolf, & Küpper, Klaus, 1954, *The genus Heterostegina in the Upper Tertiary of Europe:* Cushman Found. Foram. Research, Contrib., v. 5, pt. 3, p. 108-127, pl. 20-23, 5 text-fig., 2 tables.

(1414) Parker, F. L., 1954, *Distribution of the Foraminifera in the northeastern Gulf of Mexico:* Museum Comp. Zool. Harvard, Bull., v. 111, no. 10, p. 453-588, pl. 1-13.

(1415) Parker, W. K., 1858, *On the Miliolitidae (Agathistègues d'Orbigny) of the East Indian Seas, Part 1. Miliola:* Micro. Soc. London, Trans., new ser., v. 6, p. 53-59, pl. 5.

(1416) ———, & Jones, T. R., 1857, *Description*

of some Foraminifera from the coast of Norway: Ann. & Mag. Nat. History, ser. 2, v. 19, p. 273-303, pl. 10-11.——(1417) 1859-72, On the nomenclature of the Foraminifera: Same, (a) Pt. 1, On the species enumerated by Linnaeus and Gmelin, ser. 3, v. 3, p. 474-482 (1859); (b) Pt. 2, On the species enumerated by Walker and Montagu, ser. 3, v. 4, p. 333-351 (1859); (c) Pt. 3, ser. 3, v. 5, p. 174-183 (1860); (d) Pt. 4, ser. 3, v. 6, p. 29-40 (1860); (e) Pt. 8, Textularia, ser. 3, v. 11, p. 91-98 (1863); (f) Pt. 10, The species enumerated by d'Orbigny in the "Annales des Sciences Naturelles, vol. vii. 1826," ser. 3, v. 12, p. 429-441 (1863); (g) Pt. 15, The species figured by Ehrenberg, ser. 4, v. 10, p. 184-200 (1872). ——(1418) 1865, On some Foraminifera from the North Atlantic and Arctic Oceans, including Davis Straits and Baffin's Bay: Philos. Trans., v. 155, p. 325-441, pl. 12-19.

(1419) ——, & Jones, T. R., & Brady, H. B., 1865, On the nomenclature of the Foraminifera. Pt. 12. The species enumerated by d'Orbigny in the "Annales des Sciences Naturelles," vol. 7, 1826: Ann. & Mag. Nat. History, ser. 3, v. 16, p. 15-41, pl. 1-3.

(1420) Parkinson, James, 1811, Organic remains of a former world: v. 3, 455 p., 22 pl., Sherwood, Neely, & Jones (London).

(1421) Parr, W. J., 1932, Victorian and South Australian shallow-water Foraminifera: Royal Soc. Victoria, Proc., new ser., v. 44, pt. 1, p. 1-14, pl. 1.——(1422) 1933, The genus Pavonina and its relationships: Same, v. 45, pt. 1, p. 28-31, pl. 7.——(1423) 1935, Some Foraminifera from the Awamoan of the Medway River district, Awatere, Marlborough, New Zealand: Royal Soc. New Zealand, Trans., v. 65, p. 77-87, pl. 19, 20, text-fig. 1-2.——(1424) 1941, A new genus, Planulinoides, and some species of Foraminifera from southern Australia: Mining & Geol. Jour., v. 2, no. 5, p. 305, text-fig. a-c.——(1425) 1942, Foraminifera and a tubicolous worm from the Permian of the North-West Division of Western Australia: Royal Soc. Western Australia, Jour., v. 27 (1940-41), no. 8, p. 97-115.——(1426) 1942, New genera of Foraminifera from the Tertiary of Victoria: Mining & Geol. Jour., v. 2, no. 6, p. 361-363, fig. 1-5.——(1427) 1947, On Torresina, a new genus of the Foraminifera from eastern Australia: Royal Micro. Soc., Jour., v. 64 (1944), pt. 3-4, p. 129-135, pl. 1, text-fig. 1-3.——(1428) 1947, The lagenid Foraminifera and their relationships: Royal Soc. Victoria, Proc., new ser., v. 58, p. 116-130, pl. 6-7, 1 text-fig.——(1429) 1950, Foraminifera: B.A.N.Z. Antarctic Res. Exped. 1929-31, rept. ser. B, v. 5, pt. 6, p. 232-392, pl. 3-15.

(1430) ——, & Collins, A. C., 1930, Notes on Australian and New Zealand Foraminifera, No. 1. The species of Patellina and Patellinella, with a description of a new genus, Annulopatellina: Royal Soc. Victoria, Proc., new ser., v. 43, pt. 1, p. 89-95, pl. 4.

(1431) Pavlovskiy, E. N., & Strelkov, A. A., eds., 1955, Atlas bespozvonochnykh Dal'nevostochnykh Morey SSSR: Akad. Nauk SSSR, Zool. Inst., p. 1-243, pl. 1-66 (Moscow & Leningrad). [Atlas of invertebrates, Far Eastern Seas of the USSR.]

(1432) Payard, J.-M., 1947, Les Foraminifères du Lias supérieur de Détroit Poitevin: Thèses Faculté Sci., Univ. Paris, 255 p., 7 pl.

(1433) Penard, Eugène, 1890, Études sur les Rhizopodes d'eau douce: Soc. Phys. & Histoire Nat. Genève, Mém., v. 31, no. 2, p. 1-230, pl. 1-11. ——(1434) 1899, Les Rhizopodes de faune profonde dans le lac Léman: Revue Suisse Zool., v. 7, p. 1-142, pl. 1-9.——(1435) 1902, Faune Rhizopodique du Bassin du Léman: 714 p., 1 pl., text-fig., Henry Kündig (Genève).——(1436) 1904, Quelques nouveaux Rhizopodes d'eau douce: Archiv Protistenkunde, v. 3, p. 391-422, 11 text-fig.——(1437) 1905, Les Sarcodinés des Grand Lacs: 133 p., 57 text-fig., W. Kündig (Genève).——(1438) 1907, On some rhizopods from the Sikkim Himalaya: Royal Micro. Soc., Jour., p. 274-278, pl. 14. ——(1439) 1909, Sur quelques Rhizopodes des Mousses: Archiv Protistenkunde, v. 17, p. 258-296, text-fig.——(1440) 1910, Rhizopodes nouveaux: Revue Suisse Zool., v. 18, p. 929-940, pl. 8.——(1441) 1911, Rhizopodes d'eau douce: British Antarct. Exped. 1907-1909, v. 1, Biol., pt. 6, p. 204-257, pl. 22-23. ——(1442) 1912, Notes sur quelques Sarcodinés: Revue Suisse Zool., v. 20, no. 1, p. 1-29, pl. 1-2.

(1443) Penzig, Otto, 1898, Die Myxomyceten der Flora von Buitenzorg: 83 p., E. J. Brill (Leiden).

(1444) Pérébaskine, Victor, 1946, Note sur quelques Foraminifères nouveaux du Flysch néo-crétacé pyrénéen: Soc. géol. France, Bull., ser. 5, v. 15 (1945), no. 7-8, p. 357-360, pl. 4.

(1445) Perner, Jaroslav, 1892, Foraminifery Českého Cenomanu: Česká Akademie Císaře Františka Josefa pro Vědy, Slovesnost a Uměni v Praze (Palaeontographica Bohemiae no. 1), p. 1-65, pl. 1-10.

(1446) Peron, Alphonse, 1891-93, Fossiles nouveaux ou critiques des terrains Tertiares et Secondaires. Invertébrés fossiles des terrains Crétacés de la région sud des Hauts-Plateaux: Exploration Scientifique de la Tunisie, Illustrations de la partie paléontologique et géologique, pt. 2, pl. 12-14.

(1447) Perty, Maximilian, 1852, Zur Kenntniss

kleinster Lebensformen nach Bau, Funktionen, Systematik, mit Specialverzeichniss der in der Schweiz beobachteten: 228 p., 17 pl. (Bern).

(1447A) **Petri, Setembrino,** 1962, *Foraminiferos Cretáceos de Sergipe:* Faculdade de Filosofia, Ciências e Letras da Universidade de São Paulo, Bull. 265 (Geol. no. 20), p. 1-140, pl. 1-21, text-fig. 1-3, table 1-8.

(1448) **Petters, Victor,** 1954, *Tertiary and Upper Cretaceous Foraminifera from Colombia, S.A.:* Cushman Found. Foram. Research, Contrib., v. 5, pt. 1, p. 37-41, pl. 8.

(1449) **Pfender, Juliette,** 1933, *Sur un Foraminifère nouveau du Bathonien des Montagnes d'escreins (H.-Alpes): Kilianina blancheti, nov. gen., nov.sp.:* Univ. Grenoble, Ann, Sec. Sci. Méd., new ser., v. 10, no. 1-2, p. 243-252, pl. 1-2.——(1450) 1934, *À propos du Siderolites vidali Douvillé et quelques autres:* Soc. géol. France, Comptes Rendus, no. 6, p. 79-80.——(1451) 1935, *À propos du Siderolites vidali Douvillé et de quelques autres:* Soc. géol. France, Bull., ser. 5, v. 4, pt. 4-5 (1934), p. 225-236, 3 pl., 4 fig.——(1452) 1938, *Les Foraminifères du Valanginien provençal:* Same, ser. 5, v. 8, p. 231-242, pl. 13-16.

(1452A) **Philippi, R. A.,** 1844, *Enumeratio molluscorum Siciliae, cum viventium tum in tellure Tertiaria fossilium, quae in itinere suo observavit:* v. 2, 303 p., pl. 13-28, E. Anton (Halis Saxon).

(1452B) **Phillips, John,** 1846, *On the remains of microscopic animals in the rocks of Yorkshire:* Geol. Polytech. Soc. West Riding Yorkshire, Proc., Leeds, v. 2, p. 274-285, pl. 7.

(1453) **Phleger, F. B.,** 1951, *Displaced foraminiferal faunas:* Soc. Econ. Paleont. & Mineral., Spec. Publ. 2, p. 66-75, text-fig. 1-7.——(1454) 1960, *Ecology and distribution of Recent Foraminifera:* 297 p., 11 pl., text-fig. 1-83, Johns Hopkins Press (Baltimore).

(1455) ——, & **Parker, F. L.,** 1951, *Ecology of Foraminifera, northwest Gulf of Mexico, Pt. II. Foraminifera species:* Geol. Soc. America, Mem. 46, p. 1-64, pl. 1-20.

(1456) **Pijpers, P. J.,** 1933, *Geology and paleontology of Bonaire (Dutch West Indies):* Geog. & Geol. Meded., physiogr.-geol. ser., Utrecht, no. 8, p. 1-103, pl. 1-2, text-fig.——(1457) 1933, *Ruttenia, a new name for Bonairea Pijpers, 1933:* Cushman Lab. Foram. Research, Contrib., v. 9, pt. 2, p. 30.

(1458) **Piveteau, Jean,** 1952, *Traité de paléontologie:* v. 1, 782 p., Masson & Cie. (Paris).

(1459) **Playfair, G. I.,** 1918, *Rhizopods of Sidney and Lismore:* Linnean Soc. New S. Wales, Proc., v. 42, p. 632-675, pl. 34-41, text-fig. 1-7.

(1460) **Plessis, G. du,** 1876, *Arcellina marina, gen. et spec. nov.?, eine neue Rhizopodenform aus der Familie der Arcellideen:* Physicalisch-medicinischen Societät Erlangen, Sitzungsber., v. 8, p. 100-107.

(1461) **Plummer, H. J.,** 1927, *Foraminifera of the Midway Formation in Texas:* Univ. Texas, Bull. 2644, p. 1-206, pl. 1-15, text-fig. 1-13, chart.——(1462) 1930, *Calcareous Foraminifera in the Brownwood Shale near Bridgeport, Texas:* Same, Bull. 3019, p. 5-21, pl. 1.——(1463) 1931, *Some Cretaceous Foraminifera in Texas:* Same, Bull. 3101, p. 109-203, pl. 8-15.——(1464) 1931, *Gaudryinella, a new foraminiferal genus:* Am. Midland Naturalist, v. 12, p. 341-342, text-fig. 1.——(1465) 1932, *Ammobaculoides, a new foraminiferal genus:* Same, v. 13, p. 86-88, text-fig. 1.——(1466) 1934, *Epistominoides and Coleites, new genera of Foraminifera:* Same, v. 15, p. 601-608, pl. 24, 1 text-fig.——(1467) 1938, *Adhaerentia, a new foraminiferal genus:* Same, v. 19, no. 1, p. 242-244, text-fig. 1.——(1468) 1945, *Smaller Foraminifera in the Marble Falls, Smithwick and lower Strawn strata around the Llano uplift in Texas:* Univ. Texas, Publ. 4401, p. 209-271, pl. 15-17.

(1469) **Plunkett, O. A.,** 1934, *Contributions to the knowledge of southern California fungi, I. Myxomycetes:* Univ. Calif., Publ. Biol. Sci., v. 1, no. 2, p. 35-48.

(1470) **Poche, Franz,** 1913, *Das system der Protozoa:* Archiv Protistenkunde, v. 30, p. 125-321, 1 text-fig.

(1471) **Poignant, Armelle,** 1958, *Un nouveau genre de Foraminifères du Stampien d'Aquitaine:* Revue Micropaléont., v. 1, no. 3, p. 117-120, pl. 1.

(1472) **Pokorný, Vladimír,** 1951, *The middle Devonian Foraminifera of Čelechovice, Czechoslovakia:* Věstnik Královske Česke Společnosti Nauk Třída Matematicko-Přírodovédecka, v. 9, p. 1-29, pl. 1-2, 17 fig.——(1473) 1951, *Thalmannammina n.g. (Foraminifera) z Karpatského flyše:* Ústřed. ústavu Geol. Sborník, v. 18, p. 469-479, fig. 1-3.——(1474) 1954, *Základy zoologické mikropaleontologie:* Naklad. Česk. Akad. Věd., p. 1-651, text-fig. 1-756.——(1475) 1955, *Cassigerinella boudecensis, n.gen., n.sp. (Foraminifera, Protozoa), z oligocénu ždánického flyše:* Ústřed. ústavu Geol. Věstník, v. 30, p. 136-140, text-fig. 1-3.——(1476) 1956, *Semitextulariidae, a new family of Foraminifera:* Univ. Carolina, Geol., v. 2, no. 3, p. 279-286.——(1477) 1956, *New Discorbidae (Foraminifera) from the upper Eocene brown Pouzdřany marl, Czechoslovakia:* Same, v. 2, no. 3, p. 257-278, text-fig. 1-15.——(1478) 1958, *Grundzüge der Zoologischen Mikropaläontologie:* v. 1, 582 p., 549 text-fig. (Berlin).

(1479) **Pouchet, A.,** 1925, *Contribution à l'étude des Myxomycètes du Département du Rhône:* Soc. Linnéene Lyon, p. 42-66.

(1480) **Poyarkov, B. V.**, 1957, *O Foraminiferakh iz famenskikh i turneyskikh otlozheniy zapadnykh otrogov Tyan-shanya:* Leningrad. Univ., Vestnik 12 (geol. & geog. ser., no. 2), p. 26-41. [*On Foraminifera of the Famenian and Tournaisian deposits of the western extension of Tyan-Shan.*]

(1481) **Prever, P. L.**, 1902, *Le Nummuliti della Forca di Presta nell'Appennino centrale e dei dintorni di Potenza nell'Appennino meridionale:* Schweiz. Paläont. Gesell., Abhandl. (Soc. Pal. Suisse, Mém.), v. 29, art. 3, p. 3-121, pl. 1-8. ———(1481A) 1903, *Considerazioni sullo studio della Nummuliti:* Soc. geol. Italiana, Bull., v. 22, p. 461-487.———(1482) 1904, *Osservazioni sulla sottofamiglia della Orbitoidinae:* Rivista Italiana Paleont., v. 10, p. 111-127, 6 pl.

(1483) ———, & **Silvestri, Alfredo**, 1905, *Contributo allo studio delle Orbitolininae:* Soc. geol. Ital., Bull., v. 23 (1904), pt. 3, p. 477-486.

(1484) **Pritchard, Andrew**, 1861, *A history of Infusoria, including the Desmidiaceae and Diatomaceae, British and Foreign:* ed. 4, 968 p., 40 pl., Whittaker & Co. (London).

(1485) **Pronina, T. V.**, 1960, *Novye paraturamminidy ordovika i silura Urala:* in Novy vidy drevnikh rasteniy i bespozvonochnykh SSSR, pt. 1, Vses. Nauchno-Issledov. Geol. Inst. (VSEGEI), Minist. Geol. i Okhrany Nedr SSSR, p. 138-140, pl. 25 (Moscow). [*New Parathuramminidae from the Ordovician and Silurian of the Urals:* in New species of older plants and invertebrates of the USSR.]———(1486) 1960, *Novye vidy foraminifer iz nizhnezhivetskikh otlozheniy srednego i yuzhnogo Urala:* Akad. Nauk SSSR, Paleont. Zhurnal 1960, no. 1, p. 45-52, pl. 1. [*New species of Foraminifera from lower Givetian deposits of the central and southern Urals.*]

(1487) **Puri, H. S.**, 1954, *Contribution to the study of the Miocene of the Florida Panhandle:* Florida Geol. Survey, Bull. 36(1953), p. 1-345, pl. 1-30, 1-17.———(1488) 1957, *Stratigraphy and Zonation of the Ocala Group:* Same, Bull. 38, p. 1-248, pl. 1-15.———(1489) 1957, *Reclassification, structure and evolution of the family Nummulitidae:* Paleont. Soc. India, Jour., v. 2, p. 95-108, pl. 10-13, text-fig. 1-10.

(1490) **Purkin, M. M., Poyarkov, B. V.**, & **Rozhanets, V. M.**, 1961, *Stratigrafiya i novye vidy foraminifer Turneyskikh otlozheniy khrebta Borkoldoy (Tyan-Shan):* Akad. Nauk Kirgizskoy SSR, Izvestya, seriya estestv. & tekhn. nauk, v. 3, no. 4, p. 15-36, pl. 1-2. [*Stratigraphy and new species of foraminifers from Tournaisian deposits of the Borkoldy Range (Tyan-Shan).*]

(1490A) **Putrya, F. S.**, 1937, *K stratigrafii srednego karbona yugo-vostochnoy chasti Bolshogo Donbassa:* Azovsko-Chernomorskoye Geologicheskoye Tresta, Materialy po geologiy i poleznym iskopayemym, v. 1, p. 41-76, 2 pl. [*On the stratigraphy of the Upper Carboniferous of the southeast part of the Don Basin.*]———(1490B) 1939, *Materialy k stratigrafii verkhnego karbona vostochnoy okrainy Donetskogo basseyna:* Azovsko-Chernomorskoye Geologicheskoye Upravl., Materialy po geologiy i poleznym iskopayemym, v. 10, p. 97-156, pl. 1-5. [*Stratigraphy of the Upper Carboniferous of the eastern border of the Donets Basin.*]———(1491) 1940, *Foraminifery i stratigrafiya verkhnekamennougolnykh otlozheniy vostochnoy chasti Donetskogo basseyna:* Same, v. 11, p. 1-146, pl. 1-14. [*Foraminifers and stratigraphy of Upper Carboniferous deposits in the eastern part of the Donets Basin.*]———(1492) 1948, *Protriticites—novyy rod fuzulinid:* L'vovskogo Geol. Obshch. Gosud. Univ. Ivana Franko, Trudy, paleont. ser., no. 1, p. 89-96, pl. 1 (Lvov). [*Protriticites—a new genus of fusulinids.*]———(1493) 1948, *Pseudotriticitinae—novoe podsemeystvo fuzulinid:* Same, no. 1, p. 97-101, pl. 1. [*Pseudotriticitinae—new subfamily of fusulinids.*]———(1494) 1956, *Stratigrafiya i foraminifery srednekamennougolnykh otlozheniy vostochnogo Donbassa:* Mikrofauna SSSR, Sbornik 8, VNIGRI, Trudy, new ser., no. 98, p. 333-485, 17 pl. [*Stratigraphy and Foraminifera of the middle Carboniferous deposits of the eastern Don Basin.*]

(1495) **Quenstedt, F. A.**, 1856-58, *Der Jura:* pt. 1, p. 1-208 (April 1856); pt. 2, p. 209-368 (Sept. 1856); pt. 3, p. 369-576 (Dec. 1856); pt. 4, p. 577-842 (May 1857); Introduction and atlas, 100 pl. (1858) (Tübingen).

(1496) **Rafinesque, C. S.**, 1815, *Analyse de la nature; ou Tableau de l'univers et des corps organisés:* 224 p. (Palermo).

(1497) **Rainwater, E. H.**, 1960, *Stratigraphy and its role in the future exploration for oil and gas in the Gulf Coast:* Gulf Coast Assoc. Geol. Soc., Trans., v. 10, p. 33-75, text-fig. 1-33 (Jackson).

(1498) **Rao, S. R. Narayana**, 1940, *On Orbitosiphon, a new genus of orbitoidal Foraminifera from the Ranikot beds of the Punjab Salt Range, N.W. India:* Current Sci., v. 9, p. 414-415, 1 text-fig. (Bangalore).———(1499) 1942, *On Lepidocyclina (Polylepidina) birmanica, sp.nov., and Pseudophragmina (Asterophragmina) pagoda, subgen. nov. et sp. nov., from the Yaw stage (Priabonian) of Burma:* Geol. Surv. India, Records, v. 77, prof. paper 12, p. 1-16, 2 pl.

(1500) **Rauzer-Chernousova, D. M.**, 1937, *Rugosofusulina—novyy rod fuzulinid:* Paleont. Lab. Moskov. Gosud. Univ., Etyudy Mikropaleontologiy, v. 1, pt. 1, p. 9-26, pl. 1-3, fig. 1-2. [*Rugosofusulina, a new genus of fusulinids.*]———(1501) 1938, *Verkhnepaleozoyskiye fora-*

minifery Samarskoy Luki i Zavolzh'ya: Akad. Nauk SSSR, Trudy, Geol. Inst., v. 7, p. 69-167, pl. 1-9, fig. 1-5. [*The Upper Palaeozoic Foraminifera of the Samara Bend and the Trans-Volga Region.*]———(1502) 1948, *Rod Haplophragmella i blizkie k nemu formy:* Same, no. 62 (geol. ser. no. 19), p. 159-165, pl. 3. [*The genus Haplophragmella and forms similar to it.*]———(1503) 1948, *Rod Cribrospira Moeller:* Same, no. 62 (geol. ser. no. 19), p. 186-189, pl. 7. [*The genus Cribrospira Möller.*]
———(1504) 1948, *Nekotorye novye nizhnekamennougol'nye foraminifery Syzranskogo rayona:* Same, no. 62 (geol. ser. no. 19), p. 239-243, pl. 17. [*Certain new Lower Carboniferous Foraminifera from the Syzransky district.*]
———(1505) 1948, *Materialy k faune foraminifer kamennougol'nykh otlozheniy tsentral'nogo Kazakhstana:* Same, no. 66 (geol. scr. no. 21), p. 1-27, pl. 1-3. [*Data on the foraminiferal fauna of the Carboniferous deposits of central Kazakhstan.*]———(1506) 1960 [1961], *Reviziya shvagerin s blizkimi rodami i granitsa karbona i permi:* Voprosy Mikropaleontologii, no. 4, Akad. Nauk SSSR, Otdel. Geol.-Geog. Nauk, Geol. Inst., p. 3-32, fig. 1-6. [*Revision of Schwagerina and related genera and the limits of the Carboniferous and the Permian.*]

(1507) ———, Belyaev, G. M., & Reytlinger, E. A., 1936, *Verkhne paleozoyskie foraminifery Pechorskogo kraya:* Akad. Nauk SSSR, Trudy, Polyarnoi Komissii, no. 28, p. 159-232, pl. 1-6. [*Upper Paleozoic Foraminifera of the Pechora district.*]———(1508) 1940, *O foraminiferakh kamennougolnykh otlozheniy samarskoy Luki:* Neft. Geol.-Razved. Inst., Trudy, new ser., no. 7, 88 p., 9 pl., 18 fig., Gostoptekhizdat. [*On Foraminifera of the Carboniferous deposits of the Samara Bend.*]

(1509) ———, & Fursenko, A. V., 1959, *Osnovy Paleontologii. Obshchaya chast prosteyshie:* Akad. Nauk SSSR, p. 1-368, pl. 1-13, text-fig. 1-894. [*Principles of Paleontology. Part I, Protozoa.*]

(1509A) ———, Gryzlova, N. D., Kireeva, G. D., Leontovich, G. E., Safonova, T. P., & Chernova, E. I., 1951, *Srednekamennougolnye fusulinidy russkoy platformy i sopredelnykh oblastey:* Akad. Nauk SSSR, Inst. Geol. Nauk, Minist. Neft. Promyshlennosti SSSR, 380 p., 58 pl., 30 text-fig. [*Middle Carboniferous fusulinids of the Russian Platform and adjacent regions.*]

(1509B) ———, & Shcherbovich, S. F., 1949, *Shvageriny evropeyskoi chasti SSSR:* Akad. Nauk SSSR, Trudy, Inst. Geol., no. 105 (geol. ser. no. 35), p. 61-117, pl. 1-12. [*Schwaginidae of the European part of the USSR.*]

(1510) Redmond, C. D., 1953, *Chamber arrangement in Foraminifera:* Micropaleontologist, v. 7, p. 16-22, text-fig. 1-4.

(1511) Reichel, Manfred, 1931, *Sur la structure des Alvéolines:* Eclogae geol. Helv., v. 24, p. 289-303, pl. 13-18, fig. 1-2.———(1512) 1933, *Sur une alvéoline cénomanienne du Bassin du Beausset:* Same, v. 26, p. 269-280, fig. 1-14.
———(1513) 1936, *Bemerkungen über einige von O. Renz im zentralen Appenin gesammelte Foraminiferen:* in Renz, O., Stratigraphische und mikropalaeontologische Untersuchung der Scaglia (Obere Kreide-Tertiär) im zentralen Apennin, Same, v. 29, p. 136-142, pl. 12, 15, fig. 7, 14.———(1514) 1936-1937, *Étude sur les Alvéolines I & II:* Schweiz. Palaeont. Gesell., Abhandl. (Soc. Palaeont. Suisse, Mém.), (I) v. 57, no. 4, 93 p., 9 pl., 16 fig.; (II) v. 59, no. 3, p. 95-147, pl. 10-11, fig. 17-29.———(1515) 1941, *Sur un nouveau genre d'alvéolines du Crétacé supérieur:* Eclogae geol. Helv., v. 34, p. 254-260, pl. 15, fig. 1-2.———(1516) 1945, *Sur un miliolidé nouveau du Permien de l'île de Chypre:* Verhandl. Naturforsch. Gesell. Basel, v. 56, pt. 2, p. 521-530, text-fig. 1-2.
———(1517) 1946, *Sur quelques Foraminifères nouveaux du Permien méditerranéen:* Eclog. geol. Helv., v. 38, no. 2 (1945), p. 524-560, pl. 19.———(1518) 1947, *Multispirina iranensis, n.gen., n.sp., foraminifère nouveau du Crétacé supérieur de l'Iran:* Schweiz. Palaeont. Gesell., Abhandl. (Soc. Palaeont. Suisse, Mém.), v. 65, p. 1-13, pl. 1-4, fig. 1-5.
———(1519) 1949, *Remarques sur le genre Boreloides Cole et Bermudez:* Actes Soc. Helvet. Sci. Nat., Lausanne, p. 148.———(1520) 1949, *Alvéolines de l'Oligocene-Miocene de Cuba (abstract):* Schweiz. naturforsch. Gesell., Verhandl., 129 Vers. Lausanne, p. 148.———(1521) 1949 [1950], *Sur un nouvel Orbitoïde du Crétacé supérieur hellénique:* Eclogae geol. Helv., v. 42, no. 2, p. 480-485, text-fig. 1-10.———(1522) 1950, *Observations sur les Globotruncana du gisement de la Breggia (Tessin):* Same, v. 42, no. 2, p. 596-617, pl. 16-17, text-fig. 1-7.———(1523) 1952, *Fusarchais bermudezi, n.gen., n.sp., pénéroplidé alvéoliniforme de l'Oligo-Miocène de Cuba:* Same, v. 44, no. 2(1951), p. 458-464, 5 text-fig.———(1524) 1953 [1954], *Les caractères embryonnaires de Subalveolina:* Same, v. 46, no. 2, p. 256-262, pl. 13-14, fig. 1-4.———(1525) 1956, *Sur une trocholine du Valanginien d'Arzier:* Same, v. 48, no. 2(1955), p. 396-408, pl. 14-16, text-fig. 1-5.

(1526) Reinsch, P. F., 1877, *Notiz über die mikroskopische Fauna der mittleren und unteren fränkischen Liasschichten:* Neues Jahrb. Mineral. Geol. & Paläont., p. 176-178.

(1527) Reiss, Zeev, 1957, *Occurrence of Nezzazata in Israel:* Micropaleontology, v. 3, p. 259-262, pl. 1.———(1528) 1957, *Notes on Foraminifera from Israel:* Israel Geol. Survey; (a) *1. Remarks on Truncorotalia aragonensis cau-*

casica (Glaessner); 2. *Loxostomoides, a new Late Cretaceous and Early Tertiary genus of Foraminifera;* 3. *Sigalia, a new genus of Foraminifera,* Bull. 9, p. i-vii; (b) 4. *Occurrence and stratigraphical significance of Cuvillierina eocenica Debourle,* Bull. 10, p. 3-12, pl. A-B; (c) 5. *Studies on Victoriellidae,* Bull. 11, p. 1-9, pl. A-B, text-fig. 1.———(1529) 1957, *The Bilamellidea, nov. superfam., and remarks on Cretaceous globorotaliids:* Cushman Found. Foram. Research, Contrib., v. 8, pt. 4, p. 127-145, pl. 18-20.———(1530) 1958, *Classification of lamellar Foraminifera:* Micropaleontology, v. 4, p. 51-70, pl. 1-5.———(1531) 1959, *The wall-structure of Cibicides, Planulina, Gyroidinoides, and Globorotalites:* Same, v. 5, p. 355-357, pl. 1.———(1532) 1959, *Note zur Pseudolituonella:* Revue Micropaléont., v. 2, p. 95-98, pl. 1.———(1533) 1960, *Structure of so-called Eponides and some other rotaliiform Foraminifera:* Israel Geol. Survey, Bull. 29, p. 1-28, pl. 1-3, text-fig. 1-2.

(1534) ———, & Merling, P., 1958, *Structure of some Rotaliidea:* Israel Geol. Survey, Bull. 21, p. 1-19, pl. 1-5.

(1535) Renz, H. H., 1948, *Stratigraphy and fauna of the Agua Salada group, State of Falcón, Venezuela:* Geol. Soc. America, Mem. 32, x+219 p., 12 pl.

(1536) Resig, J. M., 1962, *The morphological development of Eponides repandus:* Cushman Found. Foram. Research, Contrib., v. 13, pt. 2, p. 55-57, pl. 14.

(1537) Reuss, A. E., 1844, *Geognostische Skizzen aus Böhmen:* v. 2, 304 p., 3 pl., C. W. Medau (Prag).———(1538) 1846, *Die Versteinerungen der böhmischen Kreideformation:* pt. 2, 148 p., pl. 14-51 (Stuttgart).———(1539) 1848, *Die fossilen Polyparien des Wiener Tertiärbeckens:* Naturwiss. Abhandl., v. 2, pt. 1, p. 1-109, pl. 1-11.———(1540) 1850, *Neues Foraminifera aus den Schichten des österreichischen Tertiärbeckens:* K. Akad. Wiss. Wien, math.-naturwiss. Cl., Denkschr., v. 1, p. 365-390, pl. 46-51.———(1541) 1851, *Ueber die fossilen Foraminiferen und Entomostraceen der Septarienthone der Umgegend von Berlin:* Deutsch. geol. Gesell., Zeitschr., v. 3, p. 49-91, pl. 3-7.

(1542) 1851, *Die Foraminiferen und Entomostraceen des Kreidemergels von Lemberg:* Haidinger's Naturwiss. Abhandl., v. 4, p. 17-52, pl. 2-6.———(1543) 1854, *Beiträge zur Charakteristik der Kreideschichten in den Ostalpen, besonders im Gosauthale und am Wolfgangsee:* K. Akad. Wiss. Wien, math.-naturwiss. Cl., Denkschr., v. 7, pt. 1, p. 1-156, pl. 1-31.

(1544) 1855, *Ein Beitrag zur genaueren Kenntniss der Kreidegebilde Mecklenburgs:* Deutsch. geol. Gesell., Zeitschr., v. 7, no. 1, p. 261-292, pl. 8-11.———(1545) 1860, *Über Lingulinopsis, eine neue Foraminiferen-Gattung aus dem böhmischen Pläner:* K. Böhm. Gesell. Wiss. Prag, math.-naturw. Cl., Sitzungsber., p. 23-24.———(1546) 1860, *Über Ataxophragmium, eine neue Foraminiferengattung aus der Familie der Uvellideen:* Same, p. 52-54.———(1547) 1860, *Über di Frondicularideen, eine Familie der Polymeren Foraminiferen:* Same, p. 77-92.———(1548) 1860, *Die Foraminiferen der Westphälischen Kreideformation:* K. Akad. Wiss. Wien, math.-naturw. Cl., Sitzungsber., v. 40, p. 147-238, pl. 1-13.———(1549) 1861, *Neuere Untersuchungen über die Fortpflanzung der Foraminiferen und über eine neue Foraminiferengattung Haplostiche:* K. Böhm. Gesell. Wiss. Prag, math.-naturw. Cl. Sitzungsber., p. 12-16.———(1550) 1861, *Beiträge zur Kenntniss der tertiären Foraminiferen-Fauna:* K. Akad. Wiss. Wien, math.-naturw. Cl., Sitzungsber., v. 42(1860), p. 355-370, pl. 1-2.———(1551) 1861, *Kurze Notiz über eine neue Foraminiferengattung Schizophora:* K. Böhm. Gesell. Wiss., Sitzungsber., v. 1861, pt. 2, p. 12-13.———(1552) 1862, *Entwurf einer systematischen Zusammenstellung der Foraminiferen:* K. Akad. Wiss. Wien, math.-naturwiss. Cl., Sitzungsber., v. 44 (1861), p. 355-396.———(1553) 1863, *Beiträge zur Kenntniss der tertiären Foraminiferen-fauna (Zweite Folge):* Same, v. 48, pt. 1, p. 36-71, pl. 1-8.———(1554) 1863, *Die Foraminiferen des norddeutschen Hils und Gault:* Same, v. 46 (1862), pt. 1, p. 5-100, pl. 1-13.———(1555) 1866, *Die Foraminiferen und Ostrakoden der Kreide am Kanara-See bei Küstendsche:* Same, v. 52 (1865), pt. 1, p. 445-470, pl. 1.———(1556) 1871, *Vorläufige Notiz über zwei neue fossile Foraminiferen-Gattungen:* Same, v. 64, pt. 1, p. 277-281.

(1557) Reyment, R. A., 1959, *The foraminiferal genera Afrobolivina, gen. nov., and Bolivina in the Upper Cretaceous and Lower Tertiary of West Africa:* Stockholm Contrib. Geol., v. 3, no. 1, p. 1-57, pl. 1-7.———(1558) 1959, *Zur Fassung der Foraminiferengattung Aragonia:* Paläont. Zeitschr., v. 33, p. 108-112, text-fig. 1-4.

(1559) Reytlinger, E. A., 1948, *Kembriyskie foraminifery Yakutii:* Moskov. Obshch. Ispyt., Prirody, Otdel Geol., Byull., v. 23, no. 2, p. 77-81, 1 pl. [*Cambrian Foraminifera of Yakutsk.*]———(1560) 1950, *Foraminifery srednekamennougolnykh otlozheniy tsentralnoy chasti Russkoy platformy (isklyuchaya semeystvo Fusulinidae):* Akad. Nauk SSSR, Geol. Inst., Trudy, no. 126 (geol. ser. no. 47), p. 1-126, pl. 1-22, text-fig. 1-15. [*Foraminifera of middle Carboniferous deposits of the central part of the Russian Platform.*]———(1561) 1954, *Devonskie foraminifery nekotorykh razrezov vostochnoy chasti Russkoy platformy:* VNIGNI, Nauchno-Issledov. Geol. Razved., Trudy, Neft.

Inst., Paleont. Sbornik 1, p. 52-81, pl. 17-22. [*Devonian Foraminifera of certain sections of the eastern part of the Russian Platform.*]——— (1562) 1956, *Novoe semeystvo Lasiodiscidae:* Voprosy Mikropaleontologii, no. 1, Akad. Nauk SSSR, p. 69-78, pl. 1-2. [*New family Lasiodiscidae.*]———(1563) 1957, *Sfery Devonskikh otlozheniy Russkoy Platformy:* Akad. Nauk SSSR, Doklady, v. 115, no. 4, p. 774-776, pl. [*Spheres from Devonian deposits of the Russian Platform.*]———(1564) 1958, *K voprosu sistematiki i filogenii nadsemeystva Endothyridea:* Voprosy Mikropaleontologii, no. 2, Akad. Nauk SSSR, p. 53-73, 4 fig. [*On the question of systematics and phylogeny of the superfamily Endothyridea.*]———(1565) 1959, *Atlas mikroskopicheskikh organicheskikh ostatkov i problematiki drevnikh tolshch Sibiri:* Akad. Nauk SSSR, Trudy, Geol. Inst., no. 25, p. 1-59, pl. 1-22. [*Atlas of microscopical organic remains and problematica of ancient strata of Siberia.*]———(1566) 1961, *Nekotorye voprosy sistematiki kvaziendotir:* Voprosy Mikropaleontologii, no. 5, Akad. Nauk SSSR, Otdel Geol. & Geog., Geol. Inst., p. 31-68, pl. 1-6, text-fig. 1-3, table. [*Certain questions of the systematics of quasiendothyrids.*]

(1567) **Rhumbler, Ludwig,** 1894, *Die Perforation der Embryonalkammer von Peneroplis pertusus Forskål:* Zool. Anzeiger, v. 17, p. 335-342, 3 fig.———(1568) 1894-95, *Beiträge zur Kenntnis der Rhizopoden:* Zeitschr. Wiss. Zool.; (a) *II. Saccammina sphaerica* M. Sars., v. 57, p. 587-617, pl. 25 (1894); (b) *III, IV, V,* v. 61, p. 38-110, pl. 4-5, 10 text-fig. (1895).——— (1568A) 1895, *Entwurf eines natürlichen Systems der Thalamophoren:* Gesell. Wiss. Göttingen, math.-physik Kl., Nachr., no. 1, p. 51-98.———(1569) 1904, *Systematische Zusammenstellung der recenten Reticulosa:* Archiv Protistenkunde, v. 3, p. 181-294, text-fig. 1-142. ———(1570) 1905, *Mitteilungen über Foraminiferen (mit Demonstrationen):* Deutsch. zool. Gesell., Verhandl., v. 15, p. 97-106, text-fig.———(1571) 1906, *Foraminiferen von Laysan und den Chatham-Inseln:* Zool. Jahresber., v. 24, no. 1, p. 21-80, pl. 2-5.———(1572) 1911-13, *Die Foraminiferen (Thalamophoren) der Plankton-Expedition:* Ergebnisse der Plankton-Exped. der Humboldt-Stiftung; (a) 1911, v. 3, Lief. c., p. 1-331, pl. 1-39 (1909); (b) 1913, *Pt. 2, Systematik: Arrhabdammidia, Arammodisclidia und Arnodosammidia,* v. 3, Lief. c., p. 332-476, 65 fig.———(1573) 1928, *Amoebozoa et Reticulosa:* Die Tierwelt der Nord- und Ostsee, Lief. 13, pt. 2, p. IIa1-IIa26, text-fig. 1-39.———(1574) 1935, *Rhizopoden der Kieler Bucht, gesammelt durch A. Remane, Teil I:* Naturwiss. ver. Schleswig-Holstein, Schrift, v. 21, p. 143-194, pl. 1-9.———(1575) 1936, *Foraminiferen der Kieler Bucht, gesammelt durch A. Remane, Teil II. (Ammodisculinidae bis einschl. Textulinidae.):* Kieler Meeresforschungen, v. 1, p. 179-242, text-fig. 127-246. ———(1576) 1938, *Foraminiferen aus dem Meeressand von Helgoland, gesammelt von A. Remane (Kiel):* Same, v. 2, p. 157-222, 64 text-fig.

(1577) **Riccio, J. F.,** 1950, *Triloculinella, a new genus of Foraminifera:* Cushman Found. Foram. Research, Contrib., v. 1, pt. 3-4, p. 90, pl. 15.

(1578) **Richarz, P. S.,** 1910, *Der geologische Bau von Kaiser Wilhelms-Land nach dem heutigen Stand unseres Wissens,* in BOEHM, G., Geologische Mitteilung aus dem Indo-Australischen Archipel: Neues Jahrb. Mineral., Geol. & Paleontol., Beil.-Bd. 29, p. 406-536, pl. 13-14, text-fig. 1-10.

(1579) **Risso, Antoine,** 1826, *Histoire naturelle des principales productions de l'Europe méridionale et particulièrement de celles des environs de Nice et des Alpes maritimes:* F.-G. Levrault (Paris & Strassburg); (a) v. 4, p. 1-439; (b) v. 5.

(1580) **Roboz, Zoltàn von,** 1884, *Calcituba polymorpha, nov.gen., nov.spec.:* K. Akad. Wiss. Wien. math.-naturw. Cl., Sitzungsber., v. 88 (1883), pt. 1, p. 420-432, pl. 1.

(1580A) **Roemer, C. F.,** 1852, *Die Kreidebildungen von Texas und ihre organischen Einschlüsse:* 100 p., 11 pl., A. Marcus (Bonn).

(1581) **Roemer, F. A.,** 1838, *Die Cephalopoden des norddeutschen tertiären Meeressandes:* Neues Jahr. Mineral., p. 381-394, pl. 3.——— (1582) 1839, *Die Versteinerungen des norddeutschen Oolithen-Gebirges. Ein Nachtrag:* Hahnschen Hofbuchhandlung (Hannover). ———(1583) 1841, *Die Versteinerungen des norddeutschen Kreidegebirges:* 145 p., 16 pl. (Hannover).

(1584) **Roissy, Felix de,** 1805, *Histoire naturelle, générale et particulière des Mollusques (Buffon et Sonnini):* v. 5, 450 p., pl. 51-56, Dufart (Paris).

(1585) **Ross, I. K.,** 1957, *Syngamy and plasmodium formation in the Myxogastres:* Am. Jour. Botany, v. 44, p. 843-850, fig. 1-19.

(1586) **Rostafiński [Rostafińskie], J. T. von,** 1873, *Versuch eines systems der Mycetozoen:* Inaugural-Dissertation der Philosophischen Facultät der Universität Strassburg im Elsass, p. i-iv, 1-21 (Strassburg). ——— (1587) 1875, *Śluzowce (Mycetozoa) Monographia:* Pamiętnik Towarzystwa Nauk Ścisłych w Paryżu, v. 5-6, p. 1-432, pl. 1-13.———(1588) 1876, *Dodatek I do Monografii Śluzowców:* Same, v. 8, p. 1-43, 4 fig.

(1588A) **Rouillier, Charles, & Vosinsky, Al.,** 1849, *Études progressives sur la géologie de Moscou:* Soc. Imper. Nat. Moscou, Bull., v. 22, p. 337-399, pl. K.

(1589) **Rouvillois, Armelle,** 1960, *Le Thanétien*

du Bassin de Paris: Museum Natl. Histoire Nat., Mem., new ser. C, v. 8, p. 1-151, pl. 1-8, 17 tables & maps.

(1590) **Rozovskaya, S. E.,** 1948, *Klassifikatsiya i sistematicheskie priznaki roda Triticites:* Akad. Nauk SSSR, Doklady, new ser., v. 59, no. 9, p. 1635-1638, fig. 1-2. [*Classification and systematic characteristics of the genus Triticites.*]————(1591) 1949, *Stratigraficheskoye raspredeleniye fuzulinid v verkhnekamennougolnykh i nizhnepermskikh otlozheniyakh yuzhnogo Urala:* Same, v. 69, p. 249-252, 1 fig. [*Stratigraphic distribution of fusulinids in Upper Carboniferous and Lower Permian deposits of the southern Urals.*]————(1591A) 1950, *Rod Triticites, ego razvitie i stratigraficheskoe znachenie:* Akad. Nauk SSSR, Trudy, Paleont. Inst., v. 26, p. 3-78, pl. 1-10. [*The genus Triticites, its development and stratigraphic significance.*]————(1592) 1950, *K sistematike semeystva Fusulinidae:* Akad. Nauk SSSR, Doklady, v. 73, no. 2, p. 375-378. [*On the systematics of the family Fusulinidae.*]————(1592A) 1952, *Fuzulinidy verkhnego karbona i nizhney permi yuzhnogo Urala:* Akad. Nauk SSSR, Trudy, Paleont. Inst., v. 40, Mater. po faune paleozoya, p. 5-50, pl. 1-6. [*Fusulinidae of the Upper Carboniferous and Lower Permian of the Southern Urals.*]————(1593) 1961, *K sistematike semeystv Endothyridae i Ozawainellidae:* Paleont. Zhurnal, 1961, no. 3, p. 19-21. [*On the systematics of the families Endothyridae and Ozawainellidae.*]

(1594) **Rütimeyer, Ludwig,** 1850, *Ueber das schweizerische Nummulitenterrain, mit besonderer Berücksichtigung des Gebirges zwischen dem Thunersee und der Emme:* Soc. Helv. Sci. Nat., Nouv. Mém., v. 11, Mém. 2, p. 1-120, pl. 1-5.

(1595) **Ruiz de Gaona, R. P. Máximo,** 1948, *Sobre un microforaminífero terciario desconocido en España:* Inst. Geol. & Minero España, Notas & Commun., no. 18, p. 77-91.

(1596) **Rutten, L. M. R.,** 1911, *On Orbitoides of the Balikpapan Bay, East Coast of Borneo:* K. Akad. Wetensch. Amsterdam, Proc., p. 1122-1139, illus.————(1597) 1913, *Studien über Foraminiferen aus Ost-Asien;* Theil 3: Geol. Reichs-Museum Leiden, Samml., ser. 1, v. 9 (1911-14), no. 3, p. 219-224, pl. 14, text-fig. 1-2.————(1598) 1914, *Foraminiferen führende Gesteine von Niederländisch Neu-Guinea:* Nova Guinea, Uitkomsten Nederland Nieuw-Guinea Exped. 1903, v. 6 (Geol.), pt. 2, p. 21-51, pl. 6-9 (Leiden).

(1599) **Rutten, M. G.,** 1935, *Larger Foraminifera of northern Santa Clara Province, Cuba:* Jour. Paleontology, v. 9, p. 527-545, pl. 59-62, 4 text-fig.

(1600) **Rzehak, Anton,** 1885, *Bemerkungen über einige Foraminiferen der Oligocän Formation:* Naturforsch. Vereins Brünn, Verhandl., v. 23 (1884), p. 123-129.————(1601) 1886, [*Ueber Foraminiferen*]: Same, v. 24, Sitzungber., p. 8.————(1602) 1888, *Die Foraminiferen der Nummulitenschichten des Waschberges und Michelsberges bei Stockerau in Nieder-Oesterreich:* K. K. Geol. Reichsanst., Verhandl., v. 1888, p. 226-229.————(1603) 1888, *Die Foraminiferen des kieseligen Kalkes von Nieder-Hollabrunn und des Melettamergels der Umgebung von Bruderndorf in Niederösterreich:* Naturhist. Hofmuseum, Wien, Ann., v. 3, p. 257-270, pl. 11.————(1604) 1891, *Die Foraminiferenfauna der alttertiären Ablagerungen von Bruderndorf in Nieder-Osterreich, mit Berüchsichtigung des angeblichen Kreidevorkommens von Leitzersdorf:* Same, v. 6, p. 1-12.————(1605) 1895, *Ueber einige merkwürdige Foraminiferen aus österreichischen Tertiär:* Same, v. 10, p. 213-30, pl. 6-7.

(1606) **Saccardo, P. A.,** 1888, *Myxomyceteae Wallr.,* in Sylloge Fungorum omnium hucusque cognitorum: Digessit, P. S. SACCARDO, v. 7, p. 323-468.

(1607) **Sacco, Federico,** 1893, *Sur quelques Tinoporinae du Miocène de Turin:* Soc. Belge. Géol. & Paléont. Hydr., v. 7 (1893-94), p. 204-207.

(1608) **Saedeleer, Henri de,** 1932, *Recherches sur les pseudopodes des Rhizopodes Testacés, Les concepts pseudopodes lobosa, filosa et granuloreticulosa:* Arch. Zool. Expér. Générale, v. 74, pt. 30, p. 597-626.————(1609) 1934, *Beitrag zur Kenntnis der Rhizopoden, morphologische und systematische Untersuchungen und ein Klassifikationsversuch:* Musée Roy. Histoire Nat. Belgique, Mem. 60, 112 p., 29 text-fig., 8 pl.

(1610) **Safonova, T. P.,** 1951. (see 1509A.)

(1611) **Sagra, Ramon de la,** 1839, *Foraminifères:* Histoire phys. pol. & nat. de l'isle de Cuba, xlviii+224 p., 12 pl.

(1612) **Sahni, M. R., & Sastri, V. V.,** 1957, *A monograph of the Orbitolines found in the Indian continent (Chitral, Gilgit, Kashmir), Tibet and Burma with observations on the age of the associated volcanic series:* Geol. Survey India, Mem., new ser., v. 33, no. 3, p. 1-44, pl. 1-6.

(1613) **Said, Rushdi,** 1950, *The distribution of Foraminifera in the northern Red Sea:* Cushman Found. Foram. Research, Contrib., v. 1, pt. 1-2, p. 9-29, text-fig. 1-4, 2 tables.————(1614) 1951, *Preliminary note on the spectroscopic distribution of elements in the shells of some Recent calcareous Foraminifera:* Same, v. 2, pt. 1, p. 11-13.————(1615) 1951, *Ecology of Foraminifera:* Micropaleontologist, v. 5, p. 12-14.

(1616) ————, **& Barakat, M. G.,** 1958, *Jurassic*

microfossils from Gebel Maghara, Sinai, Egypt: Micropaleontology, v. 4, p. 231-272, pl. 1-6, text-fig. 1-5, table 1.

(1617) **Saidova, Kh. M.**, 1960, *Raspredelenie foraminifer v donnykh otlozheniyakh Okhotskogo Morya:* Akad. Nauk SSSR, Instituta Okeanologii, Trudy, v. 32, p. 96-157, text-fig. 1-28. [*Distribution of Foraminifera in bottom sediments of the Okhotsk Sea.*]————(1618) 1961, *Ekologiya foraminifer i paleogeografiya dal'nevostochnykh Morey SSSR, i severo-zapadnoy chasti Tikhogo Okeana:* Akad. Nauk SSSR, Inst. Okeanologii, p. 1-232, pls. 1-31. [*Foraminiferal ecology and paleogeography, far eastern seas of the USSR and northwest part of the Pacific Ocean.*]

(1619) **St. Jean, Joseph, Jr.**, 1957, *A middle Pennsylvania foraminiferal fauna from Dubois County, Indiana:* Indiana Dept. Conserv., Geol. Survey, Bull. 10, p. 1-66, pl. 1-5.

(1620) **Saito, Tsunemasa**, 1962, *Eocene planktonic Foraminifera from Hahajima (Hillsborough Island):* Palaeont. Soc. Japan, Trans. & Proc., new ser., no. 45, p. 209-225, pl. 32-34.

(1621) **Sakagami, Sumio & Omata, Toshikazu**, 1957, *Lower Permian fusulinids from Shiraiwa, north-western part of Ome, Nishitama-gun, Tokyo-to, Japan:* Japanese Jour. Geol. & Geog., v. 28, no. 4, p. 247-264, pl. 19-20, fig. 1-2.

(1622) **Samoylova, R. B.**, 1940, *The genus Almaena of the lower Oligocene foraminifers of the Crimea:* Acad. Sci. U.R.S.S., Comptes Rendus, Doklady, v. 28, p. 377-378, 3 text-fig.————(1623) 1947, *O nekotorykh novykh i kharakternykh vidakh foraminifer iz verkhnego Paleogena Kryma:* Moskov. Obshch. Ispyt. Prirody, Otdel Geol., Byull., v. 22(4), p. 77-101, 3 pl. [*On certain new and characteristic species of Foraminifera from the upper Paleogene of Crimea.*]

(1624) **Sample, C. H.**, 1932, *Cribratina, a new genus of Foraminifera from the Comanchean of Texas:* Am. Midland Naturalist, v. 13, no. 5, p. 319-321, pl. 30.

(1625) **Sandahl, O.**, 1858, *Två nya former af Rhizopoder:* K. Vetenskaps.-Akad., Förhandl., Öfvers., v. 14 (1857), no. 8, p. 299-303, pl. 3.

(1625A) **Sander, N. J.**, 1962, *Aperçu paléontologique et stratigraphique du Paléogène en Arabie Séoudite Orientale:* Revue Micropaleont., v. 5, no. 1, p. 3-40, pl. 1-5, text-fig. 1-8.

(1626) **Sandon, H.**, 1927, *The composition and distribution of the protozoan fauna of the soil:* 237 p., 6 pl., Oliver & Boyd (Edinburgh & London).————(1627) 1932, *The food of Protozoa:* Egyptian Univ., Publ. Fac. Sci., no. 1, p. 1-187 (Cairo).————(1628) 1957, *Neglected animals—the Foraminifera:* New Biology, v. 24, p. 7-32, text-fig. 1-4 (London).

(1629) **Sars, Michael**, 1869, *Fortsatte Bemaerkninger over det dyriske Livs Udbredning i Havets Dybder:* Vidensk.-Selsk. Christiania, Forhandl., v. 1868, p. 246-275.

(1630) **Sars, G. O.**, 1872, *Undersøgelser over Hardangerfjordens Fauna:* Vidensk.-Selsk. Christiania, Forhandl., v. 1871, p. 246-255.

(1631) **Saunders, J. B.**, 1957, *Trochamminidae and certain Lituolidae (Foraminifera) from the Recent brackish-water sediments of Trinidad, British West Indies:* Smithsonian Misc. Coll., v. 134, no. 5, Publ. 4270, p. 1-16, pl. 1-4.————(1632) 1957, *Emendation of the foraminiferal genus Palmerinella Bermúdez, 1934, and erection of the foraminiferal genus Helenia:* Washington Acad. Sci., Jour., v. 47, no. 11, p. 370-374, fig. 1-7.————(1633) 1958, *Recent Foraminifera of mangrove swamps and river estuaries and their fossil counterparts in Trinidad:* Micropaleontology, v. 4, no. 1, p. 79-92, pl. 1-2, text-fig. 1-3.————(1634) 1961, *Helenina Saunders, new name for the foraminiferal genus Helenia Saunders, 1957, non Helenia Walcott, 1889:* Cushman Found. Foram. Research, Contrib., v. 12, pt. 4, p. 148.

(1635) **Schacko, Gustav**, 1897, *Beitrag über Foraminiferen aus der Cenoman-Kreide von Moltzow in Mecklenburg:* Verhandl. Freunde Naturg. Mecklenberg, Archiv, v. 50 (1896), p. 161-168, pl. 4.

(1636) **Schaeffer, A. A.**, 1926, *Taxonomy of the Amebas, with descriptions of thirty-nine new marine and fresh-water species:* Carnegie Inst. Washington, Papers, Dept. Marine Biol., v. 24, p. 3-116, pl. 1-12.

(1637) **Schafhäutl, K. E.**, 1851, *Geognostische Untersuchungen der südbayerischen Alpengebirges:* Liter.-Artist. Anst., p. 1-206, pl. 13.————(1638) 1863, *Süd-Bayerns Lethaea Geognostica:* 487 p., 86 pl., L. Voss (Leipzig).

(1639) **Schaub, Hans**, 1951, *Stratigraphie und Paläontologie des Schlierenflysches mit besonderer Berücksichtigung der paleocaenen und untereocaenen Nummuliten und Assilinen:* Schweiz. Palaeont., Abhandl., v. 68, p. 1-222, 9 pl., 1 table, 336 text-fig.

(1640) **Schaudinn, Fritz**, 1893, *Myxotheca arenilega, nov.gen., nov. sp., ein neuer mariner Rhizopode:* Zeitschr. Wiss. Zool., v. 57, p. 18-31, pl. 2.————(1641) 1894, *Über die systematische Stellung und Fortpflanzung von Hyalopus (Gromia dujardinii Schultze):* Gesell. naturforsch. Freunde Berlin, Sitzungsber., p. 14-22.

(1642) **Scheffelt, E.**, 1920, *Die Fauna der Chiemseemoore:* Zool. Anzeiger, v. 52, no. 3/4, p. 166-175, text-fig. 1-11.

(1643) **Scheffen, Walther**, 1932, *Zur morphologie und morphogenese der "Lepidocyclinen":*

Paläont. Zeitschr., v. 14, p. 233-256, pl. 9-10, 6 fig.

(1643A) **Scheibnerova, Viera,** 1962, *Stratigrafia strednej a vrchnej kriedy tétydnej oblasti na základe globotrunkaníd:* Geol. Sborník,, Bratislava, v. 13, no. 2, p. 197-226, text-fig. 6, 7. [*Stratigraphy of the middle and Upper Cretaceous of the Tethys region on the basis of the globotruncanids.*]

(1644) **Schellwien, Ernst,** 1898, *Die Fauna des karnischen Fusulinenkalks. Theil II, Foraminifera:* Palaeontographica, v. 44(1897), p. 237-282, pl. 17-24, text-fig. 1-7.———(1645) 1902, *Trias, Perm und Carbon in China:* Phys.-Ökon. Gesell. Königsberg, Schrift., Jahrg. 1901, v. 43, p. 59-71, pl. 3.———(1645A) 1908, *Monographie der Fusulinen; Teil I. Die Fusulinen des russisch-arktischen Meeresgebietes (nach dem Tode des Verfassers herausgegeben und Fortgesetzt von G. Dyhrenfurth und H. von Staff:* Palaeontographica, v. 55, p. 145-194, pl. 13-20.

(1646) **Schenck, H. G., & Thompson, M. L.,** 1940, *Misellina and Brevaxina, new Permian fusulinid Foraminifera:* Jour. Paleontology, v. 14, p. 584-589.

(1647) **Schepotieff, Alexander,** 1912, *Untersuchunger über niedere Organismen, II. Die Xenophyophoren des Indischen Ozeans:* Zool. Jahrbücher, v. 32, Abt. für Anatomie & Ontogenie der Tiere, p. 245-286, pl. 15, 16.

(1648) **Schlicht, E. von,** 1870, *Die Foraminiferen des Septarienthones von Pietzpuhl:* pl. 1-38 (Berlin).

(1649) **Schlotheim, E. F. von,** 1822, *Nachträge zur Petrefacktenkunde:* xi+100 p., 21 pl., Becker (Gotha).

(1650) **Schlumberger, Charles,** 1883, *Note sur quelques Foraminifères nouveaux ou peu connus du Golfe de Gascogne:* Feuille jeunes Naturalistes, v. 13(1882-83), p. 21-28, pl. 2-3, text-fig. A-C.———(1651) 1887, *Note sur le genre Planispirina:* Soc. Zool. France, Bull., v. 12, p. 105-118, pl. 7, text-fig. 1-8.———(1652) 1889, *Sur le genre Thomasinella:* Same, ser. 3, v. 17, p. 425.———(1653) 1890, *Note sur un Foraminifère nouveau de la côte occidentale d'Afrique:* Soc. Zool. France, Mém., v. 3, pt. 1, p. 211-213, pl. 7.———(1654) 1891, *Révision des Biloculines des grands fonds:* Same, v. 4, p. 542-579, pl. 9-12.———(1655) 1893, *Monographie des Miliolidées du golfe de Marseille:* Same, v. 6, p. 57-80, pl. 1-4, text-fig. 1-37.———(1656) 1893, *Note sur les genres Trillina et Linderina:* Soc. géol. France, Bull., ser. 3, v. 21, pt. 2, p. 118-123, pl. 3.———(1656A) 1894, *Note sur Lacazina wichmanni Schlumb., n.sp.:* Same, ser. 3, v. 22, pt. 5, p. 295-298, pl. 12.———(1657) 1896, *Note sur le genre Tinoporus:* Soc. zool. France, Mém., v. 9, p. 87-90, pl. 3-4.———(1658) 1898, *Note sur le genre Meandropsina Mun.-Chalm., n.g.:* Soc. géol. France, Bull., ser. 3, v. 26, no. 3, p. 336-339, pl. 8-9.———(1659) 1898, *Note sur Involutina conica n.sp.:* Feuille jeunes Naturalistes, ser. 3, v. 28(1897-98), no. 332, p. 150-151, text-fig. 1-3.———(1660) 1900, *Note sur quelques Foraminifères nouveaux ou peu connus de Crétacé d'Espagne:* Soc. géol. France, Bull., ser. 3, v. 27 (1899), p. 456-465, pl. 8-11.———(1661) 1901, *Première note sur les Orbitoïdes:* Same, ser. 4, v. 1, pt. 4, p. 459-467, pl. 7-9.———(1662) 1902, *Deuxième note sur les Orbitoïdes:* Same, ser. 4, v. 2, pt. 3, p. 255-261, pl. 6-8.———(1663) 1903, *Troisième note sur les Orbitoïdes:* Same, ser. 4, v. 3, pt. 3, p. 273-289.———(1664) 1905, *Deuxième note sur les Miliolidées Trématophorées:* Same, ser. 4, v. 5, no. 2, p. 115-134, text-fig. 1-29, pl. 2-3.———(1665) 1905, *Note sur le genre Choffatella, n.g.:* Same, ser. 4, v. 4 (1904), p. 763-764, pl. 18.

(1666) ———, **& Choffat, P.,** 1904, *Note sur le genre Spirocyclina Munier-Chalmas et quelques autres genres du même auteur:* Soc. géol. France, Bull., ser. 4, v. 4, p. 358-368, 2 pl., text-fig.

(1667) ———, **& Douvillé, Henri,** 1905, *Sur deux Foraminifères Eocènes, Dictyoconus egyptiensis Chapm. et Lituonella roberti, nov.gen. et sp.:* Soc. géol. France, Bull., ser. 4, v. 5, no. 3, p. 291-304, pl. 9.

(1668) ———, **& Munier-Chalmas, E.,** 1884, *Note sur les Miliolidées trématophorées:* Soc. géol. France, Bull., ser. 3, v. 12(1883-1884), pt. 8, p. 629-630.

(1669) **Schlumberger, P.,** 1845, *Observations sur quelques nouvelles espèces d'Infusoires de la famille des Rhizopodes:* Ann. Sci. nat., Zool., ser. 3, v. 3, p. 254-256.

(1670) **Schlüter, Clemens,** 1879, *Coelotrichium decheni, eine Foraminifere aus dem Mitteldevon:* Deutsch. geol. Gesell., Zeitschr., v. 1879, p. 668-675, text-fig. a-d.

(1671) **Schmarda, L. K.,** 1871, *Zoologie:* x+372 p., 269 text-fig., Wilhelm Braumüller (Wien).

(1672) **Schmid, E. E.,** 1867, *Ueber die kleineren organischen Formen des Zechsteinkalks von Selters in der Wetterau:* Neues Jahrb. Mineral. Geol. & Paläont., p. 576-588, pl. 6.

(1673) **Schmidt, W. J.,** 1924, *Die Bausteine des Tierkörpers in polarisierten Lichte:* 528 p., F. Cohen (Bonn).———(1674) 1929, *Rheoplasma und Stereoplasma nach Beobachtungen an einer neuen monothalamen Foraminifere Rhumblerinella bacillifera, n.g., n.sp., zugleich eine Kritik der Söderstromschen Anschauungen über die Kornchenstromung der Foraminiferen:* Protoplasma, v. 7, no. 3, p. 353-394, 7 text-fig., pl. 3-4.

(1675) **Schouteden, H.,** 1906, *Les Rhizopodes testacés d'eau douce, d'après la Monographie du*

Prof. S. Awerintzew: Ann. Biol. Lacustre, v. 1, no. 3, p. 327-382, text-fig. 1-62.

(1676) **Schröder, Olaw**, 1907, *Echinogromia multifenestrata, nov.gen., nov. spec., eine neue, zu den Rhabdamminiden gehörende Rhizopoden Art* in DRYGALSKI, E. VON, Deutsche Südpolar Exped. 1901-1903: v. 9 (Zool. v. 1), p. 343-348, pl. 26, Reimer (Berlin).

(1676A) **Schroeder, Rolf**, 1962, *Orbitolinen des Cenomans Südwesteuropas:* Paläont. Zeitschr., v. 36, no. 3/4, p. 171-202, pl. 20-21, text-fig. 1-5.

(1677) **Schröter, J. S.**, 1783, *Einleitung in die Conchylienkenntniss nach Linné:* v. 1, 860 p., 3 pl., J. J. Gebauer (Halle).————(1678) 1886, *Pilze,* in COHN, F., *Kryptogamen-Flora von Schlesien:* v. 3 (i), p. 91-135.————(1679) 1897, *Die natürlichen Pflanzenfamilien nebst ihren Gattungen und wichtigeren Arten insbesondere den Nutzpflanzen* [unter Mitwirkung zahlreicher hervorragender Fachgelehrten begründet von A. ENGLER und K. PRANTL, fortgesetzt von A. ENGLER]: pt. I, no. 1, *Acrasieae,* p. 1-4; *Phytomyxinae,* p. 5-8; *Myxogasteres (eigentliche Myxomyceten),* p. 8-35.

(1680) **Schubert, R. J.**, 1900, *Flabellinella, ein neuer Mischtypus aus der Kreideformation:* Deutsch geol. Gesell., Zeitschr., v. 52, p. 551-553, pl. 1-2.————(1681) 1902, *Neue und interessante Foraminiferen aus dem südtiroler Alttertiar:* Beiträge Paläont. & Geol. Österreich-Ungarns Orients, v. 14, p. 9-26, pl. 1.————(1682) 1902, *Ueber die Foraminiferen-"Gattung" Textularia Defr. und ihre Verwandtschaftsverhältnisse:* K. K. geol. Reichsanst., no. 3, p. 80-85.————(1683) 1906, *Heteroclypeus, eine Uebergangsform zwischen Heterostegina und Cycloclypeus:* Zentralbl. Mineral. Geol. & Paläont., p. 640-641.————(1684) 1907, *Vorläufige Mitteilung über Foraminiferen und Kalkalgen aus dem dalmatinischen Karbon:* K. K. geol. Reichsanst., Verhandl., p. 211-214.————(1685) (Same as 1684).————(1686) 1908, *Zur Geologie des österreichischen Velebit:* Same, Jahrb., v. 58, p. 345-386, pl. 16, text-fig. 1-5.————(1687) 1908, *Beiträge zu einer natürlichen Systematik der Foraminiferen:* Neues Jahrb. Mineral. Geol. & Paläont., Beil.-Bd. 25, p. 232-260, pl. 1.————(1688) 1910, in BOEHM, GEORG, *Geologische Mitteilungen aus dem Indo-Australischen Archipel, VII. RICHARZ, P. STEPH, Der geologische Bau von Kaiser Wilhelms-Land nach dem heutigen Stand unseres Wissens,* Anhang 2: Same, Beil.-Bd. 29, p. 533-534, fig. 10c.————(1689) 1911, *Die fossilen Foraminiferen des Bismarckarchipels und einiger angrezender Insel:* (a) Same, v. 2, p. 318-320; (b) K. K. geol. Reichsanst., Abhandl., v. 20, no. 4, p. 1-130, pl. 1-6.————(1690) 1912, *Über Lituonella und Coskinolina liburnica Stache sowie deren Beziehungen zu den anderen Dictyoconinen:* Same, Jahrb., v. 12, no. 2, p. 195-208, pl. 10.————(1691) 1912, *Über die Verwandtschaftsverhältnisse von Frondicularia:* K. K. geol. Reichsanst. Wien, Verhandl., p. 179-184.————(1692) 1914, *Pavonitina styriaca, eine neue Foraminifere aus dem mittelstierischen Schlier:* K. K. geol. Reichsanst., Jahrb., v. 64 (Jahrg. 1914), no. 1-2, p. 143-148, pl. 4.————(1693) 1915, *Über Foraminiferengesteine der Insel Letti: Nederlandsche Timor-Expeditie I:* Jaarb. Mijnweizen, v. 43 (1914), p. 169-183, pl. 18-20, E. J. Brill (Leiden).————(1693A) 1915, *III. Die Foraminiferen des jüngeren paläozoikums von Timor:* Paläont. von Timor, Lief. 2, p. 49-59, pl. 39-41, fig. 1-2 (Stuttgart).————(1694) 1921, *Palaeontologische daten zur Stammesgeschichte der Protozoen:* Paläont. Zeitschr., v. 3 (1920), p. 129-188.

(1695) **Schultze, M. S.**, 1854, *Ueber den Organismus der Polythalamien (Foraminiferen), nebst Bemerkungen über die Rhizopoden im Allgemeinen:* 68 p., 7 pl., Wilhelm Engelmann (Leipzig).————(1696) 1863, *Das Protoplasma der Rhizopoden und der Pflanzenzellen:* 68 p. (Leipzig).

(1697) **Schulze, F. E.**, 1875, *Zoologische Ergebnisse der Nord-seefahrt vom 21 Juli bis 9 September, 1872, I. Rhizopoden. II.:* Komm. Untersuch. deutsch. Meere in Kiel, Jahresber., v. 1872-73, p. 99-114, pl. 2.————(1698) 1875-77, *Rhizopodenstudien:* Archiv Mikro. Anat.; (a) 1875, III, v. 11, p. 94-139, pl. 5-7; (b) 1875, IV, v. 11, p. 329-353, pl. 18-19; (c) 1877, VI, v. 13, p. 9-30, pl. 2-3.————(1699) 1904, *Über den Bau und die Entwickelung gewisser Tiefsee-Organismen, welche bisher von einigen Zoologen für Hornspongien, von anderen für Foraminiferen gehalten wurden:* K. preuss. Akad. Wiss., Sitzungsber., 2 Halbbd., p. 1387.————(1700) 1905, *Die Xenophyophoren, eine besondere Gruppe der Rhizopoden:* Wiss. Ergebnisse deutschen Tiefsee-Exped. "Valdivia" 1898-99, v. 11, p. 1-55, pl. 1-8.————(1701) 1912, *Xenophyophora:* Zool. Anzeiger, v. 39, p. 38-43, 1 fig.

(1702) **Schwager, Conrad**, 1864, *Foraminifera,* in DITTMAR, A. VON, Die Contorta-Zone (Zone der Avicula contorta Portl.): p. 198-201, pl. 3, H. Manz (München).————(1703) 1866, *Fossile Foraminiferen von Kar-Nicobar:* Novara-Exped., Geol. Theil, v. 2, p. 187-268, pl. 4-7.————(1704) 1876, *Saggio di una classificazione dei Foraminiferi avuto riguardo alle loro famiglie naturali:* R. Comitato Geol. Italia, Bull., v. 7, no. 11-12, p. 475-485.————(1705) 1877, *Quadro del proposto sistema de classificazione dei foraminiferi con guscio:* Same, Bull., v. 8, no. 1-2, p. 18-27, 1 pl.————(1706) 1883, *Carbonische Foraminiferen aus China und Japan,* in VON RICHTHOFEN, F. F., China: v. 4, Palaeont. Theil, Abhandl. 7, p. 106-159, pl. 15-

(1707) 1883, *Die Foraminiferen aus den Eocaenablagerungen der Libyschen Wüste und Aegyptens* in ZITTEL, K. A. VON, Beiträge zur Geologie und Paläontologie der Libyschen Wüste und der angrenzenden Gebiete von Aegypten: Palaeontographica, v. 30, p. 79-153, pl. 24-29, 1 table.

(1708) **Scott, H. W., Zeller, E., & Zeller, D. N.,** 1947, *The genus Endothyra:* Jour. Paleontology, v. 21, p. 557-562, pl. 83-84, 2 text-fig.

(1709) **Scudder, S. H.,** 1882, *Nomenclator zoologicus:* U.S. Govt. Printing Office (Washington, D.C.); (a) *Pt. 1. Supplemental list of genera in zoology,* p. 1-376; (b) *Pt. 2. Universal index to genera in zoology,* p. 1-340.

(1710) **Sedgwick, A.,** 1898, *A students' textbook of zoology:* v. 1, 619 p., 472 fig., Swan Sonnenschein & Co. (London).

(1711) **Seguenza, Giuseppe,** 1859, *Intorno ad un nuovo genere di foraminiferi fossili del torreno miocenico di Messina:* Eco Peloritano, Giornale Sci., Lett. & Arti, ser. 2, v. 5, pt. 9, p. 1-12, 1 pl.——**(1712)** 1862, *Die terreni terziarii del distretto di Messina, Parte II. Descrizione dei foraminiferi monotalamici delle marne mioceniche del distretto di Messina:* 84 p., 2 pl., T. Capra (Messina).——**(1713)** 1880, *Le formazioni terziarie nella provincia di Reggio (Calabria):* R. Accad. Lincei, Cl. Sci. Fis. Mat. Nat., Mem., ser. 3, v. 6, p. 3-446, pl. 1-17.——
(1714) 1882, *Studi geologici e paleontologici sull Cretaceo medio dell'Italia meridionale:* R. Accad. Lincei, Cl. Sci. Fis. Mat. Nat., ser. 3, v. 12, p. 65-214, pl. 1-21.

(1715) **Seiglie, G. A.,** 1961, *Dos generos y dos especies nuevos de foraminiferos del Cretacico Superior de Cuba:* Asoc. Mexicana Geól. Petrol., Bull., v. 12 (1960), no. 11-12, p. 341-351, pl. 1-4, fig. 1-5.

(1716) **Selli, Raimondo,** 1941, *Sulla struttura della "Cristellaria" serpens Seguenza:* Giornale Geol., Ann. R. Mus. Geol. Bologna, ser. 2, v. 14 (1939-40), p. 83-92, pl. 1.——**(1717)** 1947, *Sopra alcune Dimorphina:* Soc. Ital. Sci. Nat., Atti, v. 86, p. 127-134, fig. 1-11.

(1718) **Serova, M. Ya.,** 1953, *Novye dannye o stroenii i razvitii ust'ya u foraminifer iz roda Hauerina (sem. Miliolidae):* Moskov. Obshch. Ispyt. Prirody, Byull., Ser. Geol., v. 28(2), p. 62-64, fig. 1-3. [*New data on the structure and apertural development in the foraminiferal genus Hauerina (Fam. Miliolidae).*]——
(1719) 1955, *Stratigrafiya i fauna foraminifer Miotsenovykh otlozheniy Predkarpatya* in Materialy Po Biostrat. zapadnykh oblastey Ukrainskoy SSR; Minist. Geol. Okhrany Nedr, p. 261-391, pl. 1-29, text-fig. 1-19. [*Stratigraphy and foraminiferal fauna of the Miocene deposits of the Carpathian foothills.*]——**(1720)** 1961, *Novyy pozdnetortonskiy rod Podolia (Miliolidae) zapadnoi Ukrainy:* Akad. Nauk SSSR, Paleont. Zhurnal 1961, no. 1, p. 56-60, pl. 4, text-fig. 1-4. [*A new late Tortonian genus, Podolia (Miliolidae), of the western Ukraine.*]
——**(1721)** 1961, *Taksonomicheskoe znachenie nekotorykh osobennostey mikrostruktury stenki i stroeniya kamer rakovin miliolid:* Voprosy Mikropaleontologii no. 5, Akad. Nauk SSSR, Otdel. Geol. & Geog. Nauk, Geol. Inst., p. 128-134, 2 text-fig., pl. 1-10. [*Taxonomic value of certain peculiar microstructures of the wall and composition of the chamber wall of the miliolids.*]

(1722) **Sharp, David,** 1910, *Index to names of genera and subgenera:* Zool. Record, v. 45 (for 1908), p. 1-17.

(1723) **Shchedrina, Z. G.,** 1936, *Alveolophragmium orbiculatum, nov.gen., nov. sp.:* Zool. Anzeiger, v. 114, p. 312-319, text-fig. 1-3.
——**(1724)** 1939, *Novyy rod peschanistykh foraminifer iz Arkticheskikh Morey:* Akad. Nauk SSSR, Doklady, new ser., v. 24, no. 1, p. 94-96, text-fig. 1-2. [*A new genus of arenaceous Foraminifera from the Arctic Sea.*]——
(1725) 1953, *Novye dannye po faune foraminifer Okhotskogo Morya i ee raspredeleniyu:* Akad. Nauk SSSR, Zool. Inst., Trudy, v. 13, p. 12-32. [*New data on the foraminiferal fauna of the Okhotsk Sea and its distribution.*]——
(1726) 1955, *Dva novykh roda foraminifer iz semeystva Trochamminidae (Foraminifera):* Same, v. 18, p. 5-9, text-fig. 1-3. [*Two new foraminiferal genera of the family Trochamminidae (Foraminifera).*]——**(1726A)** 1962, *Foraminifery zalivov Belogo Morya:* Biologiya Belogo Morya, Trudy, Belomorskoy biologicheskoy stantsii MGU, v. 1, p. 51-69, text-fig. 1-10. [*Foraminifera of the bays of the White Sea.*]

(1727) **Sheng, J. C.,** 1951, *Taitzehoella, a new genus of fusulinids:* Geol. Soc. China, Bull., v. 31, no. 1-4, p. 79-85, pl. 1.——**(1728)** 1955, *Some fusulinids from Changhsing Limestone:* Palaeont. Sinica, v. 3, no. 4, p. 287-308, pl. 1-4.——**(1729)** 1958, *Fusulinids from the Penchi Series of the Taitzeho Valley, Liaoning:* Same, whole no. 143, new ser. B, no. 7, 119 p., pl. 1-16.

(1730) **Sherborn, C. D.,** 1888, *A bibliography of the Foraminifera Recent and fossil, from 1565-1888; with notes explanatory of some of the rare and little known publications:* 152 p., Dulau & Co. (London).——**(1731)** 1893-96, *An index to the genera and species of the Foraminifera:* Smithsonian Misc. Coll.; (a) 1893, no. 856, p. 1-240; (b) 1896, no. 1031, p. 241-485.

(1732) ——, **& Chapman, Frederick,** 1886, *On some microzoa from the London clay exposed in the drainage works, Piccadilly, London, 1885:*

Royal Micro. Soc. London, Jour., ser. 2, v. 6, p. 737-763, pl. 14-16.

(1733) **Shifflett, Elaine,** 1961, *Living, dead, and total foraminiferal faunas, Heald Bank, Gulf of Mexico:* Micropaleontology, v. 7, no. 1, p. 45-54, text-fig. 1-3. table 1-4.

(1734) **Shirai, Takehiro,** 1960, *New genus and species of Foraminifera from the Pliocene formation, southwestern Hokkaido:* Hokkaido Univ., Jour. Faculty Sci., ser. 4, Geol. & Mineral., v. 10, no. 3, p. 537-543, pl. 1-2.

(1735) **Shmalgauzen, O. I.,** 1950, *Novyy vid foraminifery iz ozera balpash-sor (Kazakhstan):* Akad. Nauk SSSR, Doklady, v. 75, no. 6, p. 869-872. [*New species of Foraminifera from Lake Balpash-Sor (Kazakhstan).*]

(1736) **Siddall, J. D.,** 1878, *On the Foraminifera of the River Dee:* Chester Soc. Nat. Sci., Proc., no. 2, p. 42-56, 2 fig.———(1737) 1880, *On Shepheardella, an undescribed type of marine Rhizopoda; with a few observations on Lieberkühnia:* Quart. Jour. Micro. Sci., v. 20, p. 130-145, pl. 15, 16.

(1738) **Sidebottom, Henry,** 1904, *Report on the Recent Foraminifera from the coast of the Island of Delos (Grecian Archipelago):* Manchester Lit. & Philos. Soc., Mem. & Proc., v. 48, no. 5, p. 1-26, pl. 2-5.———(1739) 1905, *On Nevillina, a new genus of Foraminifera:* Same, v. 49, no. 11, p. 1-3, 1 pl.———(1740) 1907, *Report on the Recent Foraminifera from the coast of the Island of Delos (Grecian Archipelago), Pt. 4:* Same, v. 51, no. 9, p. 1-28, pl. 1-4.——— (1741) 1918, *Report on the Recent Foraminifera dredged off the east coast of Australia, H.M.S. "Dart," Station 19 (May 14, 1895), lat. 29° 22'S., long. 153° 51'E., 465 fathoms, Pteropod ooze:* Royal Micro. Soc. London, Jour., p. 121-152, pl. 3-5.

(1742) **Siebold, C. T. E. von, & Stannius, Hermann von,** 1845, *Lehrbuch der vergleichende Anatomie:* pt. 1, Wirbellose Thiere, no. 1, p. 1-679.

(1743) **Sigal, Jacques,** 1948, *Notes sur les genres de Foraminifères Rotalipora Brotzen, 1942, et Thalmanninella. Famille des Globorotaliidae:* Revue Inst. Français Pétrole et Ann. Combus. liquides, v. 3, no. 4, p. 95-103, pl. 1,2.——— (1744) 1949, *Sur quelques Foraminifères de l'Aquitanien des environs de Dax, Leur place dans l'arbre phylétique des Rotaliiformes:* Same, v. 4, no. 5, p. 155-165, pl. 1-3.———(1745) 1950, *Les genres Queraltina et Almaena (Foraminifères), Leur importance stratigraphique et paléontologique:* Soc. géol. France, Bull., ser. 5, v. 20, p. 63-71, text-fig. 1-6.———(1746) 1952, *Aperçu stratigraphique sur la micropaléontologie du Crétacé:* 19th Cong. Géol. Internatl., Mon. Région., ser. 1, Algérie, no. 26, p. 1-47, text-fig. 1-46, table.———(1747) 1956, *Notes micropaléontologiques nord-africains, 4. Biticinella breggiensis (Gandolfi) nouveau morphogenre:* Soc. géol. France, Comptes Rendus Somm. Séances, no. 3-4, p. 35-57, 1 text-fig.——— (1748) 1956, *Notes micropaléontologiques nord-africaines, 6. Sur la position systématique du genre Thomasinella Schlumberger (Foraminifères):* Same, no. 8, p. 102-105, text-fig. 1-4.———(1749) 1958, *La classification actuelle des familles de Foraminifères planctoniques du Crétacé:* Same, no. 11-12, p. 262-265.

(1750) **Silvestri, Alfredo,** 1898, *Foraminiferi pliocenici della Provincia di Siena, Parte II:* Accad. Pont. Nuovi Lincei, Mem., v. 15, p. 155-381, pl. 1-6.———(1751) 1900, *Sul genera Ellipsoglandulina:* R. Accad. Sci., Lett. & Arte degli Zelanti, Cl. Sci., Mem., new ser., v. 10 (1899-1900), p. 1-9, 1 pl.———(1752) 1901, *Sulla struttura di certe Polimorfine die dintorni di Caltagirone:* Accad. Gioenia Sci. Nat. Catania, Bull., new ser., pt. 69, p. 14-18.———(1753) 1901, *Intorno ad alcune Nodosarine poco conosciute del neogene italiano:* Accad. Pont. Nuovi Lincei, Atti, v. 54, p. 103-109.———(1754) 1902, *La Siphogenerina columellaris (Brady):* Same, v. 55(1901-02), p. 101-104, text-fig. 1-2. ———(1755) 1902, *Sulle forme aberranti della Nodosaria scalaris (Batsch):* Same, v. 55, 1901-02, p. 49-58, text-fig. 1-9.———(1756) 1903, *Linguloglanduline e lingulonodosarie:* Same, v. 56 (1902-03), p. 45-50.———(1757) 1903, *Alcune osservazioni sui protozoi fossili piemontesi:* R. Accad. Sci. Torino, Atti, v. 38, p. 206-217.———(1758) 1904, *Forme nuove o poco conosciute di Protozoi miocenici piemontesi:* Same, v. 39 (1903-04), p. 4-15, text-fig. 1-7. ———(1759) 1904, *Località Toscana del genere Chapmania Silv. et Prev.:* Rivista Ital. Sci. Nat., Boll. Nat., v. 24, p. 117-119, text-fig. 1-3.——— (1760) 1904, *Ricerche strutturali su alcune forme dei Trubi di Bonfornello (Palermo):* Accad. Pont. Nuovi Lincei, Mem., v. 22, p. 235-276.———(1761) 1905, *Lepidocyclinae ed altri fossili del territorio d'Anghiari:* Same, Atti, v. 58, p. 122-128, text-fig.———(1762) 1905, *Sul Dictyoconus aegyptiensis (Chapman):* Same, Atti, v. 58, p. 129-131.———(1763) 1905, *Notizie sommarie su tre faunule del Lazio Perugia:* Rivista Italiana Paleont., v. 11, pt. 4, p. 140-145.———(1764) 1906, *Notizie sommarie su tre faunule del Lazio. II:* Same, v. 12, p. 20-35.———(1765) 1907, *Forma italiana della "Lingulina impressa" Terquem:* Same, v. 13, p. 66-70, text-fig. 1-2.———(1766) 1907, *Probabile origine d'alcune Orbitoidine:* Same, v. 13, p. 79-81.———(1767) 1907, *Probabile origine d'alcune Orbitoidine:* Rivista Italiana Sci. Nat., Boll. Nat., Suppl., v. 27, no. 2, p. 11-12. ———(1768) 1907, *Sull' età geologica della Lepidocicline:* Accad. Pont. Nuovi Lincei, Atti, v. 60, p. 83-95, text-fig.———(1769) 1907,

Fossili dordoniani nei dintorni di Termini-Imerese (Palermo): Same, v. 60, p. 105-110.
—— **(1770)** 1908, *Sulla "Orbitulites complanata" Martelli:* Same, v. 61, p. 131-141.
—— **(1771)** 1908, *Fossili cretacei della contrada Calcasacco presso Termini-Imerese (Palermo):* Palaeont. Italica, v. 14, p. 121-170, pl. 17-20 (1-4), text-fig. 1-38.—— **(1771A)** 1910, *Lepidocicline sannoisiane di Antonimina in Calabria:* Accad. Pont. Nuovi Lincei, Mem., v. 28, p. 103-163, pl. 1, text-fig. 1-28.——
(1772) 1912, *Review of R. J. Schubert, "Die fossilen Foraminiferen des Bismarckarchipels und einiger angrenzender Inseln. Abh. K.K. Reichsanst. Wien, v. 20, p. 1-130, pl. 1-6, 1911":* Rivista Italiana Paleont., v. 18, pt. 2-3, p. 66-71.—— **(1773)** 1920, *Ortostilia e flessostilia nei Rizopodi reticolari:* Accad. Pont. Nuovi Lincei, Atti, v. 73(1919-1920), p. 50-57.——
(1774) 1923, *Lo stipite della Ellissoforme e le sue affinità:* Same, Mem., ser. 2, v. 6, p. 231-270, pl. 1.—— **(1775)** 1923, *Singolari Nodosarine dell'Eocene piemontese:* Rivista Italiana Paleont., v. 29, p. 11-24, pl. 2, text-fig. 1-12.
—— **(1776)** 1923, *Nuovi rinvenimenti di Chapmanie:* R. Accad. Nazionale Lincei, v. 32, ser. 5ᵃ, sem. 2, pt. 3-4, p. 88-92, text-fig. 1.
—— **(1777)** 1923, *Il criterio delle Alveoline:* Accad. Pont. Nuovi Lincei, Atti, v. 76, p. 115-125, fig. A-B.—— **(1778)** 1924, *Revisione di fossili della Venezia e Venezia Giulia:* Accad. Sci. Veneto-Trentino-Istriana, Atti, ser. 3, v. 14(1923), p. 7-12.—— **(1779)** 1924, *Fauna paleogenica di Vasciano presso Todi:* Soc. Geol. Italiana, Bull., v. 42(1923), pt. 1, p. 7-29, pl. 1.
—— **(1780)** 1925, *Sulla diffusione stratigrafica del genere "Chapmania" Silv. e Prev.:* Accad. Pont. Nuovi Lincei, Cl. Sci., Mem., ser. 2, v. 8, p. 31-60, pl. 1, text-fig. 1-10.—— **(1781)** 1926, *Sulla Patella cassis Oppenheim:* Rivista Italiana Paleont., v. 32(1926), p. 15-22, pl. 1.
—— **(1782)** 1927, *Sull'età di alcune rocce della Libia Italiana:* R. Liceo Sci., Ann., pt. 2, p. 223-232.—— **(1783)** 1928, *Intorno all'Alveolina melo d'Orbigny (1846):* Rivista Italiana Paleont., v. 34, p. 17-44, pl. 1-4, fig. A.
—— **(1784)** 1931, *Sul genere Chapmanina e sulla Alveolina maiellana n.sp.:* Soc. Geol. Italiana, Bull., v. 50, pt. 1, p. 63-73, pl. 1.
(1785) 1931, *Fossili Miocenici nel territorio di Rivona (Agrigento):* Rivista Italiana Paleont., v. 37, p. 29-36, pl. 4-5.—— **(1786)** 1932, *Revisione di Foraminiferi preterziarii del sud-ouest di Sumatra:* Same, v. 38, p. 75-103, pl. 2-4.—— **(1787)** 1937, *Foraminiferi dell'Oligocene e del Miocene della Somalia:* Palaeont. Italica, v. 32, suppl. 2, p. 45-264, pl. 4-22.
—— **(1787A)** 1932, *Foraminiferi del Cretaceo della Somalia:* Same, v. 32, p. 143-204, p. 9-16.
—— **(1788)** 1938-42, *Foraminiferi dell'Eocene della Somalia, Parte 1, Paleontologia della Somalia:* Same, (a) v. 32, suppl. 3, p. 49-89, pl. 3-12 (1938); (b) v. 32, suppl. 4, p. 1-102, pl. 1-12 (1939); (c) v. 32, suppl. 5, pt. 3, no. 1, p. 1-94(181-274), pl. 1-9(23-31) (1942).——
(1789) 1940, *Illustrazione di specie caratteristica del Cretaceo superiore:* Soc. Geol. Italiana, Bull., v. 58, p. 225-234, pl. 12.—— **(1790)** 1947, *La Siphonclavulina trigona A. Silv. dell'Eocene piemontese:* Same, v. 66, p. 1-3 (of reprint).
—— **(1791)** 1950, *Foraminiferi della Laguna Veneta:* Boll. Pesca, Piscicoltura & Idrobiologia, v. 26 (new ser., v. 5), pt. 1, p. 3-79, pl. 1-3.

(1792) Silvestri, Orazio, 1889, *Sopra due nuovi generi di rhizopodi (foraminifere) appartenenti al pliocene inferiore d'Italia:* Soc. Italiana Micro., Bull., v. 1, p. 51-59, pl. 3.

(1793) Singh, B. N., 1951, *Nuclear division in Amoebae and its bearing on classification:* Nature, v. 167, p. 582-584.

(1793A) Singh, S. N., 1957, *Two aberrant types of Nummulitidae from the Eocene of Rajasthan, India:* Paleont. Soc. India, Jour., v. 2, p. 209-212, pls. 25, 26.

(1794) Skinner, J. W., 1931, *Primitive fusulinids of the Mid-Continent region:* Jour. Paleontology, v. 5, p. 253-259, pl. 30.

(1795) ——, & Wilde, G. L., 1954, *The fusulinid subfamily Boultoniinae:* Jour. Paleontology, v. 28, no. 4, p. 434-444, pl. 42-45.—— **(1796)** 1954, *Fusulinid wall structure:* Same, v. 28, no. 4, p. 445-451, pl. 46-52.—— **(1797)** 1955, *New fusulinids from the Permian of West Texas:* Same, v. 29, p. 927-940, pl. 89-95.

(1798) Slama, D. C., 1954, *Arenaceous tests in Foraminifera—an experiment:* Micropaleontologist, v. 8, no. 1, p. 33-34.

(1799) Smith, D. J., 1949, *Miocene Foraminifera of the "Harang sediments" of southern Louisiana* in Pope, D. E., & Smith, D. J., The Harang fauna of Louisiana; Louisiana Geol. Survey, Geol. Bull. 26, p. 23-80, pl. 7-12.

(1800) Smith, F. D., Jr., 1955, *Planktonic Foraminifera as indicators of depositional environment:* Micropaleontology, v. 1, no. 2, p. 147-151, text-fig. 1, 2.

(1801) Smith, G. M., 1955, *Cryptogamic botany. v. 1., Algae and fungi:* ed. 2, p. 546 p., 311 fig., McGraw Hill (New York).

(1802) Smitter, Y. H., 1956, *Chitinosaccus, a new foraminiferal genus of the Allogromiidae from Santa Lucia Bay, Zululand:* South African Jour. Sci., v. 52, no. 11, p. 258-259.

(1803) Smout, A. H., 1954, *Lower Tertiary Foraminifera of the Qatar Peninsula:* British Museum (Nat. History), London, p. 1-96, pl. 1-15.
—— **(1804)** 1955, *Reclassification of the Rotaliidea (Foraminifera) and two new Cretaceous forms resembling Elphidium:* Washington Acad. Sci., Jour., v. 45, no. 7, p. 201-210, fig. 1-10.—— **(1805)** 1956, *Three new Cre-*

taceous genera of Foraminifera related to the Ceratobuliminidae: Micropaleontology, v. 2, no. 4, p. 335-348, pl. 1-2.

(1806) ———, & Eames, F. E., 1958, *The genus Archaias (Foraminifera) and its stratigraphical distribution:* Palaeontology, v. 1, pt. 3, p. 207-225, pl. 39-42.

(1807) ———, & Sugden, W., 1962, *New information on the foraminiferal genus Pfenderina:* Palaeontology, v. 4, pt. 4, p. 581-591, pl. 73-76.

(1808) Soest, J. van, 1942, *Geologie und Paleontologie des zentralen Biokovo (Dalmatien):* Dissertation, Univ. Utrecht, p. 1-42, pl. 1-4.

(1809) Soldani, Ambrogio, 1789, *Testaceographiae ac Zoophytographiae parvae et microscopicae:* Tomus Primus, xxxii+80 p., 93 pl., Rossi (Senis).———(1810) 1795, *Testaceographiae ac Zoophytographiae parvae et microscopicae:* Tomi Primi pars tertia, p. 201-289, pl. 143-179 (Senis).

(1811) Sollas, W. J., 1921, *On Saccammina carteri Brady, and the minute structure of the foraminiferal shell:* Geol. Soc. London, Quart. Jour., v. 77, pt. 3, p. 193-212, pl. 7, text-fig. 1-7.

(1812) Solovieva, M. N., 1955, *Novyy rod fuzulinid Dagmarella, ego sistematicheskoe polozhenie i geograficheskoe rasprostranenie:* Akad. Nauk SSSR, Doklady, v. 101, no. 5, p. 945-946, 1 fig. [*A new fusulinid genus Dagmarella, its systematic position and geographic occurrence.*]

(1813) Sorby, H. C., 1879, *Anniversary address of the President, Proc. Geol. Soc. London, 1878-79:* Geol. Soc. London, Quart. Jour., v. 35, appendix, p. 56-93.

(1814) Sorrentino, Stefano, 1930, *Alcune osservazioni sulla struttura interna delle Alveoline:* Soc. Geol. Italiana, Bull., v. 49, p. 170-176.———(1815) 1935, *Considerazioni sulla variabilità dei caratteri di Alveolina e Flosculina dal punto di vista del loro raggruppamento e determinazione:* Soc. Nat. Napoli, Bull., v. 46(1934), p. 121-141.

(1816) Sosnina, M. I., 1956, in Kiparisova *et al.* (see 1040.).———(1817) 1960, *K metodike issledovaniya lagenid* in Dochetvertichnaya mikropaleontologiya: Mezhdunarodnyy Geol. Congress, 21 Sess., Doklady Sovetskikh Geol. Prob. 6, p. 32-47, text-fig. 1-15, pl. 1-2. [*On research techniques in the lagenids,* in Pre-Quaternary micropaleontology.].——— (1818) 1960, *Izuchenie lyagenid metodom posledovatel'nykh prishlifovok:* Trudy Pervogo Seminara po Mikrofaune, VNIGRI, p. 88-119, text-fig. 1-30. [*Study of lagenids by the method of serial sections.*]

(1819) Sowerby, G. B., Jr., 1842, *A conchological manual:* ed. 2, 313 p., 562 fig., G. B. Sowerby (London).

(1820) Sowerby, James, & Sowerby, James de Carle, 1826, *The mineral conchology of Great Britain:* v. 6, p. 73-76, pl. 504-609, J. de C. Sowerby (London).

(1821) Spandel, Erich, 1898, *Die Foraminiferen des deutschen Zechsteins, und ein zweifelhaftes mikroskopisches Fossil ebendaher:* Verlags-Inst. "General Anzeiger," p. 1-15, text-fig. 1-11 (Nürnberg).———(1822) 1901, *Die Foraminiferen des Permo-Carbon von Hooser, Kansas, Nord Amerika:* Festschrift Nat. Gesell. Nurmberg, p. 175-194, 10 fig.———(1823) 1909, *Der Rupelton des Mainzer Beckens, seine Abteilungen und deren Foraminiferenfauna:* Offenbacher Vereins Naturkunde, Ber., no. 43-50, p. 57-230, pl. 1-2.

(1824) Speck, J., 1953, *Geröllstudien in der subalpinen Molasse am Zugersee und Versuch einer paläogeographischen Auswertung:* 175 p., 12 pl., 11 text-fig., Kalth-Zehnder (Zug).

(1825) Stache, Guido, 1865, *Die Foraminiferen der tertiären Mergel des Whaingaroa-Hafens (Prov. Auckland):* Novara Exped. 1857-59, Wien, v. 1, Geol. Theil, pt. 2, p. 159-304, pl. 21-24.———(1826) 1875, *Neue Beobachtungen in den Schichten der liburnischen Stufe:* K.K. geol. Reichsanst. Verhandl., p. 334-338.——— (1827) 1880, *Die liburnische Stufe:* Same, p. 195-209.———(1828) 1889, *Die liburnische Stufe und deren Grenz-Horizonte, eine Studie über die Schichtenfolgen der Cretacisch-Eocänen oder Protocänen Landbildungsperiode im Bereiche der Küstenländer von Österreich-Ungarn:* Same, Abhandl., v. 13, p. 1-170, pl. 1-5a. ———(1829) 1913, *Über Rhipidionina St. und Rhapydionina St.:* Same, Jahrb., v. 62(1912), no. 4, p. 659-680, pl. 26-27.

(1830) Staff, Hans von, 1909, *Beiträge zur Kenntnis der Fusuliniden:* Neues Jahrb. Min., Geol. & Paläont., Beil.-Bd. 27, p. 461-508, pl. 7-8, fig. 1-16.———(1831) 1910, *Die Anatomie und Physiologie der Fusulinen:* Zoologica, Orig.-Abh. Gesamtgebiete Zool., no. 58, 93 p., 2 pl., 62 fig.

(1832) ———, & Wedekind, Rudolf, 1910, *Der Oberkarbon Foraminiferensapropelit Spitzbergens:* Geol. Inst. Upsala, Bull., v. 10, p. 81-123, pl. 2-4.

(1833) Stainforth, R. M., 1952, *Classification of uniserial calcareous foraminifera:* Cushman Found. Foram. Research, Contrib., v. 3, pt. 1, p. 6-14, text-fig. 1.———(1834) 1952, *Ecology of arenaceous Foraminifera:* Micropaleontologist, v. 6, p. 42-44.

(1835) Stein, S. F. N. von, 1859, *Ueber die ihm aus eigener Untersuchung bekannt gewordenen süsswasser Rhizopoden:* K. Česka Společnosti Nauk, Prague (K. Böhm. Gesell. Wiss., Abhandl.), ser. 5, v. 10, p. 41-43.———(1836) 1867, *Der Organismus der Infusionsthiere, II. Abtheilung:* 355 p., 16 pl., W. Engelmann (Leipzig).

(1837) Steinmann, Gustav, 1881, Die Foraminiferengattung Nummoloculina, n.g.: Neues. Jahrb. Mineral. Geol. & Paläont., v. 1(1881), p. 31-43.

(1838) Stewart, G. A., & Lampe, Lois, 1947, Foraminifera from the Middle Devonian Bone Beds of Ohio: Jour. Paleontology, v. 21, p. 529-536, pl. 78-79.

(1839) Stewart, W. J., 1958, Some fusulinids from the upper Strawn, Pennsylvanian, of Texas: Jour. Paleontology, v. 32, p. 1051-1070, pl. 132-137, fig. 1-2.

(1840) Stöhr, Emil, 1877, Bericht über die Tripoli-Schichten auf Sicilien: Zeitschr. deutsch. geol. Gesell., v. 29, p. 638-643.

(1841) Stone, Benton, 1946, Stichocassidulina, a new genus of Foraminifera from northwestern Peru: Jour. Paleontology, v. 20, p. 59-61, text-fig. 1-3.———(1842) 1949, New Foraminifera from northwestern Peru: Same, v. 23, p. 81-83, pl. 21.

(1843) Strand, Embrik, 1928, Miscellanea nomenclatoria zoologica et palaeontologica: Archiv Naturgeschichte, v. 92, pt. A(A8) (1926), p. 30-69.———(1844) 1943, Miscellanea nomenclatorica zoologica et palaeontologica, XII: Folia Zool. Hydrobiol., v. 12, no. 1, p. 211 (Riga).

Stschedrina, Z. (see Shchedrina, Z.)

(1845) Stuart, Alexander, 1866, Ueber Coscinosphaera ciliosa, eine neue Radiolarie: Zeitschr. Wiss. Zool., v. 16, p. 328-345, pl. 18.

(1846) Subbotina, N. N., 1953, Verkhneeotsenovye Lyagenidy i Buliminidy yuga SSSR: Mikrofauna SSSR, Sbornik 6, VNIGRI, Trudy, new ser., no. 69, p. 115-255, pl. 1-13. [Upper Eocene Lagenidae and Buliminidae of southern USSR.]———(1847) 1953, Globigerinidy, Hantkeninidy i Globorotaliidy: Iskopaemye Foraminifery SSSR, VNIGRI, Trudy, new ser., no. 76, p. 1-296, 41 pl. [Globigerinidae, Hantkeninidae and Globorotaliidae: Fossil Foraminifera of the USSR.]

(1848) ———, Glushko, V. V., & Pishvanova, L. S., 1955, O vozraste nizhney vorotyshchenskoy svity predkarpatskogo kraevogo progiba: Akad. Nauk SSSR, Doklady, v. 104, no. 4, p. 605-607. [On the age of the lower Vorotyshchensky beds of the Carpathian border trough.]

(1849) Šulc, Jaroslav, 1929, Příspěvky k Poznání morfologie foraminifer: Stát. Geol. Ústavu., Vestnik, v. 5, p. 148-155, pl. 1(13).———(1850) 1936, Études sur quelques genres et espèces de Pénéroplidés: Ann. Protistologie, v. 5, p. 157-170, pl. 8-9.

(1851) Suleymanov, I. S., 1945, Some new species of small foraminifers from the Tournaisian of the Ishimbayevo oil-bearing region: Akad. Nauk SSSR, Doklady (Acad. Sci. URSS, Comptes Rendus), v. 48, no. 2, p. 124-127, fig. 1-5, 2 tables.———(1852) 1955, Novyy rod Gubkinella i dva novykh vida semeystva Heterohelicidae iz verkhnego senona yugo-zapadnykh Kyzyl-Kumov: Akad. Nauk SSSR, Doklady, v. 102, no. 3, p. 623-624, text-fig. 1-2. [A new genus, Gubkinella, and two new species of the family Heterohelicidae from the upper Senonian of the southwestern Kyzyl-Kumy.]———(1853) 1958, Novyy rod i dva novykh vida iz sem. Verneuilinidae: Akad. Nauk Uzbekskoy SSR, Doklady, 1958, no. 12, p. 19-21, text-fig. 1-2. [A new genus and two new species of the family Verneuilinidae.]———(1854) 1959, O novom rode i vide foraminifer iz sem. Ammodiscidae: Same, Doklady, 1959, no. 7, p. 19-20, text-fig. 1. [On a new genus and species of Foraminifera of the family Ammodiscidae.]———(1855) 1960, Novyy podrod i dva novykh vida iz semeystva Ammodiscidae: Same, Doklady, 1960, no. 2, p. 18-20, text-fig. 1-2. [A new subgenus and two new species of the family Ammodiscidae.]———(1856) 1960, O mikrostrukture stenki rakovin nekotorykh vidov tekstulyariid v svyazi s ikh paleoekologiey: Voprosy Mikropaleontologii, no. 3, Akad. Nauk SSSR, Otdel. Geol.-Geogr. Nauk, Geol. Inst., p. 37-40, text-fig. 1 (Moscow). [On the microstructure of the test wall of certain species of Textulariidae in relation to their paleoecology.]———(1857) 1961, K filogenii ryada Gaudryina-Gaudryinella: Voprosy Mikropaleontologii, no. 4, Akad. Nauk SSSR, Otdel. Geol.-Geogr. Nauk, Geol. Inst., p. 83-88, text-fig. 1-3. [On the phylogeny of the Gaudryina-Gaudryinella suite.]

(1858) Summerson, C. H., 1958, Arenaceous Foraminifera from the Middle Devonian limestones of Ohio: Jour. Paleontology, v. 32, p. 544-558, pl. 81-82, text-fig. 1-7.

(1859) Switzer, George, and Boucot, A. J., 1955, The mineral composition of some microfossils: Jour. Paleontology, v. 29, p. 525-533, 3 text-fig.

(1860) Sykes, W. H., 1840, A notice respecting some fossils collected in Cutch, by Capt. Walter Smee of the Bombay Army: Geol. Soc. London, Trans., ser. 2, v. 5(1834), p. 715-719, pl. 61.

(1861) Tairov, Ch. A., 1956, O dvukh novykh rodakh iz semeystv Verneuilinidae i Ammodiscidae, prinadlezhashchikh k faune foraminifer: Akad. Nauk Azerbaydzhan SSR, Doklady, v. 12, no. 2, p. 113-116, text-fig. 1-3. [On two new genera of the families Verneuilinidae and Ammodiscidae belonging to the foraminiferal fauna.]

(1862) Takayanagi, Yokichi, 1953, New genus and species of Foraminifera found in the Tonohama group, Kochi Prefecture, Shikoku, Japan: Inst. Geol. & Paleont. Sendai, Short Papers, no. 5, p. 25-36, pl. 4.———(1863) 1960, Cretaceous Foraminifera from Hokkaido, Japan: Tohoku Univ., Sci. Rept., ser. 2(Geol.), v. 32, no. 1, p. 1-154, pl. 1-11.

(1864) **Tan Sin Hok**, 1932, *On the genus Cycloclypeus Carpenter; Part 1; and an appendix on the Heterostegines of Tjimanggoe, S. Bantam, Java*: Wetensch. Meded., no. 19, p. 1-194, pl. 1-24, 4 text-fig.———(1865) 1933, *Notiz über das Basalskelett von "Verbeekina"*: Same, no. 25, p. 57-65, pl. 1.———(1866) 1936, *Zur Kenntniss der Miogypsiniden*: Ingenieur Nederland.-Indië, Mijnbouw Geol. 4, v. 3, no. 3, p. 45-61, pl. 1-2.———(1867) 1936, *Zur Kenntnis der Lepidocycliniden*: Natuurkund. Tijdschr. Nederlandsch-Indie., v. 96, p. 235-280.———(1868) 1936, *Beitrag Zur Kenntnis der Lepidocycliniden*: K. Akad. Wetensch. Amsterdam, Verh., v. 39, p. 990-999.———(1869) 1936, *Over verschillende palaeontologische criteria voor de geleding van het Tertiair*: Ingenieur Nederland.-Indië, v. 3, pt. 4 (Mijnb. Geol.), p. 173-179.———(1870) 1937, *On the genus Spiroclypeus H. Douvillé with a description of the Eocene Spiroclypeus vermicularis, nov.sp., from Koetai in East Borneo*: Same, Mijnbouw Geol. 4, v. 4, no. 10, p. 177-193, pl. 1-4, 1 fig.

(1871) **Tappan, Helen**, 1940, *Foraminifera from the Grayson Formation of northern Texas*: Jour. Paleontology, v. 14, p. 93-126, pl. 14-19.———(1872) 1943, *Foraminifera from the Duck Creek Formation of Oklahoma and Texas*: Same, v. 17, p. 476-517, pl. 77-83.———(1873) 1951, *Foraminifera from the Arctic slope of Alaska, General introduction and Part 1, Triassic Foraminifera*: U.S. Geol. Survey, Prof. Paper 236-A, p. 1-20, pl. 1-5, 2 text-fig.———(1874) 1955, *Foraminifera from the Arctic slope of Alaska, Part 2, Jurassic Foraminifera*: Same, Prof. Paper 236-B, p. 21-90, pl. 7-28.———(1875) 1957, *New Cretaceous index Foraminifera from northern Alaska*: U.S. Natl. Museum Bull. 215, p. 201-222, pl. 65-71.

(1876) **Taránek, K. J.**, 1882, *Beiträge zur Kenntniss der Süsswasser-Rhizopoden Böhmens*: K. böhm. Gesell. Wiss., Sitzungsber., v. 1881, p. 220-235.———(1877) 1882, *Monographie der Nebeliden Böhmen's. Ein Beitrag zur Kenntniss der Süsswasser-Monothalamia*: Same, Abhandl., ser. 6, v. 11, math.-nat. Cl., no. 8, p. 1-55, pl. 1-5.

(1878) **Tasch, Paul**, 1953, *Causes and paleoecological significance of dwarfed fossil marine invertebrates*: Jour. Paleontology, v. 27, p. 356-444, pl. 49, 50, 6 text-fig.

(1879) **Tatem, J. G.**, 1870, *Notes on new Infusoria*: Royal Micro. Soc., London, Trans., Monthly Micro. Jour., v. 4, p. 313-314, pl. 68.———(1880) 1877, *Mr. Archer's genus Hyalosphenia*: Same, v. 17, p. 311.

(1881) **Termier, Geneviève, & Termier, Henri**, 1947, *I. Généralités sur les invertébrés fossiles*: Service Géol. Maroc. Div. Mines & Geol., Notes & Mém., no. 69, Paléont. Marocaine, p. 1-391, pl. 1-22.———(1882) 1950, *Paléontologie Marocaine, Tome II. Invertébrés de l'Ère Primaire, Fasc. 1. Foraminifères, Spongiaires et Coelentérés*: 218 p., 51 pl., Hermann & Cie. (Paris).

(1883) **Terquem, Olry**, 1862, *Recherches sur les Foraminifères de l'Étage Moyen et de l'Étage inférieur du Lias, Mémoire 2*: Acad. Imper. Metz, Mém., v. 42 (ser. 2, v. 9, 1860-61), p. 415-466, pl. 5-6.———(1884) 1864, *Quatrième mémoire sur les Foraminifères du Lias comprenant les polymorphines des Départements de la Moselle, de la Côte-d'Or et de l'Indre*: p. 233-305, pl. 11-14, Lorette, Éditeur-Libraire, Paris (Metz).———(1885) 1864, *Mémoire sur les Foraminifères du Lias des départements de la Moselle, de la Côte-d'Or, du Rhône, de la Vienne et du Calvados, Mémoire 3*: Acad. Imper. Metz, Mém., v. 44 (ser. 2, v. 11, 1862-63), p. 361-438, pl. 7-10.———(1886) 1866, *Cinquième mémoire sur les Foraminifères du Lias des Departements de la Moselle, de la Côte-d'Or, et de l'Indre*: p. 313-454, pl. 15-18, Lorette, Éditeur-Libraire, Paris (Metz).———(1887) 1866, *Sixième mémoire sur les Foraminifères du Lias des Départements de l'Indre et de la Moselle*: p. 459-532, pl. 19-22, Lorette (Metz).———(1888) 1876, *Essai sur le classement des animaux qui vivent sur la plage et dans les environs de Dunkerque*: pt. 2, p. 55-100, pl. 7-12 (Paris).———(1889) 1878, *Les Foraminifères et les Entomostracés-Ostracodes du Pliocène supérieur de l'Isle de Rhodes*: Soc. géol. France, Mém., ser. 3, v. 1, p. 1-135, pl. 1-19.———(1890) 1882, *Les Foraminifères de l'Éocène des environs de Paris*: Soc. géol. France, Mém. 3, ser. 3, v. 2, p. 1-193, pl. 1-28.———(1891) 1883, *Cinquième mémoire sur les Foraminifères du système oolithique de la zone à Ammonites parkinsoni de Fontoy (Moselle)*: p. 339-406, pl. 38-44, The Author (Metz).———(1892) 1883, *Sur un nouveau genre de Foraminifères du Fuller's-earth de la Moselle*: Soc. géol. France, Bull., ser. 3, v. 11 (1882-83), pt. 1, p. 37-39, pl. 3.

(1893) ———, **& Berthelin, Georges**, 1875, *Étude microscopique des marnes du Lias moyen d'Essey-lès-Nancy, zone inférieure de l'assise à Ammonites margaritatus*: Soc. géol. France, Mém., ser. 2, v. 10, no. 3, p. 1-126, pl. 1-10.

(1894) **Thalmann, H. E.**, 1932, *Die Foraminiferen-Gattung Hantkenina Cushman, 1924, und ihre regional-stratigraphische Verbreitung*: Eclogae geol. Helv., v. 25, p. 287-292.———(1895) 1933, *Zwei neue Vertreter der Foraminiferen-Gattung Rotalia Lamarck 1804, R. cubana, nom.nov., und R. trispinosa, nom.nov.*: Same, v. 26, p. 248-251, pl. 12.———(1896) 1934, *Supplement to bibliography and index to* *genera and species of Foraminifera for the year 1931:* Jour. Paleontology, v. 8, p. 238-244.——(1897) 1935-58, *Bibliography and index to new genera, species and varieties of Foraminifera:* Same; (a) for 1933, v. 9, p. 715-743 (1935); (b) for 1934, v. 10, p. 294-322 (1936); (c) for 1935, v. 12, p. 177-208 (1938); (d) for 1936, v. 13, p. 425-465 (1939); (e) for 1937-38, v. 15, p. 629-690 (1941); (f) for 1942, v. 19, p. 396-410 (1945); (g) for 1945 *with supplements for 1939-44, and addenda (1942-45)*, v. 21, p. 355-395 (1947); (h) for 1948, v. 23, p. 641-668 (1949); (i) for 1949, v. 24, p. 699-745 (1950); (j) for 1951, v. 26, p. 953-992 (1952); (k) for 1952, v. 27, p. 847-876 (1953); (l) for 1955, v. 32, p. 737-762 (1958).——(1898) 1937, *Palaeontological abstracts No. 1349.* Palmer, Dorothy K.—"*New genera and species of Cuban Oligocene Foraminifera.*" Mem. Soc. Cubana Hist. Nat., 10, 123-128, 1 text-fig., pl. 5, Habana 1936: Palaeont. Zentralbl., Geol. Zentralbl., Pt. B, Palaeont., v. 10 (1937-1938), p. 351.——(1899) 1937-51, *Mitteilungen über Foraminiferen.:* Eclogae geol. Helv.; (a) *Pt. 3*, 1937, v. 30, no. 2, p. 337-356, pl. 21-23; (b) *Pt. 4*, 1939, v. 31, no. 2, p. 327-333; (c) *Pt. 5*, 1947, v. 39, no. 2, p. 309-314; (d) *Pt. 9*, 1951, v. 43, p. 221-225.——(1900) 1942, *Foraminiferal homonyms:* Am. Midland Naturalist, v. 28, p. 457-462.——(1901) 1942, *Foraminiferal genus Hantkenina and its subgenera:* Am. Jour. Sci., v. 240, p. 809-820, pl. 1.——(1902) 1950, *New names and homonyms in Foraminifera:* Cushman Found. Foram. Research, Contrib., v. 1, p. 41-45.——(1903) 1952, *New names for Foraminiferal homonyms, I:* Same, v. 3, pt. 1, p. 14.——(1904) 1954, *Pijpersia, nom.nov. for Ruttenia Pijpers, 1933, a homonoym of Ruttenia Rodhain, 1924:* Same, v. 5, pt. 4, p. 153.——(1905) 1960[1961], *An index to the genera and species of the Foraminifera, 1890-1950:* George Vanderbilt Found., Stanford Univ., Calif., p. 1-393 (1960).

(1906) ——, & Bermúdez, P. J., 1954, *Chitinosiphon, a new genus of the Rhizamminidae:* Cushman Found. Foram. Research, Contrib., v. 5, pt. 2, p. 53-54.

(1907) Thomas, A. O., 1931, *Late Devonian Foraminifera from Iowa:* Jour. Paleontology, v. 5, p. 40-41, pl. 7.

(1908) Thomas, Philippe, 1893, *Description de quelques fossiles nouveaux ou critiques des terrains tertiaires et secondaires de la Tunisie:* Exploration Scientifique de la Tunisie, p. 1-46, Imprimerie Nationale (Paris).

(1909) Thomas, Raymond, & Gauthier-Lièvre, L., 1959, *Note sur quelques Euglyphidae d'Afrique:* Soc. Hist. Nat. Afrique Nord, Bull., v. 50, no. 5-6, p. 204-221, 4 text-fig.

(1910) Thompson, M. L., 1934, *The fusulinids of the Des Moines Series of Iowa:* Univ. Iowa Studies Nat. History, new ser., no. 284, v. 16, no. 4, p. 277-332, pl. 20-23.——(1911) 1935, *The fusulinid genus Yangchienia Lee:* Eclogae geol. Helv., v. 28, no. 2, p. 511-517, pl. 17 (Bâle).——(1912) 1935, *The fusulinid genus Staffella in America:* Jour. Paleontology, v. 9, p. 111-120, pl. 13.——(1913) 1935, *Fusulinids from the Lower Pennsylvanian Atoka and Boggy Formations of Oklahoma:* Same, v. 9, p. 291-306, pl. 26.——(1914) 1936, *The fusulinid genus Verbeekina:* Same, v. 10, p. 193-201, pl. 24.——(1915) 1936, *Lower Permian fusulinids from Sumatra:* Same, v. 10, p. 587-592, fig. 1-13.——(1916) 1936, *Nagatoella, a new genus of Permian fusulinids:* Geol. Soc. Japan, Jour., v. 43, no. 510, p. 195-202, pl. 12.——(1917) 1936, *The genotype of Fusulina s.s.:* Am. Jour. Sci., v. 32, p. 287-291.——(1918) 1937, *Fusulinids of the subfamily Schubertellinae:* Jour. Paleontology, v. 11, p. 118-125, pl. 22.——(1919) 1942, *New genera of Pennsylvanian fusulinids:* Am. Jour. Sci., v. 240, p. 403-420, pl. 1-3.——(1920) 1944, *Pennsylvanian Morrowan rocks and fusulinids of Kansas:* Geol. Survey Kansas, Bull. 52, pt. 7, p. 409-431, pl. 1-2, fig. 1-2.——(1921) 1946, *Permian fusulinids from Afghanistan:* Jour. Paleontology, v. 20, p. 140-157, pl. 23-26, fig. 1.——(1922) 1948, *Studies of American fusulinids:* Univ. Kansas Paleont. Contrib., Protozoa, art. 1, 184 p., 38 pl., 7 fig.——(1923) 1949, *The Permian fusulinids of Timor:* Jour. Paleontology, v. 23, p. 182-192, pl. 34-36, 1 fig.——(1924) 1951, *Wall structures of fusulinid Foraminifera:* Cushman Found. Foram. Research, Contrib., v. 2, pt. 3, p. 86-91, pl. 9-10, 1 fig.——(1925) 1951, *New genera of fusulinid Foraminifera:* Same, v. 2, pt. 4, p. 115-119, pl. 13-14.——(1926) 1954, *American Wolfcampian fusulinids:* Univ. Kansas Paleont. Contrib., Protozoa, art. 5, 226 p., 52 pl., 14 fig.——(1927) 1957, *Northern midcontinent Missourian fusulinids:* Jour. Paleontology, v. 31, p. 289-328, pl. 21-30, fig. 1-2.——(1928) 1961, *Pennsylvanian fusulinids from Ward Hunt Island:* Same, v. 35, p. 1130-1136, pl. 135-136, 1 fig.

(1929) ——, & Foster, C. L., 1937, *Middle Permian fusulinids from Szechuan, China:* Jour. Paleontology, v. 11, p. 126-144, pl. 23-25.

(1930) ——, Pitrat, C. W., & Sanderson, G. A., 1953, *Primitive Cache Creek fusulinids from Central British Columbia:* Jour. Paleontology, v. 27, p. 545-552, pl. 57-58.

(1931) ——, Verville, G. J., & Bissell, H. J., 1950, *Pennsylvanian fusulinids of the south-central Wasatch Mountains, Utah:* Jour. Paleontology, v. 24, p. 430-465, pl. 57-64, fig. 1-2.

(1932) ——, ——, & Lokke, D. H., 1956, *Fusulinids of the Desmoinesian-Missourian con-*

tact: Jour. Paleontology, v. 30, p. 793-810, pl. 89-93, 1 fig.

(1933) ——, & **Wheeler, H. E.**, 1942, *Permian fusulinids from British Columbia, Washington and Oregon:* Jour. Paleontology, v. 16, p. 700-711, pl. 105-109, fig. 1-2.

(1934) ——, ——, & **Hazzard, J. C.**, 1946, *Permian fusulinids of California:* Geol. Soc. America, Mem. 17, 77 p., 18 pl., 4 fig.

(1935) **Tinoco, I. de Medeiros**, 1955, *Foraminíferos recentes de Cabo Frio, Estado do Rio de Janeiro:* Div. Geol. Mineral., Bull., no. 159, p. 7-43, pl. 1-4 (Rio de Janeiro).

(1936) **Tizard, Staff-Commander, & Murray, John**, 1882, *Exploration of the Faröe Channel during the summer of 1880, in Her Majesty's hired Ship "Knight-Errant":* Royal Soc. Edinburgh, Proc., v. 11, p. 638-720, pl. 6.

(1937) **Tobler, August**, 1922, *Helicolepidina, ein neues subgenus von Lepidocyclina:* Eclogae geol. Helv., v. 17, p. 380-384, text-fig.——(1938) 1927, *Verkalkung der Lateral-kammern bei Miogypsina:* Same, v. 20, p. 323-330, text-fig. 1-5.

(1939) **Todd, Ruth, & Blackmon, P.**, 1956, *Calcite and aragonite in Foraminifera:* Jour. Paleontology, v. 30, p. 217-219.

(1940) ——, & **Brönnimann, Paul**, 1957, *Recent Foraminifera and Thecamoebina from the eastern Gulf of Paria:* Cushman Found. Foram. Research, Spec. Publ. 3, p. 1-43, pl. 1-12.

(1941) **Toriyama, Ryuzo**, 1953, *New peculiar fusulinid genus from the Akiyoshi Limestone of southwestern Japan:* Jour. Paleontology, v. 27, p. 251-256, pl. 35-36.

(1942) **Torrend, C.**, 1907, *Les Myxomycètes. Étude des espèces connues jusqu'ici:* Broteria, v. 6, pt. 2, ser. Bot., p. 5-64.

(1943) **Toula, Franz**, 1915, *Über den marinen Tegel von Neudorf an der March (Dévény-Ujfalu) in Ungarn und seine Mikrofauna:* K.K. Geol. Reichsanst., Jahrb., v. 64 (1914), no. 4, p. 635-674, pl. 39, text-fig. 1.

(1944) **Toulmin, L. D.**, 1941, *Eocene smaller Foraminifera from the Salt Mountain Limestone of Alabama:* Jour. Paleontology, v. 15, p. 567-611, pl. 78-82, text-fig. 1-4.

(1945) (see 1954A.)

(1946) (see 1955.)

(1947) **Tournour, Raoul**, 1868, *Sur les lambeaux de terrain des environs de Rennes et de Dinan, en Bretagne, et particulièrement sur la présence de l'étage, des sables de Fontainebleau aux environs de Rennes:* Soc. géol. France, Bull., ser. 2, v. 25 (1867-68), pt. 3, p. 367-377.

(1948) **Trauth, Friedrich**, 1918, *Das Eozänvorkommen bei Radstadt im Pongau und seine Beziehungen zu den gleichalterigen Ablagerungen bei Kirchberg am Wechsel und Wimpassing am Leithagebirge:* K. Akad. Wiss. Wien, math.-naturwiss. Cl., Denkschr., v. 95, p. 171-278, pl. 1-5, text-fig. 1-5.

(1949) **Troelsen, J. C.**, 1950, *Contributions to the geology of northwest Greenland, Ellesmere Island and Axel Heiberg Island:* Medd. Grønland, Komm. Vidensk. Undersøg. I Grønland, v. 149, no. 7, 85 p., 17 fig., C. A. Reitzels Forlag (København).——(1950) 1954, *Studies on Ceratobuliminidae (Foraminifera):* Dansk Geol. Foren., Medd., v. 12, p. 448-478.——(1951) 1954, *Foram surgery:* Micropaleontologist, v. 8, no. 4, p. 40-41.——(1952) 1955, *On the value of aragonite tests in the classification of the Rotaliidea:* Cushman Found. Foram. Research, Contrib., v. 6, pt. 1, p. 50-51.

(1953) **Trouessart, E. L.**, 1898, *Sur un Foraminifère marin présentant le phénomène de la conjugaison:* Soc. Biol., Comptes Rendus, ser. 10, v. 50, p. 771-774.

(1954) **Trujillo, E. F.**, 1960, *Upper Cretaceous Foraminifera from near Redding, Shasta County, California:* Jour. Paleontology, v. 34, p. 290-346, pl. 43-50, 3 text-fig.

(1954A) **Tumanskaya, O. G.**, 1950, *O vysshikh fuzulinidakh iz verkhnepermskikh otlozheniy SSSR:* Moskov. Obshch. Ispyt., Prirody, Byull., Otdel. Geol., v. 25(4), p. 77-97, pl. 1-7. [*Higher fusulinids from Upper Permian deposits in the USSR.*]——(1955) 1953, *O verkhnepermskikh fuzulinidakh yuzhno-Ussuriyskogo kraya:* VSEGEI, Trudy, Minist. Geol., p. 1-56, pl. 1-15. [*Upper Permian fusulinids in the South Ussuri territory.*]——(1955A) 1962, *O nekotorykh nizhnepermskikh fuzulinidakh Urala i drugikh rayonov SSSR:* Akad. Nauk SSSR, Doklady, v. 146, no. 6, p. 1396-1398, text-fig. 1-3. [*Certain lower Permian fusulinids from the Urals and other regions of the USSR.*]

(1956) **Turnovsky, Kurt**, 1958, *Eine neue Art von Globorotalia Cushman aus dem Eozaen Anatoliens und ihre Zuordnung zu einer neuen Untergattung:* Geol. Soc. Turkey, Bull., v. 6, p. 80-86, fig. 1.

(1957) **Uchio, Takayasu**, 1951, *New species and genus of the Foraminifera of the Cenozoic formations in the middle part of the Boso Peninsula, Chiba-Ken, Japan:* Palaeont. Soc. Japan, Trans. & Proc., new ser., no. 2, p. 33-42, pl. 3.——(1958) 1952, *An interesting relation between Stomatorbina Dorreen, 1948, and Mississippina Howe, 1930, of Foraminifera:* Same, new ser., no. 7, p. 195-200, pl. 18.——(1959) 1952, *Foraminiferal assemblage from Hachijo Island, Tokyo Prefecture, with descriptions of some new genera and species:* Japanese Jour. Geol. & Geog., v. 22, p. 145-159, pl. 6-7.—— (1960) 1953, *On some foraminiferal genera in*

Japan: Same, v. 23, p. 151-162, pl. 14.——
(1961) 1960, *Ecology of living benthonic Foraminifera from the San Diego, California, area:* Cushman Found. Foram. Research, Spec. Publ. 5, p. 1-72, pl. 1-10.

(1962) **Uhlig, Victor,** 1883, *Ueber Foraminiferen aus dem rjäsan'schen Ornatenthone:* K. K. geol. Reichsanst. Wien, Jahrb., v. 33, p. 753-774, pl. 7-9.

(1963) **Ujiié, Hiroshi,** 1956, *Pseudocibicidoides, n.gen., from the sea coast of Katase, Kanagawa Prefecture, Japan:* Tokyo Kyoiku Daigaku, Sci. Repts., ser. C (Geol., Mineral. & Geog.), v. 4, no. 37, p. 263-265, pl. 13.——(1964) 1956, *The internal structures of some Elphidiidae:* Same, v. 4, no. 38, p. 267-282, pl. 14-15, text-fig. 1-2.

(1965) ——, & **Watanabe, Hikosuke,** 1960, *The Poronai Foraminifera of the northern Ishikari Coal-Field, Hokkaidō:* Tokyo Kyoiku Daigaku, Sci. Rept., ser. C, v. 7, no. 63, p. 117-136, pl. 1-3.

(1966) **Ulrich, E. O.,** & **Bassler, R. S.,** 1904, *A revision of the Paleozoic Bryozoa, Part 1, Ctenostomata:* Smithsonian Misc. Coll., v. 45, p. 256-294, pl. 65-68.

(1967) **Umbgrove, J. H. F.,** 1928, *Het genus Pellatispira in het indo-pacifische gebied:* Nederland.-Indië, Dienst. Mijnb., no. 10, p. 43-71, pls. 1-15.——(1968) 1931, *Tertiary Foraminifera:* Leidsche Geol. Meded., pt. 5, Feestbundel Prof. Dr. K. Martin, p. 35-91.——
(1969) 1936, *Heterospira, a new foraminiferal genus from the Tertiary of Borneo:* Same, pt. 8, no. 1, p. 155-157, pl. 1.——(1970) 1937, *A new name for the foraminiferal genus Heterospira:* Same, pt. 8, no. 2, p. 309.

(1971) **Upshaw, C. F.,** & **Stehli, F. G.,** 1962, *Quantitative biofacies mapping:* Am. Assoc. Petroleum Geologists, Bull., v. 46, no. 5, p. 694-699, text-fig. 1-7.

(1972) **Valkanov, A.,** 1932, *Yadrenoto dielenie i yadreniyat stroezh pri Cochliopodium i Cyathomonas kato osnova za razglezhdaneto na edin kariologichen problem:* Godishnik na Sofiyskiya Univ., II, Fiz. Mat. Fac., Book 1, v. 28, p. 153-196. [*Die Kernteilung und Kernverhältnisse bei Cochliopodium un Cyathomonas als Grundlage zur Betrachtung eines karyologischen Problems.*] [Bulgarian, German summary.]——(1973) 1938, *Über die Fortpflanzung von Gromia dujardini M. Schultze, Protistenstudien 10:* Archiv Protistenkunde, v. 90, no. 3, p. 393-395.——
(1974) 1940, *Die Heliozoen und Proteomyxien. Artbestand und sonstige Kritische Bemerkungen:* Same, v. 93, p. 225-254, 4 fig.

(1975) **Vanderpool, H. C.,** 1933, *Upper Trinity microfossils from southern Oklahoma:* Jour. Paleontology, v. 7, p. 406-411, pl. 49.

(1976) **Van Oye, Paul,** 1949, *Rhizopodes de Java:* Bijdragen Tot de Dierkunde, pt. 28, p. 327-352, pl. 28, text-fig. 1-24.——(1977) 1956, *Rhizopoda Venezuelas mit besonderer Berücksichtigung ihrer Biogeographie:* Ergebnisse der Deutschen Limnologischen Venezuela-Expedition 1952, v. 1, p. 329-360.

(1978) **Van Tieghem, Phillipe E.,** 1880, *Sur quelques Myxomycètes à plasmode agrégé:* Soc. Bot. France, Bull., v. 27 (ser. 2, v. 2), p. 317-322.——(1979) 1898, *Éléments de botanique, II. Botanique spéciale:* ed. 3, Revised, 612 p., 345 fig., Masson & Cie. (Paris).

(1980) **Van Wessem, A.,** 1943, *Geology and paleontology of central Camaguey, Cuba:* Dissertation, Univ. Utrecht, p. 1-91, pl. 1-3, 3 text-fig.

(1980A) **Varsanofeva, V. A.,** & **Reytlinger, E. A.,** 1962, *K kharakteristike verkhnedevonskikh i turneyskikh otlozheniy maloy Pechory:* Moskov. Obshch. Ispyt., Prirody, Byull., new ser., v. 67, Otdel. Geol., v. 37, no. 5, p. 36-60, pl. 1, 2. [*On the characteristic Upper Devonian and Tournaisian deposits of the Little Pechora.*]

(1981) **Vašíček, Miloslav,** 1947, *Poznámky k microbiostratigrafiy magurského flyše na Moravě:* Stat. Geol. Ústavu, Repub. Českoslov., Vestnik, Ročník 22, zvlastni otisk, p. 235-256, pl. 1-3.——(1982) 1953, *Changes in the ratio of sinistral and dextral individuals of the foraminifer Globorotalia scitula (Brady) and their use in stratigraphy:* Ústřed. Ústavu Geol., Sborník, v. 20, p. 1-76, tab. 1.——(1983) 1956, *Analysa rodu Sphaeroidina d'Orb. (Foraminifera):* Ústřed. Ústavu Geol., Rozpravy, v. 19, p. 7-162, pl. 1-7.

(1984) ——, & **Růžička, B.,** 1957, *Namurské thecamoeby z ostravsko-karvinsk'eho revíru:* Národního Musea v Praze, Sborník, Acta Musei Nationalis Pragae, v. 13, B, no. 5, p. 333-340, pl. 40, 41.——(1985) 1957, *Namurské foraminifery z ostravsko-karvinsk'eho revíru:* Same, v. 13, B, no. 5, p. 341-362, pl. 42-44.

(1986) **Vasilenko, V. P.,** 1954, *Anomalinidy. Iskopaemye foraminifery SSSR:* VNIGRI, Trudy, new ser., no. 80, p. 1-282, pl. 1-36, text-fig. 1-42. [*Anomalinidae. Fossil Foraminifera of the USSR.*]

(1987) **Vasseur, Gaston,** 1878, *Sur les terrains tertiaires de la Bretagne:* Acad. Sci. Paris, Comptes Rendus, v. 87, p. 1048-1050.

(1988) **Vaughan, T. W.,** 1924, *American and European Tertiary larger Foraminifera:* Geol. Soc. America, Bull., v. 35, p. 785-822, pl. 30-36.——(1989) 1928, *Yaberinella jamaicensis, a new genus and species of arenaceous Foraminifera:* Jour. Paleontology, v. 2, p. 7-12, pl. 4-5.——(1990) 1929, *Actinosiphon semmesi, a new genus and species of orbitoidal Foraminifera, and Pseudorbitoides trechmanni H. Douvillé:* Same, v. 3, p. 163-169, pl. 21.——

(1991) 1929, *Additional new species of Tertiary larger Foraminifera from Jamaica:* Same, v. 3, p. 373-382, pl. 39-41.———(1992) 1929, *Descriptions of new species of Foraminifera of the genus Discocyclina from the Eocene of Mexico:* U.S. Natl. Museum, Proc., v. 76, no. 2800, p. 1-18, pl. 1-7.———(1993) 1929, *Studies of orbitoidal Foraminifera: The subgenus Polylepidina of Lepidocyclina and Orbitocyclina, a new genus:* Nat. Acad. Sci., Proc., v. 15, no. 3, p. 288-295, pl.———(1994) 1936, *New species of orbitoidal Foraminifera of the genus Discocyclina from the Lower Eocene of Alabama:* Jour. Paleontology, v. 10, p. 253-259, pl. 41-43.———(1995) 1945, *American Paleocene and Eocene larger Foraminifera:* Geol. Soc. America, Mem. 9, pt. 1, p. 1-175, pl. 1-46.

(1996) ———, & Cole, W. S., 1932, *Cretaceous orbitoidal Foraminifera from the Gulf States and Central America:* Nat. Acad. Sci., Proc., v. 18, p. 611-616, pl. 1, 2.———(1997) 1938, *Triplalepidina veracruziana, a new genus and species of orbitoidal Foraminifera from the Eocene of Mexico:* Jour. Paleontology, v. 12, p. 167-169, pl. 27.———(1998) 1941, *Preliminary report on the Cretaceous and Tertiary larger Foraminifera of Trinidad, British West Indies:* Geol. Soc. America, Spec. Paper 30, 137 p., 46 pl., 2 text-fig.

(1999) Vejdovský, František, 1881, *Über die Rhizopoden der Brunnenwässer Prags:* K. böhm. Gesell. Wiss. Prag, Sitzungsber., v. 1880, p. 136-139.———(2000) 1882, *Thierische Organismen der Brunnenwässer von Prag:* 70 p., 8 pl. (Prague).

(2001) Vella, Paul, 1957, *Studies in New Zealand Foraminifera:* New Zealand Geol. Survey, Paleont. Bull. 28, p. 1-64, pl. 1-9.———(2002) 1961, *Upper Oligocene and Miocene uvigerinid Foraminifera from Raukumara Peninsula, New Zealand:* Micropaleontology, v. 7, p. 467-483, pl. 1-2.

(2003) Venglinskiy [Venglinskii], I. V., 1960, *O stroenii stenki rakoviny nekotorykh agglyutinirovannykh foraminifer:* Voprosy Mikropaleontologii, no. 3, Akad. Nauk SSSR, Otdel. Geol.-Geogr. Nauk, Geol. Inst., p. 31-36, 2 pl. [*On the wall structure of the test of certain agglutinated Foraminifera.*]

(2004) Verville, G. J., Thompson, M. L., & Lokke, D. H., 1956, *Pennsylvanian fusulinids of eastern Nevada:* Jour. Paleontology, v. 30, p. 1277-1287, pl. 133-136.

(2005) Verworn, Max, 1889, *Psycho-physiologische Protisten-Studien:* 218 p., 6 pl., text-fig. 1-27, G. Fischer (Jena).

(2005A) Viên, Le Thi, 1959, *Étude de fusulinidés du Haut-Laos, du Cambodge et du Sud Viêt-Nam:* Univ. Saigon, Ann. Faculté Sci., p. 99-120, pl. 1-2, fig. 1-2 (Saigon).

(2006) Vine, G. R., 1882, *Notes on Annelida Tubicola of the Wenlock shales from the washing of Mr. George Maw, F.G.A.:* Quart. Jour. Geol. Soc. London, v. 38, p. 377-392, pl. 15.

(2007) Vinogradov, A. P., 1953, *The elementary chemical composition of marine organisms:* Mem. Sears Found. for Marine Research, Yale Univ., Mem. 2, p. 1-647.

(2008) Vissarionova, A. Ya., 1948, *Gruppa Endothyra globulus Eichwald iz Vizeyskogo yarusa nizhnego karbona evropeyskoy chasti soyuza:* Akad. Nauk SSSR, Trudy, Geol. Inst., no. 62, geol. ser., no. 19, p. 182-185, pl. 6. [*The group of Endothyra globulus Eichwald from the Visean stage of the Lower Carboniferous in the European part of the Union.*]———(2009) 1948, *Nekotorye vidy podsemeystva Tetrataxinae Galloway iz Vizeyskogo yarusa Evropeyskoy chasti soyuza:* Akad. Nauk SSSR, Trudy, Geol. Inst., no. 62, geol. ser. 19, p. 190-195, pl. 8. [*Certain species of the subfamily Tetrataxiinae Galloway from the Visean stage of the European part of the Union.*]———(2009A) 1948, *Primitivnye Fuzulinidy iz nizhnego karbona evropeyskoy chasti SSSR:* Akad. Nauk SSSR, Trudy, Geol. Inst., no. 62 (geol. ser. 19), p. 216-226, pl. 13-14. [*Primitive Fusulinidae from the Lower Carboniferous of the European part of the USSR.*]———(2010) 1950, *Fauna foraminifer v devonskikh otlozheniyakh Bashkirii:* Bashkirskaya Neft, no. 1, p. 33-36, 1 pl. [*Foraminiferal fauna from the Devonian deposits of Bashkir.*]

(2011) Vlerk, I. M. van der, 1923, *Een overgangsvorm tusschen Orthophragmina en Lepidocyclina uit het Tertiair van Java:* Geol.-Mijnb. Genoot. Nederland Kolon., Verhandl., geol. ser., pt. 7(1923-27), stuk 2, p. 91-98, pl. 1.———(2012) 1924, *Miogypsina dehaartii, nov. species de Larat (Moluques):* Eclogae Geol. Helv., v. 18, no. 3, p. 429-432, text-fig. 1-3.———(2013) 1928, *The genus Lepidocyclina in the Far East:* Eclogae Geol. Helv., v. 21, no. 1, p. 182-211, pl. 6-23, 2 tables.———(2014) 1928, *Het genus Lepidocyclina in het Indo-Pacifische Gebied:* Wetensch. Med., no. 8, p. 7-88, pl.

(2015) Vogler, J., 1941, *Ober-Jura und Kreide von Misol (Niederländisch-Ostindien):* Palaeontographica, Suppl. 4, pt. 4, p. 246-293, pl. 19-24.

(2016) Volkonsky, M., 1931, *Hartmannella castellanii Douglas et classification des Hartmannelles (Hartmanellinae, nov. subfam., Acanthamoeba, nov.gen., Glaeseria, nov.gen.):* Arch. Zool. Expér. & Générale, v. 72, p. 317-339, pl. 2.

(2017) Vologdin, A. G., 1939, *Arkheotsiaty i*

vodorosli srednego Kembriya Yuzhnogo Urala: Paleont. Labor. Moskov. Gosudarst. Univ. SSSR, Problemy Paleontologii, v. 5, p. 209-245 (Russian), p. 245-276 (English), pl. 1-12. [*Archaeocyathids and algae from the Middle Cambrian of the southern Urals.*]———(2018) 1958, *Nizhnekembriyskie Foraminifery Tuvy:* Akad. Nauk SSSR, Doklady, v. 120, no. 2, p. 405-408, text-fig. 1-22. [*Lower Cambrian Foraminifera from Tuva.*]

(2019) Voloshinova, N. A., 1958, *O novoy sistematike Nonionid:* Mikrofauna SSSR, Sbornik 9, VNIGRI, Trudy, no. 115, p. 117-191, pl. 1-16. [*On a new systematics of the Nonionidae.*]———(2020) 1960, *Uspekhi mikropaleontologii v dele izucheniya vnutrennego stroeniya foraminifer:* Trudy Pervogo Seminara po Mikrofaune, VNIGRI, p. 48-87, pl. 1-12. [*Progress in micropaleontology in the work of studying the inner structure of Foraminifera.*]

(2021) ———, & Budasheva, A. I., 1961, *Lituolidy i trochamminidy iz tretichnykh otlozheniy ostrova Sakhalina i poluostrova Kamchatki:* Mikrofauna SSSR, Sbornik 12, VNIGRI, Trudy, no. 170, p. 169-233, pl. 1-19. [*Lituolidae and Trochamminidae from Tertiary deposits of Sakhalin Island and Kamchatka Peninsula.*]

(2022) ———, & Dain, L. G., 1952, *Nonionidy, Kassidulinidy i Khilostomellidy. Iskopaemye foraminifery SSSR:* VNIGRI, Trudy, new ser., no. 63, 151 p., 17 pl. [*Nonionidae, Cassidulinidae and Chilostomellidae. Fossil Foraminifera of the USSR.*]

(2023) Volz, P., 1929, *Studien zur Biologie der Bodenbewohnenden Thekamöben:* Archiv Protistenkunde, v. 68, p. 349-408, pl. 6, text-fig. 1-33.

(2024) Volz, Wilhelm, 1904, *Zur Geologie von Sumatra:* Geol. & Paläont. Abhandl., new ser., v. 6, no. 2, p. 87-196, text-fig. 26-45, pl. 1-12.

(2025) Voorwijk, G. H., 1937, *Foraminifera from the Upper Cretaceous of Habana, Cuba:* Royal Acad. Amsterdam, Proc., v. 40, p. 190-198, pl. 1-3.

(2026) Wade, Mary, 1955, *A new genus of the Chapmanininae from southern Australia:* Cushman Found. Foram. Research, Contrib., v. 6, pt. 1, p. 45-49, pl. 8, text-fig. 1-3.——— (2027) 1957, *Morphology and taxonomy of the foraminiferal family Elphidiidae:* Washington Acad. Sci., Jour., v. 47, no. 10, p. 330-339, text-fig. 1-4.

(2028) ———, & Carter, A. N., 1957, *The foraminiferal genus Sherbornina in southeastern Australia:* Micropaleontology, v. 3, no. 2, p. 155-164, pl. 1-3, text-fig. 1-2.

(2029) Wähner, F., 1903, *Das Sonnwendgebirge im Unterinntal, Ein typus Alpinen Gebirgsbaues, etc. Theil I:* i-xii+356 p., 19 pl. (Leipzig & Wien).

(2030) Wailes, G. H., 1927, *Rhizopoda and Heliozoa from British Columbia:* Ann. & Mag. Nat. History, ser. 9, v. 20, p. 153-156, text-fig. a-i.

(2031) ———, & Penard, Eugène, 1911, *Rhizopoda. Clare Island Survey, pt. 65:* Royal Irish Acad., Proc., v. 31, p. 1-64, pl. 1-6.

(2032) Walcott, C. D., 1899, *Pre-Cambrian fossiliferous formations:* Geol. Soc. America, Bull., v. 10, p. 199-244.

(2033) Walker, George, & Boys, William, 1784, *Testacea minuta rariora, nuperrime detecta in arena littoris Sandvicensis a Gul. Boys, arm S.A.S. multa addidit, et omnium figuras ope microscopii ampliatas accurate delineavit Geo. Walker:* 25 p., 3 pl., J. March (London).

(2034) Wallich, G. C., 1863, *Further observations on the distinctive characters, habits, and reproductive phenomena of the amoeban rhizopods:* Ann. & Mag. Nat. History, ser. 3, v. 12, p. 448-468, pl. 8.———(2035) 1864, *On the extent, and some of the principal causes, of structural variation among the difflugian rhizopods:* Same, ser. 3, v. 13, p. 215-245, pl. 15-16.———(2036) 1877, *On Rupertia stabilis, a new sessile foraminifer from the North Atlantic:* Same, ser. 4, v. 19, p. 501-504, pl. 20.

(2037) Wallroth, F. G., 1833, *Flora cryptogamica Germaniae:* Compendium florae Germanicae, Sec. II, M. J. Bluff & C. A. Fingerhuth (Norimbergae).

(2038) Wanner, Johann, 1941, *Gesteinsbildende Foraminiferen aus Malm und Unterkreide des östlichen ostindischen Archipels. Nebst Bemerkungen über Orbulinaria Rhumbler und andere verwandte Foraminiferen:* Paläont. Zeitschr., v. 22, p. 75-99, 2 pl., 37 fig.

(2039) Warren, A. D., 1957, *Foraminifera of the Buras-Scofield Bayou region, southeast Louisiana:* Cushman Found. Foram. Research, Contrib., v. 8, pt. 1, p. 29-40, pl. 3-4.

(2040) Warthin, A. S., Jr., 1930, *Micropaleontology of the Wetumka, Wewoka, and Holdenville formations:* Oklahoma Geol. Survey, Bull. 53, p. 1-95, pl. 1-7, chart.

(2041) Wedekind, P. R., 1937, *Einführung in die grundlagen der historischen geologie, Band II. Mikrobiostratigraphie die Korallen- und Foraminiferenzeit:* 136 p., Ferdinand Enke (Stuttgart).

(2042) Weijden, W. J. M., van der, 1940, *Het genus Discocyclina in Europa. Een monografie naar Aanleiding van een Heronderzoek van het Tertiair-profiel van Biarritz:* Dissertation, Rijsuniv., Leiden, p. 1-116, pl. 1-12.

(2043) Weinhandl, Rupert, 1958, *Schackoinella, eine neue Foraminiferengattung:* K.K. geol. Reichsanstalt (Bundesanst.), Verhandl., p. 141-142, text-fig. 1.

(2044) **Weinzierl, L. L., & Applin, E. R.**, 1929, *The Claiborne Formation on the coastal domes:* Jour. Paleontology, v. 3, p. 384-410, pl. 42-44.

(2045) **Wenyon, C. M.**, 1926, *Protozoology, a manual for medical men, veterinarians and zoologists:* v. 1, 778 p., 336 fig., Balliere, Tindall & Cox (London, New York).

(2046) **West, G. S.**, 1901, *On some British freshwater rhizopods and Heliozoa:* Linnean Soc. London, Jour., Zool., v. 28, p. 308-342, pl. 28-30.

(2047) **Wetzel, Otto**, 1940, *Mikropaläontologische Untersuchungen an der obersenonen Kreide von Stevns Klint-Kridtbrud auf der dänischen Insel Seeland und ihrem Feuerstein in geschiebekundlicher Hinsicht:* Zeitschr. Geschiebeforsch. & Flachlandsgeol., v. 16, p. 118-156, pl. 1-5.———(2048) 1951, *Mikroskopische Reste von Kalkorganismen als Feuersteinfossilien besonderen Aussehens:* Neues Jahrb. Geol. & Pälont. Abhandl., v. 94 (1951), no. 1, p. 112-120, pl. 14.———(2049) 1957, *Fossil "microforaminifera" in various sediments and their reaction to acid treatment:* Micropaleontology, v. 3, no. 1, p. 61-64, pl. 1.

(2050) **Weynschenk, Robert**, 1950, *Die Jura-Mikrofauna und -flora des Sonnwendgebirges (Tirol):* Schlernschriften, Univ. Innsbruck, v. 83, p. 1-32, pl. 1-3.———(2051) 1951, *Two new Foraminifera from the Dogger and Upper Triassic of the Sonnwend Mountains of Tyrol:* Jour. Paleontology, v. 25, p. 793-795, pl. 112, 3 text-fig.———(2052) 1956, *Aulotortus, a new genus of Foraminifera from the Jurassic of Tyrol, Austria:* Cushman Found. Foram. Research, Contrib., v. 7, pt. 1, p. 26-28, pl. 6, text-fig. 1-2.

(2053) **Whipple, G. L.**, 1934, *Larger Foraminifera from Vitilevu, Fiji* in LADD, H. S., Geology of Vitilevu, Fiji: Bernice P. Bishop Museum, Bull., no. 119, p. 141-153, pl. 19-23.

(2054) **White, C. A.**, 1878, *Descriptions of new species of invertebrate fossils from the Carboniferous and Upper Silurian rocks of Illinois and Indiana:* Acad. Nat. Sci. Philadelphia, Proc., p. 29-37.

(2055) **White, M. P.**, 1929, *Some index Foraminifera of the Tampico embayment area of Mexico:* Jour. Paleontology, v. 3, no. 1, p. 30-58, pl. 4-5.———(2056) 1932, *Some Texas Fusulinidae:* Univ. Texas, Bull. 3211, 104 p., 10 pl., 3 text-fig.

(2057) **Whittaker, R. H.**, 1959, *On the broad classification of organisms:* Quart. Rev. Biol., v. 34, p. 210-226.

(2058) **Wicher, C. A.**, 1952, *Involutina, Trocholina und Vidalina—Fossilien des Riffbereichs:* Geol. Jahrb., v. 66, p. 257-284, 4 text-fig. (Hannover).

(2059) **Wick, W.**, 1939, *Versuch einer biostratigraphischen Gliederung des jüngeren Tertiars auf Grund von Foraminiferen:* Preuss. geol. Landesanst., Jahrb., v. 59(1938), p. 476-512, pl. 18-23.

(2060) **Wickenden, R. T. D.**, 1949, *Eoeponidella, a new genus from the Upper Cretaceous:* Royal Soc. Canada, Trans., ser. 3, v. 42, sec. 4(1948), p. 81-82, 1 text-fig.

(2061) **Wiesner, Hans**, 1920, *Zur Systematik der Miliolideen:* Zool. Anzeiger, v. 51, p. 13-20.———(2062) 1923, *Die Milioliden der östlichen Adria:* 113 p., 20 pl., The Author (Prag-Bubenč).———(2063) 1931, *Die Foraminiferen der deutschen Sudpolar Expedition 1901-1903:* Deutsche Sudpolar Exped. 1901-03, herausgegeben von Erich von Drygalski, v. 20, Zool. vol. 12, p. 53-165, pl. 1-24.

(2064) **Williamson, W. C.**, 1848, *On the Recent British species of the genus Lagena:* Ann. & Mag. Nat. History, ser. 2, v. 1, p. 1-20, pl. 1-2.———(2065) 1858, *On the Recent Foraminifera of Great Britain:* Ray Soc. Publs., xx+107 p., 7 pl.———(2066) 1881, *On the organisation of the fossil plants of the coalmeasures, Pt. X. Including an examination of the supposed radiolarians of the Carboniferous rocks:* London, Royal Soc., Philos. Trans., v. 171 (1880), p. 493-539, pl. 14-21.

(2067) **Wingate, H.**, 1889, *Orcadella operculata Wing., a new Myxomycete:* Acad. Nat. Sci. Philadelphia, Proc., p. 280-281.

(2068) **Winter, F. W.**, 1907, *Zur Kenntniss der Thalamophoren I. Untersuchung über Peneroplis pertusus (Forskål):* Archiv Protistenkunde, v. 10, no. 1, p. 1-113, pl. 1,2, 10 text-fig.

(2069) **Witt Puyt, J. F. C. de**, 1941, *Geologische und paläeontologische Beschreibung der Umgebung von Ljubuški, Hercegovina:* Dissertation, Univ. Utrecht, p. 1-99, pl. 1-5.

(2070) **Wolańska, Henryka**, 1959, *Agathammina pusilla (Geinitz) z dolnego Cechsztynu Sudetów i gór świetokrzyskich:* Acta Palaeont. Polonica, v. 4, no. 1, p. 27-59, pl. 1-3, text-fig. 1-4.

(2071) **Wood, Alan**, 1946, *The type specimen of the genus Ophthalmidium:* Geol. Soc. London, Quart. Jour., v. 102, pt. 4, p. 461-463, pl. 29, 30.———(2072) 1948, *"Sphaerocodium," a misinterpreted fossil from the Wenlock limestone:* Geol. Assoc., Proc., v. 59, pt. 1, p. 9-22, pl. 2-5.———(2073) 1949, *The structure of the wall of the test in the Foraminifera; its value in classification:* Geol. Soc. London, Quart. Jour., v. 104, p. 229-255, pl. 13-15.

(2074) ———, **& Barnard, Tom**, 1946, *Ophthalmidium: a study of nomenclature, variation, and evolution in the Foraminifera:* Geol. Soc. London, Quart. Jour., v. 102, p. 77-113, pl. 4-10.

(2075) ——, & Haque, A. F. M. M., 1956, *The genus Cycloloculina (Foraminifera) with a description of a new species from Pakistan:* Geol. Survey Pakistan, Records, v. 7, pt. 2, p. 41-44, text-fig. A-B.

(2076) ——, & Haynes, John, 1957, *Certain smaller British Paleocene Foraminifera, Part II. Cibicides and its allies:* Cushman Found. Foram. Research, Contrib., v. 8, pt. 2, p. 45-53, pl. 5-6.

(2077) Wood, S. V., 1842, *A catalogue of shells from the Crag:* Ann. & Mag. Nat. History, ser. 1, v. 9, p. 455-462, pl.

(2078) Woodring, W. P., 1924, *Some new Eocene Foraminifera of the genus Dictyoconus* in WOODRING, W. P., BROWN, J. S., & BURBANK, W. S., Geology of the Republic of Haiti: Appendix I, p. 608-610, pl. 9, 13, Republic Haiti Geol. Survey (Port-au-Prince).

(2079) Wright, Joseph, 1875, *A list of the Cretaceous microzoa of the north of Ireland:* Belfast Nat. Field Club, Proc., new ser., v. 1 (1873-80), Appendix 3, p. 73-99, pl. 2-3. ——(2080) 1889, *Report of a deep-sea trawling cruise off the south-west coast of Ireland, under the direction of Rev. W. Spotswood Green; Foraminifera:* Ann. & Mag. Nat. History, ser. 6, v. 4, p. 447-449.

(2081) Wright, T. S., 1861, *Observations on British Protozoa and Zoophytes:* Ann. & Mag. Nat. History, ser. 3, v. 8, p. 120-135, pl. 3-5. ——(2082) 1867, *Observations on British Zoophytes and Protozoa:* Jour. Anat. & Physiol., v. 1, p. 332-338, pl. 14-15.

(2083) Yabe, Hisakatsu, 1903, *On a Fusulina-Limestone with Helicoprion in Japan:* Geol. Soc. Tokyo, Jour., v. 10, no. 113, p. 1-13, pl. 2-3.——(2084) 1918, *Notes on Operculina-rocks from Japan, with remarks on "Nummulites" cumingi Carpenter:* Tohoku Imper. Univ., Sci. Repts., ser. 2(Geol.), v. 4, no. 3, p. 104-126, pl. 17.——(2085) 1919, *Notes on a Lepidocyclina-limestone from Cebu:* Same, ser. 2(Geol.), v. 5, p. 37-51, pl. 6-7 (Sendai). ——(2086) 1946, *On some fossils from the Saling limestone of the Goemai Mountains, Palembang, Sumatra, II:* Imper. Acad. Japan, Proc., v. 22, no. 8, p. 259-264, 3 text-fig.

(2087) ——, & Asano, Kiyosi, 1937, *Contribution to the palaeontology of the Tertiary formations of west Java, Pt. I. Minute Foraminifera from the Neogene of west Java:* Tohoku Imper. Univ., Sci. Repts., ser. 2(Geol.), v. 19, p. 87-126, pl. 17-19, text-fig. 1-15.

(2088) ——, & Hanzawa, Shôshirô, 1922, *Uhligina, a new type of Foraminifera found in the Eocene of Japan and west Galicia:* Japan. Jour. Geol. & Geog., Trans. & Abstr., v. 1, no. 2, p. 71-76, pl. 12, text-fig. 1-4.——(2089) 1923, *Foraminifera from the Natsukawa-limestone, with a note on a new genus of Polystomella:* Japan. Jour. Geol. & Geog., Trans. & Abstr. v. 2, no. 4, p. 95-100.——(2090) 1925, *Nummulitic rocks of the islands of Amakusa (Kyushu, Japan):* Tohoku Imper. Univ., Sci. Repts., ser. 2 (Geol.), v. 7, p. 73-82, pl. 18-22.——(2091) 1926, *Choffatella Schlumberger and Pseudocyclammina, a new genus of arenaceous Foraminifera:* Same, ser. 2 (Geol.), v. 9, no. 1, p. 9-11, pl. 2, text-fig. 1.—— (2092) 1928, *Tertiary foraminiferous rocks of Taiwan (Formosa):* Imper. Acad. Japan, Proc., v. 4, p. 533-536, text-fig. 1-3.—— (2093) 1930, *Tertiary foraminiferous rocks of Taiwan (Formosa):* Tohoku Imper. Univ., Sci. Repts., ser. 2 (Geol.), v. 14, p. 1-46, pl. 1-16. ——(2094) 1932, *Tentative classification of the Foraminifera of the Fusulinidae:* Imper. Acad. Tokyo, Proc., v. 8, p. 40-43.

(2095) Yakovlev, V., 1891, *Opisanie neskol'kikh vidov melovykh foraminifer:* Khar'kovsk. Obshch. Ispyt., Prirody, Trudy, v. 24 (1890), p. 341-364, pl. 1-3. [*Description of some species of Cretaceous Foraminifera.*]

(2096) Yokoyamo, Matajiro, 1890, *Foraminiferen aus dem Kalksteine von Torinosu und Kompira* in NAUMANN, E., & NEUMAYR, M., Zur Geologie und Paläontologie von Japan: K. Akad. Wiss. Wien, math.-naturwiss. Cl., Denkschrift., v. 57, p. 26-27, pl. 5.

(2097) Young, John, & Armstrong, James, 1871, *On the Carboniferous fossils of the west of Scotland:* Geol. Soc. Glasgow, Trans., v. 3, suppl., p. 1-103, 1 table.

(2098) Zakharova-Atabekyan, L. V., 1961, *K revizii sistematiki Globotrunkanid i predlozhenie novogo roda Planogyrina, gen. nov.:* Akad. Nauk Armyanskoy SSR, Doklady, v. 32, no. 1, p. 49-53. [*On a revision of the systematics of the globotruncanids and proposal of the new genus, Planogyrina, gen. nov.*]

(2099) Zalessky, M. D., 1926, *Premières observations microscopiques sur le schiste bitumineux du Volgien inférieur:* Soc. Géol. Nord, Ann., v. 51, p. 65-104, pl. 2-6, 2 text-fig.

(2100) Zarnik, B., 1908, *Über eine neue Ordnung der Protozoen:* Phys.-med. Gesell. Würzburg, Sitzungsber., v. 1907, p. 72-78, 1 text-fig.

(2101) Zborzewski, Adalbert, 1834, *Observations microscopiques sur quelques fossiles rares de Podolie et de Volhynie:* Soc. Imper. Nat. Moscou, Nouv. Mém., v. 3, p. 299-312, pl. 28.

(2102) Zeller, Doris E. Nodine, 1953, *Endothyroid Foraminifera and ancestral fusulinids from the type Chesteran (Upper Mississippian):* Jour. Paleontology, v. 27, p. 183-199, pl. 27-28, 9 text-fig., 1 chart.——(2102A) 1963, *Endothyra bowmani Brown, 1843, designation of neotype:* Jour. Paleontology, v. 37, p. 502-503, text-fig. 1.

(2103) Zeller, E. J., 1950, *Stratigraphic significance of Mississippian endothyroid Foraminifera:* Univ. Kansas Paleont. Contrib., Protozoa, art. 4, p. 1-23, pl. 1-6.——
(2104) 1957, *Mississippian endothyroid Foraminifera from the Cordilleran geosyncline:* Jour. Paleontology, v. 31, p. 679-704, pl. 75-82, text-fig. 1-11.

(2105) Zittel, K. A. von, 1880, *Handbuch der Palaeontologie, Band I. Palaeozoologie:* v. 1, pt. 1, 765 p., 558 fig. (München & Leipzig).——(2106) 1913, *Text-book of Paleontology:* transl. & ed. by Eastman, C. R., ed. 2, v. 1, 839 p., 1594 text-fig., Macmillan & Co., Ltd. (London).

(2107) Zopf, W. F., 1885, *Die Pilzthiere oder Schleimpilze* in Schenk, Handbuch der Botanik; v. 3, pt. 2, 174 p., 51 fig.——(2108) 1892, *Zur Kenntniss der Labyrinthuleen, einer familie der Mycetozoen:* Beiträge Physiol. & Morphology niederer Organismen, no. 2, p. 36-48, pl. 4-5 (Leipzig).

(2109) Zulueta, A. de, 1917, *Promitosis y sindiéresis, dos modos de división nuclear coexistentes en Amebas del grupo "limax":* Museo Nacional Cienc. Nat. Madrid, Trab., Ser. Zool., no. 33, p. 1-55 (publ. Dec. 31, 1917).

SOURCES OF ILLUSTRATIONS
[Additional to those given in "References"]

(2110) Barker, R. W., new
(2111) Brown, N. K., Jr., & Brönnimann, Paul, 1947
(2112) Bykova, E. V., new
(2113) Cole, W. S.; a, 1942; b, 1949; c, new
(2114) Douglass, R. C., new
(2115) Henson, F. R. S., new
(2116) Ladd, H. S., & Hoffmeister, J. E., 1945
(2117) Loeblich, A. R., Jr., & Tappan, Helen, new
(2118) Lys, M., new
(2119) Neumann, Madeleine, 1958
(2119A) Reichel, Manfred, new
(2120) Schlumberger, Charles, 1904
(2120A) Tan Sin Hok, 1939
(2120B) Thompson, M. L., new
(2121) Van der Vlerk, I. M., & Umbgrove, J. H. F., 1927
(2122) Vaughan, T. W., 1936
(2123) ——, & Cole, W. S., 1943

INDEX

Italicized names in the following index are considered to be invalid; those printed in roman type, including morphological terms, are accepted as valid. Suprafamilial names are distinguished by the use of full capitals and authors' names are set in small capitals with an initial large capital. Page references having chief importance are in boldface type (as **C442**).

A_1 generation, **C58**, **C106**
A_2 generation, **C58**, **C106**
Abathomphalus, C138, **C663**
aboral, **C58**
Aboudaragina, **C574**
Abrardia, C309
ACALCARINÉES, C13
ACANTHARIA, C18, C39
acanthi, **C58**
Acanthospira, C782
acanthus, **C58**
Acarinina, C668
accessory apertures, **C58**
Accordiella, **C292**, **C795**
Acervoschwagerina, **C417**
Acervulina, C58, **C694**, C698
acervuline, **C58**
Acervulinida, C143, C694
Acervulinidae, C139, C145, C147, C156, **C694**
Acervulininae, C147, C694
Acipyxis, C34
ACONCHULINA, C39
ACONCHULINIDA, **C39**
Acostina, **C562**
Acrasacées, C11
ACRASIAE, C11
ACRASIALES, C11
ACRASIDA, C11
Acrasidae, C11
ACRASIDEA, C11
ACRASIEAE, C11
ACRASIÉES, C11
ACRASIEI, C11
ACRASIEOS, C11
ACRASIÈS, C11
Acrasiidos, C11
ACRASINA, C11
ACRASINEA, C11
Acruliammina, C137, **C247**
Actinocyclina, C714
Actinosiphon, C710, **C711**, C718
ACYSTOSPORÉS, C54, C164, C794
ACYSTOSPORIDIA, C54, C164, C794
ACYTTARIA, C5
ADDIFFLUENTIA, C5
Adelosina, C458
Adercotryma, **C225**
Adhaerentia, **C249**
Adhaerentina, C213
Adherentina, **C783**
Adjungentiidae, C34, C40
adventitious, **C58**
Aeolides, **C783**
Aeolostreptis, C734, C735
Aequilateralidae, C511, C559

Aethalini, C13
Aethalium, **C10**
AFFILOSIA, C39
Afghanella, C379, C380, C386, C387, **C429**
Afrobolivina, C549
Agathammina, C93, C136, **C438**
AGATHISTÈGUES, C142, C436
Agathistègues, C142, C458
agglutinated, **C58**
agglutinated, tests in foraminifers, **C89**
AGRICOLA, C55
Aguayoina, C786
Akiyoshiella, **C406**
Aktinocyclina, **C714**
Aktinorbitoides, C725
Alabamina, C60, C61, C63, C96, C138, **C748**, C750
Alabaminidae, C138, C149, C151, C161, **C748**, C750
Alanwoodia, **C600**
alar projection, **C58**
Alexandrella, C179
Alexandrellidae, C173
Alfredosilvestris, **C512**
Aljutovella, C410
Allelogromia, C177
Allelogromiini, C173
Alliatina, C96, C139, **C778**
Alliatinella, C139, **C778**, C781
Allodictya, C53
Allogromia, C59, C68, C71, C74, **C173**, C175
Allogromida, C173
Allogromidae, C173
ALLOGROMIDIACEAE, C164
ALLOGROMIIDA, C151, C164
Allogromiida, C173
Allogromiidae, C71, C119, C134, C135, C146, C147, C154, **C173**, C181
Allogromiidea, C150, C164
ALLOGROMIIDIA, C164
ALLOGROMIINA, C16, C154, **C164**
Allogromiinae, C146, C147, C173
Allogromioidea, C147, C164
Allomorphina, C137, **C743**, C745
Allomorphinella C138, **C743**
Allomorphinellinae, C146, C147, C153, C742
Allomorphininae, C146, C153, C742
Allotheca, C572

Almaena, C139, **C763**, C764
Almaeninae, C139, C152, C161, **C763**
Alocodera, C32
Altistoma, **C548**
Aluvigerina, C565
alveolar, **C58**
alveoli, C58, **C365**, **C366**
Alveolina, C506, C508, C509
Alveolinea, C503
Alveolinella, C58, C92, C503, C505, **C506**
Alveolinellidae, C146, C147, C503, C685
Alveolinellinae, C147, C503
Alveolinida, C143, C503
Alveolinidae, C58, C60, C62, C137, C145, C148, C149, C150, C152, C156, **C503**
Alveolinidea, C152, C436
Alveolinina, C503
Alveolininae, C144, C145, C503
Alveolophragmium, **C228**
alveolus, **C58**
Alveovalvulina, **C298**
Alveovalvulinella, **C298**
Amastigogenina, C6
Amaurochaetacea, C13
AMAUROCHAETACEAE, C13
Amaurochaetaceae, C13
Amaurochaetaceen, C13
Amaurochaetacées, C13
AMAUROCHAETALES, C13
AMAUROCHAETE, C13
AMAUROCHAETEAE, C13
Amaurochaeteae, C13
Amaurochaetidae, C13
Amaurochaetides, C13
AMAUROCHAETINAE, C13
Amaurochaetinae, **C13**
AMAUROCHAETINEA, C13
AMAUROCHAETINEAE, C13
Amaurochaetoinae, C13
AMAUROCHETENE, C13
Amaurosporae, C13
Amaurosporales, C13
Amaurosporeae, C12
Amaurosporées, C13
Amaurosporei, C13
AMEBEA, C5
Amébidos, C6
AMEBOIDEOS, C5
Amibiens, C6
AMIBOS, C5
Ammoasconidae, C145, C184, C194
Ammoastuta, **C238**
Ammobaculites, C89, C136, **C239**

Ammobaculoides, C137, **C241**
Ammochilostoma, C263
Ammocibicides, C247
Ammocycloloculina, **C302**
Ammodinetta, C184
Ammodiscacea, C74, C154, **C184**
Ammodiscella, **C213**
AMMODISCIDA, C151, C184,
Ammodiscida, C210
Ammodiscidae, C135, C144, C146, C147, C148, C149, C150, C151, C155, **C210**
Ammodiscidea, C151, C184
Ammodiscinae, C147, C148, C150, C151, C155, **C210, C211**
Ammodiscinea, C143, C210
Ammodiscoidea, C147, C184
Ammodiscoides, **C210**
Ammodisculinidae, C210
Ammodiscus, C129, C135, **C210, C214**
Ammoflintina, **C220**
Ammofrondicularia, C216
Ammoglobigerina, C259
Ammolagena, C135, **C214**
Ammomarginulina, **C241**
Ammomassilina, C93, **C470**
Ammonacea, C142
Ammonacés, C142
Ammoneata, C142
Ammonema, C214
Ammonia, C75, C77, C78, C139, **C607**
Ammonoïdes, C142
Ammopemphix, **C202**
Ammoscalaria, **C241**
Ammosphaeroides, C786
Ammosphaeroidina, **C259**
Ammosphaeroidininae, C146, C148, C259
Ammosphaerulina, **C227**
Ammospirata, **C251**
Ammotium, C137, **C241**
Ammovertella, **C214**
Ammovertellina, **C210**
Amoeba, C6, C58, C129
AMOEBAE, C5
AMOEBAEA, C5, C54, C164
Amoebaea, C6
Amoebaea lobosa, C6
Amoebaea reticulosa, C54, C164
Amoebea, C6
Amoebeae, C6
Amoebées, C6
AMOEBIAE, C5
AMOEBIDA, C2, **C5**, C67
Amoebida, C6
AMOEBIDA (GYMNAMOEBIDA), C5
Amoebidae, C6
AMOEBINA, C5
Amoebina, C6
AMOEBINEN, C5
AMÖBOEA, C5
Amoebogromia, **C41**
amoeboid, **C58**
AMOEBOIDEA, C5

AMOEBOIDINA, C5
amoebula, **C10**
Amorphina, C782
Amphicervicis, C135, **C202**
Amphicoryna, **C513**
Amphicoryne, C514
Amphifenestrella, **C195**
Amphigramma, C782
Amphilepidina, C721
Amphimorphina, C139, **C525**, C785
Amphimorphinella, C525
Amphisorus, C60, C139, **C496**
Amphistegina, C95, C98, **C685**, C718
Amphisteginidae, C146, C148, C149, C150, C151, C156, **C685**
Amphistegininae, C147, C685
AMPHISTOMATA, C40, C164
Amphistomata, C40
Amphistomidae, C164
Amphistomina, C47, C164
Amphistomina, C47, C164
Amphistomini, C47, C164
Amphistominidae, C47
Amphitrema, C19, **C47**, C67
Amphitrematidae, **C47**
Amphitrematides, C40
Amphitremidae, C47
Amphitreminae, C47, C146
Amphitremoida, C200
Amphitremoidea, C200
Amphizonella, C17, **C20**
Amphorina, C518
Amphycorina, C514
Ampullaria, C53
Ampullataria, **C49**
Anchispirocyclina, **C233**
ANDREAE, C607
Andromedes, C631
ANEMEAE, C14
ANEMINEA, C14
ANEMINEAE, C14
Anfistómidos, C164
Angulodiscorbis, C78, **C589**
Angulodiscus, C740
Angulogavelinella, C138, **C755**
Angulogerina, C571
Angulogerininae, C147, C565
anisogamy, **C17**
annular, **C58**
annuli, **C58**
Annulina, **C783**
Annulocibicides, C687, **C688**, C691
Annulofrondicularia, C518
Annulopatellina, **C730, C731**
Annulopatellinidae, C156, **C730**
annulus, **C58**
Anomalina, C71, C753, **C754**, C755, C757, C759
Anomalinella, **C764**
Anomalinidae, C100, C137-139, C146, C150-152, C161, C686, **C753**, C755, C757, C760, C763
Anomalininae, C146, C148,

C150, C152, C161, **C753**, C760, C763
Anomalinoides, C138, C753, **C755**
Antarcella, C22
Antenor, C520
anterior, **C58**
antetheca, **C58, C362**, C371, **C385**
Antillesina, C746
ANTONOVA, C462
ANTROPOV, C788
Anulopatellina, C730
Aoujgalia, C136, **C330-C332**
Aoujgalia, C330
apertural form, position, and modifications, **C108**
aperture, **C17, C58**
Aphrosina, C694
Apiopterina, **C783**
Apodera, **C29**
apogamic reproduction in Foraminiferida, **C85**
Apogromia, **C166**
Applinella, C666
Apterrinella, C443
Aragonella, C666
Aragonia, **C736**
aragonite, **C95**
Arammlagenum, C214
Arammodiscinia, C210
Arammodisclidia, C210
Arammodiscodum, C210
Arammodiscum, C210
Arammosphaerium, C786
Araschemonellinia, C214
Araschemonellum, C214
Arastrorhiznia, C184
Arastrorhizum, C184
Arbathysiphum, C186
Arbdelloidinum, C250
Arboderium, C166
arborescent, **C58**
Arbotellum, C191
Arbrachysiphum, C196
Arbulinarium, C501
Arcélidos, C22
Arcella, C18, **C22**, C23
Arcella, C53
Arcellacea, **C19**
Arcellida, C22
Arcellida, C17, **C22**
ARCELLINA, C19
Arcellina, C19, C22, C40
Arcellinae, C22
Arcellinea, C22
Arcellineae, C22
Arcellinés, C22
ARCELLINIDA, C2, C16, **C19**, C67
Archaecyclus, **C699**
Archaediscidae, C93, C136, C150, C153, C155, **C354**
Archaediscinae, C148, C354
Archaediscoum, C354
Archaediscus, C136, **C354**
Archaelagena, C786
Archaeochitinia, C135, **C175**, C785

Index

Archaeochitosa, C135, **C175**
Archaesphaera, C135, **C314**
Archaesphaeridae, C313
Archaia, C495
Archaiadinae, C494
Archaias, C92, C139, **C494**
Archaiasinae, C147, C156, **C494**
Archais, C495
Archaisinae, C146
Archapmanoum, C621
Archealagena, C786
ARCHER, C166
Archerella, C19, **C47**
Archiacina, C482
ARCHIMONOTHALAMIA, C184
ARCHI-MONOTHALAMIDA, C164
Archi-Monothalamidia, C164, C184, C320, C436, C598
ARCHIMYCETES, C11
Archithalamia, C184
Arcornuspira, C438
Arcornuspirinia, C438
Arcornuspirum, C438
Arcrithionum, C205
Arcyriacea, C15
ARCYRIACEAE, C15
Arcyriaceae, C15
Arcyriaceen, C15
Arcyriacées, C15
ARCYRIAE, C15
Arcyriae, C15
Arcyrieae, C15
Arcyriei, C15
Arcyriidae, C15
Arcyriinae, **C15**
Ardactylosaccum, C177
Ardendrophyrum, C192
Ardendrotubum, C177
Ardiplogromium, C177
areal aperture, **C58**
areal bulla, **C58**
Arechinogromium, C166
ARENACEA, C184
Arenacea, C184, C194, C214, C259
arenaceous, **C58**
ARENACIDAE, C145, C184
Arenácidos, C194
Arenaglobula, C281
Arenagula, C281
Arenistella, C184
Arenobulimina, **C273**
Arenodosaria, C283
Arenonina, C137, **C247**
Arenonionella, **C261**
Arenoparrella, **C262**
Arenosiphon, C186
Arenosphaera, C196
Arenoturrispirillina, **C210**
Arenovidalina, C740
Arenovirgulina, C255
areolate, **C58**
Arethusa, **C783**
ARFORAMINIFERIA, C164
argillaceous, **C58**
Argillotuba, C186

Argillotubinae, C186
Argirvanellum, C786
Arglomospirum, C212
Argynnia, C32
Arhaddonium, C248
Arhaliphysemum, C192
Arhaplostichoum, C220
Arhemidiscum, C212
Arhippocrepnia, C187
Arhippocrepum, C188
Arhomosum, C215
Arhospitellum, C177
Arhyperammum, C190
Arinvolutoum, C740
Aristeropora, **C783**
Aristerospira, C572
Arjaculum, C190
Arkalamopsum, C215
Arlagenammum, C200
Arlagunculum, C44
Arlieberkuehnium, C179
Arlituotubum, C214
Armarsipellum, C186
Armarsupium, C179
Armasonellum, C208
Armeniella, **C436**
Armenina, C427
Armiliolidia, C458
Armillettum, C23
Armorella, C184
Armyxothecnia, C164
Armyxothecum, C169
Arnaudiella, C138, **C614**
Arnodellum, C179
Arnodosammidia, C214
Arnodosaridia, C512
Arnodosaroum, C323
Arnodosinum, C323
ARNOLD, C66, C111
Arnoldia, C786
ARNUDIA, C54
Arophiotubum, C169
Arorbitolidia, C308
Arpatellinia, C602
Arpatellum, C603
Arpelosum, C200
Arperneroum, C535
Arpilulum, C201
Arplacopsinia, C247
Arplacopsum, C535
Arplagiophrum, C171
Arpolyphragmina, C248
Arpolyphragmoum, C248
Arproblematoia, C739
Arproblematoum, C740
Arproteonum, C216
Arpsammonyxum, C212
Arpsammophoum, C214
Arpsammosiphoum, C194
Arpsammosphaerum, C195
Arpseudarcelloum, C522
Arreophaxnia, C215
Arreophaxum, C216
ARRETICULARIA, C164
Arrhabdammidia, C184
Arrhabdammum, C185
Arrhabdamnia, C184
Arrhaphoscenum, C786
Arrhizammum, C186

Arrhizamnia, C186
Arrhynchogromium, C181
Arrhynchosaccum, C181
Arrogromium, C173
Arrogromnia, C173
Arrotalaridia, C605
Arsaccammum, C196
Arsaccamnia, C196
Arsagenum, C205
Arschultzellum, C173
Arshepheardellum, C182
Arsilicoum, C216
Arsorophaerum, C196
Arspirillinia, C600
Arspirillinum, C600
Arsquamulum, C444
Arstachecoum, C330
Arstorthosphaerum, C196
Arsyringammum, C192
Artechnitum, C200
Artetraxoum, C337
Artextulidia, C251
Arthalamophagum, C183
Artholosum, C205
Arthrocena, **C783**
Arthyrammum, C202
Articulina, C139, **C478**
Artrochammidia, C259
Artubinia, C477
Artubinum, C477
Arturritellum, C212
Arurnulum, C46
Arvanhoeffenum, C186
Arverrucum, C210
Arvidaloum, C440
Arwebbina, C535
Arwebbinum, C535
Arwebbum, C448
ARXENOPHYRIA, C789
ASANO, C781
Asanoina, **C607**
Asanospira, C225
Aschemonella, **C214**
Aschemonellidae, C136, C137, C145, **C214**
Aschemonellinae, C146, C148, C155, **C214**
ASCOFORAMINIFERA, C145
Ascoforaminifera, C184
ASIPHOIDEA, C164
Askopsis, C782
ASPHYCTA, C39
Aspidodexia, **C783**
Aspidospira, **C783**
Assilina, C645
Assilininae, C645
Assulina, C50
Assulinidae, C47
Astacolus, C136, **C514**
Asteriacites, C715, C796
Asteriatites, C796
Asterigerina, C96, **C592**, C777
Asterigerinacea, C572
Asterigerinata, C96, **C592**
Asterigerinella, **C592**
Asterigerinida, C592
Asterigerinidae, C142, C147, C151, C152, C156, C572, **C592**

Asterigerinoides, **C592**
Asteroarchaediscinae, C354
Asteroarchaediscus, C354
Asterocyclina, **C714, C715**
Asterocyclinidae, C712
Asterodiscina, C482
Asterodiscocyclina, C715
Asterodiscus, C693, C714, C715
Asterophragmina, **C715**
Asterorbis, **C711**
Asterorbitoides, C782
Asterorotalia, C139, **C608**
Astrammina, **C184**
Astrodiscus, C184
Astrolepidina, C721
Astrononion, **C746**
Astrorhiza, C86, C87, C89, C90, C135, **C184**, C785
Astrorhiza, C184
ASTRORHIZIDA, C151, C184
Astrorhizida, C184
ASTRORHIZIDACEAE, C184
Astrorhizidae, C135, C144-151, C154, **C184**
ASTRORHIZIDAE, C145
Astrorhizidae, C184
ASTRORHIZIDEA, C184
Astrorhizidea, C148-151, C184
ASTRORHIZINA, C145
Astrorhizina, C184
Astrorhizinae, C144-C146, C148, C154, **C184**
Astrorhizinae, C184
Astrorotalia, C668
Asymmetrina, C137, **C755**
Ataxogyroidina, C283
Ataxophragmida, C151
Ataxophragmidea, C144, C268, C283
ATAXOPHRAGMIIDA, C152, C184
Ataxophragmiidae, C136, C137, C147, C152, C155, **C268**
Ataxophragmiinae, C147, C149, C150, C152, C155, **C283**
Ataxophragmiinae, C148
Ataxophragmium, **C283**
Ataxophragmoides, C273
Atetsuella, C406
ATHALAMIA, C5, C19, C54, C164
Athalamia, C794
ATHALAMIDA, **C54**
Athecocyclina, **C715**
Atractolina, C537
Atrichae, C12, C14
Atriche, C14
Atrichées, C12, C14
attic, **C58**
Auerinella, **C215**
Aulostomella, C531
Aulotortus, C136, **C740**, C741
Auriculina, **C783**
Austrocolomia, **C514**
Austrotrillina, C139, **C474**
autogamous fertilization of amoeboid gametes in Foraminiferida, **C85**
autogamy, C58, **C85**

Averincevia, C29
AVERINTSEV, C88
Averintzia, C29
Averinzia, C29
Avganella, C429
AVNIMELECH, C89
Awerintzevia, C29
Awerintzewia, **C29**
Awerintzia, C29
axial fillings, C58, **C364, C386**
axial section, **C58, C360**
axial septulum (pl., septula), **C58, C379**
axis, **C58**
axopodia, C18
axostyle, **C58, C68, C78**
Ayalaina, C488
Azera, C747
Azoosporeae, C12, C794, C795
Azoosporés, C794
Azoosporida, C794, C795
Azoosporidae, C795

B-form, **C58**
B generation, **C106**
Bactrammina, C190
Baculogypsina, C139, **C629**
Baculogypsinidae, C150, C151, C628
Baculogypsinoides, **C629**
Baggatella, C139, **C543**
Baggatellinae, C152, C543
Baggina, C137, **C586**
Bagginae, C146, C147, C149, C150, C152, C156, **C586**
Baileya, C44
Baissunella, **C211**
Baissunellinae, C210, C211
Baituganella, C316, C796
Balanulina, C786
BANDY, C95, C96, C116, C118, C134, C253, C571, C576
BANDY & BURNSIDE, C570
Bandyella, **C730**
BANNER & BLOW, C665
Barbourina, C268
Barbourinella, **C268**
Bargoniella, **C42**
BARKER, C154
BARKER & GRIMSDALE, C718
Barkerina, **C481**
Barkerinidae, C155, **C480**
Barkerininae, C480
BARNARD, C283
Bartramella, **C406**
BARY, DE, C8
basal layer, **C58**
BASISTOMA, C511
Basistoma, C145, C320
BASSET, C574
Bathysiphon, C66, C86, C89, C90, C135, **C186**, C190
Bathysiphoninae, C186
BATSCH, C55
Bdelloidina, **C250**
BECCARIUS, C55
Beedeina, C405
Beella, C139, **C669**, C671
Begia, C481

Beisselina, C284, C291
Belaria, **C166**
Belariini, C164
Belemnites, C142
BELFORD, C595
Belorussiella, **C269**
BERMÚDEZ, C290, C580, C582, C584, C589, C600, C640, C686, C708, C742, C757, C761
Bermudezella, C707, C708
Bermudezina, **C269**
Bermudezita, C282
BERTHELIN, C56
Berthelinella, **C528**
Biapertorbis, C572
Biarritzina, C139, **C628**, C708
biconvex, **C58**
Bifarina, C138, **C654**
Bifarinella, C563
bifid, **C58**
BIFORAMINATA, C149, C511
biforaminate, **C58**
biformed, **C58**
Bifurcammina, C210
Bigenerina, C95, C96, **C254**
Bigeneropolis, C782
Biglobigerinella, C138, **C656**, C658, C665
BIGNOT & NEUMANN, C94
bilamellar walls, **C59, C99**
BILAMELLIDEA, C652, C678, C725
Bilamellidea, C100, C150, C151
Biloculina, C465
biloculine, **C59**
Biloculinella, **C467**
BILOCULINIDEA, C149, C184, C436, C511
Biloculinidea, C150, C151
Bimonilina, C255
biofacies maps, **C132**
Biomyxa, **C54**
Biomyxidae, **C54**
Biorbulina, C675
Biplanispira, C644, **C647**
Birbalina, C309
Bireophax, C217
Birrimarnoldia, C786
Bisaccium, **C746**
biserial, **C59**
Biseriammina, C136, **C338**
Biseriamminidae, C136, C152, C155, **C338**
Bisphaera, C135, **C314**, C796
Bistadiidae, C7
Biticinella, C656
Bitubulogenerina, C569
biumbilicate, **C59**
biumbonate, **C59**
Biwaella, **C418**
BLACKWELDER, C140, C141
BLAINVILLE, DE, C55, C142
Blastammina, C135, **C195**
blepharoplast, **C17, C59, C68**
BLOW & BANNER, C678
Boderia, **C166**
BOGDANOVICH, C462
Bolbodium, C547

Index

Boldia, C138, **C755**, C757
Bolivina, C137, **C549**, C553, C733, C736, C784
Bolivinella, C139, **C526**
Bolivinidae, C149, C151, C548
Bolivininae, C148, C150, C153, C548, C736
Bolivinita, C139, **C548**
Bolivinitella, C736
Bolivinitidae, C108, C111, C136, C137, C139, C153, C156, **C548**
Bolivinitinae, C146, C147, C148, C153, C548
Bolivinoides, C137, **C549**, C736
Bolivinopsis, **C251**
BOLLI, C115
BOLLI, LOEBLICH & TAPPAN, C659, C660
Bolliella, **C665**
BOLTOVSKOY, C130
Bonairea, C582
BONTE, C89
Bontourina, C714
Borelia, C506
Borelida, C503
Borelidae, C503
Borelidinae, C503
Borelis, C503, **C505**, C506
Boreloides, **C685**
Borodinia, **C694**
Borovina, C265
Bosc, C55, C508
boss, **C59**
Bostrychosaria, **C787**
Botellina, **C190**
Botellinidae, C184
Botellininae, C147, C155, **C190**
BOUCOT, C90
Boultonia, **C401**
Boultoniinae, C150, C152, C400
BOWEN, C269, C275
Brachysiphon, **C196**
BRADSHAW, C125, C127
BRADY, C56-C59, C67, C93, C97, C144, C146, C151, C215, C325, C466, C522, C569, C580, C582, C584, C600, C608, C675, C683, C784
Bradya, C786
Bradyella, C496
Bradyina, C93, C136, **C353**
Bradyinidae, C152, C342
Bradyininae, C150, C155, **C353**
Bramletteia, C222
BRAY, C96
Brefeldiaceae, C13
Brefeldiacées, C13
Brefeldiei, C13
Brefeldiidae, C13
Brevaxina, C383, **C427**
BREYN, C55
Brizalina, C136, **C552**, C734, C784
BRODERIP, C55
Broeckina, **C487**
Broeckinella, **C304**
Broekina, C487

Broekininae, C485
BRÖNNIMANN, C226, C712, C714
Bronnibrownia, C656
Bronnimannella, C656
Bronnimannia, **C574**
BROTZEN, C64, C94, C546, C574, C592, C750, C760, C761, C777
Brotzenella, C759
Brotzenia, C771
BRUGUIÈRE, C55
Bruguieria, C647
Brunsia, C136, **C355**
Brunsiella, C212
Brunsiina, C135, **C340**
buccal apparatus, **C59**
buccal aperture, **C59**
Buccella, C130, **C575**
Buccicrenata, C137, **C241**
Buccinina, C520
BUCHANAN & HEDLEY, C89, C90
Bucherina, C662
Budashevaella, **C262**
budding in foraminiferal reproduction, **C86**
Bueningia, **C589**
BÜTSCHLI, C95
Bulbobaculites, C244
Bulbophragmium, C137, **C241**
Bulimina, C138, C145, **C559**
Buliminacea, C71, C99, C137, C139, C153, C154, C156, **C543**, C734
Buliminae, C144, C145, C559
Buliminella, C95, C137, **C543**, C782
Buliminellidae, C149, C151, C543
Buliminellinae, C152, C543
Buliminellita, C139, **C544**
Buliminicae, C543
BULIMINIDA, C152, C511
Buliminida, C143, C559
Buliminidae, C64, C108, C111, C120, C138, C144-C152, C156, **C559**, C736, C782
Buliminidae, C559
Buliminidea, C144, C148, C149, C150, C151, C543, C559
Buliminidee, C559
Buliminina, C559
Bulimininae, C144-C150, C152, C156, **C559**, C736
Buliminoides, **C544**, C782
Buliminopsis, **C783**
Bulinella, C24
bulla(e), **C59**
Bullalveolina, C503, **C506**
Bullinula, C24
Bullinularia, C18, **C24**
Bullopora, **C535**
BURSCH, C238
Burseolina, **C738**
Bursullidae, C795
Bursullineae, C795
Bursullineen, C795
BYKOVA, E. V., C314, C315, C322

BYKOVA, N. K., VASILENKO, VOLOSHINOVA, MYATLYUK & SUBBOTINA, C652

Cadosina, C786, C787
Cadosinella, 786
Cadosinidae, C786
Calatharia, **C783**
CALCAREAE, C13
Calcariaceen, C13
Calcarina, C138, **C628**
CALCARINAE, C13
Calcarine, C628
CALCARINEA, C13
CALCARINEAE, C13
Calcarinées, C13
Calcarinidae, C134, C138, C139, C145, C148, C150, C156, **C628**
Calcarininae, C147, C628
Calcidiscus, C782
calcite, **C95**
"calcite eyes," **C59**
Calcitornella, C136, **C443**
Calcituba, C87, **C446**
Calcitubida, C445
Calcivertella, C136, **C443**
Calcivertellinae, C155, **C443**
Caligella, C135, **C316**
Caligellidae, C135, C151, C155, **C316**
CALKINS, C8
Calonemata, C14
CALONEMATINEAE, C14
CALONEMEAE, C14
CALONEMINEA, C14
CALONEMINEAE, C14
Caloneminées, C14
CALOTRICHEAE, C14
Camagueyia, **C284**
cameral aperture, **C59**
Camerina, C645
Camerinidae, C147-C149, C643, C714
Camerininae, C147, C148, C645
Cameroconus, **C784**
Campascus, **C54**
canal systems and stolons, **C109**
CANALICULATA, C511
canaliculate, **C59**
Canalifera, C637
Canaliferidae, C631
cancellate, **C59**
Cancellina, C379, C383, C386, C387, **C431**
Cancrininae, C150, C586
Cancris, C139, **C586**
Cancrisinae, C586
Candeina, C64, C139, **C675**, C734
Candeininae, C146, C147, C150, C152, C675
Candela, C571
Candorbulina, C675
Canepaia, **C228**
Canopus, C784
Cantharipes, C784
Cantharus, C784
Canthropes, C784

Canthropus, C784
Capidulina, C321
capillitium, **C10**
Capitellina, C518
Capsellina, **C42**
Capsulina, C786
Carbonella, C339
Carcris, C586
Caribeanella, C139, **C688**, C689, C690
carina, **C59**, **C436**
carinal band, **C59**
Carinina, C580
Carixia, **C443**
Carpenter, C57, C58, C90, C91, C647
Carpenter, Parker & Jones, C99, C143, C279, C281, C748
Carpenterella, C635, **C707**, C708
Carpenteria, C138, **C707**, C708
Carter, D. J., C131
Carter, C56, C647, C696, C786
Carteria, C319
Carterina, C95, **C764**
Carterinacea, C154, C161, **C764**
Carterinidae, C161, **C764**
Cash, C31
Caspirella, C782
Cassidella, C95, C138, **C732**, C733
Cassidulina, C71, C139, C666, **C737**, C738, C769
Cassidulina, C736
Cassidulinacea, C71, C99, C100, C154, C156, **C725**, C734, C736
Cassidulinae, C144, C736
Cassiduline, C145, C736
Cassidulineae, C736
Cassidulinella, **C738**
Cassidulinella, C768
Cassidulinida, C143, C736
Cassidulinidae, C95, C120, C139, C142, C145-C152, C161, C725, **C736**, C738
Cassidulinidea, C143, C736
Cassidulininae, C144, C146, C736
Cassidulinita, **C767**, C769, C770
Cassidulinoides, **C556**
Cassigerinella, C139, **C666**
Cassigerinellinae, C156, **C666**
Cassilamellina, C556
Cassilongina, C738
Catapsydracinae, C156, **C676**
Catapsydrax, C60, C139, **C676**
Catharia, C27, C29
Caucasina, C138. **C734**, C735
Caucasinella, C782
Caucasinidae, C95, C99, C108, C138, C156, **C731**
Caucasininae, C152, C161, C731, **C734**
Caudammina, C200
Causia, **C200**
Cavifera, **C787**
Cayeuxina, C786
Celibs, C784

Cellanthus, C139, **C635**, C643
CELLULACEA, C142, C511
CELLULACÉS, C142, C511
cellules, **C59**
Cellulia, C635
Cellulina, C786
Cenchridium, C540
Centenaria, C246
Centenarina, C246
central complex, **C59**
central section, **C59**
Centropyxidae, **C23**
Centropyxis, C18, **C23**
Ceophonus, C631
Cepekia, C340
CEPHALOPHORA, C142
CEPHALOPHORES, C142
CEPHALOPODA, C142
CEPHALOPODES FORAMINIFÈRES, C56
CEPHALOPODES SIPHONIFÈRES, C56
Cepinula, **C784**
Ceratammina, C135, **C196**
Cerataria, **C784**
Ceratestina, C175, C181
Ceratiaceae, C12
Ceratiacées, C12
Ceratiacei, C12
Ceratieae, C12
Ceratina, C453
Ceratiomyxacea, C9, **C12**
Ceratiomyxacea, C12
Ceratiomyxaceae, C12
Ceratiomyxacées, C12
Ceratiomyxales, C12
Ceratiomyxidae, **C12**
Ceratiomyxoidea, C12
Ceratobulimina, C95, C96, C138, **C766**, C769, C770, C777
Ceratobuliminidae, C96, C108, C136, C138, C148-C152, C162, **C766**, C771, C777
Ceratobuliminidea, C152, C766
Ceratobulimininae, C97, C146, C162, **C766**, C770, C777
Ceratobuliminoides, C766
Ceratocancris, C96, **C769**
Ceratolamarckina, C96, C138, **C769**
Ceratomíxidos, C12
Ceratospirulina, C478
Ceratubuliminidae, C766
Cercidina, C786
Cerelasma, **C792**, C794
Cerobertina, C96, C139, **C781**
Certesella, **C29**
Cerviciferina, C540
Chabakovia, **C787**
Chaetoproteida, C7
Chaetoproteidae, **C7**
Chaetotrochus, C782
Chaidae, C6
CHAIDEA, C5
chamber, **C59**
chamber formation in Foraminiferida, **C69**
chamber passages, **C59**

chamberlet, **C59**
chamberlets, **C371**
CHAOINEA, C5
Chaos, C6
Chaosidae, C6
Chaosina, C6
Chapman, C91, C537
Chapman & Parr, C147
Chapmania, C621
Chapmaniida, C605
Chapmaniidae, C147, C152, C605
Chapmanina, C139, **C621**
Chapmaninidae, C605, C621
Chapmanininae, C150, C156, **C621**
Charltonina, C138, **C752**, C753
Chatton, C7, C8
Chave, C92, C97, C98
Checchiaites, C508
Cheilosporites, C786
Cheilosporitidae, C786
CHEILOSTOMELLACEAE, C511
Cheirammina, C782
Cheiropsis, C783
Chelibs, **C784**
chemical composition of calcareous wall in foraminifers, **C97**
Chenella, **C432**
Chernyshinella, C136, **C352**
Chernyshinellina, C346
Chernyshinellinae, C152, C352
Chiloguembelina, C138, **C654**
Chilostomella, C138, **C742**, C743, C746
CHILOSTOMELLIDA, C511
Chilostomellida, C742
Chilostomellidae, C120, C144-C148, C150, C151, C153, C742, C748
CHILOSTOMELLIDAE, C145, C511
CHILOSTOMELLIDEA, C511
Chilostomellina, **C746**
Chilostomellinae, C146, C147, C153, C161, **C742**
Chilostomelloides, **C743**
chitin, **C59**
Chitinodendron, C135, **C175**
chitinoid wall, **C88**
Chitinolagena, C135, **C175**
Chitinosaccus, **C175**
Chitinosiphon, C179, C180
chitinous wall, **C88**
Chitinozoa, C175
Chlamydamoeba, **C19**
Chlamydomonas, C605
Chlamydophryidae, C40
Chlamydophryinae, C40
Chlamydophrys, **C42**
Choffatella, C137, **C228**
Choffatellinae, C228
choma (pl., chomata), **C59**, **C364**, **C386**
chromatin granules, **C59**
chromidia, C17, **C59**
Chrysalidina, **C279**

Chrysalidinella, C139, C562, **C563**
Chrysalidinoides, C563
Chrysalogonium, **C514**
Chrysolus, C514
Chuaria, C786
Chusenella, C415
CHYTRIDINEAE, C12
Cibicidae, C149, C685
Cibicidella, **C690**, C691
Cibicides, C59, C61, C78, C120 C128, C129, C138, C686, C687, **C688**, C690, C697, C757, C759, C760, C783
Cibicididae, C138, C139, C156, **C685**, C760
Cibicidina, **C686**, C757, C759
Cibicidinae, C146, C147, C152, C156, C685, **C687,** C688
Cibicidoides, **C757**
Cibicidoides, C757
Cibicorbis, **C678**
Cidaria, C512
Cidarollus, **C784**
Cienkowskiaceae, C13
Cimelidium, **C784**
Cincoriola, **C679**
Cingodifflugia, C35
Cingodifflugiinae, C34
Ciperozea, C569
Circus, C262
Ciry, C149
Cisalveolina, C505, **C506**
Cisseis, C715
Citharina, **C514**
Citharinella, **C516**
Clarke & Wheeler, C95, C97
classification of Foraminiferida, **C140, C141**; in *Treatise*, **C153**
Clathroptychiaceae, C14
Clathroptychiaceen, C14
Clathroptychiacei, C14
Clausulus, C505
clavate, **C59**
Clavelloides, **C565**
Clavigerinella, C139, **C665**
Clavihedbergella, C138, **C659**
Clavula, C783
Clavulina, C139, **C279**
Clavulinella, C291
Clavulinoides, C272
Clidostomum, C549
Climacammina, C136, **C333**
Climaccamina, C333
Clisiphontes, C520
Clisophontes, C520
Clypeocyclina, **C784**
Clypeolina, **C44**
Clypeorbinae, C150, C710
Clypeorbis, C711
Clyphogonium, C783
Cochlea, C514
Cochleatina, C786
Cochlidion, C514
Cochliopodidae, C19
Cochliopodiidae, **C19**
Cochliopodiinae, C19
Cochliopodium, C17, **C19**

Codonofusiella, C370, **C401**
Codonoschwagerina, C415
COELONEMEEN, C14
Coelotrochium, C786
coiling ratios, **C113**
Colania, C431
Colaniella, C136, **C328**
Colaniellidae, C135, C136, C155, **C328**
Colaniellinae, C152, C328
Cole, C128, C154, C643, C650, C710, C712, C717-C719, C725
Coleites, C138, **C757**
Collins, C582
COLLODERMACEAE, C13
Collodermaceae, C13
Collodermataceae, C13
Colloderminae, **C13**
Colomia, C95, C96, C138, **C781,** C782
Colonammina, C135, **C204**
Colpopleura, **C784**
columella, **C10**
Columella, C628
COLUMELLIFERAE, C13, C14
Compressigerina, **C565**
conchite, C92
CONCHULINA, C19
Conchulina, C23
Conicocornuspira, C438
Conicospirillina, C137, **C600**, C782
Conklin, C190
Conoglobigerina, C652
Conomiogypsinoides, C650
Conorbella, C78, C588
CONORBIDA, C149, C511
Conorbidae, C149, C151, C766
Conorbina, C137, **C575,** C769
Conorbinidae, C572
Conorbis, C769
Conorbitoides, C725
Conorboides, C137, **C769,** C777
Conorboididae, C766
Conorotalites, C138, **C752**
Conotrochammina, **C263**
contractile vacuole, **C17**
Conulina, C493
Conulites, C608
convolute, **C59**
Coprolithina, **C284**
Corbiella, C315
Corbis, C315
Corliss, C141
Cornish & Kendall, C91
Cornuloculina, **C448**
Cornuspira, C59, C92, C438
Cornuspiramia, **C447**
Cornuspirella, **C438**
Cornuspirenstamm, C145
Cornuspirida, C143, C438
Cornuspiridae, C145, C149, C152, C438
CORNUSPIRIDEA, C436
Cornuspiridea, C143, C144, C438
Cornuspirideae, C438

Cornuspirinae, C146-C148, C150, C438
cornuspirine, **C59**
Cornuspirininae, C146, C438
CORNUSPIROIDEA, C436
Cornuspiroides, **C438**
Corona, C36
Coronella, **C741**
Coronipora, C741
Corrosina, C543
Cortalus, **C784**
cortex, **C10**
Corticella, C34
Corycia, C19
Corycie, C19
Corycina, C19
Coryphostoma, C138, **C733**
Corythion, C19, **C53**
Coryzia, C19
Coscinoconus, C742, C786
Coscinophragma, C137, **C248**
Coscinophragmatinae, C155, **C248**
Coscinophragminae, C152, C248
Coscinosphaera, C675
Coscinospira, C484
Cosijn, C113
Cosinella, C783
Coskinolina, **C298**
Coskinolinella, **C304**
Coskinolinoides, **C310**
Coskinolinopsis, C308
costa, **C59**
costate, **C59**
Coxites, **C481**
Craterella, **C688**
Craterina, C173, C175
Craterininae, C173
Crateriola, C688
Craterites, **C493**
Craterularia, **C784**
Cremsia, C336
Crepidulina, C514
Crespinella, **C617**
Crespinina, **C624**
Cribáridos, C14
Criboelphydium, C635
Cribrariacea, C14
CRIBRARIACEAE, C14
Cribrariaceae, C14
Cribrariaceen, C14
Cribrariacées, C14
Cribrariacei, C14
CRIBRARIAE, C14
CRIBRARIALES, C14
Cribrarieae, C14
Cribrariidae, C14
Cribrariinae, **C14**
Cribrarioidea, C14
cribrate, **C59**
Cribratina, C137, **C220**
Cribratininae, C155, **C220**
Cribrobigerina, **C254**
Cribrobulimina, **C279**
Cribroelphidiinae, C152, C631, C643
Cribroelphidium, **C635,** C643
Cribroelphydium, C635

Cribroendothyra, C346
Cribrogenerina, C136, **C334**
Cribrogloborotalia, **C680**
Cribrogoesella, **C273**
Cribrohantkenina, C64, C139, **C666**
Cribrolinoides, **C453**
Cribrononion, **C637**
Cribroparella, C752
Cribroparrella, **C752**
Cribropullenia, **C746**
Cribropyrgo, **C470**
Cribrorobulina, **C516**
Cribrorotalia, C643
Cribrosa, C145
Cribrosphaera, C314
Cribrosphaerella, C314
Cribrosphaeroides, C135, **C314**
Cribrosphaeroides, C314
Cribrospira, C136, **C352**
CRIBROSPIRACEA, C313
Cribrospirella, C238
Cribrospiridae, C342
CRIBROSTOMACEA, C313
Cribrostomatidae, 332
Cribrostomoides, C90, **C225**
Cribrostomum, C333
Cribrotextularia, C254
Cribroturretoides, **C281**
Cribsophaeroides, C314
Crimellina, C429
Criptocanalifera, C637
Cristacea, C142, C512, C572
Cristacés, C142, 512, C572
Cristacidae, C482
Cristata, C142, C482
Cristellaria, C482, C520
CRISTELLARIACEA, 436, C511
Cristellarida, C143, C482
Cristellaridae, C482
Cristellaridea, C143, C482
Cristellarideae, C482
Cristellariidae, C482
Cristellarinae, C144, C145, C482
Cristellarioidea, C482
Cristellariopsis, C520
Cristellaroidea, C144
Cristellaroidi, C482
Crithionina, **C205,** C687
Croneisella, C200
Crossopyxis, C47
Crouch, C142
Cruciloculina, **C458**
Crustula, **C784**
Cryptasterorbis, C711
Cryptodifflugia, C18, **C37**
Cryptodifflugiacea, **C37**
Cryptodifflugiidae, **C37**
Cryptodifflugiidae, C37
Cryptomonas, C78
Cryptostegia, C143, C144, C725, C742
Ctenorbitoides, C725
Cubanina, **C285**
Cucurbitella, C17, **C35**
Cucurbitina, C784
Cumerina, C645

Cummings, C93, C324, C334
Cuneolina, **C285**
Cuneolinella, C285
cuniculus (pl., cuniculi), C59, **C372**
currents, **C130**
Cushman, C92, C93, C142, C146-C148, C247, C251, C253, C269, C272, C277, C279, C281-C283, C333, C438, C439, C466, C479, C501, C516, C520, C526, C528, C541, C547, C548, C559, C563, C569-C571, C574, C582, C584, C600, C604, C629, C635, C642, C656, C696, C728, C730, C736, C742, C748, C757, C784, C785
Cushman & Bermúdez, C708
Cushman & Todd, C582
Cushman & Warner, C93
Cushman & Waters, C325
Cushmanella, C96, C778, **C781**
Cushmania, C310
Cuvier, C55
Cuvillierina, **C614**
Cuvillierina, C614
Cuvillierininae, C156, **C614**
Cyanophyceae, C18
Cyanospira, C54
Cyclammina, C60, C63, C90, C137, **C228**
Cyclammina, C228
Cyclammininae, C155, **C228**
Cyclocibicides, C78, C687, C688, **C690**
Cycloclypeida, C643
Cycloclypeidae, C147, C643
Cycloclypeina, C643, C647
Cycloclypeinae, C144, C145, C146, C147, C156, **C647**
Cycloclypeinae, C643
Cycloclypeus, C644, C645, **C647, C649,** C694, C784
Cycloclypidae, C647
Cycloclypsinella, C302
Cyclodiscus, C572
Cyclogypsinoides, C783
Cyclogyra, C59, C92, C136, 438
Cyclogyrinae, C155, **C438,** C742
cyclogyrine, **C59**
Cyclolepidina, C721
Cyclolina, **C301**
Cyclolinineae, C155, **C301**
Cycloloceilina, C690
Cycloloculina, **C690**
Cyclomeandropsina, C485
cyclomorphosis, **C59**
Cyclopavonina, **C784**
Cyclophthalmidium, C457
Cyclopsina, C302
Cyclopsinella, **C302**
Cyclopyxis, C18, **C27**
Cyclorbiculina, **C495**
Cyclosiphon, C721
Cyclospira, C678, C680, C682

Cyclospiridae, C145, C678
CYCLOSTÈGUES, C184, C436, C511
Cyclotella, C22
Cylindria, C784
Cylindroclavulina, **C281**
Cylindropyxis, C27
Cylindrospira, C783
Cymbalopora, C138, **C698,** C699
Cymbaloporella, C139, **C699**
Cymbaloporetta, C699, **C701**
Cymbaloporettidae, C149, C151, C152, C698
Cymbaloporidae, C138, C139, C146, C148, C150, C156, **C698,** C702
Cymbaloporinae, C147, C698
Cymbella, C22
Cymbicides, C688
Cyphidium, C22
Cyphoderia, **C53**
Cyphoderiidae, **C53**
Cyphoderini, C53
Cyphoderiopsis, C32
cyst(s), **C17, C59, C68**
Cystammina, C89, **C263**
Cysteodictyina, C786
Cystidina, C26
Cystoforaminifera, C40
CYSTOFORAMINIFERA, C145
Cystoforaminifera (Vesiculata), C184
Cystophrys, **C166**
Cytharina, C515
cytoplasm, **C17, C59**

Dactyloporida, C143
Dactyloporidea, C144
Dactylosaccus, **C176**
Dagmarella, C405
Dainella, **C795**
Dainita, **C537**
Daitrona, **C205**
Daixina, C415
Danubica, C196
Darbyella, C520
Darbyellina, C520
Darvasites, **C434**
Darvasites, C434
Dasycladaceae, C742
Daucina, **C727**
Daucinoides, **C528**
Daviesina, **C617**
Daxia, C137, **C226**
Deckerella, C333
Deckerellina, C136, **C335**
Deflandre, C32, C51, C172, C181
Deflandre & Deflandre-Rigaud, C32
Deflandria, C32
Defrance, C253
Delage & Hérouard, C144, C145
Delosina, **C735, C736**
Delosinidae, C161, **C735, C736**
Dendrina, C759
dendritic, **C59**

Index

Dendritina, C139, **C482**
Dendronina, **C192**
Dendropela, C783
Dendrophrya, **C192**
Dendrophryida, C184, C192
Dendrophryidae, C145, C184, C192
Dendrophryinae, C146, C147, C150, C151, C155, **C192**
Dendrophyra, C192
Dendrotuba, **C177**
Dentalina, C136, **C516**
Dentalina trujilloi, nom. nov., **C516**
dentaline, les dentalines, C516
Dentalinella, C516
Dentalinidae, C144, C512
Dentalinoidea, C144, C512
Dentalinoides, **C516**
Dentalinopsis, **C516**
DENTATA, C149, C184, C511
Dentostomina, C93, **C458**
DEPRAT, C377, C409
Depratella, C401
depth, **C126**
DESHAYES, C508, C520
Desinobulimina, C559
deuteroconch, **C59**
deuteroforamen, **C58, C59, C108**
DEUTEROFORAMINATA, C149, C511
deuteropore, **C59**, C111
Dexiopora, **C784**
Dexiospira, C786
diagonal section, **C59**
Dianema, C15
Dianemaceae, C15
Dianemeae, C15
Dianemina, C15
Dianeminiinae, C15
diaphanotheca, **C59, C364**
Diaphorodon, **C44,** C171
diaphragm, **C17**
Dichistidae, C145
DICK, C89
Dictiostélidos, C12
Dictydiaethaliaceae, C14
Dictydiaethaliacées, C14
Dictydiaethaliidae, C14
DICTYDIINEAE, C14
Dictyoconella, C311
Dictyoconoides, C139, **C608**
Dictyoconos, C310
Dictyoconus, C137, **C310**
Dictyokathina, **C608**
Dictyopsella, **C285**
DICTYOSTELIACEAE, C11
Dictyosteliaceae, C11
Dictyosteliaceen, C11
Dictyosteliacei, C12
Dictyostelidae, C12
Dictyosteliidae, **C11**
Dictyostelinae, C12
Dicyclina, **C303**
Dicyclinidae, C137, C155, **C301**
Dicyclininae, C155, **C302**
Didimos, C13
DIDYMEAE, C13

Didymiacea, C13
DIDYMIACEAE, C13
Didymiaceae, C13
Didymiaceen, C13
Didymiacées, C13
Didymidae, C13
Didymieae, C13
Didymiées, C13
Didymiei, C13
Didymiidae, C13
Didymiinae, C9, **C13**
Didymium, C9
Dientamoebidae, C7
DIFFLUENTIA, C5
Difflugia, C18, **C34**
Difflugia, C32, C36, C47, C166
Difflugidae, C34
Diflúgidos, C34
Diflugie, C34
Difflugiella, **C38**
Difflugiida, C34
Difflugiidae, C17, **C34**
Difflugiidae, C34
Difflugiinae, C34
Difflugina, C34
Difluginae, C34
Diffusilina, **C205**
Diffusilininae, C155, **C205**
Digitina, **C274**
Dillina, C783
Dimastigamébidos, C7
Dimastigamoebidae, C7
Dimorphina, **C516**
dimorphism, **C59, C103**
Dioxeia, C202
Diplochlamys, **C20**
Diplogromia, **C177**
diploid, **C17, C59**
Diplomasta, C783
Diplophryidae, **C164**
Diplophrys, C67, **C166**
Diplostoma, **C782**
Diplotremina, C137, **C575**
Discammina, C89, **C226**
Discamminoides, **C242**
Discanomalina, C139, **C757,** C763
Dischistidae, C250, C259, C332
Discobolivina, C603
Discobotellina, C205
Discocyclina, C138, C712, **C714**
Discocyclinidae, C138, C148-C152, C156, **C712,** C714
Discocyclinidea, C678
Discocyclininae, C147, C150, C152, C712
Discogypsina, C695
Discoidina, C786
Discolita, C783
Discolites, **C498**
Discolithes, C498
Discolithus, C498
Discopulvinulina, C575
Discorbacea, C77, C99, C154, C156, **C572**
discorbes, les, C572
Discorbidae, C120, C136, C139, C148, C150, C152, C156, **C572,** C652, C750

Discorbidea, C99, C152, C572
Discorbididae, C572
Discorbidinae, C572
Discorbiidae, C572
Discorbiinae, C572
Discorbina, C572
Discorbinae, C147, C148, C150, C152, C156, **C572,** C652, C757, C763, C771
Discorbinae, C149
Discorbinidae, C572
Discorbininae, C145, C572
Discorbinella, C69, C78, **C575**
Discorbinellinae, C150, C572
Discorbis, C58, C75, C77, C78, C139, **C572,** C574, C757, C777
Discorbisinae, C146, C149, C572, C757
Discorbites, C572
Discorbitoides, C783
Discorbitura, **C575**
Discorbitus, C572
Discorbula, C607
Discorinopsis, **C281**
Discorotalia, C632, C643
Discospira, C645
Discospirina, **C457**
Discospirinia, C457
Discospiriniinae, C457
Discospirininae, C155, **C457**
Discospora, C646
Discotruncana, C584
displaced faunas, C119
distal, **C59**
Ditrema, C47
DOFLEIN, C8
Dogielina, **C470**
Dohaia, **C304**
Doliolina, C427
Doliolininae, C427
Domatocoela, C789
Dorbignyaea, **C784**
Dorothia, **C275**
DORREEN, C589, C643
dorsal, **C59**
DOUGLASS, C154, C292
DOUVILLÉ, C57, C477, C724
Dribroparella, C752
DUJARDIN, C56, C67
Dujardinia, C784
Dukhania, C290
DUNBAR, C365, C397
DUNBAR & HENBEST, C367, C372, C373, C375
DUNBAR & SKINNER, C369, C372, C375, C376
DUNBAR, SKINNER, & KING, C368, C371
Dunbarinella, **C418**
Dunbarula, **C401**
Duostomina, C137, **C577**
Duotaxis, C279
Dusenburyina, **C281**
Dutkevichella, C405
Dymia, C571
Dyocibicides, C687, C688, **C690,** C692
Dyofrondicularia, **C516**

Dyoxeia, C783

EARLAND, C181, C247, C272, C736
Earlandia, C135, **C317**
Earlandiidae, C317
Earlandiinae, C150, C155, **C317**
Earlandinella, C319
Earlandinita, C136, **C317**
Earlmyersia, **C577**
Echinogromia, **C166**
Echinopyxis, C23
Echinosteliaceae, C13
Echinosteliidae, C13
ecology, **C119**
Ectobiellidae, C795
Ectolagena, C518
ectoparasitic, **C59**
ectoplasm, C17, **C59**, C65
ectosolenian, **C59**
Ectosporeae, C12
Edentostomina, **C448**
Edomia, **C487**
Egeon, C645
Eggerella, **C275**
Eggerellidae, C268
Eggerellina, **C277**
Eggerellinae, C148, C149, C273, C795
Eggerina, **C277**
EHRENBERG, C496, C574, C600, C668, C736
Ehrenbergina, C95, C139, **C738**
Ehrenbergininae, C146, C736
Eilemammina, C783
EIMER & FICKERT, C56, C145, C333
Elaeomyxaceae, C13
elaters, **C10**
Elenis, C495
ELIAS, C330, C787
Ellipsobulimina, **C727**
Ellipsocristellaria, **C528**
Ellipsodentalina, C725
Ellipsodimorphina, **C728**
Ellipsofissurina, C540
Ellipsoglandulina, **C728**
Ellipsoidella, **C728**, C730
Ellipsoidina, C727, **C728**
Ellipsoidinidae, C146, C148, C150, C725
Ellipsoidininae, C725
Ellipsolagena, C540, C541
Ellipsolageninae, C725
Ellipsolingulina, **C728**
Ellipsomarginulina, C520
Ellipsonodosaria, C725, C728, C730
Ellipsonodosariinae, C725
Ellipsopleurostomella, C729, C730
Ellipsopolymorphina, **C729,** C730
Ellipsosiphogenerina, C569
Ellipsosiphongenerina, C569
Elliptina, **C784**
ELLIS & MESSINA, C219, C281, C535, C730, C760
Ellisina, C533

Elongobula, C544
Elphidiella, C138, **C638,** C643
Elphidiidae, C119, C120, C138, C139, C150-C152, C156, **C631**
Elphidiinae, C147, C152, C156, **C631**
Elphidioides, C139, **C594**
Elphidioides, C642
Elphidiononion, C635
Elphidium, C60, C61, C63, C64, C66, C70, C75, C83, C87, C88, C98, C104, C139, **C631,** C643
embryonic apparatus, **C59**
EMILIANI, C98, C127
ENALLOSTEGIA, C184
ENALLOSTÈGUES, C142, C184, C511
Enallostègues, C142, C250, C530, C731
Enantioamphicoryna, C520
Enantiocristellaria, C520
Enantiodentalina, C516, C533
Enantiomarginulina, C520
Enantiomorphina, **C530**
Enantiomorphinidae, C149, C150, C151, C152, C530
Enantiomorphininae, C530
Enantiovaginulina, C514
ENCLINOSTEGIA, C145, C725
Enclinostegia, C214, C511, C543
Encorycium, C537
Endamoeba, C7
Endamoebida, C7
Endamoebidae, C7
Endamoebinae, C7
endolobopodia, **C17**
endoplasm, **C17, C60**, C65
ENDOSPOREA, C14
Endosporea, C12
Endosporeae, C9
ENDOSPOREAE, C14
Endosporeae, C12, C14
Endosporeae (Myxogastres), C12, C14
ENDOSPOREEN, C14
Endosporeen, C12
ENDOSPORÉES, C14
Endosporées, C12, C14
ENDOSPOREI, C14
Endosporei, C12
ENDOSPORINEI, C14
Endosporinei, C12
Endostaffella, C343
Endothyra, C93, C126, C136, **C343,** C795
Endothyracea, C61, C93, C136, C155, **C320**
Endothyranella, C136, **C346**
Endothyranopsinae, C152, C155, **C352**
Endothyranopsis, C136, **C352**
Endothyranstamm, C145
ENDOTHYRIDA, C151, C152, C313
Endothyridae, C135, C136, C144-C150, C152, C155, **C342**
Endothyridea, C148, C151, C320
Endothyrina, C342, C350
Endothyrinae, C144-C148, C150, C152, C155, **C343**
Endothyrinae, C342
ENDOTRICHEA, C14
Endotrichea, C13
ENDOTRICHEEN, C14
Endotricheen, C13
Enerthenemaceae, C13
Enerthenemea, C13
Enerthenemeen, C13
Entamoeba, C7
Entamoebidae, **C7**
Entamoebinae, C7
ENTERIDIALES, C14
ENTERIDIEA, C12
ENTERIDIEAE, C14
Enteridioidea, C14
ENTHERIDIEAE, C14
Enthomostègues, C142, C458, C482, C503, C643, C685
Entolagena, C540
Entolingulina, **C539**
ENTOMOSTÈGUES, C142, C436
Entosalenia, C540
Entosolenia, C540
entosolenian, **C60**
Entrochus, C737
Entzia, **C264**
Eoalveolinella, C505, C508
Eoannularia, **C693**
Eoassilina, C647
Eoconuloides, **C685**
Eocristellaria, **C342**
Eocyclammina, C783
Eodictyoconus, C701
Eoendothyra, C347
Eoeponidella, C137, **C577**
Eofabiania, **C701**
Eoflabellina, C520
Eofrondicularia, C783
Eofusulina, C405
Eofusulininae, C152, C404
Eogeinitzina, C325
Eoglobigerina, C671
Eoguttulina, **C530**
Eohastigerinella, C658
Eolagena, C135, **C323**
Eolasiodiscus, C358
Eolepidina, C724
Eolituonella, C783
Eonodosaria, C323
Eoparafusulina, C420
Eoparastaffella, **C436**
Eoplacopsilina, C248
Eorupertia, C139, **C708,** C709
Eorupertiidae, C702
Eoschubertella, **C401**
Eosigmoilina, C136, **C450**
Eostaffella, C350, C396
Eostaffelloides, **C434**
Eotuberitina, C321
Eouvigerina, C137, **C556**

Index

Eouvigerinidae, C137, C139, C156, **C556**
Eouvigerininae, C146, C147, C148, C556
Eoverbeekina, C380, C383, C386, **C427**
Eovolutina, C135, **C323**
Eozawainella, **C436**
Eozoöninae, C144, C145
ephebic, **C60**
epidermal layer, **C60**
Epifusulina, C412
epipods, **C17**
Epistomaria, C139, **C592,** C598
Epistomariella, C598
Epistomariidae, C139, C156, **C592**
Epistomarioides, **C594**
Epistomaroides, **C594**
Epistomella, C592
Epistomina, C94, C96, C97, C136, **C771,** C775, C777
Epistominella, C137, **C578**
Epistominidae, C139, C149-C152, C766, C771, C776, C777
Epistomininae, C97, C137, C162, **C771**
Epistominita, C137, **C771**
Epistominites, **C784**
Epistominoides, C96, C137, **C771,** C776
epitheca, **C60**
Eponidae, C149, C678
Eponidella, C96, **C595**
Eponides, C78, C139, **C678,** C680, C682, C683, C684, C750
Eponididae, C139, C151, C156, **C678,** C680, C683
Eponidinae, C152, C678
Eponidoides, C748
Eponidopsis, C678
equatorial, **C60**
equatorial aperture, **C60**
equatorial section, **C60**
Equilateralidae, C142, C511, C559
equitant, **C60**
Erichsenella, C478
Ericson, G. Wollin, & J. Wollin, C116
Esosyrinx, **C539**
Espumáridos, C13
Estemonítidos, C13
Euarcella, C22
Eudifflugia, C34
Eudiscodina, C714
Eugenia, C46
Euglífidos, C47
Euglypha, C19, **C47**
Euglyphacea, **C47**
Euglyphida, C47
Euglyphidae, **C47**
Euglyphidae, C47
Euglyphidion, C50
Euglyphina, C27, C47
Euglyphinae, **C47**
Euglyphinae, C47

Euglyphini, C47
Eulepidina, C719, **C721**
Eulinderina, C724
EULOBOSA, C19
Eulobosa, C19
EUMYCETOZOA, C11, C12
EUMYCETOZOEN, C12
EUMYCETOZOINA, C9, **C12**
EUPLASMODIDA, C14
Euplasmodida, C13
EUPLASMODIÉS, C14
Euplasmodiés, C13
Eurycheilostoma, C137, **C578**
Eustoma, C35
Euuvigerina, C139, **C566**
Evlania, C316
evolute, **C60**
Exagonocyclina, C714
Exassula, C166
exogenous, **C60**
exolobopodia, **C17**
Exosporales, C12
Exosporea, C12
Exosporeae, C9
Exosporeae, C12
Exosporeen, C12
Exosporées, C12
Exosporei, C12
Exosporés, C12
Exosporinei, C12
Exseroammodiscus, C783
external furrow, **C60, C362**
extraumbilical aperture, **C60**
extraumbilical-umbilical aperture, **C60**

Fabiania, C139, **C701,** C702
Fabularia, C139, **C473**
Fabularidea, C143, C473
Fabulariinae, C137, C155, **C473**
Fabularina, C458
Fairliella, C202
Fallotella, C310
Fallotia, **C488**
Falsocibicides, **C691**
Falsoguttulina, **C530**
Falsopalmula, C522
Falsotetrataxis, C337
Faralieberkuehnia, C181
Fasciolites, C138, C503, C505, **C506, C508, C509**
Fasciolitidae, C503
Fascispira, C488
Faujasina, **C640,** C643
Faujasinella, C632
Faujasininae, C156, **C640,** C643
Faujasina, C640
Fauré-Fremiet, C90
Favocassidulina, **C738**
Feifel, C535
Feigl's reagent, C96
Fenestrifera, C678
Ferayina, C139, **C624**
Ferganites, C426
Feuerbornia, C53
Feurtillia, **C228**
Fichtel, von, C55
Fijiella, **C563**
filamentous, **C60**

Filoplasmodida, C794
Filoplasmodieae, C795
Filoplasmodiés, C794
Filoplasmodinae, C794
Filoplasmodinos, C794
filopodia, C17, C18
FILOSA, C39, C42, C67
Filosa, C794
FILOSA MONOSTOMATA, C39
filose, **C60**
FILOSIA, C2, C16, C18, **C39**
fimbriate, **C60**
Finlay, C277, C466, C559, C582, C642
Fisáridos, C13
Fischerina, **C441**
Fischerinella, **C443**
Fischerinidae, C136, C146, C147, C149, C155, **C438**
Fischerininae, C155, C438, **C441**
fission, **C17**
Fissistomella, C766
Fissoelphidium, C138, **C617**
fissure, **C60**
Fissurina, **C540, C541**
fistulose, **C60**
FITOMIXINOS, C12
Flabellammina, C90, C137, **C244**
Flabellamminopsis, **C244**
flabelliform, **C60**
Flabellina, C522
Flabellinella, **C516**
Flabelliporus, C650
Flabellulidae, C7
Flabellulina, C6
flagellum (pl., flagella), **C10, C17, C60**
flange, **C60**
Flectospira, C136, **C439**
Flexostili (Imperforata), C436
flexostyle, **C60,** C503
FLEXOSTYLIDIA, C436, C511, C678
Flexostylidia, C436
Flexurella, **C787**
Flint, C466, C575
Flintia, C453
Flintina, **C461**
Flintinella, **C462**
floor, **C60**
Florilus, C138, **C746**
'Flosculina,' C503, C506, C508
Flosculinella, C58, C503, **C510**
Flourensina, **C269**
fluting, **C60, C362**
Fomina, C93
foramen (pl., foramina), **C60, C364, C380**
Foraminella, C456
FORAMINIFERA, C67, C151, C164
FORAMINIFERA MONOMERA, C40, C143, C184, C436, C511
FORAMINIFERA POLYMERA, C143, C184, C436, C511

FORAMINIFERAE, C164
FORAMINIFÈRES, C142, C164
FORAMINIFERIAE, C145, C164
FORAMINIFERIDA, C16, **C55, C164,** C208, C789
FORAMINIFERIDA, economic importance, C56
FORAMINIFERIDA, living animal, **C65**
FORAMINIFERIDA, morphology and biology, **C58**
FORAMINIFERIDA, terminology, **C58**
FORAMINÍFEROS, C164
Forschia, C136, **C340**
Forschia, C340
Forschiella, C136, **C340**
Forschiidae, C338
Forschiinae, C151, C338
Forshiinae, C338
fossettes, **C60**
Fourstonella, C136, **C331**
FØYN, C71
fragmentation, in foraminiferal reproduction, **C86**
Frankeina, C245
FRANZENAU, C759, C760
Frenzelina, **C44**
FRIES, C8
Frilla, C646
FRIZZELL, C251, C624, C625
Frondicularia, **C516,** C528, 785
Frondicularidae, C145, C512
Frondicularidea, C143, C512
Frondicularideae, C512
Frondiculariinae, C147, C512
Frondiculina, C522, C528
Frondiculinita, C528
Frondilina, C135, **C324**
Frondovaginulina, C517
Frumentarium, C458
Frumentella, C412
Frustrulia, C22
Fujimotoella, C434
Fuliginoidei, C14
Fuligo, C9
Fursenkoina, C71, C138, **C731,** C733, C734, C784
Fursenkoininae, C156, **C731,** C736
Fusarchaias, **C495**
Fusiella, **C401**
fusiform, **C60**
Fusilinidae, C399
Fusulina, C95, C365, C369, C373, C374, C380, C383-C385, C389, **C404**
Fusulina Zone, C389
FUSULINACEA, C313
Fusulinacea, C64, C93, C136, C155, C358, **C394**
Fusulinacea, uncertain status, C155
Fusulinaceae, C394
fusulinacean chambers, **C369**
fusulinacean ecology, C387

fusulinacean evolutionary trends, **C381**
fusulinacean geographic distribution, **C389**
fusulinacean morphology, **C361**
fusulinacean phylogeny, **C386**
fusulinacean septa, **C371**
fusulinacean shell, **C360**
fusulinacean stratigraphic distribution, **C389**
fusulinacean test shape and size, **C382**
Fusulinella, C373, C374, C384, C385, C389, **C406**
Fusulinella Zone, C389
Fusulinellinae, C152, C404
FUSULINIDA, C151, C152, C313
Fusulinida, C399
Fusulinidae, C59, C60, C62, C145-C150, C152, C155, **C399**
Fusulinidae, C404
Fusulinidea, C150, C152, C394
FUSULININA, C154, C155, C313
Fusulinina, C399
Fusulininae, C144-C150, C152, C155, C370-C373, C375, C382-C386, **C404**
Fusulininae, C399
Fusulinoidea, C149, C394

Gabonella, C137, **C552,** C734
Galea, C450
Galeanella, **C450**
Gallionella, C22
Gallowaiina, C406
Gallowaiinella, C406
GALLOWAY, C93, C146, C298, C406, C438, C445, C453, C501, C508, C509, C516, C526, C559, C569, C582, C584, C600, C628, C654, C680, C785
GALLOWAY & HARLTON, C93
GALLOWAY & RYNIKER, C409
Gallowayina, C406, C710
Gallowayinella, C385, **C406**
gamete, **C60, C68**
gamont, **C60**
GANDOLFI, C115
Ganella, C139, **C764**
Garantella, C137, **C774,** C775
GASTEROMYCETES, C12
GASTEROMYCETES GENUINA, C12
GASTRAEADA, C184
Gastroammina, C200
GASTROMYZETES, C12
Gastrophysema, C192
Gaudryina, C90, C136, **C269**
Gaudryinella, **C269**
Gaudryna, C269
GAUTHIER-LIÈVRE, C129
Gavelinella, C138, C753, **C759,** C763
Gavelinellidae, C151, C753
Gavelinellinae, C753

Gavelinonion, C761
Gavelinopsis, **C578**
Geamphorella, C51
geological importance of thecamoebians, C18
Geinitzella, C325
Geinitzina, C325, C785
Gemellides, C759, C760
Geminaricta, C553
Geminospira, C96, C139, **C781**
Gemmulina, C254
Gemmuline, C254
generic names erroneously applied to Foraminiferida, C162
Geococcus, C37
Geophonus, C631
Geoponus, C632
Geopyxella, C38
GERKE, C210
gerontic, **C60**
Gijuella, C429
GIGNOUX, C89
GIMNAMEBOIDEOS, C5
Ginesina, C216, C217
Giraliarella, **C190**
Girtyina, C425
Girvanella, C786
Girvanellinae, C144
Glabratella, C66, C68, C69, C78, C80, C139, **C588**
Glabratellidae, C78, C139, C156, **C587**
Gladiaria, C514
Glaesneria, C783
GLAESSNER, C148, C251, C654
GLAESSNER & WADE, C680
Glandiolus, **C784**
Glandulina, C138, **C537**
Glandulinaria, C783
Glanduline, C537
Glandulinea, C537
Glandulinidae, C108, C137, C153, C156, **C537,** C729
Glandulinidea, C143, C537
Glandulininae, C156, **C537**
Glandulodimorphina, C516
Glandulonodosaria, C512
Glandulonodosariinae, C512
Glandulopleurostomella, **C530**
Glandulopolymorphina, C530
Globalternina, C783
Globanomalina, C138, **C665**
Globigenera, C669
Globigerapsis, C139, C671, **C675,** C676
Globigerina, C64, C66, C70, C87, C94, C95, C97, C138, C659, C663, **C669**-C671, C673-C676, C678, C786, C791
Globigerinacea, C87, C137, C154, C156, **C652,** C666
Globigerinae, C143, C669
Globigerinatella, C58, C139, **C676**
Globigerinatellinae, C676
Globigerinatheka, C139, **C676**
Globigerinella, C663

Index

Globigerinelloides, C138, **C656**, C665
GLOBIGERINIDA, C511
Globigerinida, C143, C669
Globigerinidae, C115, C138, C139, C144-C146, C148-C152, C156, **C669**
GLOBIGERINIDAE, C145, C511
Globigerinidae, C669
GLOBIGERINIDEA, C511, C652
Globigerinidea, C144, C152, C669
Globigerinidee, C669
Globigerínidos, C669
Globigerinina, C143, C669
Globigerininae, C144-C148, C150, C152, C156, **C669**
Globigerininae, C669
Globigerinita, C139, **C676**, C678
Globigerinitinae, C676
Globigerinoides, C139, **C670**, C675, C676, C678
Globigerinoita, C139, **C678**
Globigerinopsis, **C670**
Globivalvulina, C136, **C338**
Globivalvulinae, C338
Globivalvulinella, C258
Globivalvulininae, C150, C338
Globobulimina, C138, **C559**
Globobuliminidae, C559
Globobulimininae, C559
Globocassidulina, C139, **C738**
Globoconusa, C138, **C670**
Globoendothyra, C352
Globonota, C35
Globoquadrina, C65, C139, **C670**, C671
Globorotalia, C60, C87, C97, C115, C138, **C667**, C668, C669, C671, C680, C683
Globorotaliidae, C138, C146, C148, C150-C152, C156, **C666**
Globorotaliinae, C147, C152, C156, **C666**
Globorotalites, C63, C138, C683, **C752**
Globorotaloides, C138, C665, **C671**
Globorotaloidinae, C669
Globosiphon, C200
Globotextularia, **C273**
Globotextulariinae, C148, C155, **C273**
Globotextularinae, C146
Globotruncana, C64, C95, C115, C138, **C662**, C663
Globotruncanella, C662
Globotruncanidae, C59, C138, C156, **C662**
Globotruncaninae, C152, C662
Globotruncanita, C662
Globulina, C94, **C530**, C785
globulines, les, C531
Globulotuba, **C539**
Glomalveolina, **C509**, C510

Glomalveolina, C509
Glomerina, C259
Glomospira, C60, C135, **C212**
Glomospiranella, C340
Glomospirella, **C212**
Glomospirella, C212, C340
Glomospirellinae, C338
glomospirine, **C60**
Glomospiroides, C136, **C340**
Glomovertella, C787, **C788**
Glomulina, **C450**
Glyphostomella, C136, **C353**
Goatapitigba, **C204**, **C795**
Gocevia, **C19**
Goës, C205
Goesella, **C281**
GOLDSCHMIDT, C7
Gonatosphaera, **C528**
Goniolina, C786
Gordiammina, C212
Gordiospira, **C439**
Goupillaudina, C138, **C753**
Goupillaudina, C753
Gourisina, **C340**
Grabauina, C420
Grammobotrys, **C784**
Grammostomum, C549
granellae, C789
granellarium, C789
granular hyaline wall, **C60**
Graniliferella, C352
GRANULORETICULOSA, C54, C150
granuloreticulose pseudopodia, **C17**, **C60**
GRANULORETICULOSIA, C2, C16, C18, C39, **C54**
Gravellina, **C277**
GRELL, C66, C71, C78, C81, C83, C84, C85
GRIMSDALE, C718
Grimsdaleinella, C137, **C551**, C734
Gromia, **C40**, C59
Gromiada, C40
GROMIDA, C16, C19, **C40**, C789
Gromida, C40, C143
Gromidae, C40, C144, C146
GROMIDAE, C40, C145
Gromidea, C40, C143, C144
Gromidee, C40
Grómidos, C40
GROMIIDA, C40
Gromiidae, **C40**, C145, C175
GROMIIDEA, C40
Gromiides, C40
Gromiina, C40
Gromiinae, C40
Grominae, C40
GROZDILOVA & LEBEDEVA, C356
Grzybowskia, C650
Gubkinella, C137, **C652**
GUBLER, C375, C376
Gublerina, C138, **C654**
Gublerina, C432
GÜMBEL, C57, C777
Gümbelia, C646
Guembelina, C654

Gümbelina, C652, C655
Gümbelinidae, C148, C150, C151, C652
Guembelininae, C146, C710
Gümbelininae, C147, C148, C652
Guembelitria, C138, **C652**
Gümbelitria, C652
Guembelitriella, C138, **C652**
Guembelitriinae, C137, C156, C652
Gunteria, C139, **C702**
Guppyella, **C298**
Guttulina, **C531**
Guttulina, C11
Guttulinacea, C11
GUTTULINACEAE, C11
Guttulinaceae, C11
Guttulinacei, C11
Guttuline, C531
Guttulineae, C11
Guttulineen, C11
guttulines, les, C531
Guttulinidae, C11
Guttulínidos, C11
Guttuliniidae, C11
Guttulininae, C11
GYMNAMOEBAEA, C5
GYMNAMOEBIDA, C5
GYMNAMOEBINA, C5
Gymnesina, C455
GYMNICA, C5
Gymnococcaceae, C12
Gymnococcaceen, C12
Gymnococcidae, C12
Gymnococcinae, C12
GYMNOMYXA, C5
Gypsina, C139, C688, **C694**, C697, C698
Gypsininae, C694
Gvrammina, **C784**
Gyroidina, C138, **C750**, C753
Gyroidinella, C708, C709
Gvroidinoides, C138, C750, **C753**
Gyromorphina, C744

DE HAAN, C55
habitat of Foraminiferida, **C87**
Haddonia, **C248**
Haeckelina, C184
Haerella, C707
Haeuslerella, **C256**
Hagenowella, **C273**
Hagenowina, **C286**
HAGN, C777
Haliphysema, C192
Halkyardia, C139, C701, **C702**
Halkyardiidae, C698
Halyphysema, C86, C88, C90, **C192**
Hammonium, C606
hamulus (pl., hamuli), **C60**
Hantkenia, C647, C666
Hantkenina, C60, C139, **C666**
Hantkeninella, C666
Hantkeninidae, C138, C139, C146, C150, C151, C152, C156, **C663**, C666

Hantkenininae, C147, C148, C156, **C666**
HANZAWAI, C100, C629, C701
Hanzawaia, C139, **C759**
haploid, **C17, C60**
Haplophragmella, C136, **C350**
Haplophragmellinae, C152, C155, **C350**
HAPLOPHRAGMIACEA, C184
Haplophragmidae, C145, C225
Haplophragmiidae, C149, C225
Haplophragmiinae, C146, C148, C238
Haplophragmina, C351
Haplophragmium, **C244**
Haplophragmoides, C89, C133, C136, **C225**
Haplophragmoidinae, C155, **C225**
Haplostiche, C137, **C220**
HAQUE, C683
Hartmannellidae, C7
Hartmannellina, C6
Hartmannellinae, C7
Hasterigerininae, C663
Hastigerina, C139, **C663,** C665, C666
Hastigerinella, C64, C659, C665, C670, **C671**
Hastigerininae, C156, **C663**
Hastigerinoides, C63, C138, **C658**
Hauerina, C139, **C470**
Hauerinae, C145, C458
Hauerinella, C448
Hauerinidae, C458
Hauerinidee, C458
Hauerinina, C458
Hauerininae, C144, C147, C468
Haurania, C493
Hayasakaina, C397
HAYDEN, C375
HAYNES, C561
Hechtina, C466
HEDLEY, C40, C88
Hedbergella, C138, **C659**
Hedbergellinae, C156, **C659**
Hedbergina, **C784**
Helenia, C580
Helenina, **C580**
Helenis, C494
Heleopera, C18, **C29**
Heleoperidae, C27
Heleoperinae, C27
Helicites, C645
Helicocyclina, C724
HELICOIDEA, C143, C184, C436, C511
Helicolepidina, C139, **C724**
Helicolepidinae, C149, C724
Helicolepidinidae, C150, C151, C717
Helicolepidininae, C148, C152, C156, C717, C718, **C724**
Helicolepidinoides, C724
Helicosorina, C458, C482, C559, C643
HELICOSTEGIA, C184, C511

Helicostegia, C268
Helicostegina, C139, C685, **C724**
HELICOSTÈGUES, C142, C184, C511
Helicostègues, C142, C268, C482, C511, C559, C572, C605, C631, C669, C742, C753
Helicotrochina, C592, C631, C685
Helicoza, C635
HELIOZOIA, C18, C39
Heliostegia, C482
Hellenocyclina, C711
HELLENOIDEA, C184, C313, C436, C511
Helvetoglobotruncana, C662
Hemiarchaediscus, C355
Hemicristellaria, C520
Hemicyclammina, C137, **C229**
Hemidiscus, **C212**
Hemifusina, C404
Hemifusulina, C404
Hemifusulininae, C404
Hemigordiella, C440
Hemigordiellina, C212
Hemigordiopsis, **C439**
Hemigordius, C136, **C440**
Hemigypsina, C695
Heminwayina, C577
Hemirobulina, C524
hemisepta, **C60**
Hemisphaerammina, C68, C74, C135, **C202**
Hemisphaerammininae, C155, **C202,** C796
Hemistegina, **C784**
Hemisterea, **C784**
Hemisticta, **C784**
HENBEST, C214, C345, C346, C365, C375, C450
HENSON, C742
Hensonia, **C292**
Herion, C520
HERODOTUS, C55
HERON-ALLEN, C78
HERON-ALLEN & EARLAND, C56, C95, C186, C192, C577, C604, C728,C785
Heronallenia, **C589**
Herrmannia, C512
Heterillina, **C470**
heterocaryotic, **C66**
Heteroclypeinae, C647
Heteroclypeus, C647
Heterocosmia, C22
HETERODERMACEAE, C14
Heterodermaceae, C14
Heterodermacées, C14
HETERODERMEAE, C14
Heterodermidae, C14
Heteroglypha, **C49**
Heterogromia, **C167**
Heterogromiini, C164
HETEROHELICIDA, C151, C153, C511
Heterohelicida, C652
Heterohelicidae, C138, C146-C148, C150, C153, C156, **C652,** C710, C736
Heterohelicinae, C146-C148, C156, **C652**
Heterohelix, C138, **C652,** C654, C656
heterokaryotic, **C60, C84**
Heterolepa, C138, **C759,** C760
Heterospira, C647
Heterostegina, C129, C644, C645, C647, **C650**
Heterosteginella, C650, C783
Heterosteginae, C147, C148, C150, C152, C647
Heterosteginoides, C650
Heterostomella, **C269**
Heterostomum, C784
Hexagonocyclina, C714
Hidaella, **C406**
Hiltermannia, C775
Hippocrepina, C67, C135, **C188,** C190
Hippocrepinella, C186
Hippocrepininae, C144, C155, **C187**
Hipporina, C135, **C324**
Hippurites, C142
hispid, **C60**
Histopomphus, **C536**
Historbitoides, C725
HÖGLUND, C201, C205, C560
Hoeglundina, C95, C96, C137, C771, **C775**
HOFKER, C64, C89, C95, C99, C103, C106, C108, C111, C149, C150, C215, C269, C503, C546, C548, C549, C553, C558, C561, C563, C565, C568, C569, C572, C574, C575, C580, C582, C589, C598, C604, C625, C635, C687, C699, C702, C733, C734, C736, C750, C759, C764, C771, C777
Hofkerina, **C680**
Hofkerinella, C686, C687
Hofkeruva, C567
Hollandina, C759
Holmanella, C139, **C760**
Holocladina, C786
hologamic, **C60, C75**
Hologlypha, C50
Holosamma, **C792,** C794
holozoic nutrition, **C10**
Homoeochlamys, C53
HOMOGENEA, C5, C54
homokaryotic, **C66**
Homotrema, **C702,** C705
Homotrematidae, C138, C139, C156, **C702**
Homotrematinae, C150, C156, **C702**
Homotremidae, C146, C150, C152, C702
Homotreminae, C147
Hoogenraadia, **C25**
Hopkinsina, C139, **C567**
Hormosina, **C215,** C216
Hormosinida, C214, C215

Index

Hormosinidae, C155, **C214**
Hormosininae, C155, **C215**
HORNIBROOK, C642, C643
HORNIBROOK & VELLA, C574, C580, C589
Hospitella, **C177**, C181
Hospitellum, C177
HOTTINGER, C510
Howchinia, C136, **C358**
HOWE, C569
Hyaleina, C540
Hyalina, C687
Hyalina, C49, C50
hyaline, **C18**, **C60**
hyaline calcareous foraminiferans with perforate granular walls, **C95**
hyaline calcareous foraminifers with perforate radial walls, **C94**
hyaline calcareous monocrystalline tests in Foraminiferida, **C94**
Hyalinea, C139, **C686**, **C687**
Hyalinia, C687
Hyalodiscida, C7
Hyalodiscidae, **C7**
Hyalopus, C40
Hyalosphenia, C18, **C27**, C29
Hyalospheniidae, C27
Hyalospheniidae, **C27**
Hyalovirgulinidae, C150, C151, C559
Hybridina, C515
Hydromylina, C520
Hydromylinidae, C512
Hydromyxaceae, C795
Hydromyxales, C794
Hymenocyclus, C710
Hyperammina, C135, **C190**
Hyperammina, C190
Hyperamminella, C188, C200
Hyperamminidae, C145, C146, C147, C149-C151
Hyperammininae, C146-C148, C150, C151, C187
Hyperamminita, C196
Hyperamminoides, C189, C190
hypodermis, **C60**
Hypporina, C324

Iberina, C233
Ichthyolaria, C518
Idalina, **C470**
Illigata, C135, **C322**
Ilotes, C494
Ilyopegma, C783
Ilyoperidia, C783
Ilyosphaera, C783
Ilyozotika, C783
IMPERFORATA, C19, C40, C91, C143, C184, C436
imperforate, **C60**
IMPERFORIDA, C145, C436
IMPERFORINA, C184, C313, C436
Inaequilateralidae, C445
Inequilateralidae, C142, C445

infralaminal accessory aperture, **C60**
inframarginal sulcus, **C60**
infundibulum, **C61**
INFUSOIRES HOMOGÈNES, C5
instar, **C61**
intercameral, **C61**
intercameral foramen, **C61**
interio-areal aperture, **C61**
interiomarginal aperture, **C61**
interseptal, **C61**
intralaminal accessory aperture, **C61**
intraseptal, **C61**
intraumbilical aperture, **C61**
Involutaria, **C518**
involute, **C61**
Involutina, C136, **C740**, C741, C742
Involutinae, C739
Involutinidae, C136, C138, C149, C161, **C739**
Involutininae, C148, C152, C739
Involvina, C137, **C760**
Involvohauerina, **C472**
Iowanella, **C418**
Iraqia, C137, **C311**
Iridia, C67, C68, C71, C72, C87, C166, **C167**
Iridiella, C202
Irregularina, C135, **C315**
Irregularina, C315
Islandiella, **C556**, C738
Islandiellidae, C156, **C556**, C738
Isodiscodina, C715
isogamy, C18, **C61**
isogenotypic, **C61**
Isolepidina, C721
Isorbitoina, C723
ISRAELSKY, C131

Jabeina, C432
Jaculella, **C190**
Jadammina, **C265**
JAHN, C111
JAHN & JAHN, C140
JAHN & RINALDI, C2
Jangchienia, C412
Janischewskina, C136, **C354**
Jarvisella, **C287**
Java Sea, C125
JEPPS, C40, C75, C78, C140, C141
Jigulites, C415
JONES, C143
JONES & PARKER, C580
JONES, PARKER & BRADY, C608
Julia, C783
Jullienella, **C194**
Jungia, **C29**
Jungia, C29
juvenarium, **C61**

Kahlerina, **C346**
Kalamopsis, **C215**
Kanakaia, **C501**

Kansanella, **C418**
Kapselthierchen, C22
Kara-Kum desert, C129
Karreria, C138, C692, **C760**, C783
Karreriella, **C277**
Karrerulina, C277
karyogamy, **C10**
Katacycloclypeus, **C649**
Kathina, C137, **C608**
Kelyphistoma, C763, C764
Keramosphaera, C91, **C501**
Keramosphaeridae, C145, C146, C147, C150, C482
Keramosphaerina, C482, C786
Keramosphaerinae, C144, C145, C147, C149, C156, **C501**
Keramosphaerinae, C482
keratinous wall, **C88**
Kerionammina, C135, **C208**
keriotheca, C61, **C364**, **C366**, **C375**
Kettnerammina, C135, **C319**
Kibisidytes, **C169**
Kikrammina, C783
Kilianina, **C292**
Klubovella, C136, **C352**
Kochliopodium, C19
Kolesnikovella, C139, **C567**
Kordeella, **C788**
KRASHENINNIKOV, C94, C119, C635
KRINSLEY, C98
Krumbachina, C457
Krumbachininae, C457
KUDO, C140
KÜPPER, C629, C710
Kuglerina, C663
Kurnubia, **C292**
KUWANO, C598
Kwantoella, **C418**
Kyphamminidae, C145, C194
Kyphopyxa, **C518**

Laberintúlidos, C795
labial aperture, **C61**
Labrospira, C225
labyrinthic, **C61**
Labyrinthina, **C245**
Labyrinthochitinia, C135, **C177**
Labyrinthulales, C794
Labyrinthuleae, C794, C795
Labyrinthuleen, C795
Labyrinthulés, C794
LABYRINTHULIDA, C789, C790, **C794**
Labyrinthulida, C794, C795
Labyrinthulidae, **C795**
Labyrinthulidae, C794
Labyrinthuloidea, C794
Lacazina, **C476**
Lacazinella, **C477**
Lacazopsis, **C477**
Lachlanella, C458
Lacosteina, C137, **C547**
Lacosteininae, C150, C153, C156, **C547**
LACROIX, C89, C90, C92, C226, C253

Ladinosphaera, C786
Ladoronia, **C698**
Laevipeneroplis, C482
Laffitteina, C138, **C639**
Lagena, C59, C94, C95, C512, **C518,** C540
LAGENACEAE, C511
Lagenammina, C135, **C200**
Lagene, C512
Lagenetta, C512
Lagenicae, C511
LAGENIDA, C152, C511
Lagenida, C143, C512
LAGENIDAE, C145, C511
Lagenidae, C144, C146, C148, C149, C151, C152, C511, C512
LAGENIDEA, C511
Lagenidea, C143, C148, C149, C151, C511, C512
Lagenideae, C512
Lagénidos, C512
Lagenina, C512
Lageninae, C144, C145, C146, C147, C149, C152, C512
Lagenoglandulina, **C518**
Lagenoidea, C144, C512
Lagenonodosaria, C514
Lagenopsis, **C784**
Lagenula, C518
Lagenulina, C540
Lagunculina, **C44**
Lagynacea, C67, C87, C88, C89, C154, **C164**
Lagynida, C164
Lagynidae, C71, C134, C143, C146, C151, C154, **C164**
Lagynidea, C149, C164
Lagyninae, C146, C164
Lagynis, C151, **C166**
Laharpia, C647
LALICKER, C126, C129
LAMARCK, C35, C55, C574
Lamarckella, C776, C777
Lamarckina, C95, C96, C138, **C769,** C777
Lamarckinita, C589
lamellar, **C61**
lamellar walls in foraminifers, **C98**
Lamellodiscorbis, **C580**
Laminiuva, C567
Lampas, C520
LAMPRAMOEBAE, C19
Lamprodermaceae, C13
LAMPRODERMAE, C13
Lamprodermeae, C13
LAMPROSACRALES, C14
Lamprosporae, C14
LAMPROSPORALES, C14
LAMPROSPOREAE, C14
Lamprosporées, C14
Lamprosporei, C14
lanceolate, **C61**
LANGE, C325
LANKESTER, C58, C144, C145, C151
Lankesterina, **C518**
Lantschichites, **C432**

Laryngosigma, **C539**
Lasiodiscidae, C136, C150, C153, C155, **C358**
Lasiodiscus, C136, **C358**
Lasiotrochus, C136, **C358**
Laterostomella, **C553**
Laticarinidae, C149, C572
Laticarinina, **C580**
LATREILLE, C55
Lebedevaella, **C788**
LE CALVEZ, C66, C71, C77, C78, C80, C81, C85, 106, C111, C574, C589, C675, C688, C729, C757
LE CALVEZ, J. & LE CALVEZ, Y., C129
LECLERC, C35
Lecquereusia, C31
Lecquereusiidae, C27
Lecythium, **C44,** C171
LEE, C377
Leeina, C420
Leella, **C396**
LEIDY, C32, C35, C67, C142
Leidyella, C32
Lekithiammina, **C784**
Lekithiammina, C784
Lensarchaediscus, C357
lenticulacées, les, C512
Lenticulina, C59, C61, C65, C136, **C518**
LENTICULINACEA, C511
Lenticulinae, C512
lenticuline, **C61**
Lenticulinidae, C512
Lenticulininae, C149, C152, C512
Lenticulites, C519
LEPAMOEBAE, C19, C40
Lepidocyclina, C139, C718, C719, **C721**
Lepidocyclinae, C149, C721
Lepidocyclinidae, C139, C150, C152, C156, **C717,** C718, C721
Lepidocyclininae, C152, C156, C718, **C721**
Lepidolina, C62, C379, C380, C386, C394, **C431**
Lepidolininae, C427
Lepidorbitoides, C138, C710, **C711,** C718
Lepidorbitoididae, C150, C710
Lepidorbitoidinae, C147, C150, C152, C710
Lepidosemicyclina, C650
Lepista, C784
Leptochlamys, **C30**
Leptocystis, C22
Leptodermella, **C27**
Leptonemeae, C13
LEPTONEMINÉES, C13
Lesquereusia, C27, **C30**
LEUPOLD & BIGLER, C742
Leupoldina, C138, **C658**
LEVINE, C141
Liceacae, C14
Liceacea, C14
Liceaceae, C14

Liceaceen, C14
Liceacées, C14
Liceacei, C14
Liceae, C14
LICEALES, C14
Liceathaliaceae, C14
Liceidae, C9, **C14**
Liceidae, C14
Liceidos, C14
LICEINA, C14
Liceinae, **C14**
Liceoidea, C14
Liceoidei, C14
Lieberkuehnia, C67, **C179**
Lieberkuehniinae, C173
Lieberkuehniini, C173
Liebusella, C139, **C287**
life habits of Foraminiferida, **C87**
life history of Foraminiferida, **C70**
limax-form, **C10**
limbate, **C61**
lime-knots, **C10**
Limocaecum, C783
Linderina, C139, **C694,** C784
linellae, C790
Lingulina, C136, **C528**
Lingulinella, C528
Lingulininae, C156, **C528**
Lingulinopsis, **C528**
Linguloglandulina, C528
Lingulonodosaria, C136, **C528**
LINNÉ, C56
Linthuris, C520
Linthurus, C520
lip, **C61**
LISTER, C58, C75, C145, C151
Listerella, C15
Listerella, C282
Listerellaceae, C15
Lithocolla, C181
LITHODERMEAE, **C13**
Lituacea, C142, C214, C225, C238
Lituacés, C142, C225
Lituola, **C238**
Lituolacea, C60, C135, C155, **C214**
Lituolacea, C225
Lituolacées, C225
Lituolata, C142, C225
Lituoletta, C225
Lituolicae, C214
Lituolida, C143, C225
LITUOLIDA, C184
LITUOLIDACEAE, C184
Lituolidae, C60, C136, C137, C144, C146-C150, C152, C155, **C225**
LITUOLIDEA, C184
Lituolidea, C143, C144, C148-C152, C214, C225
Lituolideae, C225
Lituolidee, C225
LITUOLINA, C145
Lituolina, C214, C216, C225
Lituolinae, C144-C149, C152, C155, **C238**

Lituolinae, C225
Lituolites, C238
Lituolitidae, C225
Lituonella, **C298**
Lituonelloides, C292
Lituosepta, C245
Lituotuba, C135, **C214**, C440
Lituotubella, C136, **C340**
Litya, C319
Lobatula, C688
Loboforamina, **C36**
lobopodia, **C18, C61**
LOBOSA, **C67**
LOBOSA, C5
Lobosa, C6, C22, C34
LOBOSIA, **C5,** C16, C17, C18, C61
Lobularia, C784
Lockhartia, C138, **C609**
loculus, **C61**
LOEBLICH & TAPPAN, C57, C128, C154, C196, C210, C242, C254, C432, C438, C442, C448, C458, C521, C535, C629, C678, C736, C740, C785
Loeblichia, C93, C136, **C342**
Loeblichiinae, C155, **C342**
Loeblichinae, C342
Loftusia, C137, **C236**
Loftusiidae, C146, C148, C150, C225
Loftusiinae, C148, C155, **C236**
Loftusina, C225
Loftusinae, C144, C149, C236
lorica, **C18**
LORICATA, C19
lower keriotheca, **C61, C367**
lower tectorium, **C61**
Loxostoma, C736
Loxostomidae, C138, C161, **C736**
Loxostomoides, C137, **C553**, C733
Loxostomum, C138, C733, **C736**
Lugtonia, C136, **C318**
Lukaschevella, **C788**
Lunucammina, C135, **C324**
Lunucammina, C325
LYCOGALACEAE, C14
Lycogalaceae, C14, C15
Lycogalacées, C14
Lycogalactida, C14
Lycogalactidae, C14
LYCOGALALES, C14
Lycogaleen, C14
Lycogalidae, C14
Lycogalinae, **C15**
Lycophris, C645
Lyrina, **C784**

MACFADYEN, C251, C455
Macrodites, C520
Maghrebia, **C36**
main partitions, **C61**
Makarskiana, C282
MALACOZOA (MALACO-ZAIRES), C142
Mallopela, C783

maltha, C792
Mangashtia, **C305**
Manorella, C137, **C248**
Marenda, **C47**
Margarita, C15
MARGARITACEAE, C15
Margaritaceae, C15
Margaritacées, C15
MARGARITALES, C15
Margaritellina, **C15**
Margaritida, C15
Margaritidae, C15
Margaritoidea, C14
marginal chamberlets, **C61**
marginal cord, **C61**
marginal zone, **C61**
Marginolamellidae, C149, C572, C659, C662, C666
Marginopora, C92, C139, **C498**
Marginotruncana, C662
Marginulina, C136, **C520**
Marginulinae, C512
Marginulinella, C520
Marginulinellidae, C512
Marginulinidae, C512
Marginulinopsis, **C521**
MARIE, C273, C283, C487, C547, C736
Marieita, **C294**
Mariella, C537
Marsipella, C90, **C186**
Marssonella, C275
Marsupophaga, C183
Marsupulina, **C179**
Martiguesia, C137, **C231**
Martinottiella, **C282**
Maslinella, **C709**
MASLOV, C742, C786
Masonella, **C208**
Massilina, C92, **C452**
Massilininées, C458
Massillininae, C458
MASTIGAMOEBAEA, C5
Mastigamöben, C7
Mastigamoebidae, **C7**
Mastigogenina, C7
MASTIGOPHORA, C17
Matanzia, **C289**
MATHEWS, C570
Matthewina, C786
MAYER, C92, C95
Maylisoria, C135, **C179**
Maylisoriidae, C173
MAYNC, C226, C233, C238, C241, C243, C245
Mayorellida, C7
Mayorellidae, C7
Mayorellina, C6
meandrine, **C61**
Meandroloculina, **C455**
Meandropsina, **C485**
Meandropsinidae, C482, C485
Meandropsininae, C137, C149, C156, **C485**
Meandrospira, C136, **C440**
median section, **C61**
Mediocris, C396
megalospheric, **C61**
Megalostomina, C769

Megathyra, C522
Meigen's reaction, C95
meiosis, **C10, C18**
Melonia, C506, C761
Melonidae, C753
Melonis, C138, C746, **C761**
Melonisinae, C152, C753
Melonites, C505
Melossis, C761
Mendesia, C262
Meneghinia, C482
Mesoendothyra, **C232**
Mesoendothyridae, C152, C225
Mesopora, C784
Mesorbitolina, C309
Mesoschubertella, **C401**
Messina, C783
Metadoliolina, C427
METAMMIDA, C184
Metammida, C145
Metarotaliella, **C784**
Metaschwagerina, C429
Meyendorffina, **C295**
Micatuba, C186
Micatubinae, C186
MICETOZOOS, C11
Microchlamyini, C19
Microchlamys, **C20**
Microcometes, C67, **C169**
Microcometesidae, C165
Microcometidae, C165
Microcometides, C164
Microcorycia, **C19**
Microcoryciidae, **C19**
microgranular, **C61**
microgranular tests in foraminifers, **C93**
Microgromia, C166
Microgromiidae, C165
Micrometula, **C179**
microsomes, **C61**, C66
microspheric, **C61**
Migros, **C269**
Mikhailovella, C351
MIKHAYLOV, C435
MIKLUKHO-MAKLAY, K. V., C325
Mikrocoryciinae, C19
Mikrocoryciini, C19
Mikrogromia, C166
Mikrogromiidae, C165
Mikrogromiini, C165
Miliammina, C90, C129, C137, **C220**
Miliola, C139, **C468**
Miliolacea, C58, C60, C61, C68, C71, C78, C155, **C436,** C776
Milioletta, C458
Miliolicae, C436
MILIOLIDA, C152, C436
Miliolida, C143, C458
MILIOLIDACEAE, C436
Miliolidae, C87, C92, C120, C137, C142, C144-C150, C152, C155, **C458**
MILIOLIDAE, C145, C436
Miliolidae holostreptae, C448
Miliolidae opisthostreptae, C458

MILIOLIDEA, C436
Miliolidea, C143, C144, C148-C152, C436, C458
Miliolidea genuina, C143, C458
Miliolidee, C458
Miliolidina, C458
MILIOLINA, C154, C155, **C436**
Miliolina, C458, C466, C468
Miliolinae, C137, C144, C147, C155, **C468**
Miliolinae, C458
miliolinе, **C61**
Miliolinella, **C466**
Miliolinellinae, C155, **C466**
Miliolinidae, C144, C458
Miliolininae, C144, C468
Miliolites, C468, C506
Miliolithes, C468
Miliolitidae, C458
Miliospirella, C137, **C602**
Millarella, C786
MILLEPORITA, C511
Milleporita, C436
MILLER, C128, C129
Millerella, C350, C371, C374, C381, C383, C384, C389, **C396**
Millerella Zone, C389
Millettella, C23
Millettia, **C558**, C569
Millettina, C200
Mimosina, **C563**
mineralogical composition, calcareous Foraminiferida, **C95**
Miniacina, **C705**
Miniacinidae, C702
Miniuva, C565
Minoella, C429
Minojapanella, **C401**
Minouxia, **C282**
Miogypsina, C139, **C650**
Miogypsinella, C650
Miogypsinidae, C139, C148-C152, C156, **C650**
Miogypsininae, C147, C650
Miogypsinita, C650
Miogypsinoides, C139, **C650**
Miogypsinopsis, C650
Miolepidocyclina, **C650**
Mirfa, **C784**
Mirga, **C784**
Miscellanea, C138, C645, **C647**
Miscellaneidae, C150, C151
Miscellaneinae, C152
Miscellanoides, C617
Misellina, C383, C387, **C427**
Misellininae, C427
Misilus, **C785**
Mississippina, **C776**, C777
Mitosis, **C18**, **C61**
MIXOGASTEROS, C11
MIXOGASTROS, C12
MIXOMICETOS, C11
MOEBIUS, C91
MÖLLER, VON, C57, C93, C333, C397
Moellerina, C333, C427
MOLL, VON, C55

Molnaria, **C785**
Monadinae Tetraplastae, C795
Monadineae azoosporeae, C794
MONADINEAE ZOOSPORAE, C12
Monadineae Zoosporeae, C795
MONADINEN (MONADINEAE), C11
Monadinen (Monadineae), C794
Monalysidium, **C484**
Monamoebidae, C6
Monamoebina, C6
Monetulites, C645
Monobidiidae, C795
MONOCYPHIA, C5
Monocyphia, C22, C23, C34, C47, C164
Monocystis, **C785**
Monodiexodina, C420
Monogenerina, C323
monolamellar wall, **C99**
monolamellid, **C61**
MONOLAMELLIDEA, C725
Monolamellidea, C150, C151, C572
Monolepidorbis, C710
MONOSOMATIA, C164, C184, C436, C511
MONOSTEGA, C5, C19, C511
MONOSTEGIA, C5, C19
MONOSTÈGUES, C40, C142, C511
Monostègues, C142
MONOSTOMATA, C40, C164
Monostomata, C40, C47, C164
Monostomina, C40, C47, C53, C164, C173
Monostominae, C40, C164, C173
MONOTÁLAMOS, C40, C164
Monotaxinoides, C136, **C358**
Monotaxis, C358
MONOTHALAMIA, C5, C143, C164, C184, C436, C511
Monothalamia, C22
Monothalamia amphistomata, C47, C164
MONOTHALAMIA FILOSA, C39
MONOTHALAMIA LOBOSA, C5
MONOTHALAMIA MONOSTOMATA, C19
MONOTHALAMIA RHIZOPODA, C5
MONOTHALAMIEN, C164
MONTAGU, C55
MONTANARO GALLITELLI, C525, C526, C547, C654, C656
DE MONTFORT, C55
Montiparus, C425
MOORE, C387
Mooreinella, C136, **C277**
Moravammina, C135, **C319**
Moravamminidae, C135, C136, C155, **C317**
Moravammininae, C150, C155, C317, **C319**

MORET, C89
MORISHIMA, C128
morphologic terms, Foraminiferida, **C58**
morphologic terms, Thecamoebians, **C17**
MORROW, C546
Morulaeplecta, **C251**
Moscovella, C434
Moscoviella, **C434**
Mstinia, C352
mucopolysaccharide, **C88**
mucoprotein, **C88**
Mucronina, **C528**
Mucronines, C528
Multicyclina, C722
Multidiscus, C356
Multifidella, **C277**
Multilepidina, C721, C723
multilocular, **C61**
Multiloculidae, C142, C458
Multiloculidaeen, C458
Multiloculina, C458
multiple tunnels, **C61**, **C380**
Multiseptida, C135, **C328**
Multispirina, C505, **C510**
Multoloculidaeen, C458
MUNIER-CHALMAS, C472, C473
Murus reflectus, **C61**
MYATLYUK, C356
MYCETOZOA, C11, C12, C789, C794
MYCETOZOEN, C11
MYCETOZOEN (EUMYCETOZOEN), C11
MYCETOZOEN (SCHLEIMTIERE), C11
MYCETOZOIDA, C2, C8, **C11**, C790
MYCETOZOIDA, C789
Mycetozoida, C794
MYCETOZOIDEA, C11
Mychostomina, C600
MYERS, C66, C75, C78, C81, C83, C128, C134, C368, C600, C604
Myrioporina, C482
Mytilus, C97
myxamoebae, **C10**
myxoflagellula, **C10**
MYXOGASTERES, C11
MYXOGASTRALES, C11
MYXOGASTRES, C8, C11
Myxoidea, C794
MYXOMYCETALES, C11
MYXOMYCETEAE, C11
MYXOMYCETEN, C11
MYXOMYCETES, C8, C11
MYXOMYCOPHYTA, C11
MYXOMYZETEN, C11
MYXOTHALLOPHYTA, C11
Myxothallophyta, C794
Myxotheca, C71, C74, C87, C134, C166, **C169**
Myxothecinae, C144, C146, C147, C164
MYXOZOA, C11, C789
Myxozoa, C794

Index

Nadinella, **C45**
Naegleriidae, C7
NAGAPPA, C119
Nagatoella, **C418**
Nanicella, C135, **C342**
Nanicellinae, C152, C342
Nankinella, **C397**
NANTILITES, C511
Nanushukella, C769
Nautilacea, C142
Nautilacés, C142
Nautilina, C520
Nautiloculina, **C443**
Nautiloida, C143, C225, C399, C458, C482, C512, C631, C742
Nautiloidae, C142, C225, C511
NAUTILOIDEA, C436, C511
Nautiloidea, C512, C742
Navarella, C137, **C245**
neanic, **C61**
NEAVE, C796
Nebela, C18, C27, **C32,** C34
Nebela, C32
Nebelida, C27
Nebelidae, C27
Nebelina, C27
Nebelinae, C27
Nebelini, C27
Nebella, C32
Nemogullmia, C71, C74, C166, **C169**
Nemophora, C495
Neoalveolina, C506
Neoangulodiscus, C440
Neoarchaediscus, C355
Neoarchaesphaera, C783
Neobulimina, C137, **C544**
Neocarpenteria, C707, C708
Neoclavulina, C281
Neoconorbina, C66, C85, **C582**
Neocribrella, **C680**
Neodiscus, C354
Neoeponides, **C680,** C682
Neoflabellina, **C522**
Neofusulinella, **C406**
Neogeinitzina, C325
Neogyroidina, C708
NEOHELLENOIDEA, C184, C511
Neolepidina, C723
Neooperculinoides, C647
Neopeneroplis, C482
Neorotalia, C612
Neoschwagerina, C376, C379, C386, C387, C393, C394, **C428**
Neoschwagerinaceae, C394
Neoschwagerinidae, C148-C150, C152, C426
Neoschwagerininae, C62, C63, C148-C150, C155, C365, C374, C376, C379, C380, C382, C383, C385-C387, C393, **C427**
Neostaffella, **C409**
Neotrocholina, C742
Neotuberitina, C321
Neouvigerina, C571

Nephrolepidina, C721
nepionic, **C62**
NEUMAYR, C56
Neusina, C208, C792
Neusinidae, C146, C147, C149, C151, C208, C790
Neusininae, C146, C147, C790
Nevillina, **C472**
Nezzazata, **C481**
Nipponitella, C370, C418
Nodellum, **C179.** C180
Nodobacularia, C93, **C455**
Nodobaculariella, **C456**
Nodobaculariinae, C137, C146-C148, C155, **C455**
Nodocyclina, C714
Nodogenerina, C559
Nodomorphina, C528
Nodophthalmidiinae, C148, C149, C455
Nodophthalmidium, **C456**
Nodoplanulis, C786
NODOSALIDIA, C184, C313, C511
Nodosalidia, C214, C320, C511
Nodosamminidae, C214, C320
Nodosarella, C654, C728, **C730**
Nodosaretta, C512
Nodosaria, C136, C142, **C512,** C516, C528, C537, C785
Nodosariacea, C63, C71, C99, C137, C153, C154, C156, **C511**
Nodosarida, C143, C512
Nodosaridae, C144, C145, C512
Nodosaridea, C143, C512
Nodosarideae, C512
NODOSARIDIA, C313, C511
Nodosarie, C512
Nodosariella, C512
Nodosariellidae, C512
Nodosariidae, C93, C120, C136, C139, C147, C150, C151, C156, **C511,** C782
Nodosariidea, C150, C511
Nodosariinae, 146, C147, C156, **C512**
Nodosarina, C511, C512
Nodosarinae, C144, C145, C512
NODOSARIOIDEA, C511
Nodosariopsis, C512, C514
Nodosaroum, C323
Nodosinella, C135, C215, **C323**
Nodosinellida, C320
Nodosinellidae, C135, C136, C144, C146, C147, C155, **C320,** C323
Nodosinellinae, C148, C155, **C323**
Nodosinum, **C215**
Nodulina, C216
Nodulinella, C783
nomina inquirenda, C155, C436, C782
nomina nuda, C155, C162
nomina nuda, Foraminiferida, C436, C782
Nonion, C71, C129, C138, C637, **C746,** C748

Nonionella, C138, **C748**
Nonionella, C748
Nonionellina, **C748**
Nonionellinae, C152, C745
Nonionia, C746
Nonionida, C143, C742, C745
Nonionidae, C99, C120, C137, C138, C146-C148, C150-C152, C161, **C742,** C746, C748
NONIONIDEA, C725
Nonionidea, C152, C742
Nonionina, C746
Nonioninae, C147, C152, C161 **C745**
Nonioninidae, C742
Nonionineae, C742
Nonionininae, C745
nonlamellar wall, **C99**
Norcottia, C571
Normanina, **C192**
Notorotalia, C642, C643
Notorotaliinae, C640, C643
Nouria, C94, **C220**
Nouriidae, C155, **C220**
Nouriinae, C148, C220
Novella, **C346,** C350, C782
Nubecularia, C78, C120, **C445**
Nubecularida, C143, C445
Nubeculariidae, C145, C445
Nubecularidea, **C192**
Nubecularidea, C155, **C445**
Nubeculariinae, C137, C146-C150, C155, **C445,** C796
Nubecularina, C445
Nubecularinae, C144, C445
Nubeculina, C93, **C456**
Nubeculinella, **C447**
Nubeculinellinae, C445
Nubeculopsis, **C448**
nucleoconch, **C62**
nucleolus, **C62**
nucleus, C18, **C62, C66**
NUDA, C5, C143
NUDA (GYMNAMOEBA), C5
Nuditestiidae, C40, C164, C173
Nummodiscorbis, **C763**
Nummofallotia, **C488,** C490
Nummoloculina, **C468**
Nummophaga, C183
Nummulacea, C142, C643, C645
Nummulacés, C142, C643
Nummularia, C645, C783
Nummulariidae, C643
Nummulina, C645
Nummulinetta, C643
Nummulinida, C143, C643
Nummulinidae, C144, C643
NUMMULINIDEA, C511
Nummulinina, C143, C645
Nummulita, C645
NUMMULITACEA, C511
NUMMULITACEAE, C511
Nummulitella, **C785**
Nummulites, C55, C95, C138, C183, C644, **C645,** C650, C685

Nummuliti, C643
nummulitic, C139
NUMMULITIDA, C152, C511
Nummulitida, C643
Nummulitidae, C138, C144-C148, C150-C152, C156, **C643**, C714, C785
NUMMULITIDAE, C145, C511
Nummulitidae, C645
NUMMULITIDEA, C511
Nummulitidea, C143, C643
Nummulitideae, C643
Nummulitidos, C643
Nummulitina, C643
Nummulitinae, C144, C145, C147, C150, C152, C156, **C645**
Nummulitinae, C643
NUMMULITINIDEA, C511
Nummulitique, C139
Nummulitoides, C647
Nummulostegina, **C399**
Numulitinae, C144
nutrition, holophytic, **C17**
nutrition, holozoic, **C18**
nutrition of foraminifers, **C87**
nutrition of thecamoebians, **C17**
nutrition, parasitic, **C18**
nutrition, saprozoic, **C18**
Nuttallides, C139, **C595**
Nuttallina, C595
Nuttallinella, **C595**
NYHOLM, C71, C183, C687, C688, C697

OBERHAUSER, C741
oblique section, **C62, C362**
Obliquina, C540
Obruchevella, C787, **C788**
Obsoletes, C425
Occidentoschwagerina, C420
occurrence of thecamoebians, C16
Oculina, C448
Oculosa, C145
Oculosiphon, C186
Odontodictya, C34
Oinomikadoina, C688, C689
Oketaella, **C418**
Okhotsk Sea, C121
Oligostegina, C90, C501
Olssonina, **C254**
Olympina, **C338**
OMARA, C219, C481
Omphalocyclinae, C147, C148, C150, C152
Omphalocyclus, C138, **C711, C712**
Omphalophacus, C685
Oncobotrys, **C785**
Ondogordius, C440
Oolina, C60, C63, C66, C70, C71, C85, C86, C153, **C540**
Oolininae, C156, **540**
Oolitella, **C539**
Oomycétes, C794
Oopyxis, **C26**
Operculina, C95, C645
Operculinella, C647

Operculinoides, C647
Opertorbitolites, **C498**
Ophidionella, C783
Ophiotuba, **C169**
Ophtalmidiidae, C445
Ophthalmidiidae, C445
Ophthalmidiidae, C148, C149, C150, C152, C445, C741
Ophthalmidiinae, C136, C148, C149, C150, C155, **C448**
Ophthalmidium, C129, **C448**
Ophthalmina, **C450**
OPISTHO-DISCHISTIDAE, C145
Opistho-Dischistidae, C250, C332
Opistho-Trichistidae, C145, C268
Opthalmidiidae, C146, C147, C445
Opthalmidiinae, C146, C147, C448
Orbiculina, C495
Orbiculines, C482
Orbiculinida, C143, C482
Orbiculininae, C145, C494
Orbientina, C631
D'ORBIGNY, C56, C142, C283, C458, C477, C493, C506, C508, C569, C574, C575, C585, C642, C690, C748, 750, C753, C781, C785
Orbignyina, C289
Orbignyna, **C289**
Orbitammina, **C234**
Orbitella, C710
Orbitoclypeidae, C150, C712
Orbitoclypeinae, C150, C152, C712
Orbitoclypeus, C714
Orbitocyclina, C711
Orbitocyclinoides, C711
Orbitoidacea, C77, C100, C109, C137, C138, C154, C156, **C678**
Orbitoidae, C710
Orbitoidee, C710
Orbitoides, C138, **C710,** C711
Orbitoidicae, C678
Orbitoidida, C710
Orbitoididae, C138, C145-C152, C156, **C710**
Orbitoidinae, C147-C150, C152, C710
Orbitoina, C722
Orbitolina, C137, **C309**
Orbitolinella, **C308**
Orbitolinida, C308
Orbitolinidae, C59, C61-C63, C137, C146-C150, C155, **C308**
Orbitolininae, C146, C148, C308
Orbitolinoides, C309
Orbitolinopsis, C309
ORBITOLITACEA, C436
Orbitolites, C92, C98, C138, C183, **C498**
Orbitolites, C482

Orbitolithes, C498
Orbitolitidae, C144, C145, C482
Orbitolitidinae, C496
Orbitolitina, C482
Orbitolitinae, C145-C147, C149, C496
Orbitophaga, C183
Orbitophage, C183
Orbitopsella, **C308**
Orbitorotalininae, C687
Orbitosiphon, C711
Orbitulina, C309
Orbitulinidea, C496
Orbitulita, C482
Orbitulites, C498
Orbitulitida, C462, C482
Orbitulitidea, C143, C482
Orbitulitideae, C482
Orbitulitidee, C482
Orbulina, C139, **C675,** C676
Orbulinaria, C501
Orbulinida, C143, C669, C675
Orbulinidae, C147, C150, C669
Orbulininae, C146, C147, C150, C152, C156, **C675**
Orcadellaceae, C14
Orcadellacées, C14
Orcadelleae, C14
Orcadellidae, C14
Ordovicina, C135, **C200**
Oreas, C520
Oridorsalis, **C750**
Orientalia, **C277**
Orientella, **C436**
Orientoschwagerina, **C419**
orifice, **C62**
ornamentation, **C111**
Ornatanomalina, C621
Orobias, **C785**
Orthella, **C456**
Orthocera, C512
Orthoceras, C142
ORTHOCERATA, C184
Orthocerata, C142, C436, C512
Orthoceratidae, C512
Orthocérés, C142, C512
Orthocerina, **C785**
Orthocérine, C785
Orthocerinida, C512
Orthocyclina, C715
Orthokarstenia, C137, **C567**
ORTHOKLINOSTEGIA, C145, C652, C678, C725
Orthoklinostegia, C214, C320, C394, C436, C572, C605
Orthomorphina, **C522**
Orthophragmina, C714
Orthophragminidae, C712
Orthophragmininae, C712
Orthoplecta, **C556**
ORTHOSTILI, C511
ORTHOSTILI (PERFORATA), C184
Orthovertella, C136, **C440**
Oryctoderma, C136, **C208**
Oryzaria, C506
Osangularia, C61, C96, C138, **C752**

Index

Osangulariidae, C100, C138, C161, **C752**
oscillation chart depth, C131
Otostomum, **C785**
Ouladnailla, C783
Outline of classification, **C154**
Ovalveolina, C503, **C510**
Ovolina, **C785**
Ovulida, C783
Ovulina, C540
Ovulinetta, C445
Ovulinida, C537
Ovulitidea, C143
Oxinoxis, **C248**
Oxygen, effect of, **C129**
OZAWA, **C397**
Ozawaia, **C640**
Ozawaina, **C784**
Ozawainella, **C396**
Ozawainellidae, C152, C155, C371, C372, C380, C384-C386, **C394**
Ozawainellinae, C152, C394

Paalzowella, C136, **C741**
Pachyphloia, C136, **C326**
Padangia, C325
Palachemonella, C135, **C315**, C316
Palaeobigenerina, C136, **C335**
Palaeocornuspira, C783
Palaeofusulina, **C409**
PALAEOHELLENOIDEA, C313
Palaeomiliolina, **C462**
Palaeonubecularia, C136, **C332**
Palaeonummulites, C647
Palaeotextularia, C136, **C333**
Palaeotextulariidae, C136, C155, **C332**
Palaeotextulariinae, C147, C152, C332
Palaeovalvulina, C337
Palaeovalvuloria, C338
paleoecology, **C119**
Paleogaudryina, C269
Paleopolymorphina, C530
Pallaimorphina, C744
palmate, **C62**
Palmerinella, **C598**
Palmula, **C522**
Pamphagidae, C40
Pamphagus, C44
Pandaglandulina, **C522**
Pandoglandulina, C522
Pansporellidae, **C7**
PANTOSTOMATA, C5
Paraarchaediscus, C354
Paraboultonia, **C401**
Paracaligella, C316
parachomata, **C62**, **C364**, **C386**
Paracyclammina, **C236**
Paradentalina, **C533**
Paradoxiella, **C403**
Paraendothyra, C136, **C346**
Paraeofusulina, C405
Parafissurina, **C541**
Parafrondicularia, C525
Parafusulina, C369, C371, C372,
C383, C385, C393, C394, **C420**
Parafusulina Zone, C393
Paragaudryina, C272
Parageinitzina, C326
Paralieberkuehnia, **C181**
parallel section, **C62**, **C362**
Paramébidos, C8
Paramilioliidae, C149, C220
Paramillerella, C384, **C396**
Paramoeba, C8
Paramoebidae, **C8**
Paranebela, **C50**
Paranonion, C595
Parapachyphloia, C326
Parapermodiscus, C326
Paraplectogyra, C136, **C346**
Paraquadrula, **C34**
Paraquadrulidae, **C34**
Paraquadrulinae, C34
Parareichelina, **C435**
Parareichelina, C435
Pararobuloides, C342
Pararotalia, C137, **C612**
Paraschwagerina, C370, C371, **C420**
Parasites in Foraminiferida, **C70**
Paraspiroclypeus, C647
Parastaffella, C435
Parasyniella, **C788**
Paratextularia, C135, **C336**
Parathurammina, C135, **C314**, C785
Parathuramminacea, C61, C93, C155, **C313**
Parathuramminidae, C135, C151, C155, **C313**, C316 C796
Parathuramminidae, C151, C313
Paratikhinella, C135, **C319**
paratrimorphic, **C78**
Paratrocholina, C740, C741
Paratuberitina, C321
Paraverbeekina, C427
Parawedekindellina, C412
Parazellia, C420
Pareuglypha, C47
PARKER & JONES, C272, C279, C506, C525, C569, C580, C582, C584, C606, C629, C639, C680, C731, C748, C785
PARKER, JONES & BRADY, C574
Parkeriada, C143
PARKINSON, C508
Parmulina, **C21**
Paromalina, **C763**
Paronaea, C647
Paronia, **C785**
Paronia, C647
PARR, C281, C563, C584, C736
Parrella, C752
Parrellina, **C642**, C643
Parrelloides, C757
Parrelloididae, C151, C753
Parrina, **C478**
Parvicarinina, C580
Parvigenerina, C255
Parvistellites, C196
Patellina, C66, C68-C71, C81, C83, C84, C87, C104, **C603**, C731
Patellinella, **C582**
Patellininae, C137, C145, C147, C148, C150, C153, C156, **C602**
Patellinoides, **C604**, C731
Pateoris, **C462**
Patrocles, C520
Paulinella, C17, C18, **C54**
Paulinellidae, **C54**
Paulinellinae, C54
Paumotua, **C682**
Pavonina, **C563**
Pavoninidae, C145, C559, C561
Pavonininae, C139, C146, C148, C156, **C561**
Pavoninoides, **C480**
Pavonitina, **C296**
Pavonitinidae, C137, C155, **C291**, C795
Pavonitininae, C155, C291, **C295**
Pealerina, C533
Pectinaria, **C785**
peduncle, **C18**
Pegidia, C139, **C625**
Pegidia, C625
Pegidiida, C604
Pegidiidae, C147, C149, C150, C605, C625, C680
Pegidiinae, C147, C156, **C625**
Pellatispira, C644, **C647**
Pellatispirella, **C640**
pellicle, **C18**
PELOMYXACEA, **C6**
Pelomyxidae, **C6**
Pelorus, C631
Pelosina, **C200**
Pelosinella, C200
Pelosininae, C146, C147, C148, C150, C196
Pelosphaera, **C201**
PENARD, C51, C171, C181
Penardia, **C40**
Penardiidae, **C39**
Penardeugenia, **C46**
Penardiella, C29
Penardochlamys, **C22**
Penardogromia, **C181**
Peneroplida, C143, C482
Peneroplidae, C120, C146, C148, C149, C152, C482
Peneroplidea, C143, C144, C482
Peneroplideae, C482
Peneroplidee, C482
Peneropliididae, C150, C482
Peneroplidina, C482
Peneroplidinae, C144, C482
Peneroplinae, C145, C147, C156, **C482**
Peneroplinae, C482
peneropline, **C62**
Peneroplis, C61, C62, C64, C66, C70, C78, C83, C87, C88, C91, C92, **C482**, C484
Penoperculinoides, C617

Penoperculoides, **C617**
Pentagonia, C35
Pentasyderina, C783
Pentellina, C468
PERFORATA, C143, C184, C511
Perforata, C164
perforations and pore plates, **C111**
perforate, **C62**
PERFORIDA, C145, C511
PERFORINA, C184, C511
Perichaenacea, C15
PERICHAENACEAE, C15
Perichaenaceae, C15
Perichaenaceen, C15
Perichaeneae, C15
Perichaenei, C15
periembryonic chambers, **C62**
Periloculina, **C477**
Periples, C514
Perisphinctina, C520
peristome, **C62**
PERITRICHEA, C14
PERITRICHEAE, C14
PERITRICHEEN, C14
Permodiscus, C136, **C356**
Permodiscus, C356
Pernerina, C283
Petalopella, **C38**
Petalopus, C38
Petchorina, C315, **C796**
Pfenderina, **C291**, C795
Pfenderinidae, C291
Pfenderininae, C155, **C291**, C795
Phacites, C645
Phainogullmia, **C181**
Phalsopalmula, C522
Phanerostomum, **C785**
Pharamum, C519
Phenacophragma, C137, **C245**
Phialina, C518
phialine, **C62**
Phleger, C120, C124-C126, C128, C134
Phleger & Parker, C600
Phonemus, C519
Phonergates, C44
phrenothecae, **C62, C377**
Phryanella, C38
Phryganella, C18, **C38**
Phryganellidae, C38
Phyllopsamia, **C258**
Physalidia, **C587**
PHYSARACEAE, C13
Physaraceae, C13
Physaracées, C13
Physaracei, C13
PHYSARAE, C13
PHYSARALES, C13
Physarea, C13
Physareae, C13
Physareen, C13
Physarées, C13
Physarei, C13
Physarella, C9
Physaridae, C9, **C13**
PHYSARINA, C13

Physarinae, C9, **C13**
PHYSARIINEAE, C13
Physaroidea, C13
Physaroinae, C13
Physarum, C9
PHYSEMARIA, C184
Physochila, C29
Physochilini, C27
Physomphalus, **C785**
Phytomyacei, C12
Phytomyxaceae, C12
PHYTOMYXIDA, C11
Phytomyxidacées, C12
Phytomyxidae, C12
PHYTOMYXINAE, C11, C12
PHYTOMYXINEA, C11, C12
PHYTOMYXINEES, C12
PHYTOMYXINI, C11, C12
Pijpersia, C139, **C582, C757**
Pilalla, C195
Pileolina, C588
pillars, **C111**
pillars, incised, **C113**
pillars, inflational, **C113**
pillars, textural, **C113**
Pilulina, **C201**
Pilulinida, C194
Pilulinidae, C194
Pilulinina, C194
Pilulininae, C144, **C196**
PILZTIERE, C11
Pinaria, **C730**
Pirulina, C533
Pisolina, **C399**
Piveteau, C149
Placentammina, **C196**
Placentula, C678
Placentulae, C678
Placocista, **C50**
Placocysta, C50
Placopsilina, C181, **C247**
Placopsilinella, C175, **C181**
Placopsilinidae, C146, C149, C152, C181, C225
Placopsilininae, C146-C148, C152, C155, **C247**
Placopsum, C535
Plagiophryiinae, C164
Plagiophrys, **C170**
Plagiopyxidae, **C24**
Plagiopyxis, C18, **C24**
plagiostome, **C18**
Plagiostomella, C137, **C763**
Plakopodaceae, C795
Planctostoma, **C256**
Planiinvoluta, **C444**
planispiral, **C62**
Planispirillina, C137, **C601**
Planispirina, C441
Planispirinella, C99, **C443**
Planispirinellinae, C448
Planispirinoides, **C453**
Planoarchaediscus, C355
Planocamerinoides, **C647**
Planodifflugia, C34
Planodifflugiinae, C34
Planodiscorbis, **C583**
Planoelphidium, C632
Planoendothyra, **C346**

Planoglobulina, C138, **C655, C656**
Planogypsina, **C698**
Planogyrina, C659
Planomalina, C63, C138, **C656**
Planomalinidae, C63, C138, C156, **C656**
Planomalininae, C656
Planopulvinulina, **C682**
Planorbulina, C61, C66, C68, C75, C77, C78, C120, C134, C139, C688, **C693**
Planorbulinella, C139, **C694, C698**
Planorbulinidae, C139, C144, C146, C148, C150-C152, C156, C688, **C692**
Planorbulinidae, C692
Planorbulininae, C147, C148, C692
Planorbulinoides, **C691**
Planorotalia, C667, C668, C683
Planorotalites, C668
Planulacea, C142, C482
Planulacés, C142, C482
Planularia, **C522**
Planularia, C522
Planulina, C138, **C686**, C764
Planulinella, C763, C764
Planulininae, C156, **C686**
Planulinoides, **C584**
plasmagel, C2, **C18**
plasmalemma, **C18**
plasmasol, C2, **C18**
PLASMODIATA, C11
Plasmodiofóridos, C12
Plasmodiophoraceae, C12
PLASMODIOPHORALES, C12
Plasmodiophorea, C12
PLASMODIOPHOREAE, C11
Plasmodiophoreae, C12
Plasmodiophoreen, C12
Plasmodiophoridae, **C12**
PLASMODIOPHORINA, **C12**
Plasmodiophorinae, C12
plasmodium, **C10**
PLASMODROMA, C5
PLASMODROMATA, C5
plasmogamy, **C10, C18**
plastogamic reproduction, in Foraminiferida with amoeboid gametes, **C81**
plastogamic reproduction, in Foraminiferida with triflagellate gametes, **C78**
plastogamy, C60, **C62**
Platoum, C166
Platyneminées, C14
Platyoecus, **C785**
Plecanioidea, C144, C250
Plecanium, C253
Plectina, **C283**
Plectinella, C255
Plectofrondicularia, C139, **C525**
Plectofrondiculariidae, C512
Plectofrondiculariinae, C139, C147, C148, C150, C153, C156, **C525**
Plectofrondicularinae, C146

Plectofusulina, **C409**
Plectogyra, C343
plectogyral, **C62**
Plectogyridae, C342
Plectogyrina, C343
Plectogyrinae, C152, C343
Plectorecurvoides, **C258**
Plectorecurvoidinae, C155, **C258**
Plectotrochammina, **C279**
Pleiona, C517
Plesiocorine, C513
Plesiocoryna, C514
Pleurites, **C785**
Pleurophryini, C173
Pleurophrys, **C181**
Pleurostomella, C71, **C725**, C727, C730
Pleurostomellida, C725
Pleurostomellidae, C99, C108, C138, C147, C152, C156, **C725**
Pleurostomellidea, C143, C725
Pleurostomellideae, C725
Pleurostomellina, C725, C727
Pleurostomellinae, C156, **C725**, C729
Pleurostomelloides, C253
Pleurostomina, **C785**
Pleurotrema, **C785**
plicate, **C62**
Plicatilia, C458
Pliolepidina, **C722**
Pliorbitoina, C723
PLUMMER, C210, C250, C333
Plummerella, C663
Plummerinella, C136, **C444**
Plummerita, C138, **C663**
PLURILOCULINIDEA, C149, C184, C313, C436, C511
Pninaella, C759, C760
POCHE, C8
Pocheinidae, **C11**
Podolia, **C452**
podostyle, **C62, C67**
POKORNÝ, C150, C324, C526
Pokornyella, C618
Pokornyellina, C138, **C618**
POLITÁLAMOS, C164
Pollonites, C458
Pollontes, C458
Polychasmina, **C215**
POLYCYCLICA, C511
Polycyclica, C436
Polydiexodina, C369, C371, C372, C380, C381, C383, C385, C393, C394, **C420**
Polydiexodina Zone, C393
Polydiexodininae, C152, C415
Polylepidina, C718, **C724**
Polymastigamoebidae, C7
Polymorphina, C62, C138, **C530**, C531, C533, C736
POLYMORPHINACEA, C511
Polymorphinae, C145, C530
Polymorphininae, C144, C146, C147, C149, C152, C156, **C530**
polymorphine, **C62**
Polymorphinella, C514

polymorphines, les, C530
Polymorphinida, C143, C530
Polymorphinidae, C120, C136, C137, C142, C145-C152, C156, **C530**, C736, C782, C783
Polymorphinidea, C143, C144, C530
Polymorphinideae, C530
Polymorphinidée, C530
Polymorphinidées, C530
Polymorphinina, C530
Polymorphinoides, C514
polymorphism, **C62**
Polyorbitoina, C723
POLYPES À RAYONS, C164
POLYPES CORALLIGENES, C164
Polyphragma, C248
Polyphragmidae, C225
Polyphragminae, C146, C148, C248
Polysegmentina, **C472**
POLYSOMATIA, C164
POLYSTEGIA, C184, C436
Polystoma, C511, C572
Polystomata, C7, C19, C164
Polystomatium, C632
Polystomella, C632, C642
Polystomellida, C143, C631
Polystomellidae, C145, C631
Polystomellidea, C143, C631
Polystomellina, C139, **C642**, C643
Polystomellina, C143, C631
Polystomellinae, C144, C145, C146, C631
Polystomidae, C165
Polystominae, C165
Polystomini, C165
Polytaxis, C136, **C337**
Polythalama, C225, C458, C482, C503, C511, C605, C631, C643, C742
POLYTHALAMACEA, C142, C164, C511
POLYTHALAMACÉS, C142, C511
POLYTHALAMIA, C56, C143, C164
Polythalamia, C142
POLYTHALAMIEN, C164
POLYTHALAMIIS, C56, C164
polythalamous, **C62**
POLYTHALAMOUS CEPHALOPODA, C142
Polytrema, C95, C786
Polytrematidae, C702
Polytremidae, C702
polyvalence, **C70**
polyvalent individuals, **C62**
Polyxenes, C688
Pontigulasia, C17, C18, **C36**
Pontigulasiidae, C34
porcelaneous, **C62**
porcelaneous tests of foraminiferans, **C91**
PORCELLANEA, C436
Porcellanea, C145

pore plug, **C62, C111**
poreless margin, **C59**
Poritextularia, **C254**
Poritida, C482
Poroarticulina, **C480**
Poroeponides, C680, C682, **C683**, C684
Poronaia, C279
Porosia, **C32**
Porosononion, C640
Porosorotalia, C139, **C643**
Porospira, C754
portici, **C62**
Porticulasphaera, C139, **C676**
porticus, **C62**
postseptal passage, **C62**
POYARKOV, C314
Praealveolina, C503, C505, **C510**
Praeammoastuta, C238
Praebulimina, C137, **C545**
Praecosinella, C783
Praeglobobulimina, C138, **C561**
Praeglobotruncana, C62, C138, **C659**
Praelacazina, C465
Praelamarckina, C137, **C769**
Praeparafusulina, **C435**
Praepeneroplis, **C484**, C490
Praerhapydionina, C493
Praerotalininae, C586, C753, C771
Praerotalipora, C783
Praesorites, C487
Praesumatrina, **C432**
Praeuvigerina, C568
Praevirgulina, C733
Prantlitina, C18, **C36**
Prantlitinopsis, C36
Premnammina, C783
preseptal passage, **C62**
Preverina, C621
primary aperture, **C62**
primary axial septulum, **C62**
primary septulum, **C62**
primary transverse, septula, **C379**
primary transverse septulum, **C62**
Problematina, C740
Problematininae, C147, C739
Procerolagena, C518
Proemassilina, C452
Profusulinella, C61, C65, C367, C371, C374, C384, C385, C389, **C410**
Profusulinella Zone, C389
proloculi, **C62**
proloculus, **C62, C362, C367**
proloculus pore, **C62, C368**
Propermodiscus, C354
Proporocyclina, **C717**
Proroporus, C549
PROTAMMIDA, C184
Protammida, C145
proteinaceous wall, **C88**
Protelphidium, C138, **C639**
Proteana, C794
PROTEINA, C5

Proteomyx, C794
Proteomyxae, C794
Proteomyxés, C794
Proteomyxiae, C794
Proteomyxida, C794
Proteomyxidea, C794
Proteonina, C216
Proteonininae, C146, C215
protheca, **C62**
Protobotellina, **C190**
PROTOCOCCALES, C12
Protocucurbitella, **C36**
Protocyclina, C786
Protocystidae, C145, C194
Protodermaceae, C14
PROTODERMEAE, C14
Protodermiaceae, C14
PROTODERMIEAE, C14
protoforamen, **C58, C62, C108**
PROTOFORAMINATA, C149, C184, C511
Protoforaminata, C151
Protoglobobulimina, C561
Protomyxées, C794
PROTOMYXIDAE, C5
Protomyxidea, C794
Protonodosaria, C323
Protopeneroplis, **C741**, C742
Protopeneroplis, C484
protoplasm, **C18, C62**
protoplasmic body, **C65**
protoplast, C62
PROTOPLASTA, C5, C39
protopore, **C62**, C111
Protoschista, **C215**, C216
Prototrichia, C15
Prototrichiaceae, C15
Prototrichieae, C15
Prototrichiinae, C15
PROTOZOA, C144
Protriticites, C425
Protrudentiidae, C34, C47
proximal, **C62**
Psammatodendron, C192
PSAMMATOSTICHOSTEGIA, C145
Psammatostichostegia, C184
Psammechinus, C783
Psammella, C195, C792
Psammetta, **C792**
Psammina, **C792**, C794
Psamminae, C792
Psamminidae, **C792**
Psamminidae, C792
PSAMMINIDEA, C789
Psamminopelta, C137, **C221**
Psammolingulina, **C215**
Psammolychna, C783
Psammonyx, C135, **C212**
Psammopemma, C792, **C794**
Psammoperidia, C783
Psammophax, C196
Psammophis, C214
Psammophyllum, C792
Psammoplakina, C792
Psammoscene, C204
Psammosiphon, C194, C786
Psammosiphonella, C186
Psammosiphonellinae, C186

Psammosphaera, C90, C91, C135, **C195**
Psammosphaerida, C194
Psammosphaeridae, C145, C194
Psammosphaerinae, C146, C147, C148, C150, C151, C155, **C194**
Psammozotika, C783
Psecadium, C537
Pseudarcella, **C523**
Pseudastrorhiza, C135, **C196**
Pseudastrorhizula, **C785**
Pseudawerintzewia, **C32**
Pseudedomia, **C490**
Pseudobolivina, **C255**
Pseudobolivininae, C155, **C255**
Pseudobradyina, C353
Pseudobulimina, C95, C108, C139, C781, **C782**
pseudocapillitium, **C10**
pseudocarina, **C62**
pseudochambers, **C62**
pseudochitin, **C18, C62**
pseudochitinous tests, **C88**
Pseudochlamys, C17, C20
Pseudochoffatella, C137, **C233**
Pseudochoffatella, C233
Pseudochrysalidina, **C290**
Pseudocibicidoides, C688, C690
Pseudocitharina, C516
Pseudoclavulina, C279
Pseudococoscinoconus, C783
Pseudocucurbitella, C36
Pseudocyclammina, **C233**
Pseudodifflugia, C19, **C46**
Pseudodifflugiidae, C40
Pseudodifflugiinae, C40
Pseudoditrema, **C172**
Pseudodoliolina, C383, **C427**
Pseudoendothyra, **C435**
Pseudoepistominella, **C776**
Pseudoeponides, **C598**, C750
Pseudofrondicularia, C518
Pseudofusulina, C62, C371, C377, **C420**
Pseudofusulinella, **C410**
Pseudofusulininae, C415
Pseudogaudryina, **C269**
Pseudogaudryinella, **C272**
Pseudogeinitzina, C326
Pseudoglandulina, C512
Pseudogloborotalia, **C683**
Pseudoglomospira, C135, **C319**
Pseudogoesella, C290
Pseudogromiinae, C40
Pseudoguembelina, C138, **C656**
Pseudogypsina, C786
Pseudohastigerina, C665, C666
Pseudo-Heliozoa, C795
Pseudo-Heliozoaires, C794
Pseudohyperammina, **C190**
Pseudolamarckina, C137, **C769**
Pseudolepidina, C139, C721, **C724**
Pseudolepidolina, C432
Pseudolituola, C783
Pseudolituonella, **C290**
Pseudomassilina, **C463**
Pseudonebela, **C32**

Pseudonebelinae, C19
Pseudonodosaria, C136, **C523**, C785
Pseudononion, C747
Pseudonovella, **C782**
Pseudonubeculina, C455
Pseudonummulites, C647
Pseudopalmula, C135, **C337**
Pseudopalmulidae, C152, C335
Pseudoparellinae, C572
Pseudoparrella, C578
Pseudoparrellidae, C152, C572
Pseudopatellinella, **C584**
Pseudopatellinoides, **C584**
Pseudophragmina, C138, **C715**
Pseudoplacopsilina, C205
Pseudoplanulinella, C763, C764
PSEUDOPLASMODIDA, C11
Pseudoplasmodidae, C12
PSEUDOPLASMODIES, C11
pseudoplasmodium, **C10**
pseudopodia, **C18, C63, C66**
Pseudopodia, C22
pseudopodial trunk, **C18, C63**
Pseudopolymorphina, **C533**
Pseudopolymorphinoides, **C533**
Pseudopontigulasia, **C36**
Pseudorbitella, C711
Pseudorbitellinae, C710
Pseudorbitoides, **C725**
Pseudorbitoididae, C138, C150, C156, **C725**
Pseudorbitoidinae, C149, C150, C152, C725
Pseudorbitolina, **C491**
Pseudoreophax, **C269**
Pseudorotalia, C139, **C613**
Pseudoruttenia, C582
Pseudoschwagerina, C370, C379, C383, C393, **C420**
Pseudoschwagerina Zone, C393
Pseudosiderolites, C138, **C619**
Pseudosigmoilina, C783
Pseudospiroloculina, C783
Pseudospiroplectinata, **C272**
Pseudosporea, C795
Pseudosporeae, C795
Pseudosporeen, C795
pseudospores, **C10**
Pseudosporidae, **C795**
Pseudosporinae, C795
Pseudostaffella, C370, C371, C383-C385, **C410**
Pseudostaffellinae, C404
pseudostome, **C18**
Pseudosumatrina, C429
Pseudotetrataxis, C337
Pseudotextularia, C138, **C656**
Pseudotextularia, C656
Pseudotextulariella, **C295**
Pseudotriplasia, C296
Pseudotristix, C136, **C524**
Pseudotriticites, C405
Pseudotriticitinae, C404
Pseudotruncatulina, C759, C760
pseudoumbilicus, **C63, C678, C683**
Pseudouvigerina, C137, **C568**
Pseudovaginulina, C515

Index

Pseudovalvulineria, C753, C759
Pseudovermiporella, C136, **C450**
Pseudowebbinella, **C208**
Pseudowedekindellina, **C412**
Pseudowoodella, **C619**
Pseudoyabeina, **C436**
Pteroptyx, **C785**
Pterygia, C32
Ptychocladia, C136, **C328**
Ptychocladiidae, C135, C136, C149, C155, **C328**
Ptychocladiinae, C155, **C328**, C787
Ptychomiliola, **C465**
Ptygostomum, **C785**
Ptyka, C783
Pullenia, C138, **C748**
Pulleniatina, C139, **C671**
Pulleniatininae, C146, C147, C150, C152, **C669**
Pullenidae, C144, C745
Pulleninae, C745
Pullenoides, **C548**
Pulsiphonina, C96, C138, **C763**
Pulvinulina, C678
Pulvinulinella, C578, C586
Pulvinulinidae, C149, C150, C151, C678, C759
Pulvinulininae, C145, C678
Pulvinulus, C678
Punjabia, C679
Pupina, C279
PURI, C281
Pustularia, C705
Puteolina, C482
Puteolus, C482
Putrella, **C435**
Pycnochila, C34
pycnotheca, **C63**, **C373**
Pylodexia, C669
Pyramidina, **C546**
Pyramidulina, C512
Pyramis, C328
Pyrgo, C59, C92, C106, **C465**
Pyrgoella, **C465**
Pyropilus, **C702**
Pyrulina, **C533**, C782, C783
Pyruline, C533
Pyrulinella, C533
pyrulines, les, C533
Pyrulinoides, C136, **C533**, C782
Pyxidicula, C22

Qataria, **C308**
Quadratina, C539
Quadratobuliminella, **C546**
Quadrimorphina, C138, C743, **C744**, C745
Quadrula, C27, C32
Quadrulella, C27, **C32**
Quadrulellinae, C27
Quadrulidae, C27
Quadrulina, C27, C535
Quasiarchaediscus, C355
Quasiendothyra, C135, **C346**, C795
Quasiendothyrinae, C343
Quasifusulina, C380, C383, C385, **C412**

Quasifusulininae, C404
Quasifusulinoides, **C436**
Quasifusulinoides, C436
Quasituberitina, C135, **C315**
Quasituberitina, C315
QUENSTEDT, C535, C536
Queraltina, C139, **C764**
Quilostomélidos, C742
Quinqueloculina, C63, C78, C92, C93, C130, C137, **C458**
quinqueloculine, **C63**
Quinqueloculininae, C146, C155, **C458**
Quinquinella, C468

Raadshoovenia, **C477**
Rabanitina, **C482**
Racemiguembelina, C138, **C656**
Raciborskiaceae, C13
radial, **C63**
radial microstructure, **C63**
radial zone, **C63**
radiate aperture, **C63**
Radicula, **C185**
Radiocycloclypeus, **C650**
RADIOLARIA, C39
Radiolata, C142, C512, C605, C678
Radiolididae, C512, C605, C678
Raibosammina, C196
RAINWATER, C140
Ramulina, **C537**
Ramulinae, C145, C530
Ramulinella, **C537**
Ramulinidae, C530
Ramulinina, C530
Ramulininae, C144, C146, C147, C149, C152, C156, **C537**
Ranikothalia, C647
Raphanulina, **C785**
Raphidohelix, C259
Rauserella, C383, **C397**
Rauserina, C135, **C315**
Rauserites, C415
RAUZER-CHERNOUSOVA & FURSENKO, C142, C151, C322, C435, C436, C652
Receptaculitidea, C144
Rectangulina, **C788**
rectilinear, **C63**
Rectobolivina, **C553**, C654, C733
Rectobulimina, C137, **C546**
Rectochernyshinella, C346
Rectocibicidella, C690
Rectocibicides, C687, C691, **C692**
Rectocornuspira, **C440**
Rectoepistominoides, C137, **C776**
Rectoeponides, **C683**
Rectoglandulina, C523
Rectoguembelina, C654
Rectogümbelina, C654
Rectoseptaglomospiranella, C135, **C350**
Rectotrochamminoides, C783
Rectuvigerina, C139, **C569**

Recurvoidella, **C226**
Recurvoides, **C226**
Red Sea, C125
REDMOND, C103
REICHEL, C154, C508, C510, C661
Reichelina, **C397**
Reichelina, C294
Reichelininae, C394
Reinholdella, C97, C137, C774, **C776**, C777
REISS, C99, C100, C113, C150, C151, C481, C598, C614, C678, C682, C688, C699, C709, C759
Reitlingerella, **C785**
REITLINGERELLIDA, **C787**, C788, C789
Reitlingerellidae, **C787**
relict apertures, **C63**
Remaneica, **C266**
Remaneicinae, C155, **C266**
Remesella, **C290**
reniform, **C63**
Renoidea, C466
Renulina, **C484**
Renulina, C785
Renulinites, C484
Renulites, C484
Reophacella, **C272**
Reophacida, C214
Reophacidae, C146, C147, C148, C149, C150, C151, C214
Reophacidinae, C215
Reophacinae, C146, C148, C215
Reophagus, C216
Reophax, C136, C215, **C216**
Reophaxopsis, C216
reproduction of multilocular foraminifers, **C75**
reproduction of thecamoebians, **C17**
reproductive cycle of "primitive" foraminifers, **C71**
RESIG, C684
Reticella, C32
RETICULAREA, C2, C16, **C39**
RETICULARIA, C39, C143, C144, C164
Reticulariacea, C14
RETICULARIACEAE, C14
Reticulariaceae, C14, C15
Reticulariaceen, C14
Reticulariacées, C14
Reticulariacei, C14
RETICULARIEAE, C14
Reticularieae, C14
RETICULARIIDA, C164
Reticulariidae, C14, C164
Reticulariinae, **C15**
reticulate, **C63**
reticulolobopodia, **C18**
RETICULOLOBOSA, C37
Reticulophragmium, **C233**
reticulopodia, **C18**
RETICULOSA, C164
Reticulosa, C39, C54
Reticulosa (Proteomyxa), C794

Retorta, C458
retral processes, **C63**
REUSS, C56, C91, C143, C269, C277, C546, C785
Reussella, C139, C562, **C563**
Reussellinae, C147, C148, C149, C150, C152, C561
Reussia, C561, 564
Reussiinae, C146, C148, C561
Reussina, C259
Reussoolina, C540
Revolventiidae, C53
REYMENT, C549, C736
REYTLINGER, C93, C385, C440
Rhabdammina, C90, C135, **C185**
Rhabdammina, C185
Rhabdamminae, C145, C184
Rhabdamminella, C186
Rhabdamminidae, C144, C145, C184
Rhabdamminina, C184
Rhabdammininae, C144, C184
Rhabdella, **C785**
Rhabdogonium, C245
RHABDOIDEA, C143, C511
Rhabdoidea, C143, C144, C512
Rhabdoina, C512
Rhabdopleura, C190
Rhabdorbitoides, C725
Rhaphidodendron, **C785**
Rhaphidohelix, C259
Rhaphidoscene, C786
Rhapidionina, C493
Rhapydionina, **C493**
Rhapydionininae, C137, C156, **C493**
Rhenothyra, C135, **C342**
Rheophax, C216
rheoplasm, **C18, C67**
Rhinocurus, C520
Rhipidionina, **C493**
Rhipidocyclina, C714
Rhizammina, **C186**
Rhizammina, C186
Rhizamminidae, C146, C147, C149, C150, C151, C184
Rhizammininae, C144, C147, C148, C154, **C186**
Rhizo-Flagellata, C7
Rhizoflagellates, C7
Rhizomastigida, C7
Rhizomastigidae, C7
Rhizomastigina, C7
Rhizomastiginen, C7
Rhizomastix, C8
Rhizonubecula, C447
Rhizopela, C783
RHIZOPODA, C5, C39, C142, C144
RHIZOPODA FILOSA TESTACEA, C40
RHIZOPODA IMPERFORATA, C40, C164
RHIZOPODA LOBOSA TESTACEA, C19
RHIZOPODA RETICULOSA TESTACEA, C164

RHIZOPODA SPHYGMICA, C5
RHIZOPODEA, C2, **C5,** C16
RHIZOPODES, C5
Rhizopodes, C458, C512
rhizopodia, **C63**
RHIZOPODIA, C5
Rhogostoma, C42
RHUMBLER, C56, C66, C91, C103, C144, C146, C172, C175, C177, C212, C215, C501, C577
Rhumblerinella, **C173**
Rhynchogromia, **C181**
Rhynchogromiinae, C146, C173
Rhynchoplecta, **C785**
Rhynchosaccus, **C181**
Rhynchospira, C669
RICHTER, C325
Rimalina, C520
Rimelphidium, C635
Rimulina, **C529**
Ripacubana, **C493**
Riveroina, **C477**
Rizomastigidos, C7
RIZÓPODOS, C5
Robertina, C96, C139, **C777,** C782
Robertinacea, C71, C99, C154, C161, **C766**
Robertinidae, C96, C97, C108, C138, C139, C149-C152, C162, C766, **C777,** C782
Robertininae, C150, C777
Robertinoides, C96, C108, **C782**
Robulammina, C225
Robulina, C520
ROBULINACEA, C511
Robulinae, C147, C512
Robulinidae, C512
Robuloides, **C342**
Robulus, C520
Robustoschwagerina, **C420**
Robustoschwagerina, C420
Roglicia, **C770**
Rolshausenia, C607
Ropalozotika, C783
Rosalina, C85, **C584**
Rosalinella, C662
Ross, C9
Rostrolina, C530
Rotalia, C137, **C606,** C687, C742
Rotaliacea, C63, C75, C88, C99, C109, C137-C139, C154, C156, **C605**
ROTALIACEA, C511
ROTALIACEAE, C511
Rotaliammina, **C265**
Rotaliaridae, C605
ROTALIARIDIA, C184, C313, C511
Rotaliaridia, C214, C320, C394, C572, C605
Rotaliatina, **C750**
Rotalida, C143, C605
ROTALIDAE, C145, C511

Rotalidae, C144, C145, C149, C605
ROTALIDEA, C511
Rotalidea, C143, C605
Rotalideae, C605
Rotalidee, C605
Rotalidium, C607
Rotálidos, C605
Rotaliella, C85, **C604,** C784
Rotaliellidae, C156, **C604**
ROTALIFORMES, C725
Rotaliformes, C605
Rotaliicae, C605
rotaliid septa, **C99**
ROTALIIDA, C152, C511
Rotaliidae, C99, C120, C134, C137-C139, C146-C148, C150-C152, C156, **C605,** C650, C686, C742, C771
Rotaliidea, C99, C148, C150-C152, C605
ROTALIINA, C154, C156, **C511,** C699
Rotaliinae, C146, C147, C149, C156, **C605,** C742
Rotalina, C143, C605, C606
Rotalinae, C143-C146, C605
Rotalininae, C605
Rotalipora, C64, C138, **C659**
Rotaliporidae, C138, C156, **C659**
Rotaliporinae, C156, **C659**
Rotalites, **C785**
Rotamorphina, C587
Rotorbinella, C572
Rotundina, C659
Ruatoria, C569
Rubratella, C66, C68, C81, C84, C85, C769, **C770,** C771, C784
Rudigaudryina, **C272**
Ruditaxis, C337
Rugidia, **C587**
Rugofusulina, C424
Rugoglobigerina, C61, C63, C64, C115, C138, **C663**
Rugoglobigerininae, C152, C662
rugose surface, **C63**
Rugosoarchaediscus, C354
Rugosofusulina, **C424**
Rugososchwagerina, **C425**
Rugososchwagerina, C425
Rugotruncana, C662
Rupertia, C627
Rupertiidae, C146, C150-C152
Rupertiinae, C147, C148, C150, C627
Rupertina, C139, **C627**
Rupertinae, C627
Rupertininae, C156, **C627**
Russiella, **C436**
Ruttenella, C589
Ruttenia, C582
RZEHAK, C520, C656
Rzehakina, C137, **C220**
Rzehakinidae, C90, C137, C155, **C220**

Rzehakininae, C148, C150, C152, C220

Saccamina, C196
Saccammina, C135, **C196**
Saccammina, C196
Saccamminae, C145, C194
Saccamminidae, C135, C145-C151, C155, **C194,** C316, C796
Saccamminina, C194
Saccammininae, C414, C146-C148, C150, C151, C155, C194, **C196**
Saccamminis, **C204,** C796
Saccamminoides, **C201**
Saccamminoides, C204
Saccamminopsis, C135, **C319**
Saccodendron, **C205**
Saccorhina, C319
Saccorhiza, **C190**
Saccorhizidae, C145, C184
Saccularia, C516
Sacculariella, C514
Sacculinella, C196
SAEDELEER, DE, C175
Saedeleeria, **C181**
Sagenella, C205
Sagenina, **C205**
sagittal section, **C63, C362**
Sagoplecta, C136, **C533**
Sagraina, C569
Sagrina, C558, **C569**
Sagrinnodosaria, C559
SAID, C98, C119, C127-C129
SAIDOVA, C121
ST. JEAN, C345, C346
SAITO, C676
Sakesaria, **C614**
Sakhiella, C678
salinity, effect of, **C129**
Salpicola, C42
Salpingothurammina, C314
Samarina, C354
SANDON, C140
Sappiniaceae, C11
Sappiniidae, **C11**
saprozoic nutrition, **C10**
Saracenaria, **C524**
Saracenella, C524
sarcode, **C18, C63**
SARCODEA, C5
SARCODINA, C2, **C5**
SARKODINA, C5
Sarmatiella, C456
Saudia, **C308**
SAUNDERS, C239
Saxicolina, C468
Saxicoline, C468
Scarificatina, C783
Schackoina, C65, C138, **C658**
Schackoinella, **C591**
Schackoinidae, C138, C156, **C658**
SCHAUDINN, C8
Schaudinnia, C32
Schaudinnula, C53
Schellwienia, C404
Schenckiella, C282

Schizammina, C136, **C194**
Schizamminidae, C136, C155, **C192**
schizogamy, **C63**
schizont, **C63**
Schizophora, C251
Schizopyrenidae, C6
SCHIZOSTOMA, C184, C511
Schizostoma, C145
SCHLEIMPILZE, C11
Schlosserina, C139, **C777**
SCHLOTHEIM, VON, C796
SCHLUMBERGER, C57, C466, C475, C571
SCHLUMBERGER & MUNIER-CHALMAS, C472
Schlumbergerella, **C629**
Schlumbergeria, C710
Schlumbergerina, C93, **C472**
SCHMIDT, C95
SCHRÖTER, C55
SCHUBERT, C56, C145, C298, C434
Schubertella, **C400**
Schubertellidae, C152, C399
Schubertellinae, C150, C152, C155, C370-C372, C380, C383-C385, **C400**
Schubertellinidae, C399
Schubertellininae, C400
Schubertia, C558
SCHULTZE, C57, C67, C142, C151, C438
Schultzella, **C173**
Schultzia, C173
SCHULZE, C29
Schwabia, C34
SCHWAGER, C144, C376, C510
Schwagerina, C61, C65, C371, C377, C379, C393, **C415**
Schwagerinidae, C152, C399, C415
Schwagerininae, C63, C147-C150, C152, C155, C371, C373-C376, C380, C382, C383, C385, C386, **C415**
sclerotium, **C11**
Scortimus, C520
SCOTT, ZELLER & ZELLER, C345
scrobis septalis, **C63**
Scutuloris, **C468**
Seabrookia, **C540**
Seabrookiinae, C146, C147, C153, C156, **C540**
secondary apertures, **C63**
secondary axial septulum, **C63**
secondary septulum, **C63**
secondary transverse septula, **C379**
secondary transverse septulum, **C63**
Seguenza, C212
Sejunctella, **C602**
Selenostomum, C737
Semiflosculina, C508
Semiinvoluta, C136, **C742**
Seminovella, **C350**
Semirosalina, C584
Semitextularia, C135, **C335**

Semitextulariidae, C135, C150, C155, **C335**
Semivulvulina, **C254**
Semseya, **C785**
septa, **C362, C385**
Septabrunsiina, C135, **C341**
Septaglomospiranella, C341, C350
septal flap, **C63**
septal fluting, **C63**
septal foramen, **C63**
septal furrow, **C63**
septal pore, **C63, C362, C372**
Septammina, **C785**
Septatournayella, C339
Septigerina, **C258**
septula (sing., septulum), **C63, C364, C379, C386**
Serpula, C518
Serpuleidae, C145, C184
Serpulella, C213
Serpulopsis, **C214**
sessile, **C63**
Sestranophora, C683
Sestronophora, C139, C680, **C683**
Setigerella, C47
Sexangularia, **C37**
Sexloculina, C547
sexuality in foraminifers, C83
Shepheardella, **C182**
Shepheardia, C182
SHERBORN & CHAPMAN, C744
Sherbornina, **C625**
Shidelerella, C200
SHIFFLET, C130
SHLYKOVA, C251
Shuguria, C135, **C317**
Sichotenella, **C436**
SIDALL, C183
Siderina, C618
Siderolina, C630
Siderolites, C138, **C630**
Siderolithes, C630
Siderolithus, C630
Siderolitidae, C628
Siderolitinae, C150, C152, C628
Sideroporus, C630
Siderospira, **C785**
sieve-plate, **C63, C111**
SIGAL, C142, C149, C150, C219, C526
Sigalia, C654
Sigmavirgulina, C731, **C733**
sigmoid, **C63**
Sigmoidella, **C533**
Sigmoidina, C533
Sigmoilina, C63, C92, **C465**
sigmoiline, **C63**
Sigmoilopsinae, C458
Sigmoilopsis, C93, **C466**
Sigmomorpha, C531, C533
Sigmomorphina, **C533**
Sigmomorphinoides, C533
Silicina, C216
Silicinidae, C146, C147, C148, C152, C214, C220
Silicininae, C147, C215, C220
Silicoplacentina, C18, **C27**

Silicosigmoilina, C137, **C222**
Silicotextulina, C733, C734
Silicotextulinidae, C149, C731
SILVESTRI, C466, C524, C525, C624, C729, C730
Silvestria, C478
Silvestriella, C629
Silvestrina, C710
Simpalveolina, **C510**
Simplorbites, C710
Simplorbitolina, C137, **C312**
Sinzowella, **C448**
Siphogaudryina, C269
Siphogenerina, C139, **C569**
Siphogenerinoides, C137, **C558**
Siphogenerita, C567
Siphoglobulina, **C539**
siphon, **C63**
Siphonaperta, C93, **C466**
Siphonclavulina, C272
Siphonema, C786
Siphonides, C139, **C591**
Siphonina, C139, **C591**, C763
Siphoninella, C139, **C591**
Siphoninidae, C139, C151, C152, C156, **C591**
Siphonininae, C147-C150, C152, C591
Siphoninoides, **C591**
Siphonodosaria, C139, **C559**
SIPHONOFORAMINIFERA, C145
Siphonoforaminifera (Tubulata), C184
Siphotextularia, **C258**
Siphotrochammina, **C266**
Siphouvigerina, **C571**
SKINNER & WILDE, C365
Skinnerella, C420
SLAMA, C89
SMITH, C131
SMOUT, C99, C111, C480, C481, C607, C629, C639, C642
SMOUT & EAMES, C495
SMOUT & SUGDEN, C292
Smoutina, C138, **C614**
SOLDANI, C55
Soldania, C520
Soldanina, C688
SOLENOPODA, C40
SOLLAS, C93, C319, C600
Somalina, C139, **C500**
somatic nucleus, C63
SORBY, C94-C97
Sorites, C92, C139, **C496**
Soritida, C143, C482
Soritidae, C136, C137, C139, C147, C156, **C482**
Soritina, C482, C496
Soritinae, C156, **C496**
Sornayina, C137, **C236**
SOROIDEA, C143
Sorophaera, C196
SOROPHORA, C11
SOROPHORAE, C11
SOROPHOREAE, C11
SOROPHOREEN, C11
SOROPHORINA, C8, **C11**
Sorosphaera, C135, **C196**

Sorosphaeroidea, C205
SOSNINA, C328
SPANDEL, C325
Spandelina, C323, C325
Spandelinoides, C323
Sphaerammina, **C227**
Sphaerammininae, C148, C155, **C227**
Sphaeridia, **C626**
Sphaerogypsina, **C698**
Sphaeroidina, C139, **C547**
Sphaeroidinella, C60, C97, C139, **C673**, C674
Sphaeroidinellinae, C156, **C673**
Sphaeroidinellopsis, C139, **C673**
Sphaeroidinidae, C139, C156, **C547**
Sphaeroidininae, C146, C147, C153, C547
Sphaerophthalmidium, C783
Sphaeroschwagerina, **C436**
Sphaeroschwagerina, C436
Sphaerulata, C142, C458
Sphaerulina, C383, **C399**
Sphenoderia, C19, **C50**
Spherulacea, C142, C458, C512
Sphérulacés, C142, C458, C512
Spherulidae, C458
spicular tests in Foraminiferida, **C95**
Spidestomella, C466
Spincterules, C520
spinose, **C64**
spiral canals, **C64**
spiral side, **C64**
Spiralina, C484
Spirigerina, C602
Spirillina, C64, C66, C68, C71, C81-C83, C136, C137, **C600**, C742
Spirillinacea, C81, C154-C156, **C598**
Spirillinae, C145, C599
spirilline, **C64**
Spirillinida, C599
Spirillinidae, C61, C94, C136, C137, C144-C146, C150, C153, C156, **C598**, C731, C742
Spirillinidea, C143, C150, C598, C600
Spirillinina, C598
Spirillininae, C144, C146-C148, C150, C153, C156, **C600**, C742
Spirillinoidea, C147, C598
Spirillinoides, **C212**
Spirobolivina, **C547**
Spirobotrys, C693
Spirocerium, C786
Spiroclypeus, C644, C645, **C650**
Spiroculina, C453
Spirocyclina, C137, **C233**
Spirocyclinidae, C152, C225, C233
Spirocyclininae, C155, **C233**
Spirofrondicularia, **C535**

Spirolina, C139, **C484**
Spirolininae, C146, C149, C482
Spirolinites, C484
Spirolocammina, **C222**
Spiroloculina, C78, C137, **C453**
Spiroloculininae, C137, C155, **C453**
Spirolocunina, C453
Spirophthalmidium, C448
Spiroplecta, C652, C785
Spiroplectammina, C89, C136, **C251**, C253
Spiroplectammininae, C146, C155, **C251**
Spiroplectella, C241
Spiroplectina, **C785**
Spiroplectina, C272
Spiroplectinae, C146, C652
Spiroplectinata, **C272**
Spiroplectinatinae, C268
Spiroplectininae, C146, C148, C268
Spiroplectoides, C251
Spiropleurites, **C785**
Spirosigmoilina, **C466**
Spirosigmoilinella, **C222**
Spirotecta, **C748**
spirotheca, **C64**, **C364**, **C374**, **C383**
spiroumbilical, **C64**
Spirulina, C484
Spongina, C786
Sporadogenerina, **C537**
Sporadotrema, C139, **C705**
Sporamoebidae, C7
sporangium, **C11**
spore, **C11**
Sporilus, C631
Sporobulimina, C137, **C546**
Sporobuliminella, C137, **C546**
Sporohantkenina, C666
sporophore, **C11**
Spumariaceae, C13
Spumariaceen, C13
Spumariei, C13
Squamulina, **C444**
Squamulinida, C444
Squamulinidae, C155, **C444**
Squamulinidea, C143, C444
STACHE, C510
Stacheia, C136, **C330**, C331
Stacheiinae, C155, **C330**
Stacheoides, C136, C331, **C332**
Stacheya, C330
STAFF, C368, C376
Staffelininae, C150
Staffella, **C397**
Staffellidae, C155, C383, C385, **C397**
Staffellinae, C152, C397
Staffellinidae, C397
Staffellininae, C397
Staffia, C528
STAINFORTH, C133, C559, C730
Stainforthia, **C561**
Stannarium, **C790**
Stannoma, **C790**
Stannomida, C790
Stannomidae, **C790**

Stannophyllum, C208, **C792**
Stannoplegma, C790
Stegnammina, C135, **C196**
Stegnamminidae, C194
Stegnammininae, C194
STEIN, C29
Stemonitacea, C9, **C12**
STEMONITACEAE, C13
Stemonitaceae, C13
Stemonitaceae, C13
Stemonitacées, C13
Stemonitacei, C13
STEMONITAE, C13
STEMONITALES, C13
Stemonitea, C13
Stemoniteae, C13
Stemoniteen, C13
Stemonitées, C13
Stemonitei, C12, C13
Stemonitidaceae, C13
Stemonitidae, C9, **C13**
Stemonitidides, C12
STEMONITINA, C13
Stemonitinae, **C13**
Stemonitioidea, C13
Stensioina, C138, **C763**
Stephanopela, C783
stercomata, C18, **C64**, C87, C789
Stereonemeen, C13
stereoplasm, C18, **C64**, **C67**
Stetsonia, **C585**
Stichocassidulina, **C556**
Stichocibicides, C687, **C692**, C761
Sticholepis, C22
STICHOSTEGIA, C145
Stichostegia, C184, C511
STICHOSTÈGUES, C142, C511
Stichostègues, C142, C511, C537, C559
Stilostomella, **C559**
Stilostomellinae, C149, C556
Stoliczkiella, C787
stolon, **C64**
Stomasphaera, C135, **C202**
Stomatorbina, C139, C776, **C777**
Stomatostoecha, C137, **C245**
Stomiosphaera, C787
stomostyle, **C64**, **C65**
Storilus, C688
Storrsella, **C619**
Storthosphaera, **C196**
stratigraphic distribution, **C134**
Streblospira, C440
Streblus, C607
streptospiral, **C64**
striate, **C64**
Strophoconus, **C785**
Stylolina, C238
Stylonychia, C70
Subalveolina, C503, **C510**
Subbdelloidina, **C248**
Subbotina, C138, **C673**
Sublamarckella, C771
subseptate, **C64**
substrate environmental effect, **C128**

Suggrunda, **C733**, C734
Sulcoperculina, C138, C645, **C647**, C725
Sulcophax, **C217**
Sulcorbitoides, **C725**
SULEYMANOV, C90
Sumatrina, C375, C379, C380, C386, C387, **C432**
Sumatrininae, C427
supplementary apertures, **C64**
supplementary multiple areal apertures, **C64**
supraembryonic area, **C64**
sutural supplementary apertures, **C64**
suture, **C64**
Svenia, C516
Svratkina, **C750**
swarm cell, **C11**
SWITZER & BOUCOT, C90, C96
symbionts, **C70**
symbiosis, C18, **C64**
SYMPLECTOMERES, C5
syngamy, **C11**
Syniella, C787, C788, **C789**
Synspira, **C785**
Syringammina, **C192**
Syzrania, C317
syzygy, C60, **C64**, **C78**

Taberina, **C491**
Taitzehoella, **C412**
tangential section, **C64**, **C362**
TAPPAN, C269, **C434**
Tappanina, C137, **C553**
Taramellina, C496
TASCH, C126
TATEM, C29
Taurogypsina, C629
Tavajzites, C402
Tawitawia, **C258**
Tawitawiinae, C155, **C258**
TECAMEBOIDEOS, C19
Technitella, **C202**
tectine, **C64**
tectinous wall, **C88**
tectoria (sing., tectorium), **C64**, **C364**
tectum, **C64**, **C364**, **C366**, **C367**
tegillum (pl., tegilla), **C64**
Telostoma, C145, C511
temperature, C124
TEN DAM, C761
Tentifrons, **C524**
Terebralina, C137, **C602**
Terebralininae, C146, C148, **C600**
Tereuva, C567
TERQUEM, C56, C216, C635, C729, C757, C771
Terquemia, C755
Terquemina, C787
test, **C18**, **C64**, **C88**
test, chamber form and arrangement, **C100**
test form of foraminifers, relation to habitat, **C134**
test of thecamoebians, C16
test openings, **C108**

TESTACEA, C16
TESTACEA, C7, C19, C40, C143, C164
TESTACEAFILOSA, C40
TESTACEALOBOSA, C7, C19
TESTACIDA, C19
TESTAFILOSINA, C40
TESTALOBOSINA, C19
TESTAMOEBIDA, C5
Testamoebidae, C7
Testamoebiformia, C786
TESTARETICULOSINA, C37
Testulorhiza, C186
Testulosiphon, C186
Testulosiphoninae, C186
Tetragonulina, C518
Tetramitacea, **C7**
Tetramitidae, **C7**
Tetraplasia, C246
Tetrataxidae, C136, C150, C152, C155, **C337**
Tetrataxiinae, C337
Tetrataxinae, C147-C150, **C337**
Tetrataxis, C136, **C337**
Tetrataxis, C337
Teurnayellidae, C338
Textilaria, C253
Textilarida, C143, C250, C253
Textilaridae, C144, C253
Textilaridea, C143, C144, C250
Textilarideae, C250
Textilariidae, C250
Textilariinae, C253
Textilarina, C250
Textillaria, C253
Textularia, C90, C96, C136, **C253**
TEXTULARIACEA, C184
TEXTULARIACEAE, C184
TEXTULARIDA, C184
Textularida, C250
TEXTULARIDAE, C145, C184
Textulariidae, C142, C144, C250, C253
TEXTULARIDEA, C184
TEXTULARIDIA, C184
Textuláridos, C251
Textulariella, **C299**
TEXTULARIIDA, C152, C184
Textulariidae, C120, C136, C146-C150, C152, C155, **C250**
TEXTULARIIDEA, C184
TEXTULARIINA, C154, **C184**, C699
Textulariinae. C146, C147, C152, C155, **C253**
Textularina, C250
Textulariinae, C143-C145, C251, C253
Textularioides, **C255**
TEXTULINIDA, C184, C313, C511
Textulinidae, C251, C268
TEXTULINIDIA, C725
Textulinidia, C214, C320, C543
THALAMIA, C164
Thalamophaga, **C183**
THALAMOPHORA, C764

THALAMOPHOREN, C164
THALMANN, C215, C282, C466, C524, C546, C582, C608, C733
Thalmannammina, **C226**
Thalmannina, C214
Thalmanninella, C659
Thalmannita, C138, **C621**
Thecamoeba, C7
THECAMOEBAEA, C19
thecamoebian, **C18**
THECAMOEBIANS, **C16**
THECAMOEBIDA, C7, C19
Thecamoebida, C6
Thecamoebidae, **C6**
THECAMOEBINA, C19
Thecamoebina, C6
THECOLOBOSA, C19
Thekammina, C196
Thekammininae, C194
Themeon, C631
Themeone, C632
Tholosina, C135, **C205**
Thomasinella, **C217**
Thomasinella, C217
THOMPSON, C154, C372, C375, C397, C432, C434
Thurammina, C135, **C202**, C785
Thuramminoides, **C208**
Thuramminopsis, C196
Thyrammina, C202
Ticinella, C62, C138, **C661**
Tikhinella, C323
Tinogullmia, C182, C183
Tinophodella, C139, **C678**
Tinoporidae, C147, C149, C628
Tinoporidea, C144, C628
Tinoporina, C628
TINOPORINAE, C313, C511
Tinoporinae, C144, C145, C628
Tinoporininae, C628
Tinoporus, **C785**
Tintinnina, C786, C787
Tiphotrocha, **C266**
Titanopsis, C196
TITANOSTICHOSTEGIA, C145
Titanostichostegia, C511
Tobolia, **C535**
TODD & BLACKMON, C96
Tolypammina, C135, **C213**
Tolypamminella, C212
Tolypamminidae, C210
Tolypammininae, C147, C148, C151, C155, **C213**
tooth, **C64**
tooth plate, **C64**
Torinosuella, **C233**
TORIYAMA, C432
Toriyamaia, **C397**
Torreina, **C712**
Torresina, **C598**
Tortonella, **C473**
Tosaia, **C547**, C652
Tournauellinae, C338
Tournayella, C135, **C339**
Tournayellidae, C62, C64, C135, C136, C150, C151, C155, **C338**
Tournayellidea, C151, C320
Tournayellina, C341
Tournayellinae, C151, **C338**
Toxinopsis, C783
trace elements, effect of, **C129**
Tracheleuglypha, C19, **C51**
Trachelinella, C138, **C736**
Trakelina, C736
transverse septulum (pl., septula), **C64**, C379
Transversigerina, C569
Tremastegina, **C685**, C724
Trematocyclina, C233
Trematoforininae, C468
trematophore, **C64**
Trepeilopsis, C135, **C214**
Tretomphalus, C69, C75, C77, C87, **C585**, C675
Triasina, C136, **C484**
Tribrachia, **C524**
Tricarinella, C539
TRICHEAE, C15
Trichia, C10
Trichiacea, C9, **C13**
Trichiacea, C15
TRICHIACEAE, C15
Trichiaceae, C15
Trichiaceen, C15
Trichiacées, C15
Trichiacei, C15
Trichiae, C15
TRICHIALES, C15
Trichieae, C15
Trichiées, C15
Trichiei, C15
Trichiidae, C9, **C15**
Trichiides, C13
Trichiinae, **C15**
Trichinacea, C15
Trichioidea, C14
Trichioidei, C15
Trichistidae, C145, C268
Trichocisti, C15
Trichocisti (Trichioidei), C14
Trichohyalus, **C750**
Trichophorae, C12, C14
TRICHOSPERMI, C12
Trifarina, C139, **C571**
Trigenerina, C251
Trigonopsis, C27
Trigonopyxidae, **C26**
Trigonopyxis, C18, **C26**
Trigonouva, C567
Trigonulina, C540
Trillina, C466
Trilocularena, **C224**
Triloculina, C64, C78, C92, **C466**
triloculine, **C64**
Triloculinella, C466
trimorphism, **C64**
Trimosina, **C565**
Trinema, C19, **C53**
Trinematinae, **C53**
Trinème, C53
Trinemidae, C47
Trineminae, C53
Trinitella, C138, **C663**
Trioxeia, **C785**
Trioxeia, C785
Triplalepidina, C723
Triplasia, **C245**, C785
Tríquidos, C15
Trisegmentina, C441
Trisegmentininae, C448
triserial, **C65**
Tristix, C524, **C539**, C785
Tritaxia, **C272**
Tritaxilina, C291
Tritaxiopsis, C272
Tritaxis, **C266**, C337
Triticites, C61, C369, C371, C383, C385, C389, C390, **C425**
Triticites Zone, C390
Tritubulogenerina, C569
Trochamina, C259
Trochammina, C89, C129, C136, **C259**, C266, C784
Trochammina, C259
Trochamminae, C145, C259
Trochamminella, C266
Trochamminida, C259
Trochamminidae, C136, C146-C150, C152, C155, **C259**
Trochamminidea, C144, C259
Trochammininae, C144, C146-C149, C155, **C259**
Trochamminisca, C265
Trochamminita, **C226**
Trochamminoides, C136, **C227**
Trochamminula, **C266**
trochoid, **C65**
Trocholina, C136, C740, C741, **C742**
Trochonella, C742
trochospiral, **C65**
Trochospirillina, C602
Trochulina, C572
trochulines, les, C572
TROELSEN, C96, C103
Troglodytes, C44
Trophosphaera, C70
TRUJILLO, C275
Truncatulina, C688
Truncatulininae, C145, C687
Truncorotalia, C667
Truncorotaloides, C64, **C669**
Truncorotaloidinae, C156, **C668**
Trybliodiscodina, C714
Trybliolepidina, C721
Tscherdyncevella, C135, **C332**
Tschoppina, C702
Tubeporina, C135, **C322**
tuberculate, **C65**
Tuberitina, C135, **C321**
Tuberitinidae, C320, C321
Tuberitininae, C155, **C321**
Tubiferaceae, C14
Tubiferacées, C14
Tubifereae, C14
Tubiferida, C14
Tubiferidae, C14
Tubiferinae, **C14**
Tubinella, **C477**
Tubinellina, C477
Tubinellinae, C155, **C477**

Index

Tubitextularia, C654
Tubophaga, C183
Tubularina, C458
TUBULATA, C145
TUBULINACEAE, C14
Tubulinaceae, C14
Tubulinacées, C14
TUBULINAE, C14
Tubulinae. C14
Tubulinées, C14
Tubulinidae, C14
Tubulogenerina, C139, **C565**
tubulospine, **C65**
tumulus (pl., tumuli), **C65**
tunnel, **C65, C364, C380**
turbidity, **C133**
Turbienta, **C789**
Turbinacea, C142, C605, C685
Turbinacés, C142, C605, C685
Turbinida, C250, C268
Turbinoida, C143, C250, C268, C605, C736
Turbinoidae, C142, C268, C530, C559, C572, C605, C669, C685, C692, C748
TURBINOIDEA, C184, C511
Turbinoidea, C565, C685
Turbinoides, C142
Turbinolina, C584
Turbinulina, C607
turbinulines, les, C607
Turborotalia, C138, **C668,** C671
Turborotalita, C676, C677
turbulence, **C134**
Turriclavula, **C183**
Turriculacea, C142
Turriculacés, C142
Turrilina, C139, **C543**
Turrilinidae, C137, C139, C156, **C543**
Turrilininae, C146-C150, C156, **C543**
Turrispira, C319
Turrispirillina, **C602**
Turrispirillininae, C146, C148, C600, C742
Turrispiroides, C136, **C319**
Turrispirrillina, C602
Turritellella, C135, **C212**
Turritellopsis, C212
Tuvaellina, **C789**

UCHIO, C598, C777
UHLIG, C771
Uhligina, C565, C708
UJIIÉ, C635, C639
Umbella, C322
Umbellina, C135, **C322**
Umbellinae, C152, C322
Umbellininae, C155, **C322**
umbilical depression, **C65**
umbilical side, **C65**
umbilical teeth, **C65**
umbilicate, **C65**
Umbilicodiscodina, C714
umbilicus (pl., umbilici), **C65**
umbo, **C65**
Umbonaria, C32
umbonate, **C65**

Ungulatella, **C782**
Unicosiphonia, **C555**
unilocular, **C65**
Uniloculina, C458
UNILOCULINIDEA, C149, C164, C184
uniserial, **C65**
unrecognizable genera, C162, C783
upper keriotheca, **C65, C367**
upper tectorium, **C65**
UPSHAW & STEHLI, C132
Upsonella, **C785**
Uralinella, C135, **C316**
Urnula, C202
Urnulina, **C46**
Usbekistania, C212
Uslonia, **C316,** C796
Ussuriella, C346
UTERINI VERI, C12
UTRICULATA, C145
Uvellida, C143, C268, C279
Uvellidae, C268, C559
Uvellidea, C143, C250, C268
Uvellideae, C250, C268
Uvellina, C268, C530, C547, C559, C565, C572, C669
Uvellinida, C669
Uvigerina, C139, **C565,** C572
Uvigerinammina, C94, C133, **C272**
Uvigerinella, C139, **C572**
Uvigerinida, C565
Uvigerinidae, C137, C139, C147, C149, C151, C156, **C565,** C736
Uvigerininae, C146-C148, C150, C152, C565, C566, C736

Vacuoles, **C18, C65**
Vacuolispira, C647
Vaginula, C524
Vaginulina, C136, **C524**
Vaginulinella, C524
Vaginulinidea, C143, C512
Vaginulinidae, C512
Vaginulinideae, C512
Vaginulinopsis, C136, **C524**
Vaginuloglandulina, C514
Vagocibicides, C760
Vahlkampfidae, C7
Vahlkampfiidae, C7
VALKANOV, C40
Valvobifarina, **C565**
Valvopavonina, C563
Valvoreussella, C269
Valvotextularia, C277
Valvulamnina, **C283**
Valvulina, C136, C145, **C279**
Valvulinella, C337
Valvulinella, C136, **C337**
Valvulineria, C137, **C587**
Valvulineria, teuriensis, nom. nov., **C587**
Valvulinerideae, C149, C572
Valvulineriidae, C151, C572
Valvulineriinae, C586
Valvulinidae, C145, C146,

C148, C149, C268, C279, C699
Valvulininae, C136, C145, C148-C150, C152, C155, **C279**
Vampyrellae, C795
Vampyrellacea, C795
Vampyrellacées, C795
Vampyrellida, C794, C795
Vampyrellidae, **C795**
Vampyrellidea, C794
VAN BELLEN, C757
Vandenbroeckia, **C484**
Vandenbroekia, C484
Vanhoeffenella, **C186**
Variostoma, C137, **C586**
VAŠÍČEK, C115, C548
Vasicekia, C135, **C320**
VAUGHAN, C501, C712, C714
VAUGHAN & COLE, C719, C724
Vaughanina, **C725**
Velellidae, C643
VELLA, C567, C572, C642
VENGLENSKIY, C89
Venilina, C251
Ventilabrella, C654, C656
ventral, **C65**
Ventrolamina, C741, C742
Ventrolaminidae, C739
Ventrolamininae, C739
Verbeekia, C647
Verbeekina, C374, C380, C386, C393, **C427**
Verbeekina Zone, C393
Verbeekinacea, C394
Verbeekinidae, C58, C59, C62, C152, C155, C371, C385, **C426**
Verbeekinidea, C152, C394
Verbeekininae, C146-C150, C155, C374, C376, C380, C382, C383, C385-C387, C426, **C427**
Verella, **C412**
Vermiculum, C518
Verneolina, C268
Verneuilina, **C268**
Verneuilinella, C277
Verneuilinidae, C146, C148-C150, C268, C795
Verneuilininae, C136, C146-C150, C152, C155, **C268**
Verneuilinoides, C133, **C273**
Verneulina, C268
Vernonina, **C586**
Verrucina, **C210**
Vertebralina, **C456**
Vertebralinidae, C445
vertical distribution in living planktonic foraminifers, **C127**
VESICULATA, C145
Vesiculata, C40
Vicinesphaera, C314
Victoriella, C139, **C705,** C708
Victoriellidae, C149, C150, C152, C680, C702, C705
Victoriellinae, C156, **C705**
Vidalina, **C440**

VINOGRADOV, C90, C92, C97, C98, C134
Virgulina, C731, C733, C734, C784, C785
Virgulinella, **C734,** C736
Virgulinidae, C151, C731
Virgulininae, C146-C148, C152, C731
Virgulinopsis, **C572**
Virgulopsis, **C561**
VISSARIONOVA, C315
VITREA, C511
vitreous, **C65**
Vitriwebbina, **C537**
VOLOGDIN, C788
VOLOSHINOVA, C635
VOLOSHINOVA & BALAKHMATOVA, C284
VOLOSHINOVA, DAIN & REYTLINGER, C340
Voloshinovella, C284, **C291**
Volutaria, **C785**
Volvotextularia, **C785**
Volvotextularia, C785
Voorthuysenia, C771
Vorticialis, C635
Vulvulina, **C251,** C254

WADE, C635, C639
Waeringella, **C412**
WAILES, C175
Wailesella, **C38**
WALKER, C55
wall composition in foraminifers, **C88**

wall microstructure in foraminifers, **C88**
WALLROTH, C8
Wanganella, C328
WARNER, C92
WARTHIN, C354
Washitella, **C537**
Webbina, **C448,** C687
Webbina, C535
Webbinella, **C535**
Webbinellinae, C147, C148, C151, C156, **C535**
Webbinelloidea, C135, **C205**
Webbum, C448
Wechselthierchen, C6
Wedekindella, C412
Wedekindellina, C373, C380, C386, **C412**
Wedekindia, C412
Weikkoella, C135, **C210**
Wellmanella, **C466**
Wetheredella, C787
WETZEL, C89
WEYNSCHENK, C741
Wheelerella, **C730**
Wheelerellinae, C156, **C730**
WHITE, C375, C656
WHITTAKER, C140
whorl, **C65**
WICHER, C742
Wiesnerella, **C452**
WILLIAMSON, C56, C91, C604
WINTER, C78
WOOD, C92, C95, C482, C582, C600, C688, C702
WOOD & HAYNES, C688, C757

Woodella, C612
Woodringina, C138, **C652**

xanthosome, **C18, C65,** C87, C88, C789
XENOPHIOPHORAE, C789
xenophya, C789
XENOPHYOPHORA, C789
XENOPHYOPHOREN, C789
XENOPHYOPHORIDA, C208, **C789,** C790
Xenophyophoridae, C790, C792
Xenotheka, C135, **C183**

Yabeina, C62, C379-C381, C383, C394, **C432**
Yabeina Zone, C394
Yaberinella, C139, **C501**
YAKOVLEV, C251
Yangchienia, C381, **C412**
Yanischewskina, C354

Zeauvigerina, C556
Zekritia, **C308**
ZELLER, C345
Zellia, **C426**
Zonomyxa, **C22**
ZOOSPOREAE, C12
Zoosporeae, C12, C794, C795
Zoosporés, C794
Zoosporida, C794
Zoosporidae, C12
zooxanthellae, C64, C70, C87
Zotheculifida, **C259**
Zoyaella, **C453**
ZYGOSPOREAE, C11
zygote, **C18, C65**

113406

113406

QE   Joint Committee on Inverte-
770   brate ...
J6   Treatise on inverte-
pt.C   brate paleontology.
v.2